Design of Bridges
Fifth Edition

Design of Bridges

Fifth Edition

(IRC: 6-2014)

(IRC: 22-2015)

(IRC: 24-2001)

(IRC:112-2011)

N Krishna Raju BE, MSc (Engg), PhD, MIE, MI Struct. E

Emeritus Professor of Civil Engineering
MS Ramaiah Institute of Technology
Bangalore

Oxford & IBH Publishing Co. Pvt. Ltd.

New Delhi

(*A Unit of* CBS Publishers & Distributors Pvt Ltd)

CBSPD

CBS Publishers & Distributors Pvt Ltd

New Delhi • Bengaluru • Chennai • Kochi • Kolkata • Lucknow • Mumbai
Hyderabad • Jharkhand • Nagpur • Patna • Pune • Uttarakhand

Design of Bridges

Fifth Edition

ISBN-13: 978-81-204-1798-4
ISBN-10: 81-204-1798-4

© 2017, 2009, 1998, 1991, 1988   N Krishna Raju

CBS Reprint: 2018, 2019, 2021, 2023

OXFORD & IBH
New Delhi
(A Unit of CBS Publishers & Distributors Pvt Ltd)

Published by **Satish Kumar Jain** and produced by **Varun Jain** for
CBS Publishers & Distributors Pvt Ltd
4819/XI Prahlad Street, 24 Ansari Road, Daryaganj, New Delhi 110 002, India
Ph: 011-23289259, 23266861   Website: www.cbspd.com
                                    e-mail: delhi@cbspd.com

Corporate Office: 204 FIE, Industrial Area, Patparganj, Delhi 110 092, India
Ph: 011-4934 4934   Fax: 011-4934 4935   e-mail: publishing@cbspd.com;
                                              publicity@cbspd.com

Branches

- **Bengaluru:** Seema House 2975, 17th Cross, KR Road, Banasankari 2nd Stage, Bengaluru 560 070, Karnataka, India
  Ph: +91-80-26771678/79   Fax: +91-80-26771680   e-mail: bangalore@cbspd.com
- **Chennai:** 7, Subbaraya Street, Shenoy Nagar, Chennai 600 030, Tamil Nadu, India
  Ph: +91-44-26680620, 26681266   Fax: +91-44-42032115   e-mail: chennai@cbspd.com
- **Kochi:** 42/1325, 1326, Power House Road, Opp KSEB, Power House, Ernakulum Kochi 682 018, Kerala, India
  Ph: +91-484-4059061-65,67   Fax: +91-484-4059065   e-mail: kochi@cbspd.com
- **Kolkata:** 147, Hind Ceramics Compound, 1st Floor, Nilgunj Road, Belghoria, Kolkata-700056, West Bengal, India
  Ph: +033-25633055, 033-25633056   e-mail: kolkata@cbspd.com
- **Lucknow:** Basement, Khushnuma Complex, 7 Meerabai Marg (Behind Jawahar Bhawan),Lucknow-226001, UP, India
  Ph: +91-522-4000032   e-mail: tiwari.lucknow@cbspd.com
- **Mumbai:** PWD Shed, Gala no 25/26, Ramchandra Bhatt Marg, Next to JJ Hospital Gate no. 2, Opp. Union Bank of India Noorbaug, Mumbai-400009, Maharashtra, India
  Ph: 022-66661880/89   e-mail: mumbai@cbspd.com

Representatives

• Hyderabad	0-9885175004	• Jharkhand	0-9811541605	• Nagpur	0-9421945513
• Patna	0-9334159340	• Pune	0-9923910676	• Uttarakhand	0-9716462459

Printed at Chaman Enterprises, Daryaganj, New Delhi, India

*The book is dedicated to the Pioneers,
Research workers and Famous Bridge Engineers*

*Randolphe Perronet, John Rennie, Roebbling, Eugene Freyssinet
Amman, George Stephenson, Dischinger, Finister Walder, Leonhardt,
Egeman Ayna, Cusens, Rusch, Raina, Jessica Bins, Rowe, Johnson Victor
and
a host of others who toiled incessantly for the design development and
Construction of bridges in the world*

Preface to the Fifth Edition

In the past 28 years, since the book was first published in 1988, the author has been revising this book periodically to update its contents conforming to the latest developments in the field of design of bridges and the national codes governing the design principles of bridges. The fourth edition of the book published in the year 2009 has been reprinted several times establishing the popularity of the book among the civil and highway engineering students, teachers and structural designers. The Ministry of surface transport, Government of India has launched the golden quadrilateral scheme connecting the capital cities of various states with modern highways. Bridges being the basic components of highway systems, have witnessed rapid progress and innovations during the last decade.

The design philosophy of reinforced concrete bridges has been revised by the Indian Roads Congress by introducing an integrated and comprehensive code (IRC: 112-2011) for the design of concrete bridges based on the limit state philosophy. The revised edition of IRC: 6-2014 (Section-II) dealing with loads and stresses incorporates the partial safety factors for verification of structural strength and serviceability limit states. The new code lays emphasis on both strength and serviceability aspects of bridge decks and specifies the method of computation of ultimate strength, width of cracks and deflections at service loads. This new code supersedes the earlier codes IRC: 21-2000 and IRC: 18-2000. However the earlier codes IRC: 21 and 18 based on working stress method of design has been retained in the code with its use being limited to those grades of concrete and steel specified in Annexure (A-4) of the code.

The present **Fifth edition** incorporates several improvements in each of the chapters by the introduction of limit state philosophy in all the design computations in keeping with the specifications of the revised IRC: 112 Code. Review and objective type questions have been included at the end of each chapter to help the students to concentrate on the salient aspects of the subject matter and also to prepare for competitive examinations and professional interviews. Some of the chapters have been updated with additional designs of bridges in keeping with the latest developments in the field of prestressed concrete. Each chapter is provided with relevant references to help the students for further reading. Numerous figures have been included in keeping with the spirit of "**drawing is the language of the engineer**".

The author gratefully acknowledges the help rendered by his wife Pramila Raju and many of his colleagues and practicing structural and highway engineers in updating the contents of the revised edition of the book. Thanks are also due to M/S. Oxford & IBH publishers, New Delhi for their excellent co-operation in the publication of this monograph.

Bengaluru

N. Krishna Raju
2017

Preface to the First Edition

During the last thirty years, rapid developments in the field of bridge engineering has resulted in the widespread use of several new types of reinforced and prestressed concrete bridge decks, which are structurally efficient and aesthetically superior besides being economical and catering to the needs of the fast moving highway traffic.

This book 'Design of Bridges' is a modern comprehensive text meeting the requirements of senior undergraduate and post graduate students of the Civil, Structural and Transportation engineering streams and practising structural engineers. The book provides a lucid exposition of the theory and design of various types of bridge decks, like masonry arched, reinforced concrete slab culvert, Tee beam slab, steel plate girder, composite, rigid frame, box culvert, pipe culvert, steel trussed, reinforced concrete balanced cantilever, reinforced concrete continuous beam and slab decks of variable cross section, prestressed concrete bridge decks and cable stayed bridges. The design of bridge bearings is covered in one of the Chapters.

Detailed design of various bridge types conform to the latest Indian Standard Codes IS:456, IS:1343 and the codes of the Indian Roads Congress. The loading standards adopted are those recommended by the Indian Roads Congress. All the design examples are worked out in S.I. units. Working drawings showing the reinforcement details, plans, elevations and cross sections are provided. The topics covered are intended to meet the requirements of the curricula of most of the engineering institutions in India.

The book is primarily design oriented with more emphasis on the design of various types of bridge decks with minimum extent of theory presented wherever required for application in design. The various design steps are identified and provided in a logical sequence. Several examples for practice are provided at the end of each chapter to help the students preparing for university examinations.

The references provided at the end of the book have been freely used in the preparation of this text and are gratefully acknowledged. The author is grateful to his wife Pramila for extending the fullest co-operation in the preparation of the type-script. The author acknowledges the help rendered by M/s System Design 1106, II Stage Rajajinagar, Bangalore - 560 010, for Text Processing and Composing the manuscript. The author gratefully acknowledges the financial help extended by the Chairman

Shri M.R. Jayaram and Prof. Y.A. Prahlada Rao, Principal, M.S. Ramaiah Institute of Technology for the successful completion of this book writing project. Finally the author welcomes constructive criticisms and suggestions which will immensely help in up dating the contents of the book.

Bangalore

N. Krishna Raju
February 1988

List of Symbols and Notations

A : Cross sectional area

A_c : Concrete cross sectional area

A_p : Area of prestressing tendons

A_{st} : Cross sectional area of tension reinforcement

A_{sc} : Cross sectional area of compression reinforcement

$A_{s,min}$: Minimum Cross sectional area of reinforcement

A_{sw} : Cross sectional area of shear reinforcement

a : Lever arm

b : Breadth of beam, or shorter dimension of a rectangular column

b_{ef} : Effective width of slab

b_f : Effective width of flange

b_w : Breadth of web or rib

D : Overall depth of beam or slab or diameter of column; dimension Of a rectangular column in the direction under consideration

D_f : Thickness of flange

D_L : Dead load

d : Effective depth

d' : Depth of compression reinforcement from the highly compressed face

E_c : Modulus of elasticity of concrete

E_{cm} : Secant Modulus of Elasticity of concrete

$E_{c,eff}$: Effective modulus of elasticity of concrete

E_L : Earth quake load

E_s : Modulus of elasticity of steel

e : Eccentricity

F : Resisting force

f_{ck} : Characteristic cube compressive strength of concrete

f'_c : Cylinder compressive strength

f_{ctm} : Tensile strength of concretre

f_{ct} : Split tensile strength of concrete

f_{cr} : Modulus of rupture of concrete (Flexural strength of concrete)

f_{ct} : Split tensile strength of concrete

f_d : Design strength

f_y : Characteristic tensile strength of steel

f_{yk} : Characteristic tensile strength of steel

f_{ct} : Permissible Compressive stress in concrete at transfer

f_{cw} : Permissible Compressive stress in concrete at service loads

f_{tt} : Permissible tensile stress in concrete at transfer

f_{tw} : permissible tensile stress in concrete at service loads

g : Gravity load or dead load

h : Overall depth

I : Second moment of area or moment of inertia

I_{ef} : Effective moment of inertia

I_{gr} : Moment of inertia of gross section excluding reinforcement

I_r : Moment of inertia of cracked section

j : lever arm factor

K : Stiffness of member

k : Constant or coefficient or factor

L_d : Development length

LL : Live load

L : Length of a beam or column between adequate lateral restraints
Or the unsupported length of a column

L_{ef} : Effective span of beam or slab

L_x : Length of shorter side of slab

L_y : Length of longer side of slab

M : Bending moment

M_g : Dead load moment

M_q : Live load moment

M_u : Ultimate moment

M_r : Moment of resistance

m : Modular ratio

n : Neutral axis depth

n_a : Actual neutral axis depth

n_c : Critical neutral axis depth

P : Axial load on a compression member or Prestressing Force

p : Safe bearing capacity of soil or intensity of pressure

p_t : Percentage reinforcement in tension

p_c : Percentage reinforcement in compression

q : Live load

Q : Design coefficient

r : Radius

S : Spacings of stirrups

T : Torsional moment

u : Width of dispersion of wheel load in short span direction

v : Width of dispersion of wheel load in long span direction

V : Shear force

w : Distributed load per unit area

W : Total load or concentrated load

WL : Wind load

LL : Live Load

x_u : Neutral axis depth

Z : Modulus of section

δ : Displacement

γ_f : Partial safety factor for load

γ_m : Partial safety factor for material

μ : Coefficient of friction

σ_{cbc} : Permissible stress in concrete in bending compression

σ_{cc} : Permissible stress in concrete in direct compression

σ_{sc} : Permissible stress in steel in compression

σ_{st} : Permissible stress in steel in tension

σ_{sv} : Permissible tensile stress in shear reinforcement

τ_{bd} : Design bond stress

τ_c : Shear stress in concrete

$\tau_{c,max}$: Maximum shear stress in concrete with shear reinforcement

τ_v : Nominal shear stress

ϕ : Diameter of bar

ψ_{cs} : Shrinkage curvature

ε : Support to span ratio

ε_{cc} : Strain in concrete

ε_{cd} : Drying shrinkage strain

ε_{ca} : Autogenous shrinkage strain

ε_{cs} : Total shrinkage strain

ε_{sc} : Strain in steel
θ : Creep coefficient
v : Poisson's Ratio
α, β : Angles or Ratio
λ : Multiplying factor
η : Loss ratio

Contents

Introduction

1.1 HISTORICAL EVOLUTION OF BRIDGES

The history of bridge engineering is closely associated with the progress of human civilization spread over several centuries[1]. The earliest bridge on record is traced to the lake dwellers of Switzerland who pioneered the timber trestle construction for crossings of rivers around 4000 B.C. The oldest bridge still standing is a pedestrian stone slab bridge which is at least 2800 years old built across the Meles river in Smyrna, Turkey[2]. Many of the important ancient bridges were built by armies. As per Homer and Herodotus, the floating bridges were made of inflated skins (used as floats) around 800 B.C. A bridge of this kind was built in the year 556 B.C. by King Cyrus[3].

Around 320 B.C, Alexander the Great built floating bridges for the passage of his army during his great conquest of the East. India was the birth place of wooden cantilever bridges in the Himalayas with planks of wood anchored at the two banks using heavy stones and the wooden planks corbelled out progressively towards the mid stream until the gap could be spanned by a single plank.

The period between 200 B.C. and 260 A.D. witnessed the widespread use of stone arches by Romans using massive piers. After the fall of Rome, the bridge activity in Europe was mainly promoted by the religious orders. The Pont d'Avignon with 20 arch spans of about 34 m built by St. Benezet over the river Rhone in 1188 and the old London bridge[4] across the river Thames with 19 pointed arch spans of varied length built by Peter Colechurch in 1209 belong to this period. The medieval period bridges were loaded with decorative and defensive towers, chapels, statues, shops and dwellings. The Rialto Bridge in Venice, Italy built in 1591 having a single arch span of 27 m is a typical example of the many bridges built during the Renaissance period.

The first treatise on bridge engineering was published in 1714 by the French Engineer, Robert Guiter ushering the age of reason. The first engineering school in the world "The Ecole de Ponts et Chaussees" was founded at Paris in the year 1747 with Rodolphe Perronet[5], considered as the **'Father of Modern bridge building'** as the first director of the school. Perronet perfected the art of building masonry arched bridge by introducing slender piers with his best work being the Pont de la Concorde at Paris built in 1791. John Rennie designed the New London Bridge across the river

Thames using segmental masonry arches and the same was completed by his son in 1831.

In mid 19th century, demand for stronger and bigger bridges over large rivers resulted in the use of cast iron and wrought iron replacing timber and stone for bridges. The first recorded use of iron in bridges was a chain bridge built in 1734 by the German army across the Oder River in Prussia. Cast iron being brittle was not found very suitable for building large span bridges. An effective combination of cast iron for compression members and wrought iron for tension members was first used in trussed bridge around 1840 especially in railway bridges.

The development of steel by Bessemer in 1856 and the open hearth process by Sieman and Martin in 1861 paved the way for extensive use of steel and caught the imagination of bridge builders. Firth of Forth Cantilever bridge in Scotland with two main spans of 521 m built in 1899 and Roebling's Suspension bridge of 490 m were a few of the famous steel bridges of the 19th century heralding the beginning of the modern era of bridge engineering. The early part of 20th century witnessed the construction of many elegant steel bridges in America and the Europe. The giant leap came with construction of George Washington Bridge with a span of 1060 m. The great architect Le Corbusier exclaimed that, *"The Washington Bridge is the most beautiful Bridge in the World"*.

Post war years saw the emergence of reinforced concrete as a suitable material for short and medium span bridges with the added advantage of durability against aggressive environmental conditions in comparison with steel. Since the beginning of the 20th century, the use of reinforced concrete has become popular for both road and railway bridges replacing steel mainly due to low maintenance costs and durability. Reinforced concrete bridge decks having various structural configurations such as slab and box culverts, Tee beam and slab, Continuous girder, Bow string, Balanced cantilever, Open spandrel, and Rigid Frame types have been extensively used for medium to long spans.

A revolutionary and path breaking achievement in materials technology was witnessed in 1928 when Eugene Freyssinet[6], a French engineer introduced a new construction material designated as "Prestressed Concrete". In 1950, prestressed concrete came to be used mostly for bridges of ever increasing spans coupled with rapidity and ease of construction and competing in costs with other alternative types like steel and reinforced concrete. Among the 500 bridges built in Germany during 1949-53, seventy percent of them used prestressed concrete. The construction of three prestressed concrete railway bridges with spans ranging from 12.8 m to 19.2 m on the Assam rail link in 1950 heralded the use of prestressing in bridge construction in India. The Bendorf Bridge over the river Rhine in Germany with a main span of 208 m built in 1965 by Finsterwalder[7] using free cantilever method of construction is considered as a breakthrough in prestressed concrete bridge construction. The Ganga Bridge at Patna with 45 spans of 121 m each is the longest prestressed concrete bridge in Asia extending over 5.575 km built by using precast single cell segmental box girders of variable depth prestressed to form continuous spans.

Innovative efforts to reduce the depth of girders of large span bridges resulted in the development of cable stayed bridge decks in which the deck system is suspended by inclined steel cables which in turn transmit the forces through a massive tower or pylon to the foundations. The first modern cable stayed bridge built in Sweeden was the Stromsund bridge designed by Dischinger[8] and constructed in the year 1955. Subsequently more than 300 cable stayed bridges have been built throughout the world with steel, reinforced and prestressed concrete decks covering spans from 100 to 1800 m.

The development of bridge engineering from timber trestles and stone arches during the ancient period to modern cable stayed bridges of the 20th century sums up the historical evolution of bridges with the development of civilization. Notable examples of the development of various types of bridges are presented in the following sections.

1.2   TIMBER AND STONE MASONRY BRIDGES

Wooden bridges were built over the river Eupharetes during the reign of queen of Babylon in Iraq during 783 B.C. This bridge according to the legend was used by the ancient Greek epic poet Homer. Raina cites examples of early timber beam and crude suspension bridges built over Min River in China and the Himalayas as shown in Fig. 1.1. The Chinese were building stone arch bridges since 250 B.C. The Chao-Chow Bridge built around 600 A.D is perhaps the most long lived vehicular bridge to day. This stone masonry bridge with a single span of 37.6 m and a central rise of 7.2 m with a road way of 9 m is situated about 350 km south of Bejing. The secret of its longevity is attributed to the accurately dressed and matching voussoirs without any mortar in the joints.

Notable examples of a series of stone masonry arched bridges across the river Seine in Paris shown in Fig. 1.2 are unique examples of human ingenuity. The proven durability of material and the long experience in intuitive proportioning made stone masonry arched bridges the most popular form of construction in the early days of Railways until Iron and steel bridges made their way in the 17th century.

1.3   IRON AND STEEL BRIDGES

The first iron bridge comprising of five semicircular arch ribs in iron joined together side by side to form a single arch span of 30 m was built at Coalbrookdale in 1779 over the Severen in England by Abraham Darby and John Wilkinson. The Menai strait bridge which is a suspension bridge having a record breaking span of 177 m using wrought iron chains was built by Thomas Telford in Wales around 1826. The first iron railway bridge was built by George Stephenson in 1823 on the Stockton-Darlington railway. Gradually wrought iron replaced cast iron in bridge construction during the period from 1840 to 1890. Many railway bridges using various types of trusses like Pratt, Whipple, Howe, Fink and Warren were used during this period.

The development of steel by Bessemer in 1856 paved the way for extensive use of steel in road and railway bridges. The first all steel bridge was built at Glasgow,

Fibre Rope Suspension Form Over Min River in China (Total Length 552 m).

Crude Suspension Form in the Himalayas.

Fig. 1.1 Early Timber Suspension Bridges Over Min River in China and Himalayas.

South Dakota in 1878. Steel was also used in the cables and spans of the famous Brooklyn Bridge during 1869-83. The world's longest span steel cantilever bridge was built at Quebec over the St. Lawrence River having a main span of 549 m in 1917. The Howrah Bridge built in 1943 over the river Hooghly at Kolkata with a main span of 457 m is aesthetically elegant and possesses pleasing proportions among the suspended span, cantilever arms and the anchor spans.

The increased availability of steel, the earlier truss forms yielded to more efficient types such as the Baltimore, Pennsylvania and K-truss widely adopted for railway bridges. In India, many major bridges were built using steel decks in the late 19th century and early 20th century for railway tracks across the major rivers. The Upper Sone Railway Bridge built in 1899 consisting of 93 spans of approximately 30 m

Fig. 1.2 Stone Masonry Arched Bridges Across the River Siene in Paris.

each extending over a length 3.1 km is an excellent example of such a bridge. Fig. 1.3 shows a typical steel trussed bridge of 360 m span at Baltimore, U.S.A. The world's longest simple steel truss bridge span is the suspended span of J.J. Barry Bridge across Delaware River in U.S.A having a span of 251 m. In the early 20th century many steel arched bridges which were aesthetically pleasing were built, the prominent among them being the Henry Hudson bridge built in 1936 with a span of 244 m and the Rainbow bridge at Niagara falls built in 1941 with a span of 290 m. The World's longest arch bridge is the New River Gorge Bridge in West Virginia built in 1976 using weathering steel with a span of 519 m.

Fig. 1.3 Steel Trussed Bridge (360 m) at Baltimore, U.S.A.

Amman considered as the foremost bridge engineer in the world during the 20th century built Verrazano Narrows Bridge spanning 1300 m in New York which is considered as a master piece in bridge construction and opened for traffic in 1964 just a few months before he died. After the failure of the Tacoma Narrows Bridge in the state of Washington in 1940, due to aerodynamic instability, bridge engineers realized the importance of stiffening the decks of suspension bridges. Tacoma Narrows suspension bridge built with a deck thickness of 2.4 m extending over a span of 850 m was the most slender bridge with a span/depth ratio of 350. The Golden Gate Bridge in San Francisco having a central span of 1280 m extends over a total length of 2738 m. The towers soar over a height of 227 m up into the sky and the diameter of each of the two cables is 924 mm containing 27572 galvanized steel wires. The slender deck has been strengthened by 10 to 12 m deep stiffening trusses to safeguard against aerodynamic instability after the famous collapse of the Tacoma Narrows Bridge. Fig. 1.4 shows the famous Golden Gate Bridge located in San Francisco.

1.4   REINFORCED CONCRETE BRIDGES

The first reinforced concrete bridge was built by Adair in 1871 across the Waveney in England spanning 15 m[1]. The adaptability of reinforced concrete in architectural form was demonstrated by Maillart in Switzerland in building arched bridges using reinforced concrete, utilizing the integrated structural action of thin arch slabs with monolithically cast stiffening beams. Salginatobel and Schwanadbach Bridges built by Maillart in 1930 and 1933 respectively are classical examples of aesthetically beautiful and efficient use of materials coupled with economy in bridge construction. Reinforced concrete was preferred to steel as a suitable material for short and medium span bridges mainly due to the added advantage of durability against aggressive environmental conditions in comparison with steel.

Fig. 1.4 Golden Gate Bridge at San Francisco.

Reinforced concrete bridges with different types of decks have been widely used for both road and railway bridges. The most common type is the slab deck used for short spans such as culverts. For medium spans in the range of 10 to 20 m, Tee beam and slab deck (Fig. 1.5) is widely used. Bow String girder type (Fig. 1.6) bridges have been used for road bridges in the span range of 25 to 35 m. Continuous bridge decks (Fig. 1.7) with longitudinal girders of varying depth are found to be more economical in the span range of 20 to 40 m. Elegant arch bridges were built during the period from 1920 to 1950. The Dum Dum Bridge (Fig. 1.8) at Kolkata built in 1926 with two arches of 24 m span is the first major reinforced concrete arch bridge in India. The Teesta Coronation Bridge built in 1941 across Teesta river in West Bengal shown in Fig. 1.9 is an excellent example of reinforced concrete Open Spandrel Bridge with a main span of 81.71 m and a central rise of 39.6 m. In the case yielding soils where minor settlements are expected, Balanced cantilever bridges are preferred. Fig. 1.10 shows the structural elements of a typical double cantilever bridge having the cantilever and suspended spans.

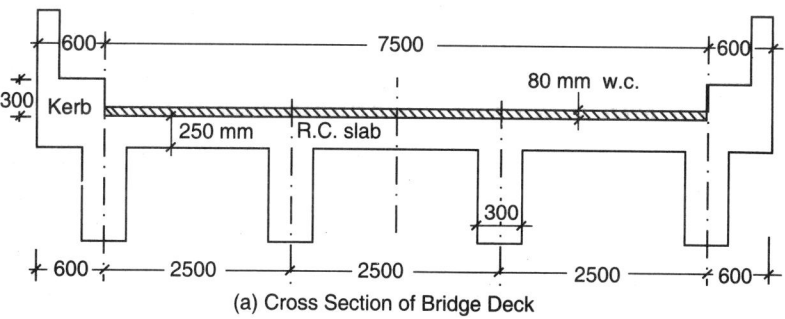

(a) Cross Section of Bridge Deck

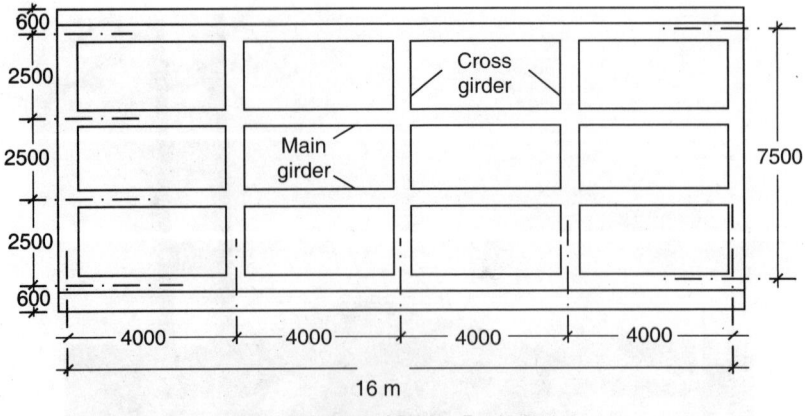

(b) Plan of Bridge Deck

Fig. 1.5 Tee Beam and Slab Bridge Deck.

Fig. 1.6 Bow String Girder Bridge.

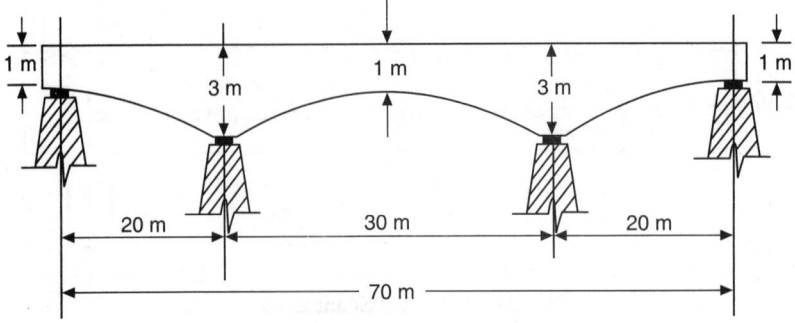

Fig. 1.7 Continuous Bridge Deck.

Fig. 1.8 Dum Dum Bridge at Kolkata.

Fig. 1.9 Teesta Coronation Bridge (Open Spandrel Type).

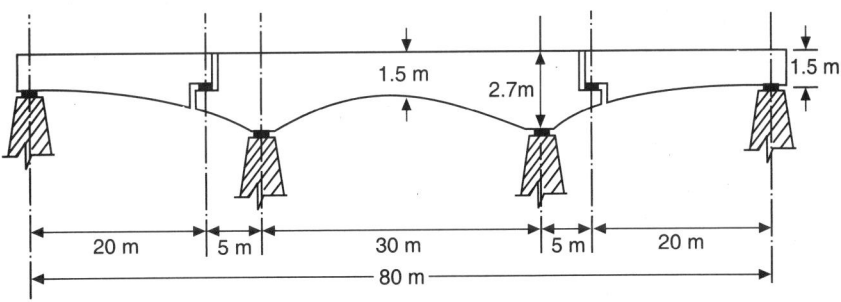

Fig. 1.10 Balanced Double Cantilever Bridge.

1.5 PRESTRESSED CONCRETE BRIDGES

A revolutionary and a path breaking achievement in materials technology was witnessed in 1928 when Eugene Freyssinet[6], a French engineer introduced a new construction material designated as '**Prestressed Concrete**', which is ideally suited for construction of long span bridges. The big boom in prestressed concrete was witnessed after the Second World War. In 1950's, prestressed concrete came to be used mostly for bridges of ever increasing spans coupled with rapidity and ease of construction and competing in costs with other alternative types like steel and reinforced concrete.

Among the 500 and odd bridges built in post war Germany during 1949-53, seventy percent of them used prestressed concrete. The use of prestressed concrete in bridges started in India in 1948 when three railway bridges of spans ranging from 12.8 m to 19.2 m were constructed in the Assam Rail link. The first prestressed concrete highway bridge built in India is the Palar Bridge near Chinglepet, built in 1954 with 23 spans of 27 m each. Since then, many prestressed concrete bridges have been successfully built in this country using innovative designs and construction techniques. In 1949 prestressed concrete was introduced in U.S.A in the construction of Magnel's Walnut Bridge.

Notable examples of prestressed concrete bridges include the Lubha bridge in Assam shown in Fig. 1.11 and the Ganga Bridge at Patna considered as the longest bridge extending over 5575 m shown in Fig. 1.12. The Bendorf Bridge over the Rhine in Germany with a main span of 208 m, the Hamana Bridge in Japan with a 240 m main span and the Boussen's Bridge in France are excellent examples of free cantilever technique of construction developed by Finsterwalder. At present the cantilever construction method is invariably used for long span prestressed concrete bridges mainly for quality control and rapidity of construction. During the last decade hundreds of fly overs built in the metropolitan cities of India have adopted the cantilever construction technique with minimal disruption of traffic.

Fig. 1.11 Lubha Bridge in Assam.

Fig. 1.12 Ganga Bridge at Patna.

1.6 CABLE STAYED BRIDGES

A revolutionary approach to bridge design and construction first conceived by the German Engineer Dischinger in 1938 and later put into practice in the construction of first modern cable stayed bridge is the Stromsund Bridge in Sweeden around 1953. This innovation paved the way for the construction of number of famous Rhine family cable stayed bridges with spans up to and exceeding 300 m.

According to Leonhardt[9], cable stayed bridges are technically, economically, aesthetically and aerodynamically superior to the classical suspension bridges for spans in the range of 700 to 1500 m. The combination of cable stays with cellular box girder prestressed concrete decks have significantly extended the span range of highway bridges. India's first cable stayed bridge is the Akkar Bridge in Sikkim completed in 1988 and extending over a length of 157 m with a single pylon of height 57.5 m shown in Fig. 1.13. Vidyasagar Sethu (Second Hooghly Bridge) at Kolkata shown in Fig. 1.14 is an excellent example of a cable stayed three lane highway bridge comprising a main span of 457 m and two side spans of 183 m. Normandie cable stayed bridge in France built using concrete and steel spans over 624 m. The Tatara Bridge[10] located on the Onomichi-Imabari highway route of the Honshu-Shikoku Bridge Project in Japan has a main span of 890 m with Y-shaped towers rising 176 m above the bridge deck. At present the longest span cable stayed bridge in the world is the Messina Straights Bridge in Italy having a main span of 1800 m.

Fig. 1.13 Akkar Bridge in Sikkim.

1.7 MODERN TRENDS IN BRIDGE ENGINEERING

Rapid developments and innovations in the field of concrete technology indicates that it is possible to produce ultra high strength concrete[11] with a characteristic compressive strength exceeding 100 N/mm^2 and high tensile steel cables of superior quality and strength required for new types of prestressed concrete and cable stayed bridges. Innovations in construction techniques coupled with rapid advances in the design philosophy of complex bridge forms has paved the way for use of longer spans and slender decks in bridges.

The evolution of long span cable stayed bridges with hybrid decks using steel girders, concrete slabs and high tensile steel cables is increasingly noticeable with the dawn of 21st century. The cantilever method of construction is the latest and most economical and popular method generally adopted for the construction of long span precast or cast *in-situ* prestressed concrete segmental bridges[12]. This method is also ideally suited for the construction of cable stayed bridges.

The Second Vivekananda Prestressed concrete cable stayed bridge across the river Hooghly just north of Kolkata is an excellent example of an extradosed bridge, comprising a hybrid structure with elements of cable stayed post tensioned prestressed concrete box girders. The nine span extradosed bridge is considered as Asia's first multispan extradosed bridge and one of only three extradosed bridges in Asia outside Japan, according to Egeman Ayna[13], principal engineer of the International Bridge Technologies (IBT) who are the design consultants for the Bridge project. The modern bridge extending over total length of 880 m comprises of seven spans of 110 m and two 55 m long spans. The extradosed bridge deck box girders of constant depth are only 2.5 m deep and are supported by a single plane of 8 stay cables of the Harp type located on the central median as shown in Fig. 1.15.

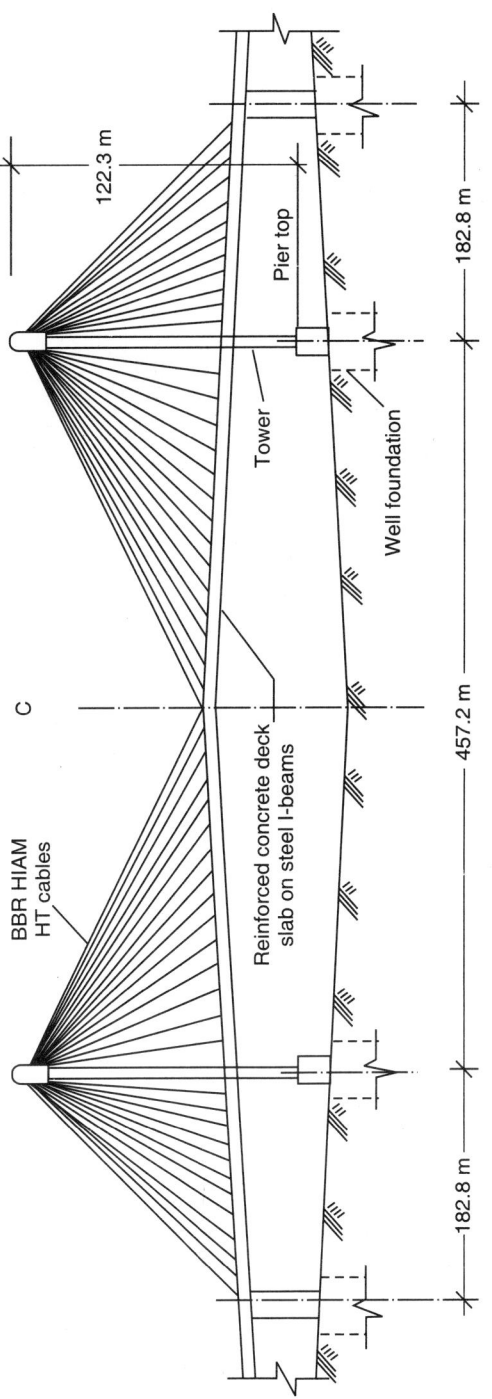

Fig. 1.14 Vidyasagar Sethu Cable Stayed Bridge at Kolkata.

Fig. 1.15 Second Vivekananda Cable Stayed Bridge Near Kolkata.

At present, the tallest and the longest cable stayed bridge is located outside the French town of Millau and France happens to be the homeland of Eugene Freyssinet, the innovator of Prestressed Concrete. The cable stayed bridge extending over a length of 2.46 km is considered as an engineering feat since some of the bridge pillars rise gracefully to a height more than 300 m. The bridge designed by the famous British architect Sir Norman Foster is currently the World's tallest and the longest cable stayed bridge shown in Fig. 1.16.

A critical survey of long span bridges constructed in various countries during the last two decades indicates that the modern trend is to adopt prestressed concrete cellular box girder decks for urban flyovers and cable stayed bridge decks for long spans since they are economical, structurally and aerodynamically efficient and aesthetically superior in comparison with other types of bridges.

1.8   STANDARD SPECIFICATIONS

Each country has formulated its own codes of practice incorporating standard specifications which have been evolved by the government organizations and professional institutions based on years of observation, development and research.

In India the highways and railways are owned and controlled by the government. Hence the bridges built on them should conform to the specifications prescribed by the designated organizations. Accordingly, the bridges built on national and state highways should adhere to the specifications of the Indian Roads Congress (IRC) codes prescribed by the Ministry of Surface Transport (Roads Wing), Government of India. The bridges built on railways should conform to the specifications of the Indian Railway Standard (IRS) Bridge Rules.

The Bureau of Indian Standards should also be followed wherever applicable for specific materials and type of work. The codes are generally revised periodically based on new materials and innovations in construction practices. The relevant Indian code provisions along with the codes of prominent countries like U.K, U.S.A, Germany, Russia, and Australia are incorporated in the text and the Bridge Engineer should be conversant with the various clauses of the codes for successful practice.

Fig. 1.16 World's Tallest and Longest Cable Stayed Bridge in France.

REFERENCES

1. Johnson Victor, D., Essentials of Bridge Engineering (Fifth Edition), Oxford & IBH Publishing Co. Pvt. Ltd., New Delhi, 2001, pp. 1–10.
2. Steinman, D.B., and Watson, S.R., Bridges and their Builders, Dover Publications, New York, 1957, 401 pp.
3. Raina, V.K., Concrete Bridge Practice, Construction, Maintenance & Rehabilitation Tata McGraw Hill Publishing Co. Ltd., New Delhi, 1988, p. 3.
4. Shirely-Smith, H., World's Great Bridges, E.L.B.S, London, 1964, 250 pp.
5. Virola, J., The World's Greatest Bridges, Civil Engineering, American Society of Civil Engineers, Vol. 38, No. 10, Oct. 1968, pp. 52–55.
6. Freysssinet, E., The Birth of Prestressing, Cement & Concrete Association, London, Translation No. CJ-59, 1956, 44 p.
7. Finsterwalder, U., Modern designs for Prestressed Concrete Bridges, Concrete and Constructional Engineering, London, Vol. 60, No. 3, 1965, pp. 99–103.
8. Raina, V.K., Concrete Bridge Practice, Analysis, design and Economics, Tata McGraw Hill Publishing Co. Ltd., New Delhi, 1991, pp. 509–530.
9. Leonhardt Fritz and Zellner, W., Cable Stayed Bridges, IABSE Surveys, S-13/80 and IABSE Periodical, 2/1980, May 1980.
10. Its, M., Fujino, Y., Miyata, T., and Narita, N., Cable Stayed Bridges-Recent Developments and their Future, Elsevier Publications, Tokyo, 1991, 438 pp.
11. Krishna Raju, N., Design of Concrete Mixes (5th Edition), C.B.S. Publishers New Delhi, 2005, pp. 1–250.
12. Cusens, A.R., Box and Cellular Girder bridges, 'A State of the Art Survey' A.C.I. Publication, SP: 26 on Concrete Bridge Design, Detroit, 1971.
13. Jessica Binns, Extradosed bridge Distinguishes Tollway project in India, Civil Engineering, Vol. 75, No. 2, February 2005, p. 20.

Bridge Loading Standards

2

2.1 EVOLUTION OF BRIDGE LOADING STANDARDS

The earliest bridge loading standards in some countries were first formulated to regulate heavy military vehicles and were generally specified by local authorities. The loadings often consisted of steam rollers and some form of traction engines. The earliest specifications of highway bridge loadings originated from the need to transport heavy military vehicles in U.K. and Europe. This resulted in the introduction of the Ministry of transport's first 'Standard Loading Train' in the U.K. in 1932 and the original loading standards of many European countries.

In the United Kingdom, these standards formed the basis for the present type HA loading of BS: 153[1]. In the U.S.A., a loading standard consisting of truck trains and equivalent loads was introduced by the American Association of State Highway Officials (AASHO) in 1935. It is significant to note that in some of the developing countries like India, the first loading standards were introduced nationally in 1937. Critical studies have been done to study the significant differences in the loading standards of various countries by Seni[2] and Rajagopalan[3]. The impact allowance was observed to vary considerably in different countries. In a recent survey made for the International Federation, Galambos[4] has reported that many countries are planning revisions of their highway bridge loading standards based on the research investigations.

The first loading standard (IRC: 6) was published in India by the Indian Roads Congress in 1958 and subsequently reprinted in 1962 and 1963. The Section-II of the code dealing with loads and stresses was revised in the second revision published in 1964. The metric version was introduced in the third revision of 1966. The fourth revion in the year 2000, including all the amemdments from time to time. The code has been throughly updated with several new provision periodically and the revised edition was released in 2014[5].

2.2 INDIAN ROADS CONGRESS BRIDGE LOADING STANDARDS

Highway bridge decks have to be designed to withstand the live loads specified by the Indian Roads Congress. The different categories of loadings were first formulated in

1958 and they have not changed in the subsequent revisions of 1964, 1966 and 2000.

The standard IRC loads specified in IRC: 6-2014 are grouped under four categories as detailed below:

2.2.1 IRC Class AA Loading

Two different types of vehicles are specified under this category grouped as tracked and wheeled vehicles. The IRC Class AA tracked vehicle (simulating an army tank) of 700 kN and a wheeled vehicle (heavy duty army truck) of 400 kN are shown in Fig. 2.1.

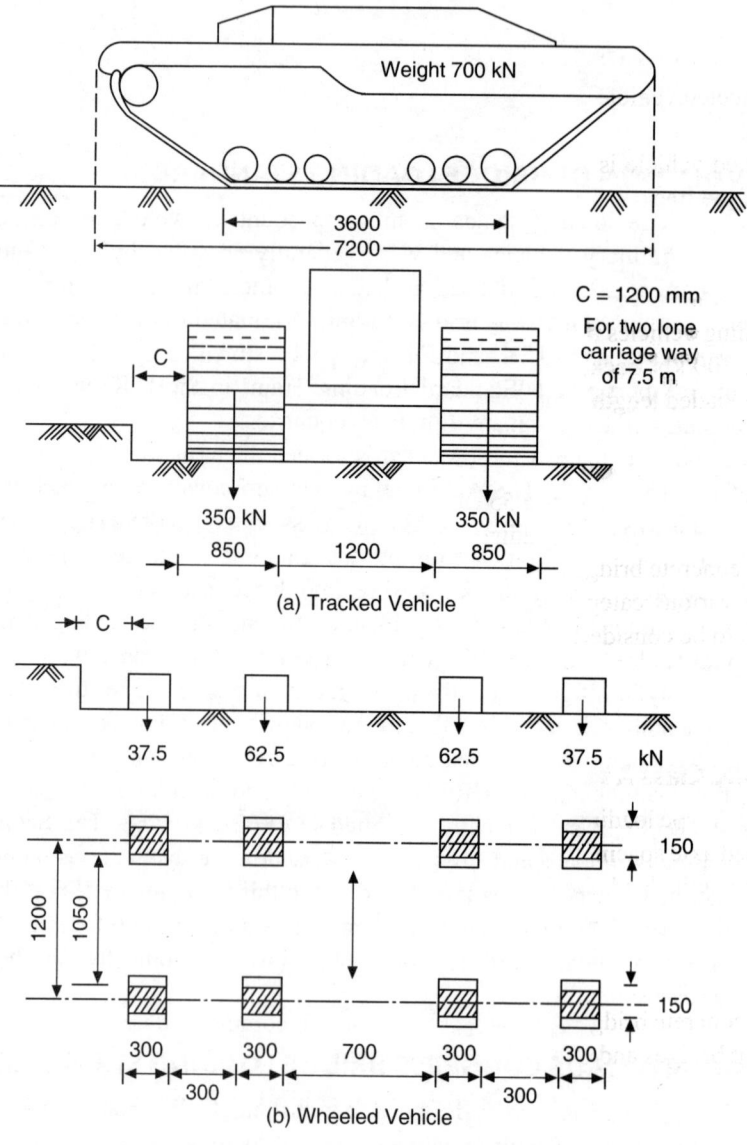

Fig. 2.1 I.R.C. Class AA Tracked and Wheeled Vehicles.

All the bridges located on National Highways and State Highways have to be designed for this heavy loading. These loadings are also adopted for bridges located within certain specified municipal localities and along specified highways. Alternatively, another type of loading designated as Class 70 R is specified instead of Class AA loading.

2.2.2 IRC Class 70 R Loading

IRC 70 R loading consists of the following three types of vehicles.

(a) Tracked vehicle of total load 700 kN with two tracks each weighing 350 kN.
(b) Wheeled vehicle comprising 4 wheels, each with a load of 100 kN totaling 400 kN.
(c) Wheeled vehicle with a train of vehicles on seven axles with a total load of 1000 kN.

The tracked vehicle is somewhat similar to that of Class AA, except that the contact length of the track is 4.87 m, the nose to tail length of the vehicle is 7.92 m and the specified minimum spacing between successive vehicles is 30 m. The wheeled vehicle is 15.22 m long and has seven axles with the loads totaling to 1000 kN. The bogie axle type loading with 4 wheels totaling 400 kN is also specified. The details of IRC Class 70 R loading vehicles are shown in Fig. 2.2.

The 700 kN tracked vehicle is common to both the classes, the only difference being the loaded length which is slightly more for the Class 70 R. The second category is the wheeled type comprising 1000 kN train of vehicles on seven axles for the Class 70 R and a 400 kN bogie axle type vehicle for the Class AA.

The Class A loading is a 554 kN train of wheeled vehicles on eight axles. Impact is to be allowed for all the loadings as per the specified formulae which is different for steel and concrete bridges.

The various categories of loads are to be separately considered and the worst effect has to be considered in design. Only one lane of Class 70 R or Class AA load is considered whereas both the lanes are assumed to be occupied by Class A loading if that gives the worst effect.

2.2.3 IRC Class A Loading

IRC Class A type loading consists of a wheel load train comprising a truck with trailers of specified axle spacing and loads as shown in Fig. 2.3. The heavy duty truck with two trailers transmits loads from 8 axles varying from a minimum of 27 kN to a maximum of 114 kN. The Class A loading is a 554 kN train of wheeled vehicles on eight axles. Impact has to be allowed as per the formulae recommended in the IRC: 6-2000. The impact factor is inversely proportional to the length of the span and is different for steel and concrete bridges. This type of loading is recommended for all roads on which permanent bridges and culverts are constructed.

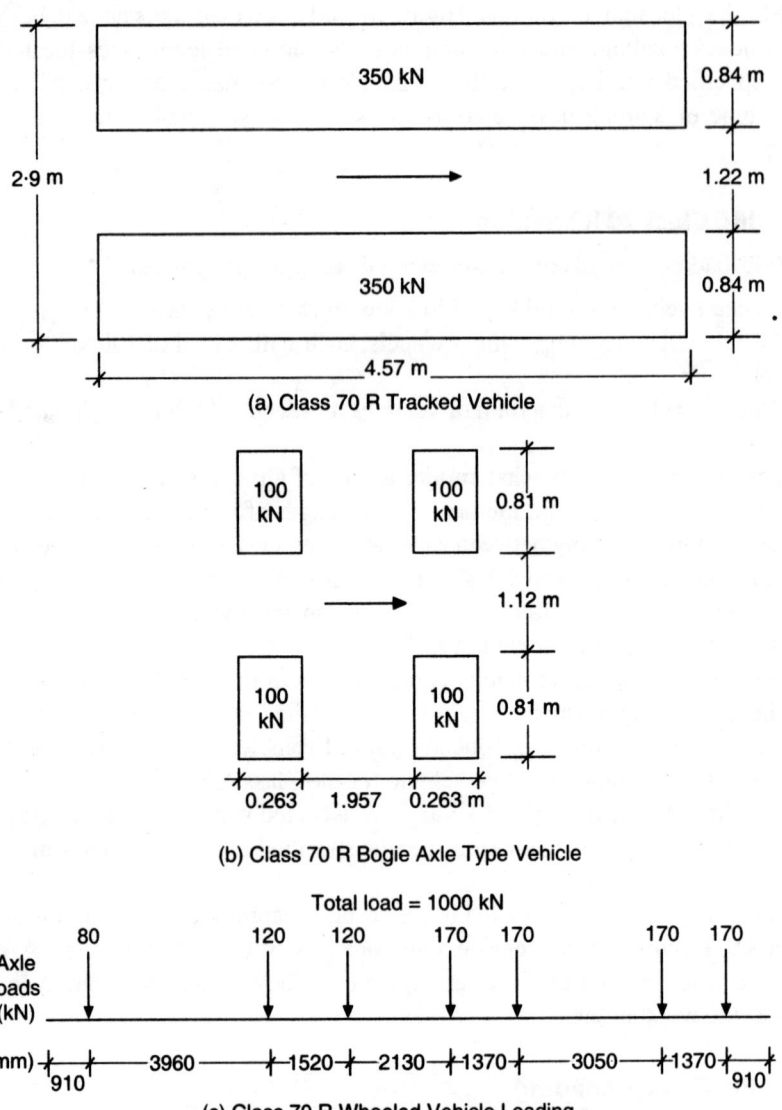

(a) Class 70 R Tracked Vehicle

(b) Class 70 R Bogie Axle Type Vehicle

(c) Class 70 R Wheeled Vehicle Loading

Fig. 2.2 I.R.C. Class 70 R Tracked and Wheeled Vehicles.

2.2.4 IRC Class B Loading

Class B type of loading is similar to Class A loading except that the axle loads are comparatively of lesser magnitude. The axle loads of Class B are a 332 kN train of wheeled vehicles on eight axles as shown in Fig. 2.3. This type of loading is adopted for temporary structures and timber bridges. Combinations of different types of live loads are recommended for the design of bridges in clause 204.3 of IRC: 6-2014. The carriageway live load combination recommended for design is compiled in Table 2.1.

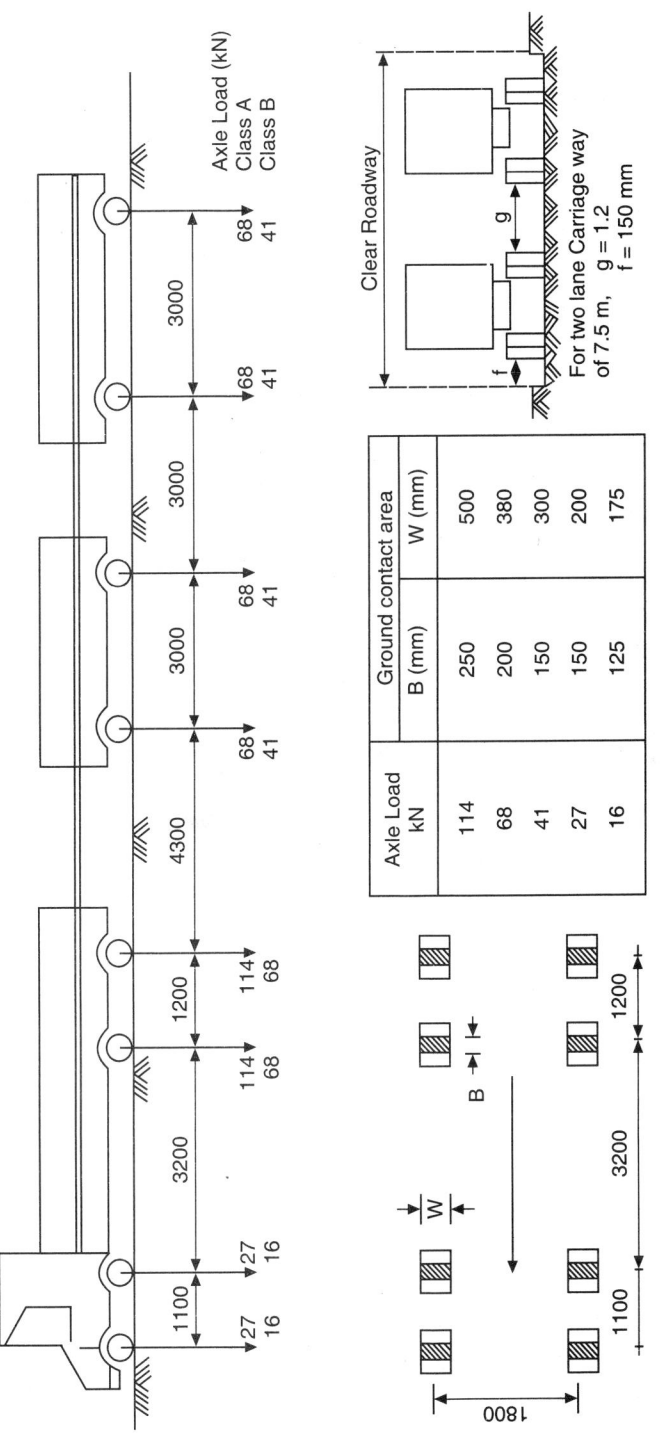

Fig. 2.3 I.R.C. Class A and B Loading Vehicle.

Axle Load kN	Ground contact area	
	B (mm)	W (mm)
114	250	500
68	200	380
41	150	300
27	150	200
16	125	175

For two lane Carriage way
of 7.5 m, g = 1.2
f = 150 mm

Axle Load (kN)	Class A	Class B

The IRC Code also provides for the reduction of the longitudinal effects on bridges accommodating more than two traffic lanes due to the low probability of all lanes not subjected to the characteristic loads simultaneously. The reduction in longitudinal effect recommended is 10 percent for three lanes and 20 percent for four lanes or more. However, it should be ensured that the reduced longitudinal effects are not less severe than the longitudinal effect resulting from simultaneous load on two adjacent lanes.

Table 2.1 Carraige Way and Live Load Combinations
(Table 2 of IRC: 6-2014)

Sl. No.	Carriage way width	Number of lanes for design purposes	Load combination
1	Less than 5.3 m	1	One lane of class A considered to occupy 2.3 m. The remaining width of carriage way shall be loaded with 5 kN/m²
2	5.3 m and above but less than 9.6 m	2	One lane of Class 70 R or Two lanes of Class A
3	9.6 m and above but less than 13.1 m	3	One lane of Class 70 R for every two lanes with one lane of Class A on the remaining lane or 3 lanes of Class A
4	13.1 m and above but less than 16.6 mm	4	One lane of Class 70 R for every two lanes with one lane of Class A for the remaining lanes, if any, or one lane of Class A for each lane.
5	16.6 m and above but less than 20.1 m	5	—————————— do ——————————
6	20.1 m and above but less than 23.6 m	6	—————————— do ——————————

Note: The width of the two lane carriage way shall be 7.5 m as per 112.1 of IRC: 5-1998

2.3 HIGHWAY BRIDGE LOADING STANDARDS OF DIFFERENT COUNTRIES

A brief survey of the highway bridge loadings prescribed in some countries based on work of Raina[6] is compiled below:

2.3.1 British Standard Loadings

The British standards prescribe two main types of loadings designated as HA and HB loading. The HA type loading is designated as the normal design loading and consists of

(a) A uniformly distributed loading carrying from 318.6 kN/m for 1 m loaded length(span) to 5.8 kN/m for 900 m loaded length(span) (as per the loading curve shown in Fig 2.4) and a knife edge load of 120 kN per lane which are inclusive of impact. There is no reduction in the intensity of HA loading for up to two lanes of traffic.

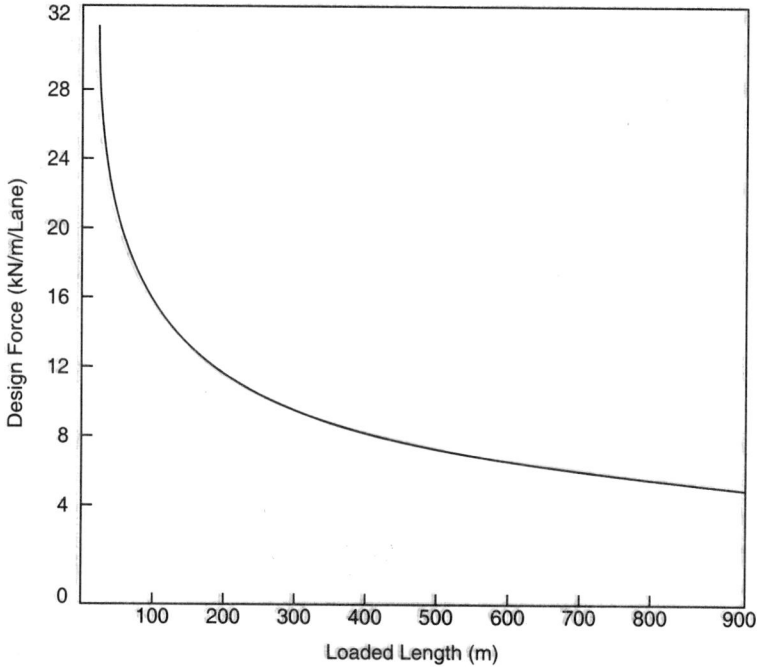

Fig. 2.4 Loading Curve for type HA-Loading (British Standard Loading-UK).

(b) And alternative axle load is also specified on which impact must be considered. This loading system consists of two loads each of 112 kN in line transversely to the direction of traffic flow speed at 0.9 m. The uniformly distributed load has a constant value of 31.5 kN·m of lane for loaded length for 6.5 m to 23 m. For span below 6.5 m, separate curers for the uniformly distributed load are specified. Two lanes are always considered as occupied by full type HA loading while all other lanes in excess of two are considered as occupied by one third of the full lane loading. The standard design lane width is 3 m. In considering the effects of the 112 kN wheel loads, an over stress 25% is permitted. An impact allowance of 25% is specified for this type of HA loading.

(c) Type HB loading is an abnormal unit loading. The number of units per axle (4 axles in all) specified in the UK for bridges carrying the heaviest class of load is 45, amounting to a total load of 1800 kN. This is idealized on four axles which allows for the weight of tractors accompanying trailers. With this loading an overstress of 25% is allowed. No allowance is to be made for impact. Only one lane is to be loaded with HB type loading, all other lanes being considered as occupied by one-third full lane HA loading only if it's presence results in worst effect. The plan view of Type HB loading is shown in Fig 2.5. These loading are followed in Malaysia, Sri Lanka, Kenya and Rhodesia.

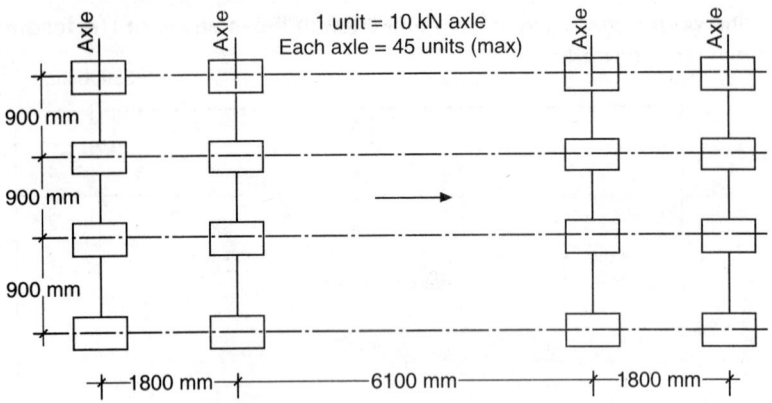

Fig. 2.5 Plan View of HB Loading (British Standard Loading-UK).

2.3.2 AASHTO (American Association of State Highway Transport Officials) Loading

The American Association of Highway and Transportation Officials (AASHTO), has prescribed the heaviest loading designated as HS 20-44. This type of loading is followed in USA, Australia, Bangladesh, Canada, Ethiopia, Philippines and Turkey. This loading comprises of a heavy tractor truck with a semi-trailer of a total load of 320.3 kN or the corresponding lane loading. The lane loading consists of a uniformly distributed load of intensity 9.3 kN/m together with a knife edge load of 80 kN for bending moment and 115.7 kN for shear force computations. In addition, impact effect is to be added for both the cases as recommended by the AASHTO recommendations. While designing bridges, both the truck and lane loading should be considered and the one which gives the worst effect is to be adopted.

When truck loading is used, only one truck is considered for each traffic lane for the whole of its length. Also there is no reduction in load intensity for up to two lanes of traffic loaded.

In USA, Canada and Australia, AASHTO loadings comprise of standard HS 20-44 truck or HS 20-44 lane loading (shown in Fig 2.6) are used.

2.3.3 French Highway Loadings

The highway loadings prescribed in France are categorized as system A and system B loads. System A loading is expressed in an empirical form given by

$$\text{U.d Load A}(L) = \left\{ 230 + \frac{3600}{L+12} \right\} \text{ kg/m}^2$$

Where L = Loaded length in meters.

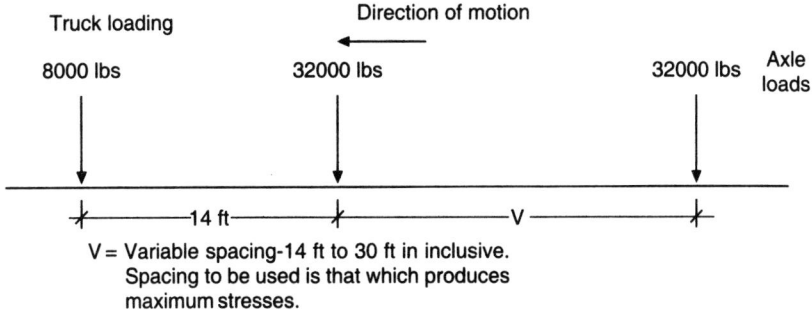

Truck loading

Direction of motion

8000 lbs   32000 lbs   32000 lbs   Axle
                                                  loads

————14 ft————   +————V————+

V = Variable spacing-14 ft to 30 ft in inclusive.
Spacing to be used is that which produces
maximum stresses.

(a) Standard HS 20-44 truck (U.S.A.)

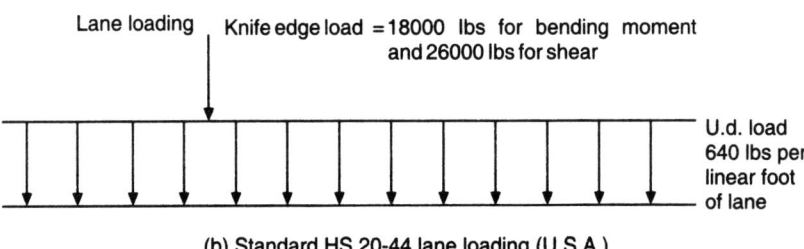

Lane loading   Knife edge load = 18000 lbs for bending moment
                              and 26000 lbs for shear

                                                  U.d. load
                                                  640 lbs per
                                                  linear foot
                                                  of lane

(b) Standard HS 20-44 lane loading (U.S.A.)

Fig. 2.6 AASHTO Loadings (U.S.A.).

The uniformly distributed load A (L) obtained from the above formula is to be multiplied by a coefficient a_1 whose value is 1.0 up to two lanes and then reduces with an increase in the number of lanes. Also $a_1 \cdot$ A (L) should be not less than $(400-0.2\ L)$ kg/m^2.

In the case of class 1 roads, if the lane width is different from the standard lane width of 3.50 m, the value of A (L) is to be multiplied by the coefficient a_2 so as to keep the total load per linear metre of lane unaltered for any loaded length.

System B loads are grouped as B_c and B_t where B_c comprises of truck loading with wheel loads (as shown in Fig 2.7(a)) and B_t is made up of tandem axle loading (as shown in Fig 2.7(b)). System A loading is inclusive of impact. For system B loading, impact factor is expressed as an empirical expression depending on the length of the element in metres, the permanent weight of the bridge and the maximum load of the truck.

2.3.4   Highway Loadings of Germany

Two types of loadings are specified in for highway bridge design in Germany. The Class 60 loading consists of a 600 kN truck together with a uniformly distributed load of 5 kN·m^2 in the portion of the lane not occupied by the truck. The substitute uniformly distributed load for the 60 t truck is 33.3 kN/m^2. The standard design lane width is 3.0 m. There is no reduction in intensity of load for up to two lanes of traffic. For the area outside the main lanes, a uniformly distributed load of 3 kN/m^2 is specified (Refer Fig 2.8). Impact factor is expressed by a formula depending upon the governing length.

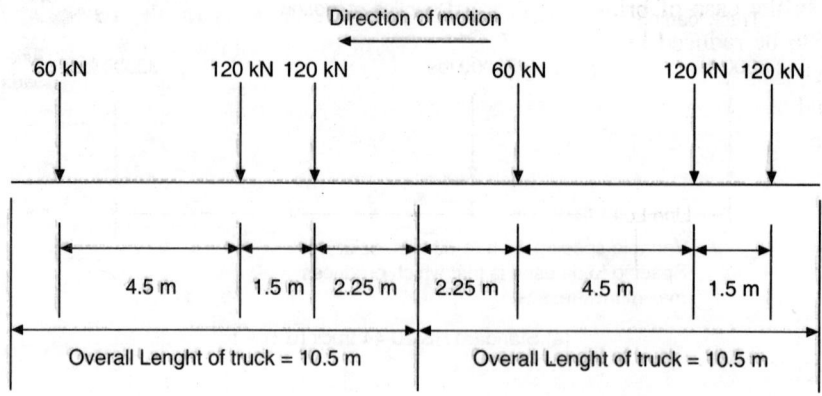

(a) System B_c Truck Loading

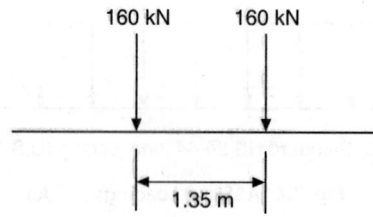

(b) System B_t Tandem Axle Loading

Fig. 2.7 Highway Loadings of France.

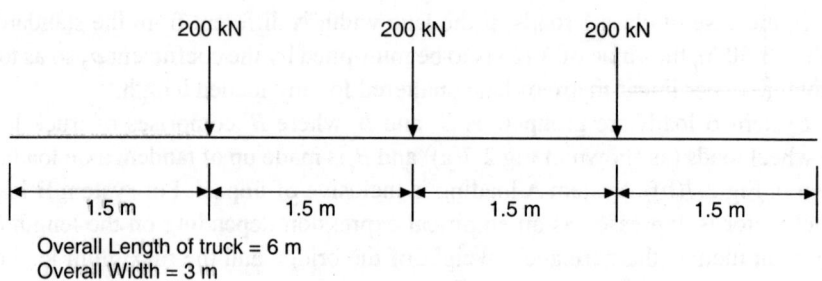

Fig. 2.8 Highway Loadings of Germany.

2.3.5 Highway Loadings of Japan

The L-20 loading consisting of a knife edge line load P of 5000 kg/m and a uniformly distributed load p which has the following values specified for a lane width of 5.5 m or less.

For $L < 80$m, $p = 350$ kg/m^2
For $L > 80$m, $p = (430 - L) \geq 300$ kg/m^2

In the case of bridges with a width of more than 5.5 m, the values of P and p are to be reduced by one-half on the portion of the road way in excess of 5.5 m. The composition of L-20 loading used in Japan is shown in Fig 2.9. Impact factor depending upon the length of the element expressed as an empirical formula should be applied for the live loads.

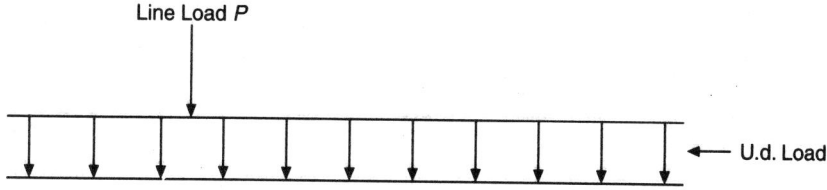

Line Load P

U.d. Load

Fig. 2.9 Highway Loadings of Japan.

2.3.6 Highway Loadings of New Zealand

The design load prescribed per lane is the same as that of AASHTO loading comprising HS20-44 truck. Alternately H20-S16-T16 truck loading (Fig 2.10) is also specified. Between the two, the loading which gives the worst effects should be considered. The standard design lane is 3 m. The impact allowance is the same as that given by the AASHTO without specifying any upper limit.

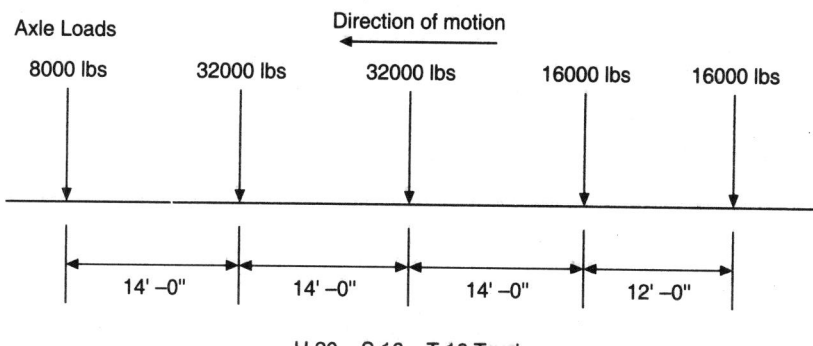

Axle Loads

Direction of motion

8000 lbs   32000 lbs   32000 lbs   16000 lbs   16000 lbs

14' –0"   14' –0"   14' –0"   12' –0"

H 20 – S 16 – T 16 Truck

Fig. 2.10 Highway Loadings of New Zealand.

2.3.7 Highway Loadings of Sweden

Two different types of highway loading are specified in Sweden. The first type consists of uniformly distributed load of 'p' t/m which depends on the loaded length and a concentrated axle load of 14 t (Fig 2.11(a)). The values of p are specified as follows:

$p = 2.4$ t/m for $L < 10$ m

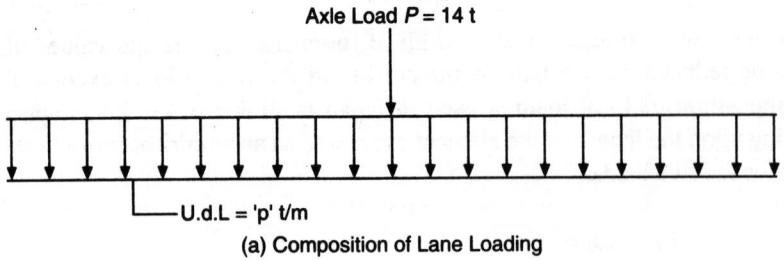

(a) Composition of Lane Loading

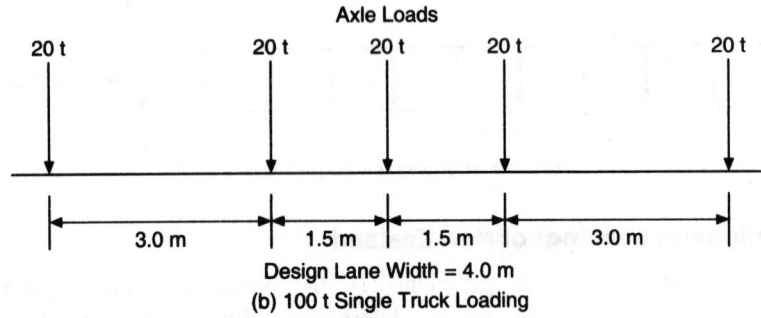

(b) 100 t Single Truck Loading

Fig. 2.11 Highway Loadings of Sweden.

$$p = 2.4 \; \frac{1.3\,(L - 10)}{80} \; \text{t/m for } 10 < L < 90 \text{ m}$$

$p = 1.1$ t/m for $L > 90$ m

Where L = loaded length in metres.

Design Lane width = 3.0 m

The second type comprises of single truck loading with axles loads spaced as shown in Fig 2.11(b). Two lane bridges are designed with lane loading in both the lanes or only with single truck loading which gives the worst results. For continuous structures there is a separate loading consisting of two axle loads and a uniformly distributed load.

2.3.8 Highway Loadings of Austria

Austrian loadings consist of tracked and truck loadings. The tracked loading with the overall length of the caterpillar being 6 m and overall width 3 m is 60 t as shown in Fig. 2.12(a). No allowance for impact is to be made for uniformly distributed load. The truck loading with axle loads is shown in Fig 2.12(b). The axle loads may be increased by 40% for impact effects. A uniformly distributed load of 0.5 t/m² is prescribed for the portion of the lane not occupied by the truck. The specifications also give the following equivalent weights of the caterpillar and truck loading which are to be used for the design of spans exceeding 30 m.

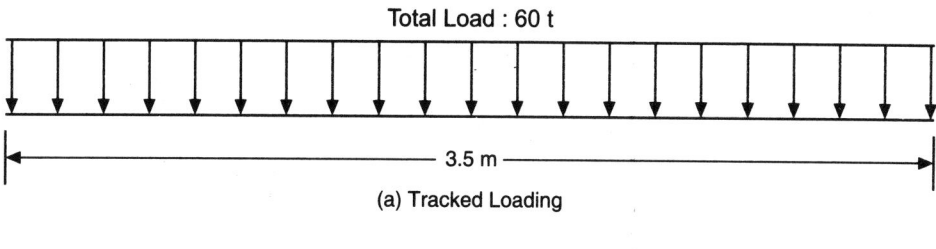

(a) Tracked Loading

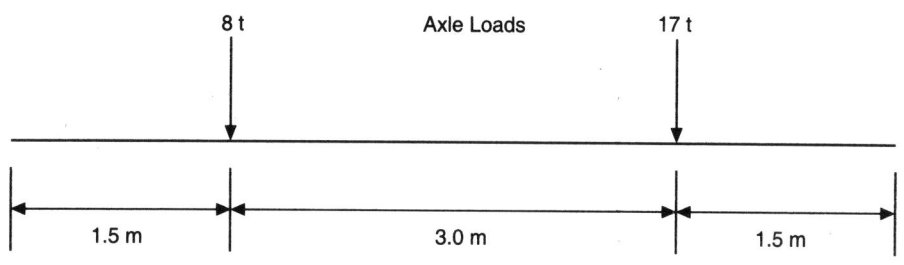

Overall Length of Truck = 6.0 m
Overall Width = 2.5 m

(b) Tracked Loading

Fig. 2.12 Highway Loadings of Austria.

$$60 \text{ t caterpillar} \text{ ——— } 3.33\text{t/m}^2$$
$$25 \text{ t truck} \text{ ——— } 1.67 \text{ t/m}^2$$

Impact allowances varying from zero to 40% corresponding to spans of 70 m to zero respectively. Separate impact factors are recommended for concrete and steel bridges as a function of the span and direct and indirect loads on the main girder.

2.3.9 Highway Loadings of Belgium

Loadings specified in Belgium are categorized as normal and heavy truck loading. The normal truck load of 32 t can have different combinations as shown in Fig 2.13(a). Heavy truck loading of 60 t comprising 3 axles is shown in Fig 2.13(b). Impact factor is expressed by a formula depending upon span, speed of vehicle, moving loads and dead weight and static deflection due to dead weight.

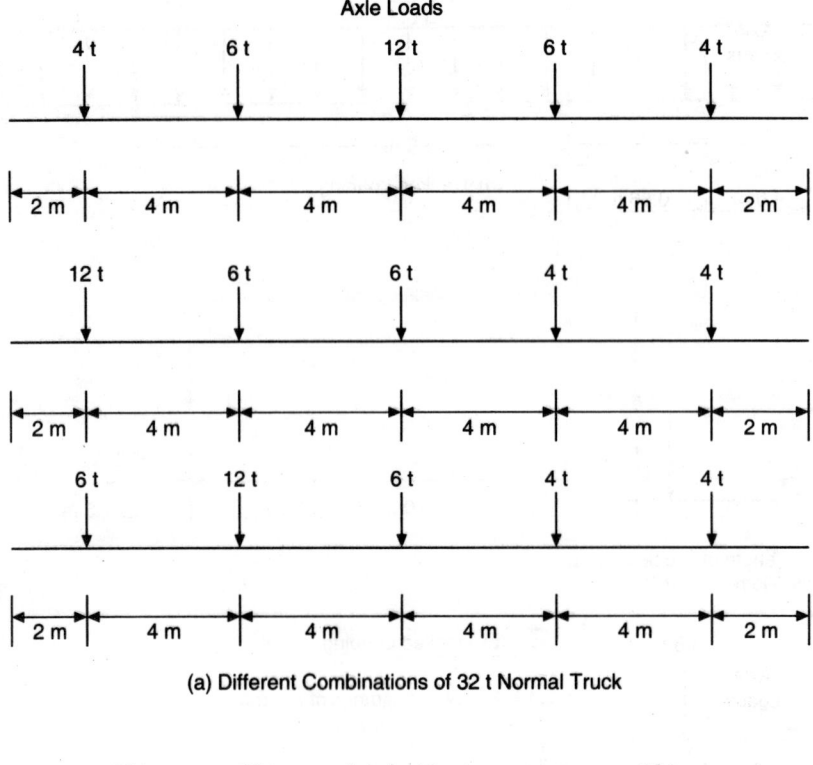

(a) Different Combinations of 32 t Normal Truck

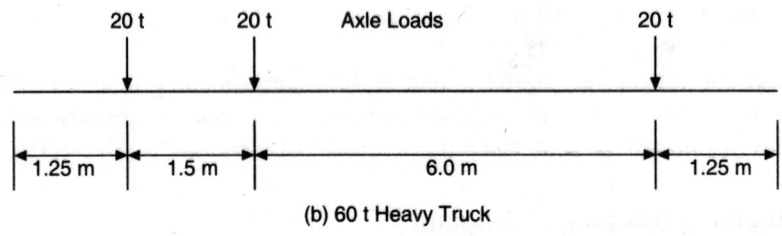

(b) 60 t Heavy Truck

Fig. 2.13 Highway Loadings of Belgium.

2.3.10 Highway Loadings of Italy

Highway bridges of Category 1 should be designed for different trains of military loads prescribed in the standards. Three different types of axles (Fig 2.14) are to be used and the worst effect should be taken for design. The total load of trains consists of 32 t, 61.5 t and 74.5 t. The width of the three types of loadings is 3.5 m. Alternately, an equivalent uniformly distributed load having different values for bending moment and shear is also specified for the design of highway bridges. Impact factor specified by a formula depend upon the span length of the bridge.

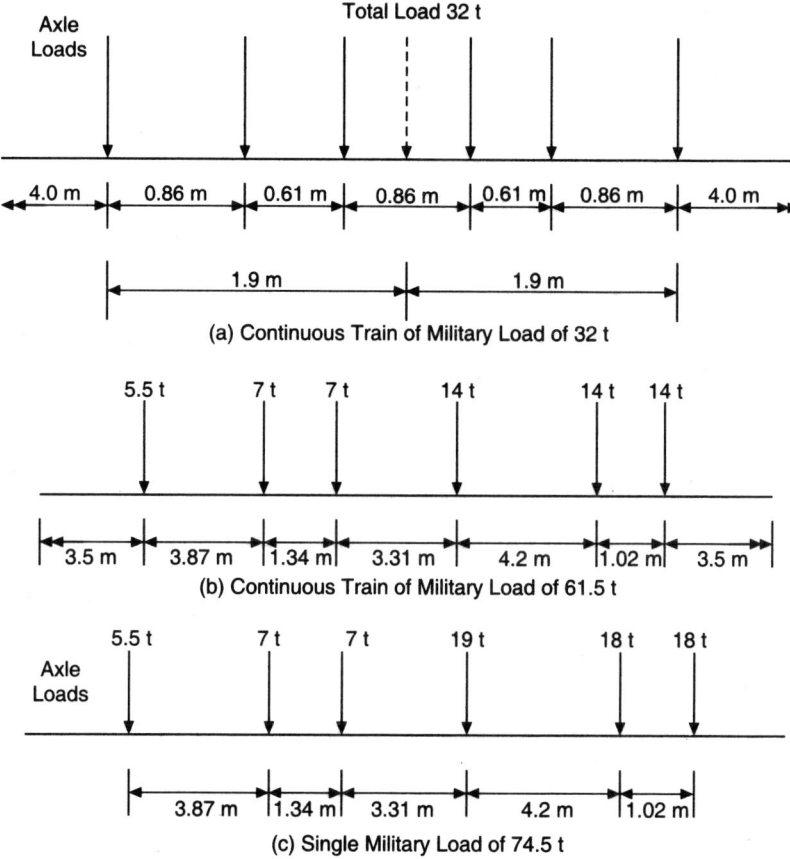

Fig. 2.14 Highway Loadings of Italy.

2.3.11 Highway Loadings of Netherlands

The highest class of loading in Netherlands is the Class 60 loading consisting of a 600 kN vehicle of three axles of 200 kN each plus a uniformly distributed load of 4 kN/m² (Fig 2.15). The impact factor depending upon the span length is expressed by an empirical formula depending upon the span length.

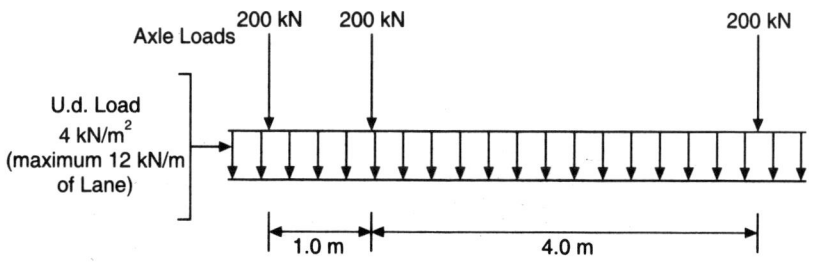

Fig. 2.15 Highway Loadings of Netherlands.

2.3.12 Highway Loadings of Norway

For the design of Class 1 type bridges, the loadings specified consists of an equivalent lane loading per lane together with a knife edge load 'A' and a uniformly distributed load 'p' (Fig 2.16). The variables 'A' and 'p' are expressed as

$$A = 12 + \left(\frac{8x}{L}\right) t$$

$$p = 0.5 + \left(\frac{35}{L+5}\right) t/m \text{ of lane}$$

L = Actual loaded length of lane in metres.
x = distance of the knife edge load from the center of span

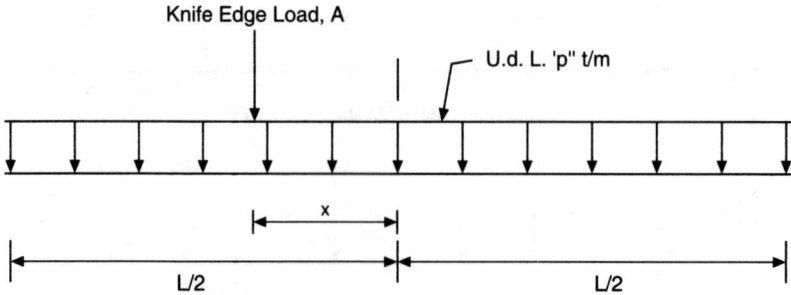

Fig. 2.16 Highway Loadings of Norway.

The loadings specified above are normally considered for lane widths varying from 3.0 to 3.75 m. For two lane bridges, the full equivalent loading is assured on both the lanes. Besides the lane loading, the structure is designed for a local loading of two axles each of 13ft. It is assumed that 38.5% impact is to be added to the heaviest axle and it is unnecessary to add any impact to the remaining axles. The values of knife edge load, 'A' and uniformly distributed load 'p' are inclusive of impact calculated on this basis. An impact of 38.5% is to be added to the axle loads.

2.3.13 Highway Loadings of USSR

Three different types of wheeled vehicle loadings are specified in USSR (Fig 2.17 a, b, c). These are designated as N-30, N-10 and NK-80 loading. Also a tracked loading designated as NG-60 caterpillar loading is specified (Fig 2.17d). The following conditions are also prescribed:

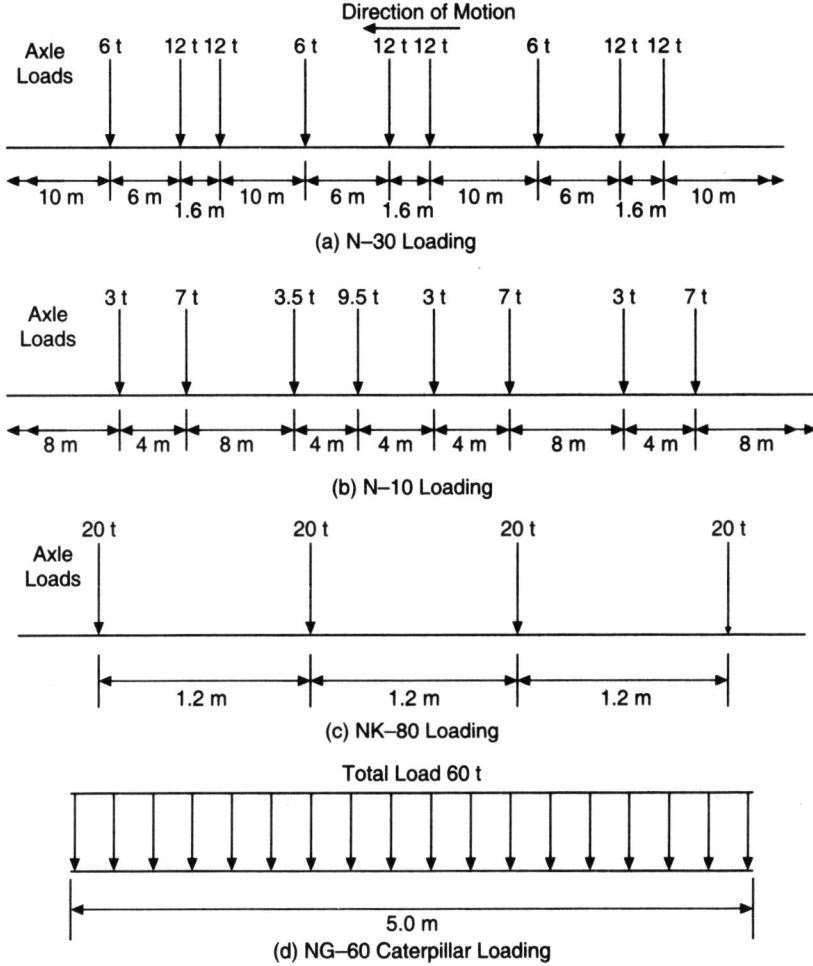

Fig. 2.17 Highway Loadings of USSR.

(i) The design load should be either (a) or (b) whichever produces the maximum effect.

(ii) Impact allowance as per AASHTO specification

(iii) For more than two lanes loaded, reduction factor on live load effect as per AASHTO specification.

2.3.14 Highway Loadings of Saudi Arabia

Highway bridge loadings of Saudi Arabia comprise of both heavy truck wheeled vehicle loading and uniformly distributed load. The truck of three axle loads comprise of 80, 260 and 260 kN; totaling 600 kN with only one truck allowed per lane in the longitudinal direction.

The uniformly distributed load is of intensity 20 kN/m per lane width with a concentrated live load per lane width of 150 kN for moment computation and 220 kN for shear force calculations. The axle loads and uniformly distributed load are shown in Fig 2.18(a) and (b) respectively.

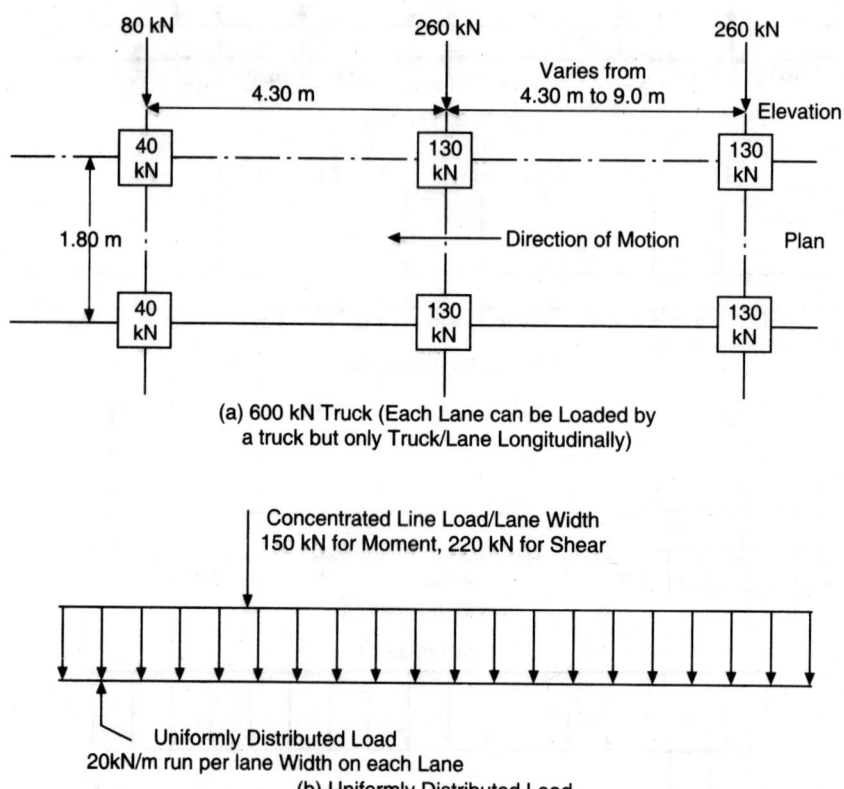

(a) 600 kN Truck (Each Lane can be Loaded by a truck but only Truck/Lane Longitudinally)

(b) Uniformly Distributed Load

Fig. 2.18 Highway Loadings of Saudi Arabia.

2.4 IMPACT FACTORS

Impact factors are generally applied to the moving wheel or distributed loads to enhance their magnitude to include their dynamic effects on the bridge deck. The impact allowance is generally expressed as a fraction of the applied live load and is expressed as an empirical expression involving constants and the span length of the bridge deck. The impact factor is always inversely proportional to the length of the span and is different for reinforced concrete and steel bridges.

The impact factors to be considered for different types of live loads of various countries are as follows:

2.4.1 Indian Standard Loadings

(a) IRC Class A loading

The impact allowance is expressed as a fraction of the applied load and is computed by the expression

$$I = \frac{A}{(B+L)} \text{ Where}$$

$I =$ Impact factor fraction

$A =$ Constant having the value of 4.5 for reinforced concrete bridges and 9.0 for steel bridges.

$B =$ constant having a value of 6.0 for reinforced concrete bridges and 13.5 for steel bridges.

$L =$ Span in metres

For spans less than 3 m, the impact factor is 0.5 for R.C. bridges and 0.545 for steel bridges. When the span exceeds 45 m, the impact factor is taken is 0.088 for R.C. bridges and 0.154 for steel bridges. The impact percentages for highway bridges can also be directly obtained from the curves of IRC code[5] given in Fig. 2.19.

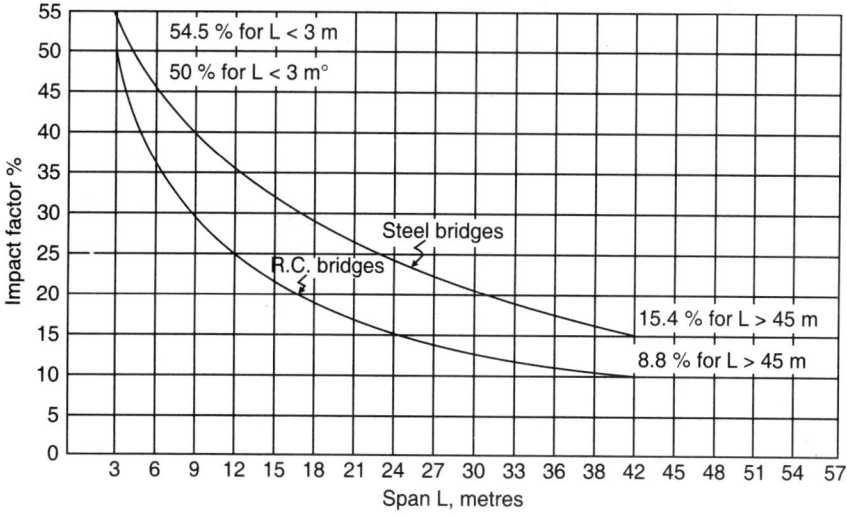

Fig. 2.19 Impact Percentage for Highway Bridges (IRC: 6–2000).

(b) IRC Class AA or 70 R loadings

For spans less than 9 m

(i) For tracked vehicle – 25% of the span up to 5 m linearly reduced to 10% for span of 9 m

(ii) For wheeled vehicle – 25%

For spans of 9 m and more,

For tracked vehicle – for R.C. Bridges, 10% up to a span of 40 m and in accordance with Fig 2.19 for spans exceeding 40 m. For steel bridges, 25% for spans up to 23 m and as per Fig 2.19 for spans exceeding 23 m.

2.4.2 British Standard Loadings

(a) For type HA loadings, an impact allowance of 25% on the heaviest axle in the train of vehicles from which the loading has been derived is provided.
(b) For type HB loading, no impact allowance has been provided since this type is considered as an abnormal loading.

2.4.3 AASHTO Loadings

$$\text{Impact allowance} = \frac{50}{L+25},$$

where L = Length in feet of the portion of the span to produce the maximum stress in the member subject to a maximum of 30%. For shear due to truck loads, the span (L) is taken as the loaded part of the span from the point being considered to the reaction, except for cantilever arms where the impact allowance is 30%.

2.4.4 French Highway Loadings

System A loading is inclusive of impact.
For system B loads, the impact factor '∂' is expressed by the formula,

$$\partial = 1 + \frac{0.4}{(1+0.2\,L)} + \frac{0.6}{1+4(G/S)}$$

Where L = length of the element in metres
G = permanent weight of the bridge
S = maximum load of the truck

2.4.5 German Highway Loading

The impact factor ϕ which has to be applied to the live load is given by the relation

$$\phi = 1.4 - 0.008\,L_\phi \geq 1.0$$

Where L_ϕ = governing length in metres

2.4.6 Highway Loading of Japan

The impact allowance 'I' is determined by the impact formula expressed as,

$$I = \left(\frac{20}{50 + L} \right)$$

Where L = Length of the element in metres.

2.4.7 Highway Loadings of New Zealand

The impact allowance is the same as that specified by the AASHTO but there is however no upper limit is prescribed.

2.4.8 Highway Loadings of Sweden

An impact allowance of 40% is to be applied for the axle loads. However no Allowance for impact is to be made for the uniformly distributed load and for the single truck loading.

2.4.9 Highway Loadings of Austria

The following impact factors are specified for concrete and steel bridges.

(i) Concrete bridges

Impact Factor for Different Spans

Span Length (m)	0	10	30	50	70
Direct Loaded Main Girder	1.40	1.30	1.20	1.10	1.00
Indirect Loaded Main Girder	1.40	1.25	1.10	1.00	1.00

Impact Factor for Floor Slab = 1.4

(ii) Steel bridges

Impact Factor for Different Spans

Span length (m)	2	6	10	20	40	60	80	100
Lane – I	1.64	1.61	1.30	1.18	1.10	1.07	1.05	1.04
Lane – II	1.32	1.20	1.15	1.09	1.05	1.03	1.02	1.02

2.4.10 Highway Loadings of Belgium

The impact factor is given by the relation

$$\phi = 1 + \frac{0.377^{v}}{\sqrt{L/\alpha}} \sqrt{1 + \frac{2Q}{P}}$$

Where v = speed in km/hour (always greater than 60)

L = span length in metres

α = (L/f_s)

f_s = static deflection in metres due to dead weight

Q = moving loads on the bridge deck in tones

P = dead weight of the bridge in tones

2.4.11 Highway Loadings of Italy

The impact factor ϕ expressed as a function of span length L is given by the Formula,

$$\phi = 1 + \frac{(100-L)^2}{100(250-L)}$$

This relation is applicable for spans up to 100 m.

For spans exceeding 100 m, the value of ϕ is unity.

2.4.12 Highway Loadings of Netherlands

The impact coefficient 'S' for bridges carrying normal traffic given by the formula,

$$S = 1 + \frac{40}{(100+L)}$$

Where L is the span in metres.

2.4.13 Highway Loadings of Norway

Impact factor of 38.5% is prescribed for the heaviest axle load and it is unnecessary to add any impact to the remaining axles. The value of the knife edge load and the uniformly distributed load are inclusive of impact calculated on this basis.

2.4.14 Highway Loadings of Saudi Arabia

The impact allowance is similar to that specified in AASHTO specifications. For more than two lanes loaded, a reduction factor on live load effect as per AASHTO specifications should be applied.

2.5 COMPARATIVE ANALYSIS OF HIGHWAY LOADING STANDARDS

(a) Type of Loads

A critical comparative analysis of highway bridge loadings prescribed in different countries has been presented by Thomas[7]. Many countries specify the same uniformly distributed load for flexure and shear except Italy which does not specify knife edge load with uniformly distributed load. Many countries specify uniformly distributed loads for the full width of the traffic lane while Sweden specifies it as a strip load in two strips of 0.6 m each running for the entire loaded length. Alternatively Sweden allows the uniformly distributed load to be applied uniformly over a width of 2.4 m. except the H.A type group, all countries which have an equivalent uniformly distributed load system, have at least an alternative truck loading which has to be considered in the design. Designs based on H.A type loading should generally be checked for H.B type loading.

In Italy separate Civil and Military loadings are specified and prominent bridges are invariably designed for the heavier military loading. Many of the countries do not explicitly specify military loadings, however IRC Class 70 R of India, the caterpillar of Austria and the NK-80 loading of Russia are based on the loading of military vehicles.

(b) Lane Width

The lane width in most countries is in the range of 2.5 to 4.0 m with 3 m as the most common value. However Japan prescribes a wider lane width of 5.5 m. In countries like India, Pakistan and Russia, where there is no standard lane loading; only the minimum widths of carriage way for different number of lanes are specified.

(c) Impact Allowance

In most of the countries, impact is related to the loaded length (span length) although the exact relationship varies considerably from country to country. Some of the countries like India and Austria specify different impact factors for concrete and steel bridges. Higher impact factors specified for steel bridges are due to the lighter structures being subjected to a larger dynamic effect.

In countries like Italy and West Germany, impact is ignored for span lengths exceeding 100 and 50 m respectively. In comparison, countries like India and Australia specify certain minimum values of impact. Belgium and France relate the impact factor to the dead load of the bridge. The impact formula of Belgium is further complicated by the inclusion of speed of the vehicle on the bridge deck.

(d) Magnitude of Loads, Bending Moment and Shear Forces

A comparative analysis of loadings specified in different countries has been reported by Thomas[7]. British loading of Type HB is the most severe followed by IRC loadings for both single and two lane simply supported spans. The loadings of different countries vary considerably both qualitatively and quantitatively.

Raina[6] has prescribed a critical comparison of bending moments and shear forces developed in simply supported bridge decks due to the live loads prescribed in various countries for spans ranging from 5 to 100 m. Tables 2.2 to 2.5 show the variation of maximum bending moment and shear force (inclusive of impact) for salient values of simply supported predominant spans of 5, 25, 50, 75 and 100 m.

Based on the comparative analysis, the following general observations are noteworthy.

(1) The maximum bending moments for a single lane loading occurs due to the British loadings of HB type for the span range of 5 to 100 m. However the

maximum shear force develops for HB type loading for spans from 5 to 75 m and IRC loadings yield the highest shear force for 100 m span.

(2) For two lanes, the West German Class 60 loading results in the maximum bending moment and shear force for all the spans from 5 to 100 m.

(3) For both single and double lanes, the AASHTO loading gives the least bending moment and shear force for all spans with magnitudes nearly half that given by the German loading.

(4) The British type HA and French loadings are almost similar in effect for spans up to about 50 m and beyond that, the latter yields slightly higher values of moments and shear forces for both single and double lanes.

(5) The New Zealand loading is marginally heavier than that of AASHTO loading for spans up to 50 m, beyond that, it gives the same values of bending moment, but in the case of shear force, it results in higher values for all spans.

The comparative analysis indicates the wide qualitative and quantitative variation in the loadings of different countries. Analysis indicates that the equivalent uniformly distributed load system appears to be the most acceptable since it is simpler for application. Due to this factor many countries have adopted the HA and AASHTO loadings for the design of bridge decks.

A critical survey of the impact factors specified by the various countries indicates the basic differences in their approach in assessing the dynamic effect of live loads on the bridge deck. Raina[6] has indicated that the impact formula prescribed by some countries is unnecessarily complicated since the effect of live load on the bridge deck is comparatively less than that of dead load for span lengths exceeding 25 m. However there is need for qualitative research in this field to investigate the behavior of the bridge structure under dynamic loads and the resulting data will be helpful in evolving rational design procedures. From considerations of simplicity of loading and ease of its application in design, type HA loading appears to be the most favorable among the various load systems.

There is significant variation in the type of highway loads specified in the standards of various countries. Raina[6] has made a critical analysis of the bending moments and shear forces developed due to the standard loadings specified by the various countries for single and double lane traffic, covering spans in the range of 5 to 100 m as shown in Tables 2.2 to 2.5. For the range of spans covered, the extreme variations in the bending moment and shear forces developed due to AASHTO loadings is only half of that due to German loadings. The loadings of the various other countries generally lie in between the American and German loadings. In view of these wide variations, there is a need for a systematic survey of vehicular loads on bridges for rationalization of Highway Bridge loading standards.

Table 2.2 Maximum Bending Moments for One Lane Simply Supported Spans of Bridge Decks due to Loads of Various Countries

Maximum Bending Moment Inclusive of Impact (kN·m)

Span (m)	UK Type H.A Loads	UK Type H.B Loads	India I.R.C. Loads	Germany Class-60 Loads	Japan L-20 Loads	Sweden	France	North America A- ASHTO HS:20-44 Loads	New Zealand
5	243	756	687	672	551	450	390	231	231
25	3156	7862	5680	4952	4083	5050	3290	2022	2485
50	8656	19029	12496	10830	11344	11300	8870	4597	5738
75	15117	30251	19683	19877	21682	17550	15848	9155	9155
100	22875	41487	33580	31268	33494	23800	24106	15184	15184

Table 2.3 Maximum Shear Forces for One Lane Simply Supported Spans of Bridge Decks due to Loads of Various Countries

Maximum Shear Force Inclusive of Impact (kN)

Span (m)	UK Type H.A Loads	UK · Type H.B Loads	India I.R.C. Loads	Germany Class-60 Loads	Japan L-20 Loads	Sweden	France	North America A- ASHTO HS:20-44 Loads	New Zealand
5	199	738	549	571	441	420	359	212	212
25	505	1451	978	807	653	820	569	369	447
50	693	1625	1090	872	908	910	710	410	501
75	806	1684	1477	1064	1156	940	845	529	606
100	915	1713	1776	1254	1340	955	964	647	758

Table 2.4 Maximum Bending Moments for Two Lane Simply Supported Spans of Bridge Decks due to Loads of Various Countries

Maximum Bending Moment Inclusive of Impact (kN·m)

Span (m)	UK Type H.A Loads	UK Type H.B Loads	India I.R.C. Loads	Germany Class-60 Loads	Japan L-20 Loads	Sweden	France	North America A- ASHTO HS:20-44 Loads	New Zealand
5	488	838	687	1224	827	640	780	462	462
25	6312	8914	5680	9904	6125	5819	6580	4044	4970
50	17132	21914	12496	21660	17016	15838	17740	9194	11476
75	30234	35290	23486	39754	32523	26250	31696	18310	18310
100	45750	49112	42880	62536	50241	37300	48212	30368	30368

Table 2.5 Maximum Shear Forces for Two Lane Simply Supported Spans of Bridge Decks due to Loads of Various Countries

Maximum Shear Force Inclusive of Impact (kN)

Span (m)	UK Type H.A Loads	UK Type H.B Loads	India I.R.C. Loads	Germany Class-60 Loads	Japan L-20 Loads	Sweden	France	North America A-ASHTO HS:20-44 Loads	New Zealand
5	398	804	594	1142	661	512	718	424	424
25	1010	1619	978	1614	980	931	1138	738	894
50	1386	1856	1174	1744	1361	1267	1420	820	1002
75	1612	1952	1572	2128	1735	1400	1690	1058	1212
100	1830	2018	1996	2508	2010	1492	1928	1294	1516

2.6 INDIAN RAILWAY BRIDGE LOADING STANDARDS

Railway bridge loadings[8,9] should conform to the specifications of the Indian Railway Standards (IRS) prescribed by the Ministry of Railways, Government of India. The various loads to be used are specified in the IRS Bridge rules. Specific recommendations are available for the design of steel. R.C.C, P.S.C, masonry and plain concrete arch bridges in the relevant bridge codes.

The railway tracks are classified according to the importance of traffic as main and branch lines. The three types of gauges used in the Indian Railways are.

(1) Broad gauge (BG): 1676 mm (5′-6″)
(2) Metre gauge (MG): 1000 mm (3′-3.375″)
(3) Narrow gauge (NG): 762 mm (2′-6″)

At present, the Indian Railways have adopted the unigauge policy with the broad gauge as the standard gauge throughout the country. Consequently many important old lines are being converted into broad gauge.

The various loads and forces to be considered in the design of bridge members are:

(1) Dead and live loads
(2) Dynamic effects
(3) Cetrifugal force due to curvature of track
(4) Temperature and frictional effects
(5) Racking force
(6) Wind and earthquake forces

IRS Bridge Rules recommends the use of equivalent uniformly distributed loads (EUDL) on each track and also the coefficient of dynamic augment (CDA) for spans varying from 1 to 130 m for both BG and MG as shown in Tables 2.6 and 2.7.

The equivalent loads specified for the computation of bending moment and shear forces can directly be used in place of the various wheel loads of the rolling stock.

Hence except in the case of special bridges like the Rigid Frame, balanced cantilever and Suspension bridges, the designer can directly use the equivalent loads in place of the basic wheel loads. The Impact Factors (CDA) listed in the Tables are for single track spans of BG and MG based on the relation:

$$CDA = 0.15 + \frac{8}{(6+L)} < 1.0 \qquad \text{where } L = \text{span}$$

For main girders of double track spans, the value specified above is multiplied by a reduction factor of 0.72.

Bridges located in the seismic zones have to be designed to resist the stresses produced due to seismic effects conforming to the recommendations in the Indian Standard Code IS: 1893–2002[10].

Table 2.6 E.U.D.L., C.D.A., Tractive Force and Braking for Modified B.G. Loading

Span (m)	Total E.U.D.L. for B.M. (kN)	Total E.U.D.L. for S.F. (kN)	C.D.A. (I.F.)	Tractive Effort (kN)	Braking Force (kN)
1	490	490	1.000	81	62
2	490	519	1.000	164	123
3	490	662	1.000	245	184
4	596	778	0.950	245	184
5	741	888	0.877	245	184
6	838	985	0.817	245	185
7	911	1068	0.765	327	221
8	981	1154	0.721	409	276
9	1040	1265	0.683	409	276
10	1101	1377	0.650	490	331
12	1377	1589	0.594	490	331
15	1631	1801	0.531	490	368
20	1964	2168	0.458	735	496
25	2356	2586	0.408	735	565
30	2727	2997	0.372	981	662
40	3498	3815	0.324	981	816
50	4253	4630	0.293	981	978
60	5051	5442	0.271	981	1140
70	5831	6254	0.255	981	1301
80	6603	7065	0.243	981	1463
90	7391	7876	0.233	981	1625
100	8201	8686	0.225	981	1787
110	9011	9496	0.219	981	1949
120	9820	10306	0.213	981	2110
130	10630	11115	0.209	981	2272

Bridges planned in the coastal areas have to be designed to withstand the effect of wind pressure. The basic wind pressure is to be obtained from the meteorological records or from the Indian Standard Code IS: 875-1987[11].

Table 2.7 E.U.D.L., C.D.A., Tractive Force and Braking for Modified M.G. Loading

Span (m)	Total E.U.D.L. for B.M. (kN)	Total E.U.D.L. for S.F. (kN)	C.D.A. (I.F.)	Tractive Effort (kN)	Braking Force (kN)
1	314	314	1.000	89	57
2	314	365	1.000	118	78
3	326	452	1.000	118	118
4	429	536	0.950	157	118
5	501	616	0.877	157	124
6	581	685	0.817	157	124
7	644	755	0.765	176	135
8	714	819	0.721	209	157
9	774	871	0.683	262	169
10	828	934	0.650	262	198
12	953	1061	0.594	314	235
15	1138	0.531	0.531	353	253
20	1421	1532	0.458	471	353
25	1677	1833	0.408	523	401
30	1991	2144	0.372	628	486
40	2589	2748	0.324	628	594
50	3099	3269	0.293	628	702
60	3625	3819	0.271	628	810
70	4178	4372	0.255	628	918
80	4727	4922	0.243	628	1026
90	5274	5470	0.233	628	1134
100	5822	6017	0.225	628	1242
110	6365	6562	0.219	628	1349
120	6908	7106	0.213	628	1457
130	7451	7649	0.209	628	1565

REFERENCES

1. Rowe, R.E., Concrete Bridge design (CR Books, London, 1962), John Wiley & Sons, New York, 1963.
2. Seni, A., Comparison of Live Loads used in High Way Bridge Design in North America with those used in Western Europe, Second International Symposium on Concrete Bridge Design, Chicago, April 1969, American Concrete Institute, Detroit, 1971, pp.1–34.
3. Rajagopalan, K.S., Comparison of Loads around the World for Design of High Way Bridges, International Symposium on Concrete Bridge Design, Chicago, April 1969, American Concrete Institute, Detroit, 1971, pp. 35–48.
4. Galambos, C.F., International Road Federation in depth study on Fatigue, Fracture and Stress Corrosion problems of High Way Bridges, World Survey. Of Current Research and Development on Roads and Road transport, International Road Federation, Washington D.C, 1972, pp. 325–365.

5. I.R.C: 6 – 2014, Standard Specifications and Code of Practice for Road Bridges, Section II, Loads and Stresses (Fourth Revision), Indian Roads Congress, New Delhi, 2000, pp. 1–84.

6. Raina, V.K., Concrete Bridge Practice, Analysis, Design and Economics, Tata McGraw Hill Publishing Co. Ltd., New Delhi, 1991, pp. 9–25.

7. Thomas, P.K., A Comparative Study of High Way Bridge Loadings in Different Countries, UK. Transport & Road Research Laboratory, Supplementary Report, 135 UC, Crowthorne, 1975, 47 pp.

8. Indian Railway Standard Code of Practice for the Design of Steel and Wrought Iron bridges carrying rail, road or pedestrian traffic. Government of India, Ministry of Railways, 1962, 87 pp.

9. Johnson Victor, D., Essentials of Bridge Engineering (Fifth Edition), Oxford & IBH Publishing Co. Pvt. Ltd., New Delhi, 2001, pp. 1–10.

10. IS: 1893-2002, Indian Standard Criteria for Earthquake Resistant design of Structures, Bureau of Indian Standards, New Delhi, 1986, 77 pp.

11. IS: 875-(Part-3)–1987, Code of Practice for Design Loads (Other than Earthquake) for Buildings and Structures, Part-3, Wind Loads (Second Revision) Bureau of Indian Standards, New Delhi, 1989.

REVIEW QUESTIONS

1. Briefly outline the basis on which the highway loading standards have been evolved for use in the design of highway bridge decks.

2. What are the various types of loadings specified in the Indian Roads Congress standards for the design of highway bridge decks in India?

3. Distinguish between tracked and wheeled vehicles specified in the IRC codes.

4. Explain the difference between IRC class A and B loadings. Specify the types of roads for which these types of loadings are used.

5. Discuss briefly the load characteristics of IRC Class 70 R tracked and wheeled vehicles specified in the IRC codes.

6. Write a brief note on the high way bridge loading standards specified in the leading countries of the world.

7. What are impact factors? In what way these factors vary with respect to the type of loading, span and types of bridges?

8. What are the main conclusions you can draw from a comparative analysis of the force components resulting from the use of highway loading standards of various countries?

9. What type of loads you would recommend for the design of railway bridges? Mention briefly the classification of railway tracks and type of lines according to the IRS bridge rules.

10. What are equivalent loads with reference to the design of railway bridges? Explain the terms a) Equivalent uniformly distributed load, b) Corresponding impact factor, specified in the IRS bridge loading standards for broad gauge and metre gauge tracks on bridges.

OBJECTIVE TYPE QUESTIONS

1. The necessity of formulating specifications of highway bridge loadings started due to the transportation of
   a) Private vehicles
   b) Military vehicles
   c) Public vehicles

2. The first loading standard in India was published by the Indian Roads Congress in the year
   a) 1963
   b) 1966
   c) 1958

3. The tracked vehicle loading specified under IRC Class AA originated due to the loading pattern of
   a) Heavy duty army trucks
   b) Army tanks
   c) Train of army truck loads

4. The worst effect to be considered in the design of bridge decks is due to the passage of
   a) IRC Class B load train
   b) IRC Class A load train
   c) IRC Class tracked vehicle

5. The maximum load transmitted by a single wheel of IRC Class A loading is
   a) 68 kN
   b) 41 kN
   c) 57 kN

6. The impact factor for reinforced concrete bridges specified in IRC code for spans exceeding 45 m is restricted to value of
   a) 25%
   b) 15.4%
   c) 8.8%

7. The impact factor used in the design of bridge decks is always inversely proportional to
   a) The width of bridge deck
   b) The span length
   c) The type of bridge deck

8. The maximum bending moment for one lane simply supported spans of bridge decks inclusive of impact develops due to the loading specified in
   a) India –IRC loads
   b) U.K. – HB type loads
   c) Germany – Class 60 loads
9. Bridges planned in the coastal areas should be designed specifically to resist the effect of
   a) Dead loads
   b) Live loads
   c) Wind loads
10. In the computation of loads in the design of railway bridges, the IRS bridge rules recommends the use of
    a) Wheel loads
    b) Tracked loads
    c) Equivalent uniformly distributed loads

Limit State Design for
Concrete Road Bridges

3.1 BASIC CONCEPTS OF LIMIT STATE DESIGN

The philosophy of limit state design was first introduced by the European Concrete Committee (CEB) in its recommendations for an International code of practice for reinforced concrete in the year 1964[1]. The basic concepts of Limit State Design[2,3] emerged during the early part of 20th century as an advancement over the traditional design philosophies like the elastic theory and ultimate load methods of design. The main deficiency in the working stress or elastic method of design is the lack of information regarding the computation of collapse or ultimate load capacity of reinforced concrete structural elements. The ultimate load method[4,5] developed during the middle of 20th century predicted various procedures to evaluate the collapse or ultimate load capacity of structural elements. A structure designed solely on the basis of ultimate loads, although having a desirable margin of safety against collapse, may not be serviceable due to excessive deflections and /or development of objectionable cracks at working loads. This type of distress is particularly noticeable in structures designed by the ultimate load method using high strength materials.

The limit state design[6] concepts slowly emerged during this period as a solution to overcome the deficiencies of the elastic and ultimate load methods. The limit state design is a method of designing structures based on the statistical concept of safety and the associated probability of failure considering both the states of serviceability and strength. The limit state method of design ensures not only the serviceability of the structure at working loads, but also satisfies the requirement of desirable strength criterion by ensuring that the structure supports the ultimate loads.

The integrated limit state method of design has been accepted worldwide and incorporated in the National Codes of various countries like USA[7], U.K[8], Germany[9], Canada[10] and India[11]. The limit state concepts have been examined thoroughly by various investigators and several research investigations have proved beyond doubt that the method can be universally adopted for the design of reinforced concrete structures. The evolution of the various design methods culminating in the limit state design is illustrated in Fig. 3.1.

1900 – 1930
Elastic Theory-Working Stress Method
1) Factor of safety applied to the yield or ultimate stress to get permissible stress
2) Structure designed to support working Loads without exceeding the permissible Stresses in Concrete and Steel
Inadequacy of the Method – Actual Safety against Ultimate Loads not known

---

1930 – 1960
Ultimate Load or Strength Method
1) Load factors applied to Working Loads to estimate the Ultimate Loads
2) Safety factors applied to Characteristic strengths to estimate design strengths
Inadequacy of the Method – Serviceability Aspects such as deflection and Cracking at service loads are not considered

---

1960 – Date
Limit State Method
1) Structure Designed to satisfy the limit states of a) Strength – Collapse
b) Serviceability – Deflection and Cracking
2) Statistical probability Concepts incorporated for Loads and Strengths
Characteristic loads and Characteristic strengths obtained by applying partial safety factors for loads and material strengths
3) Limit State method overcomes the inadequacies of working stress and Ultimate Load methods
4) Limit State method Incorporated in Most of the National Codes of Various Countries in American, European and Asian Countries

---

Fig. 3.1 Evolution of Limit State Method.

The Indian standard code of Practice for plain and reinforced concrete, first published in the year 1953 and revised in 1964 and 1978 was based on the working stress method. However the code was revised in its fourth revision by incorporating the principles of limit state design in 2000. The Indian Roads congress, entrusted with the responsibility of formulating codes of practice for road bridges, published the standard specifications for reinforced concrete road bridges in the year 1966 (IRC: 21-1966) based on the working stress method. Subsequently the code was revised in 1987 and 2000. The design criteria for prestressed concrete road bridges (IRC: 18-2000) was first published in the year 1965 and revised in the years, 1977, 1985 and 2000 adopting the working stress method.

The latest Indian roads congress code (IRC: 112-2011)[12] fulfills the need for a rationalized code for bridge structures in general, based on the limit state philosophy of design conforming to the unprecedented growth of knowledge in the field of concrete, steel and the methods of analysis and design over the past two decades. The unified code dealing with all types of concrete bridges, replaces the previous codes dealing separately with reinforced (IRC: 21) and prestressed concrete (IRC: 18) bridges. The new code is comprehensive in its recommendations as it specifies the aims of design, reliability aspects, strength, safety, serviceability, durability, design service life and economy.

3.2   APPLICATION OF CLASSICAL RELIABILTY THEORY TO LIMIT STATE DESIGN

Probabilistic concepts are explicitly incorporated in the limit state design by considering the probability density of load and strength variables of the constituent materials. Applications of classical reliability theory[13,14] to structural design require comprehensive statistical data regarding loads and strengths and their exact shapes of normal distribution curves. At present the probabilities of failure that are socially acceptable must be kept very low (1 in a million). At such low levels, the probability of failure is very sensitive to the exact shape of the normal distribution curves. To determine exact shapes of normal distribution curves, we require very large numbers of statistical data and such comprehensive data is not yet available. In particular, sufficient numbers of extreme values of the strengths of complete structures (to define accurately the shapes of the tails of the normal distribution curves) may never be available.

In a simple example, only one type of load and one-strength variables are used. For a real structure, there will in general be many types of loads and many modes of failure, normally with complex correlations between them making it very difficult to calculate the probability of failure. Hence in the limit state design, our engineering experience and judgment have been used to modify and to remedy the inadequacies of earlier design methods and partly use the probabilistic concepts. Hence, it is appropriate to designate the limit state design method currently practiced as Semi probabilistic approach[15] to structural design.

The interaction between load effects and strength is shown in Fig. 3.2 where the normal distribution curves for loads and material strengths are superposed along with the characteristic loads and strengths.

For good design the characteristic loads and strengths are expressed in terms of the standard deviation, mean strength and the probability factor as,

$$F = [F_m + 1.65\,\sigma]$$
$$f_{ck} = [f_m - 1.65\,\sigma]$$

Where F = characteristic load

F_m = mean load

f_{ck} = characteristic strength

f_m = mean strength

σ = standard deviation

3.3 LIMIT STATE CRITERIA

The new bridge code explicitly deals with the various criteria for limit states in section-5 of the code. The bridge or any of its components should be designed to ensure the achievement of an acceptable probability that the specified life of the bridge structure is not curtailed prematurely due to the attainment of an unsatisfactory condition or limit state, which covers the various forms of failure. Several limit states at which the structure ceases to function have been identified. The most prominent among them being the limit state of collapse, excessive deflection and cracking. Each of these limit states may be reached due to different types of loading configurations; However in practice, only a few of these are of primary significance in design. Some of the important criteria as identified by the British code for reinforced concrete structures to attain the limit state of collapse are listed below:

1. Failure of one or more critical sections in flexure, shear, torsion or due to their combinations
2. Bond and anchorage failure of reinforcements
3. Failure of connections between precast and cast in situ elements
4. Failure due to inelastic stability of members
5. Failure due to fatigue due to repeated loads
6. Failure due to corrosive environment and aggressive environmental conditions
7. Failure due to seismic loads on the structure
8. Failure due to fire, impact or explosions

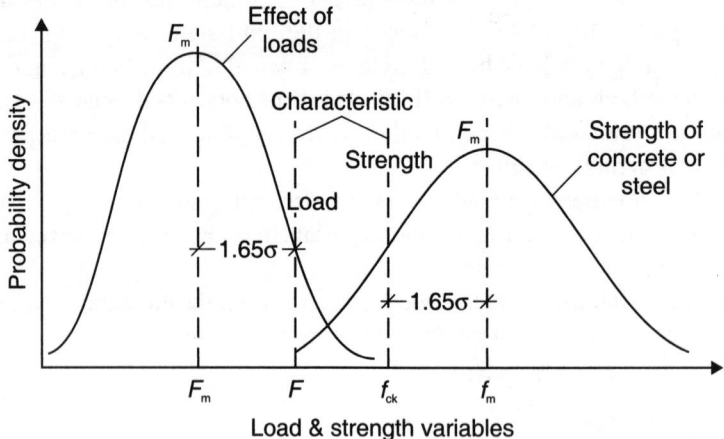

In limit state design

$F = F_m + 1.65\sigma$ ⎤ For good
$f_{ck} = f_m - 1.65\sigma$ ⎦ design

Fig. 3.2 Classical Reliability Model for Strength Design.

The bridge structure may be rendered unfit for its intended purpose due to various serviceability limit states such as:

1. Excessive internal stresses beyond the specified magnitudes stipulated in the codes.
2. Excessive deflection or displacement, adversely affecting the finishes and causing discomfort to the users of the structure.
3. Excessive local damage due to cracking or spalling of concrete which impairs the efficiency or appearance of the structure.
4. Limit state of vibration critical in foot bridges, suspension bridges under dynamic effects of wind loads.

3.4 PARTIAL SAFETY FACTORS

In limit state design, safety is expressed in terms of the probability that the structure will not become unfit for its intended function during its useful life that is the structure will not reach a limit state. The initial idea of referring to a single failure criterion has been replaced by the comprehensive concept of multiple limit states. With this concept the local or the overall behavior in all stages – elastic, cracked, inelastic and ultimate – are considered. In the limit state approach, a structure is considered as well designed if it could be shown that the probability of any limit state being attained is substantially constant for all the component members and for the structure as a whole and that consequently, the latter possesses adequate and uniform structural safety. Due to the number of variables involved, a rational determination of the safety of a structure, based on probability theory is not yet practical in the design office. Partial safety factors are therefore introduced for each limit state and these consist of γ_m, reduction factor for characteristic strength of materials and γ_f, enhancement factors for characteristic loads on the structure.

3.5 CHARACTERISTIC AND DESIGN LOADS

The characteristic load F which is independent of the limit state considered and is seldom exceeded in service is defined as

Characteristic load = [(Mean load) + (k × Standard deviation)]

where k is a factor so chosen as to ensure that the probability of the characteristic load being exceeded is small. A value of 1.64 for k ensures the probability that the characteristic load is exceeded by only five per cent during the intended life of the structure.

The statistical data required to define the characteristic loads for different types of occupancy is not readily available, since loading statistics are invariably difficult to compile as they need systematic observations and recording of data over a long period of time. In the absence of statistical observations and recording of data on loading, the nominal imposed loads provided in IRC: 6-2014[16], may be treated as characteristic loads. The characteristic values of the loads take account of expected variations but do not allow for the following:

1. Possible unusual increases in load beyond those considered in deriving the characteristic.

2. Inaccurate assessment of effects of loading and unforeseen stress distribution within the structure.

3. Variations in dimensional accuracy achieved in construction.

The design loads are obtained by enhancing the characteristic loads by suitable partial safety factors for the various limit states. The values of partial safety factors to be used for verification of equilibrium, structural strength and serviceability states are compiled in Annexure-B of the bridge code IRC: 6-2014.

Design Load = [(Characteristic load) × (Partial Safety factor)]

The basic safety factors prescribed in the code for limit states of strength and serviceability are compiled in Tables 3.1 and 3.2 for different combinations

3.6 CHARACTERISTIC AND DESIGN STRENGTHS

The variation in the properties of concrete and steel are expressed as characteristic values related to the mean values and standard variation.

Characteristic Strength (f) = [Mean strength – k × Standard Deviation]

Where 'k' is a factor chosen to ensure that the probability of the characteristic strength not being exceeded is small. Many of the national codes including the Indian standard code IS: 456-2000 and IRC: 112-2011, has recommended a value of 1.65 for k so that only 5 percent of the test results could have zstrength less than the characteristic strength. In the absence of statistical data, the characteristic strength of concrete and steel may be taken as the works cube strength and minimum proof or yield strength respectively as recommended in the current codes.

Since the materials in the structure are likely to differ in quality from those tested, design strengths are obtained by dividing the characteristic strength by γ_m, the appropriate partial safety factor for the Limit State being considered. The proposed values for the partial safety factors are as given in Table 3.3.

Design Strength = [Characteristic strength/Partial safety factor]

$$f_d = [f / \gamma_m]$$

where γ_m = partial safety factor appropriate to the material and the limit state being considered.

Table 3.1 Partial Safety Factors for Verification of Structural Strength
(Table 3.2 of IRC: 6-2014)

Loads	Ultimate Limit State		
	Basic Combination	Accidental Combination	Seismic Combination
(1)	**(2)**	**(3)**	**(4)**
Permanent Loads: Dead Load, Snow load if present, SIDL except surfacing			
a) Adding to the effect of variable loads	1.35	1.0	1.35
b) Relieving the effect of variable loads	1.0	1.0	1.0
Surfacing:			
Adding to the effect of variable loads	1.75	1.0	1.75
Relieving the effect of variable loads	1.0	1.0	1.0
Prestress and Secondary effect of prestress: (refer note no. 2)			
Back fill Weight	1.50	1.0	1.0
Earth pressure due to Back Fill:			
a) Leading Load	1.50	–	1.0
b) Accompanying Load	1.0	1.0	1.0
Variable Loads: Carriageway Live Load and associated loads (braking, tractive and centrifugal forces) and ***Pedestrian Live Load:***			
a) Leading Load	1.5	0.75	0
b) Accompanying Load	1.15	0.2	0.2
c) Construction Live Load	1.35	1.0	1.0
Wind during service and construction:			
a) Leading Load	1.50	–	–
b) Accompanying Load	0.9	–	–
Live Load Surcharge (as accompanying load)	1.2	0.2	0.2
Erection effects	1.0	1.0	1.0
Accidental Effects: i) Vehicle Collision (or) ii) Barge Impact (or) iii) Impact due to floating bodies }	–	1.0	–
Seismic Effect:			
a) During Service	–	–	1.5
b) During Construction	–	–	0.75
Hydraulic Loads (Accompanying Load):			
Water Current Forces	1.0	1.0	1.0
Wave Pressure	1.0	1.0	1.0
Hydrodynamic Effect	–	–	1.0
Buoyancy	0.15	0.15	0.15

Table 3.2 Partial Safety Factors for Verification of Serviceability Limit State
(Table 3.3 of IRC: 6-2014)

Loads	Rare Combination	Frequent Combination	Quasi-permanent Combination
(1)	(2)	(3)	(4)
Permanent Loads: Dead Load, Snow load if present, SIDL except surfacing	1.0	1.0	1.0
Back fill Weight	1.0	1.0	1.0
Prestress and Secondary effect of prestress (refer note no. 4) Shrinkage and Creep Effects	1.0	1.0	1.0
Earth pressure due to Back Fill	1.0	1.0	1.0
Settlement Effects: a) Adding to the permanent loads b) Opposing the permanent loads	1.0 0	1.0 0	1.0 0
Variable Loads: Carriageway Live Load and associated loads (braking, tractive and centrifugal forces) and **Pedestrian Live Load:** a) Leading Load b) Accompanying Load	1.0 0.75	0.75 0.2	– 0
Thermal Loads: a) Leading Load b) Accompanying Load	1.0 0.6	0.6 0.5	– 0.5
Wind: a) Leading Load b) Accompanying Load	1.0 0.60	0.60 0.50	– 0
Live Load Surcharge (Accompanying Load)	0.80	0	0
Hydraulic Loads (Accompanying Load): Water Current Forces Wave Pressure Buoyancy	1.0 1.0 0.15	1.0 1.0 0.15	– – 0.15

Table 3.3 Partial Safety Factors for Material Strengths (γ_m)
(IRC: 112-2011)

Limit State	Persistent and Transient		Accidental	
	Concrete	Reinforcing and Prestressing Steel	Concrete	Reinforcing and Prestressing Steel
Ultimate Flexure Shear Bond	1.50	1.15	1.20	1.00
Serviceability	1.00	1.00	—	—

3.7 GLOBAL FACTOR OF SAFETY

The global factor of safety concept has been proposed by Bill Moseley et al.[17] in which a combined safety factor termed as global safety factor is computed by multiplying the appropriate partial safety factors for loads and materials. The use of partial safety factors on materials and load actions provides for considerable flexibility and this can be used for special situations such as very high standards of control and construction (Ex: Construction of Nuclear reactors or in structural elements where Failure without warning may be very serious).

For example, the global factor of safety in the case of a beam failure caused by yielding of tensile reinforcement is computed as:

$(\gamma_f \times \gamma_m) = (1.50 \times 1.15) = 1.725$ For permanent (dead) loads as per IS: 456-2000 code

Alternatively, failure by crushing of concrete in the compression zone may have a global factor of safety of $(1.5 \times 1.5) = 2.25$ due to variable actions only emphasizing the fact that such failure is generally explosive without warning and may be result in serious consequences.

REFERENCES

1. C.E.B. recommendations for International Code of Practice for Reinforced Concrete, American Concrete Institute and Cement & Concrete Association, London, 1964.
2. Rowe, R.E., Cranston, W.B., and Best, B.C., New Concepts in the Design of Concrete, Structural Engineer, Vol. 43, 1965, pp. 339-403.
3. Bate, S.C.C., Why Limit State Design, Concrete, March 1968, pp. 103-108.
4. C.E.B. recommendations for International Code of Practice for Reinforced Concrete, American Concrete Institute and Cement & Concrete Association, London, 1964.
5. Hognestad, E., N.W. Hansen and D.Mc. Henry, Concrete Stress Distribution in Ultimate Strength design, Proc. of the ACI Journal, V. 52, No. 4, Dec. 1955, pp. 455-80.
6. Krishna Raju, N., Limit State Design For Structural Concrete, Proceedings of the Institution of Engineers (India), Vol. 51, No. 1, January 1971, pp. 138-143.
7. ACI: 318M-11, Building Code requirements for Structural Concrete, American Concrete Inst itute, Farmington Hills, Michigan, 2005.
8. BS EN 1992-1-12004, Design of Concrete Structures, General Rules & Rules for Buildings, British Standards Institution, 2004.
9. DIN: 1045-1988, Structural use of Concrete, Design & Construction, Din Deutsches Institute Fir Normung E.V., 1988.
10. CSA Standard A23.3-94, Design of Concrete Structures, Canadian Standards Association, Rexdale, Ontario, 1994.
11. IS: 456-2000, Indian Standard Code of Practice for Plain and Reinforced Concrete (Fourth Revision), Bureau of Indian Standards, 2000, pp. 100.

12. IRC: 112-2011, Code of Practice for Concrete Bridges, Indian Roads Congress, New Delhi, 2011, pp. 280.
13. Ranganathan, R., Reliability Analysis & Design of Structures, Tata McGraw Hill Publishing Co, Ltd, New Delhi, 1990.
14. Ellingwood, B., Reliability Basis for load and Resistance factors for R.C. design, NBS Building Science Series 110, National Bureau of standards, Washington DC, 1978.
15. Cornell, C.A., A Probability – Based Structural Code, Journal of ACI, Vol. 66, Dec. 1969, pp. 974-985.
16. IRC: 6-2014, Standard Specifications and Code of Practice for Road Bridges (Section-II, And Stresses (Revised Edition), New Delhi, 2014, pp. 1-84.
17. Bill Moseley, John Bungey & Ray Hulse., Reinforced Concrete Design to Euro Code-2, Palgrave Macmillan, London, 2012.

REVIEW QUESTIONS

1. List the main reasons for the evolution of the limit state method of design for R.C. structures.
2. Outline the difference between deterministic design and probabilistic design with reference to structural concrete members.
3. What are the various limit states to be considered in the design of structural concrete Members?
4. Differentiate between safety & serviceability with respect to structural concrete members.
5. What are the various serviceability states and why they should be considered in design?
6. Explain the terms a) Characteristic load, b) Characteristic strength c) Partial Safety factor
7. Explain clearly the concept of assigning different safety factors for different types of loads.
8. Distinguish between Design loads and Characteristic loads.
9. Explain the terms, Standard deviation, mean strength and design strength.
10. What is Global factor of safety? How do you compute the global factor of safety for flexural concrete members?

OBJECTIVE TYPE QUESTIONS

1. The inadequacy of the ultimate load/strength method is
   a) Ultimate loads cannot be estimated
   b) Collapse loads can be computed
   c) Serviceability is not ensured

2. Probabilistic concepts are incorporated in
   a) ultimate load design
   b) limit state design
   c) working stress design
3. The first method to be used in the design of structural concrete members is
   a) the ultimate load method
   b) elastic method
   c) limit state method
4. The partial safety factor specified in the Indian standard Code for the combination of live, dead and wind loads is
   a) 1.5
   b) 1.6
   c) 1.2
5. Excessive deflections in a reinforced concrete beam leads to
   a) sudden collapse
   b) damage to partitions
   c) instability of the structure
6. In the limit state design process, design loads are obtained by
   a) equating the characteristic loads
   b) enhancing the characteristic loads
   c) decreasing the characteristic loads
7. The partial safety factor used for material strength of steel at the limit state of collapse is
   a) 1.5
   b) 1.0
   c) 1.15
8. The design strength of material is
   a) directly proportional to the partial safety factor
   b) nearly equal to the characteristic strength
   c) inversely proportional to the partial safety factor
9. The partial safety factor specified in the Indian standard code for evaluating the design Strength of concrete for the limit state of deflection is
   a) 1.3
   b) 1.5
   c) 1.0
10. The global safety factor can be evaluated as the
    a) Sum of the partial safety factors
    b) Product of the partial safety factors
    c) Product of safety factor for loads and strength

Reinforced Concrete Slab Bridge Decks

4

4.1 GENERAL FEATURES

Reinforced concrete slab type decks are often referred to as culverts and are commonly used for small spans. This type of super structure is economical for spans up to 8 m. For larger spans, prestressed concrete slab decks are preferred since the thickness of the slab can be reduced. Slab decks are simpler for construction due to easier fabrication of form work, reinforcement detailing and placement of concrete. In the case of culverts the slab is supported on the two opposite sides on piers or abutments. The deck slab is designed as a one way slab to support the dead and live loads with impact. National highway bridge deck slabs are generally designed to support the IRC Class AA or A type vehicle loads whichever gives the worst effect.

In the case of reinforced concrete Tee beam and slab decks, the slab spans in two directions since it is cast integrally with main and cross girders at regular intervals. Hence the slab is designed as a two way slab for the wheel loads. The deck slab is generally designed for the worst effect of either one lane of IRC 70R/Class AA tracked vehicle loading or one lane of 70R/Class AA wheeled vehicle or two lanes of Class A load trains moving on the deck as specified in IRC: 6-2014[1]. Based on analytical investigations Victor[2] has reported that, for the computation of live load bending moment, only one loading condition need be considered, namely Class AA wheeled vehicle for spans up to 4 m and Class AA tracked vehicle for spans exceeding 4 m. For computations of maximum live load shear in two lane bridge decks, Class AA wheeled vehicle controls the design for all spans from 1 to 8 m.

The distribution or secondary transverse reinforcement in the perpendicular direction should be at least 20 percent of the main reinforcement according to the revised IRC: 112-2011 code[3].

According to the new IRC code, reinforced concrete bridge decks should be designed to conform to the philosophy of limit state design[4,5,6]. The two basic groups of limit states to be considered are:

a) Ultimate limit state or the limit state of strength in which the structural element is designed to withstand safely the ultimate design loads obtained by applying suitable partial safety factors to the service loads.

b) Serviceability limit state in which the structural element should perform its intended function satisfactorily at service loads without excessive deflection or displacement, or local cracking.

The revised new code replaces the working stress method used in the previous codes with the limit state method. However, the working stress method is included in Annexure A-4 of the code as an alternative and restricts its use for concretes of grade up to M-60. The new code envisages the use of high performance concrete classified in the range of M-30 to M-90.

The strength and deformation characteristics of normal concrete generally used for bridge construction are compiled in Table 4.1 (Table 6.5 of IRC: 112-2011). The new IRC code also permits the use of high yield strength deformed bars in the range from Fe-415 to Fe-600.

4.2 FLEXURAL STRENGTH OF REINFORCED CONCRETE BRIDGE DECKS

4.2.1 Basic Assumptions

The ultimate flexural strength of reinforced concrete sections can be determined by assuming suitable stress block in the concrete compression zone. The basic principles of estimating the ultimate strength of structural elements subjected to flexure are well established and most of the codes have specified idealized stress blocks parameters for concrete in the compression zone with the following assumptions specified in the Indian Roads Congress Code[3].

a) Plane sections normal to the axis remain plane after bending

b) The maximum strain in concrete at the outermost compression fiber is taken as 0.0035 in flexure.

c) The relation between the compressive stress distribution and strain in concrete is assumed to be a rectangular parabola which results in prediction of strength in close agreement with test results as shown in Figs. 4.1 & 4.2

d) The tensile strength of concrete is ignored.

e) The stresses in the reinforcement are derived from representative stress-strain curve for the type of steel used. Typical characteristic and design stress-strain curves are shown in Fig. 4.3. For design purposes the partial safety factors recommended in Table 3.1 are applicable,

f) The maximum strain in tension reinforcement in the section at failure shall be not less than that computed by the relation.

Table 4.1 Strength and Deformation Characteristics of Normal Concrete
(Table 6.5 of IRC: 112-2011)

Strength class		M 15	M 20	M 25	M 30	M 35	M 40	M 45	M 50	M 55	M 60	M 65	M 70	M 75	M 80	M 85	M 90
1	f_{ck} (MPa)	15	20	25	30	35	40	45	50	55	60	65	70	75	80	85	90
2	f_{cm} (MPa)	25	30	35	40	45	50	55	60	65	75	75	80	85	90	95	100
3	f_{ctm} (MPa)	1.6	1.9	2.2	2.5	2.8	3.0	3.3	3.5	3.7	4.0	4.4	4.5	4.7	4.8	4.9	5.0
4	$f_{ctk,0.05}$ (MPa)	1.1	1.3	1.5	1.7	1.9	2.1	2.3	2.5	2.6	2.8	2.9	3.0	3.1	3.2	3.3	3.3
5	$f_{ctk,0.95}$ (MPa)	2.0	2.5	2.9	3.2	3.6	3.9	4.3	4.6	4.9	5.2	5.4	5.6	5.7	5.9	6.1	6.2
6	E_{cm} (GPa)	27	29	30	31	32	33	34	35	36	37	38	38	39	40	40	41
7	ε_{c1} (‰)	1.8	1.9	2.0	2.0	2.1	2.2	2.3	2.3	2.4	2.4	2.5	2.5	2.6	2.6	2.7	2.7
8	ε_{cu1} (‰)						3.5					3.4	3.2	3.0	2.9	2.9	2.8
9	ε_{c2} (‰)						2.0					2.1	2.2	2.3	2.3	2.4	2.4
10	ε_{cu2} (‰)						3.5					3.3	3.1	2.9	2.8	2.7	2.6
11	n						2.0					1.9	1.7	1.6	1.5	1.5	1.4
12	ε_{c3} (‰)						1.8					1.8	1.8	1.9	1.9	2.0	2.1
13	ε_{cu3} (‰)						3.5					3.3	3.1	2.9	2.8	2.7	2.6

Note:

(1) Strength designation of concrete, (based on characteristic strength) and corresponding properties to be used in the design are given above. The strains are expressed in per thousand by ‰ sign. The co-relation equations used are given in Annexure A-2.

(2) The tabulated values of E_{cm} are for quartzite/granite aggregates. For other aggregates, they should be multiplied by factors as given below: limestone = 0.9, sandstone = 0.7, basalt = 1.2.

(3) Properties of materials to be used in bridge construction given in Section 18 and the acceptance criteria based on sampling theory of statistics are to be used for procurement purposes only.

$$\varepsilon_{su} = \left[\frac{f_y}{1.15E_s} + 0.002\right] = \left[\frac{0.87f_y}{E_s} + 0.002\right]$$

Where f_y = characteristic strength of steel

E_s = modulus of elasticity of steel

Characteristic Strength = f_{ck}

$$\text{Design Strength} = \left[\frac{0.67f_{ck}}{\gamma_m}\right] = \left[\frac{0.67f_{ck}}{1.5}\right] = 0.45f_{ck}$$

The stress block parameters are shown in Fig 4.2.

Area of stress block is the sum of rectangular & parabolic portion & is computed as:

$$A = (0.45 f_{ck} \times 0.42 \, x_u) + (2/3 \times 0.45 f_{ck} \times 0.58 \, x_u) = 0.36 f_{ck} \cdot x_u.$$

Where x_u = depth of Neutral Axis

f_{ck} = Characteristic Compressive Strength

Position of centre of compression from extreme compression fibre = $0.42 \, x_u$

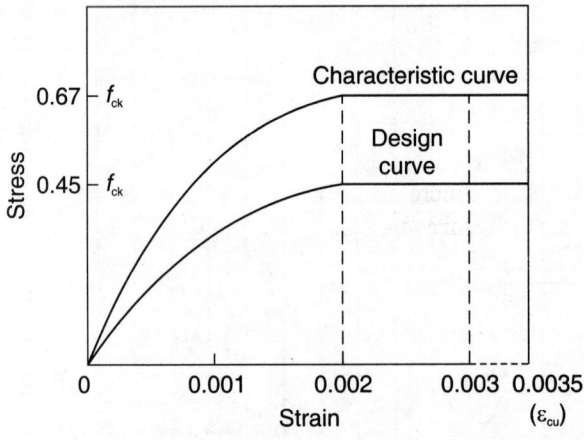

Fig. 4.1 Characterisitc and Design Strength Curves for Concrete in Compression.

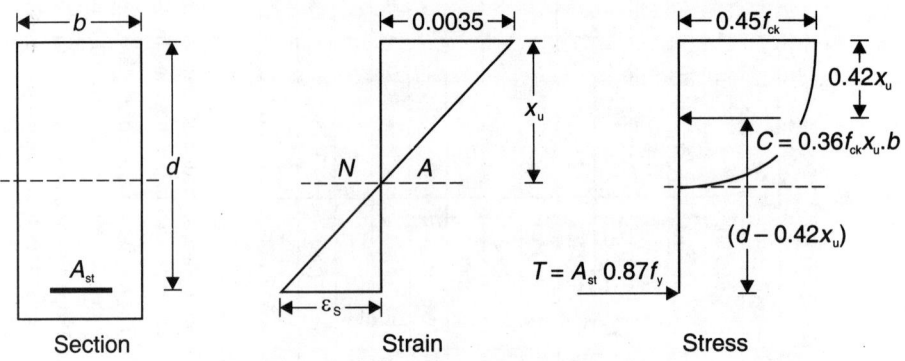

Fig. 4.2 Stress Block Parameters.

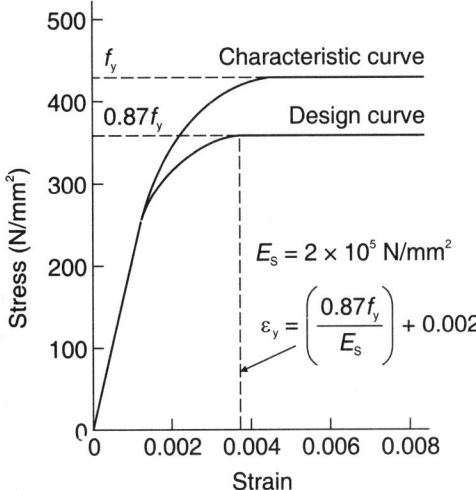

Fig. 4.3 Characteristic and Design Stress-Strain Curves for Fe-415 Grade Steel.

4.2.2 Balanced, Under Reinforced and Over Reinforced Sections

When the compressive strain in concrete reaches a value of 0.0035 as shown in Fig. 4.2, reinforced concrete sections in flexure reach the failure stage. When the sections are reinforced in such a way that the tension steel reaches the yield strain of

$$\varepsilon_y = [(0.87 f_y)/E_s + 0.002]$$

and simultaneously the concrete strain is $\varepsilon_c = 0.0035$, then the section is termed as **Balanced.** In **Under reinforced** sections, the tension steel reaches yield strain at loads lower than the load at which concrete reaches the failure strain. There will be excessive deflections and cracking with a clear indication of impending failure, when the steel yields earlier than concrete, Hence it is preferable to design beams as under reinforced since failure will take place after yielding of steel with clear warning signals like excessive deflections & cracking before the ultimate failure.

In the case of **Over reinforced** sections, concrete reaches the yield strain earlier than that of steel. Over reinforced beams fail by compression failure of concrete without much warning and with very few cracks and negligible deflections. In practice, over reinforced concrete beams are not preferred since they require large quantities of steel since they fail suddenly with explosive failures without any warning.

4.2.3 Neutral Axis Depth

Consider a rectangular beam section shown in Fig. 4.2.

Let b = width of section

d = Effective depth

A_{st} = Area of tension reinforcement

x_u = depth of neutral Axis

For equilibrium of forces at the limit state of collapse,
Total tension (T) = Total compression(C)

$$(A_{st} \cdot 0.87 f_y) = 0.36 f_{ck} \cdot b \cdot x_u$$

$$\therefore \left(\frac{x_u}{d}\right) = \left[\frac{0.87 f_y A_{st}}{0.36 f_{ck} b.d}\right] \qquad \qquad ...(4.1)$$

Limiting values of (x_u/d) to avoid brittle failure is determined from the condition that the steel strain ε_{su} at failure should be not less than the value given by

$$\varepsilon_{su} = \left[\frac{0.87 f_y}{E_s} + 0.002\right]$$

Assuming $E_s = 2 \times 10^5$ N/mm^2, the yield strain for design purposes for different grades of steel are given in Table 4.2.

From proportionality of strains, we have the relation,

$$\left(\frac{x_u}{d}\right) = \left[\frac{\varepsilon_{cu}}{\varepsilon_{cu} + \varepsilon_{su}}\right] = \left[\frac{0.0035}{0.0035 + \varepsilon_{su}}\right] \qquad ...(4.2)$$

Substituting the various values of ε_{su} for different grades of steel, the maximum limiting values of *(x_u/d)* for different grades of steel are also shown in Table 4.2.

Table 4.2 Limiting Values of ($x_{u,max}/d$) for Different Grades of Steel

Grade of Concrete	f_y	Yield strain (ε_{su})	($x_{u,max}/d$)
Fe – 250 Mild Steel	250	0.0310	0.53
HYSD bars Fe – 415	415	0.0038	0.48
HYSD bars Fe – 500	500	0.0042	0.46

4.2.4 Resisting Moment of Reinforced Concrete Sections

In the case of under reinforced sections, the moment of resistance can be computed by using the stress diagram assumed at the limit state of collapse shown in Fig.4.2 Taking moments about the centre of compression,

$$M_u = T (d - 0.42 x_u)$$

Substituting $\quad x_u = \left[\frac{0.87 f_y A_{st}}{0.36 f_{ck}.b}\right]$ from Eq. (4.1)

$$\& \ T = (0.87 A_{st} \cdot f_y)$$

Hence $\quad M_u = 0.87 A_{st} \cdot f_y[d - 0.42 (0.87 f_y A_{st} / 0.36 f_{ck} b)]$

Simplifying & rearranging,

$$M_u = 0.87 A_{st} \cdot f_y d \left[1 - \frac{A_{st} \cdot f_y}{b.d.f_{ck}} \right]$$ ...(4.3)

Eq (4.3) can be used for estimating the flexural strength of sections in which (x_u / d) is less than the limiting value given in Table 4.2. This equation is specified in ANNEX-G of IS: 456-2000.

Expressing the area of steel as a percentage of the effective area, we have

$$p = \left[\frac{A_{st}}{bd} \right] \times 100$$

Where p is the percentage of steel. Substituting for (A_{st} / bd) from the above expression in Eq. (4.3), we get

$$M_u = 0.87 f_y \left(\frac{p}{100} \right) \left[1 - \frac{f_y \cdot p}{f_{ck} \cdot 100} \right] bd^2$$

$$\left[\frac{M_u}{bd^2} \right] = 0.87 f_y \cdot \frac{p}{100} \left[1 - \left(\frac{f_y}{f_{ck}} \right) \left(\frac{p}{100} \right) \right]$$ ...(4.4)

For a given value of $[M_u / bd^2]$, f_y & f_{ck}, the value of 'p' can be computed. This is presented in SP: 16[7] as design tables. In these tables, the percentage of tension steel in the beam corresponds to the yield stress in steel when the beam fails by yielding of steel as in under reinforced sections. The design tables 1 to 4 in SP: 16 are very useful for structural designers to compute the percentage of tensile steel for known values of (M_u /bd^2) and different grades of steel & concrete. The moment of resistance of a concrete section can also be determined in terms of concrete strength by taking the moment of compression force about the tension force in steel, which yields the relation,

$$M_u = 0.36 f_{ck} b x_u (d - 0.42 x_u) = 0.36 f_{ck} \left(\frac{x_u}{d} \right) \left[1 - 0.42 \left(\frac{x_u}{d} \right) \right] bd^2$$

If $\left(\frac{x_u}{d} \right) = \left(\frac{x_{u,max}}{d} \right)$ which is the limiting value as given in Table 4.2, then the limiting values of the moment of resistance of the section is given by,

$$M_{u,lim} = 0.36 f_{ck} \left(\frac{x_{u,max}}{d} \right) \left[1 - 0.42 \left(\frac{x_{u,max}}{d} \right) \right] bd^2 = K.b.d^2$$ ...(4.5)

Where K = A constant.

The expression for M_u for different grades of steel is compiled in Table 4.3.

Table 4.3 Moment of Resistance for Limiting Values of $(x_{u, max}/d)$ for Different Grades of Steel

Grade of Steel	$\left(\dfrac{x_{u,max}}{d}\right)$	Expression for M_u
Fe – 250	0.53	$0.149 f_{ck} bd^2$
Fe – 415	0.48	$0.138 f_{ck} bd^2$
Fe – 500	0.46	$0.133 f_{ck} bd^2$

4.2.5 Reinforcements in Balanced Singly Reinforced Sections

Equating the compressive force in concrete and tensile force in steel (Fig.4.2) we have.

$$0.87 f_y A_{st} = 0.36 f_{ck} b x_u$$

Rearranging the terms,

$$\left(\frac{A_{st}}{bd}\right) = \left(\frac{0.36 x_u}{0.87 d}\right)\left(\frac{f_{ck}}{f_y}\right) = (\text{Constant})\left(\frac{f_{ck}}{f_y}\right)$$

Since (x_u/d) is constant for a given value of f_y

If p_t = Limiting percentage of tension steel.

$$p_t = \left(\frac{100 A_{st}}{bd}\right)$$

The Reinforcement Index can be expressed as

$$\left(\frac{p_t \cdot f_y}{f_{ck}}\right) = \frac{100(0.36)}{0.87}\left(\frac{x_u}{d}\right) = 41.3\left(\frac{x_u}{d}\right) \qquad \qquad \dots(4.6)$$

For different grades of steel, the reinforcement index and the limiting moment of resistance for singly reinforced rectangular sections are compiled in Table 4.4.

The British-Euro code BS EN 1992-1-1-2004[8] and the American code ACI 318-11[9] also specify simplified rectangular stress blocks for the computation of the ultimate flexural strength. While the Indian code adopts the cube strength, the American and the British codes use the cylinder compressive strength for the stress block parameters. The resulting flexural strength estimates are more or less similar in all the the three code procedures. However in contrast to the Indian code, the British and American codes are more comprehensive in their specifications.

Table 4.4 Limiting Moment of Resistance and Reinforcement Index for
Singly Reinforced Rectangular Sections
(Table-C of SP: 16)

	250	415	500
$\left(\dfrac{M_{u,\text{lim}}}{f_{ck}.b.d^2}\right)$	0.149	0.138	0.133
$\left(\dfrac{p_{t,\text{lim}}.f_y}{f_{ck}}\right)$	21.97	19.82	18.87

Balanced percentage of steel, $p_{t,\text{lim}}$, evaluated for different grades of concrete and steel
are shown in Table 4.5.

Table 4.5 Balanced Percentage of Steel, $p_{t,\text{lim}}$ for Singly Reinforced Rectangular Sections
(Table-E of SP: 16)

f_{ck} (N/mm²)	f_y (N/mm²)		
	250	415	500
15	1.32	0.72	0.57
20	1.76	0.96	0.76
25	2.20	1.19	0.94
30	2.64	1.43	1.13

4.2.6 Use of Design Charts and Tables of SP: 16 for Singly Reinforced Beams and Slabs

The Indian Standards Institution's special Publication SP:16, Design Aids for
Reinforced concrete to IS: 456-2000 contains a number of charts and tables for design
of reinforced concrete members. Based on Equations (4.3) and (4.6), the various charts
and tables have been evolved. The design tables of SP: 16 are very useful for structural
designers, since the designs of beams and slabs can be quickly worked out and checked
without using the detailed procedure of using the design equations. Using the data
given in Tables 2, 3 and 4 of SP: 16, the percentage reinforcement required in singly
reinforced sections like solid bridge deck slabs can be directly read out for a known
value of the ratio $[M_u/(bd^2)]$ for different grades of concrete such as M-20, M-25 and
M-30 using Fe-415 high yield strength deformed steel reinforcement.

4.3 SHEAR STRENGTH OF REINFORCED CONCRETE BRIDGE DECKS

4.3.1 Shear Failures in Bridge Deck Slabs

Shear failures are likely to occur near the supports of bridge decks where maximum shear forces develop due to the vehicular loads. The most common types of Shear failures are identified under the following groups:

a) Diagonal Tension

b) Flexure-Shear

c) Shear-Compression

d) Shear-Bond

e) Shear-Friction

The ultimate shear strength of a reinforced concrete beam or slab section depends upon several factors like percentage reinforcement ratio, grade of concrete and depth of slab.

The reader may refer to specialist literature[10,11,12] for theories concerning the various modes of shear failures.

Experimental studies have shown that slabs fail at loads corresponding to a nominal shear stress that is higher than that applicable for beams of usual proportions. In recognition of this Criterion, the Indian standard codes IS: 456-2000 and IRC: 112-2011 has incorporated the shear strength enhancing factor K which depends upon the depth of the solid slab. While the IS code specifies values of K ranging from 1.00 to 1.30 for slabs of depths ranging from 300 to 150 mm respectively, the IRC code specifies an empirical equation of the type,

$$K = 1 + \sqrt{\frac{200}{d}} \leq 2.00, \text{ where } d \text{ is the effective depth expressed in mm}$$

According to the IRC bridge design code, the design shear resistance ($V_{Rd.c}$) of the member without shear reinforcement is given by the expression:

$$V_{Rd.c} = [0.12K (80 \, \rho_1 f_{ck})^{0.33}] \, b_w.d \qquad \qquad ...(4.7)$$

Where $\quad \rho_1 = \left(\dfrac{A_{sl}}{b_w.d} \right) \leq 0.02$

A_{sl} = area of longitudinal reinforcement in the member

b_w = width of member in slabs and width of rib in beams

d = effective depth of the member

In the case of reinforced concrete slab decks, shear resistance being high, failure due to shear is a rare phenomenon and shear reinforcements are not generally provided in slabs. If the nominal shear stress exceeds the permissible values, the depth of the slab is increased to avoid the use of shear reinforcements.

4.4   ANALYSIS OF SLAB DECKS

Reinforced concrete slab decks used for small span culverts are generally spanning in one direction and hence the moments due to dead and live loads are critical in the longitudinal direction i.e. the direction of the moving loads. Bridge deck slabs simply supported on either side have to be designed for IRC loads specified as Class AA or A depending upon the importance and classification of the bridge. In the case of reinforced concrete Tee beam and slab bridges, the deck slab is supported along the longitudinal and lateral directions by main and cross girders. Hence the slabs in such cases have to be analyzed for moments developed in the longitudinal and lateral directions. Analysis of slabs with different support conditions are detailed under the following sub sections.

(a)   Solid Slabs Spanning in One Direction

1.   Single Concentrated Load: In the case of slabs spanning in one direction, the dead load moments are directly computed assuming the slab to be simply supported between the bearings. Live loads of vehicles transmitted through wheels are considered as concentrated loads spread over the contact area of the tyres with the deck slab. The bending moment per unit width of slab developed by concentrated loads on solid slabs may be calculated by assuming the width of slab considered as effective in resisting the bending moment due to concentrated loads. In the case of precast slabs, the term 'actual width of slab' should include the actual width of each individual precast element.

IRC: 112-2011 code specifications outline a method of computing the effective width of slab supporting a concentrated wheel load considering various parameters like the span length, dimensions of slab and the concentration area of the wheel load.

For a single concentrated load, the effective width may be calculated by the equation,

$$b_e = Kx\left[1 - \frac{x}{L}\right] + b_w$$

where b_e = effective width of slab on which the load acts

L = effective span

x = distance of centre of gravity of load from nearer support

b_w = breadth of concentration area of load, ie the dimension of the tyre or track contact area over the road surface of the slab in a direction at right angles to the span plus twice the thickness of the wearing coat or surface finish above the structural slab.

K = A constant depending upon the ratio (B/L) where B is the width of the slab

The values of the constant 'K' for different values of the ratio (B/L) is compiled in Table 4.6.

The effective width shall not exceed the actual width of the slab. Also in the case of a load near the unsupported edge of a slab, the effective width shall not exceed the above value nor half the above value plus the distance of the load from the unsupported edge.

2. Two or More Concentrated Loads in Line in the Direction of Span: When two or more concentrated loads are positioned in a line in the direction of span, the bending moment per unit width of slab shall be calculated separately for each load according to its appropriate effective width of slab as specified under the single concentrated load.

Table 4.6 Values of Constant 'K'
(IRC: 112-2011) (Annexure – B3)

B/L	K For Simply Supported Slabs	K For Continuous Slabs	B/L	K For Simply Supported Slabs	K For Continuous Slabs
0.1	0.40	0.40	1.1	2.60	2.28
0.2	0.80	0.80	1.2	2.64	2.36
0.3	1.16	1.16	1.3	2.72	2.40
0.4	1.48	1.44	1.4	2.80	2.48
0.5	1.72	1.68	1.5	2.84	2.48
0.6	1.96	1.84	1.6	2.88	2.52
0.7	2.12	1.96	1.7	2.92	2.56
0.8	2.24	2.08	1.8	2.96	2.60
0.9	2.36	2.16	1.9	3.00	2.60
1.0	2.48	2.24	2.0 & Above	3.00	2.60

3. Two or more Concentrated Loads not in Line in the direction of Span: In cases where the effective width of slab for one load overlaps the effective width of slab for an adjacent load, the resultant effective width for the two loads equals the sum of the effective widths for each load minus the width of overlap, provided that the slab so designed is tested for the two loads acting separately.

(b) Solid Cantilever Slab

The effective width of dispersion in the direction parallel to the supported edge for a single concentrated load is computed from the equation,

$$b_e = 1.2\, x + b_w$$

Where b_e = effective width.

x = distance of the centre of gravity of the concentrated load from the face of the cantilever support.

b_w = the breadth of the concentration area of the load i.e., the dimension of the tyre or track contact area over the road surface of the slab in a direction parallel to the supporting edge of the cantilever plus twice the thickness of the wearing coat or surface finish above the structural slab.

The effective width should be limited to one-third the length of the cantilever slab measured parallel to the support. When the concentrated load is placed near one of the two extreme ends of the length of the cantilever slab in the direction parallel to the support, the effective width should not exceed the above value, nor should it exceed half the above value plus the distance of the concentrated load from the nearer extreme end, measured in the direction parallel to the fixed edge.

When two or more loads act on the slab and when the effective width of slab for one load overlaps the effective width of the adjacent load, the resultant effective width for two loads should be taken as the sum of the respective effective widths for each load minus the width of the overlap.

(c)   Dispersion of Loads Along the Span

The effective length of slab in the direction of the span is computed as the sum of the tyre contact area over the wearing surface of slab in the direction of the span and twice the overall depth of the slab inclusive of the thickness of the wearing surface.

If D = depth of the wearing coat

H = depth of the slab

x = wheel load contact area along the span

v = effective length of dispersion along the span

We have the relation,

$$v = x + 2(D + H)$$

(d)   Dispersion of Loads in Slabs Spanning in Two Directions

In bridge decks comprising slab integrally cast with longitudinal and cross girders as in the case of Tee beam and slab decks, the moments develop due to wheel loads on the slab both in the longitudinal and transverse directions. These moments are computed by using the design curves developed by M. Pigeaud or Westergaard's method[2,4]. Pigeaud's method is applicable to rectangular slabs supported freely on all the four sides and the slab should be symmetrically loaded as shown in Fig. 4.4. The following notations are used in calculating the dispersion width and moments due to concentrated wheel loads on slabs.

L = Long span length

B = Short span length

u & v = Dimensions of the load spread after allowing for dispersion through the wearing coat and structural slab.

K = Ratio of short to long span of slab (B/L)

M_1 = Moment in the short span direction

M_2 = Moment in the long span direction

m_1 & m_2 = Coefficients for moments along the short and long spans

μ = Poisson's Ratio for concrete generally assumed as 0.15

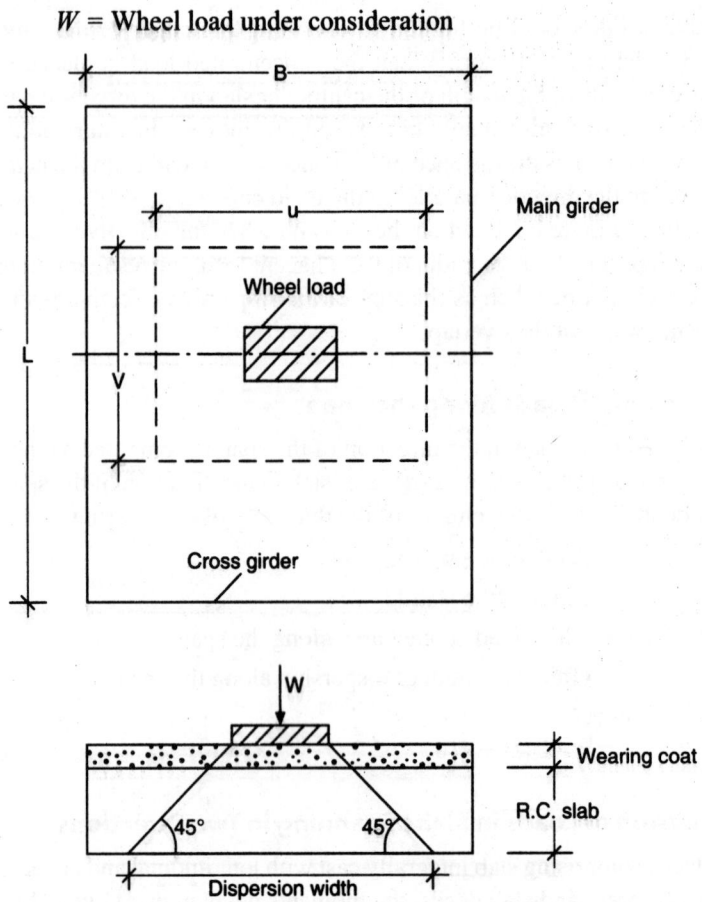

W = Wheel load under consideration

Fig. 4.4 Dispersion of Wheel Load through Wearing Coat and Deck Slab at 45°.

The dispersion of the wheel or track load may be assumed to be at 45° through the wearing coat and structural slab. Consequently the effect of contact of wheel or track load in the direction of span length is equal to the dimensions of tyre contact area over the wearing surface of the slab in the direction of the span plus twice the overall depth of the slab inclusive of the thickness of the wearing coat as shown in Fig. 4.4. The dispersion of the wheel load through the wearing coat only is shown in Fig. 4.5(a). The dispersion of wheel load through the wearing coat and the deck slab at 45° is shown in Fig. 4.5(b). According to Victor[2], the dispersion is some times assumed to be at 45 degrees through the wearing coat which is flexible and at a steeper angle through the deck slab which is rigid as shown in Fig. 4.5(c).

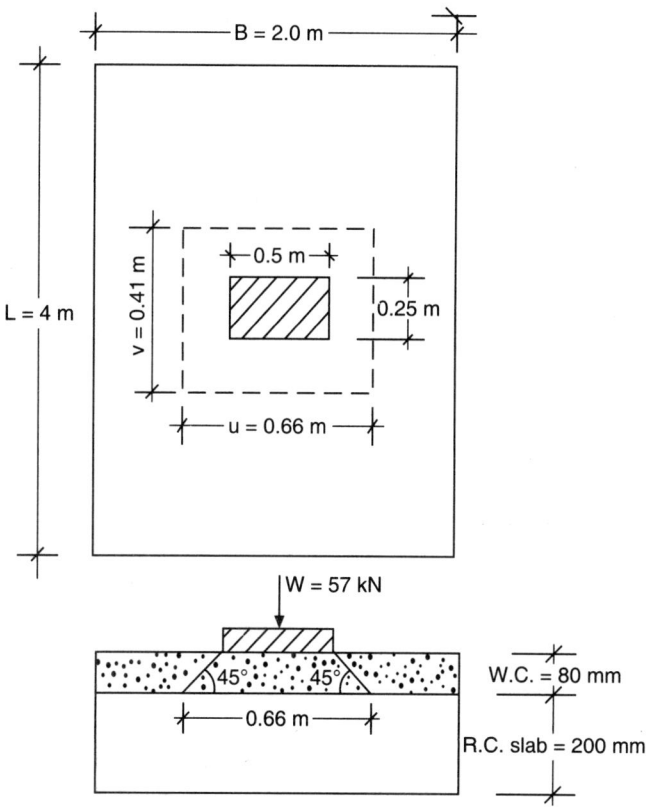

Fig. 4.5(a) Dispersion of Wheel Load through Wearing Coat.

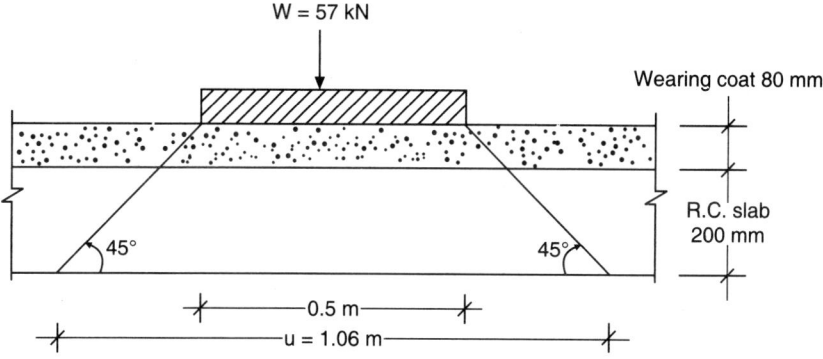

Fig. 4.5(b) Dispersion of Wheel Load through Wearing Coat and Deck Slab at 45°.

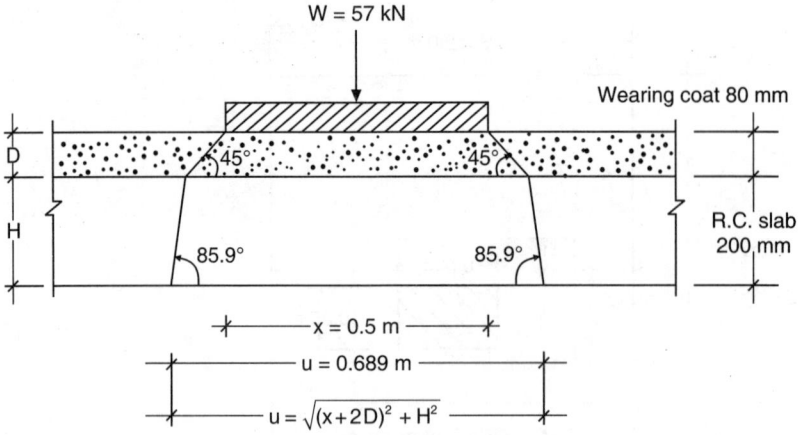

Fig. 4.5(c) Dispersion of Wheel Load at Different Angles through Wearing Coat and Deck Slab.

The bending moments in the short and long span directions are expressed as

$$M_1 = (m_1 + \mu \, m_2) \, W$$
$$M_2 = (m_2 + \mu \, m_1) \, W$$

The values of the moment coefficients m_1 and m_2 depend upon the parameters (u/B) and (v/L) and the value of $K = (B/L)$. Figures 4.6 to 4.12 are the Pigeaud's curves used for the estimation of moment coefficients m_1 and m_2 for various values of K ranging from 0.4 to 1.0. Moment coefficients m_1 and m_2 corresponding to K and $(1/K)$ for slabs supporting uniformly distributed load (dead load of slab) are obtained from Fig. 4.13.

The effect of different criteria used for determining the dispersion of wheel loads can be examined by a detailed comparative analysis. Hence a comparative study of the bending moments developed in the short and long span directions of a typical deck slab supported on main and cross girders of a Tee beam bridge, assuming different types of load distribution through wearing coat and thickness of structural slab is presented for three different types of cases for a wheel load of IRC Class A type loading using the following data:

Data: IRC Class A Wheel Load = W = 57 kN
Wheel contact dimensions = 500 mm by 250 mm
Thickness of wearing coat = D = 80 mm
Thickness of structural slab = H = 200 mm
Dimensions of slab: L = 4 m and B = 2 m
Ratio = K = (B/L) = $(2/4)$ = 0.5
Poisson's Ratio = μ = 0.15

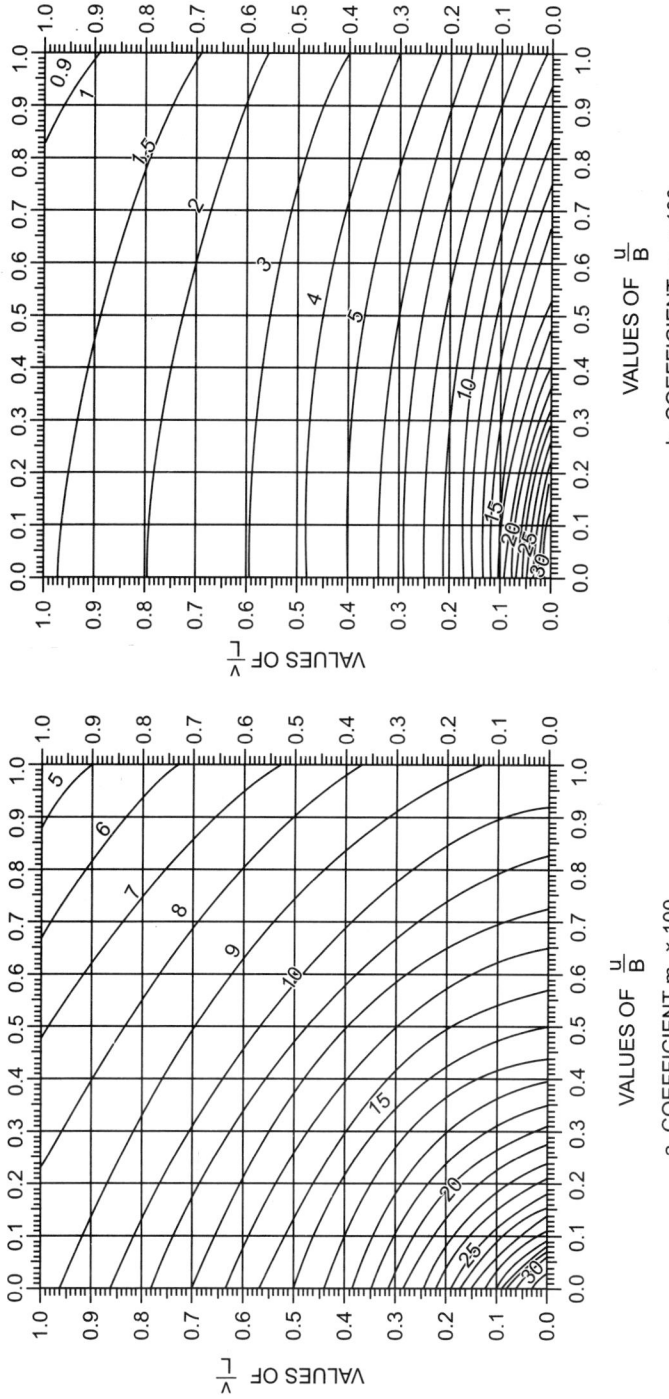

Fig. 4.6 Moment Coefficients m_1 and m_2 for $K = 0.4$.

b. COEFFICIENT $m_2 \times 100$

a. COEFFICIENT $m_1 \times 100$

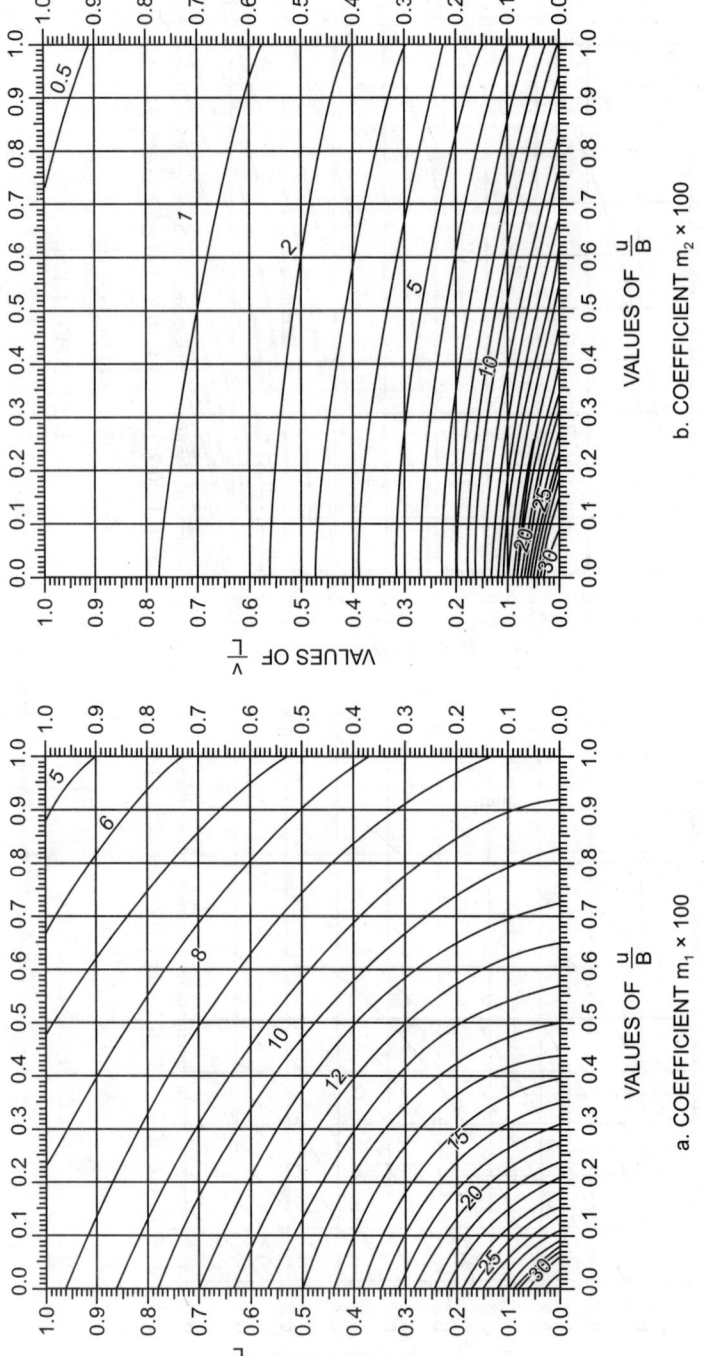

Fig. 4.7 Moment Coefficients m_1 and m_2 for $K = 0.5$.

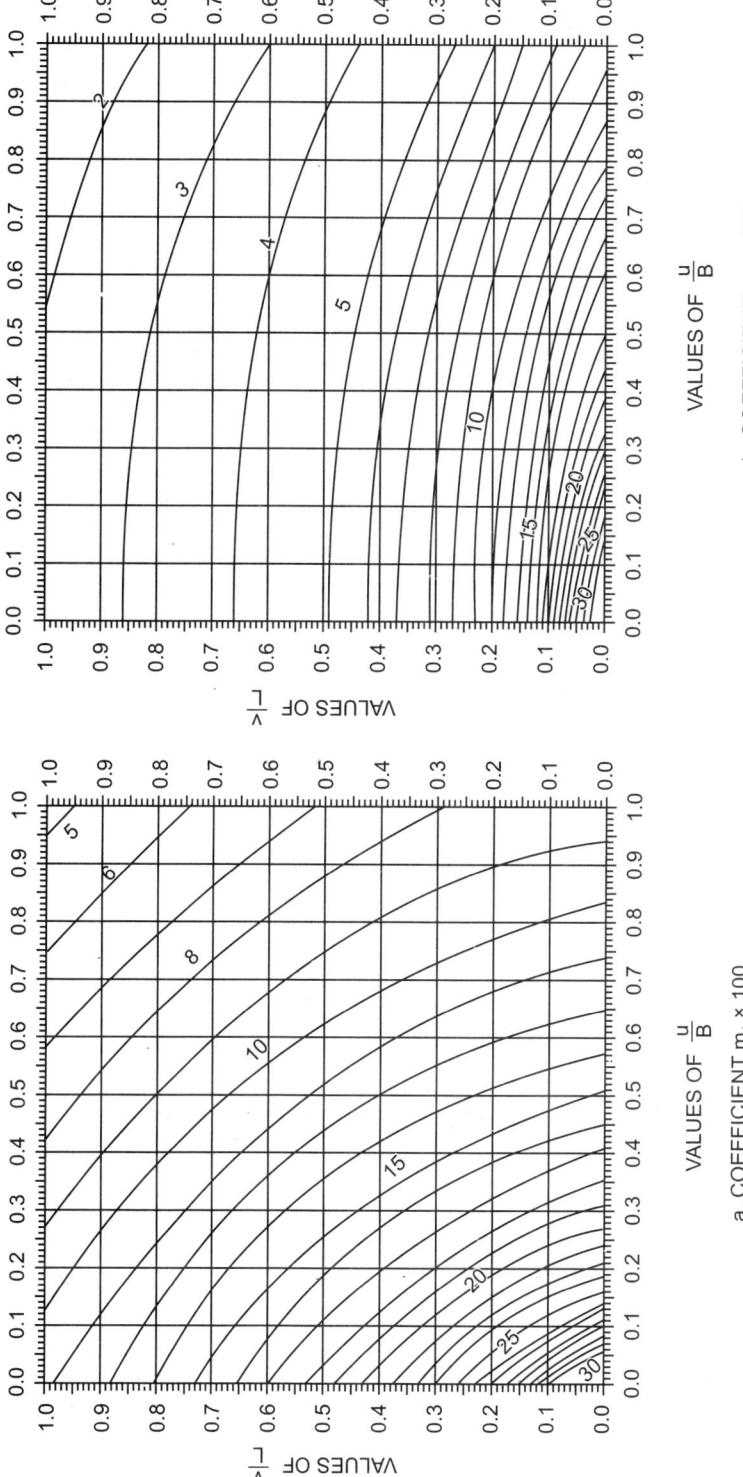

Fig. 4.8 Moment Coefficients m₁ and m₂ for K = 0.6.

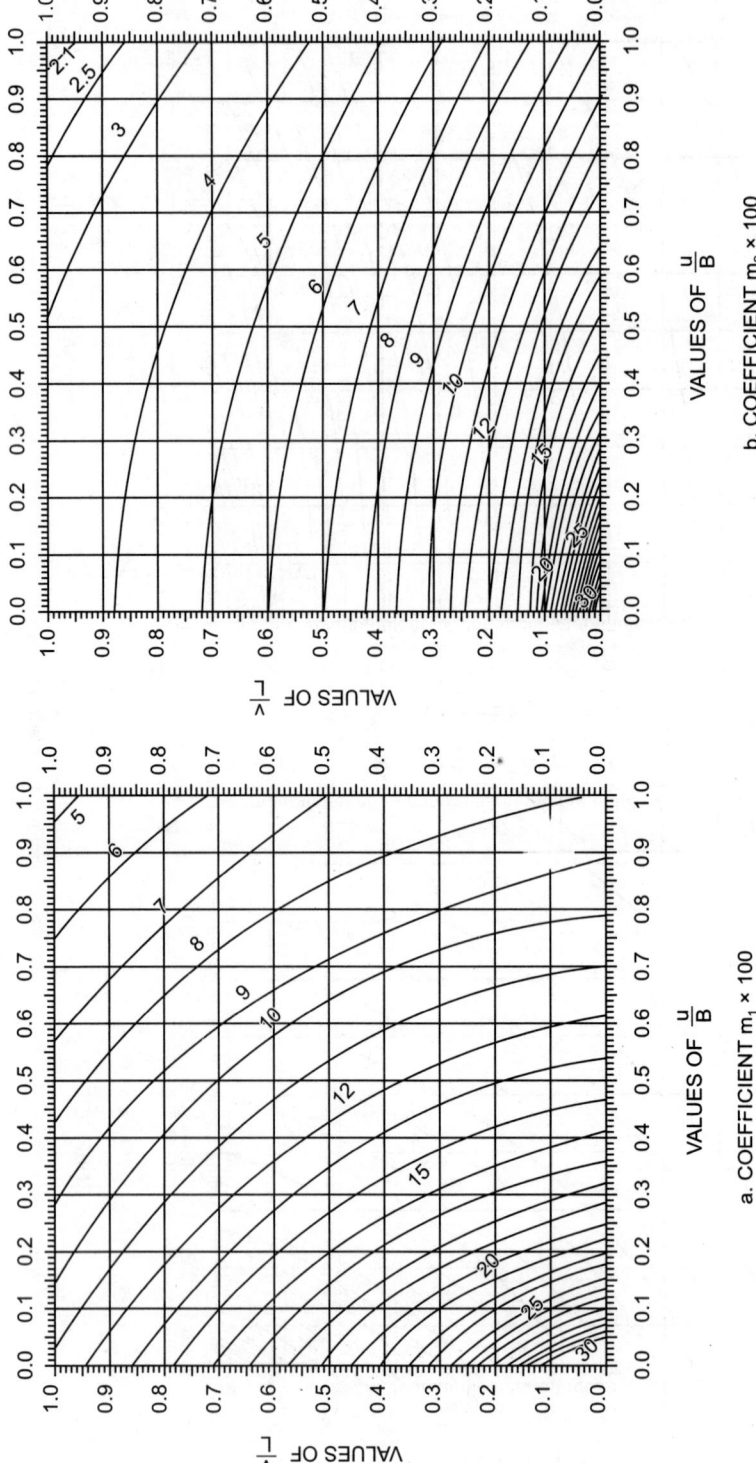

Fig. 4.9 Moment Coefficients m_1 and m_2 for $K = 0.7$.

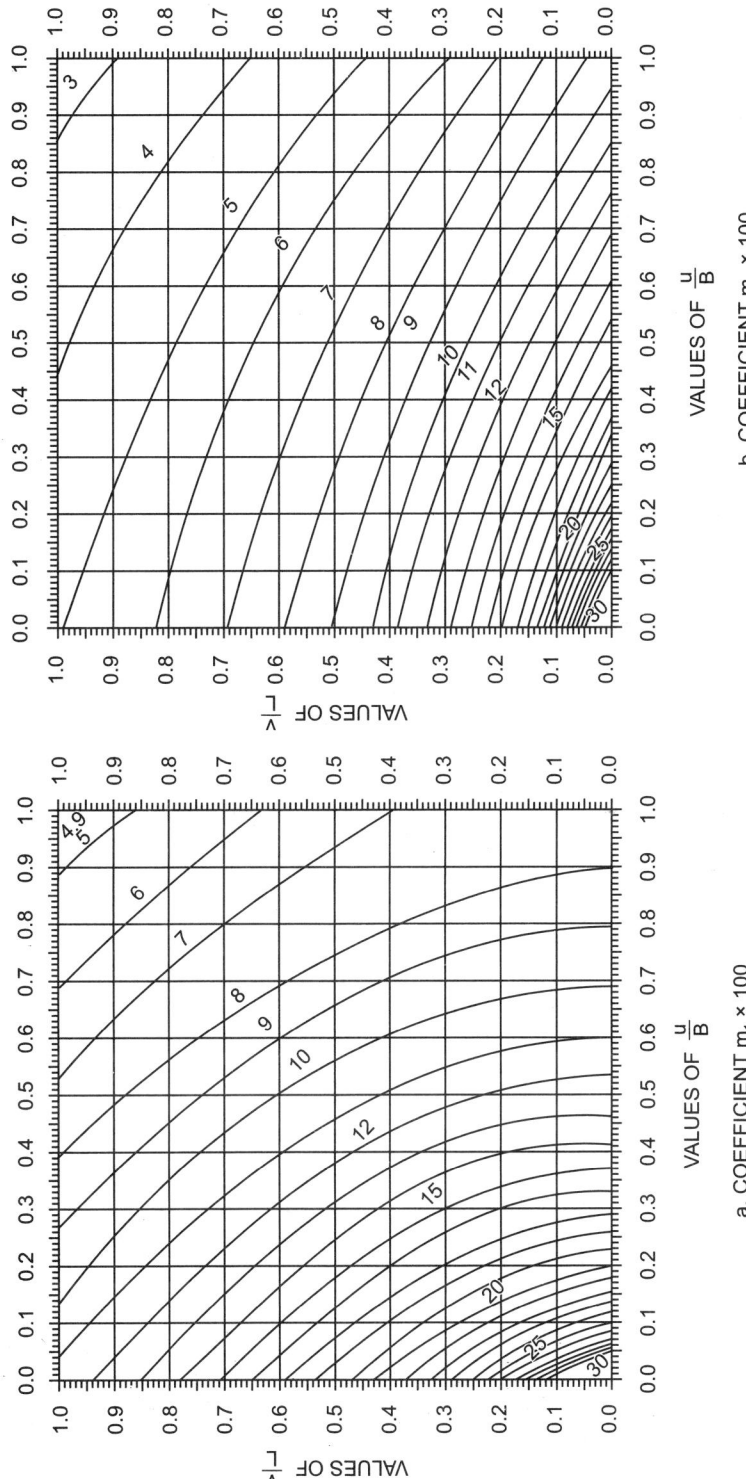

Fig. 4.10 Moment Coefficients m_1 and m_2 for $K = 0.8$.

a. COEFFICIENT $m_1 \times 100$

b. COEFFICIENT $m_2 \times 100$

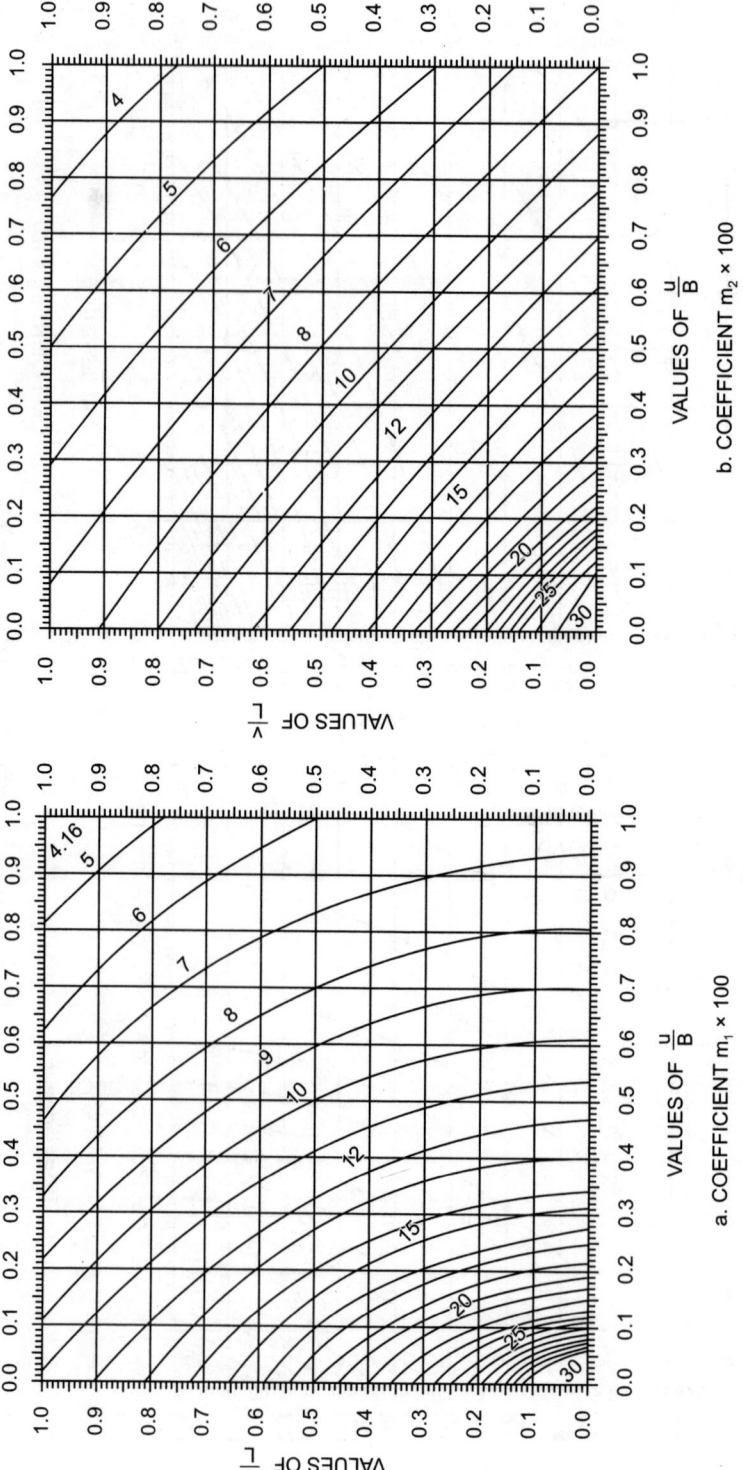

Fig. 4.11 Moment Coefficients m_1 and m_2 for $K = 0.9$.

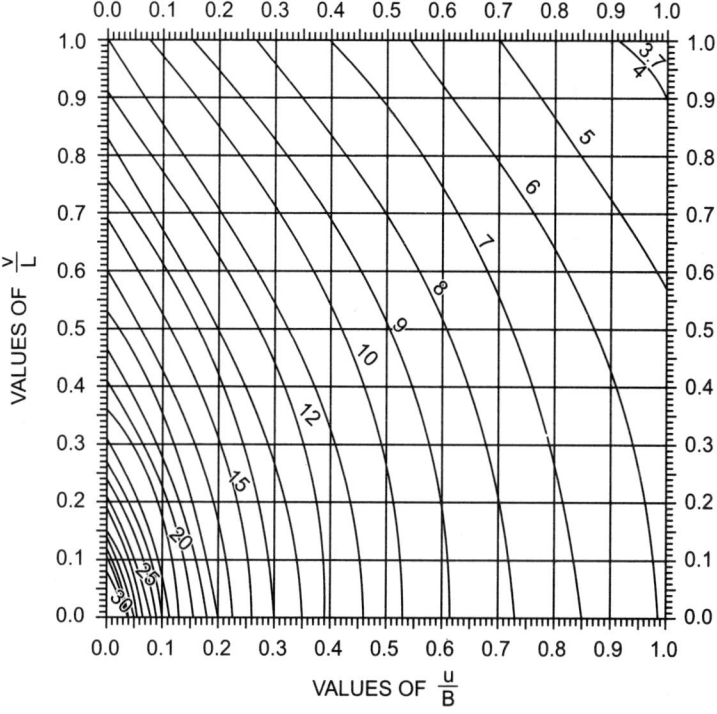

Fig. 4.12 Moment Coefficients m_1 and m_2 for $K = 0.10$.

Case 1: Dispersion of Wheel load through wearing coat only
Referring to Fig. 4.5(a),

$$u = (0.50 + 2 \times 0.08) = 0.66 \text{ m}$$
$$v = (0.25 + 2 \times 0.08) = 0.41 \text{ m}$$
$$(u/B) = (0.66/2.0) = 0.33$$
$$(v/L) = (0.41/4.0) = 0.102$$
$$K = (B/L) = (2/4) = 0.5$$

Using Pigeaud's curves corresponding to $K = 0.5$, the moment coefficients are read as

$$m_1 = 0.18 \text{ and } m_2 = 0.13$$
$$\therefore \quad M_B = W (m_1 + \mu \, m_2) = 57 (0.18 + 0.15 \times 0.13) = 11.37 \text{ kN·m}$$
$$M_L = W (m_2 + \mu \, m_1) = 57 (0.13 + 0.15 \times 0.18) = 8.94 \text{ kN·m}$$

Case 2: Dispersion of Wheel load through wearing coat and Structural Slab at 45°
Referring to Fig. 4.5(b),

$$u = (0.50 + 2 \times 0.08) = 0.66 \text{ m}$$
$$v = (0.25 + 2 \times 0.08) = 0.41 \text{ m}$$
$$(u/B) = (1.06/2.0) = 0.53$$
$$(v/L) = (0.81/4.0) = 0.202$$
$$K = (B/L) = (2/4) = 0.5$$

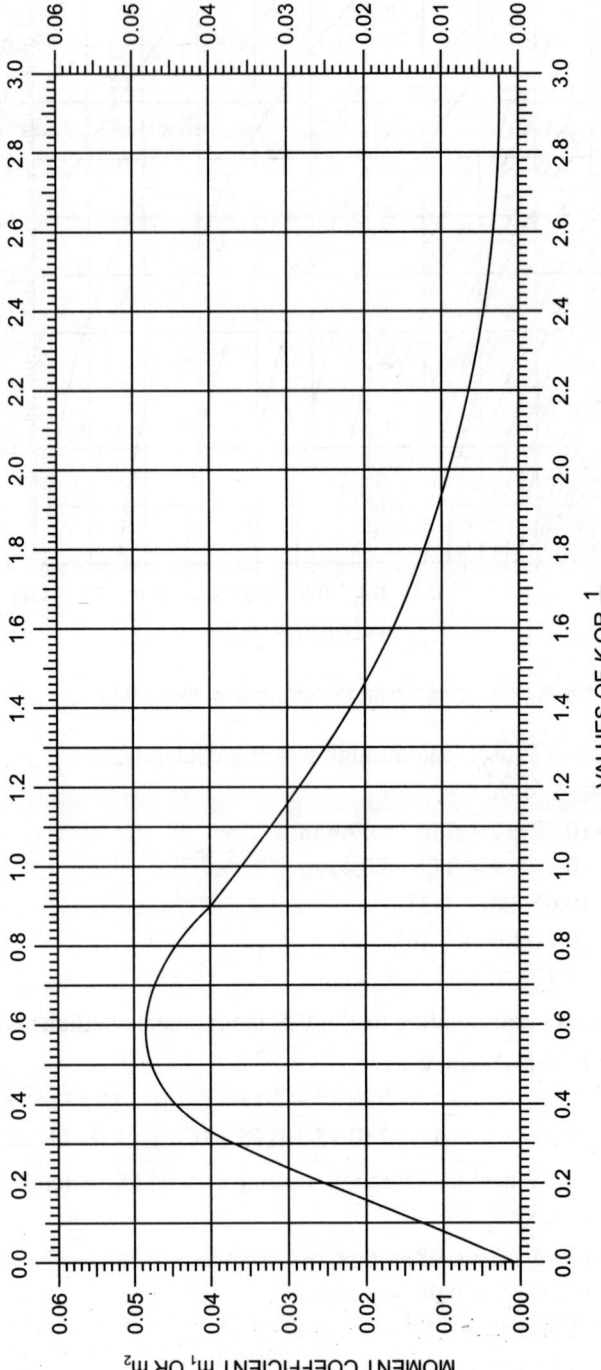

Fig. 4.13 Moment Coefficients for Slabs Completely Loaded with Uniformly Distributed Load, Coefficient is m_1 for K and m_2 for $1/K$.

Using Pigeaud's curves corresponding to $K = 0.5$, the moment coefficients are read as

$\qquad m_1 = 0.138$ and $m_2 = 0.080$

$\therefore \qquad M_B = W\,(m_1 + \mu\,m_2) = 57\,(0.138 + 0.15 \times 0.08) = 8.55$ kN·m

$\qquad\quad M_L = W\,(m_2 + \mu\,m_1) = 57\,(0.08 + 0.15 \times 0.138) = 5.74$ kN·m

Case 3: Dispersion of Wheel load through wearing coat and 85.9° through structural slab

Referring to Fig. 4.5(c),

$$u = \sqrt{(x + 2D)^2 + H^2} = \sqrt{(0.5 + 2 \times 0.08)^2 + 0.2^2} = 0.689 \text{ m}$$

$$v = \sqrt{(0.5 + 2 \times 0.08)^2 + 0.2^2} = 0.456 \text{ m}$$

Using Pigeaud's curves corresponding to $K = 0.5$, the moment coefficients are read as

$\qquad m_1 = 0.18$ and $m_2 = 0.125$

$\therefore \qquad M_B = W\,(m_1 + \mu\,m_2) = 57\,(0.175 + 0.15 \times 0.125) = 11.04$ kN·m

$\qquad\quad M_L = W\,(m_2 + \mu\,m_1) = 57\,(0.125 + 0.15 \times 0.175) = 8.62$ kN·m

A comparative study of the moments resulting from different types of dispersion of wheel load indicates that the value of moments are maximum for Case 1 in which the dispersion is assumed at 45° through wearing coat only. If the dispersion is assumed at a steeper angle of 85.9° through the stiffer structural slab, the resulting moments are only 3 percent less than Case 1. However if the dispersion is assumed at an uniform angle of 45° through both the wearing coat and slab, the resulting moments are the least with a difference of 24 percent in comparison with the results of Case 1 type load dispersion. Hence it is recommended that Case 1 type of load dispersion yielding maximum moments may as well be used for the design of two way slabs.

4.5 DESIGN AIDS AND TABLES FOR REINFORCED CONCRETE BRIDGE DECK SLABS

Pigeaud's method of computing the live load and dead load moments in concrete slabs supported on all sides assumes simple unyielding boundaries and arbitrary continuity coefficients are app. ed 'n order to account for the continuity of the slab over supports. This assumption is not valid particularly when the longitudinal and cross girders are cast monolithically with the slab so that the whole deck behaves as an integral structure under loads. In such a case the behaviour of slab can be approximated to that of a slab with fixed boundaries. This factor has been recognized in the German specifications DIN-1075 and ONORM B4204. The bridge loadings are generally specified in the form of wheel or tracked loads with the dimensions of loaded area laid down by codes of practice.

Concentrated wheel load pattern does not permit simplified solutions unlike that for uniformly distributed load covering the entire slab. Further, the multiplicity of loading conditions with several wheel loads make the designer's job very tedious and time consuming. Although Pigeaud's curves provide a ready method of analysis for simple boundaries, still a lot of computations will be necessary particularly for eccentric loads.

Based on the work of Ruesch, involving slabs with continuous boundaries, design tables have been prepared by Suryaprakasha Rao *et al.* for computation of bending moments in two way slabs with simply supported and fixed or continuous edges. The various parameters considered are:

(1) The width (B) and length (L) of slab panel

(2) Type of support at slab edges (Simply supported or fixed or continuous)

(3) Type of loading such as Class 70 R, Class AA tracked and wheeled vehicles and Class A train and uniformly distributed load.

The design tables are based on the following assumptions:

(1) The slab is analyzed as a thin plate using the elastic analysis with different boundary conditions.

(2) The dispersion of the wheel loads is taken up to the middle surface of the slab.

(3) The wearing coat thickness is assumed as 75 mm and that of the structural slab as 150 mm. In case the thickness is greater than 150 mm, the values provided will be slightly on the conservative side.

The bending moments are tabulated for two different cases which are commonly encountered in highway bridge decks. These are i) All edges are simply supported and ii) All edge fixed or continuous.

The variables considered are the slab dimensions in the perpendicular directions designated as:

L = length of slab in the longitudinal direction (direction of traffic)

B = width of slab in the transverse direction

M_{xc} and M_{yc} are the positive moments developed at the centre of slabs in the principal directions.

M_{xe} and M_{ye} are the negative moments developed at the edges in the principal directions. The range of values of slab dimensions, type of loading and the edge conditions covered in these table are as follows:

(a) The length 'L' varies from 3 to 8 m with increments of 0.25 m.

(b) The width 'B' varies from 1 to 5 m with increments of 0.25 m.

(c) The type of loading comprises of IRC Class 70 R, Class AA and Class A.

(d) A key chart indicates the plan dimensions of the deck slabs. Chart numbers varying from 1 to 285 cover the entire spectrum of the slab dimensions.

(e) Four sets of tables covering the various types of loading (Three types of Live loads and uniformly distributed load due to dead load) and two types of edges *viz.* all edges simply supported and all edges fixed or continuous are covered in the design tables with bending moments tabulated for all cases.

Tee beam slab decks generally have slabs continuous at edges enclosed by longitudinal and cross beams and hence only for this type of fixed slabs covering the slab dimensions (Length varying from 3 to 5 m and width from 2 to 2.5 m) commonly used in Indian Highway bridge decks, bending moments are tabulated.

Table 4.7 gives the live load moments for IRC Class AA tracked loading and Table 4.8 gives the dead load moments for uniformly distributed load of 10 kN/m². These design tables are very useful in the design of slabs in Tee beam and slab decks. Since the maximum bending moments in slabs occur for IRC Class AA tracked loading, only this particular case is covered in the following tables.

Table 4.7 Design Tables for Positive and Negative Moments in Slabs with Continuous Edges for IRC Class AA Tracked Loading

Slab Dimensions		Positive Moments		Negative Moments	
L (m)	B (m)	M_{xc} (kN·m)	M_{yc} (kN·m)	M_{xe} (kN·m)	M_{ye} (kN·m)
3	2	8.54	10.41	−18.52	−19.35
4	2	7.73	11.86	−16.73	−20.22
4	2.5	10.49	15.64	−22.66	−28.11
5	2.5	9.89	16.97	−22.56	−29.03
5	2	7.54	12.18	−16.40	−20.43

Table 4.8 Design Tables for Positive and Negative Bending Moments in Slab for Uniformly Distributed Load of 10 kN/m²

Slab L (m)	Dimensions B (m)	Positive M_{xc} (kN·m)	Moments M_{yc} (kN·m)	Negative M_{xe} (kN·m)	Moments M_{ye} (kN·m)
3	2.0	0.61	1.43	−2.27	−2.99
4	2.0	0.39	1.64	−2.25	−3.28
4	2.5	0.87	2.34	−3.55	−4.83
5	2.5	0.61	2.58	−3.54	−5.16
5	2.0	0.29	1.69	−2.23	−3.33

4.6 MINIMUM AND MAXIMUM REINFORCEMENTS IN SLABS

1) The minimum reinforcements in solid bridge deck slabs is governed by the specifications in Clause 16.6.1 of IRC: 112-2011. In the case of deck slabs, the effective cross sectional area of the longitudinal tensile reinforcement should be not less than that required to control cracking and not less than $A_{s,min}$ given by the relation,

$$A_{s,min} = 0.26 \left[\frac{f_{ctm}}{f_{yk}} \right] b_t \, d \text{ but not less than } 0.0013 b_t \, d \qquad \ldots(4.8)$$

Where

b_t = mean width of the tension zone in slabs and width of web in beams

f_{ctm} = mean value of axial tensile strength of concrete (Refer Table 4.1)

d = effective depth

2) The maximum reinforcement should be not greater than $0.025A_c$ at sections other than at laps.

3) Secondary transverse reinforcement of area not less than 20 percent of the main reinforcement should be provided in one way slabs.

4) The maximum spacing of main reinforcements should be the lesser of 2 times the thickness of the slab or 250 mm.

5) The maximum spacing of the secondary reinforcement in one way slabs should be the lesser of 3 times the thickness of the slab or 400 mm.

4.7 CONTROL OF CRACKING IN BRIDGE DECKS

4.7.1 Permissible Crack Widths

The IRC bridge code specifies values for the maximum permissible width of cracks in reinforced and prestressed concrete bridge decks depending upon the conditions of exposures such as moderate, severe and very severe environmental conditions as detailed in Table 4.9.

Table 4.9 Recommended Values of Maximum Crack Widths (w_{max})
(Table 12.1 of IRC: 112-2011)

Condition of Exposure[1] As per Clause 14.3.1	Reinforced members and prestressed members with un-bonded tendons	Prestressed members with bonded tendons
	Quasi-permanent load combination[2] (mm)	Frequent load combination (mm)
Moderate	0.3[2]	0.2
Severe	0.3	0.2[3]
Very Severe and Extreme	0.2	0.2[4] and decompression
(1) The condition of exposure considered applies to the most severe exposure the surface will be subjected to in service. (2) For moderate exposure class, crack width has on influence on durability and this limit is set to guarantee acceptable appearance. (3) For these conditions of exposure, in addition, decompression should be checked under the quasi-permanent combination of loads that include DL + SIDL + Prestress including secondary effect + settlement + temperature effects. (4) 0.2 applies to the parts of the member that do not have to be checked for decompression.		

The reader may refer to the Clauses 12.3.3 to 12.3.5 of the IRC: 112-2011 for the minimum reinforcements for crack control, calculation of crack width and control of shear cracks within webs respectively. However the code also recommends a simpler procedure for control of cracking without direct calculations. In this method the code recommends the use of two tables in which the maximum permissible stress in steel immediately after cracking is restricted to specified values for different bar sizes and the width of cracks. Tables 4.10 and 4.11 gives the maximum bar diameters and spacing for control of cracks widths of 0.3 mm and 0.2 mm for different stress levels in steel.

Table 4.10 Maximum Bar Diameters for Crack Control
(Table 12.2 of IRC: 112-2011)

Steel stress [MPa]	Maximum bar size [mm]	
	$W_k = 0.3$ mm	$W_k = 0.2$ mm
160	32	25
200	25	16
240	16	12
280	12	–
320	10	–

Table 4.11 Maximum Bar Spacing for Crack Control
(Table 12.3 of IRC: 112-2011)

Steel stress [MPa]	Maximum bar size [mm]	
	$W_k = 0.3$ mm	$W_k = 0.2$ mm
160	300	200
200	250	150
240	200	100
280	150	50
320	10 0	–

4.7.2 Calculation of Crack Widths

The width of cracks in the tension zone of reinforced concrete bridge depends upon the maximum spacing of bars and the mean strain in reinforcement and concrete between the cracks. IRC: 112-2011 recommends an expression for the width of cracks based on the method specified in British Code[13]. According to this method the design surface crack width is computed using the expression:

$$W_k = s_{r,max} \, (\varepsilon_{sm} - \varepsilon_{cm})$$

Where
W_k = the design surface crack width

$s_{r,max}$ = the maximum crack spacing

ε_{sm} = the mean strain in the reinforcement under the relevant combination of loads, including the effect of imposed deformations, restrained thermal shrinkage effects allowing for the effects of tension stiffening of the concrete. For prestressed members only the additional tensile strain beyond the state of zero strain of the concrete at the same level is considered.

ε_{cm} = the mean strain in the concrete between cracks

$$\left(\varepsilon_{sm} - \varepsilon_{cm}\right) = \left[\frac{\sigma_s - k_t\left(\dfrac{f_{ct,eff}}{\rho_{p,ef}}\right)\left(1 + \alpha_e \rho_{p,ef}\right)}{E_s}\right] \geq 0.6\left(\frac{\sigma_s}{E_s}\right) \qquad ...(4.9)$$

Where

σ_s = stress in tension reinforce calculated using the cracked concrete section.

k_t = a factor that accounts for the duration of loading which may be taken as 0.5

$\rho_{p,ef}$ = $[A_s/A_{c,eff}]$ is the effective reinforcement ratio based on an effective concrete tension area $(A_{c,eff})$ as shown in Fig. 4.14

$f_{ct,eff}$ = concrete tensile strength at the time of cracking and the values as listed in Table 4.1

α_e = modular ratio = $[E_s/E_{cm}]$

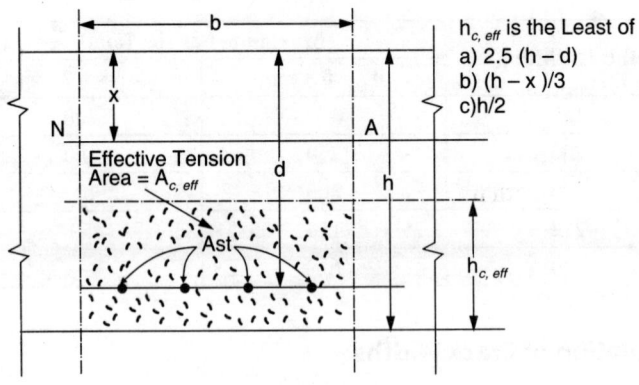

Fig. 4.14 Effective Concrete Tension Area.

The maximum crack spacing $s_{r,max}$ is given by the empirical formula

$$s_{r,max} = [3.4c + 0.425\, k_1\, k_2\, \text{Ø}/\rho_{p,ef}] \qquad ...(4.10)$$

Where

Ø = bar diameter (or average value if bars of different sizes have been used)

c = cover to the reinforcement

k_1 = 0.8 for high bond bars and

k_2 = 0.5 for flexure and 1.0 for direct tension

Substituting these values we have the expression

$$s_{r,max} = [3.4c + 0.17\text{Ø}/\rho_{p,ef}] \qquad ...(4.11)$$

This has an upper limit of $s_{r,max} = 1.3(h-x)$ where reinforcement spacing exceeds $5(c + \phi/2)$

Where h is the depth of neutral axis.

4.8   CONTROL OF DEFLECTIONS IN BRIDGE DECKS

4.8.1 Limiting Values of Deflection

The IRC: 112-2011 specifies the limiting values of deflection taking into account the nature of the super structure, bridge deck furniture and functional needs of the bridge. In the absence other criteria, the following deflection limits under live load maybe considered:

a) Vehicular..Span/800

b) Vehicular and pedestrian
    Or pedestrian alone...Span/1000

c) Vehicular on cantilever..Cantilever span/300

d) Vehicular & pedestrian and
    Pedestrian only on cantilever arms.....................Cantilever span/375

In contrast to the IS: 456-2000[14] and IS: 1343-2012[15] code of practice for reinforced and prestressed concrete in which the final deflection due to all loads including the effects of temperature, creep and shrinkage is limited to a value of span/250, the present IRC code limit of span/800 seems to be very conservative resulting in larger sizes of deck elements with more reinforcements affecting the overall cost of the deck structure.

4.8.2   Calculation of Deflection Due to Sustained Loads

The computation of deflections are considered in two parts:

a) Instantaneous or short term deflections occurring on application of the loads.

b) Long term deflections resulting from differential shrinkage and creep due to sustained loading.

In case of cracked members, appropriate value of cracked moment of inertia should be used in the computations. If actual value of cracked moment of inertia cannot be determined, the code permits the use of 70 percent of the gross moment of inertia for computations. For prestressed concrete members, fully under compression, gross moment of inertia can be used.

Deflection due Creep effects of sustained loads over a long period is calculated by using the effective modulus of elasticity for concrete using the equation,

$$E_{c,eff} = \left[\frac{E_{cm}}{1 + \emptyset(\infty, t_o)} \right] \qquad ...(4.12)$$

Where

E_{cm} = Secant modulus of elasticity

$\emptyset\ (\infty, t_0)$ = creep coefficient relevant for the load and time interval as given in Table 4.12 for different ages of loading and atmospheric conditions.

Table 4.12 Final Creep Coefficient [$\emptyset$(70 Yrs)] of Concrete
(Table 6.9 of IRC: 112-2011)

Age at loading t_o (days)	Notional Size 2A$_c$/u (in mm)					
	50	150	600	50	150	600
	Dry atmospheric conditions (RH – 50%)			Humid atmospheric conditions (RH – 80%)		
1	5.50	4.60	3.70	3.60	3.20	2.90
7	5.50	4.60	3.70	2.60	2.30	2.00
28	3.90	3.10	2.60	1.90	1.70	1.50
90	3.00	2.50	2.00	1.50	1.40	1.20
365	1.80	1.50	1.20	1.10	1.00	1.00

The deflections due to shrinkage is expressed by the relation,

$$a_{cs} = k \left(\frac{1}{r_{cs}} \right) L^2$$

where

k = a constant depending upon the support conditions having values of

= 0.5 for cantilevers

= 0.125 for simply supported members

= 0.086 for members continuous at one end and

= 0.063 for fully continuous members

$\left(\dfrac{1}{r_{cs}} \right)$ = Shrinkage curvature assessed by the relation

$$= \varepsilon_{cs}\, \alpha_e \left[\frac{S}{I} \right]$$

and ε_{cs} = free shrinkage strain = $[\varepsilon_{cd} + \varepsilon_{ca}]$

α_e = effective modular ratio = $(E_s / E_{c,eff})$

S = First moment of Area of the reinforcement about centroid of the section

I = Second moment of area of the section

ε_{cd} = drying shrinkage strain

ε_{ca} = autogenous shrinkage strain

Autogenous shrinkage strain develops during the hardening of concrete in the early days after casting and it depends on the strength of concrete and the values recommended in the IRC code are compiled in Table 4.13

Table 4.13 Autogenous Shrinkage Strain of Concrete
(Table 6.6 of IRC: 112-2011)

Grade of Concrete	M-30	M-35	M-45	M-50	M-60	M-65
Autogenous Shrinkage Strain ($\varepsilon_{ca} \times 10^6$)	35	45	65	75	95	105

The drying shrinkage strain develops slowly, since it is a function of the migration of water through the pores in the hardened concrete. The final value of the drying shrinkage strain may be taken as $k_h \varepsilon_{cd}$.

Where k_h ia a factor which depends upon the notional size h_o

and h_o is the notional size (mm) of the cross section and is equal to $2A_c/u$

where A_c is the cross sectional area of concrete and u is the perimeter of that part of the cross section which is exposed to drying.

The values of k_h and ε_{cd} are compiled in Tables 4.14 and 4.15 respectively.

Table 4.14 Values for k_h
(Table 6.7 of IRC: 112-2011)

h_o in mm	k_h
100	1.0
200	0.85
300	0.75
≥ 500	0.75

Table 4.15 Unrestrained Drying Shrinkage Values ($\varepsilon_{cd} \times 10^6$)
(Table 6.8 of IRC: 112-2011)

f_{ck}(MPa)	Relative Humidity (in %)		
	20	50	80
25	620	535	300
50	480	420	240
75	380	330	190
95	300	260	150

4.9 MINUMUM GRADE OF CONCRETE AND COVER REQUIREMENTS

Bridge structures have to be designed for a service life of at least 100 years and the new code specifies the maximum water/cement ratio, minimum cement content and grade of concrete along with the cover requirements for different types of environmental exposure conditions as shown in Table 4.16.

Table 4.16 Durability Recommendations for Service Life of at least 100 years
(20 mm aggregate) (Table 14.2 of IRC: 112-2011)

Exposure Condition	Maximum Water/ Cement Ratio	Minimum Cement Content (Kg/m³)	Minimum Grade of Concrete	Minimum Cover (mm)
Moderate	0.45	340	M-25	40
Severe	0.45	360	M-30	45
Very Severe	0.40	380	M-40	50
Extreme	0.35	400	M-45	75

4.10 DESIGN OF REINFORCED CONCRETE SLAB CULVERT FOR IRC CLASS AA LOADS

Design a reinforced concrete slab culvert for a National Highway crossing to suit the following data:

Carriage way – Two lane (7.5 m wide)
Foot paths – 1 m on ether side
Clear span = 6 m
Wearing coat = 80 mm
Width of bearing = 400 mm
Materials: M-25 Grade Concrete and Fe-415 Grade HYSD bars
Loading-IRC Class AA tracked vehicle

Design the reinforced concrete slab deck and sketch the details of reinforcements in the longitudinal and cross section of the slab. The design should conform to the specifications of IRC: 6-2014 and IRC: 112-2011.

1. Data

Clear span = 6 m
Width of bearing = 400 mm
IRC Class AA tracked vehicle loading
M-25 Grade Concrete and Fe-415 Grade HYSD bars

2. Characteristic Strength of Materials

$f_{ck} = 25$ N/mm² $\qquad E_s = 200$ GPa $\qquad \alpha_e = [E_s/E_c] = 6.66$
$f_{yk} = 415$ N/mm² $\qquad E_c = 30$ GPa

3. Depth of Slab and Effective Span

a) Based on limit state of serviceability considerations of limiting deflections
   Ratio of Span/Depth $(L/d) = 12$ to 15
   For $(L/d) = 15$, $d = $ (span/15) = (6000/15) = 400 mm
   For $(L/d) = 12$, $d = $ (span/12) = (6000/12) = 500 mm

b) For highway solid slabs bridge decks

Thickness of the slab is assumed as 80 mm/metre of span

Depth of slab = (80 × 6) = 480 mm

Based on the above considerations, adopt overall depth of slab = h = 500 mm

Assuming moderate exposure conditions, clear cover = 40 mm

Using 20 mm diameter HYSD bars as main reinforcements with a clear cover of 40 mm,

Effective depth = [500 – 40 – 10] = 450 mm

Width of bearing = 400 mm

Effective span is the least of

a) (Clear span + effective depth) = (6 + 0.45) m= 6.45 m

b) Centre to centre of bearings = (6 + 0.40) = 6.4 m

The cross section of the deck slab is shown in Fig. 4.15.

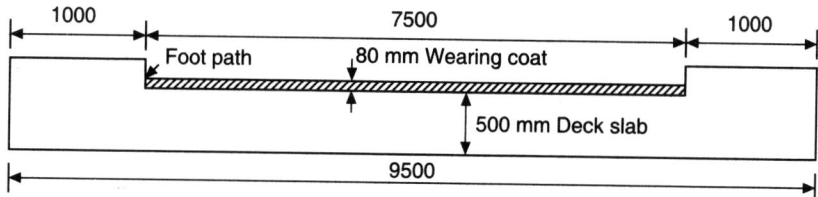

Fig. 4.15 Cross Section of Deck Slab.

4. Dead Load Bending Moments

Dead weight of slab = (0.5 × 24) = 12 kN/m²

Dead weight of W.C. = (0.08 × 22) = 1.76 kN/m²

Therefore Total load = (12 + 1.76); 14 kN/m²

Dead load bending moment = [0.125(14 × 6.4²)] = 72 kN·m

5. Live Load Bending Moments

Generally the bending moment due to live load will be maximum for IRC Class AA tracked vehicle loading. Impact factor for IRC Class AA tracked vehicle is 25% for 5 m span decreasing linearly to 10% for 9 m span. Therefore for 6.4 m span.

Impact factor = [25 – (15/4) × 96.4 – 5)] = 19.7%

The tracked vehicle is placed symmetrically on the span

Effective length of load = [3.6 + 2(0.5 + 0.08)] = 4.76 m

Effective width of slab perpendicular to span is expressed as,

$$b_e = Kx\left[1 - \frac{x}{L}\right] + b_w$$

Referring to Fig. 4.16, we have

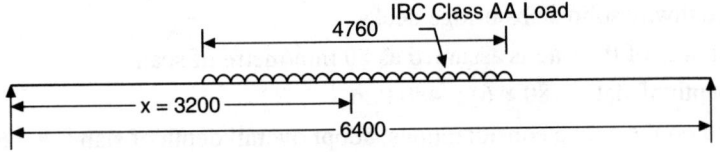

Fig. 4.16 Position of Load for Maximum Bending Moment.

$$x = 3.2 \text{ m}, L = 6.4 \text{ m}, B = 9.5 \text{ m and the ratio } (B/L) = 1.48$$
$$b_w = [0.85 + (2 \times 0.08)] = 1.01 \text{ m}$$

From Table 4.6, for $(B/L) = 1.48$ and simply supported slab the value of $K = 2.84$

$$b_e = [2.84 \times 3.2(1 - 3.2/6.4) + 1.01] = 5.56 \text{ m}$$

The tracked vehicle is placed close to the kerb with the required minimum clearance a as shown in Fig. 4.17.

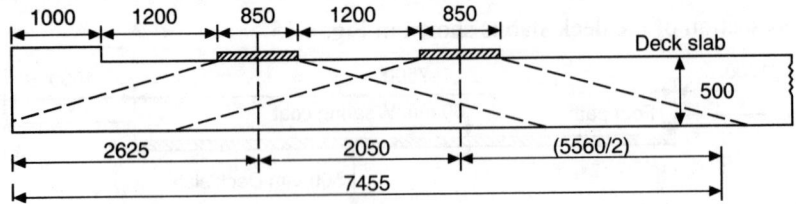

Fig. 4.17 Effective Width of Dispersion for I.R.C. Class AA Loads.

Net effective width of dispersion = 7.455 m

Total load of two tracks with impact = $(700 \times 1.197) = 838$ kN

Average intensity of load = $[838 / (4.76 \times 7.455)] = 23.61$ kN/m²

Maximum bending moment due to live load is given by,

$$M_{max} = [0.5(23.61 \times 4.76) \times 3.2] - [0.5 (23.61 \times 4.76) \times (4.76 \times 0.25)] = 113 \text{ kN·m}$$

∴ Total design bending Moment = $(113 + 72) = 185$ kN·m

Total design Ultimate moment $(M_u) = [1.35 \, M_d + 1.5 \, M_L]$
$$= [(1.35 \times 72) + (1.5 \times 113)]$$
$$= 267 \text{ kN·m/m}$$

6. Shear due to Class AA Tracked Vehicle

For maximum shear at support, the IRC Class AA tracked vehicle is arranged as shown in Fig. 4.18

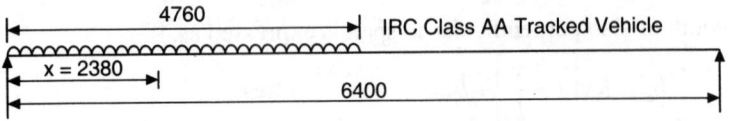

Fig. 4.18 Position of Load for Maximum Shear.

Effective width of Dispersion is given by

$$b_e = Kx\left[1 - \frac{x}{L}\right] + b_W$$

where $x = 2.38$ m $(B/L) = 1.48$

           $B = 9.5$ m $K = 2.84$

           $L = 6.4$ m

$\therefore$ $b_e = \{(2.84 \times 2.38)\,[1 - (2.38/6.4)] + 1.01\} = 5.256$ m

$\therefore$ Width of dispersion $= [2625 + 2050 + (0.5 \times 5256)] = 7303$ mm

Average intensity of load $= [838/(4.76 \times 7.303)] = 24.1$ kN/m^2

$\therefore$ Shear force $= V_A = [(24.1 \times 4.76 \times 4.02)/6.4] = 72$ kN

Dead load shear $= [0.5(14 \times 6.4)] = 45$ kN

Therefore total design shear force $= (72 + 45) = 117$ kN

Total design ultimate shear force $= [1.35\ V_d + 1.5\ V_L]$

                               $= [(1.35 \times 45) + (1.5 \times 72)]$

                               $= 169$ kN/m

7.   Design of Deck Slab

Using M-25 grade concrete and Fe-415 HYSD bars

Limiting moment of resistance for singly reinforced sections is expressed as (Refer Table 4.3)

$$M_{u,lim} = 0.138\ f_{ck}\ b\ d^2$$

$\therefore$ $d = \sqrt{\dfrac{M_u}{0.138 f_{ck}.b}} = \sqrt{\dfrac{267 \times 10^6}{0.138 \times 25 \times 1000}} = 278$ mm

Since the effective depth selected is 450 mm, the section is under reinforced.

The area of reinforcement required to resist the ultimate moment can be computed using the Relation,

$$M_u = 0.87 f_y A_{st} d\left[1 - \frac{A_{st} f_y}{b.d.f_{ck}}\right]$$

$$(267 \times 10^6) = 0.87 \times 415 \times A_{st} \times 450\left[1 - \frac{A_{st} \times 415}{1000 \times 450 \times 25}\right]$$

Solving $A_{st} = 1758$ mm^2

Using 20 mm diameter bars main reinforcement, the spacing is given by

$$S = \left[\frac{1000 a_{st}}{A_{st}}\right] = \left[\frac{1000 \times 314}{1758}\right] = 178.6\,\text{mm}$$

Provide 20 mm diameter bars at 150 mm centres

    A_{st} (provided) $= 2094$ mm^2

The distribution reinforcement should be designed to resist a moment computed as

Transverse moment = $[0.3\ M_{uL} + 0.2\ M_{uD}]$

$$M_{uL} = (1.5 \times 113) = 169.5 \text{ kN·m}$$

$$M_{uD} = (1.35 \times 72) = 97.2 \text{ kN·m}$$

Transverse moment = $[\ (0.3 \times 169.5) + (0.2 \times 97.2)] = 71$ kN·m

Area of distribution reinforcement = $[(2094/267) \times 71] = 557$ mm^2

Provide 12 mm diameter bars at a spacing of 200 mm centres.

8. Check for Ultimate Flexural Strength

$$M_u = \left(0.87 \times 415 \times 2094 \times 450\right)\left[1 - \frac{2094 \times 415}{1000 \times 450 \times 25}\right]$$

$$= (314 \times 10^6) \text{ N.mm}$$

$$= 314 \text{ kN·m} > 267 \text{ kN·m (Hence safe)}$$

9. Check for Ultimate Shear Strength

The ultimate shear strength of the reinforced concrete deck slab is checked by using the equation 4.7.

$$V_{Rd.c} = [0.12K\ (80\ \rho_1 f_{ck})^{0.33}]\ b_w.d$$

where

$$K = 1 + \sqrt{\frac{200}{d}} \leq 2.00 = \left[1 + \sqrt{\frac{200}{450}}\right] = 1.66$$

and

$$\rho_1 = \left(\frac{A_{sl}}{b_w.d}\right) \leq 0.02$$

$$= \left(\frac{2094}{1000 \times 450}\right) = 0.0046$$

$$V_{Rd.c} = [0.12 \times 1.66\ (80 \times 0.0046 \times 25)^{0.33}]\ (1000 \times 450) \text{ N}$$

$$= (188 \times 1000) \text{ N}$$

$$= 188 \text{ kN} > 169 \text{ kN (Hence safe)}$$

10. Check for Serviceability Limit States

a) Limit State of Cracking

1) Crack Width Computations by Rigorous Analysis

The width of cracks at the limit state of serviceability can be calculated using the expression.

$$W_k = s_{r,max}\ (\varepsilon_{sm} - \varepsilon_{cm})$$

$$\left(\varepsilon_{sm} - \varepsilon_{cm}\right) = \left[\frac{\sigma_s - k_t\left(\dfrac{f_{ct,eff}}{\rho_{p,ef}}\right)\left(1 + \alpha_e\,\rho_{p,ef}\right)}{E_s}\right] \geq 0.6\left(\frac{\sigma_s}{E_s}\right)$$

Substituting the relevant values in the design example, we have

$$f_{ck} = 25 \text{ N/mm}^2 \quad K_t = 0.5,$$

$f_{ct,eff} = f_{ctm} = 2.2$ N/mm^2 corresponding to the grade of concrete (Refer Table 4.1)

$$\alpha_e = [E_s/E_{cm}] = [200/30] = 6.66$$

For the cracked section using the parameters

$$h = 500 \text{ mm}, \, b = 1000 \text{ mm}, \, d = 450 \text{ mm}, \, A_s = 2094 \text{ mm}^2, \, \alpha_e = 6.66$$

Neutral axis depth 'x' is computed from the relation,

$$0.5\,b.\,x^2 = \alpha_e.A_s.(d - x)$$
$$(0.5 \times 1000 \times x^2) = 6.66 \times 2094 \times (450 - x)$$

Solving, the neutral axis depth $= x = 99$ mm

a) $2.5(h - d) = 2.5\,(500 - 450) = 125$ mm
b) $(h - x)/3 = [(500 - 99)/3] = 133.6$ mm
c) $(h/2) = (500/2) = 250$ mm

Hence the least value of $h_{c,eff} = 125$ mm

$$\rho_{p,eff} = \left[\frac{A_s}{b \times h_{c,eff}}\right] = \left[\frac{2094}{1000 \times 125}\right] = 0.0167$$

$k_1 = 0.8$ for high bond bars
$k_2 = 0.5$ for bending

For $\quad f_{ck} = 25 \text{ N/mm}^2, \, f_{cr} = 0.7\sqrt{f_{ck}} = 0.7\sqrt{25} = 3.5 \text{ N/mm}^2$

$$M_r = \left(\frac{f_{cr}I_r}{y_t}\right) = \left(\frac{3.5 \times 10.4 \times 10^9}{500}\right) = \left(72.8 \times 10^6\right) \text{N.mm}$$

$$M = 0.125\,wL^2 = [0.125 \times 14 \times 6.4^2] = 71.68 \text{ kN·m}$$

$$z = \text{Lever arm} = \left[d - \frac{x}{3}\right] = \left[450 - \frac{99}{3}\right] = 417 \text{ mm}$$

$$I_{gr} = \left[\frac{bh^3}{12}\right] = \left[\frac{1000 \times 500^3}{12}\right] = \left(10.4 \times 10^9\right) \text{mm}^4$$

$$I_r = \left(\frac{bx^3}{3}\right) + m\,A_{st}r^2 = \left(\frac{1000 \times 99^3}{3}\right) + \left(6.66 \times 2094 \times 351^2\right)$$
$$= \left(2.038 \times 10^9\right) \text{mm}^4$$

$$I_{eff} = \left[\frac{I_r}{1.2 - \left(\dfrac{M_r}{M}\right)\left(\dfrac{z}{d}\right)\left(1 - \dfrac{x}{d}\right)\left(\dfrac{b_w}{b}\right)}\right]$$

$$I_{eff} = \left[\frac{2.038 \times 10^9}{1.2 - \left(\dfrac{72.8 \times 10^6}{71.68 \times 10^6}\right)\left(\dfrac{417}{450}\right)\left(1 - \dfrac{99}{450}\right)(1)}\right] = \left(2.82 \times 10^9\right) \text{mm}^4$$

Also, $\qquad I_r \leq I_{eff} \leq I_g$

$(2.038 \times 10^6) \leq (2.82 \times 10^9) \leq (10.4 \times 10^9)$

Cover provided $= c = 40$ mm

If spacing is less than $5(c + \phi/2)$ then maximum final crack spacing may be calculated using the relation,

Maximum spacing of cracks $= s_{r,max} = [3.4c + 0.17\phi/\rho_{p,ef}]$

In the present example, $5(c + \phi/2) = 5(40 + 20/2) = 250$ mm

But spacing of main tension bars is 150 mm < 250 mm

Hence $s_{r,max} = [(3.4 \times 40) + (0.17 \times 20)/0.0167] = 339$ mm

Service load moment $= M = (72 + 113) = 185$ kN·m

$$\text{Stress in steel at service load} = \sigma_s = \left[\frac{M}{\left(d - \dfrac{x}{3}\right)A_{st}}\right] = \left[\frac{185 \times 10^6}{\left(450 - \dfrac{99}{3}\right)2094}\right] = 212 \text{ N/mm}^2$$

Using the above values of the various parameters, the width of crack is computed as

$$w_k = s_{r,max}\left[\frac{\sigma_s - k_t\left(\dfrac{f_{ct,eff}}{\rho_{p,ef}}\right)\left(1 + \alpha_e\rho_{p,ef}\right)}{E_s}\right]$$

$$= 399\left[\frac{212 - 0.4\left(\dfrac{2.2}{0.0167}\right)\left(1 + 6.66 \times 0.0167\right)}{200 \times 10^3}\right]$$

$$= 0.26\text{mm} \geq 0.6\left(\frac{\sigma_s}{E_s}\right) \geq 0.6\left(\frac{212}{200 \times 10^3}\right) \geq 0.0006 \text{ mm}$$

The maximum crack width is less than the permissible value 0.3 mm. (Hence safe)

2) Control of Cracking Without Direct Calculation

The IRC: 112-2011 code outlines a simpler method for crack control by limiting the stress in steel and the maximum bar size and spacing for permissible crack widths of 0.3 and 0.2 mm. The code recommends the use of Tables 4.10 and 4.11.

Service load moment = M_w = 185 kN·m
The stress in steel at working load is computed as

$$\sigma_s = \left[\frac{M}{\left(d - \frac{x}{3}\right)A_{st}}\right] = \left[\frac{185 \times 10^6}{\left(450 - \frac{99}{3}\right)2094}\right] = 212 \, \text{N/mm}^2$$

The slab is reinforced with 20 mm diameter bars spaced at 150 mm centers.

Referring to tables 4.10 and 4.11, the bar size and spacing are well within the safe limits for control of cracking.

b) Limit State of Deflection

1) The deflection due to shrinkage can be computed by the relation

$$a_{cs} = k\psi_{cs} L^2$$

Where, k = a constant = 0.125 for simply supported beams

$$\psi_{cs} = \text{shrinkage curvature} = \left[\varepsilon_{cs} \alpha_e \left(\frac{S}{I}\right)\right]$$

$$\varepsilon_{cs} = \text{Total shrinkage strain} = (\varepsilon_{cd} + \varepsilon_{ca})$$

Where, ε_{cd} is the drying shrinkage strain
ε_{ca} is the autogenous shrinkage strain

The development of drying shrinkage strain with time is expressed as

$$\varepsilon_{cd}(t) = [\beta_{ds}(t, t_s)k_h \varepsilon_{cd}]$$

Where, $\beta_{ds}(t, t_s) = \left[\dfrac{(t - t_s)}{(t - t_s) + 0.04\sqrt{h_o^3}}\right]$

And h_o = Notional size (mm) of the cross section = $(2A_c/u)$

A_c = cross sectional area of concrete

U = perimeter of that part of the cross section exposed to drying

t = age of the concrete in days at the time considered

t_s = age of concrete in days at the beginning of drying shrinkage, normally this is at the end of curing (28 days)

k_h = Values given Table 4.14

Substituting the numerical values, we have

$$h_o = (2A_c/u) = (2 \times 1000 \times 500)/2000 = 500 \, \text{mm}$$

$t = 365$ days, $t_s = 28$ days and $h_o = 500$

$$\beta_{ds}(t, t_s) = \left[\frac{(365 - 28)}{(360 - 28) + 0.04\sqrt{500^3}} \right] = 0.42$$

Final value of drying shrinkage strain $\varepsilon_{cd,\infty} = k_h.\varepsilon_{cd}$

From Table 4.13, the value of $k_h = 0.70$

For relative humidity of 50 percent, and M-25 grade concrete

The value of unrestrained drying shrinkage strain from Table 4.12 is extrapolated as

$$\varepsilon_{cd} = (535 \times 10^{-6})$$
$$\varepsilon_{cd}(t) = [\beta_{ds}(t, t_s)k_h\varepsilon_{cd}] = [0.42 \times 0.70 \times 535 \times 10^{-6}] = 0.000157$$

Autogenous shrinkage strain is interpolated from Table 4.13 for concrete of grade M-25 as

$$\varepsilon_{ca} = (30 \times 10^{-6})$$

Total shrinkage strain $= \varepsilon_{cs} = (\varepsilon_{cd} + \varepsilon_{ca}) = (157 + 30)\,10^{-6} = (187 \times 10^{-6})$

$$\psi_{cs} = \text{shrinkage curvature} = \left[\varepsilon_{cs}\,\alpha_e \left(\frac{S}{I} \right) \right]$$

$$\varepsilon_{cs} = (187 \times 10^{-6})$$
$$\alpha_e = 6.66$$

S = First moment of area of reinforcement about centroid of the section

$= (2904 \times 200) = (580.8 \times 10^3)$ mm^3

$I = (10.4 \times 10^9)$ mm^4

$$\left(\frac{S}{I} \right) = [(580.8 \times 10^3)/(10.4 \times 10^9)] = (55.8 \times 10^{-6})$$
$$\psi_{cs} = [(187 \times 10^{-6}) \times 6.66 \times 55.8 \times 10^{-6}] = (69494 \times 10^{-12})$$
$$a_{cs} = [k\,\psi_{cs}\,L^2] = [0.125 \times 69494 \times 10^{-12}) \times 6400^2] = 0.355 \text{ mm}$$

2) Long term deflection due to sustained (dead) loads

Total dead load $= g = 14$ kN/m $= 14$ N/mm

Effective span $= L = 6.4$ m $= 6400$ mm

Modulus of elasticity of concrete $= E_c = 30$ kN/mm^2 $= 30{,}000$ N/mm^2

Effective moment of inertia $= I_{eff} = (2.82 \times 10^9)$ mm^4

Maximum short term deflection due to dead load $= a_g = \left[\dfrac{5gL^4}{384E_cI} \right]$

$$= \left[\frac{5 \times 14 \times 6400^4}{384 \times 25000 \times 2.82 \times 10^9} \right]$$

$$= 4.34 \text{ mm}$$

For computing long term deflection, effective modulus of elasticity involving the effect of Creep has to be used.

Final Creep coefficient of concrete = Ø depends upon the notional size, age at loading and the atmospheric conditions as given in Table 4.12.

$$E_{c,eff} = \left[\frac{E_c}{1+\emptyset} \right]$$

The value of Ø depends upon the notional size of the member $(2A_c/u)$

Where, A_c = cross sectional area of the member and

              U = perimeter in contact with the atmosphere (mm)

Substituting relevant values we have

$$(2A_c/u) = [(2 \times 1000 \times 500)/2000] = 500 \text{ mm}$$

Referring to Table 4.9, for age at loading of 28 days and relative humidity of 50%, interpolating the value of Ø = 2.6

Hence, $(1 + \emptyset) = 3.6$ and $E_{c,eff} = (E_c/3.6)$

Long term deflection due to permanent loads = $(3.6 \times 4.34) = 15.6$ mm

3) Deflection due to live loads

The live load due to the IRC Class AA tracked vehicle is computed as 23.61 kN/m². Spread over a length of 4.76 m at the centre of span of 6.4 m. Assuming the entire span is loaded with an uniformly distributed load of 23.61 kN/m, a conservative estimate of the maximum deflection is computed as,

$$a_q = \left[\frac{5 \times q \times L^4}{384 E_c I_{eff}} \right] = \left[\frac{5 \times 23.61 \times 6400^4}{384 \times 30000 \times 2.82 \times 10^9} \right] = 6.08 \text{ mm}$$

Hence the total final deflection is computed as the sum of shrinkage, dead loads with creep and live load deflections.

Final deflection = $(0.355 + 15.60 + 6.08) = 22.03$mm

Maximum deflection due to live loads (6.08 mm) $= (a_q) \leq \left(\dfrac{span}{800} \right) = \left(\dfrac{6400}{800} \right) \leq 8 \text{ mm}$

Total maximum deflection $(22.03 \text{ mm}) \leq \left(\dfrac{span}{250} \right) = \left(\dfrac{6400}{250} \right) \leq 25.6 \text{ mm}$

Hence the serviceability limit state of deflection is well within the limits specified in the IRC: 112-2011 code.

11. Reinforcement Details in Deck Slab

The details of reinforcements in the deck slab are shown in Fig. 4.19.

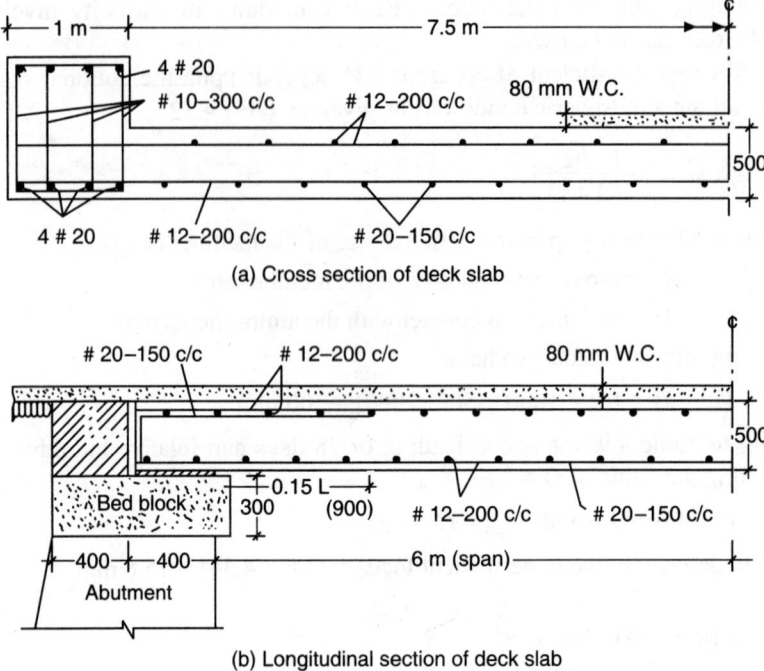

(a) Cross section of deck slab

(b) Longitudinal section of deck slab

Fig. 4.19 Details of Reinforcement in Desk Slab

4.11 DESIGN OF REINFORCED CONCRETE SLAB CULVERT FOR IRC CLASS A LOADS

Compare the design live load moments for the reinforced concrete slab culvert using IRC Class A loading. Show that the live load moments are significantly lower when compared with those resulting from IRC Class AA tracked vehicle loading on the slab culvert. Adopt the same data of example in section 4.10.

1. Data

Effective span = L = 6.4 m

Width of road way = B = 9.5 m

Wearing coat = 80 mm

M-25 Grade concrete and Fe-415 HYSD bars

Loading: IRC Class A two lane

2. Position of Live Loads for Maximum Bending Moment

The two heavier axle loads each of 114 kN (IRC Class A load train) are to be placed on the span so as to produce maximum bending moment. The position of loads for maximum bending moment is shown in Fig. 4.20.

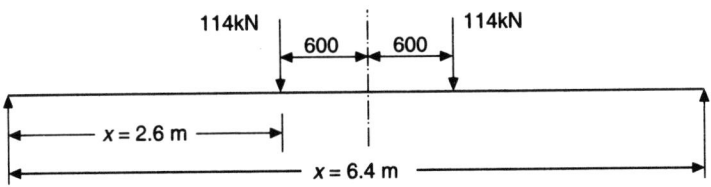

Fig. 4.20 Position of I.R.C. Class-A Loads for Maximum B.M.

3. Impact Factor

The impact factor for class A loading is given by

$$I = \left[\frac{4.5}{(6+L)}\right] = \left[\frac{4.5}{(6+6.4)}\right] = 0.363$$

4. Effective Width of Dispersion

The effective width of dispersion perpendicular to span is given by

$$b_e = Kx\left[1 - \frac{x}{L}\right] + b_w$$

Where, $x = 2.6$ m

$L = 6.4$ m   Ratio $(B/L) = 1.48$

$B = 9.5$ m

From Table 4.6, the value of $K = 2.84$

$$b_w = [0.5 + (2 \times 0.08)] = 0.66 \text{ m}$$

$$b_e = (2.84 \times 2.6)\left[1 - \frac{2.6}{6.4}\right] + 0.66 = 5.05$$

The wheels are arranged as shown in Fig. 4.21.

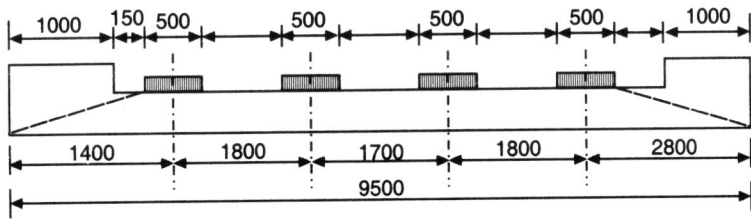

Fig. 4.21 Effective Width of Dispersion for Class-A Two Lane Wheel Loads.

The effective width of dispersion exceeds the width of the slab. Hence the effective width of dispersion is taken as 9.5 m.

The intensity of loading including impact is computed as

$$w = \left[\frac{(4 \times 114 \times 1.363)}{(2.61 \times 9.5)} \right] = 27.36 \text{ kN/m}^2$$

Width of dispersion parallel to span = $[0.25 + 2(0.5 + 0.08)] = 1.41$ m

The 114 kN wheels are placed symmetrically on the span as shown in Fig. 4.22.

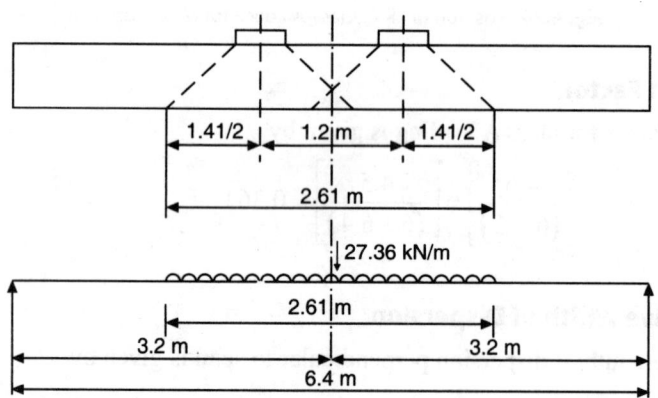

Fig. 4.22 Position of I.R.C. Class-A Wheel Loads for Maximum Bending Moment.

$$M_{max} = [(27.36 \times 0.5 \times 2.61) \times 3.2] - [(27.36 \times 0.5 \times 2.61) \times 0.25 \times 2.61]$$
$$= 91.00 \text{ kN·m}$$

The value of bending moment is significantly lower than the value of 113 kN·m obtained for IRC Class AA loading.

4.12 DESIGN OF REINFORCED CONCRETE SLAB SUPPORTED ON ALL SIDES (TEE BEAM AND SLAB DECK) FOR IRC CLASS A LOADS

The slab panel of a reinforced concrete Tee beam and slab deck is 2.5 m wide between main girders and 4 m between cross girders. Design the slab for IRC Class A loading. Adopt M-20 Grade concrete and Fe-415 grade HYSD bars.

1. Data

Two way slab panel 2.5 m wide by 4 m supported on all the four sides and continuous over main and cross girders.

Loading: IRC Class A train

Materials: M-20 Grade concrete and Fe-415 Grade HYSD bars

Thickness of slab = 200 mm

Wearing coat = 80 mm

2. Characteristic Strength of Materials

$f_{ck} = 25$ N/mm^2 $E_s = 200$ GPa $\alpha_e = [E_s/E_c] = 6.9$

$f_{yk} = 415 \text{ N/mm}^2$ $\qquad E_c = 29 \text{ GPa}$

3. Bending Moments

The arrangement of wheel loads is as shown in Fig. 4.23.

$$W_1 = W_2 = 57 \text{ kN}, L = 4 \text{ m}, B = 2.5 \text{ m}$$

B.M due to W_1

$$u = [0.5 + (2 \times 0.08)] = 0.66 \text{ m}$$
$$v = [0.25 + (2 \times 0.08)] = 0.41 \text{ m}$$
$$(u/B) = (0.66/2.5) = 0.264$$
$$(v/L) = (0.41/4) = 0.102$$
$$K = (B/L) = (2.5/4) = 0.625$$

Using Pigeaud's curve (Refer Fig. 4.8) for $K = 0.6$, the moment coefficients are read out as

$$m_1 = 0.188 \text{ and } m_2 = 0.148$$

Short span moment $= M_B = W(m_1 + 0.15\, m_2)$
$$= 57[0.188 + (0.15 \times 0.148)] = 11.99 \text{ kN·m}$$

Long span moment $= M_L = W(m_2 + 0.15\, m_1)$
$$= 57[0.148 + (0.15 \times 0.188)] = 10.04 \text{ kN·m}$$

B.M. due to load W_2 (Unsymmetrical load)

An imaginary load equal to W_2 is placed symmetrically as shown in Fig. 4.24.

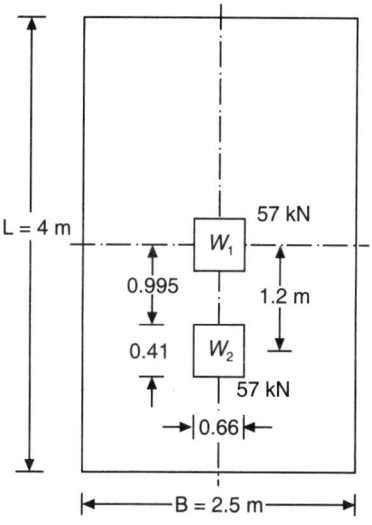

Fig. 4.23 Arrangement of Class-A Wheel Loads.

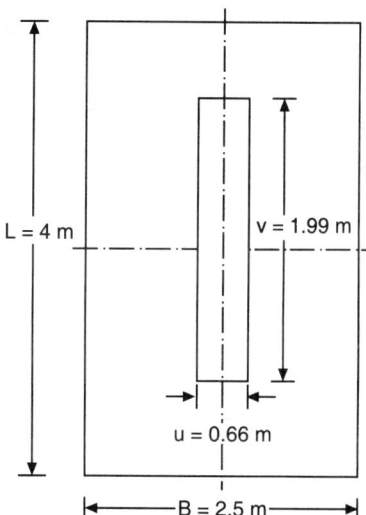

Fig. 4.24 Arrangement of Imaginary Load (Class-A) on Slab.

$$W_2 = 57 \text{ kN}$$

$$\text{Intensity of load} = \left[\frac{57}{(0.41 \times 0.66)} \right] = 210.6 \text{ kN/m}^2$$

$$(u/B) = (0.66/2.5) = 0.264$$
$$(v/L) = (2.81/4) = 0.702$$
$$K = (B/L) = (2.5/4) = 0.625$$

Using Pigeaud's curve (Fig. 4.8) for $K = 0.6$, read out the values of moment coefficients $m_1 = 0.12$ and $m_2 = 0.038$. The bending moments in the short and long span are computed as

$$M_B = W (m_1 + 0.15 \, m_2)$$
$$= [210.6 \times 2.81 \times 0.66] [0.12 + (0.15 \times 0.038)]$$
$$= 49.10 \text{ kN·m}$$
$$M_L = W (m_2 + 0.15 \, m_1)$$
$$= [210.6 \times 2.81 \times 0.66] [0.038 + (0.15 \times 0.12)]$$
$$= 21.87 \text{ kN·m}$$

Subtracting the moments due to the load as shown in Fig. 4.25,

$$(u/B) = (0.66/2.5) = 0.264$$
$$(v/L) = (1.99/4) = 0.498$$
$$K = (B/L) = (2.5/4) = 0.625$$

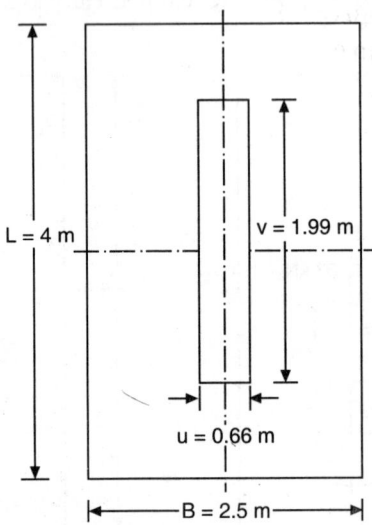

Fig. 4.25 Arrangement of Imaginary Load (Class-A) on Slab.

Using Pigeaud's curve (Fig. 4.8) for $K = 0.6$, read out the moment coefficients $m_1 = 0.142$ and $m_2 = 0.049$

$\therefore$ $\qquad M_B = W (m_1 + 0.15\ m_2)$

$\qquad\qquad = [210.6 \times 1.99 \times 0.66]\ [0.142 + (0.15 \times 0.049)]$

$\qquad\qquad = 41.30$ kN·m

$\qquad M_L = W (m_2 + 0.15\ m_1)$

$\qquad\qquad = [210.6 \times 1.99 \times 0.66]\ [0.049 + (0.15 \times 0.142)]$

$\qquad\qquad = 19.44$ kN·m

Moments due to W_2 are computed as

$\qquad M_B = 0.5[49.10 - 41.30] = 3.90$ kN·m

$\qquad M_L = 0.5[21.87 - 19.44] = 1.22$ kN·m

Applying continuity and impact factors, the total live load moments are given by

$\qquad M_B = (1.25 \times 0.8)[11.99 + 3.90] = 15.89$ kN·m

$\qquad M_L = (1.25 \times 0.8)[10.04 + 1.22] = 11.26$ kN·m

Dead weight of slab = $(0.2 \times 24) = 4.8$ kN/m²

Dead weight of wearing coat = $(0.08 \times 22) = 1.76$ kN/m²

Total dead load = 6.56 kN/m²

Referring to Pigeaud's curve (Fig. 4.13)

$\qquad (u/B) = 1, (v/L) = 1$ and $K = (B/L) = 0.625$

$\qquad m_1 = 0.049$ and $m_2 = 0.015$

Moments due to dead load are computed as

$\qquad M_B = 65.60[0.049 + (0.15 \times 0.015)] = 3.36$ kN·m

$\qquad M_L = 65.60[0.015 + (0.15 \times 0.049)] = 1.468$ kN·m

Taking continuity into effect

$\qquad M_B = (0.8 \times 3.36) = 2.688$ kN·m

$\qquad M_L = (0.8 \times 1.468) = 1.174$ kN·m

Total design moments are given by

$\qquad M_B = (15.89 + 2.688) = 18.578$ kN·m

$\qquad M_L = (11.26 + 1.174) = 12.434$ kN·m

4. Design of Slab

Total design short span Ultimate moment $(M_{Bu}) = [1.35\ M_d + 1.5\ M_L]$

$\qquad\qquad\qquad\qquad\qquad\qquad = [(1.35 \times 2.688) + (1.5 \times 15.89)]$

$\qquad\qquad\qquad\qquad\qquad\qquad = 27$ kN·m/m

Total design Long span Ultimate moment $(M_{Lu}) = [(1.35 \times 1.174) + (1.5 \times 11.26)]$

$\qquad\qquad\qquad\qquad\qquad\qquad = 18.47$ kN·m/m

Effective depth of slab required $= d = \sqrt{\dfrac{M_u}{0.138 f_{ck}.b.}} = \sqrt{\dfrac{27 \times 10^6}{0.138 \times 20 \times 1000}} = 99$ mm

Using 12 mm diameter bars and clear cover of 40 mm

Effective depth provided = 154 mm

$$\left(\frac{M_u}{b.d^2}\right) = \left(\frac{27 \times 10^6}{1000 \times 154^2}\right) = 1.13, \text{ using M-25 grade concrete and Fe-415 HYSD bars}$$

Read out the percentage of reinforcement required from Table-3 of SP: 16 Design Aids

$$p_t = 0.330 = \left(\frac{100 A_{st}}{b.d}\right)$$

Solving $\qquad A_{st} = \left(\frac{0.330 \times 1000 \times 154}{100}\right) = 508 \text{ mm}^2$

For short span, provide 12 mm diameter bars at 150 mm centers (A_{st} provided = 754 mm^2).

For long span, provide 10 mm diameter bars at 150 mm centers.

5. Check for Shear Strength

Maximum shear force developed near the support along the short span of 2.5 m between the tee beam girders should be evaluated by positioning the IRC Class A wheel loads as shown in Fig. 4.26

Dispersion of wheel load in the direction of span = [0.50 + 2(0.08 + 0.2)] = 1.06 m

Intensity of wheel load = 57 kN

Maximum live load shear force $= V_q = \left[\dfrac{57 \times 1.67}{2.2}\right] = 47.6 \text{ kN/m}$

Dead weight of slab = (1 × 1 × 0.2 × 24) = 4.8 kN/m^2

Dead weight of wearing coat = (0.08 × 22) = 1.76 kN/m^2

Total dead weight = g = (4.80 + 1.76) = 6.56 kN/m^2

Maximum dead load shear force = V_g = (6.56 × 2.2)0.5 = 7.2 kN/m

Maximum ultimate load design shear force = [1.35 V_g + 1.5 V_q]

$$= [(1.35 \times 7.2) + (1.5 \times 47.6)]$$

$$= 81.12 \text{ kN}$$

The ultimate shear strength of the reinforced concrete deck slab is checked by using the equation 4.7.

$$V_{Rd.c} = [0.12K \, (80 \, p_1 f_{ck})^{0.33}] \, b_w.d$$

Where, $\qquad K = 1 + \sqrt{\dfrac{200}{154}} = 2.13$ but limited to 2.00

and $\rho_1 = \left(\dfrac{A_{sl}}{b_w.d}\right) \leq 0.02$

$$= \left(\frac{754}{1000 \times 154} \right) = 0.0049$$

$$
\begin{aligned}
V_{\text{Rd.c}} &= [0.12 \times 2.00 \, (80 \times 0.0049 \times 25)^{0.33}](1000 \times 154) \text{ N} \\
&= (81.3 \times 1000) \text{ N} \\
&= 81.3 \text{ kN} > 81.12 \text{ kN (Hence safe)}
\end{aligned}
$$

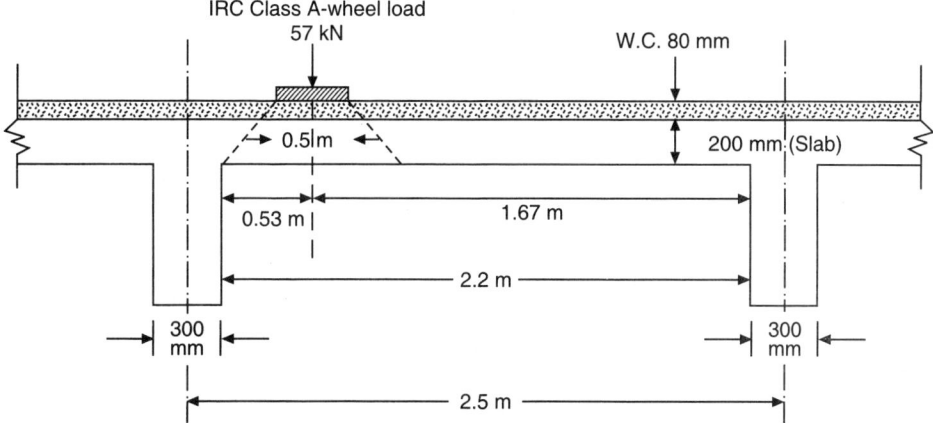

Fig. 4.26 Position of IRC Class-A Wheel Load on Slab for Maximum Shear.

6. Check for Cracking

According to IRC: 112-2011, cracking can be controlled without direct computations of the width of cracks by limiting the size of reinforcing bars, spacing of the bars, stress in steel and the permissible crack width as listed in Tables 4.10 and 4.11. This empirical method will be followed in this example to check for the serviceability limit state of cracking in the deck slab.

Depth of neutral axis (x) is determined using the relation,

$$(0.5 \times 1000 \times x^2) = [6.66 \times 754 \times (154 - x)]$$

Solving $x = 37.3$ mm

Stress in steel under working loads is computed as

$$
\sigma_s = \left[\frac{M_w}{\left(d - \dfrac{x}{3} \right) A_{st}} \right] = \left[\frac{18.578 \times 10^6}{\left(154 - \dfrac{37.3}{3} \right) 754} \right] = 174 \text{ N/mm}^2
$$

The deck slab is reinforced using 12 mm diameter bars at 150 mm centers

Referring to Table 4.10, for a stress of 174 N/mm², maximum bar diameter permissible to Limit the crack width to 0.3 is interpolated as 28 mm as against 12 mm used in the slab.

From Table 4.11, maximum spacing of bars to limit the crack width to 0.3 mm should not exceed 290 mm as against 150 mm used in the slab.

Hence the slab is safe regarding the serviceability limit stat of cracking according to the specifications of IRC: 112-2011.

7. Check for Deflection

Effective moment of inertia of the slab is computed using the relation

$$I_{eff} = \left[\frac{I_r}{1.2 - \left(\dfrac{M_r}{M}\right)\left(\dfrac{z}{d}\right)\left(1 - \dfrac{x}{d}\right)\left(\dfrac{b_w}{b}\right)} \right]$$

Where,

$$I_r = \left(\frac{bx^3}{3}\right) + mA_{st}r^2$$

$$= \left(\frac{1000 \times 37.3^3}{3}\right) + (6.66 \times 754)116.7^2$$

$$= (85.68 \times 10^6) \text{ mm}^4$$

$$f_{cr} = 0.7\sqrt{f_{ck}} = 0.7\sqrt{25} = 3.5 \text{ N/mm}^2$$

$$I_{gr} = \left(\frac{bh^3}{12}\right) = \left(\frac{1000 \times 200^3}{12}\right) = (666 \times 10^6) \text{ mm}^4$$

$$M_r = \left(\frac{f_{cr}I_{gr}}{y_t}\right) = \left(\frac{3.5 \times 666 \times 10^6}{100}\right) = (23.31 \times 10^6) \text{ N.mm}$$

$$Z = \text{lever arm} = \left(d - \frac{x}{3}\right) = \left(154 - \frac{37.3}{3}\right) = 141.57 \text{ mm}$$

$$M = \text{Service load moment} = 18.578 \text{ kN·m}$$

$$d = 154 \text{ mm and } x = 37.3 \text{ mm}$$

$$I_{eff} = \left[\frac{85.68 \times 10^6}{1.2 - \dfrac{23.31 \times 10^6}{18.587 \times 10^6}\left(\dfrac{141.57}{154}\right)\left(1 - \dfrac{37.3}{154}\right)1} \right] = (262 \times 10^6) \text{ mm}^4$$

Service dead load of slab and wearing coat = g = 6.56 kN/m²

Maximum short term deflection due to dead load is computed as

$$a_{sd} = \left[\frac{5gL^4}{384E_c I_{eff}}\right] = \left[\frac{5 \times 6.56 \times 2500^4}{384 \times 29000 \times 262 \times 10^6}\right] = 0.43 \text{ mm}$$

Creep coefficient depends upon notional size of member given by $\left(\dfrac{2A_c}{u}\right)$

$$\left(\frac{2A_c}{u}\right) = \left(\frac{2\times1000\times200}{2000}\right) = 200\,\text{mm}$$

From Table 4.12, interpolate the final creep coefficient for the given notional size of 200 mm at relative humidity of 50% and age at loading of 28 days as $\emptyset = 3.00$.

Effective modulus of elasticity of concrete $= E_{c,\text{eff}} = \left[\dfrac{E_c}{1+\emptyset}\right] = \left[\dfrac{E_c}{1+3}\right] = \left[\dfrac{E_c}{4}\right]$

Long term deflection due to dead loads $= a_{\text{dL}} = (4 \times 0.43) = 1.72$ mm

Deflection due to live load of 57 kN at centre of span of 2.5 m is given by

$$a_{\text{LL}} = \left(\frac{QL^3}{48\,E_c\,I_{\text{eff}}}\right) = \left(\frac{57\times10^3\times2500^3}{48\times30000\times262\times10^6}\right) = 2.36\,\text{mm}$$

Maximum permissible deflection due to live load $\leq \left(\dfrac{span}{800}\right) = \left(\dfrac{2500}{800}\right) = 3.1\,\text{mm} > 2.36\,\text{mm}$

Total deflection due to dead and live loads $= (1.72 + 2.36)$

$$= 4.08\,\text{mm} \leq \left(\frac{2500}{250}\right) = 10\,\text{mm}$$

Hence the slab is safe regarding the serviceability limit stat of deflection according to the specifications of IRC: 112-2011.

EXAMPLES FOR PRACTICE

1. A reinforced concrete slab culvert is required for a National highway crossing to suit the following data:

   Width of carriageway = 7.5 m
   Width of kerb = 600 mm
   Wearing coat = 80 mm
   Clear span = 5 m
   Width of bearing = 400 mm
   Foot paths 1 m on either side
   Type of loading: IRC Class AA or A whichever gives the worst effect
   Materials: M-30 Grade Concrete & Fe-415 HYSD bars

   Design the deck slab conforming to the specifications of IRC codes and draw the following views:

   (a) Half cross section of deck slab showing details of reinforcements

   (b) Half longitudinal section of deck slab showing the details of reinforcements and bearing pads.

   (Bangalore University 1985)

2. A road bridge deck consists of a reinforced concrete slab continuous over Tee beams spaced 2 m apart and cross girders spaced at 5 m centres. Thickness of wearing coat = 100 mm. Type of loading is IRC Class AA tracked vehicle. Using M-25 grade concrete & Fe-415 HYSD bars design the R.C. slab and draw the cross section and longitudinal section of the slab. The design should conform to the relevant IRC codes.

    (Bangalore University 1984)

3. Design a reinforced concrete slab culvert for a state highway crossing to suit the following data:

    Carriage way: Two lane traffic 7.5 m wide
    Foot Paths: 1 m on either side
    Clear span = 6 m
    Wearing coat = 80 mm
    Width of bearing = 500 mm
    Materials: M-30 Grade concrete & Fe-415 HYSD reinforcements
    Loading: IRC Class A load train

    Design the slab culvert and draw the cross section and longitudinal section showing the details of reinforcements.

    (Mysore University 1984)

4. The reinforced concrete slab panel of a reinforced concrete Tee beam and slab deck is 2 m wide between Tee beams and 4 m long between cross girders. Design the R.C. slab panel for IRC Class A loading using M-30 grade concrete and Fe-500 grade HYSD bars. Assume the thickness of the wearing coat as 80 mm. Sketch the details of reinforcements in the slab.

5. Design the reinforced concrete slab of a Tee beam and slab bridge deck using the following data:

    Spacings of main beams = 3 m
    Effective span of the Tee beams = 20 m
    No Cross girders have been provided
    Thickness of the wearing coat = 80 mm
    Loading: IRC Class AA tracked vehicle
    Materials: M-25 Grade concrete & Fe-415 HYSD bars

    Sketch the details of reinforcements in the slab.

6. Design a slab culvert to suit the following data:

    Effective span = 6.5 m
    Thickness of wearing coat = 80 mm
    Width of road = 7.5 m with kerbs 600 mm on either side
    Loading: IRC Class 70 R tracked vehicle
    Materials: M-30 Grade concrete & Fe-415 HYSD bars

    Design the slab to conform to IRC: 112-2011 code specifications and sketch the details of reinforcements in the deck slab.

7. Design a reinforced concrete slab for a bridge culvert for a national high way to suit the following data:

Carriage way .... Three lane road with a width of 10 m
Foot paths .... 1 m wide on either side
Clear span .... 7 m
Wearing coat.............. 75 mm
Width of bearing .... 500 mm
Materials .... M-30 grade concrete and Fe-500 HYSD bars
Loading .... RC Class AA tracked vehicle

Sketch the details of reinforcements in the slab.

The design should conform to the latest IRC bridge codes, IRC: 6-2014 and IRC: 112-2011.

8. A reinforced concrete slab culvert is to be designed for a national high way crossing Conforming to the IRC Bridge code (IRC: 112-2011). Using the following data at site design a suitable slab for the bridge deck.

Clear span between supports .... 6 m
Foot paths on either side .... 1 m wide
Width of bearing .... 400 mm
Wearing coat .... 80 mm
Materials .... M-30 grade concrete & Fe-500 HYSD bars
Type of loading .... IRC class A loads

Sketch the details of reinforcements in the cross section and longitudinal section of the deck Slab. Check for the limit states of strength and serviceability.

REFERENCES

1. IRC: 6-2014, Standard Specifications and Code of Practice for Road Bridges, Section II, Loads and Stresses (Revised Edition), Indian Roads Congress, New Delhi, 2014, pp. 1-84.

2. Victor, D.J. and Chettiar, C.G., Design Charts for Highway Bridge Slabs, Transport Communications Quarterly Review, Indian Roads Congress, New Delhi, September 1969, pp. 105-109.

3. IRC: 112-2011, Code of Practice for Concrete Road bridges, Indian Roads Congress, New Delhi, 2011, p. 181.

4. ROWE, R.E., CRANSTON, W.B., and BEST, B.C., New Concepts in the Design of Concrete, Structural Engineer, Vol.43, 1965, pp. 339-403.

5. BATE, S.C.C., Why Limit State Design, Concrete, March 1968, pp.103-108.

6. C.E.B. recommendations for International Code of Practice for Reinforced Concrete, American Concrete Institute and Cement & Concrete Association, London, 1964.

7. SP: 16-1980, Design Aids For reinforced Concrete to IS: 456-2000, Bureau of Indian Standards, New Delhi, 11th Reprint, New Delhi, 1999.

8. BS EN: 1992-1-1-2004, Euro Code – 2, Design of Concrete Structures, General Rules and Rules for Buildings, British Standard Institution, London, 2004.

9. ACI: 318M-11-2011, Building Code Requirements for Structural Concrete, American Concrete Institute, Farmington Hills, Michigan, USA, 2011.

10. ASCE-ACI Committee-426, The Shear Strength of reinforced Concrete Members, ASCE Journal, Structural Division, Vol.99, No. ST-6, 1973, pp. 1091-187.

11. Mattock, A.H and N.M. Hawkins., Shear Transfer in reinforced Concrete; Recent research, Journal of Prestressed concrete Institutue, Vol. 17, No.2, 1972, pp. 55-75.

12. Krishna Raju, N., Reinforced Concrete Structural Elements, New Age International Publishers, New Delhi, 2016, pp. 136-152.

13. BS EN: 1992-3-2006, Euro Code-2, Design of Concrete Structures, Part-3, Liquid Retaining and Containment Structures, British Standard Institution, London, 2006.

14. IS: 456-2000, Indian Standard Code of Practice for Plain and Reinforced Concrete (Fourth Revision), Bureau of Indian Standards, 2000, pp.100.

15. IS: 1343-2012, Indian Standard Code of Practice for Prestressed Concrete, Bureau of Indian Standards, New Delhi, 2012, pp. 1-54.

REVIEW QUESTIONS

1. Briefly explain the various types of Slab bridge decks generally adopted in different types of high way bridges.

2. Discuss briefly the various types of I.R.C. High way loadings specified in the design of slab bridge decks.

3. Explain the terms a) effective span, b) Impact factor, c) Dispersion of concentrated loads with reference to the design of slab bridge decks.

4. What are Pigeaud's curves? Where do you use them? Briefly specify the advantages and limitations of using these curves.

5. Explain with sketches the position of IRC class AA tracked vehicle on a deck Slab for the computation of maximum moments and shear forces.

6. How do you check for serviceability limit state of cracking in bridge slabs? Explain the Methods recommended in the IRC codes.

7. List the various factors influencing the computation of cracks in bridge deck slabs and Mention the permissible crack widths according to the IRC codes.

8. Briefly explain the method of computing the deflections in deck slabs in relation to the Serviceability limit state specified in the bridge codes.

9. How do you check for the limit state of shear strength of deck slabs? What are the factors Influencing the shear strength of slabs?

10. Discuss briefly the permissible limits of deflections for different types of loads specified in the IRC codes.

OBJECTIVE TYPE QUESTIONS

1. Reinforced concrete slab decks are economical for culverts in the span range of
   a) 12 to 15 m
   b) 10 to 12 m
   c) 6 to 8 m
2. In designing deck slabs, maximum bending moment develops when the IRC Class AA tracked vehicle Is placed
   a) Adjacent to the supports
   b) At the centre of span
   c) At quarter span
3. For computation of the effective width of dispersion of for IRC class AA loads, the depth of dispersion of the load to be considered is
   a) The wearing coat
   b) Centre of the depth of slab
   c) Complete depth of the slab
4. In the case of designing tee beam and slab bridge decks, the dispersion of the wheel loads should be considered in
   a) The long span direction
   b) Short span direction
   c) Both long and short span directions
5. In designing bridge deck slabs for IRC Class AA tracked load, maximum shear force Develops when the tracked load is
   a) At the centre of span
   b) At quarter span
   c) Adjacent to the supports
6. The notional size used for determining the drying shrinkage strain in a solid deck slab Depends upon
   a) The grade of concrete used in slab
   b) The cross sectional area of concrete
   c) The cross sectional area of reinforcement
7. The effective width of dispersion in the direction parallel to the to the supported edge For a single concentrated load on a solid cantilever deck slab depends upon the
   a) The span of the slab
   b) Distance of the centre of gravity of the load from the face of the cantilever support
   c) The thickness of he slab
8. The wearing coat specified in the Indian standard codes on reinforced concrete deck slab is in the range of
   a) 25 to 40 mm
   b) 40 to 60 mm
   c) 75 to 100 mm

9. The cross sectional area of tension reinforcement in a solid deck slab expressed as a percentage of the cross sectional area of concrete, should be not be greater than
   a) 0.4 %
   b) 0.3 %
   c) 0.25 %

10. The maximum spacing of principal reinforcement in a solid deck slab should be not more than
   a) 300 mm
   b) 250 mm
   c) 400 mm

5

Skew Slab Culvert

5.1 GENERAL FEATURES

When the alignment of a road crosses a stream at any angle other than 90 degrees, the crossing of the road is skew and the road bridge crossing the stream is classified as a skew slab culvert.

Present day traffic requirements demand straight alignment of the road in view of the fast traffic and this in turn necessitates the use of skew crossings. The analysis and design of a skew bridge deck is more complicated than that of a right bridge crossing at right angles to the stream. The span length, deck area and the length of abutments supporting the bridge deck increase in proportion to the skew angle of the bridge. With the increase in the skew angle, the stress distribution in the skew slab differs significantly in comparison with the straight slab.

A load applied on the deck slab is transmitted to the supports in proportion to the rigidity of the various possible paths; hence a major proportion of the load tends to reach the supports in a direction normal to the faces of the abutments and piers. Due to this the planes of maximum stresses are not parallel to the centre line of road way and the slab exhibits warped or twisted deformational characteristics due to the passage of wheel loads on the deck. The reactions at the obtuse angled end of the slab support are larger than the other end with its magnitude ranging from 0 to 50 percent for skew angles of 20 to 50 degrees. The bearing reactions developed are such that the acute angle supports lift up with the increase in the skew angle. Depending upon the skew angle, the following design principles have been evolved based on theoretical and experimental investigations.[1,2]

5.2 RESEARCH INVESTIGATIONS ON SKEW SLAB BRIDGE DECKS

Several research investigations on the behavior of skew slab bridge decks under loads have been reported by investigators like Deshmukh et al,[3] Khatri et al[4] and Qaqish[5] using computational methods and finite element analysis. The finite element modeling and analysis.

Of slabs in the span range of 4 to 12 m and skew angles varying from 15 to 45° resulted in the following conclusions regarding the moments and shear forces developed in the slabs:

1) The increase of shear force was significantly high for short spans of 4 to 6 m as the skew angle increased from 15 to 45°. However for long spans of 8 to 12 m the shear force was more or less constant as the skew angle increased from 15 to 45°.

2) There was an increase of 20% in the shear force when spans increased from 4 to 6 m.

3) The bending moment slightly decreased for short spans in the range of 4 to 6 m as the skew angle increased from 15 to 45°. However for long spans in the range of 8 to 12 m the bending moment increased with the skew angle from 15 to 45° by 10 percent.

4) The increase of torsional moment was similar to that of the bending moment. For skew angles exceeding 30° the torsional moments were higher requiring larger magnitude of reinforcements than for low span bridge decks.

Theoret et al[6] have reported the results of the finite element modeling of skew slab bridge decks with spans varying from 3 to 20 m and skew angles varying from 0 to 60°. The results indicated that shear force and secondary bending moments increase with skew angle while the longitudinal bending moments diminished. Also high vertical shear forces developed in the vicinity of the free edges requiring additional steel reinforcements.

5.3 ANALYSIS AND DESIGN

An approximate method of design has been evolved for smaller skew angled (< 15°) simply supported bridge decks. In this method the bending moments are evaluated as for a right bridge of span length measured centre to centre of supports and parallel to the centre line of the road way. The main reinforcements are provided in a direction parallel to the centre line of the road way. Distribution reinforcement comprising 0.2 percent of the effective cross section of the slab is placed parallel to the supports as shown in Fig. 5.1.

When the skew angles of the deck slab exceed 15°, a more rigorous analysis is required. Based on extensive experimental investigations on skew slab models made of gypsum plaster, Rusch and Hergenroder[7] have developed influence surfaces for bending and torsional moments at critical points of skew slabs under concentrated load placed anywhere on the slabs for various span/width ratios and skew angles in the range of 15 to 60°. The Ministry of Surface Transport[8] (Roads Wing) has prepared standard designs for skew bridge decks of clear spans of 5, 6 and 8 m and skew angles of 15°, 30°, 45° and 60°, on the basis of the above data applied to IRC loadings suitable for two lane traffic without foot paths on national highways. The tables are applicable for M-20 Grade concrete and Fe-415 HYSD bars with bearing width for slab assumed as 370 mm. Typical results and values of reinforcements are compiled in Table 5.1.

Fig. 5.2 shows the details of main and distribution reinforcements to be provided at the bottom and top of the slabs in a direction perpendicular and parallel to the supports. Since the main bars in such an arrangement cut the free edge at an angle and are ineffective in resisting the bending moment at the centre of the free edge for want of adequate anchorage, extra steel rods have to be provided near the free edge parallel to the edges for a width sufficient to provide anchorage for the main bars. The reinforcements at the top of the slab are needed at the obtuse angled corner and

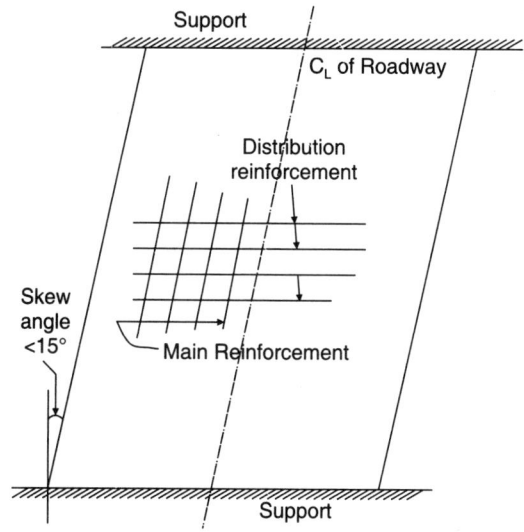

Fig. 5.1 Reinforcement Layout in Slabs (Small Skew Angles).

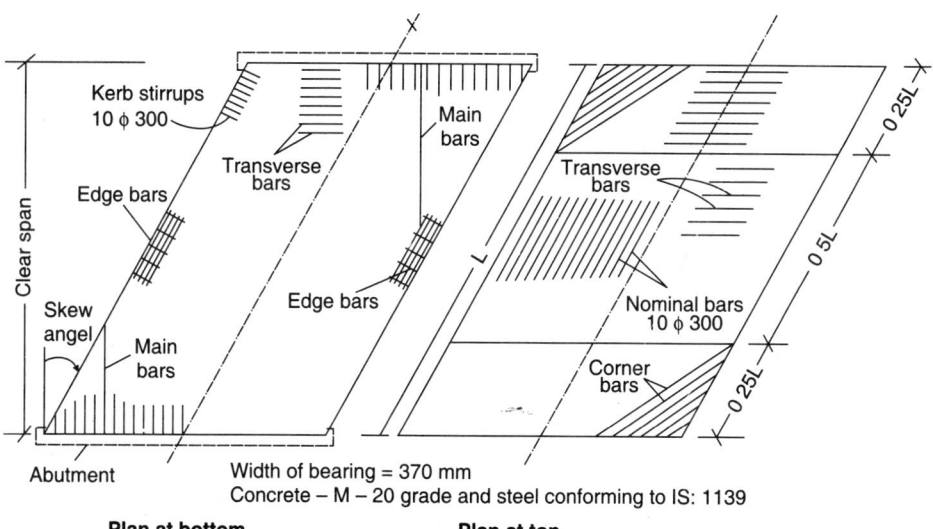

Width of bearing = 370 mm
Concrete – M – 20 grade and steel conforming to IS: 1139

Plan at bottom   **Plan at top**

Fig. 5.2 Reinforcement Details in Skew Slab Culvert.

Table 5.1 Typical Reinforcements in Skew Slab Culverts

Clear Skew Span (m)	Skew angle degrees	Slab thick-ness (mm)	Bottom Reinforcements			Top Reinforcement	
			Main bars	Transverse bars	Edge bars at each edge	Transverse bars	Corner bars
5	15	465	20 φ 130	10 φ 90	20 φ 140	10 φ 300	16 φ 300
	30	465	20 φ 130	10 φ 90	20 φ 130	10 φ 300	16 φ 300
	45	465	20 φ 130	10 φ 90	20 φ 120	10 φ 200	16 φ 190
	60	465	20 φ 120	10 φ 110	20 φ 120	10 φ 180	16 φ 110
6	15	540	20 φ 115	10 φ 100	20 φ 110	10 φ 300	16 φ 300
	30	540	20 φ 115	10 φ 100	20 φ 110	10 φ 260	16 φ 300
	45	540	20 φ 115	10 φ 100	20 φ 110	10 φ 200	16 φ 170
	60	540	20 φ 95	10 φ 160	20 φ 110	10 φ 160	16 φ 90
8	15	700	25 φ 150	10 φ 90	20 φ 150	10 φ 260	20 φ 300
	30	700	25 φ 150	10 φ 110	20 φ 140	10 φ 230	20 φ 300
	45	700	25 φ 150	10 φ 150	20 φ 130	10 φ 160	20 φ 180
	60	700	25 φ 95	10 φ 180	20 φ 120	10 φ 90	20 φ 90

the centre of the free edges. Short bars parallel to the long diagonal are provide at the obtuse angled corner. Transverse are provided parallel to the supports. Nominal reinforcements of 10 mm diameter spaced at 300 mm centres are provided parallel to the free edge over the full width of bridge deck.

DESIGN EXAMPLE

Design a skew slab culvert for a National high way crossing of a stream to suit the following data:

1. Data

Clear span............................. 6 m
Width of bearing .... 370 mm
Width of carriage way .... = 7.5 m
Overall depth of slab .... 540 mm
Wearing coat .... 80 mm
Skew angle .... 30°
Type of loading .... IRC Class AA tracked vehicle
Materials .... M-20 grade concrete and Fe-415 HYSD bars

Design the skew slab using the design tables of the Ministry of Transport (Roads Wing) and Check for the limit states of strength and serviceability according to IRC: 112-2011 code.

2. Depth of Slab and Effective Span

Referring to the design table 5.1, for the given clear span, grade of concrete and steel the minimum depth of slab is to be taken as 540 mm and width of bearing is 370 mm.

Using 20 mm diameter bars and a clear cover of 50 mm, the effective depth is computed as

Effective depth = d = (540) = 490 mm
Effective span is the least of the following:

1) [Clear span + effective depth] = [6 + 0.49] = 6.49 m
2) Centre to centre of supports = [6 + 0.37] = 6.37 m
   Hence effective span = L = 6.37 m

3. Bending Moments

Dead weight of the slab = (0.56 × 24) = 12.96 kN/m^2
Weight of wearing coat = (0.08 × 22) = 1.76 kN/m^2
                    Total dead load ≅ 15.00 kN/m^2

Dead load bending moment = M_d = [0.125 × 15 × 6.37^2] = 76 kN·m
Live load bending moment due to IRC Class AA tracked load will be almost similar to the value Calculated in example 4.10 (M_L = 113 kN·m)

Design Ultimate bending moment = $[1.35\ M_d + 1.5\ M_L)]$

$$= [(1.35 \times 76) + (1.5 \times 113)] = 272 \text{ kN·m}$$

4. Depth of Slab and Reinforcements

$$M_{u,\text{lim}} = 0.148 f_{ck}\ b\ d^2$$

$$\therefore d = \sqrt{\frac{M_u}{0.148 f_{ck} b}} = \sqrt{\frac{272 \times 10^6}{0.148 \times 25 \times 1000}} = 217 \text{ mm} < 490 \text{ mm provided (safe)}$$

Main Reinforcements of 20 mm diameter bars spaced at 115 mm centres with distribution bars of 10 mm diameter at 100 mm centers are provided from Table. 5.1 The free edges of the skew slab are strengthened by using 20 mm diameter bars at 110 mm centers.

5. Check for Strength and Serviceability Limit States

a) Ultimate Flexural Strength

Using the following parameters of the reinforced slab,

$d = 490$ mm, $A_{st} = 2700$ mm^2/m, $f_y = 415$ N/mm^2, $f_{ck} = 20$ N/mm^2

$$M_u = (0.87 \times 415 \times 2700 \times 490)\left[1 - \frac{2700 \times 415}{1000 \times 490 \times 20}\right]$$

$$= (31410^6) \text{ N.mm}$$

$$= 423 \text{ kN·m} > 272 \text{ kN·m (Hence safe)}$$

b) Crack Width Limits

Control of Cracking Without Direct Calculation

The IRC: 112-2011 code outlines a simpler method for crack control by limiting the stress in steel and the maximum bar size and spacing for permissible crack widths of 0.3 and 0.2 mm. The code recommends the use of Tables 4.10 and 4.11.

Service load moment = $M_w = (76 + 113) = 189$ kN·m, $\alpha_e = (E_s/E_c) = (200/29) = 6.89$ If x = depth of neutral axis, by taking the first moment of the areas, we have

$$(0.5 \times 1000 \times x^2) = [(6.89 \times 2700)(490 - x)$$

$$\text{Solving } x = 116 \text{ mm}$$

The stress in steel at working load is computed as

$$\sigma_s = \left[\frac{M}{(d - \frac{x}{3})A_{st}}\right] = \left[\frac{189 \times 10^6}{(490 - \frac{116}{3})2700}\right] = 155 \text{ N/mm}^2$$

The slab is reinforced with 20 mm diameter bars spaced at 115 mm centers.

Referring to tables 4.6 and 4.7, the bar size and spacing are well within the safe limits for control of cracking.

c)   *Check for Deflection*

The slab with an overall depth of 540 mm and reinforced with 20 mm bars spaced at 115 mm centers will have a service load deflection less than that computed for the example 4.10 and satisfies the serviceability limit state of deflection specified in IRC: 112-2011.

6.   Reinforcement Details

The details of reinforcements in the cross section and longitudinal section of the skew slab is shown in Fig. 5.3.

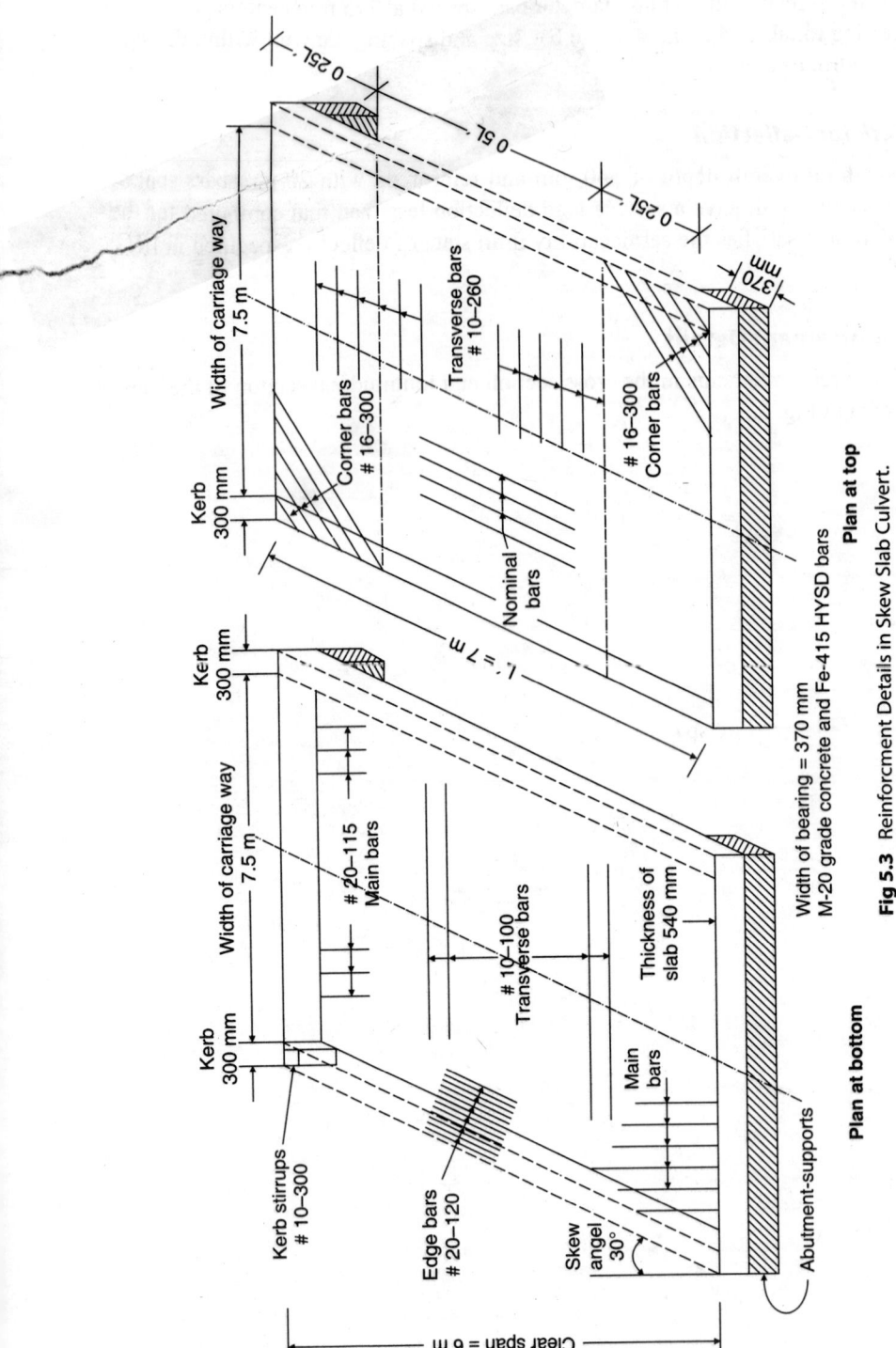

Plan at top

Plan at bottom

Width of bearing = 370 mm
M-20 grade concrete and Fe-415 HYSD bars

Fig 5.3 Reinforcment Details in Skew Slab Culvert.

EXAMPLES FOR PRACTICE

1. Design a skew slab culvert to suit the following data:

    Clear span span = 6 m
    Skew angle = 10°
    Width of carriage way = Two lane 7.5 m wide
    Loading: IRC Class AA tracked vehicle
    Materials: M-25 Grade concrete and Fe-500 Grade HYSD reinforcements

    Design the slab to conform to the specifications of IRC: 112-2011 and check for safety against strength and serviceability requirements. Sketch the details of reinforcements in the deck slab.

2. A skew slab culvert is required for a National highway crossing to suit the following data:

    Clear span = 8 m
    Skew angle = 30°
    Width of carriage way = 7.5 m
    Materials: M-20 Grade concrete and Fe-415 HYSD bars
    Loading: IRC Class AA tracked vehicle

    Design the slab to conform to the specifications of IRC: 112-2011 and check for safety against strength and serviceability requirements.

3. A slab culvert is to be designed for a district highway crossing of a stream at an angle of 15°. Design a suitable reinforced concrete skew slab deck using the limit state procedure specified in IRC: 112-2011 using the following data:

    Clear span of road way = 7.5 m
    Width of bearing = 500 mm
    Loading: IRC Class A

    Materials: M-25 Grade concrete and Fe-500 Grade HYSD reinforcements

    Design the slab to conform to the specifications of IRC: 112-2011 and check for safety against strength and serviceability requirements.

REFERENCES

1. Raina, V.K., Concrete Bridge Practice, Analysis, Design and Economics, Tata McGraw-Hill Publishing Co. Ltd, New Delhi, 1991, pp. 90-96.
2. Johnson Victor, D., Essentials of Bridge Engineering (Fifth Edition), Oxford & IBH Publishing Co, Ltd, New Delhi, 2001, pp. 99-130.
3. Deshmukh, N.V. and Waghe, U.P., Analysis and Design of Skew Bridges, International Journal of Science and Research (IJSR), Vol. 4, No. 4, April 2015, pp. 399-402.
4. Khatri, P.R. et al., Analysis of Skew Bridges using Computational Methods, International Journal Computational Engineering research, May-June 2012, Vol. 2, No. 3, pp. 628-636.

5. Qaqish, M.S., Effect of Skew Angle on Distribution of Bending Moments in Bridge Slabs, Journal of Applied Science, Vol. 6, No. 2, 2006, pp. 366-372.

6. Theoret, P., Massicitte, B. and Conciatori, D., Analysis and Design of Straight and Skewed Slab Bridges, Journal of Bridge Engineering (ASCE), Vol. 10, 2012, pp. 289-301.

7. Rusch, H. and Hergenroder, A., Influences Surfaces for Moments in Skew Slabs, cement and Concrete Association, London, 1964.

8. DRg. No. BD/12-74, and BD/13-74, Ministry of Shipping and Transport (Roads Wing), Govt. of India, July 1974.

REVIEW QUESTIONS

1. Briefly outline the necessity of using skew slab bridge decks for National high way crossings.

2. Bring out the salient differences between the right bridge decks and skewed bridge decks with respect to national high way crossings.

3. In what way the structural behavior of skew slab bridge decks differ from right bridge decks in supporting moving high way live loads?

4. What are the significant parameters influencing the structural behavior of skew slab bridges while resisting the dead and live loads?

5. Explain briefly the method of design of skew slab bridge decks when the skew angle of the high way crossing does not exceed 15 degrees.

6. Briefly illustrate the method adopted for arranging the reinforcements in skew slab bridge decks when the skew angle is less than 15 degrees?

7. In what way the skew angle influences the variation of bending and torsional moments in the decks slab due to moving loads?

8. Briefly explain the variation of shear forces at supports of a skew slab bridge deck under loads due to the variation of span and skew angle.

9. Explain the basis of Ministry of Transport (Roads Wing) recommendations for the design of skew slab bridge decks and their limitations.

10. Explain with sketches the detailing of reinforcements in a typical skew slab bridge culvert with Skew angle of 30° as specified in the Ministry of Transport (Road Wing) (Govt. of India) recommendations.

OBJECTIVE TYPE QUESTIONS

1. Skew slab culverts are necessary when the alignment of the high way crosses a stream at
   a)   Right angles
   b)   The bed level of stream
   c)   an angle other than 90 degrees

2. In the structural design of skew slab decks, approximate methods can be used if the angle of road crossing is
   a) Less than 15 degrees
   b) Greater than 30 degrees
   c) 90 degrees
3. Under the action of wheel loads on a skew slab deck, the support reactions will be
   a) Similar at all corners
   b) Larger at the acute angled corners
   c) Larger at the obtuse angled corners
4. In skew slab decks with skew angle less than 15 degrees, the main reinforcements can be arranged
   a) Perpendicular to the supports
   b) Inclined to the supports
   c) Parallel to the centre lime of the road way
5. In the analysis of skew slab decks with skew angles exceeding 15 degrees
   a) Rigorous methods have to be used
   b) Method of design followed for right bridge decks can be used
   c) Method of design as per IRC codes can be used
6. The main reinforcements in skew slab decks having skew angles exceeding 15 degrees should be arranged
   a) Parallel to the centre line of the road way
   b) Perpendicular to the supports
   c) Parallel to the supports
7. The distribution reinforcements in slabs with skew angles less than 15 degrees should be provided in a direction
   a) Perpendicular to the main reinforcements
   b) Parallel to the supports
   c) Perpendicular to the unsupported edges
8. In the case of slabs with skew angles greater than 15 degrees, the unsupported edges should be provide with
   a) Nominal reinforcements
   b) Reinforcements similar to that at the centre of the slab
   c) Additional reinforcements for strengthening the edges
9. In comparison with right slab bridge decks, skew slab decks require
   a) The same thickness and reinforcements
   b) Smaller thickness and less reinforcements
   c) Larger thickness and more reinforcements
10. Skew slab bridge decks require additional reinforcements near
    a) Acute angled corners
    b) Obtuse angled corners
    c) Mid portion of supports

6

Pipe Culvert

6.1 INTRODUCTION

Reinforced concrete pipes are widely used as cross drainage structural elements for a road or railway embankment when the discharge in the stream is small. During the last three decades Precast R.C.C. pipes have replaced the steel pipes for many works due to the initial and maintenance costs. Reinforced concrete pipes[1,2] are commonly used for various types of applications like water supply and sanitary systems, bridge structures, gravity mains for carrying water under hydrostatic pressure. The main advantage of reinforced concrete pipes is its initial negligible maintenance costs in comparison with the traditional steel pipes which are costly and prone to corrosion damage. When the earth fill over the pipe is small, return type wing walls are provided at the ends of the road formation width to retain the earth. If the earth fill is large the length of the pipe culvert is increased so that the embankment with its natural side slopes is retained by splayed wing walls.

In coastal regions where the structure has to withstand severe environmental conditions, reinforced concrete pipes are invariably used due to its superior durability[3,4] characteristics. Rapid improvements in the manufacturing process of precast pipes by spinning process[5] has reduced the costs and also significantly improved the structural quality of reinforced concrete pipes. For the last few decades, concrete pipes have more or less replaced steel pipes due to their superior durability and reduction in costs.

The application of prestressing techniques[6,7] has further improved the structural performance of concrete pipes for large scale water supply systems using large diameter pipes to convey water under hydrostatic pressure. Prestressed concrete pipes have been used for the water supply systems of metropolitan areas in many countries. Reinforced concrete hume pipes have been used for culverts of bridges and for drainage of storm water at city road crossings. The reader may refer to investigative research reports of Rao[8] and Frank[9] and other prominent codes[10] for design aspects of reinforced concrete pipes. The various tests to be conducted for certification of R.C.C. pipes can be obtained from Indian standard code IS: 3597-1966[11]. Specifications for laying of concrete pipes is codified in IS: 783-1985[12].

6.2 CLASSIFICATION OF R.C.C. PIPES

According to the Indian Standard Code IS: 458-2003, reinforced concrete pipes are classified as non pressure and pressure pipes with their applications as shown in Table 6.1. Non pressure pipes NP-2, NP-3 and NP-4 are used for culverts.

Table 6.1 Classification of Reinforced Cement Concrete Pipes
(IS: 458-2003)

	Pipe Designation	Conditions Where Used
NP-1	Unreinforced concrete non pressure pipe	For drainage and irrigation use above ground or in shallow trenches
NP-2	Reinforced concrete light duty non pressure pipe	For drainage and irrigation use for culverts carrying light traffic
NP-3	R.C. heavy duty non pressure pipe	For drainage and irrigation use and for culverts carrying heavy traffic
NP-4	R.C. heavy duty non pressure pipe	For drainage and irrigation use and for culverts carrying very heavy traffic such as Railway loadings
P-1	R.C. pressure pipes tested to a hydraulic pressure of 0.2 N/mm² (20 m head)	For use in gravity mains, the design pressure not exceeding 2/3 of test pressure
P-2	R.C. pressure pipes tested to a hydraulic pressure of 0.4 N/mm² (40 m head)	For use in pumping mains, the design pressure not exceeding half the test pressure
P-3	R.C. pressure pipes tested to a hydraulic pressure of 0.6 N/mm² (60 m head)	For use in gravity mains, the design pressure not exceeding half the test pressure

6.3 REINFORCEMENTS IN PIPES

The longitudinal and hoop reinforcements are designed for the loads but the minimum quantity of steel reinforcements is specified for different classes of pipes in IS: 458-2003. The typical reinforcement, design and strength test requirements for pipes of class NP-2, NP-3 and NP-4 generally used for high way and railway embankment culverts specified in IS: 458-2003 are shown in Table 6.2, 6.3 and 6.4. The precast pipes should conform to the serviceability and strength requirements specified in the code with regard to the cracking and ultimate loads.

6.4 DESIGN PRINCIPLES

The cross sectional area of the pipe depends upon the amount of flow in the stream crossing the road way.

If Q = discharge to be carried in the pipe
A = cross sectional area of the pipe
v = velocity of flow
d = diameter of the pipe

Table 6.2 Reinforced Concrete, Light Duty, Non Pressure Pipes of Class NP-2
(Table-2 of IS: 458-2003)

Internal Diameter of Pipes	Barrel Wall Thickness	Reinforcements		Spirals, Hard Drawn Steel	Strength Test Requirements for Three Edge Bearing Test	
		Longitudinal, Mild Steel or Hard Drawn Steel			Load to Produce 0.25 mm Crack	Ultimate Load
mm	mm	Minimum number	kg/linear metre	kg/linear metre	kN/linear metre	kN/linear metre
(1)	(2)	(3)	(4)	(5)	(6)	(7)
80	25	6	0.59	0.16	10.05	15.08
100	25	6	0.59	0.18	10.05	15.08
150	25	6	0.59	0.24	10.79	16.19
200	25	6	0.59	0.38	11.77	17.66
225	25	6	0.59	0.46	12.26	18.39
250	25	6	0.59	0.58	12.55	18.83
300	30	8	0.78	0.79	13.48	20.22
350	32	8	0.78	1.13	14.46	21.69
400	32	8	0.78	1.49	15.45	23.18
450	35	8	0.78	1.97	16.18	24.27
500	35	8	0.78	2.46	17.16	25.74
600	45	8	0.78	3.47	18.88	28.32
700	50	8	1.22	4.60	30.35	30.53
800	50	8	1.22	6.71	21.57	32.36
900	55	8	1.22	9.25	22.80	34.20
1000	60	8	1.76	10.69	24.27	36.41
1100	65	8	1.76	12.74	25.50	38.25
1200	70	8	1.76	15.47	26.97	40.46
1400	75	12	2.64	20.57	29.42	44.13
1600	80	12 or 8 + 8	3.52	25.40	32.15	48.18
1800	90	12 or 8 + 8	3.52	32.74	35.06	52.59
2000	100	12 + 12	5.28	45.14	37.76	56.64
2200	110	12 + 12	5.28	56.17	40.21	60.32

Notes:
1. If mild steel is used for spiral reinforcement, the weight specified under col 5 shall be increased to 140/125.
2. Soft grade mild steel lwire for spirals may be used for pipes of internal diameters 80 mm, 100 mm and 150 mm only, by increasing weight to 140/84.
3. If the longitudinal reinforcement given in this table is valid for pipes up to 2.5 m effective length for internal diameter of pipe up to 250 mm and up to 3 m effective length for higher diameter pipes.
4. Total mass of longitudinal reinforcement shall be calculated by multiplying the values given in col 4 by the length of the pipe and then deducting for the cover length provided at the two ends.

Table 6.3 Reinforced Concrete, Medium Duty, Non Pressure Pipes of Class NP-3
(Table-3 of IS: 458-2003)

Internal Diameter of Pipes	Barrel Wall Thickness	Reinforcements			Strength Test Requirements for Three Edge Bearing Test	
		Longitudinal, Mild Steel or Hard Drawn Steel		Spirals, Hard Drawn Steel	Load to Produce 0.25 mm Crack	Ultimate Load
mm	mm	Minimum number	kg/linear metre	kg/linear metre	kN/linear metre	kN/linear metre
(1)	(2)	(3)	(4)	(5)	(6)	(7)
80	25	6	0.59	0.16	13.00	19.50
100	25	6	0.59	0.22	13.00	19.50
150	25	6	0.59	0.46	13.70	20.55
200	30	6	0.59	0.81	14.50	21.75
225	30	6	0.59	1.03	14.80	22.20
250	30	6	0.59	1.24	15.00	22.50
300	40	8	0.78	1.80	15.50	23.25
350	75	8	0.78	2.95	16.77	25.16
400	75	8	0.78	3.30	19.16	28.74
450	75	8	0.78	3.79	21.56	32.34
500	75	8	0.78	4.82	23.95	35.93
600	85	8 or 6 + 6	1.18	7.01	28.74	43.11
700	85	8 or 6 + 6	1.18	10.27	33.53	50.30
800	95	8 or 6 + 6	2.66	13.04	38.32	57.48
900	100	6 + 6	2.66	18.30	43.11	64.67
1000	115	6 + 6	2.66	21.52	47.90	71.85
1100	115	6 + 6	2.66	27.99	52.69	79.00
1200	120	8 + 8	3.55	33.57	57.46	86.22
1400	135	8 + 8	3.55	46.21	67.06	100.60
1600	140	8 + 8	3.55	65.40	76.64	114.96
1800	150	12 + 12	9.36	87.10	86.22	129.33
2000	170	12 + 12	9.36	97.90	95.80	143.70
2200	185	12 + 12	9.36	133.30	105.38	158.07
2400	200	12 + 12	14.88	146.61	114.96	172.44
2600	215	12 + 12	14.88	175.76	124.54	186.81

Notes:
1. If mild steel is used for spiral reinforcement, the weight specified under col 5 shall be increased to 140/125.
2. The longitudinal reinforcement given in this table is valid for pipes up to 2.5 m effective length for internal diameter of pipe up to 250 mm and up to 3 m effective length for higher diameter pipes.
3. Total mass of longitudinal reinforcement shall be calculated by multiplying the values given in col 4 by the length of the pipe and then deducting for the cover length provided at the two ends.
4. Concrete for pipes shall have a minimum compressive strength of 35 N/mm² at 28 days.

Table 6.4 Reinforced Concrete, Heavy Duty, Non Pressure Pipes of Class NP-4
(Table-6 of IS: 458-2003)

Internal Diameter of Pipes	Barrel Wall Thickness	Reinforcements		Spirals, Hard Drawn Steel	Strength Test Requirements for Three Edge Bearing Test	
		Longitudinal, Mild Steel or Hard Drawn Steel			Load to Produce 0.25 mm Crack	Ultimate Load
mm	mm	Minimum number	kg/linear metre	kg/linear metre	kN/linear metre	kN/linear metre
(1)	(2)	(3)	(4)	(5)	(6)	(7)
80	25	6	0.59	0.24	22.1	33.15
100	25	6	0.59	0.36	22.1	33.15
150	25	6	0.59	0.74	23.3	34.95
200	30	6	0.59	1.30	24.6	36.90
225	30	6	0.59	1.64	25.2	37.80
250	30	6	0.59	1.98	25.5	38.25
300	40	8	0.78	2.71	26.4	39.60
350	75	8	0.78	3.14	29.8	44.70
400	75	8	0.78	3.52	33.9	50.90
450	75	8	0.78	3.88	36.9	55.30
500	75	8	0.78	5.96	40.0	61.20
600	85	8 or 6 + 6	2.34	9.63	46.3	69.40
700	85	8 or 6 + 6	3.44	14.33	52.2	78.30
800	95	8 or 6 + 6	3.44	21.20	59.3	89.10
900	100	6 + 6	3.44	27.13	66.3	99.40
1000	115	8 + 8	6.04	35.48	72.6	108.90
1100	115	8 + 8	6.04	43.76	80.4	120.60
1200	120	8 + 8	6.04	53.07	88.3	132.40
1400	135	8 + 8	9.36	77.62	104.2	156.40
1600	140	12 + 12	9.36	108.97	119.6	179.50
1800	150	12 + 12	14.88	150.22	135.3	203.00
2000	170	12 + 12	14.88	151.79	135.3	203.00
2200	185	12 + 12	14.88	160.90	142.2	213.30
2400	200	12 + 12	14.88	216.90	155.0	232.50
2600	215	12 + 12	14.88	258.93	166.7	250.00

Notes:
1. If mild steel is used for spiral reinforcement, the weight specified under col 5 shall be increased to 140/125.
2. The longitudinal reinforcement given in this table is valid for pipes up to 2.5 m effective length for internal diameter of pipe up to 250 mm and up to 3 m effective length for higher diameter pipes.
3. Total mass of longitudinal reinforcement shall be calculated by multiplying the values given in col 4 by the length of the pipe and then deducting for the cover length provided at the two ends.
4. Concrete for pipes shall have a minimum compressive strength oi 35 N/mm^2 at 28 days.

Then using the relation,

$$A = Q/v = (\pi d^2)/4$$

$$d = \sqrt{(4Q)/(\pi v)}$$

The structural design of the pipe involves the computations of the three-edge bearing strength of pipe, the weight of earthfill over the pipe and the load on pipe due to a surface concentrated live load, each associated with a strength factor generally taken as 1.5.

The type of non-pressure pipe and bedding are so chosen that under the worst combination of field loading, a factor of safety of 1.5 is available as given by the equation,

$$\begin{bmatrix} \dfrac{\text{Three - edge bearing}}{\text{strength} (\text{kN/m})} \\ \dfrac{}{\text{Factor of}} \\ \text{safety} (1.5) \end{bmatrix} = \begin{bmatrix} \dfrac{W \text{ due to filling}}{\text{material} (\text{kN/m})} + \dfrac{W \text{ due to surface}}{\text{load} (\text{kN/m})} \\ \text{Corresponding} \quad \text{Strength factor} \\ \text{strength factor} \quad\quad (1.5) \end{bmatrix}$$

The load acting on the pipe due to the soil in embankment is computed from the equation,[5,6]

$$W = C_e \cdot w \cdot D^2$$

Where, W = Vertical external load in kN/m of pipe due to embankment material as shown in Table 6.5

C_e = Coefficient depending on the ratio of height of embankment H to the external diameter of the pipe

w = density of the embankment material in kN/m³

D = External diameter of pipe (m)

The load on the pipe due to a concentrated highway wheel load 'P' is obtained from the equation,

$$W = 4 \cdot C_s \cdot I \cdot P$$

Where, W = Vertical external load in kN/m due to concentrated surface load

C_s = Influence coefficient depending upon D and H as compiled in Table 6.6.

H = Vertical depth of top of pipe below the surface (m)

D = External diameter of pipe (m)

P = Concentrated wheel load (kN)

I = Impact factor (1.5 for highways)

In the case of Railway loading, the load is uniformly distributed because of the sleepers and ballast. The load on a buried pipe in a railway embankment is given by the equation,

$$W = 4 \cdot C_s \cdot U \cdot D$$

Table 6.5 Load on Pipe Due to Earthfill

Pipe Size (d) (mm)	Outer Diameter (D) (mm)	Embankment loading on pipe in kN/m for various depths H in metres									
		1	2	3	4	5	6	7	8	9	10
NP-3											
500	650	16.8	34.9	54.0	65.3	99.0	118.5	150.0	171.0	191.0	212.0
600	760	18.7	42.6	62.4	75.5	103.0	133.0	160.5	186.0	197.0	230.0
700	860	20.0	48.0	68.0	86.5	118.5	154.5	174.0	209.0	228.0	246.0
800	980	24.2	55.3	79.5	107.0	136.5	163.0	202.0	230.0	260.0	288.0
900	1100	28.3	58.7	91.5	122.0	150.0	185.0	209.0	257.0	297.0	338.0
1000	1200	28.6	59.5	101.0	132.0	163.0	202.0	228.0	259.0	324.0	357.0
1200	1430	33.4	74.0	115.0	159.0	200.0	226.0	274.0	315.0	352.0	427.5

Table 6.6 Influence Coefficient C_s for Concentrated Surface Load for Highways

Pipe Size (d) (mm)	Outer Diameter (D) (mm)	C_s for various Depths in metres									
		0.1	0.2	0.3	0.4	0.6	0.8	1.0	2.0	3.0	4.0
NP-3											
500	650	0.246	0.228	0.198	0.169	0.117	0.083	0.060	0.017	0.008	0.005
600	760	0.247	0.234	0.210	0.182	0.131	0.094	0.068	0.022	0.010	0.006
700	860	0.247	0.236	0.215	0.186	0.140	0.102	0.075	0.024	0.010	0.006
800	980	0.249	0.240	0.220	0.196	0.149	0.110	0.083	0.027	0.013	0.007
900	1100	0.249	0.241	0.225	0.202	0.156	0.117	0.089	0.029	0.014	0.008
1000	1200	0.249	0.242	0.228	0.205	0.162	0.123	0.095	0.032	0.015	0.010
1200	1430	0.249	0.242	0.230	0.209	0.171	0.131	0.104	0.036	0.020	0.011

Where, W = Load on pipe in kN/m

C_S = Influence coefficient depending on the length of the sleeper, distance between two axles and depth of top of pipe below surface.

U = Uniformly distributed load in kN/m² on the surface directly above the pipe = $[(PI/4AB) + 2. W_t. B]$

P = Axle load in kN (229 kN for B.G.)

A = Half the length of the sleeper in m (1.35 m for B.G.)

B = Half the distance between the two driving axles in m (0.92 m for B.G.)

W_t = Weight of track structure in kN/m (generally 3 kN/m)

D = Outside diameter of the pipe in m for broad gauge loading, the equation reduces to

$$W = 339 \cdot C_s \cdot D$$

The values of the coefficient C_s is compiled in Table 6.7.

Table 6.7 Influence Coefficient C_s for Broad Gauge Railway Loading

H (m)	C_s	H (m)	C_s
0.1	0.250	1.0	0.183
0.2	0.249	2.0	0.094
0.3	0.245	3.0	0.052
0.4	0.240	4.0	0.032
0.5	0.233	5.0	0.021
0.6	0.224	6.0	0.015
0.7	0.218	7.0	0.011
0.8	0.205	8.0	0.009
0.9	0.193	9.0	0.007
		10.0	0.005

6.5 DESIGN EXAMPLE

Design a suitable R.C.C. pipe culvert to suit the following data:

Discharge through pipe culvert = 1.57 m³/s
Velocity of flow through pipe = 2 m/s
Width of road (two lane) = 7.5 m
Top width of embankment = 1.5 : 1
Bed level of stream = 100.00
Top of embankment = 103.00

Loading: I.R.C. class AA wheeled vehicle with a maximum wheel load of 62.5 kN
Draw the longitudinal section, plan and end view of the pipe culvert.

1. Diameter of Pipe Culvert

Discharge $Q = Av$

$$A = (Q/v) = (1.57/2) = 0.785 \text{ m}^2$$

If $\quad d$ = diameter of the pipe

$$\pi\, d^2/4 = 0.785 \qquad \therefore d = 1.00 \text{ m}$$

Adopt NP-3, R.C.C. heavy duty non-pressure pipe for carrying heavy road traffic. From the Indian standard code IS: 458-1971, for a pipe of internal diameter 1 m, the external diameter D = 1.2 m.

2. Load Due to Earthfill

Height of embankment over pipe = 2 m

From Table 5.3, for d = 1000 mm and H = 2 m,

Load due to earthfill = 59.5 kN/m

3. Load Due to I.R.C. Class AA Wheel Load

Assuming I.R.C. Class AA wheel load of 62.5 kN to be directly above the pipe, from Table 5.4,

Loading on pipe $= 4\, C_S. I. P. = (4 \times 0.032 \times 1.5 \times 62.5)$

$$= 12 \text{ kN/m}$$

4. Check for Strength Factor

The type of non-pressure pipe and bedding should be so chosen that under the worst combination of field loading, a factor of safety of 1.5 is available.

This is computed as follows:

$$\left[\frac{\text{Three - edge bearing strength (kN/m)}}{\text{F.S.(i.e., 1.5)}} \right] = \left[\frac{\text{Load on pipe due to earthfill in kN/m}}{\text{Corresponding strength factor}} + \frac{\text{Load on pipe due to wheel load (kN/m)}}{\text{Strength factor}} \right]$$

From IS: 458-1971 code,

Three-edge bearing strength for NP-3 class 1000 mm diameter pipe is 111 kN/m

$$(111/1.5) = (59.5/\text{S.F.}) + (12/1.5)$$

solving, $\quad$ S.F. = 0.90

The strength factor for first class bending is 2.3 and for concrete cradle bedding shown in Fig. 6.1, is 3.7. Any of these two beddings can be provided for the pipe culvert.

5. Reinforcements in pipe

The minimum reinforcements in the pipe according to IS: 485-1971 are as follows:

Spiral reinforcement of hard drawn steel wire with a permissible stress of 140 N/mm²
= 44 kg/m.

Longitudinal reinforcement of mild steel with a permissible stress of 126.5 N/mm²
= 5.80 kg/m

Using 12 mm diameter bars at 60 mm centres as spiral reinforcement, Average diameter
of spiral = 1.1 m

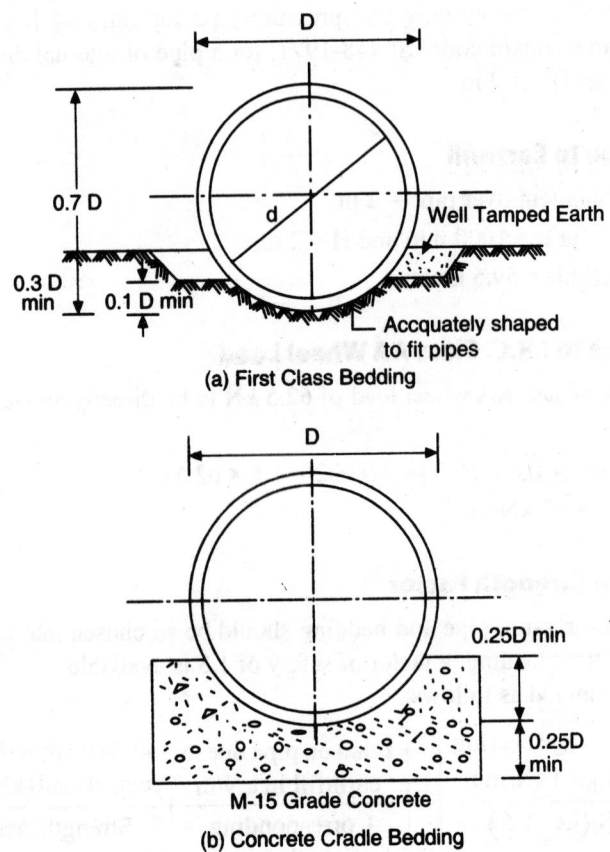

(a) First Class Bedding

(b) Concrete Cradle Bedding

Fig. 6.1 Bedding for Concrete Pipes in Pipe Culverts.

Weight of one spiral of 12 mm diameter
$$= (\pi \times 1.1 \times 0.88) = 3.045 \text{ kg}$$
Number of spirals in 1 m = (1000/60) = 16.66

Weight of spiral reinforment per meter length of pipe
$$= (3.045 \times 16.66) = 50.7 \text{ kg/m}$$

The quantity of spiral steel provided is greater than the minimum of 44 kg/m specified
in the code.

Providing 6 mm diameter mild steel bars as longitudianal reinforcement,

Weight of each bar = $(\pi \times 0.0062/4) \times 7800 = 0.22$ kg/m

Number of bars required = $(5.80/0.22) = 26.36$

Spacing = $(\pi \times 1100/26.36) = 131$ mm

Adopt 130 mm spacing for the longitudinal reinforcements.

The details of the pipe culvert are shown in Fig. 6.2.

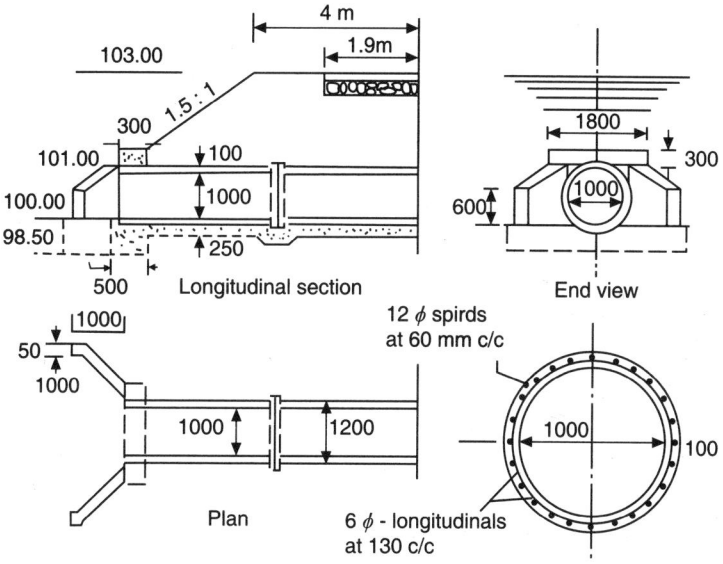

Fig. 6.2 R.C.C. Pipe Culvert.

EXAMPLES FOR PRACTICE

1. Design a R.C.C. pipe culvert to suit the following data:

    Bed level of the stream = 100.00
    Top of road level = 102.68
    Road width = 7.5 m
    Maximum flood discharge = 0.5 m³/s
    Velocity of flow in pipe = 2 m/s

    Type of loading—I.R.C. class AA wheeled vehicle. Non-pressure R.C.C. pipes of classification NP-3 (heavy duty) are available for use in diameters from 500 to 1200 mm. Design a suitable pipe for the culvert and check the structural adequacy of the pipe designed.

2. Design a suitable pipe culvert to carry a discharge of 1 m³/s with a velocity of 1.5 m/s. The depth of earthfilling over the pipe is 3 m. Adopt I.R.C. class AA loading with M-20 grade concrete and steel conforming to IS: 432. Sketch the details of reinforcements and bedding for the pipe.

3. Design a R.C.C. pipe culvert to suit the following data:

   Bed level of stream = 100.00
   Top of road level = 105.00
   Road width (two lane) = 7.5 m
   Maximum flood discharge at site = 0.75 m³/s
   Velocity of flow in pipe = 1.5 m/s
   Side slopes of embankment = 1.5 : 1

   Type of loading—I.R.C. class AA vehicle. Design a suitable pipe for the culvert and check for the structural safety of pipe.

   Draw the following views of the pipe culvert.

   (a) Longitudinal section through the centre of pipe.

   (b) Half plan at top and half sectional plan through the centre of pipe.

   (c) Half cross-section through centre of pipe culvert and half elevation.

4. Design a R.C.C. hume pipe and culvert for the following specifications:

   Bed level of stream = 100.00
   Top of Roafc level = 108.00
   Road width = 7.5 m
   Maximum flood discharge at site = 1.58 m³/s
   Velocity of flow in pipe = 1.4 m/s
   Side slopes of the embankment = 2 : 1

   Type of loading—I.R.C. class AA vehicle. Design a suitable pipe for the culvert and check for the structural safety of the pipe.

   Draw the following views of the pipe culvert.

   (a) Longitudinal section through the centre of pipe.

   (b) Half plan at top and half sectional plan.

   (c) Half cross section through the centre of pipe and half elevation.

5. Check the adequacy of NP3 Class for concrete pipe with first class bedding for a pipe culvert having a diameter of 600 mm and located under 1 m depth of cover across a road.

6. Determine the Class of pipe and the type of bedding required for a concrete pipe of 1200 mm diameter to be laid with its invert level at 7 m under a broad gauge railway embankment.

REFERENCES

1. IS: 458-2003, Indian Standard Specification For Concrete Pipes (With and Without Reinforcement), Second Revision, V-Reprint, Bureau of Indian Standards, New Delhi, April 1983, pp. 1-30.

2. Bacher. A.E, Banke, A.N, and Kirkland, D.E., Reinforced Concrete Pipe Culverts, Design Summary and implementation, 61st Annual Meeting of the Transportation Research Board Washington, USA, 1982, pp. 83-92.
3. ACI Committee-201, Guide to Durable Concrete, Journal of the American Concrete Institute, Vol. 74, 1977, pp. 573-609.
4. Unnikrishna Pillai. S and Devada Menon., Reinforced Concrete Design (Third Edition), McGraw-Hill education Pvt. Ltd, New Delhi, 2009, p. 62.
5. Reinforced Cement Concrete Spun/Hume Pipe manufacturing Co, IS and ISO: 9001-2000, (Web Portal) Rajasthan, 2000.
6. Joshi, N.G., Prestressed Concrete Pipes, State of the Art, International Symposium on Prestressed Concrete Pipes, Poles, Pressure Vessels and Sleepers, Proceedings, Vol. 2, Madras, 1972, pp. p/3-1 to p/3-40.
7. Krishna Raju, N., Prestressed Concrete (Fifth Edition), McGraw-Hill education Pvt. Ltd, New Delhi, 2012, pp. 488-521.
8. Rao, V.V.S., Structural design of Concrete Pipes in accordance with IS: 783-1959, Journal of The Institution of Engineers (India), Vol. 50, No. 5 Part CI-3, Jan. 1970, pp. 99-107.
9. Frank. H.J., Structural design Method for Precast R.C.C. Pipes, 61st Annual Meeting of the Transportation Research Board, Washington, USA, 1982, pp. 93-100.
10. A.124-1962, Australian Standard for Concrete Pressure Pipes, Australian Water Resources Council Technical Paper No. 9, Canberra, 1975, pp. 105-110.
11. IS: 3597-1966, Indian Standard Code of Practice for Test on R.C.C. Pipes, Bureau of Indian Standards, New Delhi, 1966, pp. 1-5.
12. IS: 783-1985 (First Revision), Indian Standard Code of Practice for laying of Concrete Pipes, Bureau of Indian Standards, New Delhi, 1985, pp. 1-77.

REVIEW QUESTIONS

1. Under what situations you would prefer to adopt Reinforced concrete pipes in place of Steel pipes for bridge culverts?
2. What are the advantages of Reinforced concrete pipes in comparison with steel, Cast iron and G.I. pipes?
3. Discuss briefly the classification of reinforced concrete pipes according to the Indian Standard codes, mentioning the situations under which they are used.
4. Briefly outline the design principles to be followed, while designing reinforced concrete Non Pressure pipes for bridge culverts according to the specifications of the Indian Standard codes.
5. What are the various types of reinforcements used in concrete pipes? Mention the structural purpose of using these reinforcements.
6. What are the minimum cover requirements adopted in reinforced concrete pipes designed according to the I.S. Codes?

7. Briefly mention the serviceability and strength tests specified for non pressure pipes used for culverts in the Indian standard codes.
8. Briefly explain the method of considering the loads on pipe due to earth fill and Concentrated loads due to vehicles on highway and loading due to railways.
9. Explain with sketches the first class and concrete cradle bedding used for pipe culverts using reinforced concrete non pressure pipes.
10. Explain with sketches a typical NP-2 class reinforced concrete pipe showing the details of reinforcements along with the bedding in a bridge culvert.

OBJECTIVE TYPE QUESTIONS

1. Reinforced concrete Pipes are preferred to steel and other types of metallic pipes mainly due to
   a) Lighter weight
   b) Durability aspects
   c) Faster construction
2. Reinforced concrete circular pipes are generally made
   a) at the site
   b) by casting using vertical forms
   c) by spinning process
3. Reinforced concrete pipes of different diameter are generally made in lengths of
   a) 1 to 2m
   b) 5 to 8 m
   c) 3 to 4 m
4. The thickness of the concrete pipe is designed by the considerations of
   a) flexural stresses developed
   b) limiting the tensile stress from cracking considerations
   c) depth of soil above the pipe.
5. The minimum clear concrete cover required for spun concrete pipes of thickness more than 75 mm according to the specifications of Indian standard code is
   a) 10 mm
   b) 30 mm
   c) 18 mm
6. The minimum quantity of spiral hard drawn wire at permissible stress of 140 N/mm^2 Specified in the I.S. Code for pipes of Class NP-3 having 1000 mm diameter is
   a) 3.8 kg/m
   b) 50 kg/m
   c) 21.52 kg/m
7. In the structural design of non pressure reinforced concrete pipes, used in bridge culverts with high embankments should be tested for safety using
   a) Hydrostatic test
   b) Absorption test
   c) Three edge bearing test

8. In the case of beddings used to support the pipes in culverts, well tamped earth is used in
   a) Concrete cradle bedding
   b) First class bedding
   c) Continuous concrete raft bedding

9. The minimum grade of concrete used for pipe culverts using Class NP-3 and NP-4 reinforced concrete pipes specified in the Indian standard code is
   a) M-15
   b) M-35
   c) M-25

10. In the serviceability test prescribed for reinforced concrete pipes used in bridge culverts the Indian standard code specifies the load required per linear metre to produce a crack of
    a) 0.1 mm
    b) 0.3 mm
    c) 0.25 mm

7

Box Culvert

7.1 GENERAL FEATURES

Reinforced concrete rigid frame box culverts consisting of two horizontal and vertical slabs built monolithically are ideally suited for a road or railway bridge crossing with high embankments, crossing a stream with limited flow. Box culverts[1] of square or rectangular vent spans of up to 4 m are commonly used for crossing small rivulets. The height of the vent rarely exceeds 3 m. The box culvert generally comprises the following structural components:

1) Solid barrel or box section of sufficient length to accommodate the road width of the carriage way along with kerbs and foot paths.

2) In the case of deep embankments, wing walls splayed at 45 degrees are used to guide the flow of water in the stream though the box culvert.

Box culverts are economical due to their rigidity and monolithic action and separate foundations are not required since the bottom slab resting directly on the soil serves as a raft slab foundation for the culvert. For small discharges, single celled box culvert is used and for larger flow, multi celled box culverts can be employed. The barrel of the box culvert should be of sufficient length to support the entire width of the carriage way.

7.2 DESIGN PRINCIPLES

a) Hydraulic Design

In the hydraulic design, the vent way required to carry the discharge in the stream is computed by examining the discharge records over a period of time at bridge site. Except in the case of buried barrel, the maximum flood level is generally fixed below the bottom of top slab allowing for a vertical clearance. In the case of buried barrel, the design vent way will be similar to that for a pipe culvert. In general, the ratio of span to height of the vent way lies between 1 : 1 and 1.5 : 1. The top of soffit slab is generally fixed at the bed level of the stream.

b) Structural Design

The structural design of a reinforced concrete box culvert[2,3] comprises the detailed analysis of the rigid frame for moments, shear forces and thrusts developed in the various structural elements of the box culvert due to the various type of loading conditions outlined listed below:

1. Concentrated Loads

In cases where the top slab forms the deck of the bridge, concentrated loads due to the wheel loads of the I.R.C. class AA or A type loading have to be considered.

If $\quad$ W = Concentrated load on the slab

$\quad\quad\quad P$ = Wheel load

$\quad\quad\quad I$ = Impact factor

$\quad\quad\quad e$ = Effective width of dispersion

Then $\quad\quad W = (P.I/e)$

The soil reaction on the bottom slab is assumed to be uniform. The notations used for the box culvert and the type of loadings to be considered are shown in Fig. 7.1 (a) to (f).

2. Uniform Distributed Load

The weight of embankment, wearing coat, and, deck slab and the track load are considered to be uniformly distributed loads on the top slab with the uniform soil reaction on the bottom slab.

3. Weight of Side Walls

The self weight of two side walls acting as concentrated loads are assumed to produce uniform soil reaction on the bottom slab.

4. Water Pressure Inside Culvert

When the culvert is full with water, the pressure distribution on side walls is assumed to be triangular with a maximum pressure intensity of $p = wh$ at the base, where

$\quad\quad\quad w$ = density of water and 'h' is the depth of flow.

5. Earth Pressure on Vertical Side Walls

The earth pressure on the vertical side walls of the box culvert is computed according to the Coloumb's theory.[4,5] The distribution of earth pressure on the side walls is shown in Fig. 7.1(e).

6. Uniform Lateral Load on Side Walls

Uniform lateral pressure on vertical side walls has to be considered due to the effect of live load surcharge. Also trapezoidal pressure distribution on side walls due to embankment loading can be obtained by combining the cases (5) and (6).

7.3 DESIGN MOMENTS, SHEARS AND THRUSTS

The box culvert is analysed for moments, shear forces and axial thrusts developed due to the various loading conditions by any of the Classical methods such as moment distribution, slope deflection or Column analogy procedures. Alternatively coefficients for moments, shears and thrusts compiled by Victor (Ref–1) are very useful in the computation of the various force components for the different loading conditions.

The fixed end moments developed for the six different loading cases are compiled in Table 7.1. The moment, shear and thrust co-efficients for the various loading cases are shown in Table 7.2, for two different ratios of $(L/H) = 1$ and 1.5.

Where L = span of the culvert

             H = height of the culvert

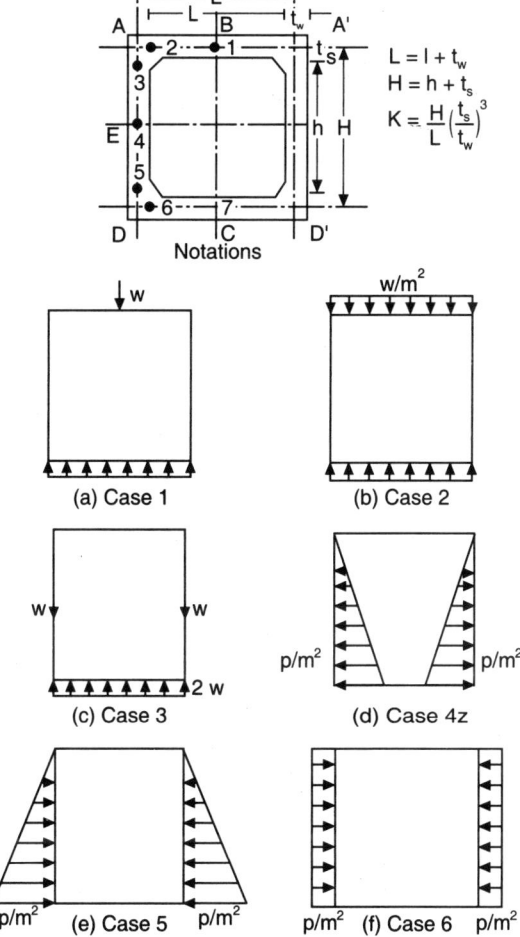

Fig. 7.1 Types of Loadings for Box Culverts.

Table 7.1 Fixed End Moments in Box Culvert

Loading case	Fixed end moments	
	$M_A = M_{A'}$	$M_D = M_{D'}$
1	$- [WL/12] (2K + 4.5)/[(K + 3)(K + 1)]$	$- [WL/24] (K + 6)/[(K + 3)(K + 1)]$
2	$- WL^2/[12(K + 1)]$	$- WL^2/[12(K + 1)]$
3	$+ [WL'6][K/(K + 3) + (K + 1)]$	$- [WL/6][(3 + 2K)/(K + 3) + (K + 1)]$
4	$+ [pH^2/60] [K(2K + 7)/(K + 3) (K + 1)]$	$+ [pH^2/60][K(3K + 8)/(K + 3) (K + 1)]$
5	$- [pH^2/60] [K(2K + 7)/(K + 3) (K + 1)]$	$- [pH^2/60][K(3K + 8)/(K + 3)(K + 1)]$
6	$- p.KH^2/[12(K + 1)]$	$- p.KH^2/[12(K + 1)]$

Note: Positve moment indicates on insides face.

Table 7.2 Coefficients for Moment, Shear and Trust

L:H	Section	Loading Case					
		1	2	3	4	5	6
	Factor for						
	M	WL	wL^2	WL	pL^2	pL^2	pL^2
	N	W	wL	WL	pL	pL	pL
	V	W	wL	W	pL	pL	pL
1:1 B 1	M	+0.182	+0.083	+0.021	+0.019	−0.019	−0.042
	N	0	0	0	−0.167	+0.167	+0.500
A 2	M	−0.068	+0.042	+0.021	+0.019	−0.019	−0.42
	N	0	0	0	−0.167	+0.167	−0.500
	V	+0.500	+0.500	0	0	0	0
A 3	M	−0.068	−0.042	+0.021	+0.019	−0.019	−0.042
	N	0	0	0	+0.167	−0.167	−0.500
	V	0	0	0	+0.167	−0.167	−0.500
E 4	M	−0.052	−0.042	−0.042	−0.043	−0.043	+0.083
	N	+0.500	+0.500	+0.500	0	0	0
D 5	M	−0.036	−0.042	−0.004	+0.023	−0.023	−0.042
	N	+0.500	+0.500	+1.000	−0.333	+0.333	0
	V	0	0	0	0	0	+0.500
D 6	M	−0.036	−0.042	−0.104	+0.023	−0.023	−0.042
	N	0	0	0	0	0	+0.500
	V	−0.500	−0.500	−1.000	−0.333	+0.333	0
C 7	M	+0.088	+0.083	+0.146	+0.023	−0.023	−0.042
	N	0	0	0	−0.333	+0.333	0
1.5:1B1	M	+0.170	+0.075	+0.018	+0.015	−0.015	−0.033
	N	0	0	0	−0.167	+0.167	+0.500
A 2	M	−0.079	−0.050	+0.018	+0.015	−0.015	−0.033
	N	0	0	0	−0.167	+0.167	+0.500
	V	+0.500	+0.500	0	0	0	0
A 3	M	−0.079	+0.050	+0.018	+0.015	−0.015	−0.033
	N	+0.500	+0.500	0	0	0	0
	V	0	0	0	+0.167	−0.167	−0.500
E 4	M	−0.062	−0.052	−0.050	−0.047	+0.047	+0.092
	N	+0.500	+0.500	+0.500	0	0	0
D 5	M	−0.045	−0.050	−0.118	+0.018	+0.018	−0.033
	N	+0.500	+0.500	+0.1000	0	0	0
	V	0	0	0	−0.333	+0.333	+0.500
D 6	M	−0.045	−0.050	−0.118	+0.018	−0.018	−0.033
	N	0	0	0	−0.033	+0.033	+0.500
	V	−0.500	−0.500	−0.1000	0	0	0
C 7	M	−0.079	−0.075	−0.132	−1.018	−0.018	−0.033
	N	0	0	0	−0.033	+0.033	+0.500

Refer Fig. 7.1 for detail and notations.

Note: 1. Positive moment indicate tension on inside face.
2. Positive shear indicates that the summation of force at the left of the section acts outwards when viewed from within.
3. Positive thrust indicates compression on the section.

7.4   DESIGN OF CRITICAL SECTIONS

The maximum design moments resulting from the combination of the various loading cases are determined. The moments at the centre of span of top and bottom slabs and the support sections and at the centre of the vertical walls are determined by suitably combining the different loading patterns. The maximum moments generally develop for the following loading conditions.[6]

1. When the top slab supports the dead and live load and the culvert is empty.

2. When the top slab supports the dead and live loads and the culvert is running full.

3. When the sides of the culvert do not carry the live load and the culvert is running full.

The slabs of the box culvert is reinforced on both faces with fillets at the inside corners.

The critical sections of the box culvert have to be designed for the limit states of strength and serviceability according to the specifications of IRC: 112-2011 and this is illustrated by the following typical example.

7.5   DESIGN EXAMPLE

Design a reinforced concrete box culvert having a clear vent way of 3 m by 3 m. The superimposed dead load on the culvert is 12.8 kN/m². The live load is estimated as 50 kN/m². Density of soil at site is 18 kN/m³. Angle of repose = 30°. Adopt M-20 grade concrete and Fe-415 HYSD bars. Sketch the details of reinforcements in the box culvert. The design should conform to the specifications of IRC: 112-2011.

1.   Data

Clear span = L = 3 m

Height of vent = h = 3 m

Dead load = 12.8 kN/m²

Live load = 50 kN/m²

Density of soil = 18 kN/m³

Angle of repose = 30°

Grade of concrete = M-20

Grade of steel = 415 HYSD bars

2.   Material Parameters

$$f_{ck} = 20 \text{ N/mm}^2$$
$$f_{yk} = 415 \text{ N/mm}^2$$
$$E_c = 29 \text{ GPa}$$
$$E_s = 200 \text{ GPa}$$
$$\alpha_e = (E_s/E_c) = 6.89$$

3.   Dimensions of Box Culvert

Adopting the thickness of slab as 100 mm/m span, we have

Thickness $= t_s = t_w = 300$ mm

Effective span $= 3300$ mm

4.   Loads

Self weight of top slab $= (0.3 \times 24) = 7.2$ kN/m^2

Superimposed dead load .... $= 12.8$

Live load .... $= 50.0$

Total load $= w$ .... $= 70$ kN/m^2

Weight of vertical side walls $= (0.3 \times 3.3 \times 24) = W = 24$ kN

Angle of repose $= \emptyset = 30°$ and height of soil fill $= h = 3.3$ m

Soil pressure $= p = \left[wh \left(\dfrac{1 - \sin \emptyset}{1 + \sin \emptyset} \right) \right] = \left[18 \times 3.3 \left(\dfrac{1 - \sin 30}{1 + \sin 30} \right) \right] = 20$ kN/m^2

Uniform lateral pressure due to the effect of superimposed dead load and live load surcharge is calculated as,

$$p = \left[50 + 12.8 \right] \left[\left(\dfrac{1 - \sin \emptyset}{1 + \sin \emptyset} \right) \right] = \left[62.8 \times (1/3) \right] = 21 \text{ kN/m}^2$$

Uniform lateral pressure due to the effect of superimposed dead load surcharge only is evaluated as,

$$p = \left[12.8 \right] \left[\left(\dfrac{1 - \sin \emptyset}{1 + \sin \emptyset} \right) \right] = \left[12.8 \times (1/3) \right] = 4.26 \text{ kN/m}^2$$

Intensity of water pressure $= p = wh = (10 \times 3.3) = 33$ kN/m^2

5.   Analysis of Moments, Shears and Thrusts

The various loading patterns considered are shown in Fig. 7.2. The moments, shears and Thrusts corresponding to the different cases of loading (Case 1 to 6) are evaluated using the coefficients given in Table 7.2, are compiled in Table 7.3. The service load design forces resulting from the combination of various cases yielding maximum moments and forces at the supports and mid span sections are shown in Table 7.4.

The maximum positive moments develop at the centre of bottom and top slabs for the condition that the sides of the culvert not carrying the live load and the culvert is running full with water. The maximum negative moments develop at the support sections of the bottom slab for the condition. Culvert is empty and the top slab carries the dead and live loads.

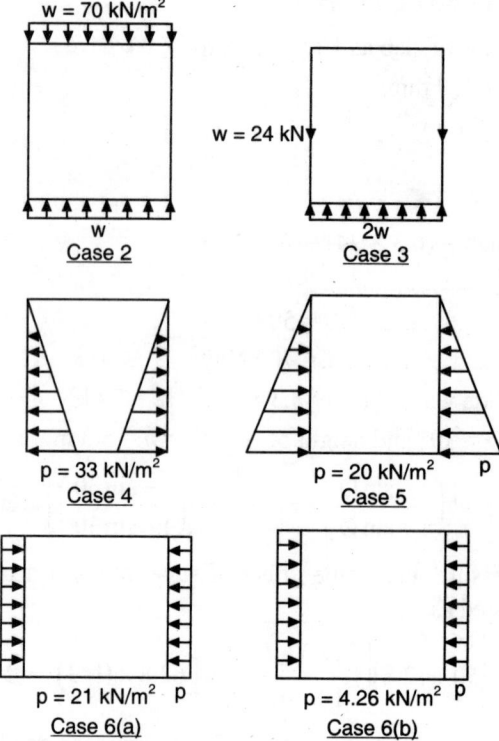

Fig. 7.2 Loading Cases Considered for Box Culvent.

6. Design of Reinforcements

a) Section C-7 (Mid Span of Bottom Slab)

Service load moment = M_w = 76.10 kN·m

Design Ultimate moment = M_u = (1.5 × 76.10) = 114.15 kN·m

Service load Thrust = N = −7.43 kN (tension)

Ultimate thrust N_u = (1.5 × 7.43) = −11.14 kN (tension)

Compute the parameters $(N_u/f_{ck}.b.D)$ and $(M_u/f_{ck}.b.D^2)$ for using the design charts of SP-16[7]

$(N_u/f_{ck}.b.D)$ = [(11.14 × 10³)/(20 × 1000 × 300) = 0.0018

$(M_u/f_{ck}.b.D^2)$ = [(114.15 × 10⁶)/(20 × 1000 × 300²) = 0.0634

Adopting effective cover as 50 mm, D = 300 mm and d' = 50 mm, Ratio (d'/D) = 0.16

Refer Design Chart-70 of SP-16 for tension and bending and read out (p/f_{ck}) corresponding to the two parameters computed above as,

(p/f_{ck}) = 0.05 and adopting effective cover as 50 mm, d = 250 mm and d' = 50 mm

Hence $\quad p = (100A_{st}/bd) = (0.05 \times 20) = 1.0$

$A_{st} = (bd/100) = [(1000 \times 250)/100] = 2500 \text{ mm}^2/\text{m}$ (equally distributed on both faces)

Provide 16 mm diameter bars at 150 mm centres on each face as main reinforcement (A_{st} provided = 2680 mm^2)

Provide 10 mm diameter bars at 150 mm centers on both faces as distribution reinforcement. The serviceability limit state of cracking is easily satisfied since the diameter and spacing of bars are 16 mm and 150 mm respectively and the stress in steel at working loads will be less than 240 N/mm^2 for a permissible crack width of 0.3 mm, according to tables 4.9 and 4.10.

Table 7.3 Force Components for Different Cases of Loading

Section	Forces	Loading case					
		Case 2	Case 3	Case 4	Case 5	Case 6(a)	Case 6(b)
B-1	M	63.2	1.66	6.82	−4.13	−9.6	−1.92
	N	0	0	−18.18	+11.0	+34.65	+6.93
A-2	M	−31.6	1.66	6.82	−4.13	−9.6	−1.92
	N	0	0	−18.18	+11.0	+34.65	−6.93
	V	115.5	0	0	0	0	0
A-3	M	−31.6	1.66	6.82	−4.13	−9.6	−1.92
	N	115.5	0	0	0	0	0
	V	0	0	18.18	−11.0	−34.65	−6.93
E-4	M	−31.6	−3.32	−15.45	−9.39	+19.2	+3.84
	N	115.5	+39.6	0	0	0	0
D-5	M	−31.6	−0.317	8.26	−5.00	−9.6	−1.92
	N	115.5	+79.2	−36.26	+21.9	0	0
	V	0	0	0	0	+34.65	+6.93
D-6	M	−31.6	−8.23	8.26	−5.00	−9.6	−1.92
	N	0	0	0	0	+34.65	+6.93
	V	−115.5	−79.2	−36.26	+21.9	0	0
C-7	M	63.2	11.56	8.26	−5.00	−9.60	−1.92
	N	0	0	−36.26	+21.9	+34.65	+6.93

Note: Moment are in kN·m
    Shear Force and Thrust are in kN

Table 7.4 Service Load Design Moments and Forces in Box Culvert

Section	Loading Combination Cases	Moment (M) (kN·m)	Thrust (N) (kN)	Shear Force (V) (kN)
D–6	2 + 3 + 5 + 6 (a)	−54.43	+34.65	−172.8
A–2	2 + 3 + 5 + 6 (a)	−43.67	−23.65	+115.5
B–1	2 + 3 + 4 + 5 + 6 (b)	65.63	−1.23	0
C–7	2 + 3 + 4 + 5 + 6 (b)	76.10	−7.43	0
E–4	2 + 3 + 4 + 5 + 6 (b)	−55.89	+155.1	0

b) Section D-6 (Support Section)

Service load moment = $M_w = -54.43$ kN·m

Design Ultimate moment = $M_u = (1.5 \times 54.43) = -81.64$ kN·m

Service load Thrust = $N = 34.65$ kN (Compression)

Ultimate thrust $N_u = (1.5 \times 34.65) = 51.97$ kN (Compression)

Compute the parameters $(N_u/f_{ck}.b.D)$ and $(M_u/f_{ck}.b.D^2)$ for using the design charts of SP-16

$$(N_u/f_{ck}.b.D) = [(51.97 \times 10^3)/(20 \times 1000 \times 300) = 0.0086$$
$$(M_u/f_{ck}.b.D^2) = [(81.64 \times 10^6)/(20 \times 1000 \times 300^2) = 0.045$$

Referring to Chart-33 of SP-16 with parameters of $f_y = 415$ N/mm^2, $(d'/D) = 0.15$

Read out the value of $(p/f_{ck}) = 0.03$

Hence, $\quad p = (100A_{st}/bd) = (0.03 \times 20) = 0.6$

$\qquad A_{st} = (0.6bd/100) = [(0.6 \times 1000 \times 250)/100] = 1500$ mm^2

Provide 12 mm diameter bars at 150 mm centres on both faces with 10 mm diameter at 150 mm centers as distribution reinforcement.

Both strength and serviceability requirements are easily satisfied with the detailing of reinforcements mentioned above.

c) Section E-4 (Vertical Side Wall)

Service load moment = $M_w = -55.89$ kN·m

Design Ultimate moment = $M_u = (1.5 \times 55.89) = -83.83$ kN·m

Service load Thrust = $N = 155.1$ kN (Compression)

Ultimate thrust $N_u = (1.5 \times 155.1) = 232.65$ kN (Compression)

Compute the parameters $(N_u/f_{ck}.b.D)$ and $(M_u/f_{ck}.b.D^2)$ for using the design charts of SP-16

$$(N_u/f_{ck}.b.D) = [(232.65 \times 10^3)/(20 \times 1000 \times 300) = 0.038$$
$$(M_u/f_{ck}.b.D^2) = [(83.83 \times 10^6)/(20 \times 1000 \times 300^2) = 0.046$$

Referring to Chart-33 of SP-16 with parameters of $f_y = 415$ N/mm^2, $(d'/D) = 0.15$

Read out the value of $(p/f_{ck}) = 0.025$

Hence, $\quad p = (100A_{st}/bd) = (0.025 \times 20) = 0.5$

$\qquad A_{st} = (0.5bd/100) = [(0.5 \times 1000 \times 250)/100] = 1250$ mm^2

Provide 12 mm diameter bars at 150 mm centres on both faces with 10 mm diameter at 150 mm centers as distribution reinforcement.

Both strength and serviceability requirements are easily satisfied with the detailing of reinforcements mentioned above.

7. Reinforcement Details

The details of reinforcements in the box culvert are shown min Fig. 7.3

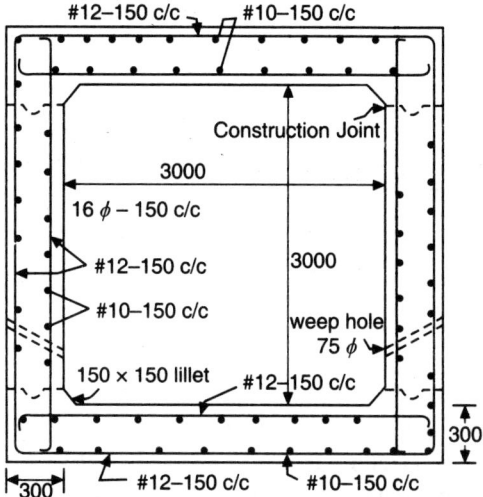

Fig. 7.3 Reinforcement Details in Box Culvert.

EXAMPLES FOR PRACTICE

1. Design a reinforced concrete box culvert witn inside dimehsions 3 m height and 4.5 width. The box culverts has to carry a superimposed load of 10 kN/m² and a live load of 50 kN/m². The density of the earth is 18 kN/m³. Angle of repose of the soil is 30°. Adopt M-20 grade concrete and Fe-415 grade tor steel. Sketch the details of reinforcement in the box culvert.

2. A reinforced concrete box culvert is required for a national highway crossing. The clear ventway of the box culvert is 4 m by 4 m. Design the box culvert assuming a superimposed dead load of 12 kN/m² and a live load of 50 kN/m². The density of the soil is 16 kN/m³ Angle of repose of the soil is 30°. Adopt M-20 grade concerte and Fe-415 grade tor steel. Sketch the details of reinforcements in the box culvert.

3. A reinforced concrete box culvert of prismatic form with a clear ventway of 3.5 m by 3.5 m is required for a road crossing. The box culvert has to support a superimposed dead load of 8 kN/m² and a lice load of 50 kN/m². Density of the soil is 18 kN/m³ and the angle of repose of the soil is 30°. Adoption M-20 grade concrete and Fe-415 grade tor steel design the box culvert and sketch the details of reinforcement in the box culvert.

4. A reinforced concrete box culvert of rectangular water way 3.75 m wide by 2.5 m deep is required for a roads crossing. The box culvert has to support a superimposed dead load of 10 kN/m² and a live load of 40 kN/m². Density of soil at site is 16 kN/m² and the angle of repose of the soil is 30°.

Adopting M-25 grade concrete and Fe-415 grade HYSD bars, design the box culvert and sketch the details of reinforcements.

REFERENCES

1. Johnson Victor, D., Essentials of Bridge Engineering, (Fifth Edition), Oxford & IBH Publishing Co. Pvt. Ltd., New Delhi, 2001, pp. 122–130.
2. Reynolds, C.E and Steedman, J., Reinforced Concrete Designers Hand Book, Concrete Publications Ltd., London, 1974.
3. Wang, P.C., Numerical and Matrix Methods in Structural Mechanics, John Wiley & Sons, New York, 1966.
4. Dunham, C.W., Foundations of Structures, McGraw-Hill Book Co. Ltd., New York, Second Edition, 1962, 722 pp.
5. Peck, R.B, Hanson, W.E and Thornburn, T.H., Foundation Engineering, Asia Publishing House, Bombay, First edition, 1959, 410 pp.
6. Krishna Raju, N. and Gururaja, D.R., Advanced Mechanics of Solids and Structures, Narosa Publishing House, New Delhi, 1997.
7. SP: 16-1980, Design Aids For Reinforced Concrete to IS: 456-1980, Bureau of Indian Standards, New Delhi, 1980, pp. 1–232.

REVIEW QUESTIONS

1. Briefly explain the situations under which you would recommend the use of reinforced concrete box culverts in bridge structures.
2. Mention the various advantages of reinforced concrete box culverts in comparison with other types of cross drainage works adopted in high way structures.
3. How do you fix the vent way in a box culvert? Discuss briefly the hydraulic design of Reinforced concrete box culverts used in bridge structures.
4. Briefly explain the various types of loads to be considered in the structural design of R.C.C. box culverts.
5. What are the critical loading conditions to be considered in evaluating the maximum design moments and forces in a box culvert?
6. What type of analysis you would adopt to calculate the design moments and forces in an indeterminate structure like rigid box culvert?
7. Specify the critical locations of the sections in the box culvert for which the design moments and forces are generally maximum under the given system of loads acting on box culverts.
8. Explain the method of designing typical sections of a box culvert subjected to flexure and axial forces.
9. How do you check for the limit states of strength and serviceability of the typical section of a box culvert?
10. Explain with sketches the detailing of reinforcements in a typical reinforced concrete box culvert.

OBJECTIVE TYPE QUESTIONS

1. Reinforced concrete box culverts are ideally suited in high way structures
   a) Criss crossing in urban fly overs
   b) Crossing small streams
   c) Crossing of major rivers
2. Box culverts are economical since
   a) They require simple footings as foundations
   b) Separate foundations are not required
   c) They are easy to construct
3. Box culverts can be classified for analysis under the category of
   a) Determinate structures
   b) Pin jointed structures
   c) Indeterminate structures
4. The number of various types of loading cases to be considered in determining the Maximum design moments and forces in the centre of the bottom slab of box culvert is
   a) 3
   b) 4
   c) 5
5. In a box culvert subjected to various types of loads, the maximum positive bending moment develops at the
   a) Centre of the top slab
   b) Centre of the bottom slab
   c) Centre of the side slab
6. In the design of box culverts subjected to various forces, the maximum thrust under the combination of various loads occurs at
   a) The support of top slab
   b) The support of bottom slab
   c) The centre of vertical side slab
7. Uniform lateral pressure on the side walls of the box culvert is due to
   a) Earth pressure on the side walls
   b) Water pressure flowing through the culvert
   c) Live load surcharge
8. The various structural elements of the box culvert should be designed using
   a) Elastic theory
   b) Ultimate load method
   c) Limit state method
9. The various sections of the box culvert should normally be reinforced
   a) On one side only
   b) Both sides
   c) On the side of the maximum moments

10. Suitable construction joints in box culverts are normally provided at
    a) The centre of horizontal slabs
    b) The centre of vertical slabs
    c) The junctions of vertical and horizontal slabs

Tee Beam and Slab Bridge Deck

8

8.1 GENERAL FEATURES

Tee beam and slab decks are the most common type of super structures generally adopted in most of the national high ways in the country. A typical tee beam deck generally comprises the longitudinal reinforced concrete girders with an integral continuous deck slab between the tee beams and cross girders to provide lateral rigidity to the bridge deck. The longitudinal girders are normally spaced at intervals of 2 to 3 m and cross girders are provided at 4 to 5 m intervals along the span. Reinforced concrete tee beams are ideally suited for spans in the range of 10 to 25 m. For larger span lengths, the depth of the tee girder being large, the total dead loads are abnormal with larger magnitudes of reinforcements in the tee girders. However Victor[1] has reported the example of a simply supported tee beam bridge deck constructed in Goa with a single span of 35 m. For spans more than 20 m it is economical to use prestressed tee beam decks for high way crossings in national high ways.

8.2 TYPES OF TEE BEAM SLAB DECKS

Basically there are three types of tee beam and slab decks gradually developed for use as high way bridge crossings. The distinctive features of these three types are outlined below:

1) Girder and slab type in which the beams and slab are cast monolithically without any cross girders. In this case the deck slab is designed as a one way slab spanning between the girders. This type of deck developed in the early stages does not possess torsional rigidity and hence not currently used.

2) Girder, slab and diaphragm type where in the slab is cast monolithically with the girders. Diaphragms connecting the girders are provided at supports and a few intermediate locations without extending up to the deck slab. This type is marginally better in resisting loads due to improved torsional rigidity in comparison with the first type.

3) Girder, slab and cross beams are cast monolithically to form an integrated bridge deck possessing superior flexural and torsional rigidity. This type evolved after several research investigations is the most commonly used system used at present in high way bridge decks.

Experimental investigations reported by Victor et al[2] using micro concrete models of the three different tee beam systems have conclusively proved that the third type of girder, slab and cross beams system cast integrally is significantly superior in resisting the ultimate loads.

Fig. 8.1 shows the cross section and plan of a typical T-beam bridge deck generally used in various highways of India.

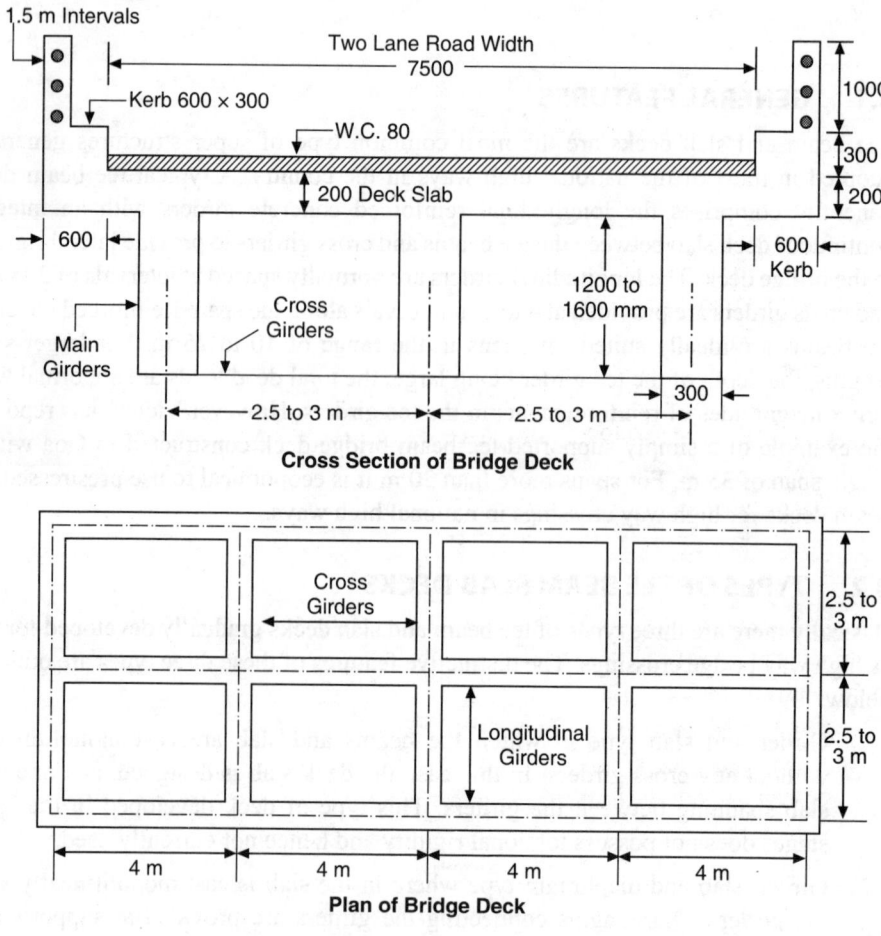

Fig. 8.1 Typical Tee Beam and Slab Deck.

8.3   DESIGN PRINCIPLES OF TEE BEAM BRIDGE DECKS

The three main structural elements of a tee beam bridge deck comprise the slab, longitudinal beams and the cross girders. The deck slab supported on all the sides by longitudinal and cross girders is designed using the moment coefficients of Pigeaud's curves outlined in section 4.12 for different types of IRC loads. The design of longitudinal beams involves the evaluation of the live load distribution among the number of beams used in the deck. With three or more girders, the load distribution is estimated using any of the three rational methods given below:

a)   Courbon's Method[3]

b)   Guyon-Massonet Method[4,5]

c)   Hendry- Jaegar Method[6]

By using any of the three methods, the maximum reaction factors are determined for the intermediate and end longitudinal girders and these are used for the computation of bending moments and shear forces.

The cross beams continuous over supports are designed to resist the maximum dead load and live load moments resulting from the critical positioning of IRC live loads. The reinforcements in the various structural elements are designed conforming to limit state criteria specified in the Indian Roads congress code IRC: 112-2011.

8.4   COURBON'S METHOD

Among these methods, Courbon's method is the simplest and is applicable when the following conditions are satisfied:

(a) The ratio of span to width of deck is greater than 2 but less than 4.

(b) The longitudinal girders are interconnected by at least five symmetrically spaced cross girders.

(c) The cross girder extends to a depth of at least 0.75 times the depth of the longitudinal girders.

Courbon's method is popular due to the simplicity of computations as detailed below:

When the live loads are positioned nearer to the kerb as shown in Fig. 8.2, the centre of gravity of live load acts eccentrically with the centre of gravity of the girder system. Due to this eccentricity, the loads shared by each girder is increased or decreased depending upon the position of the girders. This is calculated by Courbon's theory by a reaction factor given by,

$$R_x = (\Sigma W/n) \, [1 + (\Sigma I / \Sigma d_x^2 \cdot I) d_x \cdot e]$$

Where R_x = Reaction factor for the girder under consideration

I = Moment of inertia of each longitudinal girder

d_x = Distance of the girder under consideration from the central axis of the bridge

W = Total concentrated live load

n = Number of longitudinal girders

e = Eccentricity of live load with respect to the axis of the bridge

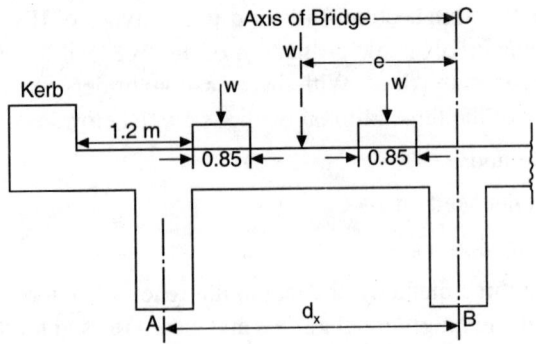

Fig. 8.2 Position of Live Loads for Maximum BM in Girder A.

The live load bending moments and shear forces are computed for each of the girders. The maximum design moments and shear forces are obtained by adding the live load and dead load bending moments. The reinforcements in the main longitudinal girders are designed for the maximum moments and shears developed in the girders.

An approximate method may be used for the computation of the bending moments and shear forces in cross girders. The cross girders are assumed to be rigid so that the reactions due to dead and live loads are assumed to be equally shared by the cross girders. This assumption will simplify the computation of bending moments and shear forces in the cross girders.

8.5 GUYON-MASSONET METHOD

Guyon-Massonet method is based on the application of orthotropic plate theory to the bridge deck system. Morice and Little[7] have successfully applied this theory to the analysis of bridge deck systems. The method has the advantage of using a single set of distribution coefficients for the two extreme cases of no torsion grillage and a full torsion slab thus enabling the determination of the load distribution behaviour of any type of bridge deck.

The longitudinal bending moments at various points along the cross-section are obtained by multiplying the mean longitudinal bending moment by the appropriate distribution coefficients for these points. The mean longitudinal bending moment is the bending moment developed by considering the total load on the span as, uniformly spread over the whole width of the bridge. Hence the mean bending moment per girder can be expressed as

$$M_{\text{mean}} = (M/n)$$

Where M = Total mean longitudinal bending moment

n = Number of girders

The design bending moment is then computed as

Design B.M.= $(1.10 \times K \times M_{mean} \times I.F.)$

Where K = Distribution coefficient

I.F. = Impact factor

The factor 1.10 is used to compensate for the error involved in using only the first term of the Fourier series in finding the distribution coefficients, as suggested by Rowe[8] based on experiments.

The distribution coefficient 'K' depends on the flexural and torsional parameters expressed as,

Flexural parameter $\theta = (b/2a) (i/j)^{0.25}$ (1)

Torsional parameter $\alpha = \left[G(i_0 + j_0) / (2E\sqrt{ij}) \right]$ (2)

Where $2a$ = Span of the bridge

$2b$ = Effective width of bridge

i = Second moment of area per unit transverse width

j = Second moment of area per unit longitudinal width

$G . i_0$ = Torsional stiffness per unit width

$G . j_0$ = Torsional stiffness per unit length

The values of distribution coefficient Ka is calculated from the interpolation formula.

$$K_\alpha = K_0 + \left(K_1 - K_0\sqrt{\alpha}\right)$$ (3)

Where K_0 and K_1 refers to the distribution coefficients corresponding to $\alpha = 0$ and $\alpha = 1$. Morice[9] has presented the values of K_0 and K_1 for five reference stations (0, $b/4$, $b/2$, $3b/4$ and b) and for various load positions and for values of θ from 0 to 3.0 in a graphical form. The values of K_0 and K_1 for range of θ between 0.2 to 0.8 have been presented in a tabular form for ready use in design office by Sarkar[10]. These are compiled in Table 8.1.

The reference stations and load positions for maximum distribution are shown in Fig. 8.3.

The maximum transverse moment occurs when an internal line of wheels coincides with the longitudinal centre line of the bridge, the maximum moment being at the centre of the bridge at the reference station O. The equation of transverse moment for a concentrated load 'W' at a distance 'u' from the left support is given by,

Table 8.1 Values of K_0 and K_1 for Various Values of θ

					$\theta = 0.20$				
Ref. pt	$-b$	$-3b/4$	$-b/2$	$-b/4$	0	$b/4$	$b/2$	$3b/4$	b
Load at									
					K_0				
0	0.94	0.99	0.97	1.02	1.06	1.02	0.97	0.99	0.94
$b/4$	0.25	0.42	0.63	0.84	1.02	1.19	1.35	1.56	1.73
$b/2$	−0.53	−0.15	0.25	0.63	0.97	1.35	1.72	2.10	2.49
$3b/4$	−1.20	−0.66	−0.15	0.42	0.99	1.56	2.10	2.70	3.27
b	−1.90	−1.20	−0.53	0.25	0.94	1.73	2.49	3.27	4.00
					K_1				
0	0.96	0.99	1.00	1.00	1.03	1.00	1.00	0.99	0.96
$b/4$	0.91	0.93	0.97	0.98	1.00	1.03	1.03	1.03	1.03
$b/2$	0.86	0.90	0.93	0.97	1.00	1.03	1.07	1.10	1.13
$3b/4$	0.80	0.85	0.90	0.93	0.99	1.03	1.10	1.16	1.23
b	0.75	0.80	0.86	0.91	0.96	1.03	1.13	1.23	1.35
					$\theta = 0.225$				
					K_0				
0	0.92	0.98	0.98	1.03	1.07	1.03	0.98	0.98	0.92
$b/4$	0.235	0.415	0.63	0.845	1.03	1.195	1.35	1.55	1.715
$b/2$	−0.53	−1.15	0.245	0.63	0.98	1.35	1.72	2.10	2.48
$3b/4$	−1.185	−0.65	−0.15	0.415	0.98	1.55	2.10	2.705	3.275
b	−1.185	−1.185	−0.53	0.235	0.92	1.715	2.48	3.275	4.00
					K_1				
0	0.96	0.985	1.00	1.01	1.035	1.01	1.00	0.985	0.96
$b/4$	0.891	0.92	0.965	0.975	1.01	1.04	1.04	1.04	1.035
$b/2$	0.835	0.88	0.92	0.965	1.00	1.04	1.085	1.115	1.145
$3b/4$	0.775	0.825	0.88	0.92	0.985	1.04	1.115	1.19	1.265
b	0.72	0.775	0.835	0.895	0.96	1.035	1.145	1.265	1.405
					$\theta = 0.225$				
					K_0				
0	0.90	0.97	0.985	1.04	1.08	1.04	0.985	0.97	0.90
$b/4$	0.22	0.41	0.63	0.85	1.04	1.20	1.35	1.54	1.70
$b/2$	−0.53	−0.15	0.24	0.63	0.985	1.35	1.72	2.10	2.47
$3b/4$	−1.17	−0.64	−0.15	−0.41	−0.97	1.54	2.10	2.71	3.28
b	−1.85	−1.17	−0.53	−0.22	0.90	1.70	2.47	3.28	4.00
					K_1				

Ref. pt / Load at	$-b$	$-3b/4$	$-b/2$	$-b/4$	0	$b/4$	$b/2$	$3b/4$	b
0	0.96	0.98	1.00	1.02	1.04	1.02	1.00	0.98	0.96
$b/4$	0.88	0.91	0.96	0.97	1.02	1.05	1.05	1.05	1.04
$b/2$	0.81	0.86	0.91	0.96	1.00	1.05	1.10	1.13	1.16
$3b/4$	0.75	0.80	0.86	0.91	0.98	1.05	1.13	1.22	1.30
b	0.69	0.75	0.81	0.88	0.96	1.04	1.16	1.30	1.46
$\theta = 0.275$									
K_0									
0	0.88	0.96	0.98	1.045	1.09	1.045	0.98	0.96	0.88
$b/4$	0.21	0.405	0.63	0.86	1.045	1.21	1.355	1.535	1.69
$b/2$	−0.535	−0.155	0.24	0.63	0.98	1.355	1.725	2.10	2.465
$3b/4$	−1.16	−0.635	−0.155	0.405	0.96	1.535	2.10	2.72	3.295
b	−1.82	−1.16	−0.535	0.21	0.88	1.69	2.465	3.295	4.05
K_1									
0	0.95	0.975	1.00	1.02	1.045	1.02	1.00	0.975	0.95
$b/4$	0.865	0.90	0.95	0.97	1.02	1.055	1.055	1.05	1.05
$b/2$	0.79	0.84	0.90	0.95	1.00	1.055	1.115	1.15	1.185
$3b/4$	0.725	0.775	0.84	0.90	0.975	1.05	1.15	1.255	1.34
b	0.66	0.725	0.79	0.865	0.95	1.05	1.185	1.34	1.525
$\theta = 0.300$									
K_0									
0	0.86	0.95	0.97	1.05	1.10	1.05	0.97	0.95	0.86
$b/4$	0.20	0.40	0.63	0.87	1.05	1.22	1.36	1.53	1.68
$b/2$	−0.54	−0.16	0.24	0.63	0.97	1.36	1.73	2.10	2.46
$3b/4$	−1.15	−0.63	−0.16	0.40	0.95	1.53	2.10	2.73	3.31
b	−1.79	−1.15	−0.54	0.20	0.86	1.68	2.46	3.31	4.10
K_1									
0	0.94	0.97	1.00	1.02	1.05	1.02	1.00	0.97	0.94
$b/4$	0.85	0.89	0.94	0.97	1.02	1.06	1.06	1.05	1.06
$b/2$	0.77	0.82	0.89	0.94	1.00	1.06	1.13	1.17	1.21
$3b/4$	0.70	0.75	0.82	0.89	0.97	1.05	1.17	1.29	1.38
b	0.63	0.70	0.77	0.85	0.94	1.06	1.21	1.38	1.59
$\theta = 0.325$									
K_0									
0	0.83	0.94	0.975	1.065	1.125	1.065	0.975	0.94	0.83
$b/4$	0.185	0.395	0.63	0.88	1.065	1.235	1.37	1.515	1.65
$b/2$	−0.54	−0.165	0.24	0.63	0.975	1.37	1.74	2.10	2.445
$3b/4$	−1.13	−0.615	−0.165	0.395	0.94	1.515	2.10	2.74	3.325
b	−1.745	−1.13	−0.54	0.185	0.83	1.65	2.445	3.325	4.15
K_1									

Ref. pt Load at	$-b$	$-3b/4$	$-b/2$	$-b/4$	0	$b/4$	$b/2$	$3b/4$	b
0	0.94	0.965	1.00	1.03	1.055	1.03	1.00	0.965	0.94
$b/4$	0.83	0.87	0.93	0.97	1.03	1.07	1.07	1.06	1.06
$b/2$	0.74	0.795	0.87	0.93	1.00	1.07	1.19	1.19	1.23
$3b/4$	0.675	0.725	0.795	0.87	0.965	1.06	1.32	1.32	1.42
b	0.595	0.675	0.74	0.83	0.94	1.06	1.42	1.42	1.655
$\theta = 0.350$									
K_0									
0	0.80	0.93	0.98	1.08	1.15	1.08	0.98	0.93	0.80
$b/4$	0.17	0.39	0.63	0.89	1.08	1.25	1.38	1.50	1.62
$b/2$	-0.545	-0.17	0.24	0.63	0.98	1.38	1.75	2.10	2.43
$3b/4$	-1.11	-0.60	-0.17	0.39	0.93	1.50	2.10	2.75	3.34
b	-1.70	-1.11	-0.545	0.17	0.80	1.62	2.43	3.34	4.20
K_1									
0	0.94	0.96	1.00	1.04	1.06	1.04	1.00	0.96	0.94
$b/4$	0.81	0.85	0.90	0.97	1.04	1.08	1.08	1.07	1.06
$b/2$	0.71	0.77	0.85	0.92	1.00	1.08	1.17	1.21	1.25
$3b/4$	0.65	0.70	0.77	0.85	0.96	1.07	1.21	1.35	1.46
b	0.56	0.65	0.71	0.81	0.94	1.06	1.25	1.46	1.72
$\theta = 0.375$									
K_0									
0	0.76	0.90	0.99	1.10	1.18	1.10	0.99	0.90	0.76
$b/4$	0.15	0.39	0.64	0.86	1.10	1.27	1.38	1.48	1.60
$b/2$	-0.54	-0.16	0.23	0.64	0.99	1.38	1.75	2.09	2.40
$3b/4$	-1.09	-0.60	-0.16	0.39	0.90	1.48	2.09	2.77	3.36
b	-1.67	-1.09	-0.54	0.15	0.76	1.60	2.40	3.36	4.30
K_1									
0	0.91	0.96	1.00	1.04	1.07	1.04	1.00	0.96	0.91
$b/4$	0.79	0.84	0.91	0.96	1.04	1.10	1.09	1.09	1.07
$b/2$	0.68	0.75	0.83	0.91	1.00	1.1	1.19	1.24	1.29
$3b/4$	0.60	0.67	0.75	0.85	0.96	1.09	1.24	1.40	1.52
b	0.52	0.60	0.68	0.79	0.91	1.07	1.29	1.53	1.81
$\theta = 0.400$									
K_0									
0	0.71	0.90	0.99	1.11	1.2	1.11	0.99	0.90	0.71
$b/4$	0.12	0.36	0.64	0.91	1.11	1.29	1.40	1.47	1.56
$b/2$	-0.55	-0.17	0.23	0.63	0.99	1.37	1.76	2.10	2.40
$3b/4$	-1.07	-0.58	-0.17	0.36	0.90	1.47	2.10	2.77	3.38
b	-1.65	-1.07	-0.55	0.12	0.71	1.56	2.40	3.38	4.30
K_1									

Ref. pt / Load at	$-b$	$-3b/4$	$-b/2$	$-b/4$	0	$b/4$	$b/2$	$3b/4$	b
0	0.90	0.95	1.00	1.05	1.08	1.05	1.00	0.95	0.90
$b/4$	0.77	0.83	0.90	0.96	1.05	1.10	1.10	1.09	1.07
$b/2$	0.66	0.73	0.81	0.90	1.00	1.10	1.20	1.26	1.30
$3b/4$	0.58	0.65	0.73	0.83	0.95	1.09	1.26	1.41	1.55
b	0.50	0.58	0.66	0.77	0.90	1.07	1.30	1.55	1.88

$\theta = 0.425$

K_0

	$-b$	$-3b/4$	$-b/2$	$-b/4$	0	$b/4$	$b/2$	$3b/4$	b
0	0.67	0.875	0.995	1.13	1.22	1.13	0.995	0.875	0.67
$b/4$	0.10	0.35	0.64	0.925	1.13	1.31	1.41	1.455	1.50
$b/2$	−0.545	−0.17	0.23	0.635	0.995	1.375	1.77	2.095	2.375
$3b/4$	−1.045	−0.57	−0.17	0.35	0.875	1.455	2.095	2.785	3.405
b	−1.60	−1.045	−0.545	0.10	0.67	1.53	2.37	3.405	4.04

K_1

	$-b$	$-3b/4$	$-b/2$	$-b/4$	0	$b/4$	$b/2$	$3b/4$	b
0	0.89	0.95	1.00	1.055	1.09	1.055	1.00	0.95	0.89
$b/4$	0.75	0.81	0.885	0.96	1.055	1.12	1.12	1.095	1.08
$b/2$	0.63	0.70	0.785	0.885	1.00	1.12	1.225	1.28	1.325
$3b/4$	0.54	0.615	0.70	0.81	0.95	1.095	1.28	1.455	1.61
b	0.47	0.54	0.63	0.75	0.39	1.08	1.325	1.61	1.94

$\theta = 0.450$

K_0

	$-b$	$-3b/4$	$-b/2$	$-b/4$	0	$b/4$	$b/2$	$3b/4$	b
0	0.63	0.85	1.00	1.15	1.25	1.15	1.00	0.85	0.63
$b/4$	0.08	0.34	0.64	0.94	1.15	1.34	1.42	1.44	1.50
$b/2$	−0.54	−0.17	0.23	0.64	1.00	1.38	1.78	2.09	2.35
$3b/4$	−1.02	−0.56	−0.17	0.34	0.85	1.44	2.09	2.08	3.43
b	−1.55	−1.02	−0.54	0.08	0.63	1.50	2.35	3.43	4.50

K_1

	$-b$	$-3b/4$	$-b/2$	$-b/4$	0	$b/4$	$b/2$	$3b/4$	b
0	0.88	0.95	1.00	1.06	1.10	1.06	1.00	0.95	0.88
$b/4$	0.73	0.79	0.87	0.96	1.06	1.14	1.14	1.10	1.09
$b/2$	0.60	0.67	0.76	0.87	1.00	1.14	1.25	1.30	1.35
$3b/4$	0.50	0.58	0.67	0.79	0.95	1.10	1.30	1.50	1.67
b	0.44	0.50	0.60	0.73	0.88	1.09	1.35	1.67	2.00

$\theta = 0.475$

K_0

	$-b$	$-3b/4$	$-b/2$	$-b/4$	0	$b/4$	$b/2$	$3b/4$	b
0	0.59	0.82	1.00	1.18	1.285	1.18	1.00	0.82	0.59
$b/4$	0.04	0.32	0.635	0.95	1.18	1.37	1.43	1.42	1.45
$b/2$	−0.54	−0.17	0.225	0.635	1.00	1.39	1.79	2.085	2.325
$3b/4$	−0.99	−0.55	−0.17	0.32	0.82	1.42	2.085	2.82	3.465
b	−1.49	−0.99	−0.54	0.04	0.59	1.45	2.325	3.465	4.65

K_1

Ref. pt Load at	$-b$	$-3b/4$	$-b/2$	$-b/4$	0	$b/4$	$b/2$	$3b/4$	b
0	0.865	0.935	1.00	1.065	1.115	1.065	1.00	0.935	0.865
$b/4$	0.705	0.775	0.865	0.96	1.065	1.15	1.145	1.11	1.09
$b/2$	0.575	0.65	0.745	0.865	1.00	1.145	1.275	1.325	1.37
$3b/4$	0.475	0.555	0.65	0.775	0.935	1.11	1.325	1.54	1.715
b	0.41	0.475	0.575	0.705	0.865	1.90	1.37	1.715	1.075

$$\theta = 0.500$$

$$K_0$$

0	0.55	0.79	1.00	1.21	1.32	1.21	1.00	0.79	0.55
$b/4$	0.00	0.30	0.63	0.96	1.21	1.40	1.44	1.40	1.40
$b/2$	−0.54	−0.17	0.22	0.63	1.00	1.40	1.80	2.08	2.30
$3b/4$	−1.96	−0.54	−0.17	0.30	0.79	1.40	2.08	2.84	3.50
b	−1.43	−0.96	−0.54	0.0	0.55	1.40	2.30	3.50	4.80

$$K_1$$

0	0.85	0.92	1.00	1.07	1.13	1.07	1.00	0.92	0.85
$b/4$	0.68	0.76	0.86	0.96	1.07	1.16	1.15	1.12	1.09
$b/2$	0.55	0.63	0.73	0.86	1.00	1.15	1.30	1.35	1.39
$3b/4$	0.45	0.53	0.63	0.76	0.92	1.12	1.35	1.58	1.76
b	0.38	0.45	0.55	0.68	0.85	1.09	1.39	1.76	2.15

$$\theta = 0.525$$

$$K_0$$

0	0.485	0.765	1.01	1.24	1.36	1.24	1.01	0.765	0.485
$b/4$	−0.05	0.275	0.63	0.97	1.24	1.425	1.45	1.375	1.33
$b/2$	−0.535	−0.175	0.215	0.63	1.01	1.415	1.82	2.075	2.275
$3b/4$	−0.925	−0.52	−0.175	0.275	0.765	1.375	2.075	2.855	3.600
b	−1.365	−0.925	−0.535	−0.05	0.485	1.33	2.275	3.60	4.95

$$K_1$$

0	0.83	0.91	1.00	1.08	1.14	1.08	1.00	0.91	0.83
$b/4$	0.665	0.735	0.85	0.96	1.08	1.17	1.16	1.13	1.09
$b/2$	0.525	0.605	0.71	0.85	1.00	1.16	1.325	1.375	1.415
$3b/4$	0.425	0.505	0.605	0.735	0.91	1.13	1.375	1.615	1.815
b	0.355	0.425	0.525	0.665	0.83	1.09	1.415	1.815	2.24

$$\theta = 0.550$$

$$K_0$$

0	0.42	0.74	1.02	1.27	1.40	1.27	1.02	0.74	0.42
$b/4$	−0.10	0.25	0.63	0.98	1.27	1.45	1.46	1.35	1.26
$b/2$	−0.53	−0.18	0.21	0.63	1.02	1.43	1.84	2.07	2.25
$3b/4$	−0.89	−0.50	−0.18	0.25	0.74	1.35	2.07	2.87	3.70
b	−1.30	−0.89	−0.53	−0.10	0.42	1.26	2.25	3.70	5.10

$$K_1$$

Ref. pt / Load at	$-b$	$-3b/4$	$-b/2$	$-b/4$	0	$b/4$	$b/2$	$3b/4$	b
0	0.81	0.90	1.00	1.09	1.15	1.09	1.00	0.90	0.81
$b/4$	0.65	0.71	0.84	0.96	1.09	1.18	1.17	1.14	1.09
$b/2$	0.50	0.58	0.69	0.84	1.00	1.17	1.35	1.40	1.44
$3b/4$	0.40	0.48	0.58	0.71	0.90	1.14	1.40	1.65	1.87
b	0.33	0.40	0.50	0.65	0.81	1.09	1.44	1.87	2.33

$$\theta = 0.575$$

$$K_0$$

Ref. pt / Load at	$-b$	$-3b/4$	$-b/2$	$-b/4$	0	$b/4$	$b/2$	$3b/4$	b
0	0.35	0.70	1.02	1.33	1.46	1.33	1.02	0.70	0.35
$b/4$	−0.13	0.22	0.62	1.00	1.33	1.50	1.48	1.34	1.10
$b/2$	−0.53	−0.18	0.21	0.62	1.02	1.48	1.86	2.08	2.22
$3b/4$	−0.84	−0.49	−0.18	0.22	0.70	1.34	2.08	2.90	3.80
b	−1.16	−0.84	0.53	0.13	0.35	1.10	2.22	3.80	5.30

$$K_1$$

Ref. pt / Load at	$-b$	$-3b/4$	$-b/2$	$-b/4$	0	$b/4$	$b/2$	$3b/4$	b
0	0.80	0.89	1.00	1.11	1.17	1.11	1.00	0.89	0.80
$b/4$	0.60	0.70	0.81	0.95	1.11	1.21	1.20	1.14	1.08
$b/2$	0.47	0.55	0.66	0.81	1.00	1.20	1.38	1.44	1.45
$3b/4$	0.37	0.45	0.55	0.70	0.89	1.14	1.44	1.72	1.92
b	0.30	0.37	0.47	0.60	0.80	1.08	1.45	1.92	2.42

$$\theta = 0.600$$

$$K_0$$

Ref. pt / Load at	$-b$	$-3b/4$	$-b/2$	$-b/4$	0	$b/4$	$b/2$	$3b/4$	b
0	0.31	0.66	1.02	1.35	1.50	1.35	1.02	0.66	0.31
$b/4$	−0.17	0.21	0.62	1.02	1.35	1.53	1.47	1.31	1.03
$b/2$	−0.52	−0.18	0.20	0.62	1.02	1.47	1.87	2.06	2.19
$3b/4$	−0.80	−0.47	−0.18	0.21	0.66	1.31	2.06	2.92	3.08
b	−1.05	−0.80	−0.52	−0.20	0.31	1.10	2.19	3.08	5.45

$$K_1$$

Ref. pt / Load at	$-b$	$-3b/4$	$-b/2$	$-b/4$	0	$b/4$	$b/2$	$3b/4$	b
0	0.80	0.89	1.00	1.12	1.19	1.12	1.00	0.89	0.80
$b/4$	0.58	0.67	0.80	0.95	1.12	1.23	1.20	1.15	1.08
$b/2$	0.44	0.52	0.63	0.80	1.00	1.20	1.40	1.45	1.46
$3b/4$	0.34	0.41	0.52	0.67	0.89	1.15	1.45	1.75	1.96
b	0.28	0.34	0.44	0.58	0.80	1.08	1.46	1.96	2.50

$$\theta = 0.625$$

$$K_0$$

Ref. pt / Load at	$-b$	$-3b/4$	$-b/2$	$-b/4$	0	$b/4$	$b/2$	$3b/4$	b
0	0.23	0.635	1.023	1.39	1.54	1.39	1.023	0.63	0.23
$b/4$	−0.22	0.18	0.615	1.03	1.39	1.57	1.49	1.29	0.965
$b/2$	−0.52	−0.18	0.02	0.615	1.023	1.49	1.89	2.06	2.16
$3b/4$	−0.755	−0.455	−0.18	0.18	0.635	1.29	2.06	2.935	3.045
b	−0.925	−0.755	−0.52	−0.235	0.23	1.01	2.16	3.045	5.25

$$K_1$$

Ref. pt / Load at	$-b$	$-3b/4$	$-b/2$	$-b/4$	0	$b/4$	$b/2$	$3b/4$	b
0	0.775	0.87	0.99	1.13	1.21	1.13	0.99	0.87	0.775
$b/4$	0.515	0.655	0.785	0.95	1.13	1.25	1.22	1.15	1.07
$b/2$	0.42	0.495	0.615	0.785	0.99	1.22	1.425	1.48	1.48
$3b/4$	0.32	0.385	0.495	0.655	0.87	1.15	1.48	1.845	2.01
b	0.26	0.32	0.42	0.555	0.775	1.07	1.48	2.01	2.775
$\theta = 0.650$									
K_0									
0	0.15	0.61	1.025	1.42	1.58	1.42	1.025	0.61	0.15
$b/4$	−0.26	0.15	0.61	1.04	1.42	1.60	1.51	1.26	0.90
$b/2$	−0.52	−0.18	0.20	0.61	1.025	1.51	1.91	2.06	2.13
$3b/4$	−0.71	−0.44	−0.18	−0.15	0.61	1.26	2.06	2.95	3.01
b	−0.80	−0.71	−0.52	−0.27	0.15	0.92	2.13	3.01	5.07
K_1									
0	0.75	0.85	0.98	1.14	1.23	1.14	0.98	0.85	0.75
$b/4$	0.55	0.64	0.77	0.95	1.14	1.27	1.24	1.15	1.06
$b/2$	0.40	0.47	0.60	0.77	0.98	1.24	1.45	1.50	1.50
$3b/4$	0.30	0.36	0.47	0.64	0.85	1.15	1.50	1.84	2.06
b	0.24	0.30	0.40	0.53	0.75	1.06	1.50	2.06	2.65
$\theta = 0.675$									
K_0									
0	0.055	0.57	1.027	1.47	1.63	1.465	1.027	0.57	0.055
$b/4$	−0.315	0.13	0.605	1.05	1.47	1.65	1.53	1.235	0.785
$b/2$	−0.51	−0.185	0.19	0.605	1.027	1.53	1.935	2.055	2.08
$3b/4$	−0.64	−0.42	−0.185	0.13	0.57	1.235	2.055	2.975	3.15
b	−0.64	−0.64	−0.51	−0.32	0.055	0.825	2.08	3.51	5.5
K_1									
0	0.73	0.842	0.98	1.155	1.255	1.155	0.98	0.842	0.73
$b/4$	0.525	0.615	0.755	0.945	1.155	1.30	1.25	1.15	1.05
$b/2$	0.365	0.45	0.575	0.755	0.98	1.25	1.48	1.525	1.51
$3b/4$	0.27	0.37	0.45	0.615	0.842	1.15	1.525	1.385	2.11
b	0.20	0.27	0.365	0.51	0.73	1.05	1.51	2.11	2.75
$\theta = 0.700$									
K_0									
0	−0.04	0.53	1.03	1.52	1.68	1.51	1.03	0.53	−0.04
$b/4$	−0.37	0.11	0.00	1.06	1.51	1.70	1.55	1.21	0.67
$b/2$	−0.50	−0.19	0.18	0.60	1.03	1.55	1.96	2.05	2.03
$3b/4$	−0.57	−0.40	−0.19	0.11	0.53	1.21	2.05	3.00	4.01
b	−0.48	−0.57	−0.50	−0.37	−0.04	0.73	2.03	4.01	6.03
K_1									

Ref. pt Load at	$-b$	$-3b/4$	$-b/2$	$-b/4$	0	$b/4$	$b/2$	$3b/4$	b
0	0.71	0.835	0.98	1.17	1.28	1.17	0.98	0.835	0.71
$b/4$	0.50	0.59	0.74	0.94	1.17	1.33	1.27	1.15	1.04
$b/2$	0.33	0.43	0.55	0.74	0.98	1.27	1.51	1.55	1.52
$3b/4$	0.24	0.32	0.43	0.59	0.835	1.15	1.55	1.93	2.16
b	0.18	0.24	0.33	0.49	0.71	1.04	1.52	2.16	2.85

$$\theta = 0.725$$

$$K_0$$

0	−0.125	0.495	1.025	1.55	1.725	1.55	1.025	0.495	−0.125
$b/4$	−0.40	0.08	0.585	1.07	1.55	1.735	1.57	1.18	0.585
$b/2$	−0.495	−0.185	0.175	0.585	1.025	1.570	1.98	2.04	1.99
$3b/4$	−0.505	−0.375	−0.185	0.08	0.495	1.18	2.04	3.025	3.60
b	−0.39	−0.505	−0.495	−0.40	0.125	0.645	1.99	3.60	6.365

$$K_1$$

0	0.685	0.817	0.98	1.185	1.305	1.185	0.98	0.817	0.685
$b/4$	0.475	0.57	0.73	0.94	1.185	1.35	1.285	1.15	1.03
$b/2$	0.315	0.41	0.53	0.73	0.98	1.285	1.54	1.575	1.535
$3b/4$	0.225	0.30	0.41	0.57	0.817	1.15	1.575	1.92	2.205
b	0.165	0.225	0.315	0.47	0.685	1.03	1.535	2.205	2.925

$$\theta = 0.750$$

$$K_0$$

0	−0.21	0.46	1.02	1.58	1.77	1.58	1.02	0.46	−0.21
$b/4$	−0.43	0.05	0.57	1.08	1.58	1.77	1.59	1.15	0.50
$b/2$	−0.49	−0.18	0.17	0.57	1.02	1.59	2.00	2.04	1.95
$3b/4$	−0.44	−0.35	−0.18	0.05	0.46	1.15	2.04	3.05	3.20
b	−0.30	−0.44	−0.49	−0.43	−0.21	0.56	1.95	3.20	6.7

$$K_1$$

0	0.66	0.80	0.98	1.20	1.33	1.20	0.98	0.80	0.66
$b/4$	0.45	0.55	0.72	0.94	1.20	1.37	1.30	1.15	1.02
$b/2$	0.30	0.39	0.51	0.72	0.98	1.30	1.57	1.60	1.55
$3b/4$	0.21	0.28	0.39	0.55	0.50	1.15	1.60	2.01	2.25
b	0.15	0.21	0.30	0.45	0.66	1.02	1.55	2.25	3.00

$$\theta = 0.775$$

$$K_0$$

0	−0.28	0.425	1.02	1.62	1.825	1.62	1.02	0.425	−0.28
$b/4$	−0.46	0.035	0.56	1.09	1.62	1.825	1.607	1.125	0.415
$b/2$	−0.485	−0.18	0.16	0.56	1.02	1.607	2.03	2.03	1.885
$3b/4$	−0.39	−0.325	−0.18	0.35	0.425	1.125	2.03	3.075	3.61
b	−0.23	−0.39	−0.485	−0.455	−0.28	0.475	1.885	3.61	6.86

$$K_1$$

Ref. pt / Load at	$-b$	$-3b/4$	$-b/2$	$-b/4$	0	$b/4$	$b/2$	$3b/4$	b
0	0.645	0.79	0.98	1.21	1.355	1.21	0.98	0.79	0.645
$b/4$	0.425	0.53	0.70	0.935	1.21	1.40	1.32	1.145	1.01
$b/2$	0.275	0.365	0.49	0.70	0.98	1.32	1.60	1.62	1.55
$3b/4$	0.185	0.255	0.365	0.53	0.79	1.145	1.62	2.055	2.29
b	0.135	0.185	0.275	0.425	0.645	1.00	0.155	2.29	3.10

$$\theta = 0.800$$

K_0	$-b$	$-3b/4$	$-b/2$	$-b/4$	0	$b/4$	$b/2$	$3b/4$	b
0	−0.35	0.39	1.025	1.66	1.88	1.66	1.025	0.39	−0.35
$b/4$	−0.49	0.02	0.55	1.10	1.66	1.88	1.64	1.10	0.33
$b/2$	−0.48	−0.18	0.15	0.55	1.025	1.64	2.06	2.03	1.82
$3b/4$	−0.34	−0.30	−0.18	0.02	0.39	1.10	2.03	3.10	4.02
b	−0.16	−0.34	−0.48	−0.48	−0.35	0.39	1.82	4.02	7.02

K_1	$-b$	$-3b/4$	$-b/2$	$-b/4$	0	$b/4$	$b/2$	$3b/4$	b
0	0.63	0.78	0.98	1.22	1.38	1.22	0.98	0.78	0.63
$b/4$	0.40	0.51	0.68	0.93	1.22	1.43	1.34	1.14	1.00
$b/2$	0.25	0.34	0.47	0.68	0.98	1.34	1.63	1.64	1.55
$3b/4$	0.16	0.23	0.34	0.51	0.78	1.14	1.64	2.10	2.33
b	0.12	0.16	0.25	0.40	0.63	0.98	1.55	2.33	3.20

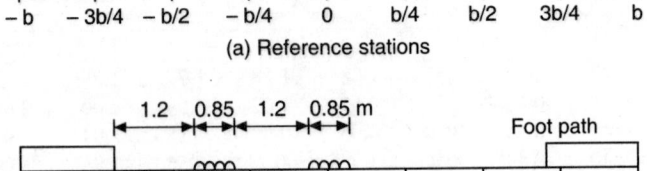

(a) Reference stations

(b) Position of Class AA Loads for Maximum Distribution

Fig. 8.3 Reference Station and Position of Class AA Loads for Maximum Distribution.

$$M_y = (Wb/a) \left[\mu_\theta \sin(\pi u/2a) - \mu_{3\theta} \sin(3\mu u/2a) + \mu_{5\theta} \sin(5\pi u/2a) + ... \right] \qquad 4(a)$$

If there is a uniformly distributed load 'p' acting over a distance '2c' then,

$$M_y = (4pb/\pi) \left[\mu_\theta \sin(\pi c/2a) + (1/3)\,\mu_{3\theta} \sin(3\pi c/2a) \right.$$
$$\left. + (1/5)\,\mu_{5\theta} \sin(5\pi c/2a) + ... \right] \qquad 4(b)$$

Where μ_θ, $\mu_{3\theta}$, $\mu_{5\theta}$, are the distribution coefficients corresponding to the flexural parameters θ, 3θ, and 5θ respectively. Coefficient 'μ' is analogous to the distribution coefficient 'K' for longitudinal moments, 'μ_0' represents the distribution coefficient for $\alpha = 0$ and μ_1 for $\alpha = 1.0$. The value of m corresponding to any other intermediate value of a can be evaluated using the interpolation relationship.

$$\mu_\alpha = \mu_0 + \left(\mu_1 - \mu_0\right)\sqrt{\alpha}$$

The coefficients μ_0 and μ_1 are determined for values of θ, 3θ, and 5θ, from the charts shown in Fig. 8.4 and 8.5 for the reference station 0, where the maximum transverse moment will occur for position of loads shown in Fig. 8.6. Graphs of these functions are plotted and values of 'μ' for actual load positions are determined. Then M_{y0} and M_{y1} are calculated for μ_0 and μ_1 respectively using the equations 4(a) or 4(b). The transverse moment M_y at the centre of the bridge is given by,

$$M_y = M_{y0} + (M_{y1} - M_{y0})\sqrt{\alpha} \tag{5}$$

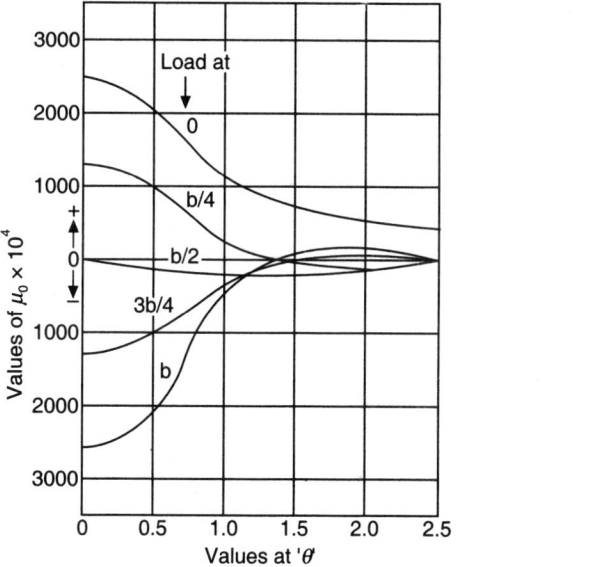

Fig. 8.4 Transverse Moment Coefficient μ_0 at Reference Station 0 for Various Load Eccentricities.

It has been found from several computations and from tests by Best, Rowe and Gifford, that transverse mild steel reinforcement of 1030 mm²/metre placed near the top of bottom flange or about 150 mm to 180 mm from the soffit of the sections is adequate for reinforced concrete cross girders. Additional mild steel reinforcement of 240 mm²/m should be provided in the transverse direction at the top of *in-situ* concrete slab with a cover of about 20 mm to cater for any small transverse hogging moments and transverse shrinkage stresses. The provision of these reinforcements will obviate the rigorous computation of transverse moments.

8.6 HENDRY-JAEGAR METHOD

In the analysis proposed by Hendry and Jaegar, the cross beams can be replaced by a uniform continuous transverse medium of equivalent stiffness. According to this method, the load distribution in an interconnected bridge deck system depends upon three dimensionless parameters given by,

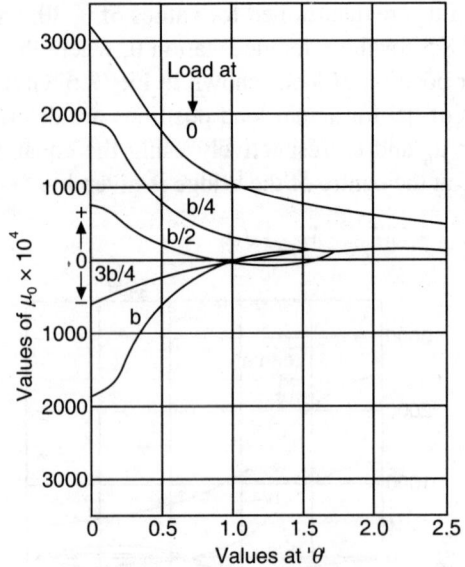

Fig. 8.5 Transverse Moment Coefficient μ_1 at Reference Station 0 for Various Load Eccentricities.

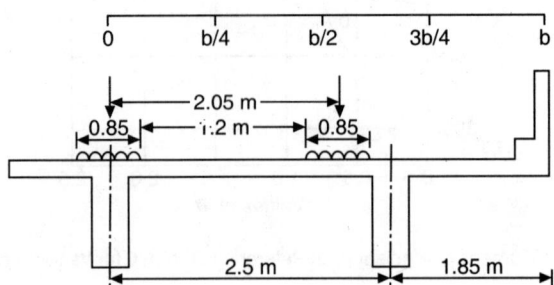

Fig. 8.6 Position of I.R.C. Class AA Loads for Maximum Transverse Moment in Cross Girders.

$$A = (12/\pi^4) (L/h)^3 (n.EI_r/EI) \tag{6}$$
$$F = (\pi^2/2n) (h/L) (GJ/EI_r) \tag{7}$$
$$C = (EI_1/EI_2) \tag{8}$$

Where
L = The span of the bridge deck

h = Spacing of longitudinal girders

n = Number of cross beams

EI = Flexural rigidity of one longitudinal girder

GJ = Torsional rigidity of one longitudinal girder

EI_1EI_2 = Flexural rigidities of the outer and inner longitudinal girders, where these are different

EI_r = Flexural rigidity of one cross beam

For slab bridges without cross beams, nEI_r in equations 6 and 7 is to be replaced by LEI_p, which represents the total flexural rigidity of the slab deck. In general the flexural rigidities of the outer and inner longitudinal girders of a Tee beam deck are nearly equal in magnitude.

The first parameter 'A' represents a function of the ratio of span to the spacing of longitudinal girders and the ratio of transverse to longitudinal flexural rigidity. The second parameter 'F' is a measure of the ratio of torsional to flexural rigidity of longitudinal and crosse girders respectively. The parameter is difficult to be evaluated due to the uncertainties in computations of torsional rigidity values for practical girders sections. In the case of T-beam bridges having three or four longitudinals with a number of cross beams, it is permissible to employ the distribution coefficients for $F = \infty$. The torsional rigidity of the transverse system is neglected in the analysis.

Hendry and Jaegar have presented graphs giving the values of the distribution coefficients (m) for different number of longitudinal girders (two to six) and for the two extreme value of $F = 0$, $F = \infty$. Coefficients for intermediate values of F may be obtained by interpolation from the equation,

$$m_F = m_0 + (m_\infty - m_0)\sqrt{F\sqrt{A}/(3 + F\sqrt{A})} \qquad (9)$$

Where m_F = required distribution coefficient

               m_0 = coefficient for $F = 0$

               m_∞ = coefficient for $F = \infty$

Typical graphs for distribution coefficients for a three girder system for $F = 0$ and $F = \infty$ are given in Figs. 8.7, and 8.8.

8.7   DESIGN EXAMPLE

Design a R.C.C. Tee beam girder bridge to suit the following data:

1.   Data

Clear width of carriage way = 7.5 m
Span (centre to centre of bearings) = 16 m
Kerbs on either side = 600 by 300 mm
Live load: IRC Class AA tracked vehicle
Thickness of the wearing coat = 80 mm
Materials: M-25 Grade concrete and Fe-415 HYSD reinforcements

Using the Courbon's method, compute the design moments and shear forces and design the deck slab, main girders and cross girders conforming to the specifications of IRC: 6-2014 and IRC: 112-2011. Sketch the details of reinforcements in deck slab, main and cross girders.

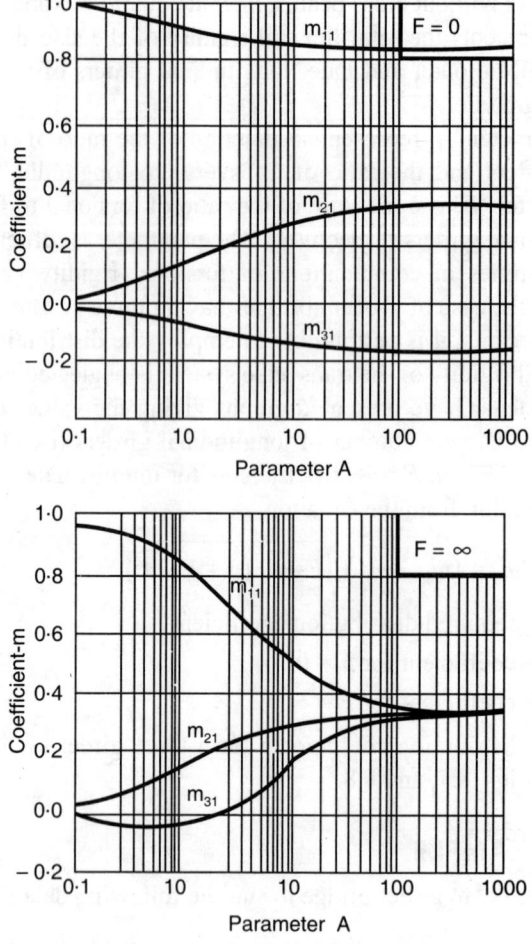

Fig. 8.7 Distribution Coefficients for Three Girder Bridge with Load on Girder No. 1.

2.   Characteristic Strength of Materials

f_{ck} = 25 N/mm² E_c = 30 GPa α_e = $[E_s/E_c]$ = [200/30] = 6.66

f_{yk} = 415 N/mm² E_s = 200 GPa

3.   Cross-section of Deck

Three main girders are provided at 2.5 m centres,

Thickness of deck slab = 200 mm

Wearing coat = 80 mm

Width of main girders = 300 mm

Kerbs 600 mm wide by 300 deep, are provided. Cross girders are provided at every 4 m intervals.

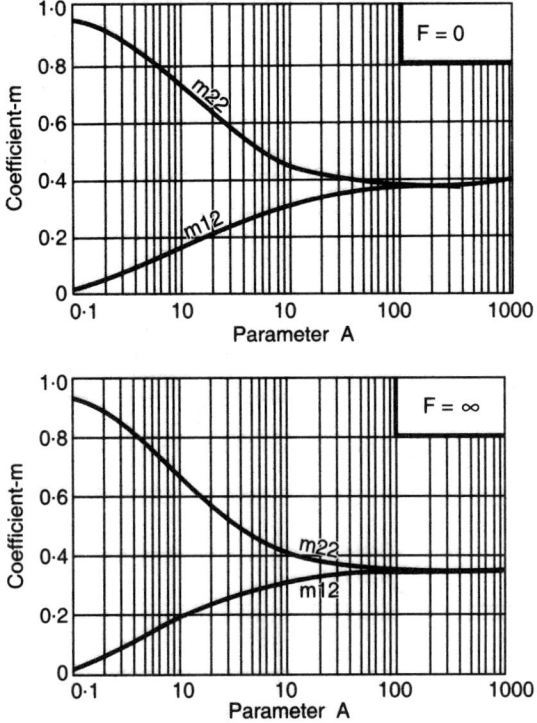

Fig. 8.8 Distribution Coefficient for Three Girder Bridge with Load on Girder No. 2.

Breadth of cross girder = 300 mm

Depth of main girder = 160 cm at the rate of 10 cm per metre of span

The depth of cross girder is taken as equal to the depth of main girder to simplify the computations. The cross-section of the deck and the plan showing the spacing of cross girders are shown in Fig. 8.9

4. Design of Interior Slab Panels

(a) Bending Moments

Dead weight of slab = $(1 \times 1 \times 0.2 \times 24) = 4.80$ kN/m^2

Dead weight of W.C. = $(0.08 \times 22) = 1.76$

Total dead load = 6.56 kN/m^2

Live load is class AA tracked vehicle. One wheel is placed at the centre of panel as shown in Fig. 8.10.

$$u = (0.85 + 2 \times 0.08) = 1.01 \text{ m}$$
$$v = (3.60 + 2 \times 0.08) = 3.76 \text{ m}$$
$$(u/B) = 1.01/2.5 = 0.404$$

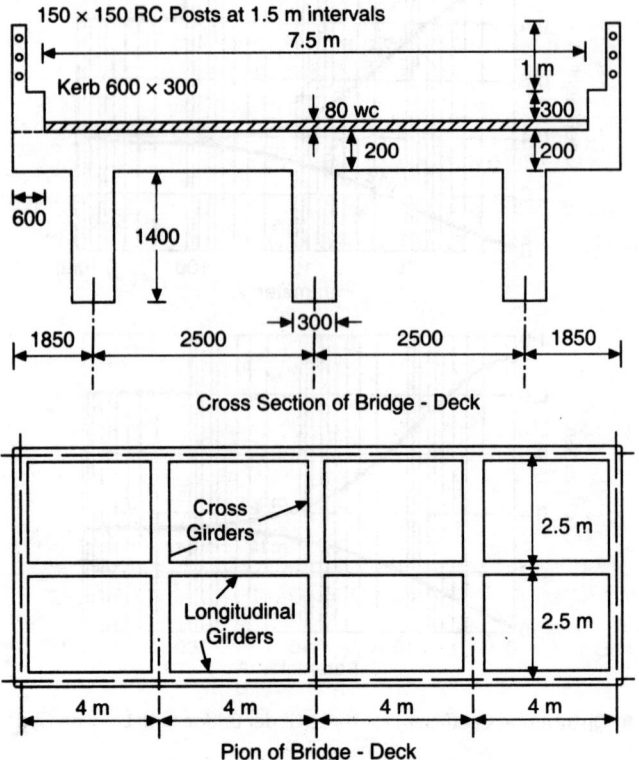

Cross Section of Bridge - Deck

Pion of Bridge - Deck

Fig. 8.9 Tee Beam and Slab Bridge Deck.

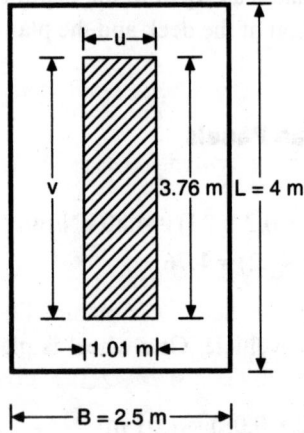

Fig. 8.10 Position of Wheel Load for Maximum Bending Moment.

(v/L) 3.76/4.0 = 0.94

$$K = (B/L) = (2.58/4.0) = 0.625$$

Referring to Pigeaud's curves (refer Fig. 3.4)

$$m_1 = 0.085 \text{ and } m_2 = 0.024$$

$$M_B = W(m_1 + 0.15\ m_2)$$

$$= 350(0.085 + 0.15 \times 0.024)$$

$$= 31.01 \text{ kN·m}$$

As the slab is continuous design B.M. = 0.8 M_B

Design B.M. including impact and continuity factor given by

M_B (short span) = (1.25 × 0.8 × 31.01) = 31.01 kN·m

Similarly, M_L = 350 (0.024 + 0.15 × 0.085)

$$= 12.845 \text{ kN·m}$$

M (long span) = (1.25 × 0.8 × 12.845) = 12.845 kN·m

(b)   Shear Forces

Dispersion in the direction of span = 0.85 + 2 (0.08 + 0.2) = 1.41 m

For maximum shear, load is kept such that the whole despersion is in span. The load is kept at 1.41/2 = 0.705 m from the edge of beam as shown in Fig. 8.11.

Effective width of slab = $K x (1 - x/L) + b_w$

Breadth of cross girder = 30 cm

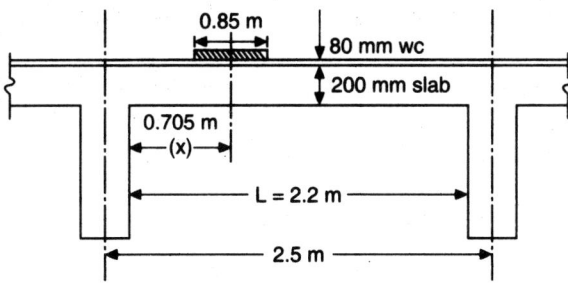

Fig. 8.11 Position of Wheel Load for Maximum Shear.

Clear length of panel = L = 3.7 m

$$(B/L) = (3.7/2.2) = 1.68$$

From Table 3.1, 'K' for a continuous slab is obtained as K = 2.52. Effective width of a slab

$$= [2.52 \times 0.705 (1 - 0.705/2.2) + 3.6 + (2 \times 0.08)] = 5 \text{ m}$$

Load per metre width = (350/5) = 70 kN

Shear force = 70 (2.2 – 0.705)/2.2 = 47.60 kN

Shear force with impact = (1.25 × 47.60) = 59.50 kN

(c)   Dead Load Bending Moments and Shear Forces

Dead load = 6.56 kN/m^2

Total load on panel = (4 × 2.5 × 6.56) = 65.6 kN

$(u/B) = 1$ and $(v/L) = 1$ as panel is loaded with uniformly distributed load

$$K = (B/L) = (2.5/4) = 0.625 \text{ and } (1/K) = 1.6$$

From Pigeaud's curve (Refer Fig. 3.9)

$$m_1 = 0.049 \text{ and } m_2 = 0.015$$
$$M_B = 65.6 (0.049 + 0.15 \times 0.015) = 3.36 \text{ kN·m}$$

taking continuity into effect

$$m_b = (0.8 \times 3.36) = 2.688 \text{ kN·m}$$
$$M_L = 65.6 (0.015 + 0.15 \times 0.049) = 1.468 \text{ kN·m}$$

taking continuity into effect

$$M_L = (0.8 \times 1.468) = 1.174 \text{ kNm}$$

Dead load shear force = (6.5 × 2.2)/2 = 7.216 kN

(d)   Design Ultimate Moments and Shear Forces

Service Dead and live load moments and shear forces for short span are computed as,

For short span: $M_{BD} = 2.688$ kN·m and $M_{BL} = 31.01$ kN·m

For Long span: $M_{LD} = 1.174$ kN·m and $M_{LL} = 12.845$ kN·m

For short span: $V_{BD} = 7.216$ kN and $V_{BL} = 59.5$ kN

Design load Ultimate moments and shear forces for short and long spans are computed as

Short span moment = $M_{BU} = [1.35\ M_{BD} + 1.5\ M_{BL}] = [(1.35 \times 2.688) + (1.5 \times 31.01)] = 50.13$ kN·m

Long span moment = $M_{LU} = [1.35\ M_{LD} + 1.5\ M_{LL}] = [(1.35 \times 1.174) + (1.5 \times 12.85)] = 20.85$ kN·m

Short span shear = $V_{BU} = [1.35\ V_{BD} + 1.5\ V_{BL}] = [(1.35 \times 7.216) + (1.5 \times 59.5)]$
$= 99.0$ kN

(e)   Design of Section

Using M-25 grade concrete and Fe-415 HYSD bars

Limiting moment of resistance for singly reinforced sections is expressed as (Refer Table 4.3)

$$M_{u,\lim} = 0.138 f_{ck}\ b\ d^2$$

$$\therefore \qquad d = \sqrt{\frac{M_u}{0.138 f_{ck} b}} = \sqrt{\frac{50.13 \times 10^6}{0.138 \times 25 \times 1000}} = 120.5 \text{ mm}$$

Using 40 mm clear cover and 16 mm diameter bars, effective cover is taken as 48 mm

Hence effective depth = $d = (200 - 48) = 152$ mm

Compute the parameter ratio $(M_u/b.d^2) = [(50.13 \times 10^6)/(1000 \times 152^2)] = 2.2$ N/mm^2
Refer SP: 16 Design Table 3 and read out the reinforcement percentage corresponding to

$$f_{ck} = 25 \text{ N/mm}^2 \text{ and } f_y = 415 \text{ N/mm}^2 \text{ as}$$
$$p_t = (100 \, A_{st}/b.d) = 0.689$$
$$\therefore \quad A_{st} = [(0.689 \times 1000 \times 152)/100] = 1047 \text{ mm}^2$$

Provide 16 mm diameter bars at a spacing of 150 mm (A_{st} provided = 1340 mm^2)

For long span, the moment being comparatively small, provide 10 mm diameter bars at a spacing of 150 mm.

(f)   Check for Ultimate Shear Strength

$$V_{Rd.c} = [0.12K \, (80 \, \rho_1 f_{ck})^{0.33}]b_w.d$$

where

$$K = 1 + \sqrt{\frac{200}{d}} \le 2.00 = \left[1 + \sqrt{\frac{200}{150}}\right] = 2.15 \, (\text{taken as } 2.0)$$

and

$$\rho_1 = \left(\frac{A_{sl}}{b_w.d}\right) \le 0.02$$

$$= \left(\frac{1340}{1000 \times 152}\right) = 0.009$$

$$V_{Rd.c} = [0.12 \times 2.00 \, (80 \times 0.0088 \times 25)^{0.33}](1000 \times 152) \text{ N}$$
$$= (98.5 \times 1000) \text{ N}$$
$$= 98.5 \text{ kN} \cong 99.0 \text{ kN (Hence safe)}$$

(g)   Check for Serviceability Limit State of Cracking

According to IRC: 112-2011, cracking can be controlled without direct computations of the width of cracks by limiting the size of reinforcing bars, stress in steel and the permissible crack width as listed in Tables 4.6 and 4.7. This empirical method will be followed in this example to check for the serviceability limit state of cracking in the deck slab.

Depth of neutral axis (x) is determined using the relation,

$(0.5 \times 1000 \times x^2) = [6.66 \times 1340 \times (152-x)]$

Solving, $x = 44$ mm and service load moment = $M_w = 31.01$ kN·m
Stress in steel under working loads is computed as

$$\sigma_s = \left[\frac{M_w}{\left(d - \dfrac{x}{3}\right)A_{st}^*}\right] = \left[\frac{31.01 \times 10^6}{\left(152 - \dfrac{44}{3}\right)1340}\right] = 168 \text{ N/mm}^2$$

The deck slab is reinforced using 16 mm diameter bars at 150 mm centers.

Referring to Table 4.7, for a stress of 168 N/mm², maximum bar diameter permissible to Limit the crack width to 0.3 is interpolated as 28 mm as against 16 mm used in the slab.

From Table 4.8, maximum spacing of bars to limit the crack width to 0.3 mm should not exceed 290 mm as against 150 mm used in the slab.

Hence the slab is safe regarding the serviceability limit stat of cracking according to the specifications of IRC: 112-2011.

5. Design of Longitudinal Girders

(a) Reaction Factors

Using Courbon's theory, the I.R.C. class AA loads are arranged for maximum eccentricity as shown in Fig. 8.12.

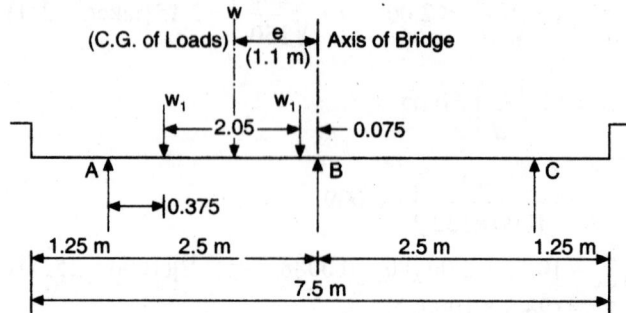

Fig. 8.12 Transverse Position of I.R.C. Class AA Tracked Vehicle.

Reaction factor for outer girder is given by

$$R_A = (2W_1/3)\,[1 + (3I \times 2.5 \times 1.1)/(2I \times 2.5^2)] = 1.107\ W_1$$

Reaction factor for inner girder is

$$R_B = (2W_1/3)(1 + 0) = (2W_1/3)$$

If W = axle load = 700 kN

$W_1 = 0.5\ W$

$R_A = (1.107 \times 0.5\ W) = 0.5536\ W$

$R_B = (0.667 \times 0.5\ W) = 0.3333\ W$

(b) Dead Load from Slab for Girder

Dead load of deck slab is calculated with reference to Fig. 8.13.

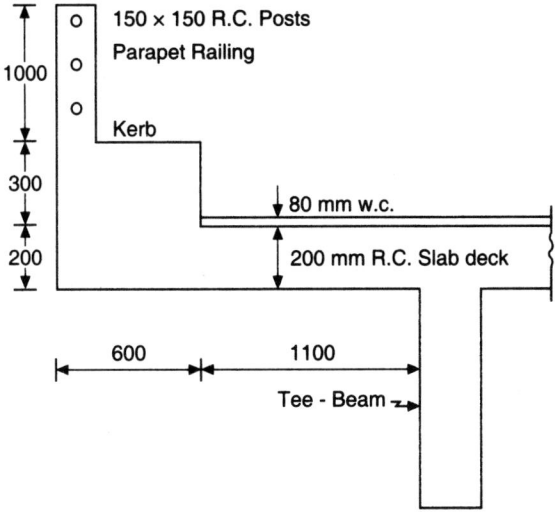

Fig. 8.13 Details of Deck Slab Kerb and Parapet.

Weight of

1. Parapet railing ....0.700 kN·m
2. Wearing coat = (0.08 × 1.1 × 22)....................1.936 kN·m
3. Deck slab = (0.2 × 1.10 × 24).....................5.280 kN·m
4. Kerb = (0.5 × 0.6 × 1 × 24) ....7.200 kN·m

 15.116 kN·m

Total deal load of deck = (2 × 15.116) + (6.56 × 5.3) = 65 kN·m
It is assumed that the dead load is shared equally by all girders
∴ Dead load/girder = (65/3) = 21.66 kN·m

(c)   Live Load Bending Moments in Girder

Span of girder = 16 m

Impact factor (For class — AA Loads) = 10%

The live load is placed centrally on the span as shown in Fig. 8.14

Bending moment = 0.5 (4 + 3.1) 700 = 2485 kN·m

Bending moment including impact and reaction factor for outer girder is = (2485 × 1.1 × 0.5536) = 1513 kN·m. Bending moment including impact and reaction factor for inner girder is (2485 × 1.1 × 0.3333) = 912 kN·m.

(d)   Live Load Shear

For estimating the maximum live load shear in the girders, the I.R.C. class AA loads are placed as shown in the Fig. 8.15

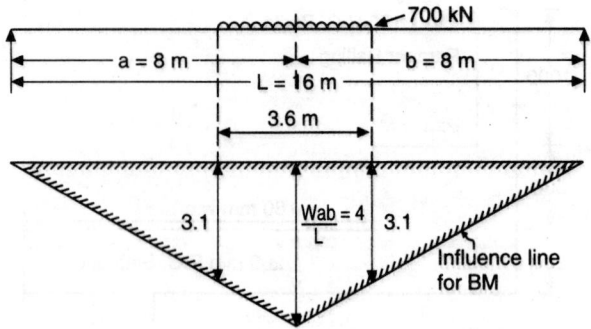

Fig. 8.14 Influence Line for Bending Moment in Girder.

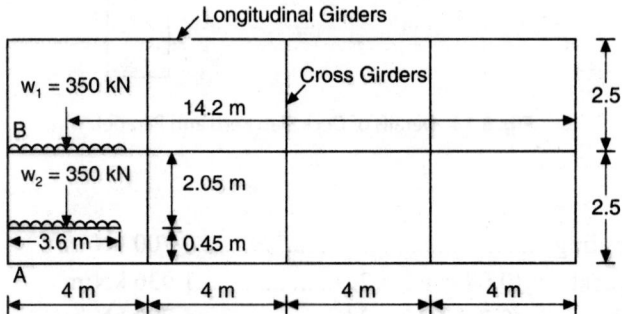

Fig. 8.15 Position of I.R.C. Class AA Loads for Maximum Shear.

Reaction of W_2 on girder B = (350 × 0.45)72.5 = 63 kN

Reaction of W_2 on girder A = (350 × 2.05)/2.5 = 287 kN

Total load on girder B = (350 + 63) = 413 kN

Maximum reaction in girder B = (413 × 14.2)716 = 366 kN

Maximum reaction in girder A = (287 × 14.2)716 = 255 kN

Maximum live load shears with impact factor in

Inner girder = (366 × 1.1) = 402.6 kN

Outer girder = (255 × 1.1) = 280.5 kN

(e)   Dead Load Bending Moments and Shear Forces in Main Girder

The depth of the girder is assumed as 1600 mm (100 mm for every metre of span)

Depth of rib = 1.4 m

Width = 0.3 m

Weight of rib/m = (1 × 0.3 × 1.4 × 24) = 10.08 kN·m

The cross girder is assumed to have the same cross-sectional dimensions of the main girder.

Weight of cross girder = 10.08 kN·m

Reaction on main girder = $(10.08 \times 2.5) = 25.2$ kN

Reaction from deck slab on each girder = 21.66 kN·m

Total dead load/m on girder = $(21.66 + 10.08)$

$$= 31.74 \text{ kN·m}$$

Referring to Fig. 8.16, the maximum bending moments are computed.

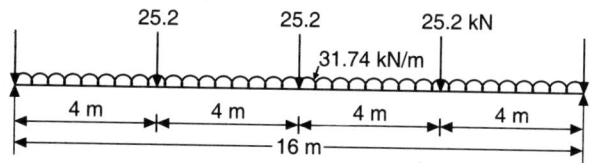

Fig. 8.16 Dead Load on Main Girders.

Maximum bending moment at centre of span is obtained as,

$$M_{\text{max}} = (31.74 \times 16^2)/8 + (25.2 \times 16)/4 + (25.2 \times 16)/4 = 1218 \text{ kN·m}$$

Dead load shear at support

$$(31.74 \times 16)/2 + 25.2 + (25.2/2) = 292 \text{ kN}$$

(f) Design bending Moments and Shear Forces

The design service load and ultimate load bending moments and shear forces are compiled in Table 8.2.

Table 8.2 Design Moments and Shear Forces in Girders

B.M	Service load D.L.B.M	B.M L.L.B.M	Design Ultimate load B.M. Total design B.M = [1.35 DLBM + 1.5 LLBM]	Units
Outer Girder	1218	1513	3914	kN·m
Inner Girder	1218	912	3012	kN·m
S.F	D.L.S.F	L.L.S.F	Total design S.F = [1.35 DLSF + 1.5 LLSF]	
Outer Girder	292	280.1	814.4	kN
Inner Girder	292	402.6	998	kN

(g) Design of Reinforcements for Flexure

Maximum design B.M = M_u = 3914 kN·m

Maximum design S.F = V_u = 998 kN

Using 25 mm diameter bars in four rows with an effective cover of 150 mm on the tension side,

Hence effective depth = $d = (1600 - 150) = 1450$ mm

Using the following parameters of the tee beam:

b_f = 2500 mm, f_{ck} = 25 N/mm², f_y = 415 N/mm², d = 1450 mm, D_f = 200 mm, M_u = 3914 kN·m

If x_u = depth of neutral axis

Assuming the neutral axis to be within the flange depth ($x_u < D_f$), the value of x_u can be determined by using the equation,

$$M_u = 0.36 f_{ck} b_f x_u (d - 0.42 x_u)$$
$$(3914 \times 10^6) = (0.36 \times 25 \times 2500 \times x_u)(1450 - 0.42 x_u)$$

Solving, the value of x_u = 125 mm < 200 mm (hence the neutral axis falls within the flange)

The area of tensile steel (A_{st}) required to resist the moment M_u can be determined by the Equation,

$$M_u = 0.87 f_y A_{st} d \left[1 - \frac{A_{st} f_y}{b \, d f_{ck}} \right]$$

Substituting the numerical values, we have

$$(3914 \times 10^6) = 0.87 \times 415 \times A_{st} \times 1450 \left[1 - \frac{A_{st} \times 415}{2500 \times 1450 \times 25} \right]$$

Solving, the area of steel A_{st} = 7753 mm²

Provide 16 bars of 25 mm diameter providing an area of 7854 mm² at an effective depth of 1450 mm arranged in 4 rows.

Alternatively, the area of steel can be obtained using SP: 16 design Table No. 3 for the parameters, f_{ck} = 25 N/mm² and f_y = 415 N/mm² corresponding to the value of the ratio

$$(M_u/b \, d^2) = [(3914 \times 10^6)/(2500 \times 1450^2)] = 0.74$$

From the design tables, interpolate the value of the percentage reinforcement as

$$p_t = \left(\frac{100 A_{st}}{b.d} \right) = 0.21$$

$$\therefore \quad A_{st} = \left(\frac{0.21 \times 2500 \times 1450}{100} \right) = 7613 \text{ mm}^2$$

(h) Design of Reinforcements for Shear

Maximum design ultimate shear force = V_u = 998 kN

According to IRC: 112-2011, the shear strength of concrete section is computed using the equation 4.7 given by,

$$V_{Rd.c} = [0.12 K (80 \, p_1 f_{ck})^{0.33}] b_w.d$$

where

$$K = 1 + \sqrt{\frac{200}{d}} \leq 2.00 = \left[1 + \sqrt{\frac{200}{1450}} \right] = 1.37 < 2.0$$

and

$$p_1 = \left(\frac{A_{st}}{b_w.d} \right) = \left(\frac{7854}{300 \times 1450} \right) = 0.018 \leq 0.02$$

$$V_{Rd.c} = [0.12 \times 1.37 \, (80 \times 0.018 \times 25)^{0.33}](300 \times 1450) \text{ N}$$
$$= (2788 \times 1000) \text{ N}$$

$$= 278 \text{ kN} < 998 \text{ kN}$$

Hence, shear reinforcements have to be designed to resist the balance shear force.

Balance shear force $= V_{\text{Rd.s}} = (998 - 278) = 720$ kN

Using 10 mm diameter, 4-legged stirrups, the spacing of stirrups is evaluated as

$$S_v = \left(\frac{087 \, f_y \, A_{sv} \, z}{V_{\text{Rd.s}}} \right)$$

where z = lever arm to be taken as 0.9d for RCC sections

$$S_v = \left(\frac{087 \times 415 \times 4 \times 79 \times 0.9 \times 1450}{720 \times 10^3} \right) = 206 \text{ mm}$$

Provide 10 mm diameter 4-legged stirrups at 200 mm centers.

6. Design of Cross Girders

Self-weight of cross girder = 10.08 kN·m

Referring to Fig. 8.17.

Dead load from slab = $(2 \times 1/2 \times 2.5 \times 1.25 \times 6.56)$

$$= 20.5 \text{ kN}$$

Uniformly distributed load = $(20.5/2.5) = 8.2$ kN·m

Total load on cross girder = $(10.08 + 8.2)$

$$= 18.28 \text{ kN·m}$$

Assuming the cross girder to be rigid, reaction on each cross girder

$$= (18.28 \times 5)73 = 30.47 \text{ kN}$$

For maximum bending moment in the cross girder, the loads of I.R.C. class AA should be placed as shown in Fig. 8.18.

Load coming on cross girder = $[350 \, (4 - 0.9)/4] = 271.25$ kN

Assuming the cross girder as rigid, reaction on each longitudinal girder is

$$= [(2 \times 271.25)/3] = 180.83 \text{ kN}$$

Maximum B.M. in cross girder under the load

$$= (1 \, 80.83 \times 1.475) = 266.7 \text{ kN·m}$$

L.L.B.M. including impact = $(1.1 \times 266.7) = 293.37$ kN·m

Dead load B.M. at 1.475 m from support

$$= (30.47 \times 1.475 - 18.28 \times 1.4752/2) = 25.10 \text{ kN·m}$$

Total design B.M. = $(293.37 + 25.10) = 318.47$ kN·m

Live load shear including impact = $(2 \times 271.25/3) \times 1.5 = 198.917$ kN

Dead load shear = 30.47 kN

Total design shear = $(198.917 + 30.47) = 229.39$ kN

Design ultimate load bending moment = $[(1.35 \times 25.10) + (1.5 \times 293.37)] = 474$ kN·m

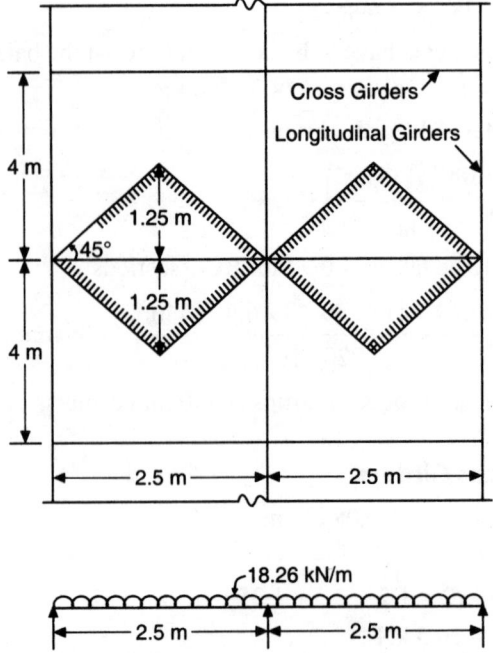

Fig. 8.17 Loads on Cross Girders.

Design ultimate load shear force = $[(1.35 \times 30.47) + (1.5 \times 198.917)] = 337$ kN
The moments and shear forces being comparatively of smaller magnitude, provide 4 bars of 25 mm diameter at top and bottom with 10 mm diameter stirrups at a spacing of 200 mm.

The details of reinforcements in the cross section and longitudinal section of the tee beam bridge deck are shown in Figs. 8.19 and 8.20.

8.8 COMPARATIVE ANALYSIS OF COURBON, GUYON-MASSONET AND HENDRY-JAEGAR METHODS

Analyse the tee beam bridge deck of Example 8.7 using the Guyon-Massonet and Hendry-Jaegar methods and compare the design moments in the exterior girder with those resulting from Courbon's method.

a) Guyon-Massonet Method

1. Cross-sectional Properties of Girders

The cross-section of the deck together with the cross-sections of the main and cross girders are shown in Fig. 8.21.

The cross-sectional properties are as follows:

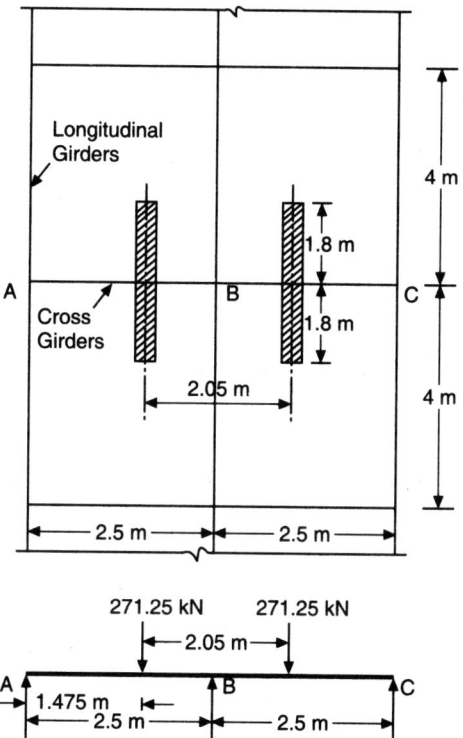

Fig. 8.18 Position of Live Loads for Maximum B.M in Cross Girder.

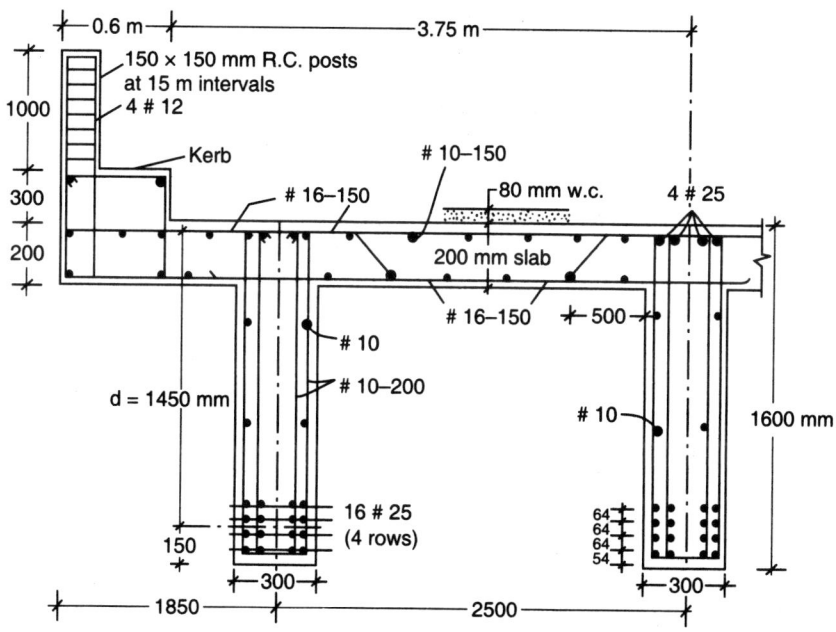

Fig. 8.19 Cross Section of Tee Beam and Slab Deck.

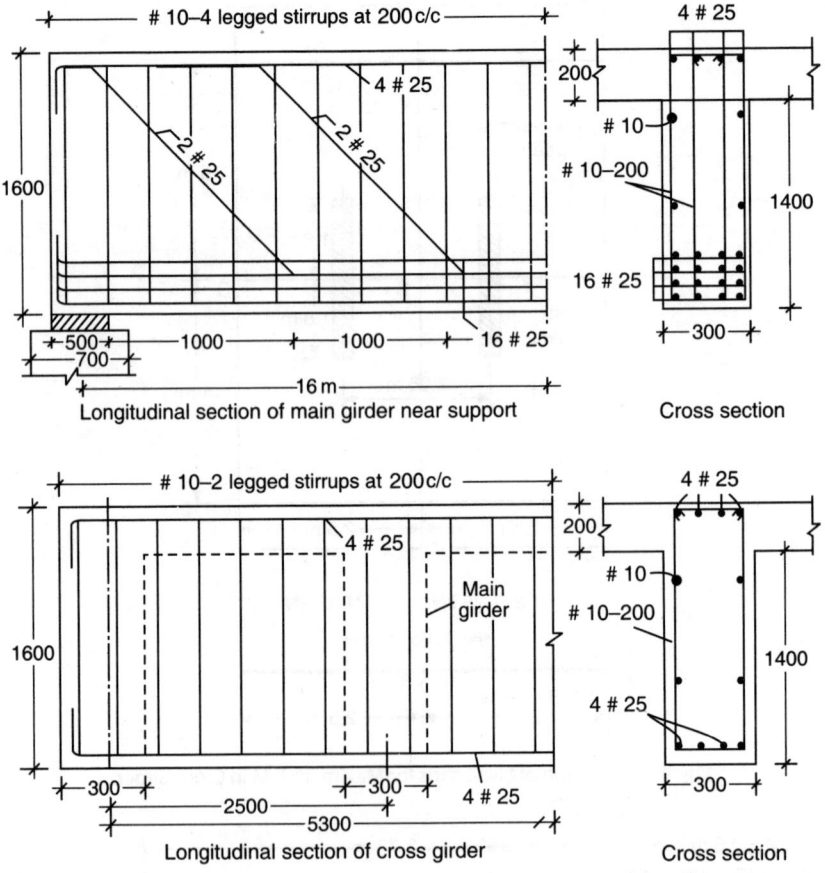

Fig. 8.20 Reinforcement Details in Longitudinal and Cross Girders.

Main Girder

$$I = 21.62 \times 10^{10} \text{ mm}^4$$
$$i = (I/B) = (21.62 \times 10^{10})/2500 = 0.864 \times 10^8 \text{ mm}^4/\text{mm}$$
$$Z_t = (I/Y_t) = (21.62 \times 10^{10})/465 = 4.64 \times 10^8 \text{ mm}^3$$
$$Z_b = (I/Y_b) = (21.62 \times 10^{10})/1135 = 1.90 \times 10^8 \text{ mm}^3$$

Cross Girder

$$J = 24.74 \times 10^{10} \text{ mm}^4$$
$$j = (J/B) = (21.74 \times 10^{10})/4000 = 0.618 \times 10^8 \text{ mm}^4/\text{mm}$$
$$Z_t = (24.74 \times 10^{10})/375 = 6.59 \times 10^8 \text{ mm}^3$$
$$Z_b = (24.74 \times 10^{10})/1225 = 2.01 \times 10^8 \text{ mm}^3$$

Torsional Inertia of Girders

$$I_0 \text{ or } J_0 = R.a^3.b$$

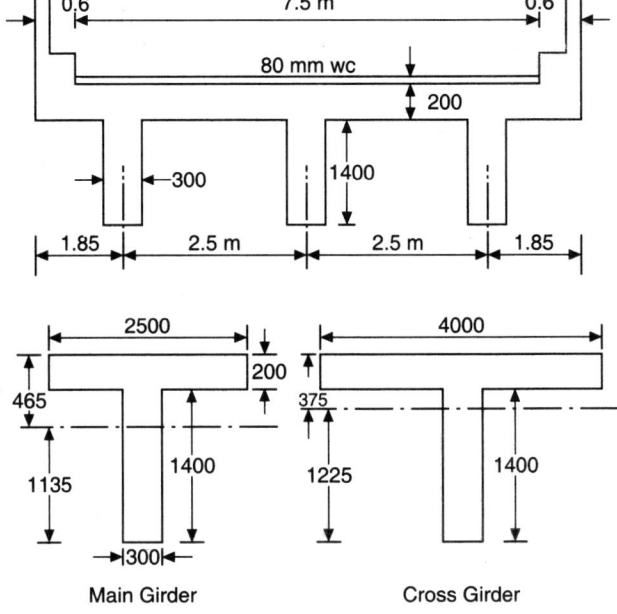

Fig. 8.21 Cross-section of Deck, Main, and Cross Girders.

Where a and b are shorter and longer sides of a rectangle and R is a constant given in Table 8.3.

Table 8.3 Values of Torsion Coefficient R

(b/a)	R	(b/a)	R
1.0	0.141	3.0	0.263
1.2	0.166	4.0	0.281
1.5	0.196	5.0	0.291
2.0	0.229	10.0	0.312
2.25	0.240	Above 10	0.333
2.50	0.249		

Main Girder

$$(b/a) = (2500/200) = 12.5 \quad \therefore R = 0.333$$
$$(b/a) = (1400/300) = 4.66 \quad \therefore R = 0.287$$
$$I_0 = (0.333 \times 200^3 \times 2500) + (0.287 \times 300^3 \times 1400)$$
$$= 1.75 \times 10^{10} \text{ mm}^4$$
$$\therefore \quad i = (I_0/B) = (1.75 \times 10^{10})/2500 = 0.07 \times 10^8 \text{ mm}^4/\text{mm}$$

Cross Girder

$$(b/a) = (4000/200) = 20 \quad \therefore R = 0.333$$
$$(b/a) = (1400/300) = 4.66 \quad \therefore R = 0.287$$

$$J_0 = (0.333 \times 200^3 \times 4000) + (0.287 \times 300^3 \times 1400)$$
$$= 1.15 \times 10^{10} \text{ mm}^4$$
$$j = (J_0/B) = (1.15 \times 10^{10})/4000$$
$$= 0.028 \times 10^8 \text{ mm}^4/\text{mm}$$

2. Distribution Coefficients

$$\theta = (b/2a)\,(i/j)^{0.25}$$

Where $\quad 2b$ = Effective width of bridge = 8.7 m

$\qquad 2a$ = Span of bridge = 16 m

$\qquad \theta = (4.35/1\ 6)[(0.864 \times 10^8)/(0.618 \times 10^8)]^{025} = 0.3$

$\qquad \alpha = (G/2E)(i_0 + j_0)/\sqrt{ij}$

Assuming the value of $G = 0.4\,E$

$$\alpha = [0.2\,(0.07 \times 10^8)$$

$$\alpha = \left[0.2\,(0.07 \times 10^8) + (0.028 \times 10^8)/\sqrt{(0.864 \times 10^8)\,(0.618 \times 10^8)}\right]$$

$$= 0.026$$

$\therefore \qquad \sqrt{\alpha} = 0.161$

The values of K_0 and K_1 for $\theta = 0.3$ are compiled in Table 8.4.

Table 8.4 Values of K_0 and K_1 for $\theta = 0.3$

Value of K_0									
Ref. pt	$-b$	$-3b/4$	$-b/2$	$-b/4$	0	$b/4$	$b/2$	$3b/4$	b
Load at									
0	0.86	0.95	0.97	0.5	1.10	1.05	0.97	0.95	0.86
$b/4$	0.20	0.40	0.63	0.87	1.05	1.22	1.36	1.53	1.68
$b/2$	−0.54	−0.63	0.24	0.63	0.97	1.36	1.73	2.10	2.46
$3b/4$	−1.15	−0.63	−0.16	0.40	0.95	1.53	2.10	2.73	3.31
b	−1.79	−1.15	−0.54	0.20	0.86	1.68	2.46	3.31	4.10
Value of K_0									
0	0.94	0.97	1.00	1.02	1.05	1.02	1.00	0.97	0.94
$b/4$	0.85	0.89	0.94	0.97	1.02	1.06	1.06	1.05	1.06
$b/2$	0.77	0.82	0.89	0.94	1.00	1.06	1.13	1.17	1.21
$3b/4$	0.70	0.75	0.82	0.89	0.97	1.05	1.17	1.29	1.38
b	0.63	0.70	0.77	0.85	0.94	1.06	1.21	1.38	1.59

3. Calculation of Weighting Factors

The loading position for absolute maximum distribution for I.R.C. class AA tracked vehicle is shown in Fig. 8.22.

These weighting factors for track load and kerb load are used in the computation of distribution coefficients.

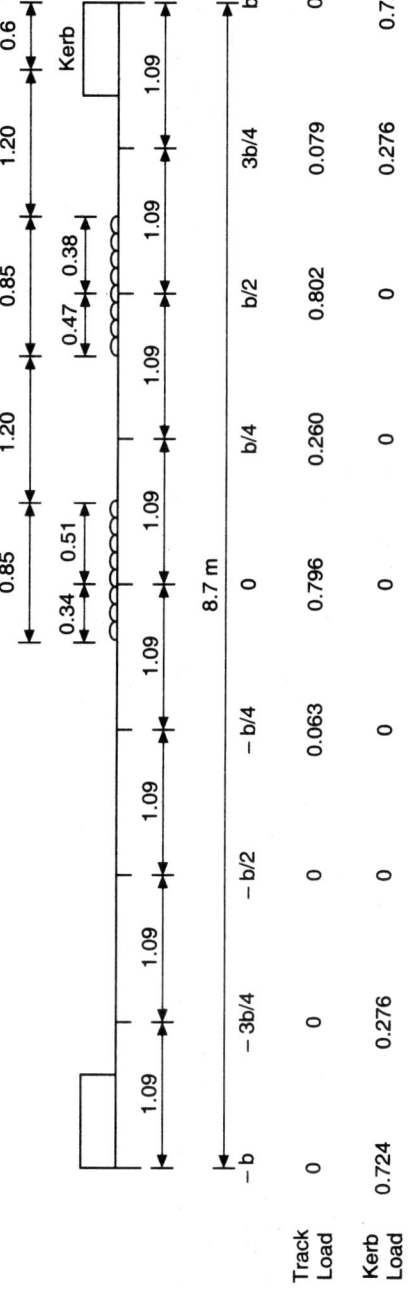

Fig. 8.22 Weighting Factors for Track and Kerb Loading.

4. Distribution Coefficients for Footpath Loading

The distribution coefficients for footpath loading are compiled in Table 8.5. The values of distribution coefficients for different load positions marked in Fig. 8.22, are computed as λK_0 and λK_1, and the distribution coefficient for footpath loading at the centre of end beam is computed as shown in Table 8.5.

5. Distribution Coefficient for Class AA Loading

The distribution coefficients for I.R.C. class AA loading are compiled in Table 8.6. The values of distribution coefficients for different load positions shown in Fig. 8.22 are computed as λK_0 and λK_1 and the distribution coefficient for class AA loading is computed as shown in Table 8.6. The maximum distribution coefficient occurs at the centre of end beam.

6. Maximum Moments in Longitudinal Girders

The maximum moments in longitudinal girders are computed for the position of live loads and dead loads shown in Fig. 8.23.

(a) Dead Loads

Dead weight of slab = (0.2×24) = 4.80 kN/m²
Dead weight of W.C. = (0.08×22) = 1.76 kN/m²

6.56 kN/m²

Weight of kerb = $(0.3 \times 0.6 \times 24)$ = 4.32 kN/m
Weight of railing posts etc. = 1.68 kN/m

6.00 kN/m

Self-weight of girder = $(0.3 \times 1.4 \times 24)$
= 10.08 kN/m

Weight of cross girder = $(0.3 \times 1.4 \times 24)$
= 10.08 kN/m

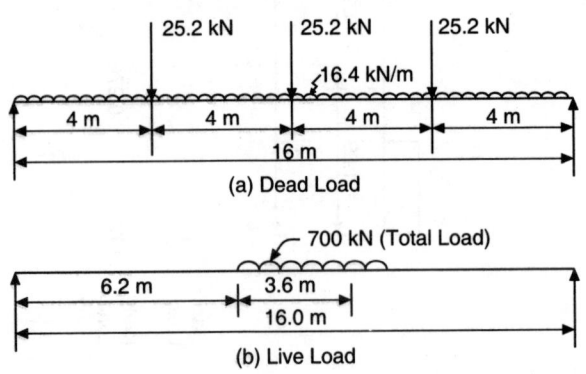

Fig. 8.23 Dead and Live Load Positions for Maximum Bending Moment.

Table 8.5 Distribution Coefficients for Foot Path Loading

Load at	Weighting factor	$-b$	$-3b/4$	$-b/2$	$-b/4$	0	$b/4$	$b/2$	$3b/4$	b
					Value of λK_0					
$3b/4$	0.276	−0.327	−0.173	−0.044	0.110	0.262	0.422	0.579	0.753	0.913
b	0.724	−1.295	−0.832	−0.390	0.144	0.622	1.216	1.781	2.396	2.968
$-3b/4$	0.276	0.913	0.753	0.579	0.422	0.262	0.110	−0.044	−0.173	−0.317
$-b$	0.724	2.968	2.396	1.781	1.216	0.622	0.144	−0.390	−0.832	−1.295
$\Sigma\lambda K_0$	=	2.269	2.144	1.926	1.892	1.768	1.892	1.926	2.144	2.269
$\Sigma\lambda K_0/2$	=	1.134	1.072	0.963	0.946	0.884	0.946	0.963	1.072	1.134
					Value of λK_1					
$3b/4$	1.276	0.193	0.207	0.226	0.245	0.267	0.289	0.322	0.356	0.380
b	0.724	0.456	0.506	0.557	0.615	0.630	0.767	0.876	0.999	1.151
$-3b/4$	0.276	0.380	0.356	0.322	0.289	0.267	0.245	0.226	0.207	0.193
$-b$	0.724	1.151	0.999	0.876	0.767	0.680	0.615	0.557	0.506	0.456
$\Sigma\lambda K_1$		= 2.180	2.068	1.981	1.916	1.894	1.916	1.981	2.068	2.180
$\Sigma\lambda K_1/2$		= 1.090	1.034	0.991	0.958	0.947	0.958	0.991	1.034	1.090
$[\Sigma\lambda K_1/2 - \Sigma\lambda K_0/2]\sqrt{\alpha}$		= −0.007	−0.006	0.004	0.002	0.010	0.002	0.004	−0.006	−0.007
$[\Sigma\lambda K_0/2 + [\Sigma\lambda K_1/2 - \Sigma\lambda K_0/2]]\sqrt{\alpha}$		= 1.127	1.066	0.967	0.948	0.894	0.948	0.967	1.066	1.127

Maximum distribution coefficient for foot path loading at the centre of end beam is given by

$$DK_F = 0.967 + [(1.066 - 0.0967)/0.09]\,0.33 = 0.996$$

Table 8.6 Distribution Coefficients for Class AA Loading

Load at	Weighting factor	$-b$	$-3b/4$	$-b/2$	$-b/4$	0	$b/4$	$b/2$	$3b/4$	b
				Value of λK_0						
0	0.796	0.684	0.756	0.772	0.835	0.875	0.835	0.772	0.756	0.684
$b/4$	0.260	0.052	0.104	0.163	0.226	0.273	0.317	0.353	0.397	0.436
$b/2$	0.802	-0.433	-0.128	0.192	0.505	0.777	1.090	1.387	1.684	1.972
$3b/4$	0.079	-0.090	-0.049	-0.012	0.031	0.075	0.120	0.165	0.215	0.261
$-b/4$	0.063	0.105	0.096	0.085	0.076	0.066	0.054	0.039	0.025	0.012
$\Sigma \lambda K_0 =$		0.318	0.779	1.200	1.673	2.066	2.416	2.716	3.077	3.365
$\Sigma \lambda K_0/2 =$		0.159	0.389	0.600	0.837	1.033	1.208	1.358	1.539	1.683
				Value of λK_1						
0	0.796	0.748	0.772	0.796	0.811	0.835	0.811	0.795	0.772	0.748
$b/4$	0.260	0.221	0.231	0.244	0.252	0.265	0.275	0.275	0.273	0.275
$b/2$	0.802	0.617	0.657	0.713	0.753	0.802	0.850	0.906	0.938	0.970
$3b/2$	0.079	0.055	0.059	0.064	0.070	0.076	0.082	0.092	0.101	0.109
$-b/4$	0.063	0.066	0.066	0.066	0.066	0.064	0.061	0.059	0.056	0.053
$\Sigma \lambda K_1 =$		1.707	1.785	1.883	1.952	2.042	2.079	2.129	2.129	2.155
$\Sigma \lambda K_1/2 =$		0.854	0.893	0.942	0.976	1.021	1.039	1.064	1.070	1.078
$[(\Sigma \lambda K_1/2 - \Sigma \lambda K_0/2)]\sqrt{\alpha} =$		0.111	0.081	0.055	0.022	-0.002	0.027	-0.046	-0.075	-0.097
$[\Sigma \lambda K_0/2 + [\Sigma \lambda K_1/2 - \Sigma \lambda K_0/2)]\sqrt{\alpha} =$		0.270	0.470	0.655	0.859	1.031	1.181	1.312	1.464	1.586

The maximum distribution coefficient occurs at the centre of end beam and is given by

$$DK_w = 1.312 + [(1.464 - 1.312)/1.09]\,0.33 = 1.358$$

Reaction on main girder:

Due to Weight of cross girders

$$= (10.08 \times 2.5) = 25.2 \text{ kN}$$

Reaction from deck slab on each girder

$$= (6.56 \times 2.5) = 16.4 \text{ kN/m}$$

Maximum B.M. at centre due to dead loads

$$= [(16.4 \times 16^2)/8 + (25.2 \times 16)/4 + (25.2 \times 16)/4] = 726.4 \text{ kN·m}$$

(b) B.M. Considering Load Distribution Effect

Live load of footpath $= [P' - (40L - 300)/9] \text{ kg/m}^2$

$$= [400 - (40 \times 16 - 300)/9] = 363 \text{ kg/m}^2 = 3.63 \text{ kN/m}^2$$

Footpath live load moment $= (3.63 \times 0.6 \times 2 \times 16^2)/8 = 139.39 \text{ kN·m}$

Mean moment due to footpath dead loads

$$M_{mean} = [(6 \times 2 \times 16^2)/8] = 384 \text{ kN·m}$$
$$M_{FDL} = [(1.1 \times D_{KF} \times M_{mean})/3]$$
$$= [(1.1 \times 0.996 \times 384)/3] = 140.23 \text{ kN·m}$$
$$M_{FLL} = [(1.1 \times 0.996 \times 139.39)/3] = 50.90 \text{ kN·m}$$

(c) Live Load Moments

Total moment due to live loads (Referring to Fig. 8.23) is given by,

$$M_{mean} = [(350 \times 8) - (350 \times 0.9)] = 2485 \text{ kN·m}$$
$$M_{LL} = [(1.1 \times DK_W \times M_{mean})/3] = (1.1 \times 1.358 \times 2485)/3$$
$$= 1237.36 \text{ kN·m}$$

∴ Total design moment in the exterior girder is computed as

$$M_{max} = (726.40 + 140.23 + 50.90 + 1237.36) = 2155 \text{ kN·m}$$

7. *Bending Moment in Cross Girder*

The values of θ, 3θ, and 5θ are computed for the given problem and the influence curves for μ_0 and μ_1 shown in Figs. 8.24 and 8.25 are derived for load position shown in Fig. 8.6 on the cross girder. From the influence curves the values of μ_0 and μ_1 for different values of q are tabulated in Table 8.7.

$$(\mu_\alpha)_\theta = 0.265 + (0.280 - 0.265)\, 0.161 = 0.2674$$
$$(\mu_\alpha)_{3\theta} = 0.1\,03 + (0.088 - 0.1\,03)\, 0.161 = 0.1006$$
$$(\mu_\alpha)_{5\theta} = 0.050 + (0.0505 - 0.0500)\, 0.161 = 0.050$$

Transverse moment per metre width of cross girder is given by the equation 4(b) as,

$$M_y = 4pb/\pi\ [\mu_\theta \sin(\pi c/2a) + 1/3\ \mu_{3\theta} \sin(3\pi c)/2a + 1/5\ \mu_{5\theta} \sin(5\pi c)/2a + ...]$$

Where $\quad p = (700/3.6) = 194.4 \text{ kN/m}$

$$c = (3.6/2) = 1.8 \text{ m}$$

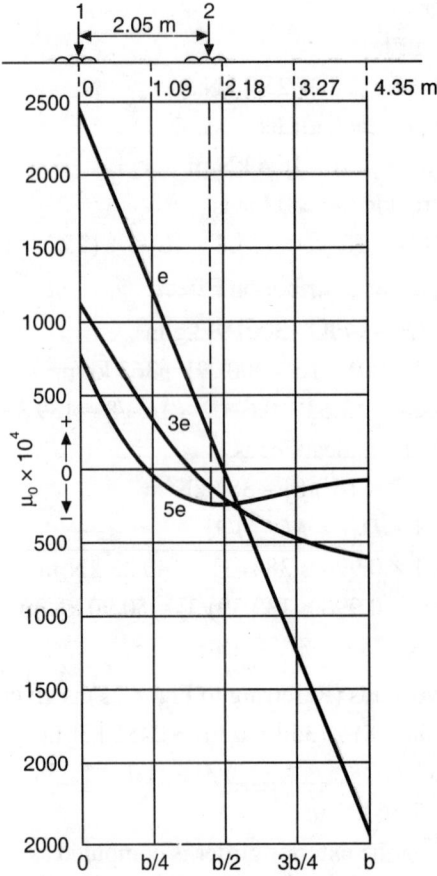

Fig. 8.24 Influence Curve for μ_0.

$$2b = 8.7 \text{ m} \quad \therefore b = (8.7/2) = 4.35 \text{ m}$$
$$2a = 16 \text{ m} \quad \therefore a = (16/2) = 8 \text{ m}$$
$$M_y = (4 \times 194.4 \times 4.35)/\pi \, [0.2674 \sin (\pi \times 1.8)/(2 \times 8)]$$
$$+ 1/3 \times 0.1006 \times \sin [3\pi \times 1.8)/(2 \times 8)]$$
$$+ 1/5 \times 0.05 \times \sin [(5\pi \times 1.8)/(2 \times 8)]$$
$$= 142 \text{ kN·m/m}$$

Total transverse moment = $(142 \times 4) = 568$ kN·m

b) Hendry-Jaegar Method

1. Design Parameters

The main design parameters A and F are computed as,

$$A = 12/\pi^4 \, (L/h)^3 \, (nEI_r/EI)$$

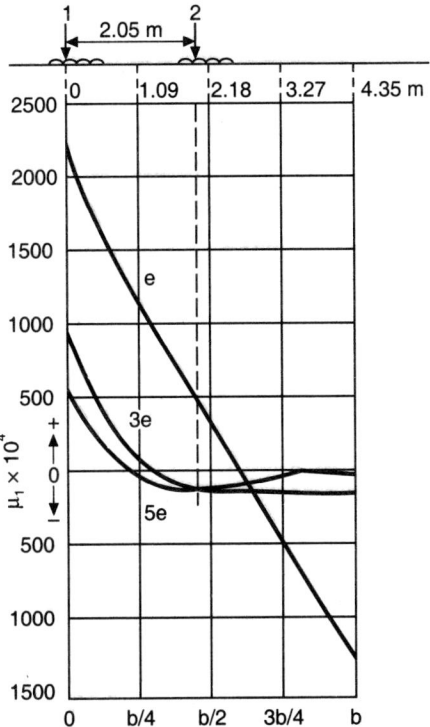

Fig. 8.25 Influence Curve for μ_1.

$$= (12/\pi^4)\,(16/2.5)3\,(5 \times 28 \times 24.74 \times 10^{10})/(28 \times 21.62 \times 10^{10})$$
$$= 184.84$$

$$F = (\pi^2/2n)\,(h/L)\,(CJ/EI_r)$$
$$= (\pi^2/2 \times 5)\,(2.5/16)\,(0.4E \times 1.75 \times 10^{10})/(E \times 24.74 \times 10^{10})$$
$$= 0.004$$

$$m_F = m_0 + (m_\infty - m_0)\sqrt{(F\sqrt{A})/(3 + F\sqrt{A})}$$

From Fig. 8.7 and 8.8, the following values of m_0 and m_∞ are obtained.

For exterior girder $m_0 = 0.82$ $m_\infty = 0.35$
For interior girder $m_0 = 0.35$ $m_\infty = 0.35$

Since F is very nearly equal to zero

$$m_F = m_0$$

For exterior girder, $m_F = 0.82$
For interior girder, $m_F = 0.35$

Design Moments in Exterior Girder:

Dead load moment $= 1218$ kN·m

Table 8.7 Values of μ_0 and $\mu_1 \times 10^4$ for Reference Station '0'

Value of q	Values of $\mu_0 \times 10^4$ for reference station '0' Load at					Average Value under Wheel Positions		
	0	b/4	b/2	3b/4	b	1	2	$\Sigma\mu_0 \times 10^4$
$\theta = 0.300$	2400	1200	–20	–1200	–2400	2400	250	2650
$3\theta = 0.900$	1150	400	–190	–400	–600	1150	–120	1030
$5\theta = 1.500$	750	–20	–250	–100	100	750	–250	500
	Values of $\mu_1 \times 10^4$ for reference station '0'							$\Sigma\mu_0 \times 10^4$
$\theta = 0.300$	2250	1150	300	–375	–1125	2250	550	2800
$3\theta = 0.900$	1000	250	–100	–120	–120	1000	–120	880
$5\theta = 1.500$	625	50	–100	0	0	625	–120	505

Live load moment $\quad = (2485 \times 1.1 \times D.F)$

$\qquad\qquad\qquad = (2485 \times 1.1 \times 0.82) = 2241 \text{ kN·m}$

Total design moment $= (1218 + 2241) = 3459 \text{ kN·m}$

The design moments obtained in the exterior girder by various methods are compared in Table 8.8.

Table 8.8 Comparison of Design Moments in Outer Girder by Different Methods

Design Method	Design Moment (kN·m)
Courbon	2731
Guyon-Massonet	2155
Hendry-Jaegar	3459

c) Comparative Analysis

The results of the various methods indicate that the Guyon-Mssonet method results in the least value of moments in the end girder while the Hendry-Jeagar's method gives the highest value. The magnitude of moments resulting from the Courbon's method lies in between the above two values. The high values resulting in the Hendry-Jaegar's method is attributed to the assumption of zero values for the ratio of torsional to flexural rigidity in the analysis. In the Courbon's method the interaction between the flexural and torsional rigidities of the girders are not considered. Among the three methods, Guyon-Massonet method is more rational since it considers both the flexural and torsional rigidities of the girders which influence the distribution coefficients used in the analysis. Hence this method is considered to be more accurate than the other methods and results in realistic estimates of the moments in the girders in a tee beam bridge deck.

8.9 DESIGN EXAMPLE

Determine the maximum live load bending moment in the exterior girder of the Tee beam and slab of Example 7.5 for two trains of I.R.C. Class A loading using Courbon's method and compare the live load bending moment with that of I.R.C. Class AA tracked loading.

The cross section of the Tee beam and slab deck with two trains of I.R.C. Class A loading positioned to achieve maximum eccentricity of the centroid of the loading system is shown in Fig. 8.26.

In this case, $\Sigma W = 4W$, $n = 3$, $e = 0.7 \text{ m}$

Second moment of area is same for all girders.

Reaction Factors (Courbon's Method)

$$R_x = \frac{\Sigma W}{n}\left[1 + \frac{\Sigma I}{\Sigma(d_x^2 \cdot I)}(d_x \cdot e)\right]$$

$$R_A = \frac{4W}{3}\left[1 + \frac{(3I \times 2.5 \times 0.7)}{(2I \times 2.5^2)}\right] = 1.893\,W$$

$$R_B = \frac{4W}{3}[1+0] = 1.333\,W$$

$$R_C = \frac{4W}{3}[1-0.42] = 0.77\,W$$

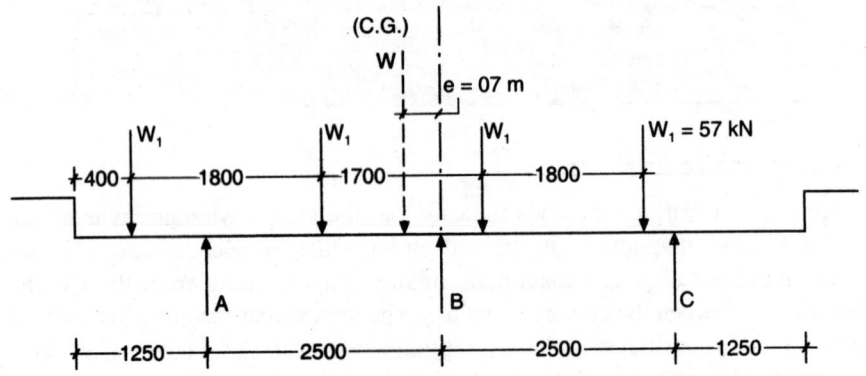

Fig. 8.26 Arrangement of I.R.C. Class A Wheel Loads for Maximum Moment in Girder A.

The wheel loads of I.R.C. Class A loading is arranged on the exterior girder A such that the heavier wheel load and the centre of gravity of the load system are equidistant from the center of the girder as shown in Fig. 8.27.

The centre of gravity of the load system is at a distance of 6.42 m from the leading wheel.

Maximum bending moment occurs under the 4th wheel 'load from left.

Span of the girder = 16 m

Wheel load = (Train load) × (Reaction factor) × (Impact factor)

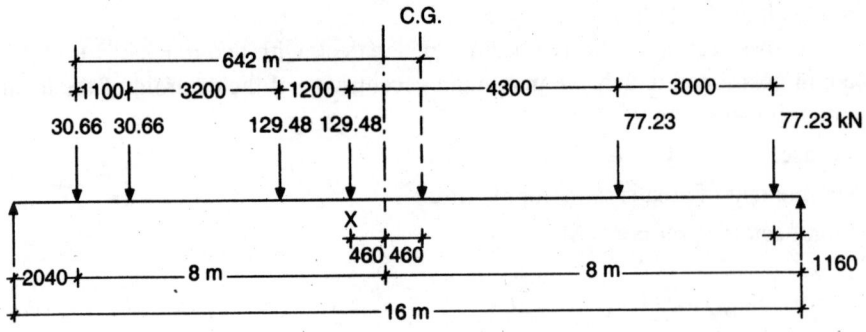

Fig. 8.27 I.R.C. Class A Wheel Load position for maximum Bending Moment in Exterior Girder A.

$$\text{Impact Factor} = \left[\frac{A}{B+L} \right] = \left[\frac{4.5}{6+16} \right] = 0.20$$

First and Second wheel load = (27/2) = 13.5 kN

Load with R.F and I.F. = (13.5 × 1.893 × 1.20) = 30.66 kN

Third and Fourth wheel load = (114/2) = 57 kN

Load with R.F. and I.F = (57 × 1.893 × 1.20) = 129.48 kN

Fifth, Sixth, Seventh and Eighth' wheel load = (68/2) = 34 kN

Load with R.F and I.F. = (34 × 1.893 × 1.20) = 77.23 kN

Maximum bending moment at X (under 4th wheel load from left) is computed as,

$$M_{max} = (223.72 \times 7.54) - 30.66 (5.50 + 4.40) - (129.48 \times 1.20)$$
$$= 1228 \text{ kN·m}$$

Comparison of live load bending moments resulting from I.R.C. Class AA tracked vehicle and I.R.C. Class A wheel loads are shown in Table 8.9.

Table 8.9 Comparison of Live Load Bending Moments

Type of Loading	Live load B.M
I.R.C. Class AA tracked load	1513 kN·m
I.R.C. Class A Wheel loads	1228 kN·m

8.10 DESIGN EXAMPLE

Design a R.C.C. Tee beam and slab deck to suit the following data:

Effective span of girders	=	16 m
Clear width of Road way	=	7.5 m
Width of Kerbs	=	600 mm
Thickness of wearing coat	=	80 mm
Number of Main Girders	=	4
Spacing of Main Girders	=	2.5 m
Spacing of Cross Girders	=	4 m

Type of loading: I.R.C. Class 70R tracked vehicle

Materials: M-20 Grade concrete and Fe-415 Grade HYSD bars

Design the deck slab and the exterior girder for flexure only and sketch the details of reinforcements.

1. Data

Effective span of Tee beam	=	16 m
Width of Carriageway	=	7.5 m
Thickness of wearing coat	=	80 mm

M-20 Grade Concrete and Fe-415 Grade HYSD bars.

2. Characteristic Strength of Materials

$f_{ck} = 20 \text{ N/mm}^2$ $\quad E_c = 29 \text{ GPa} \quad \quad \propto_e = (E_s/E_c) = 7$
$f_{yk} = 415 \text{ N/mm}^2 \quad E_s = 200 \text{ GPa}$

3. Cross Section of Deck

Four main girders are provided at 2.5 m centres
Adopt thickness of deck slab = 250 mm
Width of main girder = 300 mm
Kerbs 600 mm wide by 300 mm deep are provided
Cross girder*, are provided at every 4 m intervals
Width of cross girder = 300 mm
Depth of main girder = 1600 mm at the rate of 100 mm per metre span. The depth of cross girder is taken as equal to that of main girder.

The cross section of the deck and the plan showing the spacing of main girders and cross girders is showing in Fig. 8.28

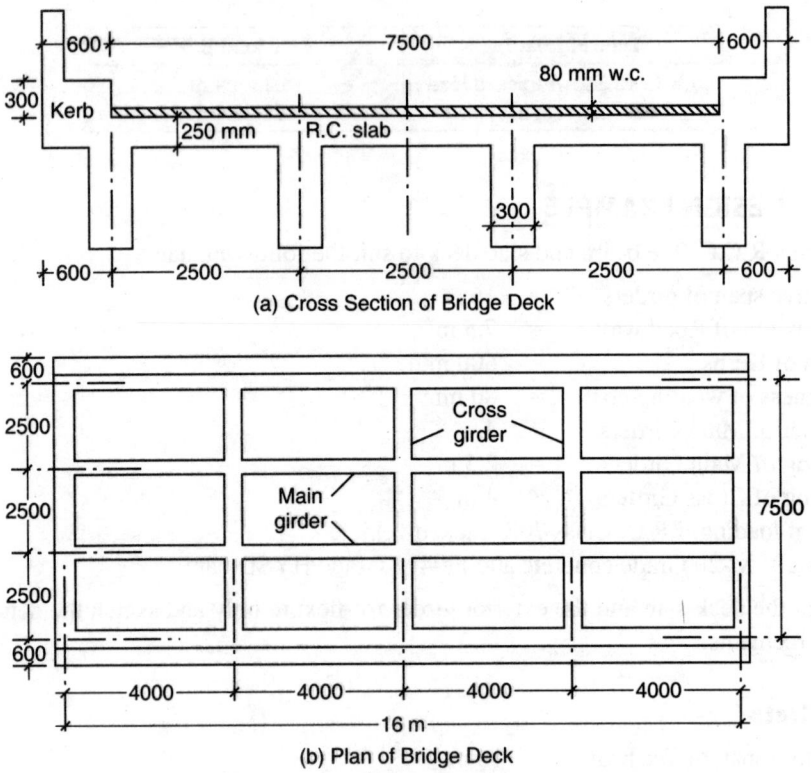

(a) Cross Section of Bridge Deck

(b) Plan of Bridge Deck

Fig. 8.28 Tee Beam and Slab Bridge Deck.

4. Design of Interior Slab Panel

(a) Bending Moment

Dead weight of slab $= (1 \times 1 \times 0.25 \times 24) = 6.00$ kN/m^2
Dead weight of W.C. $= (0.08 \times 22)$ $\qquad = 1.76$ kN/m^2

Total dead load $\qquad\qquad\qquad\qquad = 7.76$ kN/m^2

Live load is IRC Class 70 R tracked vehicle.
One wheel is placed at the centre of the panel as shown in Fig. 8.29.

$$u = (0.84 + 2 \times 0.08) = 1.00 \text{ m}$$
$$v = (4.57 + 2 \times 0.08) = 4.73 \text{ m but limited to 4 m which is the spacing}$$
$$\text{of cross girders.}$$
$$(u/B) = (1.00/2.5) = 0.40$$
$$(v/L) = (4.00/4.00) = 1.00$$
$$K = (B/L) = (2.5/4.00) = 0.625$$

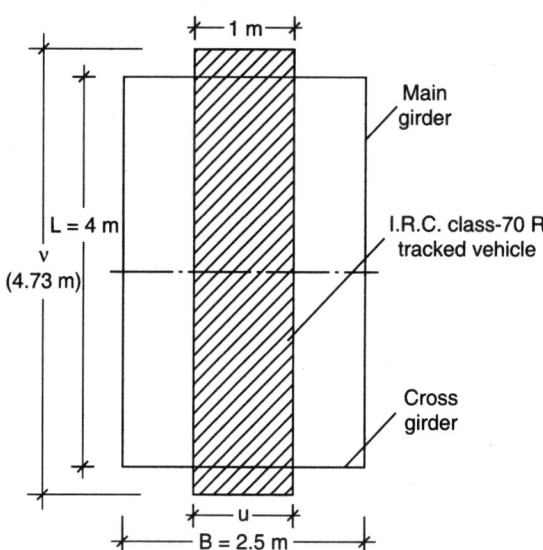

Fig. 8.29 Position of Wheel Load for Maximum Moment.

Referring to Pigeaud's curves (Refer Fig. 4.7) corresponding to $K = 0.6$, the values of moment coefficients are read out as

$$m_1 = 0.081 \text{ and } m_2 = 0.022$$

Total load per track including impact $= (1.25 \times 350) = 437.5$ kN

$W =$ Effective load on the span $= 437.5 (4.00/4.73) = 370$ kN

Moment along shorter span $= M_B = W(m_1 + \mu m_2)$
$$= 370 (0.081 + 0.15 \times 0.022)$$
$$= 31.19 \text{ kN·m}$$

Moment along longer span $= M_L = W(m_2 + \mu m1)$
$$= 370 \, (0.022 + 0.15 \times 0.081)$$
$$= 12.63 \text{ kN·m}$$

As the slab continuous, the design bending moments are obtained by applying the continuity factor as
$$M_B = 0.8 \, (31.19) = 24.95 \text{ kN·m}$$
$$M_L = 0.8 \, (12.63) = 10.10 \text{ kN·m}$$

Dead load bending moments are computed using Pigeaud's curves shown in Fig. 4.13.
Dead load $= 7.76 \text{ kN/m}^2$
Total dead load on Panel $= (4 \times 2.5 \times 7.76) = 77.6 \text{ kN}$

$$(u/B) = (v/L) = 1 \text{ as panel is loaded with uniformly distributed load.}$$
$$K = (B/L) = (2.5/4.00) = 0.625 \text{ and } (1/K) = 1.6$$

From Pigeaud's Curves, read out
$$m_1 = 0.049 \text{ and } m_2 = 0.015$$

Taking continuity into effect, the design moments are
$$M_B = (0.8 \times 77.6) \, (0.049 + 0.15 \times 0.015) = 3.18 \text{ kN·m}$$
$$M_L = (0.8 \times 77.6) \, (0.015 + 0.15 \times 0.049) = 1.38 \text{ kN·m}$$

Hence the service load design moments in the slab are obtained as
$$M_B = (24.95 + 3.18) = 28.13 \text{ kN·m}$$
$$M_L = (10.10 + 1.38) = 11.48 \text{ kN·m}$$

Design Ultimate bending moments in the slab are evaluated by applying load factors as,
$$M_B = [(1.35 \times 3.18) + (1.5 \times 24.95)] = 44 \text{ kN·m}$$
$$M_L = [(1.35 \times 1.38) + (1.5 \times 10.10)] = 30 \text{ kN·m}$$

(b) Design of Reinforcements in Deck Slab

Effective depth $= d = \sqrt{\dfrac{M_u}{0.138 \times f_{ck} \times b}} = \sqrt{\dfrac{44 \times 10^6}{0.138 \times 20 \times 1000}} = 126 \text{ mm}$

Assumed depth of the slab $= 250 \text{ mm}$ and allowing for a clear cover of 40 mm and using 12 mm diameter bars, provide an effective depth, $d = 200 \text{ mm}$

Refer Table 2 of SP: 16 and interpolate the percentage of reinforcement corresponding to the parameter $(M_u/b \, d^2) = [(44 \times 10^6)/(10^3 \times 200^2)] = 1.1$

$$p_t = 0.322 = (100 \, A_{st}/b.d)$$

Hence, $\quad A_{st} = [(0.322 \times 1000 \times 200)/100] = 644 \text{ mm}^2$

Provide 12 mm diameter bars at a spacing of 150 mm for short span and 10 mm diameter bars at a spacing of 150 mm along the long span

5. Design of Longitudinal Girders

(a) Reaction Factors

Using Courbon's theory, the IRC Class 70 R tracked vehicle loads are arranged for maximum eccentricity as shown in Fig. 8.30. Reaction factor for outer girder A is given by

$$R_A = \left(\frac{2W_1}{4}\right)\left[1 + \frac{4I \times 3.75 \times 1.1}{(21 \times 3.75^2) + (2I \times 1.25^2)}\right] = 0.764\,W_1$$

If W = axle load = 700 kN

           $W_1 = 0.5\,W$

∴ $R_A = (0.764 \times 0.5\,W) = 0.382\,W$

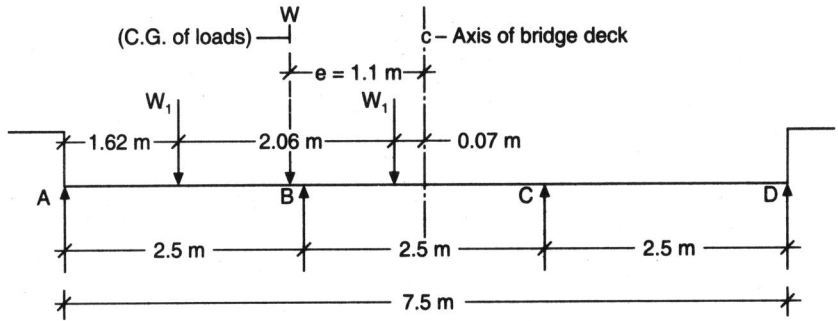

Fig. 8.30 Transverse Position of I.R.C. Class 70 R Tracked Vehicle.

(b) Dead Load from Slab per Girder

Dead load of deck slab is calculated with reference to Fig. 8.31.

Weight of

  (1) Parapet railing (lumpsum).....................................= 0.700 kN/m
  (2) Kerb and Deck slab = (0.55 × 0.6 × 1 × 24).........= 7.920 kN/m

       Total Load ....= 8.620 kN/m

Total dead load of deck = [(2 × 8.62) + (7.5 × 7.76)] = 75.44 kN/m

It is assumed that the dead load of deck is shared equally by all the four girders.

∴ Dead load per girder = (75.44/4) = 18.86 kN/m

(c) Live load Bending Moment in Girder

Effective span of girder = 16 m

Impact factor (for Class 70 R loading) = 10%

The live load is placed centrally on the span as shown in Fig. 8.32.

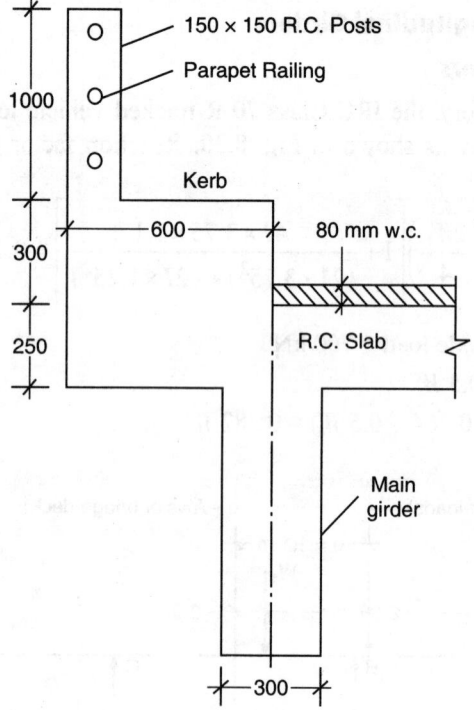

Fig. 8.31 Details of Deck Slab-Kerb and Parapet.

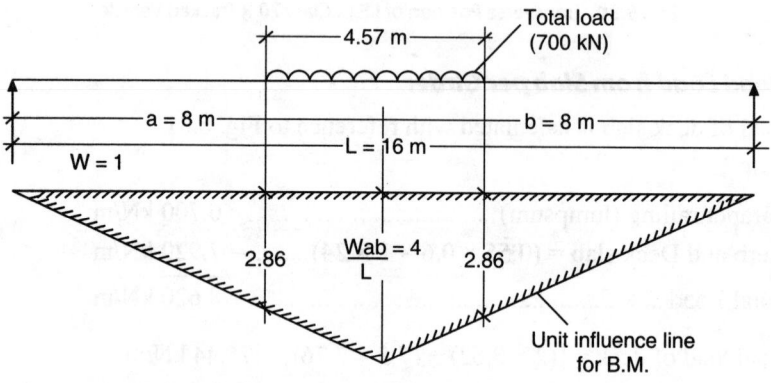

Fig. 8.32 Influence Line for Bending Moment at Centre of Span of Girder.

Total bending moment = 0.5 (4 + 2.86) 700 = 2401 kN·m

Bending moment including impact and reaction factors in outer girder A is computed as

$$M = (2401 \times 1.01 \times 0.382) = 1009 \text{ kN·m}$$

(d) Dead Load Bending Moment in Girder-A

Overall depth of girder = 1600 mm
Depth of rib = (1600 – 250) = 1350 mm
Width of rib = 300 mm
Self weight of rib = (1 × 0.3 × 1.35 × 24) = 9.72 kN/m

The cross girder is assumed to have the same cross sectional dimensions of the main girder.

Weight of cross girder = 9.72 kN/m
Reaction on main girder = (9.72 × 2.5) = 24.3 kN
Reaction from deck slab on each girder = 18.86 kN/m
∴ Total dead load on girder = (18.86 + 9.72) = 28.58 kN/m

Referring to Fig. 8.33, the maximum bending moment in the exterior girder A is computed as

$$M_{max} = [928.58 × 16^2)/8] + [(24.3 × 16)/4] + [(24.3 × 16)/4] = 1109 \text{ kN·m}$$

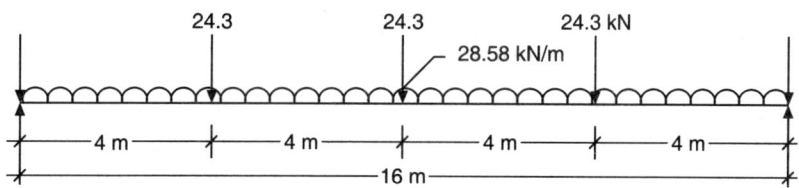

Fig. 8.33 Dead Loads on Main Girder.

(e) Design Moments in Girder A

Service load moments are computed as
Dead load moment = M_D = 1109 kN·m
Live load moment = M_L = 1009 kN·m
Design Ultimate bending moment is obtained by applying the load factors as
$$M_u = [1.35 M_D + 1.5 M_L) = [(1.35 × 1109) + (1.5 × 1009)] = 3011 \text{ kN·m}$$
As the moment is of larger magnitude, the beam section will be designed as doubly reinforced.

Providing an effective cover of 150 mm for tension steel and using 25 mm diameter HYSD bars, the effective depth = d = (1600 – 150) = 1450 mm

Providing an effective cover of 50 mm to compression reinforcement, d' = 50 mm
Hence the ratio (d'/d) = (50/1450) = 0.03
Compute the parameter $(M_u/b\, d^2)$ = (3011 × 10⁶)/(300 × 1450²)] = 4.77
Refer Table 50 of SP: 16 corresponding to the parameters f_{ck} = 20 N/mm², f_y = 415 N/mm² and (d'/d) = 0.05, interpolate the percentage steel as,

$$p_t = 1.550 \text{ and } p_c = 0.620$$
$$p_t = (100 \, A_{st}/b.d) \text{ and } p_c = (100 \, A_{sc}/b.d)$$

Hence,
$$A_{st} = [(1.550 \times 300 \times 1450)/100] = 6743 \text{ mm}^2$$
$$A_{sc} = [(0.620 \times 300 \times 1450)/100] = 2697 \text{ mm}^2$$

Provide 16 bars of 25 mm diameter in four rows on the tension side and 8 bars of 25 mm on the compression side to resist negative moments developed near supports.

Provide 10 mm diameter 4-legged stirrups at 200 mm centers to resist the shear forces. The reinforcement details in the cross section are shown in Fig. 8.34.

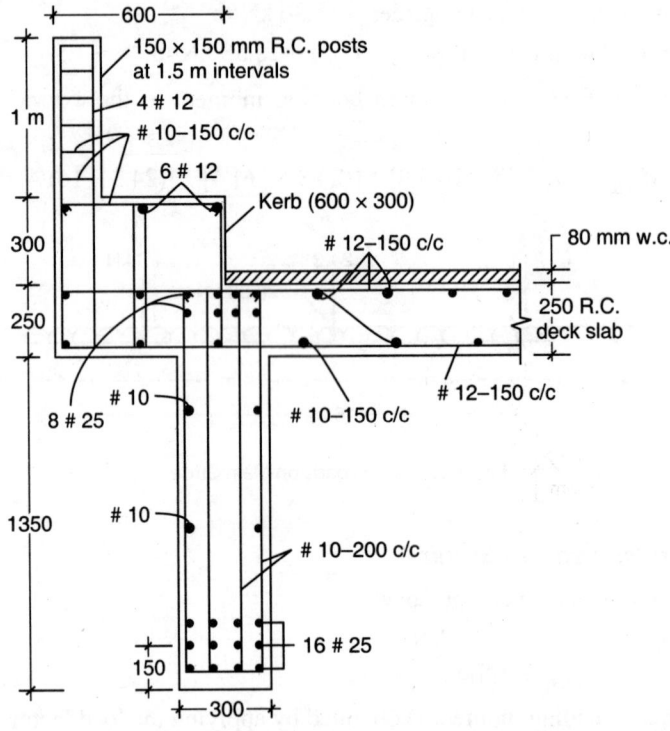

Fig. 8.34 Cross Section of Exterior Tee Beam and Slab.

8.11 DESIGN EXAMPLE

Design the cantilever slab of a Tee beam and slab bridge deck using the following data:

Width of Roadway = 7.5 m
Width of Kerb = 600 mm
Depth of Kerb = 300 mm
Number of Longitudinal Girders = 3
Width of Girder = 300 mm
Spacing of Longitudinal Girders = 2.5 m
Thickness of Wearing Coat = 80 mm

Type of Loading: IRC Class A wheel loads
Materials: M-20 Grade concrete
Fe-415 Grade HYSD bars

Design the cantilever slab and sketch the details of reinforcements conforming to the relevant Indian roads congress standards.

1.   Data

Figure 8.35 shows the cantilever portion of the Tee beam and slab bridge deck with the dimensional details of cantilever projection, kerb. R.C. posts and hand rails.

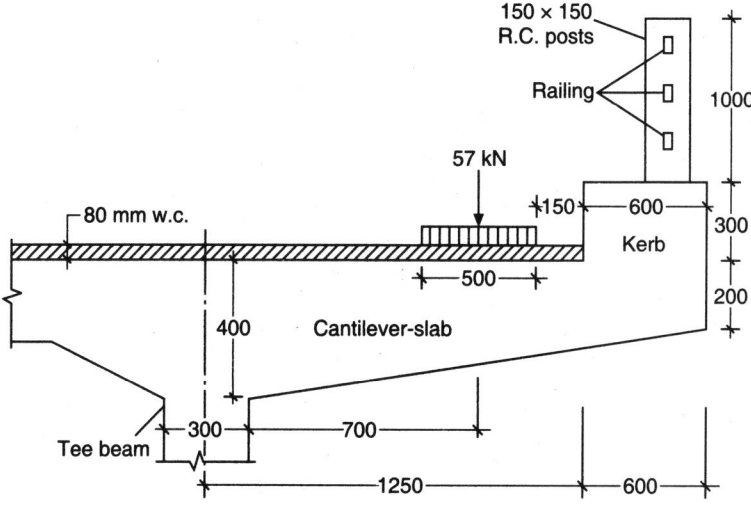

Fig. 8.35 Cantilever Slab with Class A Wheel Load.

Clear projection of cantilever slab = (1250 – 150 + 600) = 1700 mm = 1.7 m R.C.C. posts 150 mm × 150 mm × 1 m are provided at every 1.5 m intervals.

Thickness of wearing coat = 80 mm

M-20 Grade concrete and Fe-415 Grade HYSD vars.

2.   Characteristic Strength of Materials

$f_{ck} = 20$ N/mm^2 $E_c = 29$ GPa
$f_{yk} = 415$ N/mm^2 $E_s = 200$ GPa

3.   Dead Load Moment

Considering one metre width of cantilever slab, the dead load moment at the fixed end of the cantilever is computed considering the self weight of slab, kerb, parapet and railings.

Sl. No.	Dimensions of Structural Element	Load (kN)	Lever arm (m)	Moment (kN·m)
1.	Hand Rails (lump sum)	2.00	1.575	3.150
2.	R.C.C. posts			
	$(0.15 \times 0.15 \times 1 \times 24)$	0.54	1.575	0.850
3.	Kerb $(0.6 \times 0.3 \times 24)$	4.32	1.400	6.048
4.	Wearing Coat			
	$(1.1 \times 0.08 \times 22)$	1.94	0.550	1.067
5.	R.C. slab			
	$(0.3 \times 1.7 \times 24)$	12.24	0.85	10.404
Total dead Load Moment (M_g)				21.519

4. Live Load Moment

The heaviest wheel (57 kN) of IRC Class-A loads is placed with its edge 150 mm from the kerb as shown in Fig. 8.35. Effective width of dispersion (b_e) perpendicular to span is given by

$$b_e = 1.2\,x + b_w$$

Where, $x = 0.70$ m

$$b_w = (0.25 + 2 \times 0.08) = 0.41 \text{ m}$$

∴ $b_e = [(1.2 \times 0.7) + 0.41] = 1.25$ m

Live load per metre width including impact $= [(57 \times 1.5)71.25] = 68.4$ kN

Design live load moment $(M_q) = (68.4 \times 0.7) = 47.88$ kN·m

5. Design Ultimate Load Moment

$$M_u = [1.35M_D + 1.5M_L) = [(1.35 \times 21.519) + (1.5 \times 47.88)] = 101 \text{ kN·m}$$

6. Design of Reinforcements

Effective depth required $= d = \sqrt{\dfrac{M_u}{0.138 \times f_{ck} \times b}} = \sqrt{\dfrac{101 \times 10^6}{0.138 \times 20 \times 1000}} = 365$ mm

Provide an overall depth of 420 mm and using 16 mm diameter bars, and clear cover of 42 mm, Effective depth $= (420 - 42 - 8) = 370$ mm

Compute the parameter $(M_u/b\ d^2) = (101 \times 10^6)/(1000 \times 370^2)] = 0.73$

Refer Design Table-2 of SP: 16 and interpolate the percentage reinforcement as,

$$p_t = (100\ A_{st}/b.d) = 0.212$$

Hence, $A_{st} = [(0.212 \times 1000 \times 370)/100] = 785$ mm^2

Provide 16 mm diameter bars at a spacing of 150 mm as main reinforcement and 12 mm bars at a spacing of 150 mm as distribution reinforcements. The details of reinforcements are shown in Fig. 8.36.

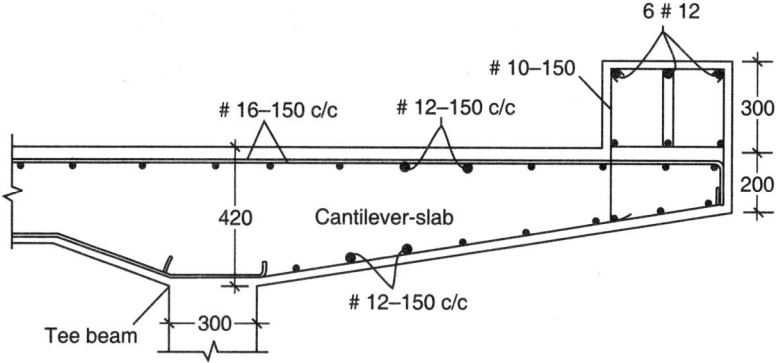

Fig. 8.36 Reinforcement Details in Cantilever Slab.

EXAMPLES FOR PRACTICE

1. Design a R.C.C. Tee beam girder for a national highway bridge to suit the following data:

   Clear width of Roadway = 7.5 m, width of kerbs = 600 mm
   Effective span = 20 m
   Live load = I.R.C. Class-AA tracked vehicle
   Thickness of wearing coat = 80 mm, Number of main girders = 4
   Concrete = M-20 grade
   Steel = Fe-415 grade tor steel
   Spacings of cross girders = 4 m, Spacings of main girders = 2.5 m

   Design the deck slab, main girders and one of the cross girders, using Courbon's method. Sketch the typical details of reinforcements.

2. Design a R.C.C. Tee beam girder deck for a bridge crossing using the following data:

   Effective span = 18 m
   Clear width of Roadway = 7.5 m, width of kerbs = 600 mm
   Number of main girders = 4
   Foot paths = one metre on either side
   Live load = I.R.C. Class-AA or A whichever produces the worst effect
   Thickness of wearing coat = 100 mm
   Concrete = M-20 grade
   Steel = Fe-415 grade tor steel
   Spacings of cross girders = 4.5 m, spacings of main girders = 2.5 m

   Design the deck slab, main girders and cross girders using Guyon-Massonet theory. Sketch the details of reinforcements in the slab and beams.

3. An R.C.C. Tee beam and slab girder deck is required for the crossings of a National Highway. The data available is as follows:

Clear width of Roadway = 15 m, Foot paths 1 m on either side, Effective span = 20 m

Live load = I.R.C. Class-AA or A whichever gives the worst effect

Thickness of wearing coat = 100 mm, Number of main girders = 8, Concrete = M-20 grade

Steel = HYSD Fe-415 tor steel bars

Spacing of cross girders = 4 m, Spacings of main girders = 2 m

Design one of the interior panels of deck slab and one of the exterior girders and sketch the details of reinforcements.

4. Design an R.C.C. Tee beam and slab bridge deck to suit the following data:

Effective span of girders = 16 m

Foot path = 1 m on either side

Live Load = I.R.C. Class AA tracked Vehicle

Thickness of wearing coat = 80 mm

Number of main girders = 4

Spacings of main girders = 2.5 m

Spacings of cross girders = 4 m

Concrete = M-20 Grade and Steel = Fe-415 grade HYSD bars

Design the deck slab, exterior girder and cross girder using Guyon-Massonet method of load distribution.

Sketch the typical details of reinforcements in slab and girders.

5. A reinforced concrete tee beam and slab bridge deck is to be designed for a major national high way crossing of a river. Design the typical structural elements of the bridge deck to suit the following data:

Clear span of the girders = 16 m

Width of carriage way = 3 standard lanes of 3.5 m each with foot paths 1.5 m wide

Thickness of the wearing coat = 80 mm

Number of main girders = 4

Cross girders at every 4 m intervals

Live load = IRC Class AA tracked vehicle

Materials = Concrete M-25 grade and Fe-500 HYSD bars

Design the deck slab, longitudinal and cross girders using limit state design concepts conforming to the specifications of IRC: 112-2012. Sketch the details of reinforcements in each of the typical structural elements.

6. Design the typical slab and girder of a tee beam and slab reinforced concrete bridge deck to suit the following data:

Effective span = 15 m

Width of road way = 7.5 m

Kerbs = 600 mm on both sides

Spacings of main girders = 2.5 m

Number of cross girders = 5
Live load = IRC Class A loading
Materials = M-25 grade concrete and Fe-500 HYSD bars

Design the slab and girders using limit state design, conforming to the specifications of the Indian Roads congress code IRC: 112-2012. Check the slab and girders for the limit states of strength and serviceability.

REFERENCES

1. Johnson Victor. D., Essentials of Bridge Engineering, Fifth Edition, Oxford & IBH, New Delhi, 2001, pp. 131-174.
2. Victor, D.J and Lakhmanan, N., Reactions in Three girder Girder Bridge Decks, journal of The Indian Roads Congress, Vol. 36-1, October 1975, paper No. 299, pp. 41-68.
3. Courbon, J., Application de la Resistance des Materius au Calculdes Fonts, Dunod Publications, Paris, 1950.
4. Raina, V.K., Concrete Bridges, Practice, Analysis, Design and Economics, Tata McGraw-Hill Publishing Co, Ltd, New Delhi, 1991, pp. 209-211.
5. Massonet. C., Method of Calculation of Bridges with Several longitudinal Beams taking into Account their Torsional Resistance, Proceedings of the International Association For Bridges & Structural Engineering, Vol. 10,1950, pp. 147-182.
6. Hendry, A.W and Jaegar, L.G., The Analysis of Grid Frame Works and Related Structures, Chatto and Windus, London, 1958, 308 pp.
7. Morice, P.B and Little, G., Analysis of Right bridge Decks Subjected to Abnormal Loading, Cement and Concrete Association, London, July 1956, 43 pp. August 1955, 29 pp.
8. Rowe, R.E., Concrete Bridge Design, Applied Science Publishers, London, 1962, p. 372.
9. Morice, P.B, Little, G and Rowe, R.E., Design Curves for the Effects of Concentrated Loads On Concrete Bridges, Cement and Concrete Association Publication No. 42-202, London, August 1955, p.29.
10. Sarkar, S, Kapla, M.S, Prasada Rao, A.S and Chhauda, J.N., Hand Book for Prestressed Concrete Bridges, Structural Engineering research Centre Publication, Roorke (U.P), 1969, pp. 15-23.

REVIEW QUESTIONS

1. Under what situations you would prefer to adopt reinforced concrete Tee-beam and slab Bridge decks for high way crossings?
2. Briefly discuss the typical structural elements of a reinforced concrete tee beam and slab bridge deck and their functions.
3. What are the various methods of analysis generally used for determining the moments and Shear forces in the girders of a Tee beam and slab bridge deck?

4. Briefly explain the method of determining the live load bending moments in the girders of a Tee beam and slab bridge deck using Courbon's method.

5. Explain briefly the Guyon-Massonet method of determining the moments in the main Girders of Tee beam and slab bridge deck and its advantages.

6. Explain the terms, a) Flexural rigidity b) Torsional Rigidity c) Distribution Coefficient with Respect to the analysis of moments in the girders of a Tee beam and slab bridge deck.

7. Briefly explain the method of analysis adopted in Hendry-Jaegar's method to determine the Moments in the girders of a Tee beam and slab bridge deck.

8. Specify the position of the IRC Class AA tracked vehicle loads for which the maximum Bending moment and shear forces occur in the main girders of a tee beam and slab bridge deck.

9. Explain briefly the method of checking the serviceability limit state of deflection and cracking specified in the Indian Roads Congress code IRC: 112-2011.

10. Describe the simplified empirical method specified in the IRC code for checking the limit state of cracking without detailed calculations of the width of cracks.

OBJECTIVE TYPE QUESTIONS

1. Reinforced concrete Tee beam and slab type bridge decks are generally adopted for high way crossings in the span range of
   a)   6 to 10 m
   b)   12 to 20 m
   c)   20 to 30 m

2. The spacings of the cross girders in a tee beam and slab bridge deck is generally in the range of
   a)   1 to 2 m
   b)   2 to 3 m
   c)   4 to 6 m

3. According to Courbon's method of analysis, the ratio of the depth of cross to main girders in a tee beam slab bridge deck should be at least
   a)   0.50
   b)   0.75
   c)   1.00

4. For the application of Courbon's method to determine the maximum bending moments in the longitudinal girders of a tee beam and slab bridge deck, the IRC. Class AA loads have to be arranged for maximum
   a)   Concentricity
   b)   Eccentricity
   c)   Distance from girders

5. Among the various methods of analysis of moments in the girders of a tee beam and slab bridge deck, the method which considers the maximum number of influencing parameters is proposed by
   a) Courbon
   b) Guyon-Massonet
   c) Hendry-Jaegar
6. In a reinforced concrete tee beam slab bridge deck having 6 longitudinal girders, the maximum bending moment under the action of IRC Class AA tracked vehicle, occurs in the
   a) Central girder
   b) Intermediate girder
   c) Exterior girder
7. In a tee beam and slab type bridge deck with three main girders with several cross girders, the maximum bending moment in the cross girder develops when the IRC Class AA tracked vehicle loads are positioned
   a) To have maximum eccentricity to the centre of the road
   b) To be equidistant from the central main girder
   c) To be nearest to the exterior girder
8. The introduction of cross girders in a tee beam slab bridge deck with main girders results in the load resistance of the deck by improvement of
   a) Flexural rigidity
   b) Torsional rigidity
   c) Shear strength
9. The width cracks developed in a reinforced concrete beam of a tee beam and slab deck depends upon the
   a) The stress in concrete
   b) The stress in steel
   c) The difference between the mean strains between the steel and concrete
10. The long term deflection of reinforced concrete beams in a tee beam slab bridge deck is significantly influenced by
    a) The loads on the deck
    b) Creep of concrete
    c) Grade of concrete

9
Plate Girder Bridges

9.1   GENERAL FEATURES

Plate girder bridges are the most common type of steel bridges generally used for railway and high way crossings from the latter half of 18th century when Henry Bessemer of England invented and patented the process of making steel in 1855[1,2]. The earliest forms of steel bridges constructed happen to be the plate girder type of bridges due to the simplicity of the structural form and their elegant aesthetics. In general, Plate girders are economical for railway and high way bridges of spans in the range of 15 to 40 m. They may be very competitive for much longer spans, when they are continuous in the range of 50 to 200 m.

   Most of the plate girder bridges constructed after 1960 are shop welded, replacing the riveted type of fasteners and they are built up using two flange plates and a web plate to form an I-shaped cross section. The longest plate girder bridge in the world is a three span continuous bridge constructed over the Save river at Belgrade[3], Yugoslavia with two end spans of 75 m and a central span of 260 m. It has a double box girder cross section varying in depth from 4.5 m at mid span to 9.6 m at the pier. Plate girders are also used for the construction of Industrial buildings like factories, air craft hangers and ware houses. The great advantage of the plate girder choice being its rapid assembly and lesser construction time in comparison with the concrete bridges. However in the long run, maintenance costs of steel bridges are abnormally high due to their vulnerability to rusting and deterioration especially in coastal areas due to aggressive exposure conditions.

   In the case of railway bridges[4,5], the plate girders support the sleepers over which the steel rails are fastened. Each rail is supported on a plate girder so that the wheel loads are transmitted directly to the plate girder without the effects of torsion. The twin plate girders are braced laterally at the level of the bottom flange to provide lateral stability. Cross bracings consisting of angles are provided at the ends and at intervals of 4 to 5 m. The lateral bracings and the end cross frames resist the lateral loads on the plate girder.

9.2   STRUCTURAL ELEMENTS OF PLATE GIRDER

The typical structural elements used to assemble a plate girder are shown in Fig. 9.1. The following three types of plate girders are generally used in railway and high way crossings.

1. Fig. 9.1(a) shows the simplest type built using cover plates to function as flanges and a web plate connecting the flange plates.

2. Fig. 9.1(b) shows a plate girder used for longer spans built by using more cover plates, to strengthen the flanges and a web plate.

3. Fig. 9.1(c) shows a built up plate girder generally used for railway bridges of long spans using two plate girders connected by lateral bracings.

Plate girders used for longer span lengths require deeper web plates and to prevent buckling, intermediate stiffeners comprising rolled steel angles are used at regular intervals along the span and at the supports. If the span length is large, the web and flange plates are connected by splices for structural integrity in resisting external loads.

9.3   DESIGN PRINCIPLES

The main structural elements in a plate girder to be designed are the web and flange plates together with the intermediate and end bearing stiffeners conforming to the specifications of IRC: 24-2001 and IS: 800-1987[6]. The various design steps involved are listed below:

1. Compute the dead load and live load moments and the shear forces. The dead loads may be obtained by empirical estimations such as

   a. Self weight of girder = $(0.2L + 1)$ kN/m where L = span of the girder

   b. Self weight of the girder = $w = (W/200)$ where w = self weight expressed in kN/m and W is the total factored load applied to the girder in kN.

2. The design moments and shear forces are computed by applying impact factors to the live load moments and shears. The impact factors for steel bridges prescribed in the relevant IRC and IRS bridge codes outlined in Chapter 2.

3. The depth of girder is estimated using the following empirical expressions:

   a. Approximate depth of the girder = 1/10 to 1/15 of span

   b. Economical depth of the plate girder[7] = $\left(\dfrac{Mk}{f_y}\right)^{0.33}$

   Where M = design moment

   $$K = (d/t_w) \le 200\epsilon \text{ where } \epsilon = (250/f_y) \text{ and } f_y = \text{yield stress of steel}$$

   c. Economical depth[8] = $5\sqrt[3]{\dfrac{M}{f_{cb}}}$ where M = design moment and

   f_{cb} = permissible bending stress in steel as specified in IRC: 24-2001

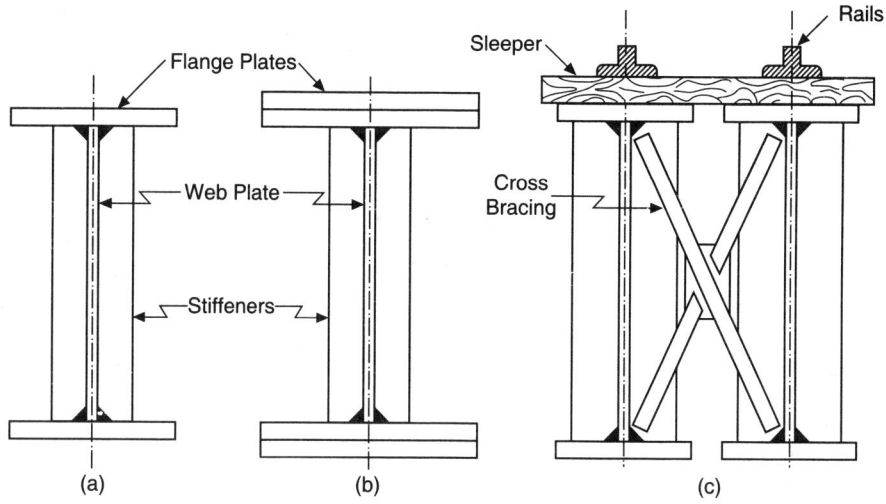

Fig. 9.1 Types of Plate Girders.

4. Minimum thickness of the web from corrosion considerations $= t_w = 6$ to 8 mm
   From handling considerations in large bridge depth girders, $t_w = 8$ to 12 mm
   From serviceability considerations, $t_w \geq \left(\dfrac{d}{200}\right)$ where $d =$ depth of web
   From shear buckling considerations, $(d/t_w) > 67\varepsilon$ which results in thick webs eliminating the use of intermediate stiffeners.

5. The thickness and the width of flange plates are determined using the following considerations:
   Approximate width of flange plates $= b_f = 0.3$ times the depth of web
   According to an empirical relation, $b_f = (L/40)$ to $(L/45)$ where $L =$ span of girder
   Approximate flange area $= A_f = [(M/f_y\, d) - (A_w/6)]$

6. The flange plates should be so proportioned that it should satisfy the requirements of plastic/compact section. For this criterion the ratio of (b_f/t_f) should be less than 8.4ε or 9.4ε for plastic and compact sections respectively where $\varepsilon = \sqrt{250/f_y}$

7. The cross section of the plate girder is finalized using the various empirical formulas and other considerations.

8. The plastic moment capacity[9] of the plate girder section is checked by using the relation:
   $M_d = [Z_p f_y/\gamma_{mo}]$ where $\gamma_{mo} =$ partial safety factor against yield stress and buckling $= 1.1$ and $Z_p =$ plastic section modulus.

9. Nominal plastic Shear resistance of the section is checked by the relation,

$$V_p = \left(\frac{A_w f_{yw}}{\sqrt{3}} \right) \text{ where } A_w = \text{Area of the web} = (d\, t_w)$$

10. Design of welded connections between flange and web
11. Design of Intermediate stiffeners when thin webs are used
12. Design of end bearing stiffeners
13. Design of Lateral Bracing
14. Design of Cross frames

The use of these design principles is illustrated by the following design example

9.4 DESIGN EXAMPLE

The cross-section of a stream is as shown in the Fig. 9.2.

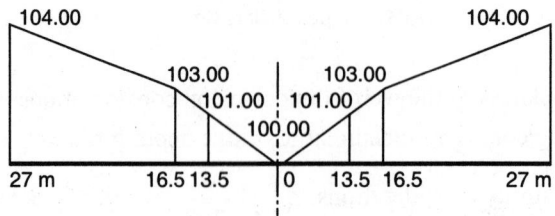

Fig. 9.2 Cross Section of Stream.

1. Data

Effective span of the plate girder = L = 30 m

Broad gauge rail track (1676 mm), main Line, Single track

Grade of steel = f_y = 250 N/mm²

Each rail supported by a plate girder with cross bracings at intervals of 6 m

2. Dead Loads

Dead load of track[10] (open floor) = 7.5 kN/m

Self weight of plate girder = (0.2L + 1) kN/m = [(0.2 × 30) + 1] = 7 kN/m

Total dead load = g = 14.5 kN/m

3. Live Loads

Referring to Table 2.6 (Chapter 2)

Equivalent Total live load for B.M per track (30 m span B.G) = 2727 kN

Since two plate girders share this load,

Total live load per girder = q = (2727/2) = 1363.5 kN

Equivalent Total live load for S.F per track (30 m span B.G) = 2927 kN
Total live load per girder = q = (2927/2) = 1463.5 kN

4. Impact Factors

The impact factor for steel bridges (From Table 2.6) is expressed as the Coefficient of dynamic augment (CDA) = 0.372

5. Design Bending Moments

B.M due to dead load = [(14.5 × 30²)/8] = 1632 kN·m
B.M due to live load = [1363.5 × 30)/8] = 5113 kN·m
B.M due to impact of live loads = (0.372 × 5113) = 1902 kN·m
Total design B.M = M = 8647 kN·m

6. Design Shear Forces

Shear force due to dead load = [14.5 × 30)/2] = 217.5 kN
Shear force due to live load = (1463.5/2) = 731.75 kN
Shear force due to impact on live load = (0.372 × 731.75) = 272.21 kN
Total design shear force = V = 1222 kN

7. Web Plate Dimensions

Economical depth of plate girder $= d = \left(\dfrac{Mk}{f_y}\right)^{0.33}$

where $k = (d/t_w) = 180$ for thin webs and $f_y = 250$ N/mm²

substituting the values, $d = \left(\dfrac{8647 \times 10^6 \times 180}{250}\right)^{0.33} = 1800$ mm

Web depth based on shear considerations, assuming 12 mm thick plate and spacing of stiffeners as equal to the depth of web plate [Refer Table 9.3 of IRC: 24 and interpolate τ for the ratio of $(d/t_w) = 133$] is computed as,

$$d = \left[\frac{V}{(\tau \times 12)}\right] = \left[\frac{1222 \times 10^3}{(95 \times 12)}\right] = 1072 \text{ mm}$$

Adopt a web plate of size 1600 mm depth and thickness of 12 mm

8. Flange Plates

Approximate area of flange plates is evaluated as,

$$A_f = \left[\left(\frac{M}{\sigma_b.d}\right)\right] \text{ where } \sigma_b = \text{permissible bending stress in girders as per IRC: 24-2011}$$

And

$$\sigma_b = 0.62 f_y = (0.62 \times 250) = 155 \text{ N/mm}^2$$

$$A_f = \left[\left(\frac{8647 \times 10^6}{155 \times 1600}\right)\right] = 34866 \text{ mm}^2$$

Flange width = $B = (L/40)$ to $(L/45) = (30,000/400$ to $(30,000/45) = 750$ to 670 mm

Adopt flange width = 700 mm

Thickness of plate required = $(A_f/B) = [34866/700] = 49.8$ mm

Adopt flange plates of size 700 mm by 50 mm

The section selected is shown in Fig. 9.3

9. Check for Maximum Flexural Stress

The second moment of area of the cross section of the girder about axis XX and YY are computed to determine the permissible stresses.

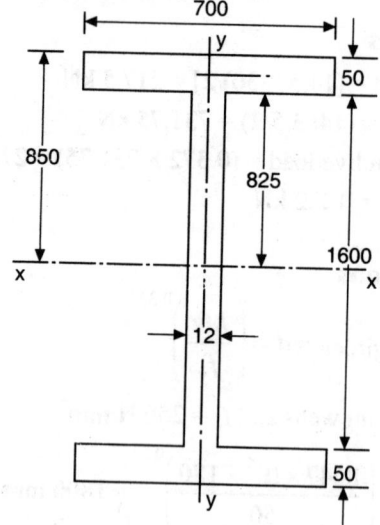

Fig. 9.3 Cross-section of Plate Girder.

$$I_{xx} = \left[\left(t_w \cdot \frac{D^3}{12}\right) + (2 \times A_f \cdot r^2)\right]$$

$$= \left[\left(12 \times \frac{1600^3}{12}\right) + (2 \times 700 \times 50 \times 825^2)\right]$$

$$= (5181 \times 10^7) \text{ mm}^4$$

$$I_{yy} = \left[\frac{(2 \times 50 \times 700^3)}{12} + (1600 \times 12^3)/12\right]$$

$$= (285 \times 10^7) \text{ mm}^4$$

$$A = [(2 \times 50 \times 700) + (1600 \times 12)] = 89200 \text{ mm}^2$$

Minimum radius of gyration $= r_y = \sqrt{(I_{yy} / A)} = \sqrt{\left[\dfrac{258 \times 10^7}{89200} \right]} = 178$

According to IRC: 24-2011, the permissible bending stress (σ_{bc}) is interpolated from Table 9.1 (Table 8.2 of IRC: 24-2001) corresponding to the values of f_{cb} and f_y, and the equation,

$$f_{cb} = k_1 \, (X + k_2 \, Y)(c_2/c_1) \text{ MPa}$$

where $\qquad X = Y \sqrt{L + \left(\dfrac{1}{20} \right) \left[(LT) / (r_y . D) \right]^2} \text{ MPa}$

and $\qquad Y = [(2650000)/(L/r_y)^2]$

$\qquad\qquad f_{cb}$ = elastic critical stress

$\qquad c_1$ and c_2 are lesser and greater distances from neutral axis to extreme fibers

$\qquad\qquad D$ = overall depth of girder

$\qquad\qquad T$ = mean thickness of the compression flange

$\qquad\qquad L$ = effective length of compression flange (Between cross bracings at 6 m intervals)

$\qquad\qquad r_y$ = radius of gyration

$\qquad\qquad k_1$ = coefficient to allow for reduction in thickness or breadth of flanges between points of effective lateral restraint

$\qquad\qquad k_2$ = coefficient to allow for unequal flanges

In the present case, c_1, c_2, k_1 and k_2 have values of 1.0

The simplified equation for the value of elastic critical stress is given by

$$f_{cb} = (X + Y) \text{ MPa}$$

However, the code provides for a simpler method for evaluating the value of X and Y which involves the computation two parameters such as $(D/T) = (1700/50) = 34$ and $(L/r_y) = (6000/178) = 33.7$

Using these values, X and Y can be interpolated from Table 9.2 (Table 8.5 of IRC: 24-2001 code) for known values of the ratio (D/T) and (L/r_y).

Using this method, the value of $X = 1712$ and $Y = 1656$ and

$$f_{cb} = (1712 + 1656) = 3368 \text{ N/mm}^2$$

Using the values of f_{cb} and $f_y = 250 \text{ N/mm}^2$, the permissible compressive stress σ_{bc} can be interpolated from Table 9.1 as $\sigma_{bc} = 161.6 \text{ N/mm}^2$.

Actual maximum bending stress is given by

$$\sigma_{bc} = \sigma_{bt} = \left(\dfrac{M_y}{I} \right) = \left(\dfrac{8647 \times 10^6 \times 850}{5181 \times 10^7} \right) = 141.8 \text{ N/mm}^2 < 161.6 \text{ N/mm}^2 \text{ (Hence safe)}$$

Table 9.1 Permissible Bending Stress (σbc) in Plate Girders
(Table 8.2 of IRC: 24-2001)
(All values in N/mm²)

$f_y \rightarrow$ $f_{cb} \downarrow$	250	340	400
20	13	13	13
30	19	19	19
40	25	26	26
50	31	31	32
60	36	37	38
70	41	43	44
80	46	48	49
90	51	54	55
100	55	59	60
110	60	64	65
120	64	68	70
130	67	73	75
140	71	77	80
150	74	81	84
160	78	85	89
170	81	89	93
180	84	93	97
190	87	97	102
200	89	100	105
210	92	103	109
220	94	106	112
230	96	110	116
240	99	113	119
250	101	115	122
260	103	118	126
270	104	121	129
280	106	123	132
290	108	126	135
300	110	128	137
310	111	130	140
320	113	133	143
330	114	135	145
340	115	137	148
350	117	139	150
360	118	141	152
370	119	143	155
380	120	144	157
390	121	146	159
400	122	148	161
420	124	151	165
440	126	154	169

Contd...

Table 9.1 Contd.

$f_y \rightarrow$ $f_{cb} \downarrow$	250	340	400
460	128	157	172
480	130	159	175
500	131	162	178
520	133	164	181
540	134	166	184
560	135	168	187
580	136	170	189
600	137	172	192
620	138	174	194
640	139	175	196
660	140	177	198
680	141	178	200
700	142	180	202
720	143	181	204
740	143	182	205
760	144	184	207
780	145	185	208
800	145	186	210
850	147	188	213
900	148	191	216
950	149	193	219
1000	150	195	222
1050	151	196	224
1100	152	197	226
1150	152	199	228
1200	153	200	230
1300	154	203	233
1400	155	205	236
1500	156	206	237
1600	157	208	240
1700	157	209	242
1800	158	210	243
1900	158	211	245
2000	159	212	246
2200	160	213	248
2400	160	215	250
2600	161	2126	251
2800	161	216	252
3000	161	217	253
3500	162	218	255
4000	163	219	257
4500	163	220	258
5000	163	221	259
5500	163	221	259
6000	164	222	260

Table 9.2 Values of X and Y for calculating fcb
(Table 8.5 of IRC: 24-2001)

D/T → L/r_y →	X															Y
	8	10	12	14	16	18	20	25	30	35	40	50	60	80	100	
40	2484	2222	2066	1965	1897	1849	1814	1759	1728	1709	1697	1683	1675	1667	1663	1656
45	2103	1856	1709	1612	1546	1409	1465	1411	1380	1362	1349	1335	1327	1319	1216	1309
50	1822	1590	1449	1357	1293	1248	1214	1161	1131	1113	1101	1086	1078	1070	1067	1060
55	1607	1389	1234	1166	1105	1060	1028	976	947	929	917	902	894	886	883	876
60	1437	1232	1104	1020	961	918	886	835	806	788	776	762	754	746	743	736
65	1301	1107	985	904	847	806	775	726	697	679	657	653	645	637	634	627
70	1188	1005	889	811	757	717	687	637	610	592	581	567	559	551	547	541
75	1094	920	810	735	682	644	615	567	540	522	511	467	489	481	478	471
80	1014	849	743	672	621	584	556	509	482	465	454	440	432	424	421	414
85	945	788	687	618	570	533	506	461	434	417	406	392	385	377	373	349
90	886	735	639	573	526	491	464	420	394	377	366	353	345	337	373	327
95	833	689	597	534	488	454	428	385	360	343	332	319	311	304	300	294
100	787	649	560	499	455	423	398	346	331	314	304	290	283	275	272	265
110	708	582	499	443	402	371	347	307	283	268	257	244	237	229	226	219
120	644	527	451	398	359	330	308	270	247	232	222	209	202	194	191	184
130	591	482	414	361	325	298	277	240	218	204	194	181	174	167	163	157
140	546	444	379	331	267	271	251	217	195	181	172	160	153	145	142	135
150	508	412	350	306	274	249	230	197	17	163	154	142	135	145	124	118
160	474	385	326	284	254	230	212	181	161	148	139	137	121	113	110	104
170	445	360	306	265	236	214	197	167	148	135	126	115	109	102	95	92
180	420	339	286	249	221	200	184	155	137	125	116	105	98	92	88	82
190	397	320	270	235	208	188	172	145	127	115	107	96	90	83	80	73
200	476	304	256	222	197	177	162	136	119	107	99	89	83	76	78	66

Contd...

Table 9.2 Contd.

$D/T \rightarrow$ $L/r_y \downarrow$	8	10	12	14	16	18	20	25	30	35	40	50	60	80	100	Y
							X									
210	358	238	243	210	186	168	153	128	112	101	93	82	76	70	66	60
220	341	275	231	200	177	159	145	121	105	94	87	77	71	64	61	55
230	326	262	220	191	169	152	138	115	99	89	82	72	66	60	56	50
240	312	251	211	183	161	145	132	109	94	84	77	67	62	55	52	46
250	229	241	202	175	151	139	126	104	90	80	73	64	58	53	49	42
260	288	231	194	167	148	133	121	99	85	76	68	60	5	48	45	39
270	277	222	186	161	142	127	115	95	82	72	66	57	52	46	42	36
280	267	214	180	155	137	122	111	91	78	69	63	54	49	43	40	34
290	257	207	173	149	132	118	107	88	75	66	60	52	46	41	38	32
300	249	200	167	144	127	114	103	84	72	64	57	49	44	38	35	30

10. Check for Maximum Shear Stress

Shear stress depends upon the ratio of $(d/t) = (1600/12) = 133$

Using stiffener spacing $= c = 0.33$ to $1.5\ d$

Adopting, $c = 1500$ mm, $(d/t) = (1500/12) = 125$ and spacing of stiffeners at $0.9\ d$

Referring to Table 9.3 (Table 8.6 of IRC: 24-2001), interpolate the average permissible shear stress as $\tau_{va} = 98$ N/mm^2.

Actual average shear stress $= \left(\dfrac{V}{d.t_w}\right) = \left(\dfrac{1222 \times 10^3}{1600 \times 12}\right) = 63.6$ N/mm$^2 < 98$ N/mm^2
(Hence safe)

11. Connections Between Flanges and Web

Maximum shear force at the junction of web and flange is given by

$$\tau = \left(\frac{VAy}{I}\right)$$

Where
$\qquad V = 1222$ kN
$\qquad A = (700 \times 50) = 35000$ mm^2
$\qquad y = 825$ mm
$\qquad I = 5181 \times 10^7$ mm^4

$$\tau = \left(\frac{VAy}{I}\right) = \left(\frac{1222 \times 10^3 \times 35000 \times 825}{5181 \times 10^7}\right) = 681 \text{ N/mm}$$

Refer Table 9.4 and read out the size of weld required as 6 mm with a strength of 793 N/mm. Provide 6 mm continuous fillet welds on either side.

12. Intermediate Stiffeners

Since the ratio of $(d/t) = (1600/12) = 133 > 85$, vertical stiffeners are required

Adopting the spacing of stiffeners $= c = 1500$ mm

Greater unsupported panel dimension of the web $= 1500$ mm $< 270\ t_w < (270 \times 12) = 3240$ mm

The Intermediate stiffeners are designed to have a minimum moment of inertia specified as

$$I = \left[\frac{1.5d^3t^3}{c^2}\right] = \left[\frac{1.5 \times 1600^3 \times 12^3}{1500^2}\right] = (471 \times 10^4) \text{ mm}^4$$

Using 10 mm thick plate, outstand of stiffener should be not greater than
$\qquad 12\ t = (12 \times 10) = 120$ mm

Adopt a plate 10 mm by 120 mm, having,
$\qquad I = (10 \times 120^3)/3) = (576 \times 10^4) > (471 \times 10^4)$ mm^4

Table 9.3 Permissible Average Shear Stress (τ_v)
(Table 8.6 of IRC: 24-2001)

Stress τ_{vd} (Mpa) for different distances between stiffeners

	0.1d	0.3d	0.4d	0.5d	0.6d	0.7d	0.8d	0.9d	1.0d	1.1d	1.2d	1.3d	1.4d	1.5d
90	100	100	100	100	100	100	100	100	100	100	100	100	100	100
95	100	100	100	100	100	100	100	100	100	100	100	100	100	99
100	100	100	100	100	100	100	100	100	100	100	100	99	99	98
105	100	100	100	100	100	100	100	100	100	100	99	98	97	96
110	100	100	100	100	100	100	100	100	100	99	98	96	95	94
115	100	100	100	100	100	100	100	100	100	98	96	95	94	93
120	100	100	100	100	100	100	100	100	98	96	95	93	92	91
125	100	100	100	100	100	100	100	98	97	95	93	92	91	90
130	100	100	100	100	100	100	99	97	96	94	92	90	89	88
135	100	100	100	100	100	100	98	96	94	92	90	89	87	86
140	100	100	100	100	100	99	96	95	93	91	89	87	86	85
150	100	100	100	100	97	94	92	90	88	85	84	83	81	
160	100	100	100	98	94	92	89	88	85	83	81	80	78	
170	100	100	100	96	92	89	87	85	82	80	78	76	75	
180	100	100	98	94	90	87	84	82	80	77	75	73	72	
190	100	100	97	92	88	84	82							
200	100	100	95	90	86	82	81							
210	100	99	93	88	83	81								
220	100	98	91	86	81	80								
230	100	96	90	84	79			Non-applicable zone						
240	100	95	88	83	77									
250	100	93	86	82	74									
260	100	92	85	81										
270	100	90	84	81										

Table 9.4 Design Capacity of Fillet Welds

Leg length, s (mm)	Design capacity per unit length, R_{nw} (kN/mm)			
	Fe 410 with E43 and E51 electrodes		Fe 540, shop welded	
	Shop welded, $f_{wd} = 189$ MPa	Site welded, $f_{wd} = 158$ MPa	With E43, $f_{wd} = 189$ MPa	With E51, $f_{wd} = 235$ MPa
4	0.529	0.442	0.529	0.658
5	0.661	0.553	0.661	0.822
6	0.793	0.663	0.793	0.987
8	1.058	0.884	1.058	1.316
10	1.323	1.106	1.323	1.645
12	1.587	1.327	1.587	1.974
15	1.984	1.659	1.984	2.467
18	2.381	1.990	2.381	2.961
20	2.646	2.212	2.646	3.290
22	2.910	2.433	2.910	3.619
25	3.307	2.765	3.307	4.112

13. Connections of Vertical Stiffener to Web

Shear on welds connecting stiffener to web = $(126 \, t^2/h)$

Where t = web thickness (mm)

h = outstand of stiffener (mm)

Shear on welds = $[(126 \times 12^2)/120] = 151$ N/mm

Size of weld = $s = [151/(0.7 \times 158)] = 1.36$

Effective length of weld should be not less than $10t = (10 \times 12) = 120$ mm

Provide 160 mm long, 5 mm fillet welds alternately on either side.

14. End Bearing Stiffener

The end bearing stiffener is designed as a column having the ratio (h/t) not greater than 12.

Where, h = outstand and t = thickness

If $h = 300$ mm, $t = (300/12) = 25$ mm

Use 300 mm by 25 mm plate for the end bearing stiffener

Permissible bearing stress = $0.8 \, f_y = (0.8 \times 250) = 200$ N/mm^2

Bearing area required = $[(1222 \times 10^3)/200] = 6110$ mm^2

If two plates are used, total are provided = $(300 \times 25 \times 2) = 15000$ mm$^2 > 6110$ mm^2

The length of web plate which acts along with stiffener plates in bearing reaction is specified as $20t = (20 \times 12) = 240$ mm

Referring to Fig. 9.4

$$I = \left[\frac{25 \times 612^3}{12} + (2 \times 240 \times 12^3)12 \right] = (479 \times 10^6) \, \text{mm}^4$$

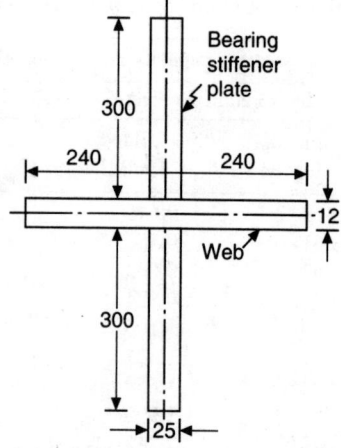

Fig. 9.4 End Bearing Stiffener.

Area of section $= A = [(612 \times 25) + (480 \times 12)] = 20880$ mm^2

Slenderness ratio $= \lambda = (L/r)$

Radius of gyration $= r = \sqrt{\dfrac{I}{A}} = \sqrt{\dfrac{479 \times 10^6}{20880}} = 151$ mm

Effective length of stiffener $= L = (0.7 \times 1600) = 1120$ mm

Hence, $\lambda = (1120/151) = 7.4$

Refer Table 9.5 (Table 11.1 of IRC: 24-2001) and read out the permissible compressive stress.

Corresponding to the slenderness ratio of 7.4 as $\sigma_{ac} = 150$ N/mm^2

Area required $= [(1222 \times 10^3)/150] = 8147$ mm$^2 < 20880$ mm^2

Table 9.5 Permissible Axial Compressive Stress (σ_{ac}) for Steels with Various Yield Stress
(Table 11.1 of IRC: 24-2001)

$\lambda = L/r \downarrow$	Yield stress (f_y) MPa		
	250	340	400
10	150	204	239
20	148	201	235
30	145	194	225
40	139	183	210
50	132	168	190
60	122	152	168
70	112	135	147
80	101	118	127
90	90	103	109
100	80	90	94
110	72	79	82
120	64	69	71
130	57	61	62
140	51	54	55
150	45	48	49
160	41	43	43
170	37	38	39
180	33	34	35
190	30	31	32
200	28	218	28
210	25	26	26
220	23	24	24
230	21	22	22
240	20	20	20
250	18	18	19

15. Connections Between Bearing Stiffener and Web

Length available for welding using alternate intermittent welds

$$= 2(1600 - 40) = 3120 \text{ mm}$$

Required strength of weld = $[(1222 \times 10^3)/3120] = 391.5$ N/mm

Size of weld = $[391.5/(0.7 \times 158)] = 3.5$ mm

Use 6 mm fillet welds of 160 mm length intermittently on both sides.

16. Lateral Bracing and Cross Frames

For resisting, wind, racking and centrifugal forces, lateral bracing is provided at intervals of 6 m along the span. End cross frames and intermediate cross frames are provided for spans greater than 20 m.

Wind load = 1.5 kN/m^2

Depth of girder = 1.7 m

Coefficient for wind load on leeward girder = 0.25

Wind load on windward girder = $(1.5 \times 1.7 \times 30) = 76.6$ kN

Wind load on leeward girder = $(0.25 \times 76.6) = 19.15$ kN

∴ Total wind load = $(76.6 + 19.15) = 95.75$ kN

Lateral load due to racking forces = 6 kN/m

Total racking force = $(6 \times 30) = 180$ kN

Total lateral load = $(180 + 95.75) = 276$ kN

Referring to Fig. 9.5

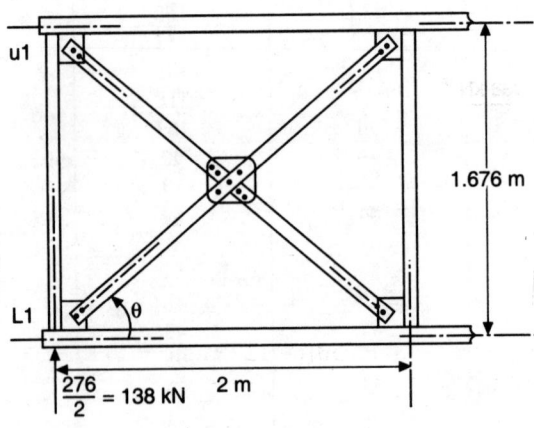

Fig. 9.5 Lateral Bracing.

Maximum tension in the diagonal = $(276/2)$ cosec θ

$$= \left[(276/2) \times \sqrt{(2^2 + 1.676^2)/1.676} \right] = 180 \text{ kN}$$

Area required = $[(180 \times 10^3)/(0.6 \times 250)] = 1194$ mm

Use ISA-8050 with 10 mm thickness

Area provided = 1202 mm^2

Maximum compressive force in $U_1L_1 = 138$ kN
Length of member = 1.676 m
Effective length = 0.65 L = (0.65 × 1.676) = 1.089 m
Try an angle ISA-10075 with 10 mm thickness
Area $\qquad A = 1650$ mm^2 $\quad r = 21.6$ mm
$\qquad\qquad \lambda = (L/r) = (1089/21.6) = 50.4$
From Table 9.5 (IRC:24)
Permissible stress $\sigma_{ac} = 132$ N/mm^2
Safe load on member = [(1650 × 132)/1000]
$\qquad\qquad\qquad = 217.8$ kN > 138 kN

16. Design of Cross Frames

Referring to Fig. 9.6

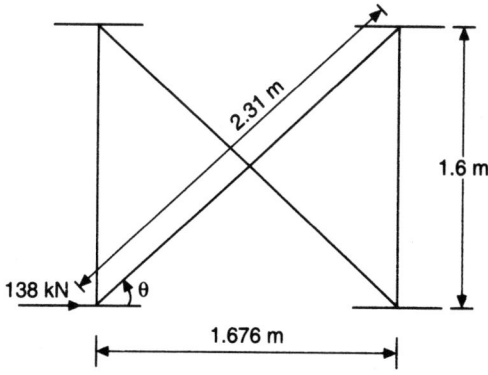

Fig. 9.6 End Cross Frame.

Lateral load to be resisted by one frame = 138 kN
Tension in diagonal = 138 sec θ = (138 × 2.31)/1.676 = 191 kN
Area required = [(191 × 10^3)/(0.6 × 250)] = 1275 mm^2
Use ISA-9060, 10 mm thick
Area provided = 1401 mm^2
The cross frames are provided at 6 m intervals.

The plan, elevation, cross-section and details of lateral Bracing and stiffeners are shown in Figs. 9.7, 9.8 and 9.9.

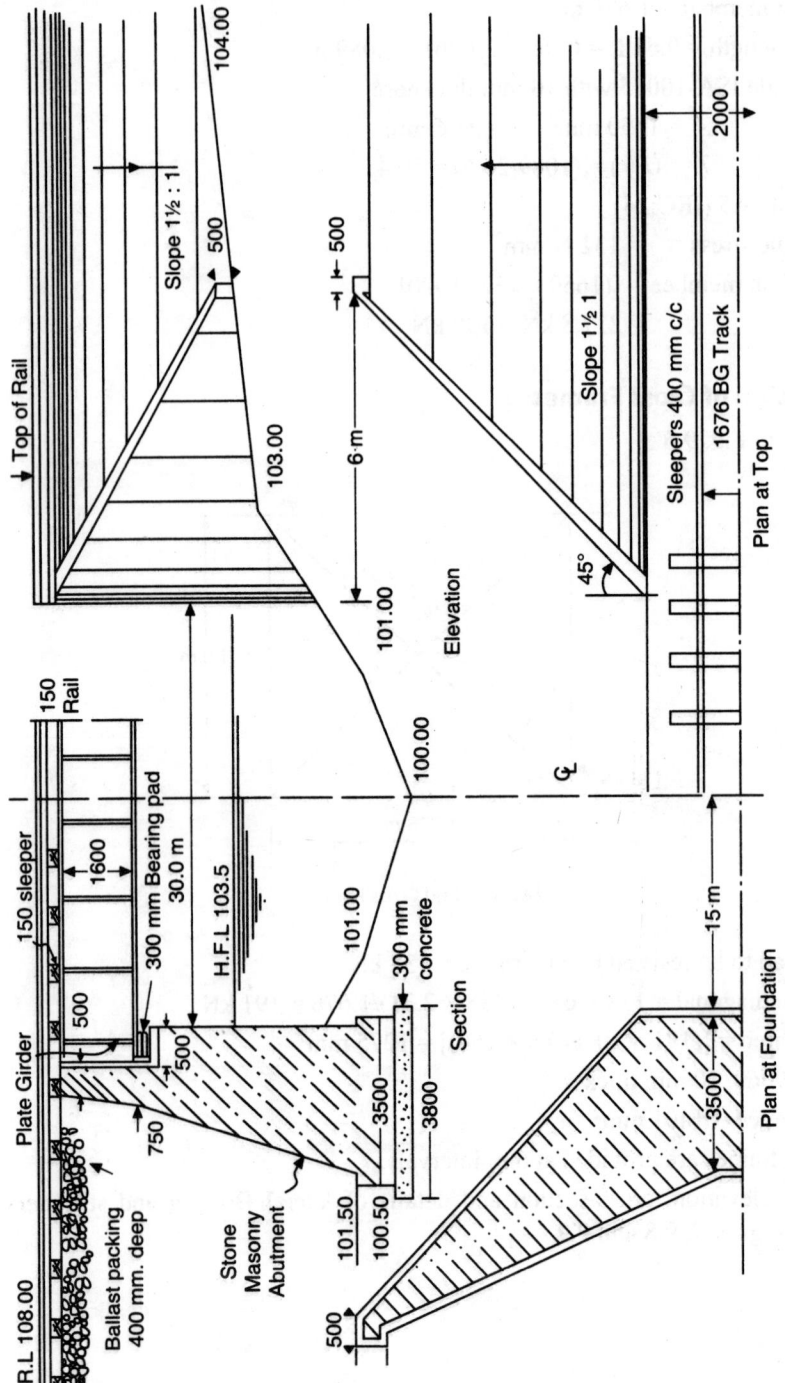

Fig. 9.7 Longitudinal Section and Elevation of Plate Girder Bridge.

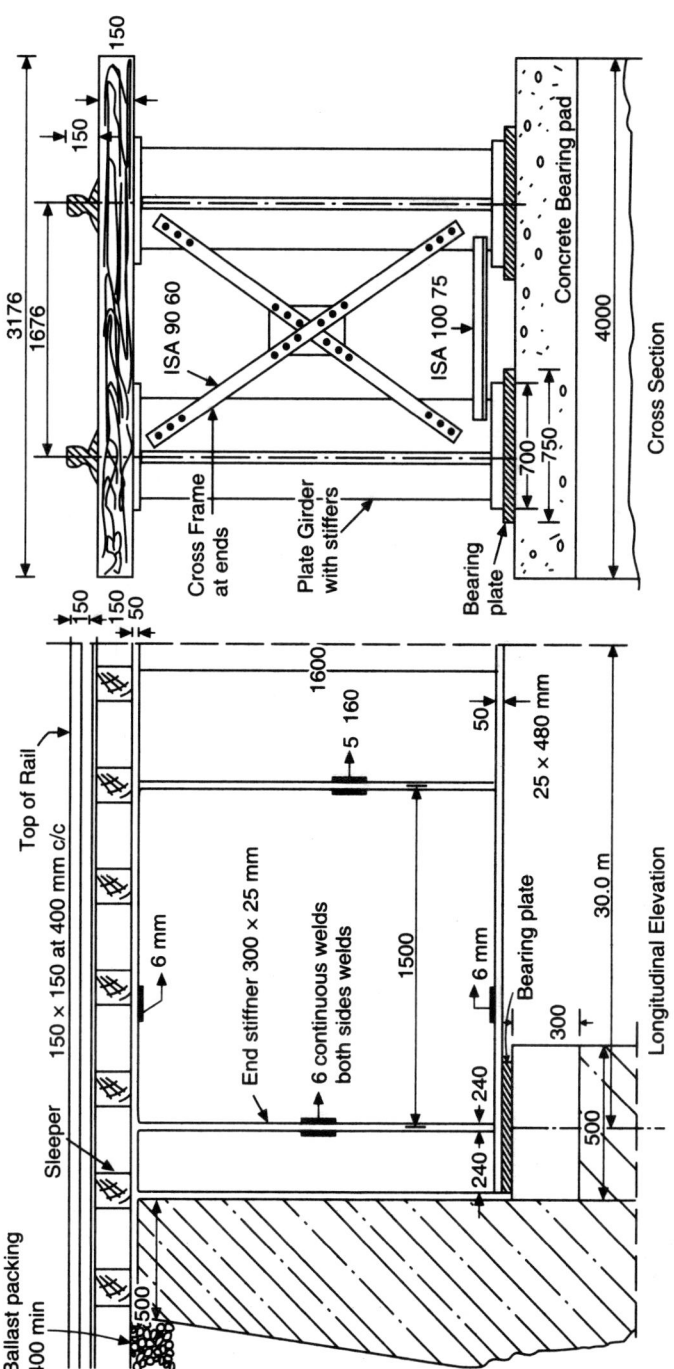

Fig. 9.8 Details of Plate Girder.

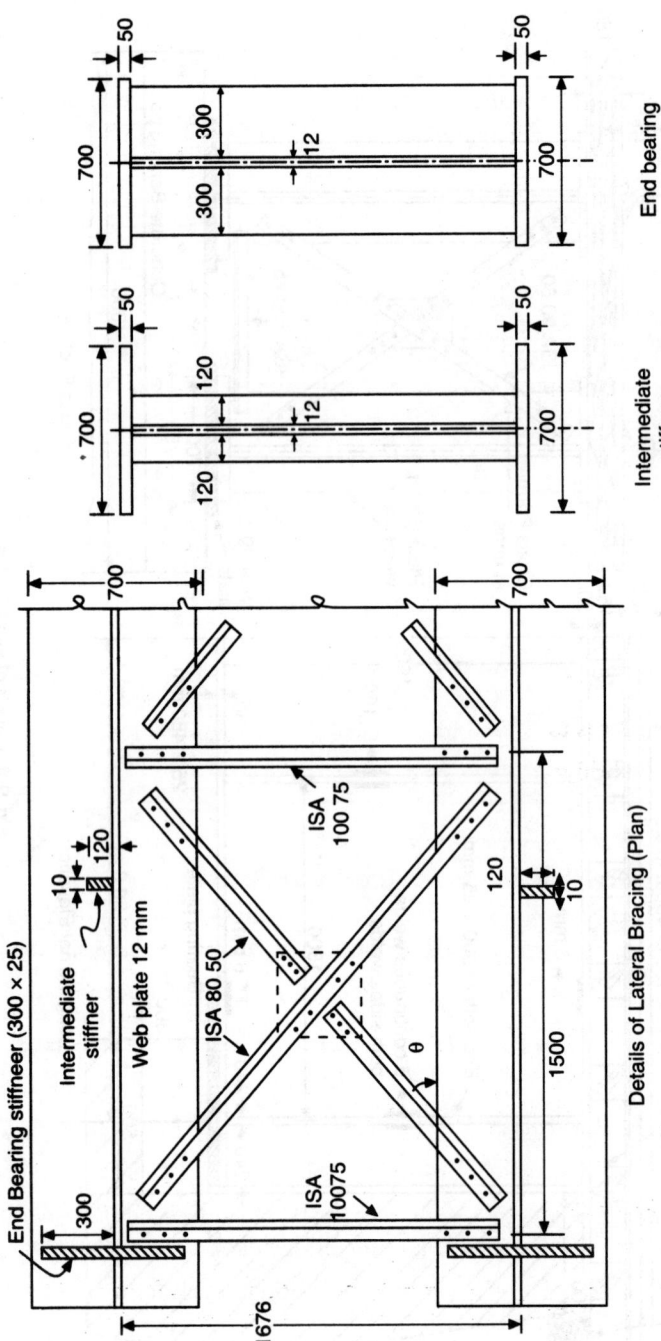

Fig. 9.9 Lateral Bracing.

EXAMPLES FOR PRACTICE

1. A deck type welded steel plate girder bridge of single span 25 m is required for a B.C. main line, singly track (gauge length 1676 pm) crossing over, a stream. The following data is available. Dead load/track (open floor)

   = 7.5 kN·m

   E.U.L.L. for B.M. calculations/track = 2356 kN

   E.U.L.L. for shear calculations/track = 2586 kN

   Side slopes of embankment = 1.5:1

   Average bed level = 100.00

   High flood level = 103.00

   Top of rail level = 107.00

   Hard rock available at 98.50, Width of embankment = 4 m.

   (a) Design the main plate girder with intermediate and bearing stiffeners and lateral bracings and cross frame at ends.

   (b) Draw the following views:

   1. Half longitudinal section and half longitudinal elevation
   2. Half plan at foundations and half plan at top
   3. Cross-sections showing the plate girder and end cross frame
   4. Longitudinal elevation of plate girder near support showing the bearing and intermediate stiffeners and weld details.

   (Bangalore University 1985)

2. A plate girder is to be designed for a B.G. track to suit the following data:

   Span of the bridge = 20 m

   Dead load of track = 7.5 kN/m

   E.U.L.L. for B.M. calculations/track = 1964 kN

   E.U.L.L. for shear calculations/track = 2168 kN

   Design the plate girder and sketch the details of the longitudinal and cross-sections showing the details of cross bracing and welded connections.

3. A plate girder bridge deck is to be designed for a B.G. track to suit the following data:

   Effective span of the girder = 15 m

   Dead load of sleepers, rails and fittings = 10 kN/m

   E.U.L.L. for B.M. calculations/track = 1631 kN

   E.U.L.L. for S.F. calculations/track = 1801 kN

   Design the plate girder to conform to the IRS loadings and IRC specifications and sketch the typical cross section of the bridge deck.

4. Design a welded deck type plate girder bridge deck for a B.G track to suit the Following data:

   Effective span = 40 m

   Dead load of track = 10 kN/m

EULL for B.M calculations/track = 3498 kN
EULL for Shear Force calculations/track = 3815 kN

Design the plate girder and sketch the details of cross bracing and welded connections. The design should conform to the relevant IRC and IRS code specifications.

5. Design a welded plate girder for a simply supported bridge deck beam with an effective span of 20 m to support the following loads:

Dead load including self weight = 20 kN/m
Imposed load = 10 kN/m
Two moving loads = 150 kN each spaced 2 m apart.

Assume the top compression flange of the plate girder is restrained laterally and prevented from rotating. Use mild steel with f_y = 250 N/mm^2. Design as an unstiffened plate girder with thick webs.

6. The preliminary dimensions selected for a bridge girder of span 20 m are as follows:

Overall depth of the girder = 1270 mm
Depth of web plate = 1200 mm
Thickness of web plate = 12 mm
Thickness of flange plates = 35 mm
Width of flange plates = 450 mm
Spacing of stiffeners = depth of web plate

The plate girder has to resist a design moment of 4275 kN·m and a shear force of 880 kN Check the girder for its moment and shear resistance according to the IRC codes.

REFERENCES

1. Shirley-Smith, H., The World's Great Bridges, English language Book Company, London, 1977, 250 pp.
2. Steel Box girder bridges, Institution of Civil Engineers, London, 1973, 315 pp.
3. Subramanian, N., Design of Steel Structures, Oxford University Press, New Delhi, 2008, p. 932.
4. IRC: 24-2001, Standard Specifications and Code of Practice for Road bridges, Section-V, Steel Road Bridges, Indian Roads Congress, New Delhi, 2001, pp. 157.
5. Indian Railway Standard Code of Practice for the Design of Steel and Wrought Iron Bridges carrying, Rail, Road or Pedestrian traffic, Ministry of railways, Govt. of India, New Delhi, 1962, pp. 87.
6. IS: 800-1987, Indian Standard Code of Practice for Steel Construction, Bureau of Indian Standards, New Delhi, 1987.

7. Martin, L.H and J.A. Purkis., Structural design of Steel Work to BS: 5950, Edward Arnold Publishers, London, 1992, 468 pp.
8. Salmon. C.G and J.E. Johnson., Steel Structures – Design and Behavior, 4th Edition, Harper Collins College Publishers, New York, 1996, 1024 pp.
9. Kulak, G.L and G.Y. Grondin., Limit State Design in Structural Steel, 7th Edition, Canadian Institute of Steel Construction, Toronto, Ontario, Canada, 2002.
10. Johnson Victor, D., Essentials of Bridge Engineering (Fifth Edition), Oxford & IBH Publishing Co. Ltd, 2001, p. 231.

REVIEW QUESTIONS

1. Briefly explain the necessity of using plate girders for long span bridge decks. What are its advantages?
2. List the various structural components of a steel plate girder mentioning their functions in resisting the external loads.
4. What is the advantage of using unstiffened webs in plate girders? How do you design an unstiffened plate girder?
5. Explain the terms, a) Economical depth b) Plastic moment resistance c) Plastic Shear capacity with respect to a plate girder.
6. List the various types of stiffeners used in plate girders mentioning their structural functions.
7. Distinguish between intermediate stiffener and end bearing stiffener in a plate girder and bring out the salient differences between them.
8. How do you design a plate girder bridge deck for lateral stability against wind loads?
9. Explain the method of designing the welded connections between the web and stiffeners and flanges and web of a plate girder.
10. Explain the functions of lateral bracings and cross frames used in plate girders. How do you design them in a typical plate girder?

OBJECTIVE TYPE QUESTIONS

1. Plate girders are ideally suited for the construction of bridge decks in the span range of
   a) 8 to 10 m
   b) 20 to 30 m
   c) 40 to 80 m
2. The construction time required for a highway or railway bridge deck is the least in the case of
   a) Reinforced concrete
   b) Steel girders and concrete deck
   c) Steel plate girders

3. In the case of bridge decks with unstiffened plate girders, the thickness of the web plate required will be
   a) Less
   b) Large
   c) Normal

4. The approximate depth of a steel plate girder selected for a railway bridge deck will be in the range of
   a) 1/5 to 1/10 span
   b) 1/10 to 1/15 span
   c) 1/15 to 1/20 span

5. For large depth plate girders, the minimum thickness required for the web plate should be in the range of
   a) 6 to 8 mm
   b) 10 to 15 mm
   c) 15 to 20 mm

6. According to the IRC: 24-2001, intermediate stiffeners are not required in a plate girder when the ratio of depth of web to its thickness is less than
   a) 85
   b) 120
   c) 360

7. The main function of flange plates in a steel plate girder is to resist
   a) Flexure
   b) Shear
   c) torsion

8. End bearing stiffeners are generally used in a plate girder bridge decks at
   a) The centre of span
   b) The supports
   c) Quarter span points

9. The main function of lateral bracings in a steel plate girder bridge deck is to resist
   a) The moments due to moving loads
   b) The impact forces of moving loads
   c) Lateral loads due to wind and racking forces

10. The welded connections to fasten the flange plates and web of a steel plate girder is designed to resist the
    a) The moment
    b) The shear forces
    c) The torsion

10

Composite Bridges

10.1 GENERAL ASPECTS

Composite bridge decks comprising of a reinforced concrete continuous deck slab cast on steel plate girders are generally adopted for bridges in the medium span rang of 10 to 20 m. This type of bridge deck is not only economical but also reduces the construction time considerably and facilitates early resumption of traffic on the high way. In a composite bridge deck comprising steel and reinforced concrete, the individual materials are utilized efficiently since concrete is strong in compression and steel in tension. The speedy erection of the prefabricated steel girders serves as supports for cast-*in-situ* concrete deck with minimum form work. Also the savings in the overall depth of the longitudinal beams leads to savings in lengths of bridge approaches in the case of embankments.

Victor[1] has reported that the flexural stiffness of a composite beam will be about 2 to 4 times that for a corresponding steel beam resulting in reduced deflections and vibrations under moving loads. Economy in the quantity of steel to the extent of 10 to 60 percent against pure steel beam construction has been reported by Yan[2]. Composite bridge decks can also be constructed using precast prestressed concrete girders[3] supporting a cast-*in-situ* reinforced concrete continuous deck slab.

10.2 SHEAR CONNECTORS

Shear connectors are the most important structural elements in a composite bridge deck, provided at the junction of the concrete slab and longitudinal steel girders. The main function of the shear connector is to prevent the separation between the steel girder and the *in-situ* concrete slab by transferring the horizontal shear force along the contact surface without slip. In the case of composite girder decks, the deflections are comparatively less than that of non composite girder decks due to the increased moment of inertia of the composite section.

The most common types of shear connectors used in composite bridge decks are

(a) the rigid connector[4] comprising short length bars, stiffened angles, tees or channels welded on the flange of the steel girders. In order to prevent the separation of the *in-situ* slab from the prefabricated steel girder in the direction perpendicular to the contact surface, mechanical devices such as U-type hoops are welded to the shear connectors to provide a rigid connection as shown in Fig. 10.1(a).

(b) the flexible connectors[5] consisting of studs or angles or channels or tees welded on the flange plates of prefabricated units as shown in Fig. 10.1(b).

(c) Anchorage type shear connectors shown in Fig. 10.1(c) are provided for composite sections comprising of precast prestressed concrete girders and cast *in-situ* reinforced concrete slab.

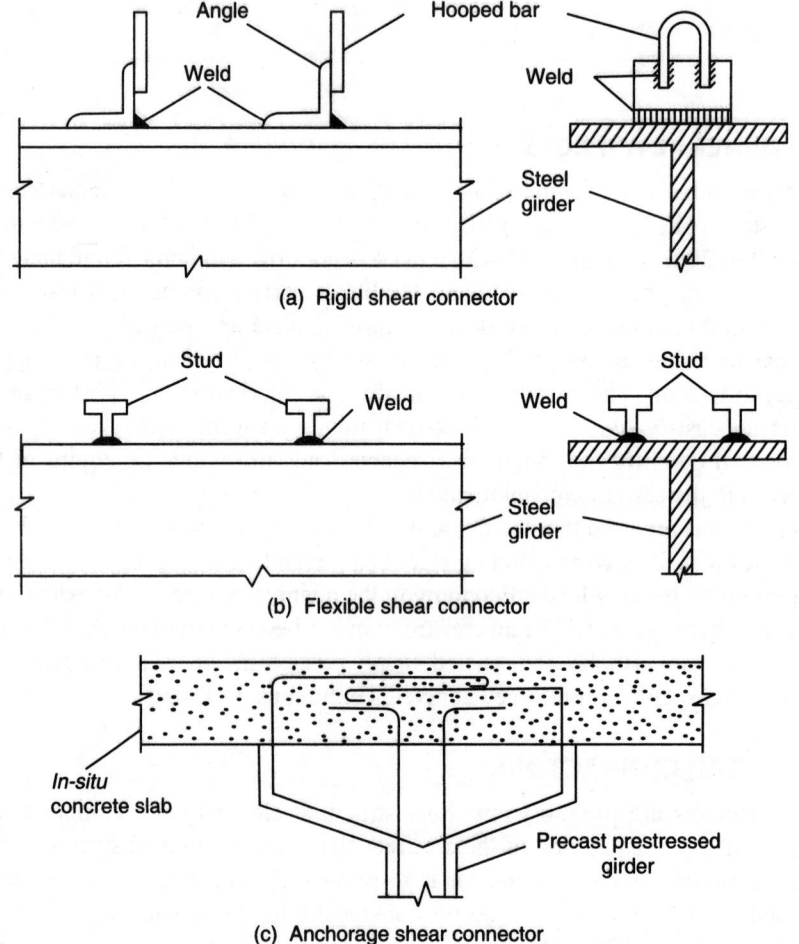

Fig. 10.1 Types of Shear Connectors.

The shear strength of shear connectors depend upon the type of steel, the cross sectional dimensions of the connector, compressive strength of concrete in the slab.

According to IRC: 22-2015[6], the safe shear resistance of high tensile steel connectors is computed by empirical relations specified in the code depending upon the type of connectors.

(i) For mild steel shear connectors, the safe shear resisted by each connector is computed by the following empirical relation:

For channel, Tee or angle shear connectors of mild steel:

$$(f_{st} = 420 \text{ to } 500 \text{ N/mm}^2 \text{ and } f_y = 230 \text{ N/mm}^2),$$

$$Q = 107\left(h_f + 0.5t\right)L\sqrt{f_{ck}}$$

(ii) For welded stud connectors of mild steel with f_u = 420 N/mm^2 and f_y = 350 N/mm^2 and having a ratio of (h/d) less than 4.2,

$$Q = 48\,hd\sqrt{f_{ck}}$$

For welded stud connectors of mild steel having a ratio of (h/d) equal to or greater than 4.2,

$$Q = 196\,d^2\sqrt{f_{ck}}$$

where Q = Safe shear resistance of one shear connector (N)

   f_{ck} = Characteristic compressive cube strength of concrete (N/mm^2)

   h_f = Maximum thickness of flange measured at the faces of the web (mm)

   L = Length of shear connector (mm)

   h = Height of stud connector (mm)

   d = Diameter of stud (mm)

   t = Thickness of web of shear connector (mm).

(iii) When anchorage type shear connectors are used to connect the concrete slab deck with precast prestressed concrete girders, the ultimate shear resistance of one connector is given by the empirical relation,

$$Q_u = A_s \cdot \sigma_u \cdot 10^{-3}$$

where Q_u = Ultimate shear resistance of each connector (kN)

   A_s = Cross sectional area of each connector (mm^2)

   σ_u = Ultimate tensile strength of steel of the anchorage connector (N/mm^2).

The ultimate bond stress at the interface should not exceed 2.1 N/mm^2 and the interface should be roughened for effective bonding.

The spacing of the shear connector is computed by the relation,

$$p = \left(\frac{\Sigma Q}{V_L}\right) \text{ or } \left(\frac{\Sigma Q_u}{V_{Lu}}\right)$$

where p = Spacing of the shear connector (mm)

Q = Safe shear resistance of one connector (kN)

Q_u = Ultimate shear resistance of one connector (kN)

V_L = Longitudinal working shear per unit length

V_{Lu} = Ultimate longitudinal shear per unit length.

The longitudinal shear (working or ultimate) is computed using the equations,

$$V_L = \left(\frac{VA_C\bar{Y}}{I}\right) \text{ and } V_{Lu} = \left(\frac{V_u A_C\bar{Y}}{I}\right)$$

where V = Vertical shear due to dead load placed after composite section is effective and working live load with impact

V_u = Vertical shear due to ultimate loads computed with load factors of 1.5 for dead load and 2.5 for live load

A_C = Transformed compressive area of concrete above the neutral axis of the composite section

$\bar{Y}$ = Distance from neutral axis to the centroid of the area A_C.

I = Second moment of area of the whole transformed composite section.

The design of a composite bridge deck using steel plate girders with cast *in-situ* reinforced concrete slab using stud connectors is presented in the following example.

10.3 DESIGN EXAMPLE

Design a composite bridge deck with reinforced concrete slab and steel plate girders to cover a span of 18 m.

Clear width of road way = 7.5 m

Footpath: 1 m on either side

Spacing of main girders = 2 m

Materials: concrete M-20 grade and Fe-415 grade tor steel, rolled steel sections with yield stress of 236 N/mm².

Design the reinforced concrete deck slab and the steel plate girders with shear connectors.

Draw the following views to a suitable scale:

(a) The cross-section of the deck slab continuous over steel girders and cross-section of the steel girders.

(b) Longitudinal elevation of the steel girders showing the details of the shear connectors.

1.   Cross-section the Deck

The cross-sectional details of the deck slab assumed are as shown in Fig. 10.2

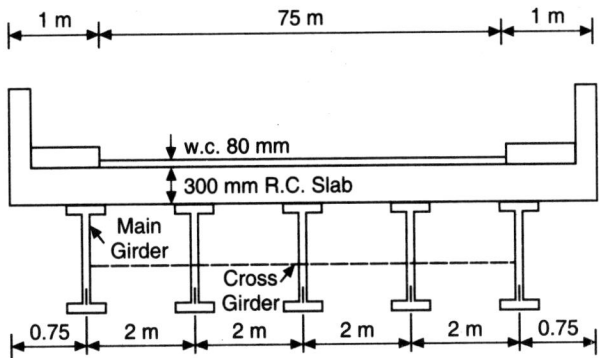

Fig. 10.2 Cross-section of Deck Slab.

2. Design of Deck Slab

Panel dimensions = 2 m by 4.5 m

Dead weight of slab = (0.3×24) = 7.20 kN/m^2

Dead weight of W.C. = (0.08×22) = 1.76

Total load ....= 9.00 kN/m^2

3. Live Load B.M.

Live load is I.R.C. Class AA tracked vehicle

Referring to Fig. 10.3

$$u = (0.85 + 2 \times 0.08) = 1.01 \text{ m}$$
$$v = (3.6 + 2 \times 0.08) = 3.76 \text{ m}$$
$$(u/B) = (1.01/2.0) = 0.50$$
$$(v/L) = (3.76/4.50) = 0.83$$
$$K = (B/L) = (2.0/4.5) = 0.45$$

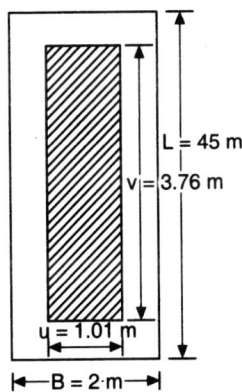

Fig. 10.3 I.R.C. Class A Wheel Load.

From Pigeaud's curves (Refer Fig. 4.7)

for $K = 0.5$ read out the values of

$$m_1 = 0.085 \text{ and } m_2 = 0.017$$

Short span moment $M_B = W(m_1 + 0.15 \, m_2)$

$$= 350 \,(0.085 + 0.15 \times 0.017) = 31 \text{ kN·m}$$

B.M. including impact and continuity factor

$$= (1.25 \times 0.8 \times 31) = 31 \text{ kN·m}$$

Long span moment $M_L = W(m_2 + 0.015 \, m_1)$

$$= 350 \,(0.017 + 0.015 \times 0.085) = 10.5 \text{ kN·m}$$

B.M. including impact and continuity factor

$$= (1.25 \times 0.8 \times 10.5) = 10.5 \text{ kN·m}$$

4. Dead Load B.M.

Dead load of deck slab $= 9 \text{ kN/m}^2$

Total dead load/panel $= (9 \times 2 \times 4.5) = 81 \text{ kN}$

$$(u/B) = 1 \text{ and } (v/L) = 1$$

$$K = (B/L) = (2/4.5) = 0.445 \,(1/K) = 2.25$$

Using Pigeaud's curves (Refer Fig. 3.9), read out

$$m_1 = 0.047 \text{ and } m_2 = 0.006$$

$$M_B = 81 \,(0.047 + 0.15 \times 0.006) = 3.94 \text{ kN·m}$$

Taking continuity into account

$$M_B = (0.08 \times 3.94) = 3.15 \text{ kN·m}$$

$$M_L = 81 \,(0.006 + 0.15 \times 0.047) = 1.05 \text{ kN·m}$$

Taking continuity into account

$$M_L = (0.8 \times 1.05) = 0.84 \text{ kN·m}$$

5. Design Moments

The revised IRC: 22-2015[6] code introduces the principles of limit state design for Composite concrete road bridges which requires the structure to satisfy both the limit states of serviceability and strength. Accordingly the design ultimate moments due to dead and live loads are obtained by applying the partial safety factors to the service load moments according to the IRC: 6-2014[7] code.

The ultimate design moments in the short and long span directions are given by

Short span moment $= M_{BU} = [(1.35 \times 3.15) + (1.5 \times 31)] = 51 \text{ kN·m}$

Long span moment $= M_{LU} = [(1.35 \times 0.84) + (1.5 \times 10.5)] = 17 \text{ kN·m}$

6. Design of Section

For M-20 grade concrete and Fe-415 HYSD bars, the effective depth of slab is given by

$$d = \sqrt{\frac{M_u}{0.138 f_{ck}.b}} = \sqrt{\frac{51 \times 10^6}{0.138 \times 20 \times 1000}} = 136 \text{ mm}$$

Overall depth = 300 mm and effective depth = (300 – 50) = 250 mm

Compute the parameter ratio $(M_u/b.d^2) = [(51 \times 10^6)/(1000 \times 250^2)] = 0.816$ N/mm²

Refer SP: 16 Design Table 2 and read out the reinforcement percentage corresponding to

$$f_{ck} = 20 \text{ N/mm}^2 \text{ and } f_y = 415 \text{ N/mm}^2 \text{ as}$$
$$p_t = (100 A_{st}/b.d) = 0.239$$
$$\therefore \quad A_{st} = [(0.239 \times 1000 \times 250)/100] = 598 \text{ mm}^2/\text{m}$$

Provide 12 mm diameter bars at a spacing of 150 mm (A_{st} provided = 754 mm²)

For long span, provide 10 mm diameter bars at 150 mm centers.

7. Design of Steel Plate Girder

The steel plate girder should be designed according to the specifications prescribed in the Indian Standard codes, IRC: 24-2001[8] and IS: 800-1987[9], dealing with steel construction in bridges and general types of structures respectively.

Spacings of main girders	= 2m
Spacings of cross girders	= 4.5 m
Dead load on girder = (9 × 2)	= 18 kN/m
Self-weight of main girder (0.2L + 1) kN/m	= 4 kN/m
Total load = W	= 22 kN/m

Self-weight of cross girders (assumed as 1 kN/m)

$$= (2 \times 1) = 2 \text{ kN}$$

(a) Dead Load Moments

Referring to Fig. 10.4

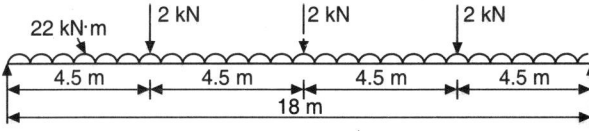

Fig. 10.4 Dead Loads on Plate Girder.

The maximum dead load moment is computed as

$$M_{max} = [(22 \times 18^2)/8 + (2 \times 18)/4 + (2 \times 4.5)] = 909 \text{ kN·m}$$

(b) Live Load Moments

Referring to Fig. 10.5

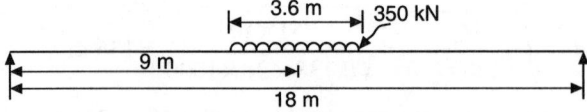

Fig. 10.5 Live Load on Plate Girder.

The maximum live load moment is computed as

$$M_{max} = [(350 \times 9)/2 - (350 \times 0.9)/2] = 1418 \text{ kN·m}$$

Impact factor = 10 percent

Live load B.M = (1418 × 1.1) = 1560 kN·m

Dead load B.M = 909

Design B.M = 2470 kN·m

(c) Shear Forces

Dead load shear = [(22 × 18)/2 + 2 + 2/2] = 201 kN

Live load shear with impact factor

$$= 1.1[(350 \times 16.2 \times 18] = 347 \text{ kN}$$

Total design shear = V = (201 + 347) = 548 kN

(d) Proportioning of Trial Section of Web Plate

Approximate depth of girder = 1/8 to 1/10 span = 18/10 = 1.8 m

Economical depth

$$= 5\sqrt[3]{M/\sigma_b} = 5\sqrt[3]{(2470 \times 10^6)/165} = 1230 \text{ mm}$$

$$\text{or}$$

$$= 5(M/\sigma_b)^{0.33} = 5\left[(2470 \times 10^6)/165\right]^{0.33} = 1230 \text{ mm}$$

Web depth based on shear considerations assuming 10 mm thick plate is,

$$d = V/(\tau \times 0) = [(548 \times 10^3)/(85 \times 10)] = 644.7 \text{ mm}$$

Try web 1000 mm by 10 mm.

(e) Flange Plates

Approximate flange area required

$$A_f = [(M/\sigma_o \cdot d) - (A_w/6)] = [(2470 \times 10^6)/(165 \times 1000) - (10 \times 1000)/6]$$
$$= 13303 \text{ mm}^2$$

Flange width B = L/40 to L/45 = 450 mm to 400 mm

Thickness of plate = (13303/500) = 26.6 mm

Adopt flange plates 500 mm by 30 mm.

The section selected is shown in Fig. 10.6

(f) Check for Maximum Stresses

$$I = [(10 \times 1000^3)/12 + 2 (30 \times 500) 515^2]$$
$$= 879 \times 10^7 \text{ mm}^4$$

Bending tensile stress $= \sigma_b = (My/I)$
$$= (2470 \times 10^6 \times 530)/(879 \times 10^7)$$
$$= 149 \text{ N/mm}^2 < 165 \text{ N/mm}^2$$

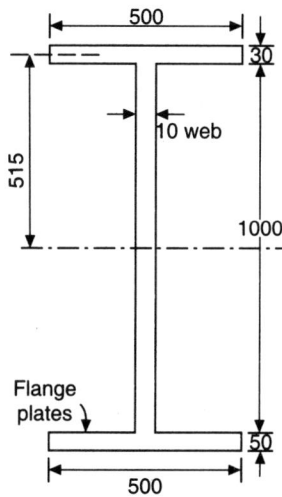

Fig. 10.6 Cross-section of Plate Girder.

Average shear stress $= (548 \times 10^3)/(1000 \times 10) = 55 \text{ N/mm}^2$

Permissible average shear stress depends upon the ratio of $(d/t) = (1000/10)$
$$= 100$$

Using stiffener spacing $c = 1000 \text{ mm} = d$

From Table 9.3, allowable average shear stress is 100 N/mm^2

Hence the average shear stress is within safe permissible limits.

(g) Connection between Flange and Web

Maximum shear force at the junction of web and flange is given by,

$$\tau = (Va\overline{y}/I)$$

where $V = 548 \times 10^3 \text{ N}$

$a = (500 \times 30) = 15{,}000 \text{ mm}^2$

$I = 879 \times 10^7 \text{ mm}^4$

$\overline{y} = 515 \text{ mm}$

$\tau = [(548 \times 10^3 \times 15 \times 10^3 \times 515)/(879 \times 10^7)] = 483 \text{ N/mm}$

Assuming continuous weld on either side, strength of weld of size 's' is
$$= (2 \times 0.7 \times s \times 102.5) = 145 \, s$$

$\therefore$　　　$145\,s = 483$, $s = 3.33$ mm

Use 5 mm fillet weld, continuous on either side,

(h) Intermediate Stiffeners

Since　$(d/t) = (1000/10) = 100 < 85$

Vertical Stiffeners are required.

Spacing of Stiffeners $= 0.33\,d$ to $1.5\,d$

$$= (0.33 \times 1000) \text{ to } (1.5 \times 1000)$$

$$= 333 \text{ mm to } 1500 \text{ mm}$$

Adopt 1000 mm spacing. Hence $c = 1000$ mm

The intermediate stiffeners are designed to have a minimum moment of inertia of,

$$I = [(1.5\ d^3 t^3)/c^2] = [(1.5 \times 1000^3 \times 10^3)/1000^2] = 15 \times 10^5 \text{ mm}^4$$

Using 10 mm thick plate

Maximum width of plate not to exceed $12t$ for flats

Use a plate 10 mm × 80 mm $h = 80$ mm

$$I = [(10 \times 80^3)/3] = 17 \times 10^5 \text{ mm}^4 > 15 \times 10^5 \text{ mm}^4$$

(i) Connection of Vertical Stiffener to Web

Shear on welds connecting stiffener to web

$$= [(125 \times t^2)/h] \text{ kN/m}$$

where　　t = web thickness (mm)

　　　　h = outstand of stiffener (mm)

Shear on welds $= [(125 \times 10^2)/80] = 156.25$ kN/m $= 156.25$ N/mm

Size of welds $= [156.25/(0.7 \times 102.5)] = 2.17$ mm

Use 5 mm minimum size intermittent welds.

Effective length of weld $\not< 10\ t \not< (10 \times 10) = 100$ mm

Use 100 mm long, 5 mm fillet welds alternately on either side.

(j) End Bearing Stiffener

Maximum shear force $= 548$ kN

The end bearing stiffener is designed as a column

$$(h/t) \not> 12$$

where　　h = outstand

　　　　t = thickness

If　　　$h = 180$ mm, $t = (180/12) = 15$ mm

Use 180 mm by 15 mm size plate

Permissible bearing stress

$$= \sigma_p = 189 \text{ N/mm}^2 \text{ (IRC: 24)}$$

$\therefore$ Bearing area required $= [(548 \times 10^3)/189] = 2900$ mm^2

If two plates are used

Total area provided = $(2 \times 180 \times 15)$

$\qquad = 5400 \text{ mm}^2 > 2900 \text{ mm}^2$

The length of web plate which acts along with stiffener plates in bearing the reaction = $20t = (20 \times 10) = 200$ mm

$$I = [(15 \times 370^3)/12 + (2 \times 200 \times 10^3)/12]$$
$$= 633^4 \times 10^4 \text{ mm}^4$$

Area $\quad A = [(360 \times 15) + (400 \times 10)] = 9400 \text{ mm}^2$

$\qquad \lambda$ = Slenderness Ratio = (L/r)

$$r = \sqrt{I/A} = \sqrt{\left(6334\,\pi\,10^4\right)/9400} = 82 \text{ mm}$$

Effective Length of Stiffener = $(0.7 \times 1000) = 700$ mm

$\qquad \lambda = (700/82) = 8.53$

From Table 9.5 (Table 11.1 of IRC: 24-2001),

Permissible stress sac in axial compression is obtained as 150 N/mm²

∴ Area required = $[(548 \times 10^3)/150] = 3653 \text{ mm}^2 < 9400 \text{ mm}^2$

(k) Connection between Bearing Stiffener and Web

Length available for welding using alternate intermittent welds

$\qquad = 2(1000 - 40) = 1920$ mm

Required strength of weld = $[(548 \times 10^3)/1920] = 286$ N/mm

Size of weld = $[286/(0.7 \times 102.5)] = 3.98$ mm

Use 5 mm fillet weld

Length of weld ≮ 10 t ≮ $(10 \times 10) = 100$ mm

Use 100 mm long 5 mm welds alternately.

(l) Properties of the Composite Section[6]

Referring to Fig. 10.7

$\qquad A_{ce} = [(2000 \times 300)/13] = 46154 \text{ mm}^2$

Modular ratio = $m = 13$

The centroid of the composite section is determined by first moment of the areas about the axis XX.

$$A\bar{y} = [(46154 \times 1210) + (500 \times 30 \times 1045) + (1000 \times 10 \times 530)$$
$$+ (500 \times 30 \times 15)]$$
$$= 77046340$$
$$A = [46154 + (500 \times 30) + (1000 \times 10) + (500 \times 30)]$$
$$= 86154 \text{ mm}^2$$

∴ $\qquad \bar{y} = (77046340/86154) = 894$ mm

$$I_{comp} = [(46154 \times 316^2) + (500 \times 1060^3)/12 - (400 \times 1000^3)/12$$
$$+ (40000 \times 364^2)]$$
$$= 1.957 \times 10^{10} \text{ mm}^4$$

Maximum shear force at junction of slab and girder is given by

$$\tau = (V\,a\,\bar{y}/I)$$

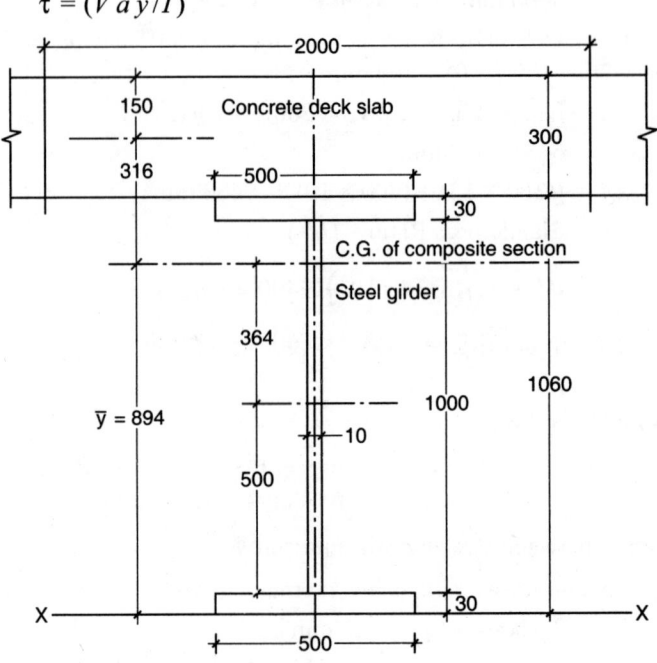

Fig. 10.7 Properties of Composite Section.

where $V = 548$ kN

           $a = 46154$ mm^2

           $I = 1.957 \times 10^{10}$ mm^4

           $\bar{y} = 316$ mm

∴ $\tau = [(548 \times 10^3 \times 46154 \times 3160)/(1.957 \times 10^{10})]$

           $= 408$ N/mm

Total shear force at junction $= (408 \times 500) = 204000$ N

Using 20 mm diameter mild steel studs, capacity of one shear connector is given by

$$Q = 196\,d^2\sqrt{f_{ck}}$$

where $H = 5d - (5 \times 20) = 100$ mm

           $d = 20$ mm

           $f_{ck} = 20$ N/mm^2

           $Q = 196 \times 20^2\sqrt{20} = 350615$ N

Number of studs required in one row $= (204000/350615) < 1$

Provide a minimum of 2 mild steel studs in a row

(m) Pitch of shear connectors $= p = [(NQ/(F\tau)]$

    where $N =$ Number of shear connectors in a row

$$Q = \text{Capacity of one shear connector}$$

$$\tau = \text{Horizontal shear per unit length}$$

$$F = \text{Factor of safety} = 2$$

$$\therefore \quad p = [(2 \times 343460)/(2 \times 408)] = 841 \text{ mm}$$

Maximum permissible pitch is the least of

    (i) three times the thickness of slab = (3 × 300) = 900 mm

    (ii) 4 times the height of the stud = (4 × 100) = 400 mm

    (iii) 600 mm

Hence adopt a pitch of 400 mm in the longitudinal direction.

The arrangement of shear connectors is shown in Fig. 10.8

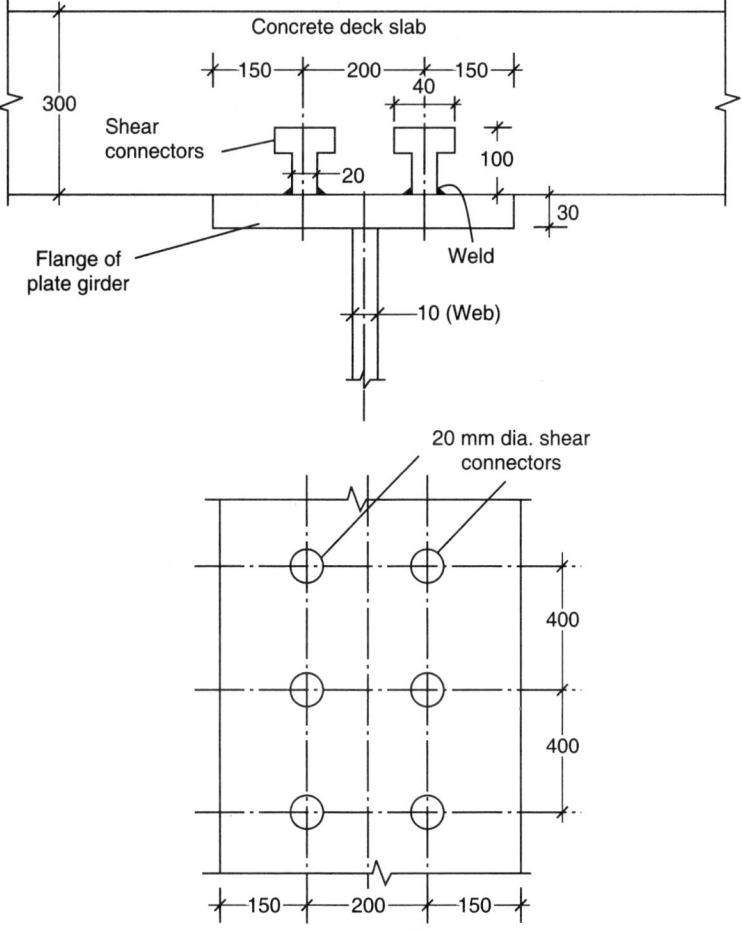

Fig. 10.8 Details of Shear Connectors.

The cross section and longitudinal section of the composite girder are shown in Figs. 10.9 and 10.10.

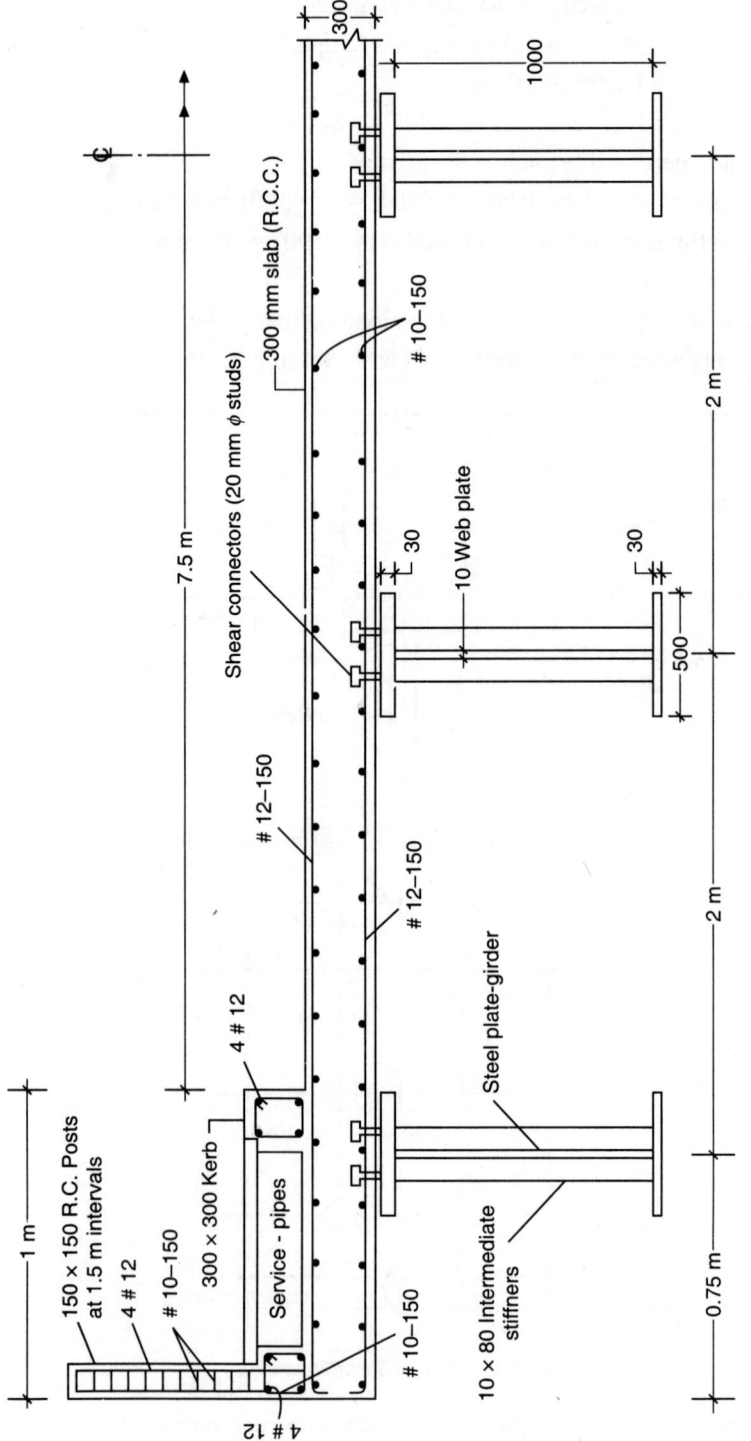

Fig. 10.9 Cross Section of Deck Slab.

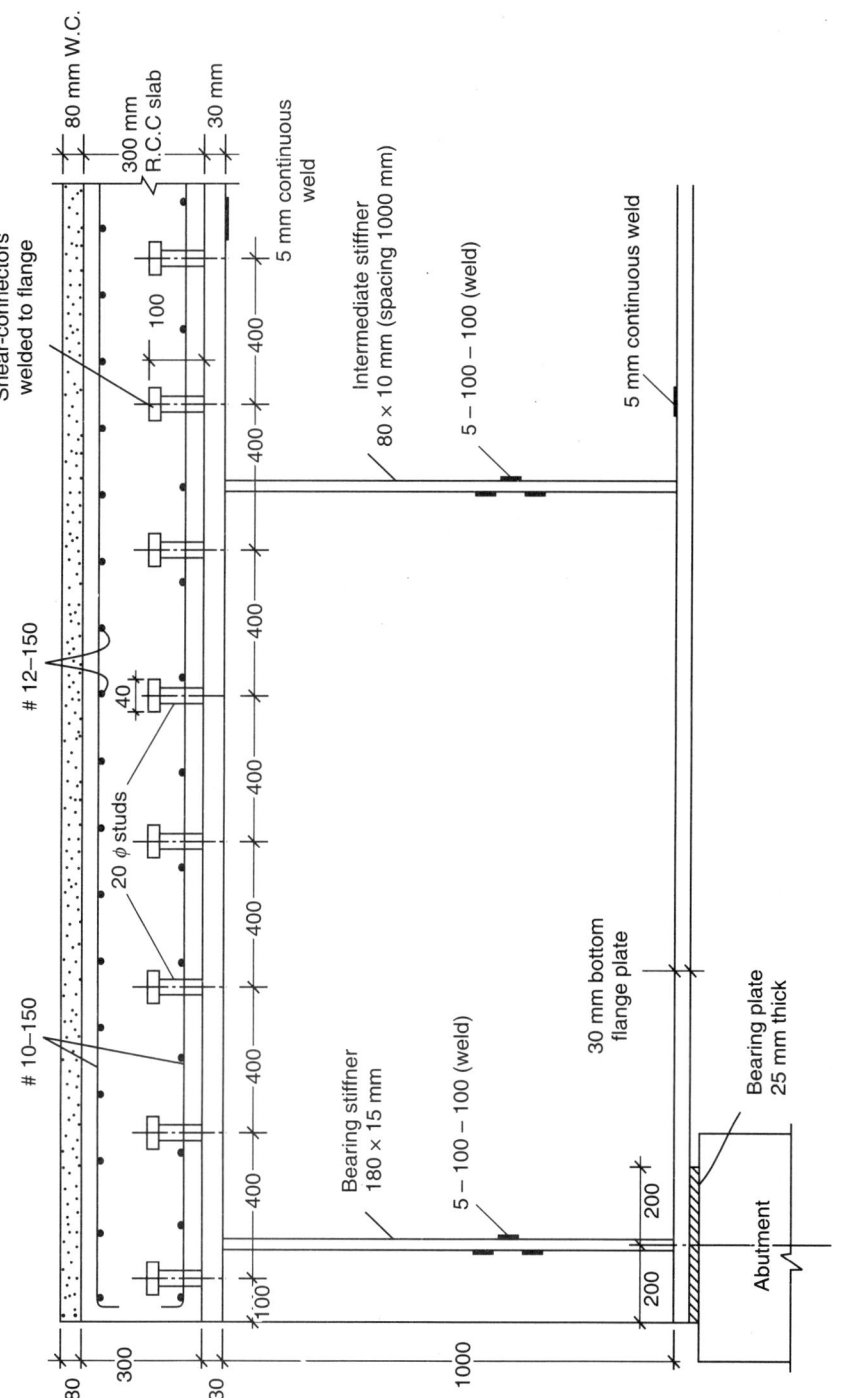

Fig. 10.10 Longitudinal Elevation of Composite Beam Showing Details of Shear Connectors.

8. Design Drawings

The structural detailing of reinforcements in the deck slab and the connections between the steel plate girder and the concrete deck slab should conform to the specifications of the special publication SP: 34-1987[10]. The cross section and longitudinal section of the composite bridge deck is shown in Figs. 10.9 and 10.10.

EXAMPLES FOR PRACTICE

1. A composite bridge deck consisting of a reinforced concrete slab and steel girders is required for a national highway of span 20 m. The following data is available:

   Clear width of roadway = 7.5 m
   Footpath = 1 m on either side
   Spacings of longitudinal girders = 2 m
   Spacings of cross girders = 4 m
   Thickness of the wearing coat =100 mm
   Loading: I.R.C. class AA or A whichever gives the worst effect.
   Materials M-300 concrete and Fe-415 grade HYSD bars.

   Design the continuous reinforced concrete deck slab, one of the steel plate girders with intermediate and bearing stiffeners and suitable shear connectors.

   Draw the following views:

   (a) Half cross-section of the deck slab supported on plate girders, showing the reinforcement details and shear connectors.

   (b) Longitudinal section of the composite bridge deck near supports, showing the details of shear connectors. Reinforcements in deck slab, bearing and intermediate stiffeners and weld details.

   (Bangalore University Nov. 1984)

2. A single span composite steel girder and an R.C.C. deck slab bridge is proposed for a state highway across a stream. The span of the bridge may be taken as 16 m. The width of the road is 6 m. the R.C.C. deck slab is supported by 5 numbers of rolled steel joists placed longitudinally symmetrically. Design the R.C.C. deck slab for an equivalent live load of 1200 kg/m^2 and an impact factor of 0.5. Also design an intermediate girder. Adopt M-15 grade concrete and tor steel bars.

   (Mysore University Dec. 1983)

3. A composite bridge deck using R.C.C. deck slab and plate girders is required for a highway bridge of span 12 m.

   Clear width of roadway = 7.5 m
   Spacings of longitudinal girders = 1.75 m
   Spacings of cross girders = 3.5 m
   Thickness of wearing coat = 80 mm
   Loading: I.R.C. class AA tracked vehicle
   Materials: M-3Q grade concrete and Fe-415 grade HYSD bars.

Design the R.C.C. deck slab and one of the main girders and sketch the longitudinal and cross-sections of composite deck bridge.

4. A composite bridge deck system comprising steel plate girders and reinforced concrete slab deck is to be designed for a National Highway crossing. Using the following data, design the composite bridge deck and the shear connectors and sketch the longitudinal and cross sections of the bridge deck.

Effective span of the girders = 25 m
Clear width of Road way = 7.5 m
Foot paths 1.5 m on either side
Spacings of longitudinal girders = 2 m
Spacings of cross girders = 5 m
Thickness of wearing coat = 100 mm
Loading: IRC Class AA tracked vehicle
Materials: M-25 Grade concrete and Fe-415 Grade HYSD bars and rolled steel plates and sections with an yield stress of 236 N/mm^2.

REFERENCES

1. Johnson Victor, D., Essentials of Bridge Engineering (Fifth Edition), Oxford & IBH Publishing Co. Ltd, 2001, pp. 259-265.
2. Yan. H.T., Composite Construction in Steel and Concrete, Orient Longmans, Calcutta, 1965, 210 pp.
3. Krishna Raju, N., Prestressed Concrete, McGraw Hill Education (Fifth Edition), New Delhi, 2012, pp. 409-437.
4. Viest, I.M, Fountain, R.S and Singleton, R.C., Composite Construction in Steel and concrete for Bridges and Buildings, McGraw-Hill Publishing Co, Ltd, New York, 1958, pp. 1-176.
5. Antia, K.F, Composite Construction with Prestressed Concrete, Journal of the Institution of Engineers (India), Vol. 40, No.1, March 1960, pp. 421-450.
6. IRC: 22-2015, Standard Specifications and Code of Practice for Road Bridges, Section VI, Composite Construction (Limit State Design) (Third Revision), Indian Roads Congress, New Delhi, 2015.
7. IRC: 6-2014, Standard Specifications and Code of Practice for Road Bridges, Section II, Loads and Stresses (Revised Edition), Indian Roads Congress, New Delhi, 2014, pp. 1-84.
8. IRC: 24-2001, Standard Specifications and Code of Practice for Road bridges, Section-V, Steel Road Bridges, Indian Roads Congress, New Delhi, 2001, pp. 157.
9. IS: 800-1987, Indian Standard Code of Practice for Steel Construction, Bureau of Indian Standards, New Delhi, 1987.
10. SP: 34-1987, Hand Book of Concrete Reinforcement and Detailing, Bureau of Indian Standards, New Delhi, 1987.

REVIEW QUESTIONS

1. Under what situations you would prefer to use composite bridge decks? What are the specific advantages of composite bridge decks?
2. What are shear connectors? Explain the main purpose of the shear connectors.
3. Explain with sketches the various types of shear connectors and their specific use in composite bridge decks.
4. Briefly explain the method of designing stud type shear connectors
5. What are the various factors influencing the magnitude of the shear resistance of the shear connectors?
6. How do you compute the maximum shear force at the junction of the cast-*in-situ* concrete slab and the steel girder?
7. Explain the method of determining the ultimate design load moments in a composite girder deck comprising the cast-*in-situ* concrete slab cast over steel plate girders when it is supporting IRC Class AA tracked vehicle loads.
8. Briefly explain the method of computing the second moment of area of the composite girder section and its use.
9. How do you determine the pitch of the shear connectors? What are the various factors influencing the pitch of shear connectors?
10. Explain with sketches the typical structural details of a composite girder deck bridge showing the details of the shear connectors.

OBJECTIVE TYPE QUESTIONS

1. For faster construction of high way bridge decks, it is better to adopt
   a) Reinforced concrete tee beam and slab
   b) Post tensioned Prestressed concrete decks
   c) Composite bridge decks
2. In a comparative analysis of the flexural stiffness of several types of bridge decks designed for a major national high way crossing, the type of bridge deck having the highest flexural stiffness is
   a) Reinforced concrete deck
   b) Composite deck
   c) Prestressed concrete deck
3. Shear connectors are invariably used in the construction of
   a) Prestressed concrete bridge decks
   b) reinforced concrete bridge decks
   c) Composite bridge decks
4. In the case of welded stud shear connectors of mild steel having the ratio of height to the diameter of the stud greater than 4.2, the safe shear resistance of the shear connector is depends upon the
   a) Height the shear connector
   b) Yield stress of the material
   c) Square of the diameter of the connector

5. Rigid shear connectors used in composite bridge decks comprise of
   a) Studs
   b) Steel angles
   c) Steel bars projecting from the main girders
6. In the case of anchorage type shear connectors, the ultimate shear resistance of the shear connector depends upon the
   a) Diameter of the shear connector
   b) Height of the shear connector
   c) Cross sectional area of the connector
7. In the case of composite bridge decks comprising reinforced concrete slab supported by steel girders, the spacing of the girders is in the range of
   a) 1 to 1.5 m
   b) 2 to 2.5 m
   c) 3 to 4 m
8. The maximum shear force at the junction of the deck slab and steel girder is inversely proportional to the
   a) Shear force at the section
   b) Second moment of area of the composite section
   c) Equivalent area of the deck slab
9. Anchorage type shear connectors are generally adopted when the supporting girders are
   a) Steel plate girders
   b) Prestressed concrete girders
   c) Reinforced concrete girders
10. The pitch of shear connectors in a composite bridge deck is directly proportional to the
    a) Horizontal shear per unit length
    b) Diameter of the shear connector
    c) Number of shear connectors in a row

Prestressed Concrete Bridges

11

11.1 GENERAL FEATURES

Prestressed concrete developed by Freyssinet[1] in 1935 is ideally suited for the construction of medium and long span bridges. According to Leonhardt[2], the earliest applications of prestressed concrete after the second world war during the period from 1945 to 1960 could be seen in long span bridges in Germany, France and the USA. Prestressed concrete has almost replaced steel for the construction of bridges due to its inherent advantages of superior durability, ease of maintenance in comparison with steel with its basic disadvantages of corrosion under aggressive atmospheric conditions and inhibitive initial costs.

Solid slabs are used for the span range of 10 to 20 m while Tee beam slab decks are suitable for spans in the range of 20 to 40 m. Single or multicell box girders are preferred for larger spans of the order of 30 to 70 m. Prestressed Concrete[3] is ideally suited for long span continuous bridges in which precast box girders of variable depth are used for span exceeding 50 m. Prestressed Concrete, has been widely used throughout the world for simply supported, continuous, balanced cantilever, suspension, hammer head and bridle chord type bridges in the span range from 20 to 500 m.

11.2 ADVANTAGES OF PRESTRESSED CONCRETE BRIDGES

Prestressed Concrete made up of high strength concrete[4] and high tensile steel[5] has distinct advantages when used for bridge construction. The salient benefits resulting from the use of prestressed concrete in bridges are out lined as follows:

1. The use of high strength concrete and high tensile steel results in slender sections which are aesthetically superior coupled with overall economy.

2. Prestressed concrete bridges can be designed as class 1 type structures without any tensile stresses under service loads resulting in a crack free structure.

3. In comparison with steel bridges, prestressed concrete bridges require very little maintenance.

4. Prestressed concrete is ideally suited for composite bridge construction in which precast prestressed girders support the cast *in-situ* slab deck. This type of construction is very popular since it involves minimum disruption of traffic.

5. Post tensioned prestressed concrete finds extensive applications in long span continuous girder bridges of variable cross section resulting in sleek structures and with considerable savings in the overall cost of construction.

6. In recent years, partially prestressed concrete[6] (type 3 structure) is preferred for bridge construction with considerable savings in the quantity of costly high tensile steel used in the girder.

11.3 PRE-TENSIONED PRESTRESSED CONCRETE BRIDGES

Pretensioned prestressed concrete bridge decks generally comprise precast pretensioned units used in conjunction with cast *in-situ* concrete resulting in composite bridge decks ideally suited for small and medium spans in the range of 20 to 30 m. In general, pretensioned girders are provided with straight tendons. The use of seven wire strands have been found to be advantageous in comparison with plain or indented wires. Deflected strands are employed in larger girders in U.S.A.

In U.K., the precast prestressed I and inverted T-beams have been standardized by the Cement and Concrete Association for the use in the construction of bridge decks of spans varying from 7 to 36 m. Standard I and T units are widely employed in high way bridge beams in U.S.A. Recently in U.K., Y-beams have been developed to replace the M-beams introduced in 1960. The design and development of the Y-beams which are superior to M-beams are ideally suited for medium spans of 15 to 30 m.

The typical cross section of the standard inverted Y-beams[7] developed by the research group in U.K. is shown in Fig 11.1 and the section properties of the Y-beam are compiled in Table 11.1. The salient features of composite bridge decks with precast pretensioned standard beams are shown in Fig. 11.2.

Table 11.1 Section Properties of Standard Y-Beams (U.K)

Section	Depth	Area	Height of Centroid above Soffit Y_b	Section Modulus Top fibre Z_t	Section Modulus Bottom fibre Z_b	Approximate Self Weight
	(mm)	(mm^2)	(mm)	(mm$^3 \times 10^6$)	(mm$^3 \times 10^6$)	kN/m
Y-1	700	309202	255.24	24.85	43.40	7.42
Y-2	800	339882	298.68	35.02	58.78	8.14
Y-3	900	373444	347.12	47.88	76.27	8.95
Y-4	1000	409890	399.71	63.53	95.41	9.82
Y-5	1100	44920	455.72	82.06	116.02	10.78
Y-6	1200	491433	514.50	103.58	138.00	11.78
Y-7	1300	536530	575.54	128.15	161.31	12.86
Y-8	1400	584708	638.54	155.98	186.01	14.02

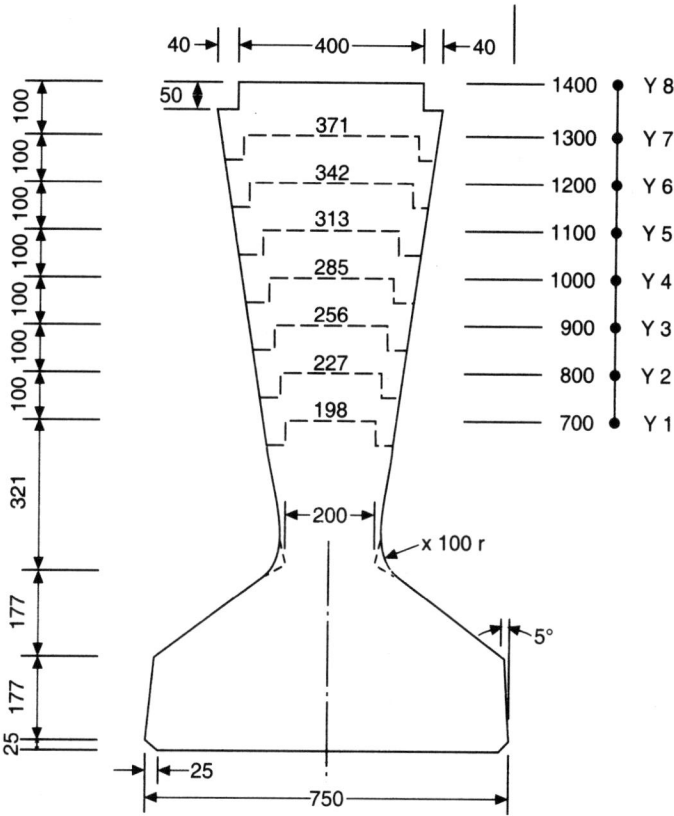

Fig. 11.1 Cross Section of Standard Y-Beams (U.K.).

11.4 POST-TENSIONED PRESTRESSED CONCRETE BRIDGE DECKS

Post tensioned bridge decks are generally adopted for longer spans exceeding 20 m. Bridge decks with precast post tensioned girders of either Tee type or box type in conjunction on with a cast *in-situ* slab is commonly adopted for spans exceeding 30 m. Post tensioning facilitates the use of curved cables which improve the shear resistance of the girders.

Post tensioning is ideally suited for prestressing long span girders at the site of construction without the need for costly factory type installations like pretensioning beds. Segmental construction[8] is ideally suited for post tensioning work. In this method a number of segments can be combined by prestressing, resulting in an integrated structure. In India a large number of long span bridges have been constructed using the cantilever method of construction.[9] Some of the notable examples being the Barak bridge at Silchar built in 1960 with a main span of 130 m and the Lubha bridge in Assam with a span of 130 m between the bearings. Long span continuous prestressed concrete bridges are invariably built up of multicelled box girder segments of variable depth using the post tensioning system. Typical cross sections of post tensioned prestressed

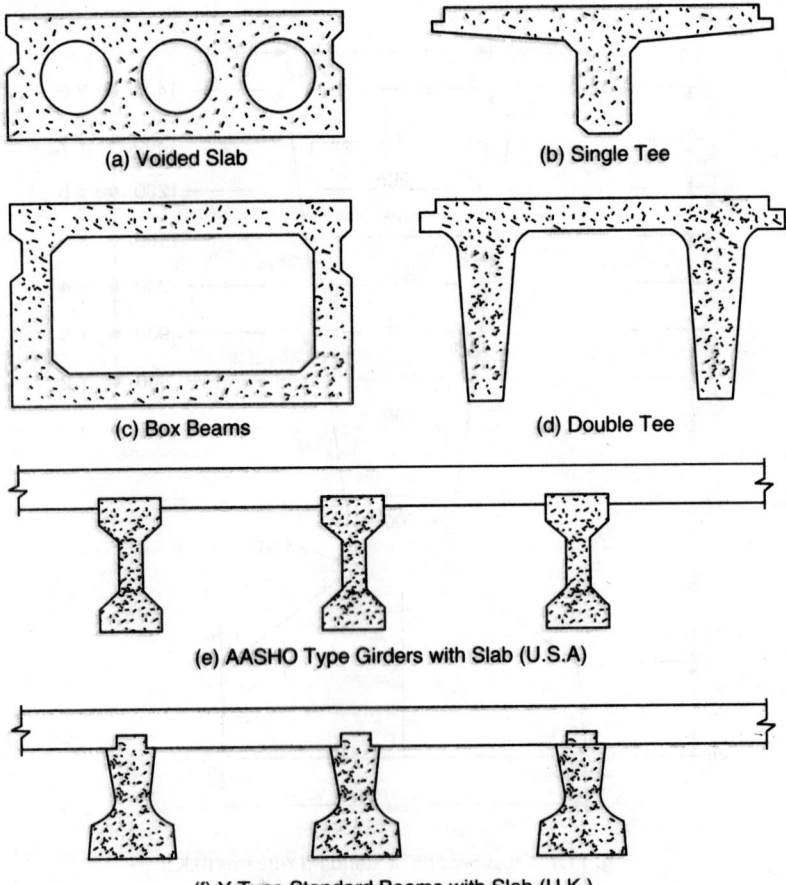

(a) Voided Slab

(b) Single Tee

(c) Box Beams

(d) Double Tee

(e) AASHO Type Girders with Slab (U.S.A)

(f) Y Type Standard Beams with Slab (U.K.)

Fig. 11.2 Typical Cross Sections of Pretensioned Prestressed Concrete Bridge Decks.

concrete bridge cocks are shown in Fig. 11.3. The salient features of cantilever construction method using cast *in-situ* segments and precast concrete elements are shown in Figs. 11.4(a) and (b).

11.5 DESIGN OF POST-TENSIONED PRESTRESSED CONCRETE SLAB BRIDGE DECK

Design a post tensioned prestressed concrete slab bridge deck for a National highway crossing to suit the following data:

1. Data

Clear span	10 m
Width of bearing	400 mm
Clear width of Road way	7.5 m

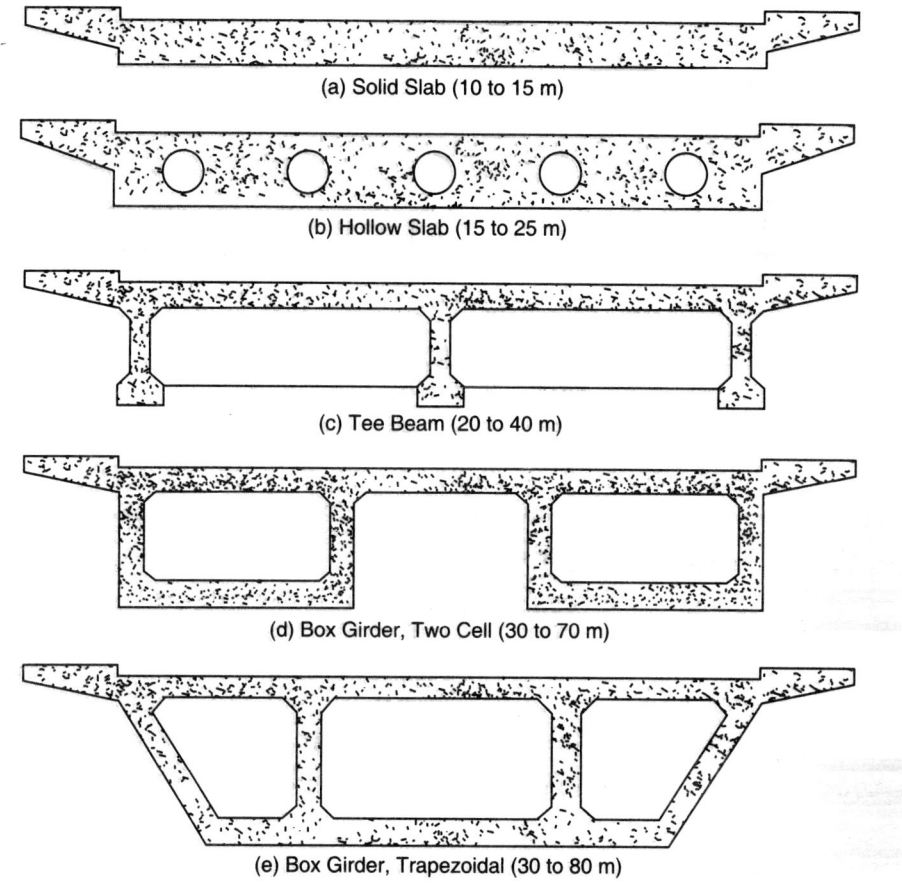

(a) Solid Slab (10 to 15 m)

(b) Hollow Slab (15 to 25 m)

(c) Tee Beam (20 to 40 m)

(d) Box Girder, Two Cell (30 to 70 m)

(e) Box Girder, Trapezoidal (30 to 80 m)

Fig. 11.3 Typical Cross Sections of Post-tensioned Prestressed Concrete Bridge Decks.

Foot paths	1 m on either side
Kerbs	600 mm wide by 300 mm deep
Thickness of wearing coat	80 mm
Live load	IRC Class AA tracked vehicle
Type of structure	Class 1 type
Materials	M-40 Grade concrete and 7 mm diameter high tensile wires with an ultimate tensile strength of 1500 N/mm^2 housed in cables with 12 wires and anchored by Freyssinet anchorages of 150 mm diameter.

For supplementary reinforcement, adopt Fe-415 Grade HYSD bars

Assume compressive strength of concrete at transfer as 35 N/mm^2 and Loss ratio as 0.8.

The design should conform to the specifications of the codes IRC: 6-2014[10], IRC: 112-2011[11] and IS: 1343-2012[12].

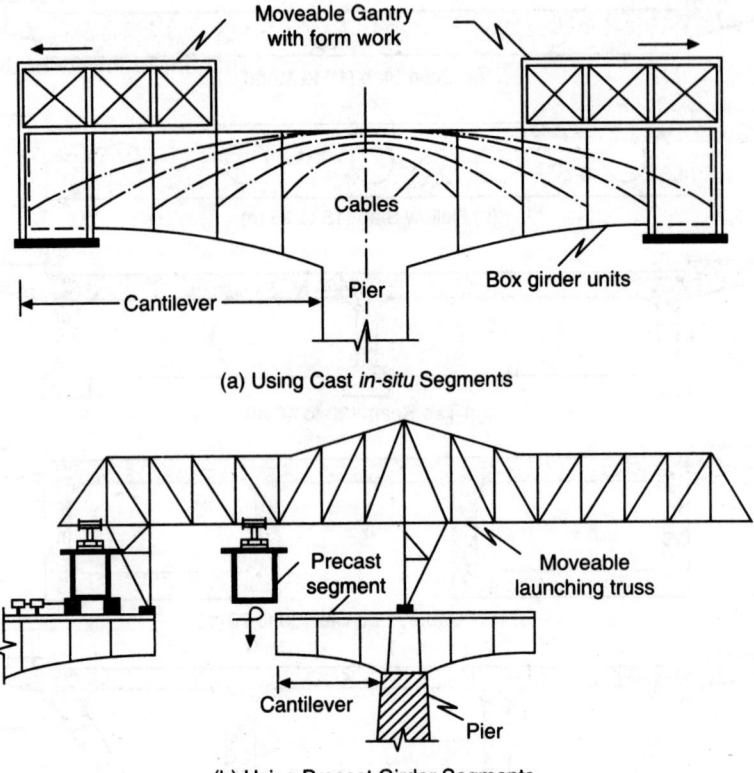

Fig. 11.4 Cantilever Method of Construction of Prestressed Concrete Bridges.

2. Maximum Permissible Stresses in Concrete and Steel

According to the given data, we have the

Compressive strength of concrete $= f_{ck} = 40$ N/mm^2

Compressive strength of concrete at transfer $= f_{ci} = 35$ N/mm^2

The permissible compressive stresses in concrete at transfer and service loads as recommended in IS: 1343 code are as follows:

Assuming that the compressive stresses are not likely to increase (Zone-I) (Post tensioned work)

Compressive stress at transfer $= f_{ct} = 15$ N/mm$^2 < 0.50\,f_{ci} = (0.50 \times 35) = 17.5$ N/mm^2

Compressive stress at service loads $= f_{cw} = 12$ N/mm$^2 < 0.39\,f_{ck} = (0.39 \times 40) = 15.6$ N/mm^2

Permissible tensile stress (Class 1 type structure) $= f_{tt} = f_{tw} = 0$

Ultimate tensile strength of 12 wires of 7 mm diameter high tensile cables $= f_p = 1500$ N/mm^2

Maximum permissible stress in the H.T. cable $= 1200$ N/mm^2

3. Depth of Slab and Effective Span

Assuming the thickness of slab at 50 mm per metre of span for high way bridge decks, overall thickness of slab = (10 × 50) = 500 mm.

Width of bearing = 400 mm

Effective span = 10.4 m

The cross section of deck slab is shown in Fig. 11.5.

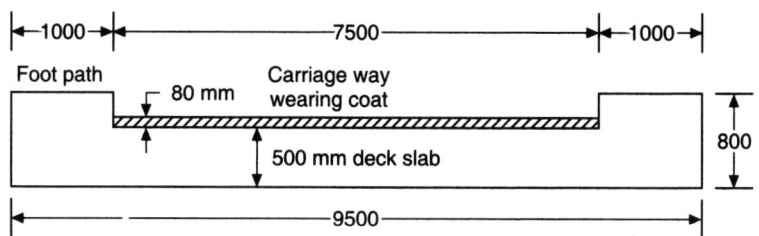

Fig. 11.5 Cross Section of Deck Slab.

4. Dead Load Bending Moments

Dead weight of slab = (0.5 × 24) = 12 kN/m^2

Dead weight of W.C. = (0.08 × 22) = 1.76 kN/m^2

Total dead load = 14.00 kN/m^2

Dead load bending moment (M_g) = (14 × 10.4^2)/8 = 190 kN·m

5. Live Load Bending Moments

Generally the bending moment due to live load will be maximum for I.R.C. Class AA tracked vehicle. Impact factor for class AA tracked vehicle is 25% for 5 m span, decreasing linearly to 10% for 9 m span.

∴ Impact factor = 10 percent for a span of 10.4 m.

The tracked vehicle is placed symmetrically on the span.

Effective length of load = [3.6.+ 2 (0.5 + 0.08)] = 4.76 m

Effective width of slab perpendicular to span is expressed as,

$$b_e = k \cdot x \,(1 - x/L) + b_w$$

Referring to Fig. 11.6.

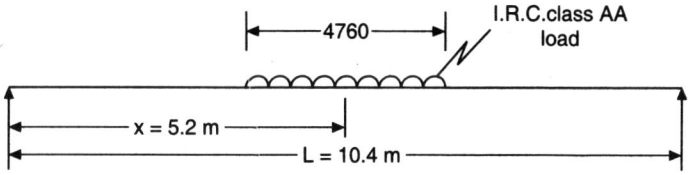

Fig. 11.6 Position of Load for Maximum Bending Moment.

$$x = 5.2 \text{ m}, \quad L = 10.4 \text{ m}, \quad B = 9.5 \text{ m}$$
$$(B/L) = (9.5/0.4) = 0.913$$
$$b_w = (0.85 + 2 \times 0.08) = 1.01 \text{ m}$$

From Table 4.6 for $(B/L) = 0.913$, simply supported slabs, $k = 2.37$

$$\therefore \qquad b_e = 2.37 \times 5.2 \,(1 - 5.2/10.4) + 1.01 = 7.172 \text{ m}$$

The tracked vehicle is placed close to the kerb with the required minimum clearance as shown in Fig. 11.7.

Net effective width of dispersion = 8.261 m
Total load of two tracks with impact = $(700 \times 1.10) = 770$ kN
Average intensity of load = $770/(4.76 \times 8.261) = 19.58$ kN/m^2
Maximum bending moment due to live load is given by

$$M_q = [(19.58 \times 4.76)\,0.5 \times 5.2] - [(19.58 \times 4.76)\,0.5 \times 0.25 \times 4.76]$$
$$= 187 \text{ kN·m}$$

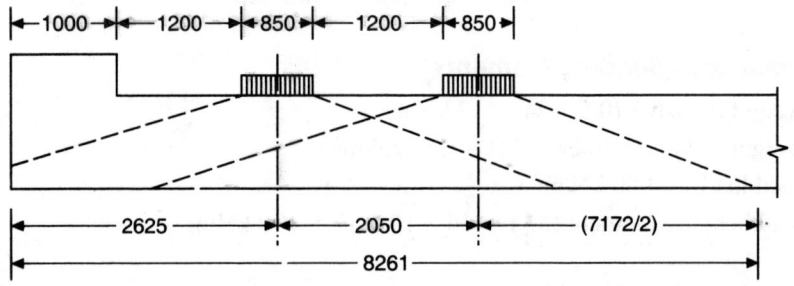

Fig. 11.7 Effective Width of Dispersion for IRC Class AA, Tracked Vehicle.

6. Shear due to Class AA Tracked Vehicle

For maximum shear force at support section; the I.R.C. Class AA tracked vehicle is arranged as shown in Fig. 11.8.

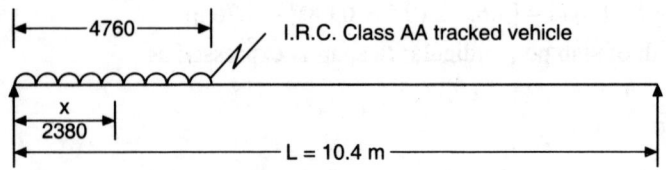

Fig. 11.8 Position of Load for Maximum Shear.

Effective width of dispersion is given by

$$b_e = k.x \,(1 - x/L) + b_w$$

where $\qquad x = 2.38 \text{ m}, L = 10.4 \text{ m}, B = 9.5 \text{ m}, b_w = 1.01 \text{ m}$

$$(B/L) = (9.5/10.4) = 0.913$$

$\therefore$ From table 3.5 for $(B/L) = 0.913$ the value of $k = 2.37$

$\therefore$ $b_e = [2.37 \times 2.38\,(1 - 2.38/10.4) + 1.01] = 5.364$ m

Referring to Fig. 11.7

width of dispersion for two tracks

$$= [2625 + 2050 + (5364/2)]$$
$$= 7357 \text{ mm}$$

$\therefore$ Intensity of load $= [770/(4.76 \times 7.357)]$
                         $= 22$ kN·m^2

$\therefore$ Shear force V_A $= (22 \times 4.76 \times 8.02)/10.4$
                         $= 80.75$ kN

Dead load shear $= (0.5 \times 14 \times 10.4)$
                         $= 72.8$ kN

$\therefore$ Total Design shear $= (80.75 + 72.80)$
                         $= 153.55$ kN

7. Check for Minimum Section Modulus

Dead load Moment $M_g = 190$ kN·m

Live load moment $M_q = 1\ 87$ kN·m

Section Modulus $= Z_t = Z_b = Z = \left(\dfrac{1000 \times 500^2}{6}\right) = 41.66 \times 10''$ mm^3

The permissible stresses in concrete at transfer (f_{ct}) is obtained from IRC – 18

$$f_{ct} = 15.0 \text{ N/mm}^2, f_{cw} = 12.0 \text{ N/mm}^2, f_{tw} = 0,$$
$$\eta = \text{Loss Ratio} = 0.8, f_{br} = (\eta f_{ct} - f_{rw}) = (0.8 \times 15 - 0)$$
$$= 12.0 \text{ N/mm}^2$$

The minimum section Modulus is given by

$$Z_b \geq \left[\frac{M_q + (1 - \eta)M_g}{f_{br}}\right]$$

$$\geq \left[\frac{187 \times 10^6 + (1 - 0.8)190 \times 10^6}{12}\right]$$

$$\geq 18.75 \times 10^6 \text{ mm}^3 < 41.66 \times 10^6 \text{ mm}^3$$

Hence the section selected is adequate to resist the service loads without exceeding the permissible stresses.

8. Minimum Prestressing Force

The minimum prestressing force required is computed using the relation

$$P = \left[\frac{A(f_{\inf} \cdot z_b + f_{\sup} \cdot z_t)}{z_b + z_t}\right]$$

Where
$$f_{sup} = \left(f_{tt} - \frac{M_g}{z_t} \right) = \left(0 - \frac{190 \times 10^6}{41.66 \times 10^6} \right) = -4.56 \text{ N/mm}^2$$

$$f_{inf} = \left(\frac{f_{tw}}{\eta} + \frac{M_q + M_g}{\eta z_b} \right) = \left[0 + \frac{(187 + 190)10^6}{0.8 \times 41.66 \times 10^6} \right]$$

$$= 11.31 \text{ N/mm}^2$$

$\therefore$
$$p = \left[\frac{1000 \times 500 \times 41.66 \times 10^6 (11.31 - 4.56)}{2 \times 41.66 \times 10^6} \right]$$

$$= 1687.5 \times 10^3 \text{ N}$$

$$= 1687.5 \text{ kN}$$

Using Freyssinet cables containing 12 wires of 7 mm diameter stressed to 1200 N/mm², Force in each cable =

$$(12 \times 38.5 \times 1200)/1000 = 554 \text{ kN}$$

$\therefore$ Spacings of cables $= \left(\dfrac{1000 \times 554}{1687.5} \right) = 328 \text{ mm}$

9. Eccentricity of Cables

The eccentricity of the cables at the centre of span is obtained from the relation:

$$e = \left[\frac{z_t z_b (f_{inf} - f_{sup})}{A(f_{sup} z_t + f_{inf} z_b)} \right]$$

$$= \left[\frac{(41.66)^2 \times 10^{12} (11.31 + 4.56)}{1000 \times 500 \times 41.66 \times 10^6 (-4.56 + 11.31)} \right]$$

$$= 195 \text{ mm}$$

The cables are arranged in a parabolic profile with a maximum eccentricity of 195 mm at centre of span reducing to zero eccentricity (Concentric) at supports.

10. Check for Stresses at Service Loads

$$P = 1687.5 \text{ kN}, \ e = 195 \text{ mm}$$
$$A = (1000 \times 500) = 5 \times 10^5 \text{ mm}^2$$
$$Z_t = Z_b = Z = 41.66 \times 10^6 \text{ mm}^3$$
$$M_g = 190 \text{ kN·m} \quad M_q = 187 \text{ kN·m}$$
$$(P/A) = (1687.5 \times 10^3/5 \times 10^5) = 3.375 \text{ N/mm}^2$$
$$(P_e/Z) = (1687.5 \times 10^3 \times 195/41.66 \times 10^6) = 7.89 \text{ N/mm}^2$$
$$(M_g/Z) = (190 \times 10^6/41.66 \times 10^6) = 4.56 \text{ N/mm}^2$$
$$(M_q/Z) = (187 \times 10^6/41.66 \times 10^6) = 4.48 \text{ N/mm}^2$$

Stresses at transfer

At top of slab = $(3.375 - 7.89 + 4.56) = 0.045$ N/mm^2

At bottom of slab = $(3.375 + 7.89 - 4.56) = 6.705$ N/mm^2

Stresses at working loads

At top of slab = $0.8 (3.375 - 7.89) + 4.56 + 4.48 = 5.428$ N/mm^2

At bottom of slab = $0.8 (3.375 + 7.89) - 4.56 - 4.48$

$$= - 0.028 \text{ N/mm}^2$$

The actual stresses developed are within the permissible limits.

11. Check for Ultimate Flexural Strength

The moment of resistance of rectangular sections is evaluated by using the expression specified in IS: 1343-2012 code as

$$M_u = f_{pb} A_{ps} (d - 0.42 x_u)$$

Where f_{pb} = tensile stress in the tendons at failure

f_{pe} = effective prestress in tendons

A_{ps} = area of prestressing tendons in the tension zone

d = effective depth to the centroid of the steel area

x_u = neutral axis depth

For pretensioned and post tensioned members with effective bond, the values of f_{pb} and x_u are interpolated using the values given in Table 11.2

Table 11.2 Conditions at the Ultimate Limit State for Rectangular Beams with Pre-tensioned Tendons or Post-tensioned Tendons Having Effective Bond

(Table 11 of IS: 1343-2012)

Sl. No.	$\dfrac{A_{ps}.f_{pu}}{bd.f_{ck}}$	Stress in Tendon as Proportion of the Design Strength $\dfrac{f_{pb}}{0.87 f_{pu}}$		Ratio of the Depth of Neutral Axis to that of the Centroid of the Tendon in the Tension Zone x_u/d	
		Pre-tensioning	Post-tensioning with Effective Bond	Pre-tensioning	Post-tensioning with Effective Bond
(1)	(2)	(3)	(4)	(5)	(6)
i)	0.025	1.0	1.0	0.054	0.054
ii)	0.05	1.0	1.0	0.109	0.109
iii)	0.10	1.0	1.0	0.217	0.217
iv)	0.15	1.0	1.0	0.326	0.316
v)	0.20	1.0	0.95	0.435	0.414[D]
vi)	0.25	1.0	0.90	0.542	0.488[D]
vii)	0.30	1.0	0.85	0.655	0.558[D]
viii)	0.40	0.9	0.75	0.783	0.653[D]

[D] The netural axis depth in these cases is too low to provide the necessary elongation for developing $0.87 f_{pu}$ stress level. Hence, it is essential that the strength provided exceeds the required strength by 15 percent for these cases.

The values of the various parameters are computed as,

$$A_{ps} = [(12 \times 38.5 \times 1000)/328] = 1408 \text{ mm}^2$$
$$d = 445 \text{ mm}$$
$$f_{pu} = 1500 \text{ N/mm}^2$$
$$f_{ck} = 40 \text{ N/mm2}$$
$$b = 1000 \text{ mm}$$

Compute the ratio $\left(\dfrac{A_{ps} f_{pu}}{b.d.f_{ck}} \right) = \left(\dfrac{1408 \times 1500}{1000 \times 445 \times 40} \right) = 0.118$

Refer Table 11.2 and interpolate the values of the ratios $\left(\dfrac{f_{pb}}{0.87 f_{pu}} \right)$ and $\left(x_u/d \right)$ as

$$\left(\dfrac{f_{pb}}{0.87 f_{pu}} \right) = 1.0 \text{ and } \left(x_u/d \right) = 0.267$$

Hence $\qquad f_{pb} = (1.0 \times 0.87 \times 1500) = 1305 \text{ N/mm}^2$

And $\qquad x_u = (0.267 \times 445) = 118.8 \text{ mm}$

∴ $\qquad M_u = f_{pb} A_{ps} (d - 0.42 x_u)$
$$= [1305 \times 1408)(445 - 0.42 \times 118.8)] = (727.8 \times 10^6) \text{ N.mm}$$
$$= 727.8 \text{ kN·m}$$

According to IRC: 6-2014, the required moment of resistance is given by

$$M_u = [(1.35 M_g + 1.5 M_q)]$$
$$= [(1.35 \times 190) + (1.5 \times 187)]$$
$$= 537 \text{ kN·m} < 727.8 \text{ kN·m}$$

Hence the ultimate moment capacity of the designed section is greater than the required ultimate moment.

12. Check for Ultimate Shear Strength

Ultimate Shear Force $= [(1.35 V_g + 1.5 V_q)]$
$$= [(1.35 \times 72.8) + (1.5 \times 87.72)]$$
$$= 130 \text{ kN}$$

The design shear resistance of the support section is calculated by using the equation specified in IRC: 112-2011 clause 10.3 as

$$V_{Rd.c} = \left(\dfrac{I.b_w}{S} \right) \sqrt{(f_{ctd})^2 \mp k_1 \sigma_{cp} f_{ctd}} + \eta P \sin \theta$$

Where $\qquad I = $ second moment of area of gross cross section

$\qquad S = $ first moment of area between centroidal axis and compression fiber about centroidal axis

$\qquad f_{ctd} = $ design value of concrete tensile strength $= (f_{ct}/\gamma_m)$

K_1 = Constant depending upon transmission length and has a value of 1 for post tensioned beams

σ_{cp} = mean compressive stress at centroidal axis = $(\eta P/A_c)$

η = loss ratio

P = prestressing force

θ = slope of the cable at support section

Computing the numerical values of the various parameters, we have

$$I = \left(\frac{bD^3}{12}\right) = \left(\frac{1000 \times 500^3}{12}\right) = (10.4 \times 10^9) \text{ mm}^4$$

$$S = \left(\frac{1000 \times 250 \times 50}{2}\right) = (31.25 \times 10^6) \text{ mm}^3$$

$$f_{ctd} = (f_{ct}/\gamma_m) = (3/1.5) = 2 \quad \text{since } f_{ct} = 3 \text{ N/mm}^2 \text{ for } f_{ck} = 40 \text{ N/mm}^2$$

$$\sigma_{cp} = (\eta P/A_c) = [(0.8 \times 1687.5 \times 10^3)/(1000 \times 500)] = 2.7 \text{ N/mm}^2$$

$$\theta = (4e/L) = [(4 \times 195)/(10.4 \times 1000)] = 0.075$$

$$\therefore \quad V_{Rd.c} = \left(\frac{I.b_w}{s}\right)\sqrt{\left(f_{ctd}\right)^2 \mp k_1 \sigma_{cp} f_{ctd}} + \eta P \sin\theta$$

$$= \left(\frac{(10.4 \times 10^9)1000}{31.25 \times 10^6}\right)\sqrt{(2)^2 \mp 1 \times 2.7 \times 2}$$

$$+ \left(0.8 \times 1687.5 \times 10^3 \times 0.075\right) \text{N}$$

$$= (493.69 \times 10^3) \text{ N}$$

$$= 493.69 \text{ kN} > 130 \text{ kN}$$

The shear resistance of the support section is greater than the required ultimate shear force.

13. Supplementary Reinforcement

According to Clause 16.5.1 of IRC: 112-2011, the minimum longitudinal reinforcement should be not less than that given by the relation,

$$A_{s.min} = 0.26 \ (f_{ctm}/f_{yk}) \ b_t \ d \text{ but not less than } 0.0013 \ b_t \ d$$

Where b_t = mean width of the tension zone

$$A_{s.min} = [0.26 \ (3/415) \ (1000 \times 450)] \text{ but not less than } (0.0013 \times 1000 \times 450)$$

$$= 846 \text{ mm}^2/\text{m} \quad \text{or} \quad 585 \text{ mm}^2/\text{m}$$

Provide 10 mm diameter Fe-415 HYSD bars at a spacing of 200 mm both at top and bottom faces of the slab in the longitudinal and transverse directions.

14. Check for Serviceability Limit States

a) Limit State of Deflection

At service loads, deflections due to prestressing force, dead load and live load is computed as given below:

Dead load = g = 14 kN/m = 0.014 kN/mm

Live load spread over a length of 4.76 m is assumed as a concentrated load at centre of span is computed as

$$Q = (4.76 \times 19.58) = 93.2 \text{ kN}$$

Effective prestressing force after losses = $\eta P = (0.8 \times 1687.5) = 1350$ kN

$$E_c = 33 \text{ KN/mm}^2$$
$$I = (10.4 \times 10^9) \text{ mm}^4$$

Upward deflection due to prestressing force

$$a_p = \left(\frac{5PeL^2}{48E_cI}\right) = \left(\frac{5 \times 1350 \times 195 \times 10400^2}{48 \times 33 \times (10.4 \times 10^9)}\right) = 8.69 \text{ mm(upwards)}$$

Down ward deflection due to dead weight

$$a_p = \left(\frac{5gL^4}{384E_cI}\right) = \left(\frac{5 \times 0.014 \times 10400^4}{384 \times 33 \times (10.4 \times 10^9)}\right) = 6.2 \text{ mm(downwards)}$$

Down ward deflection due to live load

$$a_p = \left(\frac{QL^3}{48E_cI}\right) = \left(\frac{93.2 \times 10400^3}{48 \times 33 \times (10.4 \times 10^9)}\right) = 6.36 \text{ mm(downwards)}$$

Maximum deflection due to prestress + self weight + live loads

$$a_r = (a_p + a_g + a_Q) = (-8.69 + 6.2 + 6.36) = 5.67 \text{ mm}$$

Long term deflection considering the effect of creep is limited to (span/250)

Notional size of cross section = $(2A_c/u) = [(2 \times 1000 \times 500)/2000] = 500$ mm

From Table 4.12, interpolate the final creep coefficient for the given notional size of 500 mm at relative humidity of 50% and age at loading of 28 days as Ø = 2.50.

Effective modulus of elasticity of concrete = $E_{c,\text{eff}} = \left[\dfrac{E_c}{1+\text{Ø}}\right] = \left[\dfrac{E_c}{1+2.5}\right] = \left[\dfrac{E_c}{3.5}\right]$

Maximum long term deflection = $(3.5 \times 5.67) = 19.84$ mm < $(10400/250) = 41.6$ mm

Maximum permissible deflection due to loads only

$$\leq \left(\frac{\text{span}}{800}\right) = \left(\frac{10400}{800}\right) = 13 \text{ mm} > 6.36 \text{ mm}$$

b) Limit State of Cracking

The deck slab has been designed as a Class-1 type structure without any tensile stresses at service loads. Hence the serviceability limit state of cracking is automatically satisfied.

Hence the slab is safe regarding the serviceability limit state of deflection and cracking according to the specifications of IRC: 112-2011.

15. Design of End Block Reinforcement

At the support section, concentric cables carrying a force of 554 kN are spaced at intervals of 328 mm. The end block has to be designed for bursting tension due to the anchorage force.

The bursting tensile force is computed using the Table 11.3 as recommended in IRC: 112-2011

Table 11.3 Design Bursting Tensile Force in End Blocks
(Table 13.1 of IRC: 112-2011)

(Y_{po}/Y_o)	0.3	0.4	0.5	0.6	0.7
(F_{bst}/P_k)	0.26	0.23	0.19	0.16	0.12

In the present design problem

$$P_k = 554 \text{ kN}$$
$$2\,Y_{po} = 150 \text{ mm}$$
$$2\,Y_o = 328 \text{ mm}$$

Ratio $(Y_{po}/Y_o) = (150/328) = 0.457$

Interpolating the value of (F_{bst}/P_k) for $(Y_{po}/Y_o) = 0.457$ from Table 11.3

$$(F_{bst}/P_k) = 0.185$$

$$\therefore \qquad F_{bst} = (0.185 \times 554) = 103 \text{ kN}$$

Using 10 mm diameter Fe-415 HYSD bars as end block reinforcement

$$\text{Area of steel required} = \left(\frac{103 \times 10^3}{0.87 \times 415} \right) = 285 \text{ mm}^2$$

Provide 10 mm diameter bars at 100 mm centers in the vertical and horizontal direction as a mesh in front of the anchorages at 100 and 200 mm respectively.

16. Reinforcement Details in the Deck Slab

The reinforcement details in the cross section and longitudinal section of the deck slab are shown in Figs. 11.9 and 11.10.

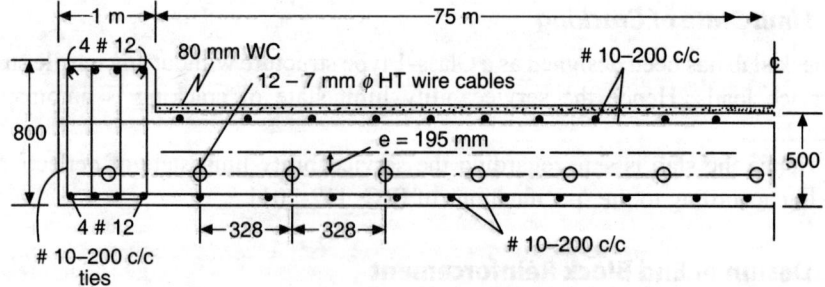

Fig. 11.9 Cross Section of Deck Slab at Centre of Span.

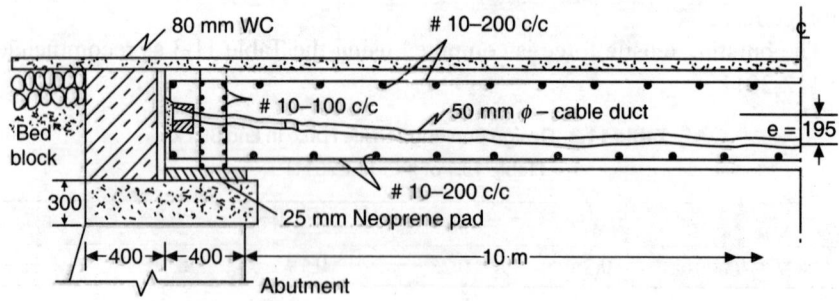

Fig. 11.10 Longitudinal Section of Deck Slab.

11.6 DESIGN OF POST-TENSIONED PRESTRESSED CONCRETE TEE BEAM AND SLAB BRIDGE DECK

Design a post tensioned prestressed concrete Tee beam slab bridge deck for a National High way crossing to suit the following data:

1. Data

Effective span = 30 m
Width of road = 7.5 m
Kerbs = 600 mm on each side
Foot path = 1.5 m wide on each side
Thickness of wearing coat = 80 mm
Live load = I.R.C. Class AA tracked vehicle
For deck slab, adopt M-20 grade concrete

For prestressed concrete girders, adopt M-50 grade concrete with cube strength at transfer as 40 N/mm².

Loss ratio = 0.85
Spacings of cross girders = 5 m

Adopt Fe-415 grade HYSD bars and strands of 15.2 mm-7 ply conforming to IS: 6006-1983 are available for use.

Design the girders as Class I type member.

Permissible stress in concrete at transfer = 18 N/mm²

Permissible stress in concrete at service loads = 16 N/mm²

The design should conform to the specifications of the codes IRC: 6-2014[10], IRC: 112-2011[11] and IS: 1343-2012[12].

2. Stresses in Concrete and Steel

For M-20 grade concrete and Fe-415 HYSD bars adopt the following parameters.

$$f_{ck} = 20 \text{ N/mm}^2 \text{ and } f_y = 415 \text{ N/mm}^2$$

$$M_u = 0.138 f_{ck} b d^2$$

For M-50 grade concrete and High tensile steel cables

$$f_{ck} = 50 \text{ N/mm}^2$$

$$f_{ct} = 18 \text{ N/mm}^2$$

$$f_{cw} = 16 \text{ N/mm}^2$$

$$E_c = 35 \text{ kN/mm}^2$$

Freyssinet system H.T cables of Type 7K-15 (7 strands of 15.2 mm diameter) in 65 mm cable ducts conforming to IS: 6006-1983[13]

3. Cross Section of Deck

4 main girders are provided at 2.5 m intervals.

Thickness of deck slab = 250 mm

Wearing coat = 80 mm

Kerbs 600 mm wide by 300 mm deep are provided

The cross section of the deck is shown in Fig. 11.11

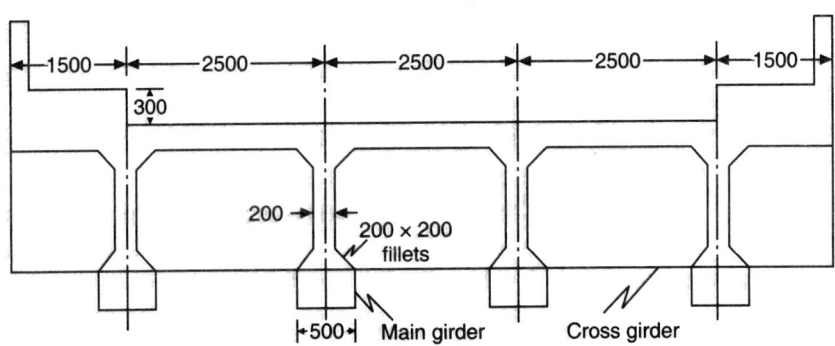

Fig. 11.11 Cross Section of Bridge Deck.

The main girders are precast and the slab connecting the girders is cast *in-situ*.

Spacing of cross girders = 5 m

Spacing of main girders = 2.5 m

4. Design of Interior Slab Panel

(a) Bending Moments

Dead weight of slab $= (1 \times 1 \times 0.25 \times 24) = 6.00 \text{ kN/m}^2$
Death weight of W.C. $= (0.08 \times 22) \qquad\qquad = 1.76$
Total dead load $\qquad\qquad\qquad\qquad\qquad\quad = \overline{7.76 \text{ kN/m}^2}$

Live load is IRC Class AA tracked vehicle, one wheel is placed at the centre of panel as shown in Fig. 11.12.

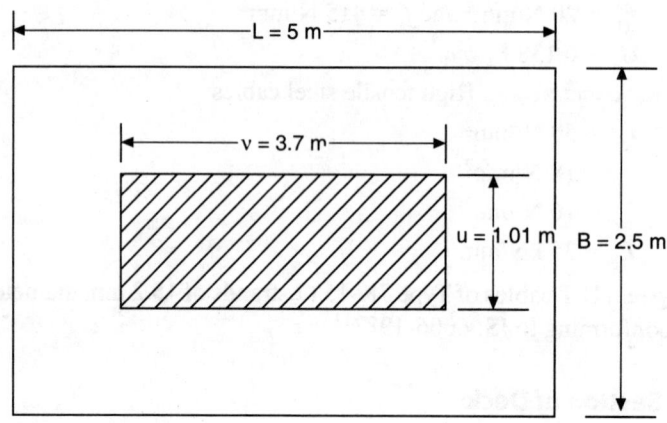

Fig. 11.12 Position IRC Class AA Wheel Load for Maximum Bending Moment.

$$u = (0.85 + 2 \times 0.08) = 1.01 \text{ m}$$
$$v = (3.60 + 2 \times 0.08) = 3.76 \text{ m}$$
$$(u/B) = (1.01/2.5) = 0.404$$
$$(v/L) = (3.76/5.0) = 0.752$$
$$K = (B/L) = (2.5/5.0) = 0.5$$

Referring to Pigeaud's curves (Fig. 11.13)

$$m_1 = 0.098 \text{ and } m_2 = 0.02$$
$$M_B = W (m_1 + 0.15 \, m_2)$$
$$= 350 (0.098 + 0.15 \times 0.02) = 35.35 \text{ kN·m}$$

As the slab is continuous, design B.M $= 0.8 \, M_B$. Design B.M, including impact and continuity factor is given by

M_B (Short span) $= (1.25 \times 0.8 \times 35.35) = 35.35 \text{ kN·m}$
Similarly $\qquad M_L = W (m_2 + 0.15 \, m_1)$
$$\qquad\qquad = 350 (0.02 + 0.15 \times 0.098) = 12.14 \text{ kN·m}$$
M_L (Long span) $= (1.25 \times 0.8 \times 12.14) = 12.14 \text{ kN·m}$

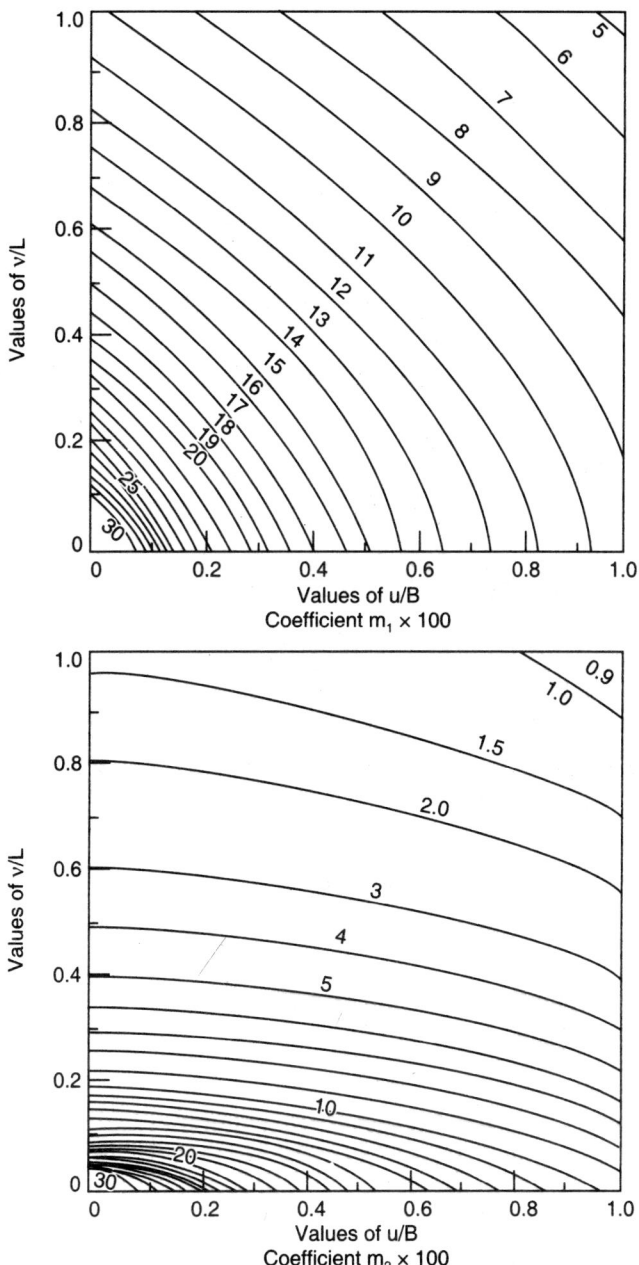

Fig. 11.13 Moment Coefficients m_1 and m_2 for $K = 0.5$ (Pigeaud's curves).

(b) Shear Forces

Dispersion in the direction of span

$$= [0.85 + 2 (0.08 + 0.25)] = 1.51 \text{ m}$$

For maximum shear, load is kept such that the whole dispersion is in span. The load is kept at $(1.51/2) = 0.755$ m from the edge of the beam as shown in Fig. 11.14

Effective width of slab = $kx [1 - (x/L)] + b_W$

Breadth of cross girder = 200 mm

Clear length of panel = $(5 - 0.2) = 4.8$ m

$\therefore \qquad (B/L) = (4.8/2.3) = 2.08$

From table 4.6, k for continuous slab is obtained as 2.60

Effective width of slab = $2.6 \times 0.755 [1 \times (0.755/2.3)] + 3.6 + (2 \times 0.08)$

$\qquad\qquad\qquad\qquad = 5.079$ m

Load per metre width $\quad = (350/5.079) = 70$ kN

Shear force/metre width $= 70 (2.3 - 0.755)/2.3 = 47$ kN

Shear force with impact $= (1.25 \times 47) = 58.75$ kN

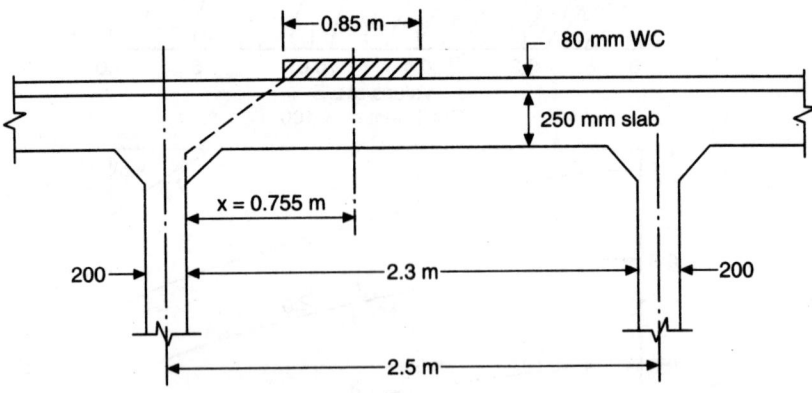

Fig. 11.14 Position of Wheel Loads of Maximum Shear.

(c) Dead Load Bending Moments and Shear Forces

Dead load = 7.76 kN/m^2

Total load on panel = $(5 \times 2.5 \times 7.76) = 97$ kN

$(u/B) = 1$ and $(v/L) = 1$, as panel is loaded with uniformly distributed load.

$$k = (B/L) - (2.5/5) = 0.5 \text{ and } (1/k) = 2.0$$

From Pigeaud's curve (refer Fig. 11.15)

$$m_1 = 0.047, m_2 = 0.01$$
$$M_B = 97 (0.047 + 0.15 \times 0.01) = 4.70 \text{ kN·m}$$
$$M_L = 97 (0.01 + 0.15 \times 0.047) = 1.65 \text{ kN·m}$$

Design B.M including continuity factor,

$$M_B = (0.8 \times 4.7) = 3.76 \text{ kN·m}$$
$$M_L = (0.8 \times 1.65) = 1.32 \text{ kN·m}$$

Dead load shear force = $(0.5 \times 7.76 \times 2.3) = 8.924$ kN

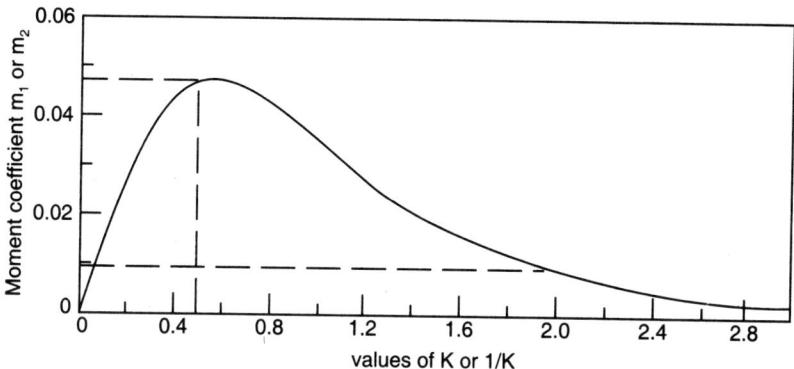

Fig. 11.15 Moment Coefficients for Slabs Completely Loaded with Uniformly Distributed Load, Coefficient is m_1 for K and m_2 for $1/K$.

(d) Design Moments and Shear Forces

Short span moment = $M_B = (35.35 + 3.76) = 39.11$ kN·m

Long span moment = $M_L = (12.14 + 1.32) = 13.46$ kN·m

Shear Force = $V = (V_g + V_L) = (8.92 + 58.75) = 67.67$ kN

Design Ultimate Load Moments & Shear Forces are computed by applying appropriate load factors to the service load moments.

Total Design Short Span Ultimate Moment (M_{Bu}) = $[1.35 \, M_d + 1.5 \, M_L]$

$$= [(1.35 \times 3.76) + (1.5 \times 35.35)]$$
$$= 58.1 \text{ kN·m/m}$$

Total Design Long Span Ultimate Moment (M_{Lu}) = $[(1.35 \times 1.32) + (1.5 \times 12.14)]$

$$= 20 \text{ kN·m/m}$$

Total Design Ultimate Shear Force = $V_u = [(1.35 \times 8.92) + (1.5 \times 58.75)] = 90.26$ kN

(e) Design of Deck Slab and Reinforcements

Effective depth of slab required = $d = \sqrt{\dfrac{M_u}{0.138 f_{ck}.b}} = \sqrt{\dfrac{58.1 \times 10^6}{0.138 \times 20 \times 1000}} = 145$ mm

Adopt effective depth, $d = 200$ mm and overall depth of 250 mm

Using 12 mm diameter bars

Effective depth provided = 200 mm

$$\left(\frac{M_u}{b.d^2} \right) = \left(\frac{58.1 \times 10^6}{1000 \times 200^2} \right) = 1.45, \text{ using M-25 grade concrete and Fe-415 HYSD bars}$$

Read out the percentage of reinforcement required from Table 2 of SP: 16 Design Aids

$$p_t = 0.443 = \left(\frac{100 A_{st}}{b.d} \right)$$

Solving $A_{st} = \left(\dfrac{0.443 \times 1000 \times 200}{100}\right) = 886\,\text{mm}^2$

For short span, Provide 12 mm diameter bars at 120 mm centers (A_{st} provided = 942 mm^2). For long span, provide 10 mm diameter bars at 150 mm centers.

(f) Check for Ultimate Flexural Strength

$$M_u = (0.87 \times 415 \times 942 \times 200)\left[1 - \frac{942 \times 415}{1000 \times 200 \times 20}\right]$$

$$= (61.42 \times 10^6)\,\text{N.mm}$$
$$= 61.42\,\text{kN·m} > 58.1\,\text{kN·m (Hence safe)}$$

(g) Check for Ultimate Shear Strength

The ultimate shear strength of the reinforced concrete deck slab is checked by using the equation 4.7.

$$V_{Rd.c} = [0.12K\,(80\rho_1 f_{ck})^{0.33}]b_w.d$$

where $\quad K = 1 + \sqrt{\dfrac{200}{d}} \le 2.00 = \left[1 + \sqrt{\dfrac{200}{200}}\right] = 2.00$

and $\quad \rho_1 = \left(\dfrac{A_{sl}}{b_w.d}\right) \le 0.02$

$$= \left(\frac{942}{1000 \times 200}\right) = 0.0047$$

$$V_{Rd.c} = [0.12 \times 2.00(80 \times 0.0047 \times 20)^{0.33}](1000 \times 200)\,\text{N}$$
$$= (96 \times 1000)\,\text{N}$$
$$= 96\,\text{kN} > 90.26\,\text{kN (Hence safe)}$$

5. Design of Longitudinal Girders

(a) Reaction Factors

Using Courbon's theory, the I.R.C. Class AA loads are arranged for maximum eccentricity as shown in Fig. 11.16.

Reaction factor for outer girder A is

$$R_A = \frac{2W_1}{4}\left[1 + \frac{4I \times 3.75 \times 1.1}{(2I \times 3.75^2) + (2I \times 1.25^2)}\right] = 0.764\,W_1$$

Reaction factor for inner girder B is

$$R_B = \frac{2W_1}{4}\left[1 + \frac{4I \times 1.25 \times 1.1}{(2I \times 3.75^2) + (2I \times 1.25^2)}\right] = 0.588\,W_1$$

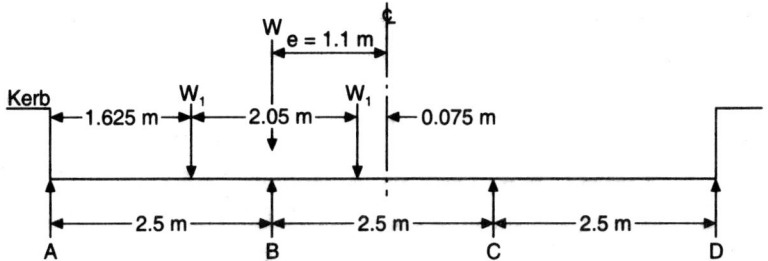

Fig. 11.16 Transverse Disposition of IRC Class AA Tracked Vehicle.

If $\qquad W$ = Axle load = 700 kN

$\qquad\qquad W_1$ = 0.5 W

∴ $\qquad\quad R_A$ = (0.764 × 0.5 W) = 0.382 W

$\qquad\qquad R_B$ = (0.588 × 0.5 W) = 0.294 W

(b) *Dead Load from Slab Per Girder*

The dead load of deck slab is calculated with reference to Fig. 11.17.

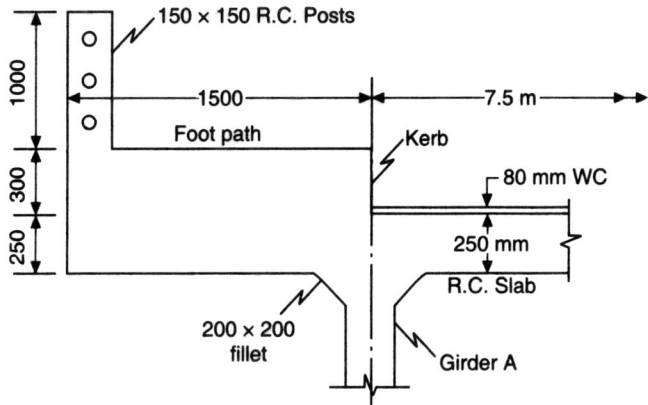

Fig. 11.17 Details of Foot Path, Kerb, Parapet and Deck Slab.

Weight of

1. Parapet railing (Lumpsum) $\qquad\qquad\qquad$ 0.92 kN/m
2. Foot path and kerb = (0.3 × 1.5 × 24) $\qquad$ 10.08
3. Deck slab = (0.25 × 1.5 × 24) $\qquad\qquad$ 9.00
$\qquad\qquad\qquad\qquad\qquad\qquad\qquad\qquad\qquad$ ——————
$\qquad\qquad\qquad\qquad\qquad\qquad\qquad\qquad\qquad$ 20.00 kN/m

Total dead load of deck = [(2 × 20) + (7.76 × 7.5)] = 98.2 kN/m

It is assumed that the deck load is shared equally by all the four girders

Dead load/girder = (98.2/4) = 24.55 kN/m

(c)   Dead Load of Main Girder

The overall depth of the girder is assumed as 1800 mm at the rate of 60 mm for every metre of span.

Span of the girder = 30 m
Overall depth = (60 × 30) = 1800 mm

The bottom flange is selected so that four to 6 cables are easily accommodated in the flange. The section of the main girder selected in shown in Fig. 11.18.

Dead weight of rib = (1.15 × 0.2 × 24)   = 5.52 kN/m
Dead weight of bottom flange = (0.5 × 0.4 × 24)   = 4.80
                                                 ───────────
                                                 10.32 kN/m

Weight of cross girder = (0.2 × 1.25 × 24) = 6 kN/m

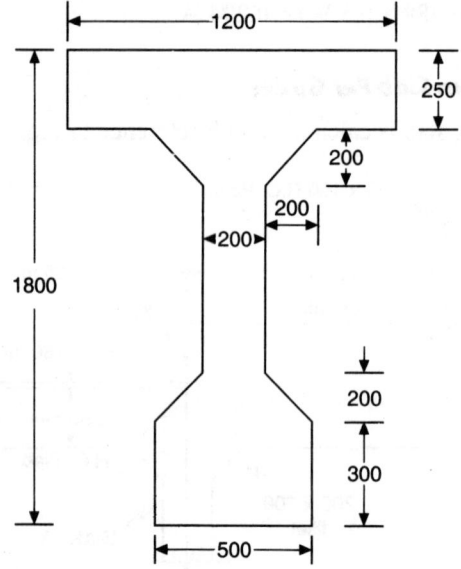

Fig. 11.18 Cross Section of Prestressed Concrete Girder.

(d)   Dead Load Moments and Shears in Main Girder

Reaction from deck slab on each girder = 24.55 kN/m
Weight of cross girder = 6 kN/m
Reaction on main girder = (6 × 2.5) = 15 kN/m
Self weight of main girder = 10.32 kN/m
Total dead load on girder = (24.55 + 10.32) = 34.87 kN/m

The maximum dead load bending moment and shear force is computed using the loads shown in Fig. 11.19 thus

$$M_{max} = [(0.125 \times 34.87 \times 30^2) + (0.25 \times 15 \times 30) + (15 \times 10) + (15 \times 5)]$$
$$= 4261 \text{ kN·m}$$

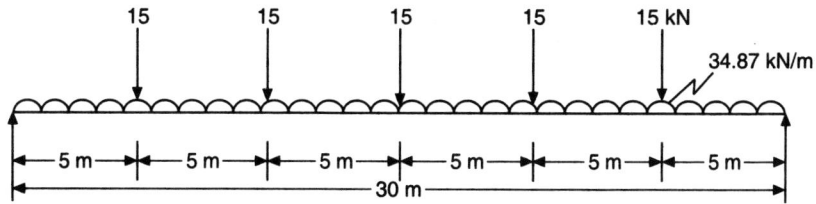

Fig. 11.19 Dead Load on Main Girder.

Dead load shear at support

$$V_{max} = [(0.5 \times 34.87 \times 30) + (0.5 \times 75)]$$
$$= 561 \text{ kN}$$

(e) Live Load Bending Moments in Girder

Span of the girder = 30 m
Impact factor (Class AA) = 10%
The live load is placed centrally on the span as shown in Fig. 11.20.

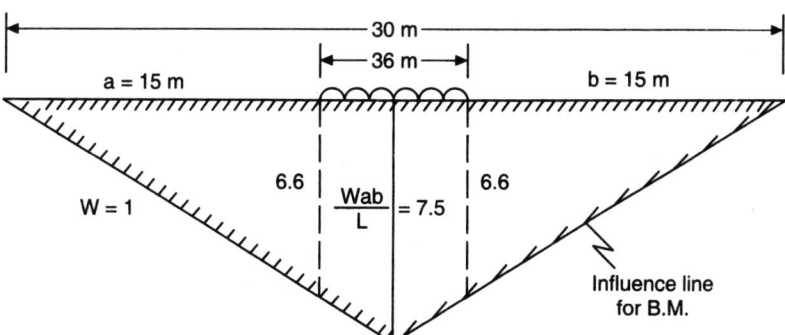

Fig. 11.20 Influence Line for Bending Moment in Girder.

Bending Moment at centre of span

$$= 0.5 (6.6 + 7.5) 700 = 4935 \text{ kN·m}$$

B.M including impact and reaction factors for outer girder is,
Live load B.M = $(4935 \times 1.1 \times 0.382) = 2074$ kN·m
For Inner girder, B.M = $(4935 \times 1.1 \times 0.294) = 1596$ kN·m

(f) Live Load Shear Forces in Girders

For estimating the maximum live load shear in the girders, the I.R.C. Class AA loads
are placed as shown in the Fig. 11.21.

Reaction of W_2 on girder $B = (350 \times 0.45)/2.5 = 63$ kN
Reaction of W_2 on the girder $A = (350 \times 2.05)/2.5 = 287$ kN
Total load on girder $B = (350 + 63) = 413$ kN

Maximum reaction in girder $B = (413 \times 28.2)/30 = 388$ kN

Maximum reaction in girder $A = (287 \times 28.2)/30 = 270$ kN

Maximum live load shear with impact factor in inner girder

$$= (388 \times 1.1) = 427 \text{ kN}$$

Outer girder $= (270 \times 1.1) = 297$ kN

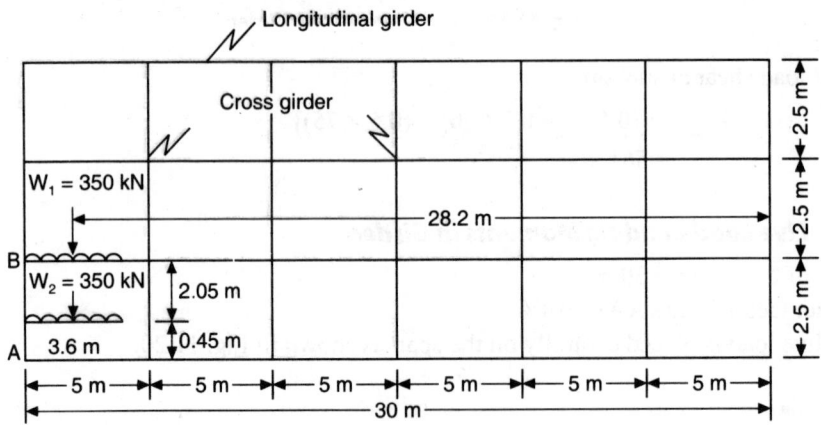

Fig. 11.21 Position of IRC Class AA Loads for Maximum Shear.

(g) Design Bending Moments and Shear Forces

The design moments and shear forces are compiled in Table 11.4.

Table 11.4 Abstract of Design Moments and Shear Forces in Main Girders

Bending Moment	D.L.B.M	L.L.B.M	Total B.M	Units
Outer Girder	4261	2074	6335	kN·m
Inner Girder	4261	1596	5857	kN·m
Shear Force	D.L.S.F	L.L.S.F.	Total S.F.	Units
Outer Girder	561	297	858	kN
Inner Girder	561	427	988	kN

(h) Properties of Main Girder Section

The main girder section is as shown in Fig. 11.22 for computational purposes. The properties of the section are:

$$A = 73 \times 10^4 \text{ mm}^2$$
$$y_1 = 750 \text{ mm}, y_b = 1050 \text{ mm}, I = 2924 \times 10^8 \text{ mm}^4$$
$$z_1 = (I/y_t) = (2924 \times 10^8)/750 = 3.89 \times 10^8 \text{ mm}^4$$
$$z_b = (I/y_b) = (2924 \times 10^8)/1050 = 2.78 \times 10^8 \text{ mm}^3$$

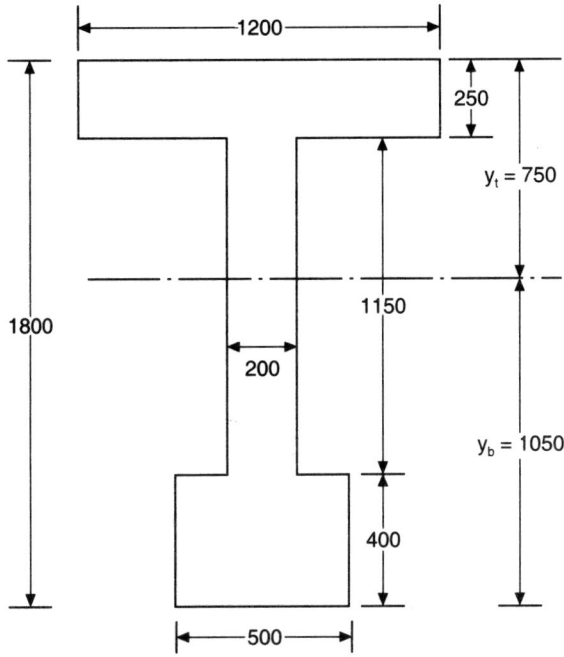

Fig. 11.22 Cross Section of Main Girder.

(i) Check for Minimum Section Modulus

$f_{ck} = 50$ N/mm² $\qquad \eta = 18$ N/mm²

$f_{ct} = 18$ N/mm² $\qquad M_g = 4261$ kN·m

$f_{ci} = 40$ N/mm² $\qquad M_q = 2074$ kN·m

$f_{tt} = f_{tw} = 0 M_d = (M_g + M_q) = 6335$ kN·m

$f_{cw} = 16$ N/mm²

$f_{br} = (\eta f_{ct} - f_{tw}) = (0.85 \times 18 - 0) = 15.3$ N/mm²

$f_{tr} = (f_{cw} - \eta f_{tt}) = 16$ N/mm²

$f_{inf} = (f_{tw}/\eta) + (M_d/\eta z_b)$

$\qquad = 0 + (6335 \times 10^6)/(0.85 \times 2.78 \times 10^8)$

$\qquad = 26.80$ N/mm²

$$z_b = \left[\frac{M_q + (1-\eta) M_g}{f_{br}} \right]$$

$$= \left[\frac{(2074 \times 10^6) + (1 - 0.85) 4261 \times 10^6}{15.3} \right]$$

$$= 1.77 \times 10^8 \text{ mm}^3 < 2.78 \times 10^8)$$

Hence the section provided is adequate.

(j) Prestressing Force

Allowing for two rows of cables, cover required = 200 mm

Maximum possible eccentricity $e = (1050 - 200) = 850$ mm

Prestressing force is obtained as

$$P = (A.f_{\text{inf}}.Z_b)/(Z_b + A.e)$$
$$= [(0.73 \times 10^6 \times 26.80 \times 2.78 \times 10^8)/(2.78 \times 10^8) + (0.73 \times 10^6 \times 850)]$$
$$= 6053 \times 10^3 \text{ N}$$
$$= 6053 \text{ kN}$$

Using Freyssinet system, anchorage type 7K-15 (7 strands of 15.2 mm diameter) in 65 mm cables ducts, (IS: 6006-1983)

Force in each cable $= (7 \times 0.8 \times 260.7) = 1459$ kN

∴ Number of cables $= (6053/1459) \approx 5$

Area of each strand $= 140 \text{ mm}^2$

Area of 7 strands in each cable $= (7 \times 140) = 980 \text{ mm}^2$

Area of strands in 5 cables $= A_p = (5 \times 980) = 4900 \text{ mm}^2$

The cables are arranged at centre of span section as shown in Fig. 11.23.

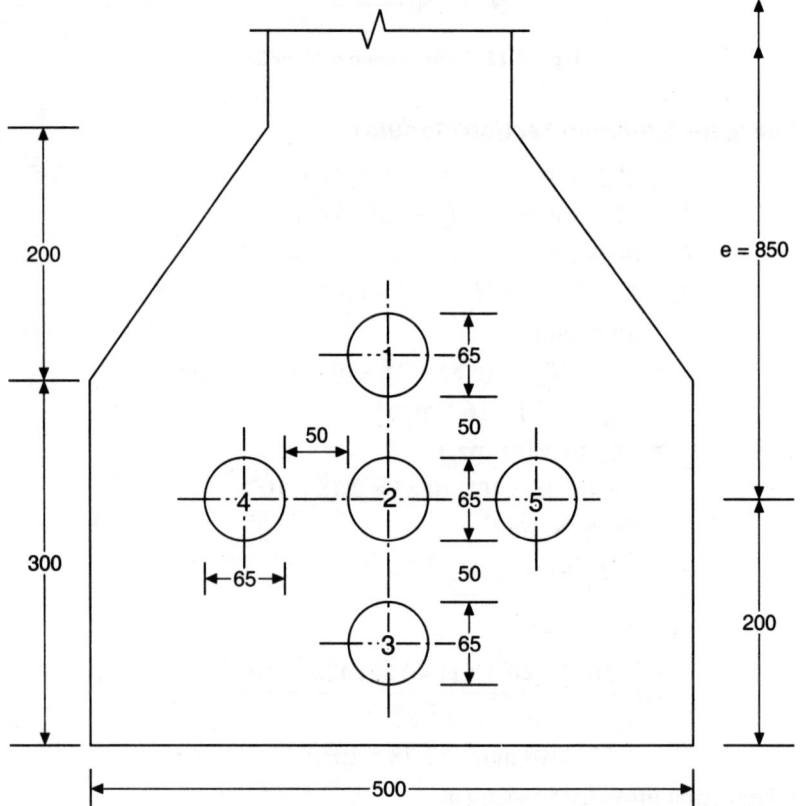

Fig. 11.23 Arrangement of Cable at Centre of Span Section.

(k) Permissible Tendon Zone

At support section,

$$e \leq (Z_b \cdot f_{ct}/P) - (Z_b/A)$$
$$\leq (2.78 \times 10^8 \times 18)/(6053 \times 10^3) - (2.78 \times 10^8)/(0.73 \times 10^6)$$
$$\leq 445 \text{ mm}$$

and

$$e \geq (Z_b \cdot f_{tw}/\eta P) - (Z_b/A)$$
$$\geq 0 - (2.78 \times 10^8)/(0.73 \times 10^6)$$
$$\geq -380 \text{ mm}$$

The 5 cables are arranged to follow a parabolic profile with the resultant force having an eccentricity of 180 mm towards the soffit at the support section. The position of cables at support section is shown in Fig. 11.24.

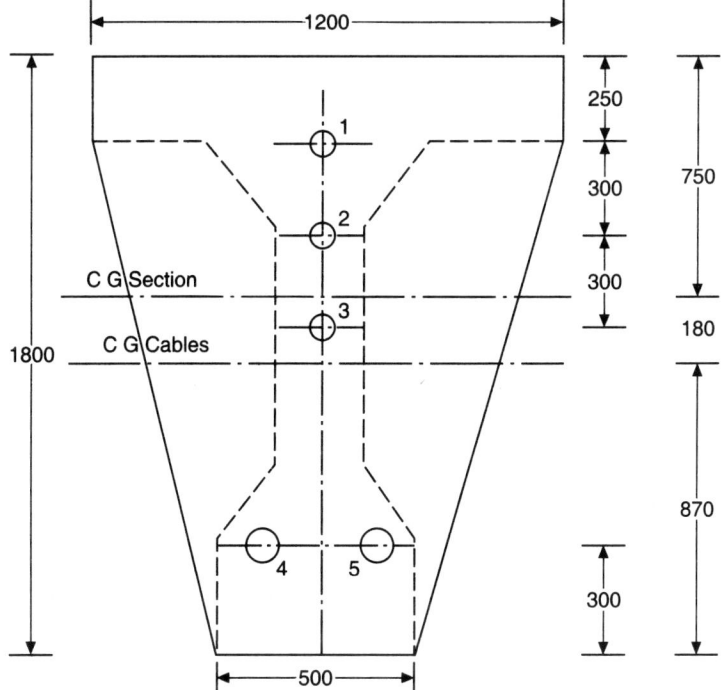

Fig. 11.24 Arrangement of Cables at Support Section.

6. Check for Stresses

For the centre of span section, we have

$P = 6053$ kN	$z_t = 3.89 \times 10^8$ mm^3
$e = 850$ mm	$h = 0.857? = 0.85$
$A = 0.73 \times 10^6$ mm^2	$M_g = 4261$ kN·m
$Z_b = 2.78 \times 10^8$ mm^3	$M_q = 2074$ kN·m

$$(P/A) = (6053 \times 10^3)/(0.73 \times 10^6) = 8.29 \text{ N/mm}^2$$
$$(Pe/z_t) = (6053 \times 10^3 \times 850)/(3.89 \times 10^8) = 13.22 \text{ N/mm}^2$$
$$(Pe/Z_b) = (6053 \times 10^3 \times 850)/(2.78 \times 10^8) = 18.50 \text{ N/mm}^2$$
$$(M_g/Z_t) = (4261 \times 10^6)/(3.89 \times 10^8) = 10.95 \text{ N/mm}^2$$
$$(M_g/z_b) = (4261 \times 10^6)/(2.78 \times 10^8) = 15.32 \text{ N/mm}^2$$
$$(M_q/Z_t) = (2074 \times 10^6)/(3.89 \times 10^8) = 5.33 \text{ N/mm}^2$$
$$(M_q/Z_b) = (2074 \times 10^6)/(2.78 \times 10^8) = 7.46 \text{ N/mm}^2$$

At transfer stage:

$$\sigma_t = [(P/A) - (Pe/Z_t) + (M_g/Z_t)]$$
$$= (8.29 - 13.22 + 10.95)$$
$$= 6.02 \text{ N/mm}^2$$
$$\sigma_b = [(P/A) + (Pe/Z_b) - (M_g/Z_b)]$$
$$= [8.29 + 18.50 - 15.32]$$
$$= 11.47 \text{ N/mm}^2$$

At working load stage:

$$\sigma_t = [\eta (P/A) - \eta (Pe/Z_b) - (M_g/Z_t) + (M_q/Z_t)]$$
$$= [0.85 (8.29 - 13.22) + 10.95 + 5.33]$$
$$= 12.09 \text{ N/mm}^2 \text{ (Compression)}$$
$$\sigma_b = [\eta (P/A) - \eta (Pe/Z_b) - (M_g/Z_b) + (M_q/Z_b)]$$
$$= [0.85 (8.29 + 18.50) - 15.32 - 7.46]$$
$$= -0.01 \text{ N/mm}^2 \text{ (Tension)}$$

All the stresses at top and bottom fibres at transfer and service loads are well within the safe permissible limits.

7. Check for Ultimate Flexural Strength

For the centre of span section

$$A_p = (5 \times 7 \times 1400) = 4900 \text{ mm}^2$$
$$b = 1200 \text{ mm}$$
$$d = 1600 \text{ mm}$$
$$b_w = 200 \text{ mm}$$
$$f_{ck} = 50 \text{ N/mm}^2$$
$$f_p = 1862 \text{ N/mm}^2$$
$$D_f = 250 \text{ mm}$$

According to the specifications of IRC: 6-2014, the design ultimate moments and shear forces in the girder are calculated by applying the partial safety factors for dead and live loads as given below:

The required design ultimate bending moment in the outer girder is evaluated as

$$M_u = [1.35 M_d + 1.5 M_L]$$

$$= [(1.35 \times 4261) + (1.5 \times 2074)]$$
$$= 8864 \text{ kN·m}$$

According to IS: 1343-1980, the ultimate flexural strength of the centre span section is computed as follows:

$$A_p = (A_{pw} + A_{pf})$$
$$A_{pf} = 0.45 f_{ck} (b - b_w) (D_f/f_p)$$
$$= 0.45 \times 50 (1200 - 200) (250/1862)$$
$$= 3021 \text{ mm}^2$$

∴ $\qquad A_{pw} = (4900 - 3021) = 1879 \text{ mm}^2$

Ratio $\left(\dfrac{A_{pw} \cdot f_p}{b_w \cdot d \cdot f_{ck}} \right) = \left(\dfrac{1879 \times 1862}{200 \times 1600 \times 50} \right) = 0.218$

From Table 11 of IS: 1343, we have for post tensioned beams with effective bond,

$$(f_{pu}/0.87 f_p) = 0.93 \text{ and } (x_u/d) = 0.43$$

∴ $\qquad f_{pu} = (0.93 \times 0.87 \times 1862) \text{ and } x_u = (0.43 \times 1600)$
$$= 1506 \text{ N/mm}^2 = 688 \text{ mm}$$

∴ $\qquad M_u = [f_{pu} A_{pw} (d - 0.42 x_u) + 0.45 f_{ck} (b - b_w) D_f (d - 0.5 D_f)]$
$$= [1506 \times 1879 (1600 - 0.42 \times 688)]$$
$$+ 0.45 \times 50 \times 1000 \times 250 (1600 - 0.5 \times 250)]$$
$$= 12006 \times 10^6 \text{ N.mm}$$
$$= 12006 \text{ kN·m}$$

$\qquad M_u = 12006 \text{ kN·m} > 8864 \text{ kN·m (Hence safe)}$

8. Check for Ultimate Shear Strength

Ultimate Shear Force $= [(1.35 V_g + 1.5 V_q)]$
$$= [(1.35 \times 561) + (1.5 \times 427)]$$
$$= 1398 \text{ kN}$$

The design shear resistance of the support section is calculated by using the equation specified in IRC: 112-2011 clause 10.3 as

$$V_{Rd.c} = \left(\frac{I \cdot b_w}{S} \right) \sqrt{(f_{ctd})^2 \mp k_1 \sigma_{cp} f_{ctd}} + \eta P \sin \theta$$

Computing the numerical values of the various parameters, we have

$$I = \left(\frac{b_w D^3}{12} \right) = \left(\frac{200 \times 1800^3}{12} \right) = (972 \times 10^8) \text{ mm}^4$$

$$S = \left(\frac{200 \times 750 \times 750}{2} \right) = (56.2 \times 10^6) \text{ mm}^3$$

$$f_{ctd} = (f_{ct}/\gamma_m) = (3.5/1.5) = 2.33 \text{ since } f_{ct} = 3.5 \text{ N/mm}^2 \text{ for } f_{ck} = 50 \text{ N/mm}^2$$

$$\sigma_{cp} = (\eta P/A_c) = [(0.85 \times 6053 \times 10^3)/(73 \times 10^4)] = 7.04 \text{ N/mm}^2$$

$$\theta = (4e/L) = [(4 \times 670)/(30 \times 1000)] = 0.089$$

$$\therefore \quad V_{Rd.c} = \left(\frac{I.b_w}{s}\right)\sqrt{(f_{ctd})^2 \mp k_1\sigma_{cp}f_{ctd}} + \eta P \sin\theta$$

$$= \left(\frac{(972 \times 10^8)\,200}{56.2 \times 10^6}\right)\sqrt{(2.33)^2 \mp 1 \times 7.04 \times 2.33}$$

$$+ \left(0.8 \times 6053 \times 10^3 \times 0.089\right) \text{N}$$

$$= (2021 \times 10^3) \text{ N}$$

$$= 2021 \text{ kN} > 1398 \text{ kN}$$

Provide nominal stirrups of 10 mm diameter 2-legged stirrups of Fe-415 HYSD bars at a maximum spacing of 300 mm throughout the span according to the specifications of IRC: 112-2011.

9. Supplementary Reinforcements

According to Clause, 16.51 of IRC: 112-2011, minimum Longitudinal reinforcements of not less than 0.13 percent of gross cross sectional area are to be provided to safeguard against shrinkage cracking.

$$A_{SL} = (0.0013 \times 0.73 \times 10^6) = 949 \text{ mm}^2$$

20 mm diameter bars are provided in the compression flange as shown in Fig.11.25.

10. Design of End Block

Solid end blocks are provided at end supports over a length of 1.5 m. Typical equivalent prisms on which the anchorage forces are considered to be effective are detailed in Fig. 11.26.

In the horizontal plane, we have the data,

$$P_K = 1459 \text{ kN}, \quad 2Y_{po} = 225 \text{ mm} \quad \text{and} \quad 2Y_o = 900 \text{ mm}$$

Hence the ratio $(Y_{po}/Y_o) = (112.5/450) = 0.25$

Interpolating from Table 11.3 , the bursting tension is computed as

$$F_{bst} = (0.26 \times 1459) = 380 \text{ kN}$$

Area of steel required to resist this tension is obtained as,

$$A_s = [(380 \times 10^3)/(0.87 \times 415)] = 1052 \text{ mm}^2$$

Provide 10 mm diameter bars at 100 mm centres in the horizontal direction. In the vertical plane, the ratio of (Y_{po}/Y_o) being higher, the magnitude of bursting tension is smaller. However the same reinforcements are provided in the form of a mesh both in the horizontal and vertical directions as shown in Fig. 11.26.

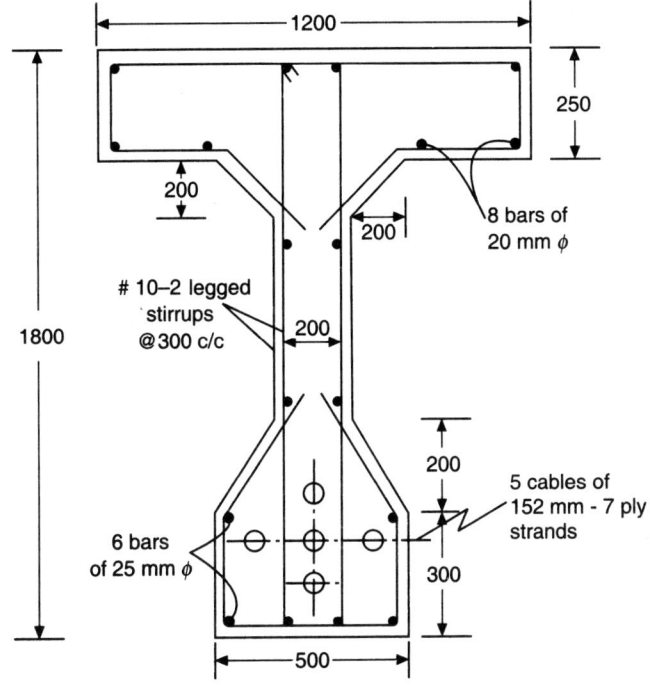

Fig. 11.25 Reinforcement Details at Centre of Span Section.

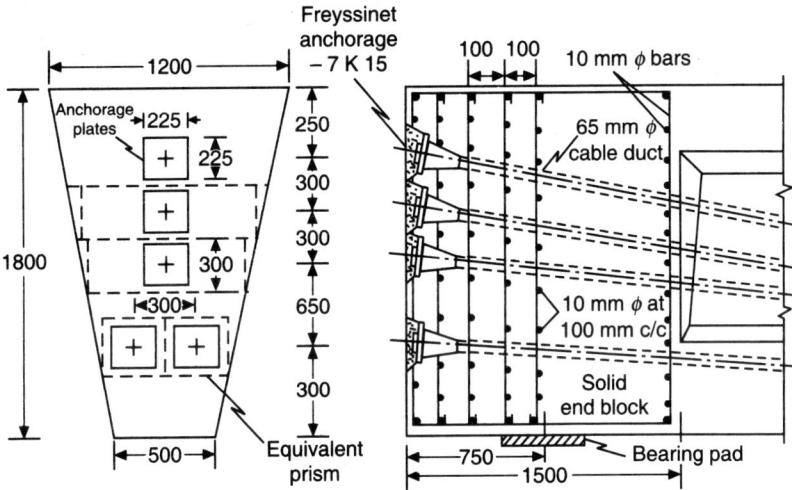

Fig. 11.26 Equivalent Prisms and Anchorage Zone Reinforcement.

11. Cross Girders

Cross girders of width 200 mm and depth 1250 mm are provided at intervals of 5 m along the span of the main girders. Nominal reinforcements of 0.15 percent of the cross section, consisting of 12 mm diameter bars spaced two at top, two at mid depth and two at bottom are provided in the cross girders. Nominal stirrups made up of 10 mm diameter two legged links are provided at 200 mm centres. Two straight H.T cables each consisting of 12 high tensile wires of 7 mm diameter are positioned at mid third points along the depth of the cross girders.

11.7 DESIGN OF POST-TENSIONED PRESTRESSED CONCRETE CONTINUOUS TWO SPAN BEAM AND SLAB BRIDGE DECK

Design a post tensioned prestressed concrete continuous beam and slab and bridge deck for a National high way crossing to suit the following data:

1. Data

Width of carriage way = 7.5 m
Two continuous spans of 40 m each
Kerbs: 600 mm wide on each side
Wearing coat thickness = 80 mm
Live load: IRC Class AA tracked vehicle

For prestressed concrete girders, adopt M-60 grade concrete with compressive strength of concrete at transfer as 40 N/mm²

For cast-*in-situ* deck slab, adopt M-20 Grade concrete

Spacings of cross girders = 5 m
Spacing of main girders = 2.5 m
Loss ratio = 0.8

High tensile strands of 15.2 mm diameter conforming to IS: 6006-1983[13] and Fe-415 HYSD bars are available for use.

Design the bridge deck as Class-1 type structure conforming to the codes IRC: 6-2014, IRC: 112-2011 and IS: 1343-2012.

2. Maximum Permissible Stresses in Concrete and Steel

According to the given data, we have for M-60 grade concrete

Compressive strength of concrete $= f_{ck} = 60$ N/mm²
Adopt permissible stress in concrete at transfer $= f_{ct} = 20$ N/mm²
Permissible tensile stress (Class 1 type structure) $= f_{tt} = f_{tw} = 0$
Ultimate tensile strength of 12 wires of 7 mm diameter high tensile cables $= f_p = 1500$ N/mm²
Maximum permissible stress in the H.T. cable = 1200 N/mm²
Modulus of elasticity of concrete $= E_c = 37$ kN/mm²

For M-20 grade concrete and Fe-415 HYSD bars adopt the following parameters.

$$f_{ck} = 20 \text{ N/mm}^2 \text{ and } f_y = 415 \text{ N/mm}^2$$
$$M_u = 0.138 f_{ck} b d^2$$

3. Cross Section of Deck

4 main girders are provided at 2.5 intervals
Thickness of deck slab = 250 mm
Wearing coat = 80 mm
Kerbs 600 mm wide by 300 mm deep are provided at each end.
Spacing of cross girders = 5 m
The overall depth of main girders is assumed at 50 mm per metre of span
∴ overall depth of girder = $(50 \times 40) = 2000$ mm
Thickness of top and bottom flanges = 400 mm
Thickness of web = 200 mm

The section properties of the main girder are as follows:

Cross sectional area $A = 0.88$ m^2
Second Moment of area $I = 0.447$ m^4
$$Y_b = Y_t = 1 \text{ m}$$
Section Modulus = $Z_t = Z_b = Z = 0.447$ m^3

The main girders are precast and the deck slab is cast *in-situ*
The cross section of the bridge deck and girders are shown in Figs. 11.27 and 11.28 respectively.

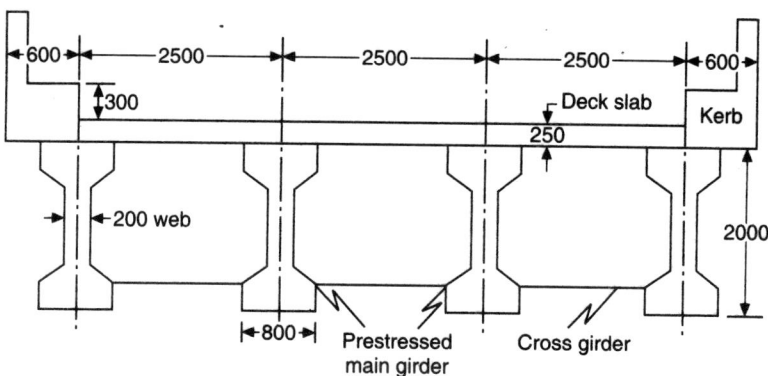

Fig. 11.27 Cross Section of Bridge Deck.

4. Design of Interior Slab Panel

The slab panel 2.5 m by 5 m is supported on all the four sides. The design is similar to that pressented in Section 11.6.

5. Design of Continuous Longitudinal Girders

(a) Reaction Factors

The spacing of the girders being the same as that in example 11.6 the reaction factor for the outer girder $R_A = 0.382$ W

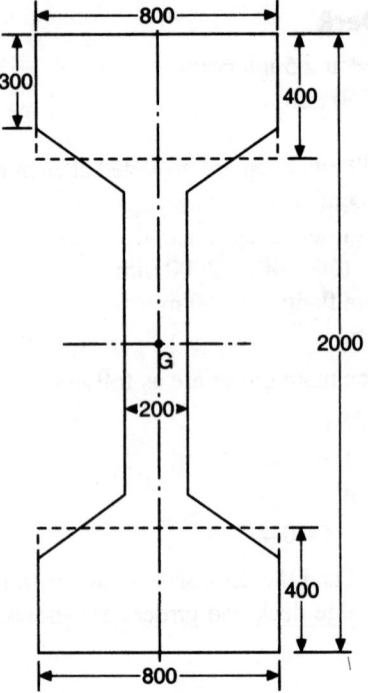

Fig. 11.28 Cross Section of Main Girder.

(b) Loads Acting on Main Girder

Self weight of slab $=(0.25 \times 25) = 6.25$ kN/m^2
Weight of W.C $=(0.80 \times 22) = 1.76$
Total load $\overline{8.00}$ kN/m^2

Load from slab and W.C $= (8 \times 2.5) = 20$ kN/m
Self weight of main girder $= (0.88 \times 25) = 22$ kN/m
Weight of cross girders assumed to act as uniformly distributed load $= 5$ kN/m
$\therefore$ Total dead load on main girder (g) $= 47$ kN/m

(c) Dead Load Moments and Shear Forces

Dead load moment at mid support section

$$M_{gB} = 0.125 \, gL^2 = (0.125 \times 47 \times 40^2)$$
$$= 9400 \text{ kN·m}$$

Dead load moment at mid span section

$$M_{gD} = 0.071 \, gL^2$$
$$= (0.071 \times 47 \times 40^2)$$
$$= 5340 \text{ kN·m}$$

Dead load shear is maximum near mid support section and is computed as follows

$$V_g = 0.62\, gL$$
$$= (0.62 \times 47 \times 40)$$
$$= 1166\ \text{kN}$$

(d) Live Load Bending Moments in Girder

Referring to the Influence Line for bending moments[14] at mid span section D shown in Fig. 11.29 the maximum live load moment at mid span in computed as

$$M_D = \left(\frac{7.3 + 8.12}{2}\right) 700 = 5397\ \text{kN·m}$$

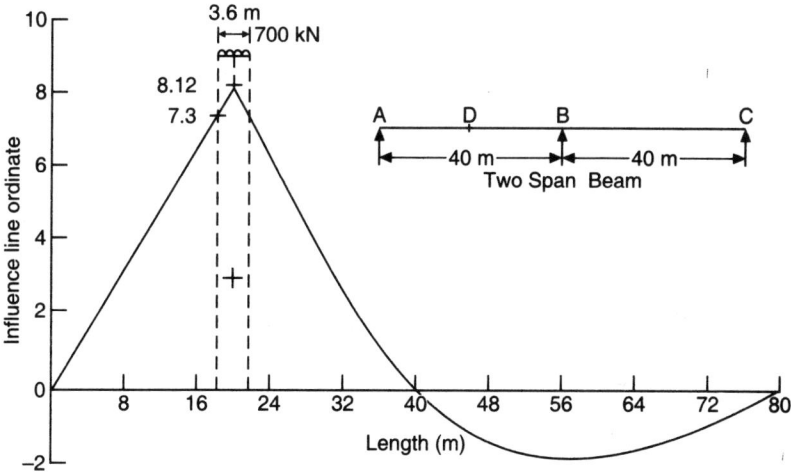

Fig. 11.29 Influence Line for Bending Moment at Mid Span D.

Similarly from Fig. 11.30, using the influence line for bending moment at mid support, the live load bending moment at support B is computed as

$$M_B = (3.76 \times 700) = 2632\ \text{kN·m}$$

The live load bending moments including the Reaction factor and Impact factors for the exterior girder are,

$$M_{qD} = (5397 \times 1.1 \times 0.382) = 2268\ \text{kN·m}$$
$$M_{qB} = (2632 \times 1.1 \times 0.382) = 1106\ \text{kN·m}$$

(e) Live Load Shear Forces in Girder

The maximum live load shear develops in the interior girders when the I.R.C class AA loads are placed near the mid support as shown in Fig. 11.31.

Reaction of W_2 on girder $B = (350 \times 0.45)/2.5 = 63$ kN

Reaction of W_2 on girder $A = (350 \times 2.50)/2.5 = 287$ kN

∴ Total load on girder B = (350 + 63) = 413 kN

Maximum reaction in girder B = (413 × 38.2)/40 = 394 kN

Maximum live loads shear with impact factor in inner girder

= (394 × 1.1) = 434 kN

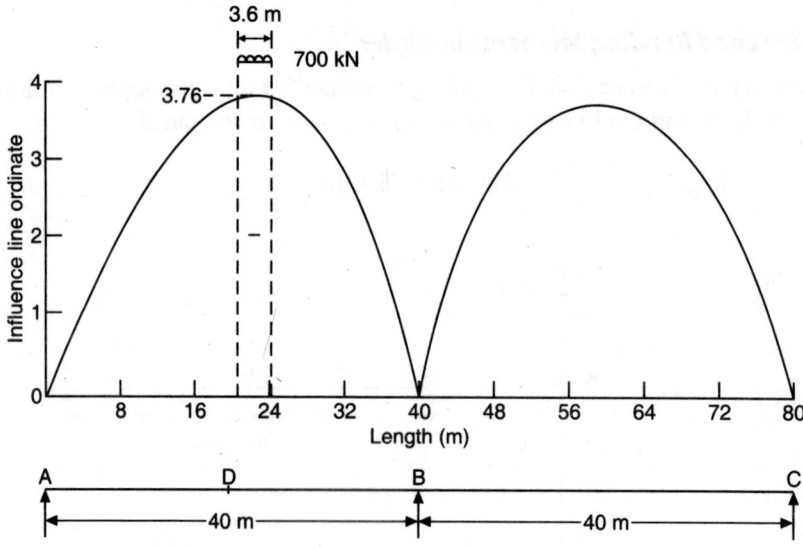

Fig. 11.30 Influence Line for Bending Moment at Mid Support B.

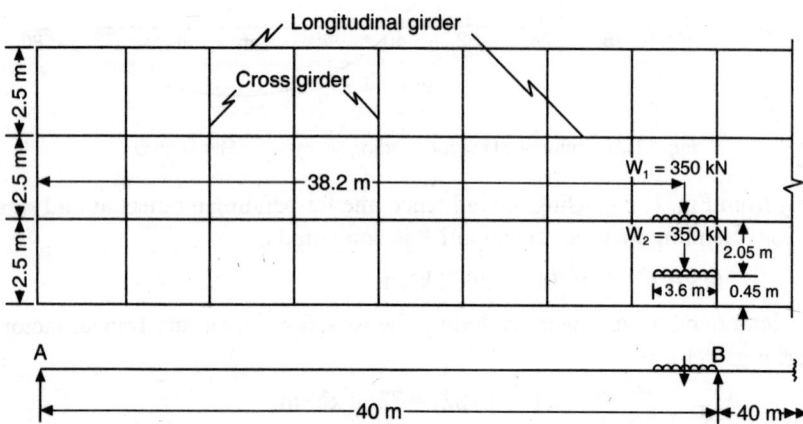

Fig. 11.31 Position of IRC Class AA Loads for Maximum Shear.

(f) Design Bending Moments and Shear Forces

The design bending moments and shear forces at service loads and ultimate loads as stipulated IRC: 6-2014 are compiled in Table 11.5.

Table 11.5 Service Load and Ultimate Load Moments and Shear Forces in Longitudinal Girders

(a) Bending Moments (Outer Girder)				
Section	Dead Load B.M (M_g)	Live Load B.M (M_q)	Service Load B.M $(M_g + M_q)$	Ultimate Load B.M $(1.35\,M_g + 1.5\,M_q)$	Units
Mid Span At (D)	5340	2268	7608	10611	kN.m
Mid Support At (B)	9400	1106	10506	14349	kN.m
(b) Shear Forces (Inner Girder)				
	Dead Load S.F (V_g)	Live Load S.F (V_q)	Service Load S.F $(V_g + V_q)$	Ultimate Load S.F $(1.35\,V_g + 1.5\,V_q)$	Units
Middle Support Section	1116	434	1550	2158	kN

(g) Check for Minimum Section Modulus

At mid support section B

$$M_g = 9400 \text{ kN·m}$$
$$M_q = 1106 \text{ kN·m}$$
$$M_d = (M_g + M_q) = 10506 \text{ kN·m}$$
$$f_{br} = (\eta f_{ct} - f_{tw}) = (0.8 \times 20 - 0) = 16 \text{ N/mm}^2$$
$$f_{inf} = (f_{tw}/\eta) + (M_d/\eta Z_b)$$
$$= 0 + (10506 \times 10^6)/(0.8 \times 0.447 \times 10^9)$$
$$29.37 \text{ N/mm}^2$$

$$Z_b \geq \left[\frac{M_g + (1-\eta)M_g}{f_{br}} \right]$$

$$\geq \left[\frac{(1106 \times 10^6) + (1-0.8)\,9400 \times 10^6}{16} \right]$$

$$\geq 0.186 \times 10^9 \text{ mm}^3 < 0.447 \times 10^9 \text{ mm}^3$$

Hence the section provided is adequate.

(h) Prestressing Force

For the two continuous spans AB and BC, a concordant cable profile is selected such that the secondary moments are zero. The cable profile selected is shown in Fig. 11.32. The maximum possible eccentricity at mid support section B is determined by providing suitable cover to house the cables. Assuming a cover of 250 mm,

Maximum possible eccentricity $e = (1000 - 250) = 750$ mm

Prestressing force is obtained from the relation,

$$P = [(A \cdot f_{inf} \cdot Z_b)/(Z_b + A \cdot e)]$$

$$= \left[\frac{(0.88 \times 10^6 \times 29.37 \times 0.447 \times 10^9)}{(0.447 \times 10^9) + (0.88 \times 10^6 \times 750)} \right]$$

$$= 10434000 \text{ N}$$

$$= 10434 \text{ kN}$$

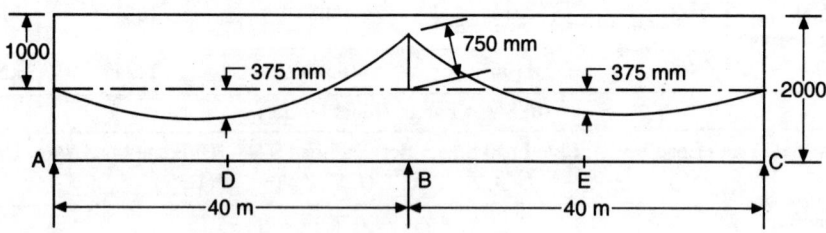

Fig. 11.32 Concordant Cable Profile.

Using Freyssinet system anchorage type 19K-15 (19 strands of 15.2 mm diameter) in 95 mm cable ducts. (IS: 6006-1983).

Force in each cable = $(19 \times 0.8 \times 260.7) = 3962$ kN

Provide 3 cables carrying an initial prestressing force

$$P = (3 \times 4000) = 12000 \text{ kN}$$

Area of each strand of 15.2 mm diameter = 140 mm²

Area of 19 strands in each cable = $(19 \times 140) = 2660$ mm²

Total Area in 3 cables = $A_p = (3 \times 2660) = 7980$ mm²

The cables are arranged in a parabolic concordant profile so that their centroid has an eccentricity of 750 mm towards the top fibre at mid support B and an eccentricity of 375 mm towards the soffit at mid span section D. The centroid of the cables is concentric at the end supports A and C. The selected cable profile is shown in Fig. 11.33.

6. Check for Stresses

(i) Centre of Span Section

$$P = 12000 \text{ kN} \qquad\qquad \eta = 0.80$$

$$e = 375 \text{ mm} \qquad\qquad M_g = 5340 \text{ kN·m}$$

$$A = 0.88 \times 10^6 \text{ mm}^2 \qquad M_q = 2268 \text{ kN·m}$$

$$Z = 0.447 \times 10^9 \text{ mm}^3$$

$$(P/A) = (12000 \times 10^3)/(0.88 \times 10^6) = 13.63 \text{ N/mm}^2$$

$$(Pe/Z) = (12000 \times 10^3 \times 375)/(0.447 \times 10^9) = 10.06 \text{ N/mm}^2$$

$$(M/Z) = (5340 \times 10^6)/(0.447 \times 10^9) = 11.94 \text{ N/mm}^2$$

$$(M/Z) = (2268 \times 10^6)/(0.447 \times 10^9) = 5.07 \text{ N/mm}^2$$

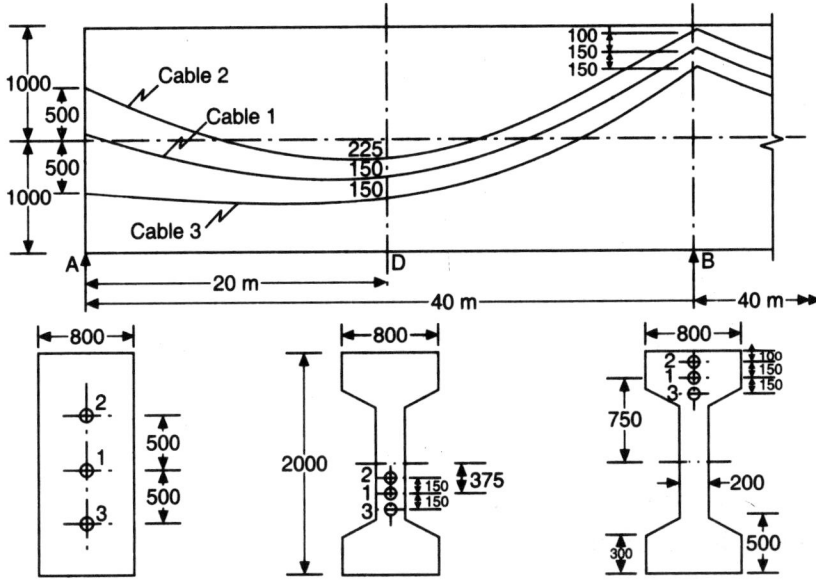

Fig. 11.33 Cable Lay Out in Main Girder.

At the stage of transfer:

$$\sigma_t = [(P/A) - (Pe/Z) + (M_g/Z)]$$
$$= [13.63 - 10.06 + 11.94] = 15.51 \text{ N/mm}^2$$
$$\sigma_b = [(P/A) + (Pe/Z) - (M/Z)]$$
$$= [13.63 + 10.06 - 11.94]$$
$$= 11.75 \text{ N/mm}^2$$

At the service load state:

$$\sigma_t = [\eta(P/A) - \eta(Pe/Z) + (M_g/Z)]$$
$$= [0.8 \ (13.63 - 10.06) + 11.94 + 5.07]$$
$$= 19.86 \text{ N/mm}^2 < 20 \text{ N/mm}^2$$
$$\sigma_b = [\eta(P/A) + \eta(Pe/Z) - (M_g/Z) - (M_q/Z)]$$
$$= [0.8 \ (13.63 + 10.06) - 11.94 - 5.07]$$
$$= 1.94 \text{ N/mm}^2$$

(ii) *Mid Support Section*

$$P = 12000 \text{ kN} \qquad\qquad \eta = 0.80$$
$$e = 750 \text{ mm} \qquad\qquad M_g = 9400 \text{ kN}$$
$$A = 0.88 \times 10^6 \text{ mm}^2 \qquad M_q = 1106 \text{ kN·m}$$
$$Z = 0.447 \times 10^9 \text{ mm}^3$$
$$(P/A) = 13.63 \text{ N/mm}^2$$
$$(Pe/Z) = (12000 \times 10^3 \times 750)/(0.447 \times 10^9) = 20.13 \text{ N/mm}^2$$

$$(M_g/Z) = (9400 \times 10^6)/(0.447 \times 10^9) = 21.02 \text{ N/mm}^2$$
$$(M_q/Z) = (1106 \times 10^6)/(0.447 \times 10^9) = 2.47 \text{ N/mm}^2$$

At the stage of transfer:

$$\sigma_t = [13.63 + 20.13 \times 21.02] = 12.74 \text{ N/mm}^2$$
$$\sigma_b = [13.63 - 20.13 + 21.02] = 14.52 \text{ N/mm}^2$$

At the service load stage:

$$\sigma_b = [0.8(13.63 + 20.13) - 21.02 - 2.47]$$
$$= 3.52 \text{ N/mm}^2$$
$$\sigma_b = [0.8(13.63 - 20.13) + 21.02 + 2.47]$$
$$= 18.29 \text{ N/mm}^2$$

The stresses in general are within the maximum permissible limit of 20 N/mm^2.

7. Check for Ultimate Flexural Strength

(a) Centre of Span Section

The ultimate flexural strength of the Tee section girder is computed using the specifications of IS: 1343-2012.

A_p = 7980 mm^2 D_f = 200 mm f_p = 1862 N/mm^2

b = 800 mm d = 1375 mm M_u (Required) = 10611 kN·m

b_w = 200 mm f_{ck} = 60 N/mm^2

$A_p = (A_{pw} + A_{pf})$

$$A_{pf} = 0.45 f_{ck} (b - b_w)\left(\frac{D_f}{f_p}\right) = \left[0.45 \times 60 (800 - 200)\left(\frac{200}{1862}\right)\right] = 1740 \text{ mm}^2$$

$A_{pw} = (7980 - 1740) = 6240 \text{ mm}^2$

$$\text{Ratio}\left[\frac{A_{pw} f_p}{b_w d f_{ck}}\right] = \left[\frac{6240 \times 1862}{200 \times 1375 \times 60}\right] = 0.70$$

From Table 11 of IS: 1343, interpolate the values of the ratio $(f_{pb}/0.87 f_{pu})$ and (x_u/d) corresponding to the above ratio of 0.70.

$$\left(\frac{f_{pb}}{0.87 f_{pu}}\right) = 0.75 \text{ and } (x_u/d) = 0.653$$

$$f_{pb} = (0.75 \times 0.87 \times 1862) = 1214.9 \text{ N/mm}^2$$
$$x_u = (0.653 \times 1375) = 897.8 \text{ mm}$$

∴
$$M_u = \{f_{pb} A_{pw} (d - 0.42 x_u) + 0.45 f_{ck} (b - b_w) D_f (d - 0.5 D_f)\}$$
$$= \{1214.9 \times 4500(1375 - 0.42 \times 897.8\}$$
$$+ \{0.45 \times 60(800 - 200) 400(1375 - 0.5 \times 400)\}$$

$$= (13070 \times 10^6) \text{ N.mm}$$
$$= 13070 \text{ kN·m} > 10611 \text{ kN·m (Hence safe)}$$

(b) Mid Support Section

All the parameters are the same except the effective depth, $d = 1750$ mm and M_u (Required) = 14349 kN·m

$$\text{Ratio}\left[\frac{A_{pw} f_p}{b_w d f_{ck}}\right] = \left[\frac{4500 \times 1862}{200 \times 1750 \times 60}\right] = 0.40$$

From Table 11 of IS: 1343, interpolate the values of the ratio $(f_{pb}/0.87 f_{pu})$ and (x_u/d) corresponding to the above ratio of 0.40.

$$\left(\frac{f_{pb}}{0.87 f_{pu}}\right) = 0.75 \text{ and } (x_u/d) = 0.653$$

$$f_{pb} = (0.75 \times 0.87 \times 1862) = 1214.9 \text{ N/mm}^2$$
$$x_u = (0.653 \times 1750) = 1142.75 \text{ mm}$$

∴
$$M_u = \{f_{pb} A_{pw} (d - 0.42 \, x_u) + 0.45 f_{ck} (b - bw) D_f (d - 0.5 \, D_f)\}$$
$$= \{1214.9 \times 4500(1750 - 0.42 \times 1142.75)\}$$
$$+ \{0.45 \times 60(800 - 200) \, 400(1750 - 0.5 \times 400)\}$$
$$= (16987 \times 10^6) \text{ N.mm}$$
$$= 16987 \text{ kN·m} > 14349 \text{ kN·m (Hence safe)}$$

Hence the mid span and support sections satisfy the limit state of ultimate strength.

8. Check for Ultimate Shear Strength

The mid support section is checked for the ultimate shear strength.

Shear Strength required = 2158 kN

The design shear resistance of the support section is calculated by using the equation specified in IRC: 112-2011 clause 10.3 as

$$V_{Rd.c} = \left(\frac{I.b_w}{S}\right) \sqrt{(f_{ctd})^2 \mp k_1 \sigma_{cp} f_{ctd}} + \eta P \sin \theta$$

Computing the numerical values of the various parameters, we have

$$I = \left(\frac{b_w D^3}{12}\right) = \left(\frac{200 \times 2000^3}{12}\right) = (1333 \times 10^8) \text{ mm}^4$$

$$S = \left(\frac{200 \times 1000 \times 1000}{2}\right) = (100 \times 10^6) \text{ mm}^3$$

$$f_{ctd} = (f_{ct}/\gamma_m) = (4.0/1.5) = 2.66 \text{ since } f_{ct} = 4.0 \text{ N/mm}^2 \text{ for } f_{ck} = 60 \text{ N/mm}^2$$
$$\sigma_{cp} = (\eta P/A_c) = [(0.8 \times 12000 \times 10^3)/(88 \times 10^4)] = 10.9 \text{ N/mm}^2$$

$$\theta = (4e/L) = [(4 \times 750)/(40 \times 1000)] = 0.075$$

$$\therefore \quad V_{Rd.c} = \left(\frac{I.b_w}{s}\right)\sqrt{(f_{ctd})^2 \mp k_1 \sigma_{cp} f_{ctd}} + \eta P \sin\theta$$

$$= \left(\frac{(1333 \times 10^8) \, 200}{100 \times 10^6}\right)\sqrt{(2.66)^2 \mp 1 \times 10.9 \times 2.33}$$

$$+ (0.8 \times 12000 \times 10^3 \times 0.089) \, N$$

$$= (2371 \times 10^3) \, N$$

$$= 2371 \, kN < 2158 \, kN$$

Nominal shear reinforcements of 10 mm diameter two legged stirrups are designed using the ralation

$$S_v = \left(\frac{0.87 f_y A_{sv}}{0.4b}\right) = \left(\frac{0.87 \times 415 \times 2 \times 79}{0.4 \times 200}\right) = 713 \, mm$$

The stirrups are provided at a maximum spacing of 300 mm throughout the span as per IRC: 112-2011 Specifications.

9. Supplementary Reinforcements

According to Clause, 16.51 of IRC: 112-2011, minimum Longitudinal reinforcements of not less than 0.13 percent of gross cross sectional area are to be provided to safeguard against shrinkage cracking.

$$A_{SL} = \{0.0013 \times 88 \times 10^4) = 1144 \, mm^2$$

The details of reinforcements and cables provided at mid span and support sections are shown in Fig. 11.34.

10. Design of End Block

Solid end blocks are provided at end supports over a length of 2 m. Typical equivalent prisms on which the anchorage forces are considered to be effective are detailed in Fig. 11.35.

In the horizontal plane, we have the data,

$$P_K = 4000 \, kN, \quad 2 \, Y_{po} = 340 \, mm \quad \text{and} \quad 2 \, Yo = 800 \, mm$$

Hence the ratio $(Y_{po}/Y_o) = (340/800) = 0.425$

Interpolating from Table 11.3 , the bursting tension is computed as

$$F_{bst} = (0.22 \times 4000) = 880 \, kN$$

Area of steel required to resist this tension is obtained as,

$$A_s = [(880 \times 10^3) / (0.87 \times 415)] = 2437 \, mm^2$$

Provide 16 mm diameter bars at 150 mm centres in the horizontal plane distributed in the region $0.2\ Y_o$ to $2\ Y_o$ (80 to 800 mm0 as shown in Fig. 11.35. In the vertical plane the ratio of (Y_{po}/Y_o) being larger, the magnitude of bursting tension is less. However the same reinforcements are provided in the vertical plane in the form of a mesh to resist bursting tension.

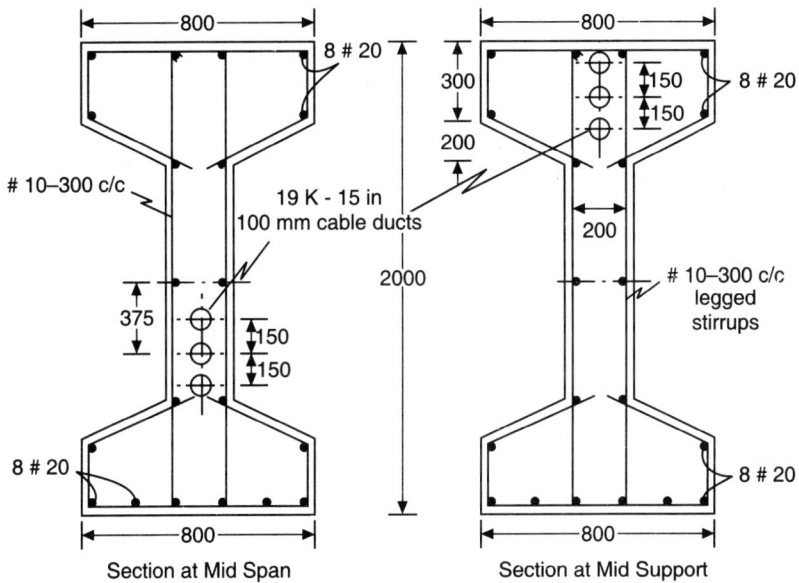

Section at Mid Span   Section at Mid Support

Fig. 11.34 Reinforcement Details at Mid Span and Mid Support Sections.

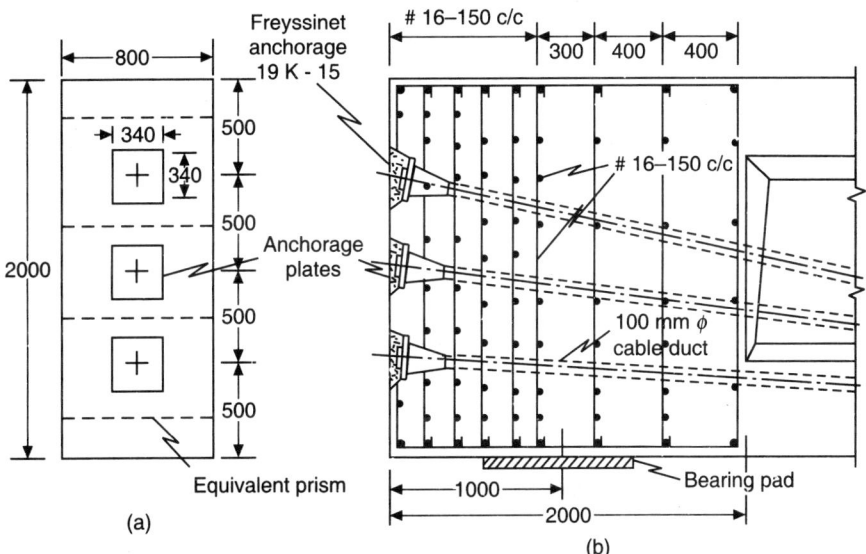

Fig. 11.35 Anchorage Zone Reinforcement in End Block.

11. Cross Girders

Cross girders of 200 mm width and 1600 mm depth, are provided with nominal reinforcements of 0.15 percent of the cross section comprising 12 mm diameters spaced two at top and bottom and two each at distances of 700 mm from top and bottom of the girders respectively. Nominal stirrups of 10 mm diameter two legged links are also provided at 200 mm centres. Two cables consisting of 12 numbers of 7 mm high tensile wires are positioned at mid third points along the depth with nominal prestress to provide lateral stiffness to the bridge deck system.

11.8 DESIGN OF PRESTRESSED CONCRETE CELLULAR BOX GIRDER BRIDGE DECK

A cellular multi celled prestressed concrete box girder deck is to be designed for a National highway crossing. The proposed bridge deck is made up of two continuous spans each of 50 m. The road width is 7.5 m with foot paths 1.25 m on each side. The box girder is proposed to have 4 cells 2 m wide by 2 m deep and should support IRC Class AA tracked vehicle loading. Design the cellular bridge deck adopting M-60 Grade concrete, Fe-415 HYSD bars and high tensile steel strands of 15.2 mm diameter conforming to the relevant Indian standards.

1.   Data

Span = 50 m
Cross section: Multi celled box girder
Cell dimensions = 2 m wide by 2 m deep
Road width = 7.5 m
Foot Paths: 1.25 m wide on either side of road way
Wearing coat = 80 mm
Thickness of web = 300 mm to house 27 K-15 Freyssinet type anchorages (27 strands of 15.2 mm diameter in 110 mm diameter cables)
(Refer Appendix 3 of Reference 3, p. 746 for details of Freyssinet Anchorages)
Thickness of top and bottom slabs = 300 mm
Concrete Grade M-60
Loss ratio = 0.80
Type of tendons: High tensile strands of 15.2 mm diameter conforming to IRC: 6006 – 2000[13]
Type of Supplementary reinforcements: Fe-415 HYSD bars
Design the bridge deck as Class-1 type structure conforming to the Codes
IRC: 6 – 2014[10], IRC: 112-2011[11] and IS: 1343-2012[12].

2.   Maximum Permissible Stresses in Concrete and Steel

According to the given data, we have for M-60 grade concrete
Compressive strength of concrete $= f_{ck} = 60$ N/mm^2
Adopt permissible stress in concrete at transfer $= f_{ct} = 20$ N/mm^2

Permissible tensile stress (Class 1 type structure) $= f_{tt} = f_{tw} = 0$
Modulus of elasticity of concrete $= E_c = 37$ kN/mm^2
Loss ratio $= 0.8$
High tensile strands of 15.2 mm diameter conforming to IS: 6006-1983 and Fe-415 HYSD bars are available for use.
Type of supplementary reinforcement = Fe-415 HYSD bars having $f_y = 415$ N/mm^2

3. Cross Section of Box Girder

Overall depth of the box girder $= \left(\dfrac{\text{Span}}{25} \right) = \left(\dfrac{50}{25} \right) = 2$ m

Width of road way $= 7.5$ m
Width of foot paths $= (2 \times 1.25) = 2.5$ m
Total width of box girder at road level $= (7.5 + 2.5) = 10$ m
Spacing between webs $= 2$ m
4 celled box girder is adopted
Thickness of web as per Clause 9.3.2.1 of IRC: 18-2000 is computed as

$\qquad t_w = [200 + \text{diameter of cable duct for housing 27 K-15 strands}]$
$\qquad\qquad = [200 + 100] = 300$ mm

At end supports where anchorages are located, web thickness increased to 600 mm
Thickness of top and bottom slabs $= 300$ mm
The multi celled box girder section selected is shown in Fig. 11.36.

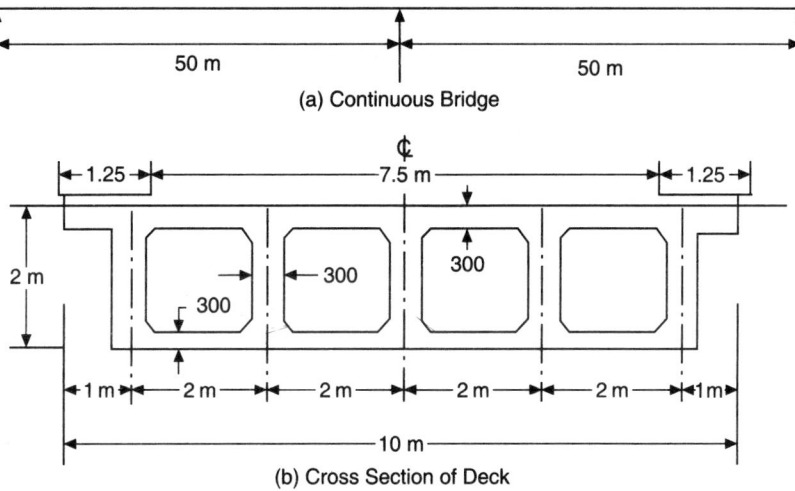

Fig. 11.36 Two Span Box Girder Bridge Deck.

The multi celled box girder section selected is shown in Fig. 11.36.
Section properties of the symmetrical I-girder shown in Fig. 11.37 are as follows:

Cross sectional area $= A = 1.62$ m^2
Second moment of area $= I = 0.94$ m^4

Distance of the extreme fibre from centroid $= y = y_t = y_b = 1$ m

Section Modulus $= Z = Z_t = Z_b = (I/y) = 0.94$ m^3

4. Design of Slab Panel

(a) Dead Load Bending Moments

Dead weight of slab $= (1 \times 1 \times 0.3 \times 24) = 7.20$ kN/m^2

Dead weight of W.C $= (0.08 \times 22)$.......... $= 1.76$

Total Dead load $= g$.................................$\cong 9.00$ kN/m^2

Referring to the bending moment coefficients compiled in a separate monograph by the author[14] and the Fig. 11.38.

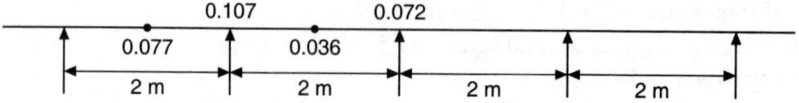

Fig. 11.38 Dead Load Bending Moment Coefficients in four Span Continuous Slab.

Maximum negative bending moment due to dead load at supports

$$= (0.107 \times gL)$$
$$= (0.107 \times 9 \times 2) = 1.93 \text{ kN·m/m}$$

Maximum positive bending moment at centre of span

$$= (0.077 \times gL)$$
$$= (0.077 \times 9 \times 2) = 1.38 \text{ kN·m/m}$$

Maximum shear force $= (0.60 \times gL) = (0.60 \times 9 \times 2) = 10.8$ kN

(b) Live Load bending Moments

The slab panel is continuous over webs in the transverse direction and free in the longitudinal direction. The slab spanning in the transverse direction is designed for IRC Class AA tracked loading using the procedure specified in IRC: 21-2000. When IRC Class AA tracked vehicle traverses on the deck, maximum bending moment in the transverse direction of the slab will develop when one tracked wheel occupies the centre of slab as shown in Fig. 11.39.

The effective width of dispersion of the wheel through the wearing coat is computed as

$$u = [0.85 + (2 \times 0.08)] = 1.01 \text{ m}$$
$$v = [3.60 + (2 \times 0.080)] = 3.76 \text{ m}$$

Average intensity of wheel load with Impact Factor $= \left[\dfrac{1.25 \times 350}{3.76 \times 1.01} \right] = 115.20$ kN/m^2

Concentrated load acting at the centre of span in the transverse direction is computed as

$$Q = (115.20 \times 1.01) = 116.4 \text{ kN}$$

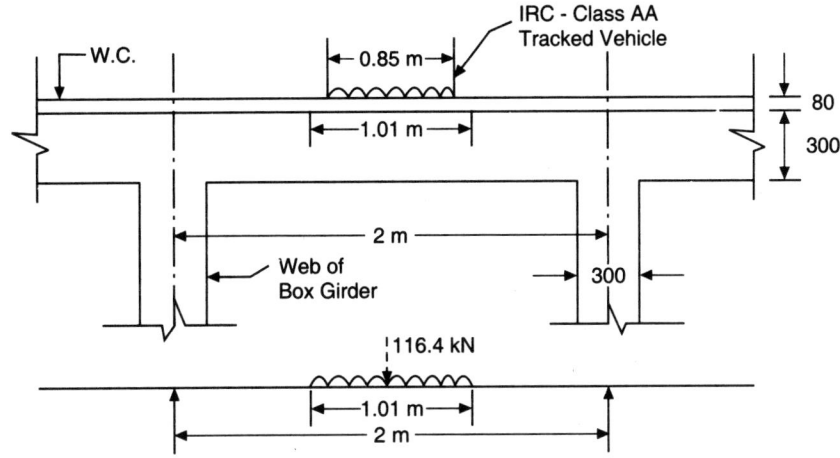

Fig. 11.39 Position of IRC Class AA Load for Maximum B.M. in Slab.

Referring to the bending moment and shear force coefficients[13] compiled in Fig. 11.40.

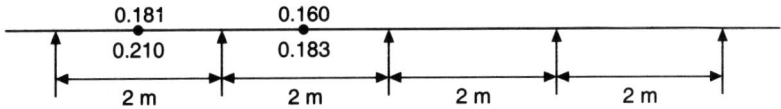

Fig. 11.40 Live Load Bending Moment and Shear Force Coefficients in Four Span Continuous Slab.

Maximum Positive B.M at middle of end span = $[0.210 \, QL]$
$$= [0.210 \times 116.4 \times 2] = 48.9 \text{ kN·m}$$

Maximum Negative B.M at penultimate support = $[0.181 \, QL]$
$$= [0.181 \times 116.4 \times 2] = 48.13 \text{ kN·m}$$

Maximum Shear Force = $[0.60 \, Q] = [0.60 \times 116.4] = 69.8 \text{ kN}$

(c) Design Ultimate Bending Moments and Shear Forces

The design ultimate bending moments are obtained by applying the appropriate safety factors to the service load bending moments and shear forces as detailed below:

Total Positive Bending Moment = $M_{up} = [1.35 \, M_d + 1.5 \, M_L]$
$$= [(1.35 \times 1.38) + (1.5 \times 48.9)]$$
$$= 74.2 \text{ kN·m}$$

Total Negative Bending Moment = $M_{un} = [1.35 \, M_d + 1.5 \, M_L]$
$$= [(1.35 \times 1.93) + (1.5 \times 42.13)]$$
$$= 65.8 \text{ kN·m}$$

Total Maximum Shear Force = $[1.35 \, V_g + 1.5 \, V_q]$
$$= [(1.35 \times 10.8) + (1.5 \times 69.8)]$$
$$= 119.3 \text{ kN}$$

(d) Design of Deck Slab and Reinforcements

Effective depth of slab required $= d = \sqrt{\dfrac{M_u}{0.138 f_{ck}.b.}} = \sqrt{\dfrac{74.2 \times 10^6}{0.138 \times 60 \times 1000.}} = 94.7\,\text{mm}$

Overall depth adopted = 300 mm

Adopt effective depth, $d = 250$ mm and overall depth of 300 mm

The area of tension reinforcement required to resist the moment is calculated by the equation,

$$M_u = 0.87 f_y A_{st} d \left[1 - \frac{A_{st} f_y}{b.d.f_{ck}} \right]$$

$$\left(74.2 \times 10^6\right) = \left(0.87 \times 415 \times A_{st} \times 250\right)\left[1 - \frac{A_{st} \times 415}{1000 \times 250 \times 60} \right]$$

Solving, $A_{st} = 2074\,\text{mm}^2$

Provide 20 mm diameter bars at 150 mm centers ($A_{st} = 2094\,\text{mm}^2$) as main reinforcements and 12 mm diameter bars at 150 mm centres as distribution reinforcements.

(e) Check for Ultimate Shear Strength

The ultimate shear strength of the reinforced concrete deck slab is checked by using the equation 4.7.

$$V_{Rd.c} = [0.12K\,(80\rho_1 f_{ck})^{0.33}]b_w.d$$

where $\quad K = 1 + \sqrt{\dfrac{200}{d}} \le 2.00 = \left[1 + \sqrt{\dfrac{200}{250}} \right] = 1.89$

and $\quad \rho_1 = \left(\dfrac{A_{sl}}{b_w.d}\right) \le 0.02$

$\qquad = \left(\dfrac{2094}{1000 \times 250}\right) = 0.008$

$V_{Rd.c} = [0.12 \times 1.89\,(80 \times 0.008 \times 60)^{0.33}](1000 \times 250)\,\text{N}$

$\qquad = (191.6 \times 1000)\,\text{N}$

$\qquad = 191.6\,\text{kN} > 119.3\text{kN}$ (Hence safe)

5. Design of Web Girder

(a) Dead Load Bending Moments and Shear Forces

The continuous box girder is treated as an assemblage of I-sections with web serving the function of a main girder and flanges of symmetrical size as shown in Fig. 11.37.

Self weight of Flanges = (2 × 0.3 × 24) = 14.40 kN/m²

Self weight of W.C. ... = (1 × 1 × 22)   = 1.76

Total Load...= 16.16

Self weight of web = $(1.4 \times 0.3 \times 24)$ = 10.08

Total load on each I-girder = g = $[(2 \times 16.16) + 10.08] = 43$ kN/m

The dead load bending moment coefficients[14] for a two span continuous beam is shown in Fig. 11.41. The dead load bending moments at mid support and mid span sections are computed as:

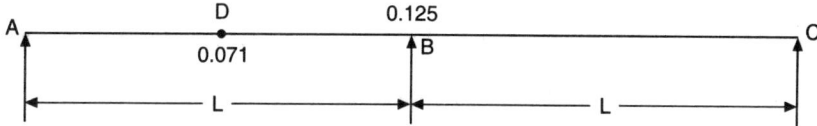

Fig. 11.41 Dead Load Bending Moment Coefficients.

$$M_{gB} = 0.125\ gL^2 = (0.125 \times 43 \times 50^2) = 13438\ \text{kN·m}$$
$$M_{gD} = 0.071\ gL^2 = (0.071 \times 43 \times 50^2) = 7633\ \text{kN·m}$$

Dead load shear is maximum near the mid support section and is computed as

$$V_g = 0.62\ gL = (0.62 \times 43 \times 50) = 1333\ \text{kN}$$

(b) Live load Bending Moments in Continuous Web Girder

Maximum live load reaction occurs in the web girder when the transverse disposition of the IRC Class AA tracked vehicle load is arranged to have the maximum eccentricity with respect to the centre of the bridge deck as shown in Fig. 11.42. Maximum reaction due to live loads in girder B is computed as

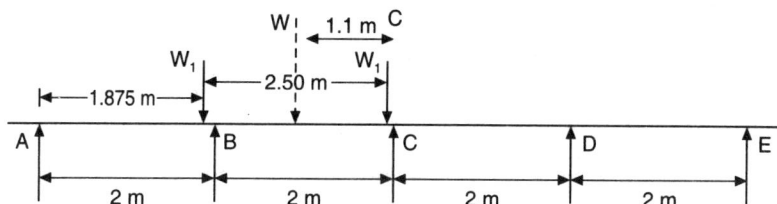

Fig. 11.42 Position of IRC Class AA Live Loads for Maximum Reaction in Girder.

$$R_B = \left[\frac{W \times 1.1}{2}\right] = 0.55\,W = (0.55 \times 700) = 385\,\text{kN}$$

Hence the concentrated load = Q = 385 kN

This load acting over a length of 3.6 m in the longitudinal direction is positioned at the centre of span of the two span continuous beam as shown in Fig. 11.43 to compute the maximum positive and negative moments.

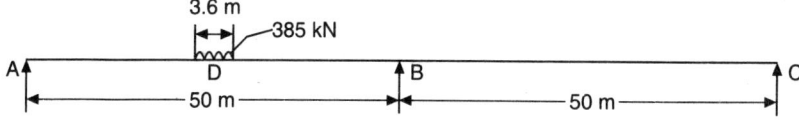

Fig. 11.43 Position of Live Loads for Maximum Moments in two Span Continuous Beam.

The live load bending moment coefficients[14] for maximum positive and negative moments in a two span continuous beam are shown in Fig. 11.44.

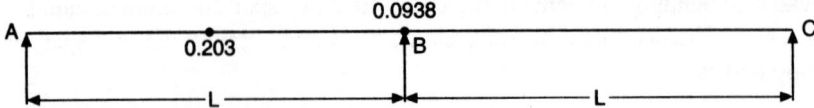

Fig. 11.44 Live Loads Bending Moment Coefficients for a two Span Continuous Girder.

Maximum positive live load bending moment with impact factor at centre of span is computed as

$$M_{max} \text{ (Positive)} = \text{(I.F)}(0.203 \ QL)$$
$$= (1.10) \, (0.203 \times 385 \times 50)$$
$$= 4298 \text{ kN·m}$$

Maximum Negative live load bending moment with impact factor at mid support is computed as

$$M_{max} \text{ (Negative)} = \text{(I.F)}(0.0938 \ QL)$$
$$= (1.1)(0.0938 \times 385 \times 50)$$
$$= 1986 \text{ kN·m}$$

(c) Live load Shear Force in Girder

The maximum live load shear force develops in the interior webs when the IRC Class AA loads are placed near the mid support as shown in Fig. 11.45.

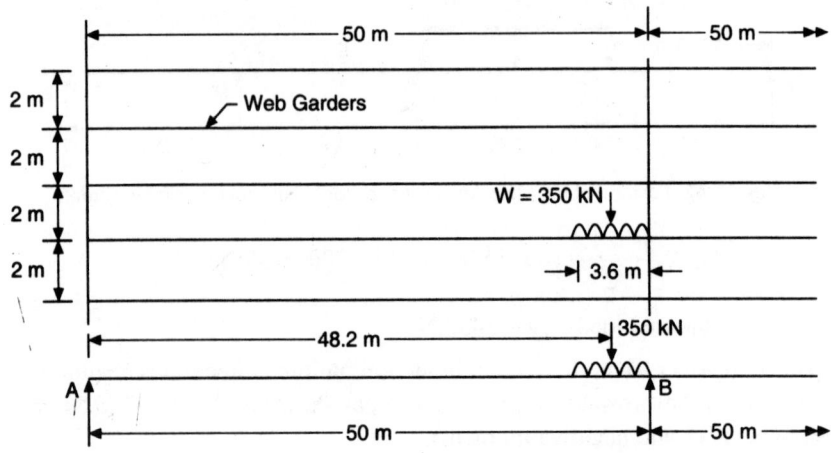

Fig. 11.45 Position of IRC Class AA Loads for Maximum Shear Force in Web Girder.

Reaction of load W on interior girder $= \left(\dfrac{350 \times 48.2}{50} \right) = 338 \text{ kN}$

Maximum live load shear force with impact $= (338 \times 1.1) = 372 \text{ kN}$

(d) Design Bending Moments and Shear Forces

The design bending moments and shear forces at service and ultimate loads are compiled in Table 11.6.

Table 11.6 Service Load and Ultimate Load Moments and Shear Forces in Web Girders

(a) Bending Moments (Outer Web Girder)					
Section	Dead Load B.M (M_g)	Live Load B.M (M_q)	Service Load B.M $(M_g + M_q)$	Ultimate Load B.M $(1.35\,M_g + 1.5\,M_q)$	Units
Mid Span At (D)	7633	4298	11931	16751	kN·m
Mid Support At (B)	13438	1986	15429	19730	kN·m
(b) Shear Forces (Inner Girder)					
	Dead Load S.F (V_g)	Live Load S.F (V_q)	Service Load S.F $(V_g + V_q)$	Ultimate Load S.F $(1.35\,V_g + 1.5\,V_q)$	Units
Middle Support Section	1333	372	1705	2357	kN

(e) Check for Minimum Section Modulus at Service Loads

At the mid support section B, the dead and live load moments are listed as

$$M_{gB} = 13438 \text{ kN·m}$$
$$M_{QB} = 1986 \text{ kN·m}$$
$$M_{dB} = [M_{gB} + M_{QB}] = [13438 + 1986] = 15424 \text{ kN·m}$$
$$f_{br} = [\eta f_{ct} - f_{tw}] = [(0.8 \times 20) - 0] = 16 \text{ N/mm}^2$$
$$f_{inf} = \left[\frac{f_{tw}}{\eta} + \frac{M_g}{\eta Z_b} \right] = \left[0 + \frac{13438 \times 10^6}{0.8 \times 0.94 \times 10^9} \right] = 17.86 \text{ N/mm}^2$$
$$Z_b \geq \left[\frac{M_q + (1-\eta) M_g}{f_{br}} \right]$$
$$\geq \left[\frac{(1986 \times 10^6) + (1 - 0.8)13438 \times 10^6}{16} \right]$$
$$\geq (0.292 \times 10^9) \text{ mm}^3 < (0.94 \times 10^9) \text{ mm}^3 \text{ (section provided)}$$

Hence section provided is adequate.

(f) Prestressing Force

For the two continuous spans AB and BC, a concordant cable profile is selected such that the secondary moments are zero. The cable profile selected with eccentricity at mid support twice that at mid spans is shown in Fig. 11.46.

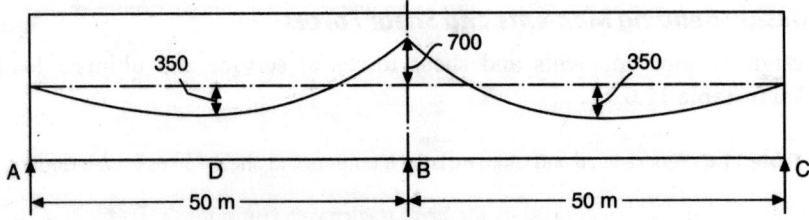

Fig. 11.46 Concordant Cable Profile.

Providing an effective cover = 300 mm,

Maximum possible eccentricity at support = $e = (1000 - 300) = 700$ mm

Prestressing force is computed from the relation,

$$P = \left[\frac{Af_b Z_b}{Z_b + Ae}\right] = \left[\frac{(1.62 \times 10^6)(17.86)(0.94 \times 10^9)}{(0.94 \times 10^9) + (1.62 \times 10^6 \times 700)}\right]$$

$$= 13109932 \text{ N}$$

$$= 13110 \text{ kN}$$

Using Freyssinet system with anchorage type 27K-15 (27 strands of 15.2 mm diameter) in 110 mm diameter cable ducts (Refer Appendix 3 of Reference 3),

Force in each cable = $(27 \times 0.8 \times 265) = 5724$ kN

Provide three cables carrying an initial prestressing force of

$$P = (3 \times 5000) = 15000 \text{ kN}$$

Area of each strand of 15.2 mm diameter tendon = 140 mm²

Area of 27 strands in each cable = $(27 \times 140) = 3780$ mm²

Total area in 3 cables = $A_p = (3 \times 3780) = 11340$ mm²

The cables are arranged in a parabolic concordant profile so that the centroid of the group of cables has an eccentricity of 700 mm towards the top fibre at mid support section B and an eccentricity of 350 mm towards the soffit at mid span section D. The centroid of the cables is concentric at the end supports A & C. The selected cable profile is shown in Fig. 11.47.

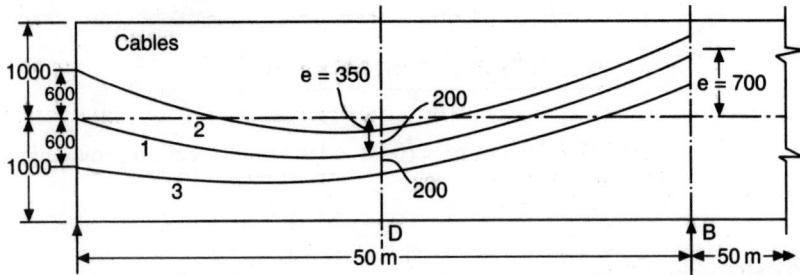

Fig. 11.47 Profiles of Individual Cables in Span.

(g) Check for Stresses at Service Loads

(1) Centre of Span Section

$$P = 15000 \text{ kN} \qquad \eta = 0.80$$
$$e = 350 \text{ mm} \qquad M_g = 7633 \text{ kN·m}$$
$$A = (1.62 \times 10^6) \text{ mm}^2 \qquad M_q = 4298 \text{ kN·m}$$
$$Z = (0.94 \times 10^9) \text{ mm}^3$$

$$\left(\frac{P}{A}\right) = \left[\frac{15000 \times 10^3}{1.62 \times 10^6}\right] = 9.25 \text{ N/mm}^2$$

$$\left(\frac{Pe}{Z}\right) = \left[\frac{15000 \times 10^3 \times 350}{0.94 \times 10^9}\right] = 5.58 \text{ N/mm}^2$$

$$\left(\frac{M_g}{Z}\right) = \left[\frac{7633 \times 10^6}{0.94 \times 10^9}\right] = 8.12 \text{ N/mm}^2$$

$$\left(\frac{M_q}{Z}\right) = \left[\frac{4298 \times 10^6}{0.94 \times 10^9}\right] = 8.12 \text{ N/mm}^2$$

At transfer stage, the stresses at extreme fibres are computed as

$$\sigma_t = \left[\frac{P}{A} - \frac{Pe}{Z} + \frac{M_g}{Z}\right] = [9.25 - 5.58 + 8.12] = 11.79 \text{ N/mm}^2$$

$$\sigma_b = \left[\frac{P}{A} - \frac{Pe}{Z} + \frac{M_g}{Z}\right] = [9.25 + 5.58 - 8.12] = 6.71 \text{ N/mm}^2$$

At Service load stage, the stresses at extreme fibres are computed as

$$\sigma_t = \left[\frac{\eta P}{A} - \frac{\eta Pe}{Z} + \frac{M_g}{Z} + \frac{M_q}{Z}\right]$$

$$= [0.8\,(9.25 - 5.58) + 8.12 + 4.57]$$

$$= 15.62 \text{ N/mm}^2 < 20 \text{ N/mm}^2$$

$$\sigma_b = \left[\frac{\eta P}{A} + \frac{\eta Pe}{Z} - \frac{M_g}{Z} - \frac{M_q}{Z}\right]$$

$$= [0.8\,(9.25 + 5.58) - 8.12 - 4.57]$$

$$= -0.8 \text{ N/mm}^2 \text{ (Negligible tension)}$$

(2) Mid Support Section

$$P = 15000 \text{ kN} \qquad \eta = 0.80$$
$$e = 700 \text{ mm} \qquad M_g = 13438 \text{ kN·m}$$
$$A = (1.62 \times 10^6) \text{ mm}^2 \qquad M_q = 1986 \text{ kN·m}$$
$$Z = (0.94 \times 10^9) \text{ mm}^3$$

$$\left(\frac{P}{A}\right) = \left[\frac{15000 \times 10^3}{1.62 \times 10^6}\right] = 9.25 \text{ N/mm}^2$$

$$\left(\frac{Pe}{Z}\right) = \left[\frac{15000 \times 10^3 \times 700}{0.94 \times 10^9}\right] = 11.16 \text{ N/mm}^2$$

$$\left(\frac{M_g}{Z}\right) = \left[\frac{13438 \times 10^6}{0.94 \times 10^9}\right] = 14.29 \text{ N/mm}^2$$

$$\left(\frac{M_q}{Z}\right) = \left[\frac{1986 \times 10^6}{0.94 \times 10^9}\right] = 2.11 \text{ N/mm}^2$$

At transfer stage, the stresses at extreme fibres are computed as

$$\sigma_t = [9.25 + 11.16 - 14.29] = 6.12 \text{ N/mm}^2$$
$$\sigma_b = [9.25 - 11.16 + 14.29] = 12.38 \text{ N/mm}^2$$

At Service load stage, the stresses at extreme fibres are computed as

$$\sigma_t = [0.8 (9.25 + 11.16) - 14.29 - 2.11] = -0.072 \text{ N/mm}^2$$
$$\sigma_b = [0.8 (9.25 - 11.16) + 14.29 + 2.11] = 14.87 \text{ N/mm}^2$$

All the stresses are well within the maximum permissible limits of 20 N/mm² And no tensile stresses develop at transfer and service load stages.

(h) Check for Ultimate Flexural Strength

(1) Centre of Span Section

$$A_p = 11340 \text{ mm}^2 \quad D_f = 300 \text{ mm} \quad\quad f_p = 1862 \text{ N/mm}^2$$
$$b = 2000 \text{ mm} \quad\quad d = 1350 \text{ mm} \quad M_u \text{ (Required)} = 16751 \text{ kN·m}$$
$$b_w = 300 \text{ mm} \quad\quad f_{ck} = 60 \text{ N/mm}^2$$
$$A_p = (A_{pw} + A_{pf})$$

$$A_{pf} = 0.45 f_{ck} (b - b_w) \left(\frac{D_f}{f_p}\right)$$

$$= \left[0.45 \times 60 (2000 - 300) \left(\frac{300}{1862}\right)\right] = 7395 \text{ mm}^2$$

$$A_{pw} = (11340 - 7395) = 3945 \text{ mm}^2$$

$$\text{Ratio} \left[\frac{A_{pw} f_p}{b_w d f_{ck}}\right] = \left[\frac{3945 \times 1862}{300 \times 1350 \times 60}\right] = 0.30$$

From Table 11 of IS: 1343, interpolate the values of the ratio $(f_{pb}/0.87 f_{pu})$ and (x_u/d) corresponding to the above ratio of 0.30.

$$\left(\frac{f_{pb}}{0.88 f_{pu}}\right) = 0.85 \text{ and } (x_u/d) = 0.558$$

$$f_{pb} = (0.85 \times 0.87 \times 1862) = 1377 \text{ N/mm}^2$$
$$x_u = (0.558 \times 1375) = 787.25 \text{ mm}$$

$$\therefore \quad M_u = \{f_{pb} A_{pw} (d - 0.42 x_u) + 0.45 f_{ck} (b - b_w) D_f (d - 0.5 D_f)\}$$
$$= \{1377 \times 3945(1375 - 0.42 \times 787.25\}$$
$$+ \{0.45 \times 60(2000 - 300) \, 300(1375 - 0.5 \times 300)\}$$
$$= (22541 \times 10^6) \text{ N.mm}$$
$$= 22541 \text{ kN·m} > 16751 \text{ kN·m (Hence safe)}$$

(2)   Mid Support Section

$$A_p = 11340 \text{ mm}^2 \quad D_f = 300 \text{ mm} \qquad\qquad f_p = 1862 \text{ N/mm}^2$$
$$b = 2000 \text{ mm} \qquad d = 1700 \text{ mm} \quad M_u \text{ (Required)} = 19730 \text{ kN·m}$$
$$b_w = 300 \text{ mm} \qquad f_{ck} = 60 \text{ N/mm}^2$$

$$A_{pf} = \left[0.45 f_{ck} (b - b_w) \left(\frac{D_f}{f_p}\right)\right] = \left[0.45 \times 60(2000 - 300) \left(\frac{300}{1862}\right)\right] = 7395 \text{ mm}^2$$

$$A_{pw} = (11340 - 7395) = 3945 \text{ mm}^2$$

$$\text{Ratio} \left[\frac{A_{pw} f_p}{b_w d f_{ck}}\right] = \left[\frac{3945 \times 1862}{300 \times 1700 \times 60}\right] = 0.24$$

From Table 11 of IS: 1343, interpolate the values of the ratio $(f_{pb}/0.87 f_{pu})$ and (x_u/d) corresponding to the above ratio of 0.24.

$$\left(\frac{f_{pb}}{0.87 f_{pu}}\right) = 0.91 \text{ and } (x_u/d) = 0.472$$

$$f_{pb} = (0.91 \times 0.87 \times 1862) = 1474 \text{ N/mm}^2$$
$$x_u = (0.472 \times 1700) = 802.4 \text{ mm}$$

$$\therefore \quad M_u = \{f_{pb} A_{pw} (d - 0.42 x_u) + 0.45 f_{ck} (b - b_w) D_f (d - 0.5 D_f)\}$$
$$= \{1474 \times 3945(1700 - 0.42 \times 802.4\}$$
$$+ \{0.45 \times 60(2000 - 300) \, 300(1700 - 0.5 \times 300)\}$$
$$= (29269 \times 10^6) \text{ N.mm}$$
$$= 29269 \text{ kN·m} > 19730 \text{ kN·m (Hence safe)}$$

The ultimate flexural strength of centre of span and mid support sections are greater than the required design ultimate moment. Hence the design satisfies the limit state of collapse as specified in IRC: 112-2011.

(i) Check for Ultimate Shear Strength

The mid support section is checked for the ultimate shear strength.

Shear Strength required = 2357 kN

The design shear resistance of the support section is calculated by using the equation specified in IRC: 112-2011 clause 10.3 as

$$V_{Rd.c} = \left(\frac{I.b_w}{s}\right)\sqrt{(f_{ctd})^2 \mp k_1 \sigma_{cp} f_{ctd}} + \eta P \sin\theta$$

Computing the numerical values of the various parameters, we have

$$I = \left(\frac{b_w D^3}{12}\right) = \left(\frac{300 \times 2000^3}{12}\right) = (2000 \times 10^8)\,\text{mm}^4$$

$$S = \left(\frac{300 \times 1000 \times 1000}{2}\right) = (150 \times 10^6)\,\text{mm}^3$$

$f_{ctd} = (f_{ct}/\gamma_m) = (4.0/1.5) = 2.66$ since $f_{ct} = 4.0$ N/mm^2 for $f_{ck} = 60$ N/mm^2

$\sigma_{cp} = (\eta P/A_c) = [(0.8 \times 15000 \times 10^3)/(1.62 \times 10^6)] = 7.4$ N/mm^2

$\theta = (4e/L) = [(4 \times 700)/(50 \times 1000)] = 0.056$

$$\therefore \quad V_{Rd.c} = \left(\frac{I.b_w}{s}\right)\sqrt{(f_{ctd})^2 \mp k_1 \sigma_{cp} f_{ctd}} + \eta P \sin\theta$$

$$= \left(\frac{(2000 \times 10^8)300}{150 \times 10^6}\right)\sqrt{(2.66)^2 \mp 1 \times 7.4 \times 2.66}$$

$$+ (0.8 \times 15000 \times 10^3 \times 0.056)\,\text{N}$$

$$= (2740 \times 10^3)\,\text{N}$$

$$= 2740\,\text{kN} < 2357\,\text{kN}$$

Nominal shear reinforcements of 10 mm diameter two legged stirrups are designed using the relation,

$$S_v = \left(\frac{0.87 f_y A_{sv}}{0.4b}\right) = \left(\frac{0.87 \times 415 \times 2 \times 79}{0.4 \times 300}\right) = 475\,\text{mm}$$

The stirrups are provided at a maximum spacing of 300 mm throughout the span as per IRC: 112-2011 Specifications.

(j) Supplementary Reinforcements

According to Clause, 16.51 of IRC: 112-2011, minimum Longitudinal reinforcements of not less than 0.13 percent of gross cross sectional area are to be provided to safeguard against shrinkage cracking.

$$A_{SL} = \{0.0013 \times 1.62 \times 10^6) = 2106\,\text{mm}^2$$

12 mm diameter bars are distributed in the cross section as shown in top and bottom flanges and web of the girder as shown in Fig. 11.48.

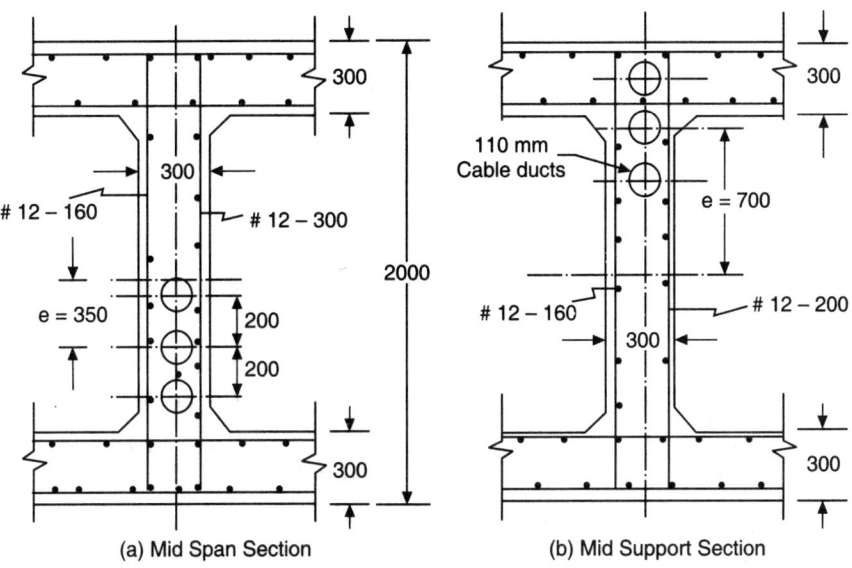

(a) Mid Span Section   (b) Mid Support Section

Fig. 11.48 Details of Reinforcements and Cables at Mid Span and Support Sections.

(k) Design of End Blocks

The width of webs are increased to 600 mm near the end supports to house the anchorages. Typical equivalent prisms on which the anchorage forces are considered to be effective are detailed in Fig. 11.49.

In the horizontal plane, we have the data,

$$P_K = 5000 \text{ kN}, \quad 2 Y_{po} = 400 \text{ mm} \quad \text{and} \quad 2 Yo = 600 \text{ mm}$$

Hence the ratio $(Y_{po}/Y_o) = (200/300) = 0.66$

Interpolating from Table 11.3, the bursting tension is computed as

$$F_{bst} = (0.14 \times 5000) = 700 \text{ kN}$$

Area of steel required to resist this tension is obtained as,

$$A_s = [(700 \times 10^3)/(0.87 \times 415)] = 1938 \text{ mm}^2$$

Provide 16 mm diameter bars at 150 mm centres in the horizontal direction. In the vertical plane, the ratio of (Y_{po}/Y_o) being higher, the magnitude of bursting tension is smaller. However the same reinforcements are provided in the form of a mesh both in the horizontal and vertical directions as shown in Fig. 11.49.

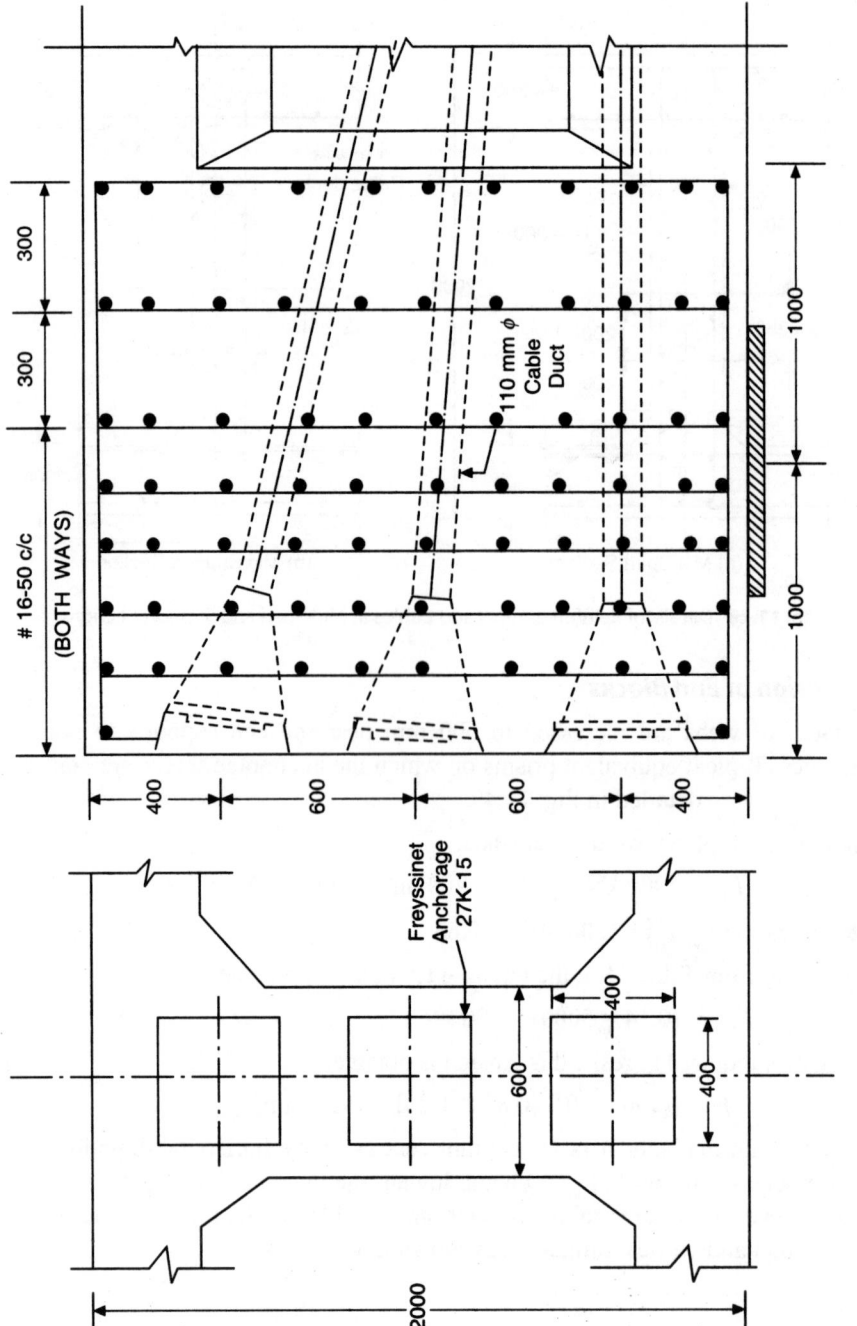

Fig. 11.49 Anchorage Zone Reinforcements in End Block.

EXAMPLES FOR PRACTICE

1. A prestressed concrete slab 400 mm thick with parallel post tensioned cables is provided for a road bridge of effective span 8 m. The live load analysis indicates an equivalent live load of 40 kN·m². The force at transfer in each of the cables is 400 kN. If the compressive stress permissible in concrete at transfer is 16 N/mm², design the slab as Class-1 type member and determine the spacing of the cables and their eccentricity at mid span. Assume a loss ratio of 0.8.

2. A box girder of a prestressed concrete bridge of span 40 m has overall dimensions of 1200 mm width and 1800 mm depth. The uniform thickness of the walls being 200 mm. The live load analysis indicates a maximum live load moment of 2000 kN·m. The compressive strength of concrete at transfer is 16 N/mm². The loss ratio is 0.8. Cables consisting of 12 high tensile wires of 8 mm diameter initially stressed to 1000 N/mm² are available for use. Design the box girder as Class-1 type structure and determine the number of cables required at the centre of span section and their position from the centroidal axis.

3. Design a post tensioned prestressed concrete Tee beam and slab bridge deck to suit the following data:

   Effective span = 24 m
   Width of carriage way = 7.5 m
   Kerbs 600 mm wide on either side of the road
   Spacing of main girders = 2 m
   Spacing of cross girders = 4 m
   Loading is IRC Class AA tracked vehicle

   Adopt M-45 Grade concrete and High tensile steel strands conforming to IS: 6006 and supplementary reinforcement comprising Fe-415 Grade HYSD bars. Permissible stresses are as specified in IS: 1343-2012, Loss ratio = 0.85.

4. A post tensioned prestressed concrete continuous beam and slab bridge deck is to be designed for a national highway crossing to suit the following data:

   Two continuous spans of 30 m each
   Width of carriage way = 15 m
   Kerbs: 600 mm wide on either side
   Thickness of wearing coat =100 mm
   Live Load: IRC Class A A tracked vehicle
   For R.C.C. slab adopt M-30 Grade concrete & Fe-415 HYSD bars
   For prestressed concrete girders, adopt M-50 Grade concrete and 12K–15 type Freyssinet cables conforming to IS: 6006-1983
   Loss ratio = 0.85
   Spacings of cross girders = 5 m

   Design the deck slab and the continuous girder as Class 1 type members conforming to IRC: 6-2014, IRC: 112-2011 & IS: 1343-2012.
   Sketch the details of reinforcements and cables at critical sections.

5. A cellular prestressed concrete box girder bridge deck is to be designed for a National highway crossing. Design the bridge deck to suit the following data:

Effective span of the Bridge = 40 m
Road width = 7.5 m
Foot Paths: 1 m wide on either side
Wearing Coat = 80 mm
Type of Loading: IRC Class AA tracked vehicle
Grade of Concrete: M-50
Type of High Tensile Steel: H.T Strands of 15.2 mm diameter conforming to IS: 6006 with Freyssinet anchorages.
Type of Supplementary reinforcement: Fe-415 HYSD Bars

Design the bridge deck conforming to the Indian Standard IRC: 6, IRC: 112 codes. Sketch the details of cable profile at salient sections and the details of reinforcements in the anchorage zone.

6. A prestressed concrete multi cell continuous box girder bridge deck is proposed for a National High way crossing of a river 120 mm wide. Design the bridge deck continuous over two spans of 60 m each to suit the following data:

Road width = 15 m
Foot Paths: 1 m on either side
Wearing Coat: 100 mm
Type of loading: IRC Class AA tracked vehicle
Grade of Concrete: M-60
High tensile steel strands conforming to IS: 6006 in conjunction with Freyssinet K-Range anchorages are available for use. Fe-415 HYSD bars are available for use in deck slab and as supplementary reinforcement in webs. It is proposed to have box cells of 2 m by 2 m vents with 300 mm thick cell walls.

Design the bridge deck conforming to the Indian standards IRC: 6-2014, IRC: 112-2011 and IS: 1343-2012. Sketch the details of cable profile along the span and details of reinforcements at critical sections and the anchorage zone.

REFERENCES

1. Freyssinet, E., The Birth of Prestressing, Cement & Concrete Association, No. CJ. 59, London, 1956, p. 44.
2. Leonhardt, F., New Trends in design and Construction of Long Span Bridges and Viaducts, Preliminary Publication of Eighth Congress of International Association for Bridge & Structural Engineering (IABSE), New York, 1968.
3. Krishna Raju, N., Prestressed Concrete (Fifth Edition), Tata McGraw Hill Publishing Co. Ltd., New Delhi, 2012.
4. Krishna Raju, N., Design of Concrete Mixes (Fifth Edition), C.B.S. Publishers & Distributors, New Delhi, 2014.

5. IS: 1785 (Part II)–1983, Indian Standard Specification for Plain & hard Drawn Steel Wire for Prestressed Concrete (Second Revision), Bureau of Indian Standards, New Delhi, 1983.
6. Abeles, P.W., An Introduction to Prestressed Concrete, Vol. II, Concrete Publications Ltd., London, 1965, p. 555.
7. Taylor, H.P.J, Clark, L.A and Banks, C.C., The Y-beam; A Replacement for the M-beam and Slab Bridge Decks, Journal of the Institution of Structural Engineers London, Vol. 68, No. 23, Dec. 1990, pp. 459–465.
8. Finsterwalder, U., Free cantilever Construction of Prestressed Concrete Bridges and Mushroom shaped Bridges, First International Symposium on Concrete Bridge Design, ACI Publication SP: 23, American Concrete Institute, Detroit, 1969, pp. 467–494.
9. Mathivat, J., The Cantilever Construction of Prestressed Concrete Bridges, John Wiley & Sons, New York, 1983, 341 pp.
10. IRC: 6-2014, Standard Specifications and Code of Practice for Road Bridges, Section II, Loads and Stresses (Revised Edition), Indian Roads Congress, New Delhi, 2014, pp. 1-84.
11. IRC: 112-2011, Code of Practice for Concrete Road bridges, Indian Roads Congress, New Delhi, 2011, p. 181.
12. IS: 1343-2012, Indian Standard Code of Practice for Prestressed Concrete (Second Revision), BIS, New Delhi, 2012.
13. IS: 6006-1983, Indian Standard Specification for uncoated stress relieved strand for Prestressed Concrete (First Revision), BIS, New Delhi, 1983, pp. 1-12.
14. Krishna Raju, N., Advanced Reinforced Concrete Design (IS: 456-2000, BSEN: 1992-1-1-2004, ACI-318 M-2011), Third Edition, C.B.S. Publishers, New Delhi, 2016, p. 463.

REVIEW QUESTIONS

1. List the various advantages of Prestressed Concrete bridges in comparison with the steel, reinforced concrete and composite bridges construction.
2. Explain the various types of prestressed concrete bridges generally used in bridge construction mentioning their suitability for various span ranges.
3. Briefly explain the various types of pretensioned prestressed concrerte bridge decks used for national high way crossings.
4. Explain with sketches the typical cross sections of post tensioned prestressed concrete bridge decks suitable for low, medium and long span bridge decks.
5. What is cantilever method of construction of prestressed concrete bridges? What are its advantages and where do you adopt this type of construction?
6. Briefly explain the method of determining the maximum moments and shear forces in slab bridge decks due to the standard high way wheel loads.
7. Explain with sketches the typical cross section of a post tensioned prestressed concrete tee beam and slab bridge deck suitable for a two lane national high way crossing. Mention the various structural elements of ther bridge deck.

8. Briefly discuss the limit state design principles and specifications enshrined in the recent Indian Roads Congress codes with relevance to prestressed concrete bridges.

9. Explain the IRC code methods prescribed to check the limit states of serviceability of prestressed concrete bridge decks.

10. When do you resort to continuous prestressed concrete bridges? Explain the typical cross sections generally adopted for continuous long span prestressed concrete bridges.

OBJECTIVE TYPE QUESTIONS

1. Use of prestressed concrete in bridge construction results in
   a) Large size structural elements
   b) Slender structures
   c) Savings in overall cost of construction

2. For Prestressed concrete bridge decks in the span range of 10 to 15 m, it is economical to use
   a) Tee beam and slab type structure
   b) Cellular box type structure
   c) Solid or hollow slab type structure

3. In tee beam and slab type bridge decks, the maximum moments due to IRC wheel loads develops in the
   a) Interior girders
   b) Middle girder
   c) Exterior girder

4. In the design of prestressed concrete deck slabs the design live load moments developed in the slabs due to wheel loads are generally determined by using
   a) Courbon's method
   b) Pigeaud's method
   c) Moment coefficients

5. In the determination of design moments in the outer girder of a tee beam and slab bridge deck due to IRC wheel loads, the method which gives the least value is due to
   a) Hendry-Jaegar
   b) Courbon
   c) Guyon-Massonet

6. For long span continuous prestressed concrete bridge decks in the span range of 30 to 60 m, it is economical to use
   a) Slab decks
   b) Tee beam and slab decks
   c) Cellular box type decks

7. The minimum thickness of web in a prestressed concrete girder required to house cables of 65 mm diameter should be not less than
   a) 100 mm
   b) 200 mm
   c) 300 mm

8. According to the IRC: 112-2011 specifications regarding the serviceability limit state of deflection, the maximum deflection of the deck under vehicular live loads is limited to a value of
   a) Span/250
   b) Span/1000
   c) Span/800

9. The ultimate shear resistance of a prestressed concrete slab deck without design shear reinforcements depends upon
   a) Prestressing force
   b) Depth of deck slab
   c) Tensile strength of concrete

10. In prestressed concrete bridge decks with bonded tendons located in zones of moderate environmental exposure conditions, the maximum permissible crack width is limited to
    a) 0.1 mm
    b) 0.2 mm
    c) 0.3 mm

12

Rigid Frame Bridges

12.1 GENERAL FEATURES

In National high way crossings carrying multilane traffic, reinforced concrete rigid frame bridge decks are often preferred to other types mainly due to the rigidity and the aesthetic appearance of the portal frame structure. Reinforced concrete rigid frame bridges[1] comprising of single or double portal frames are generally used in the span ranges of 10 to 20 mm. Prestressed concrete rigid frame bridges are often preferred for larger span ranges mainly to reduce the cross sectional dimensions of the transom and the vertical legs resulting in considerable reduction in dead load of the structure.

Rigid frame bridge structure generally comprises of a horizontal member referred to as transom serving as the deck and it is supported at the ends by columns rigidly connected at the junction resulting in a statically indeterminate structure. The column supports are generally hinged so that the cross section required is smaller since there are no moments at the hinged ends. A typical rigid frame bridge of the solid barrel or slab type is shown in Fig. 12.1(a). For larger spans in the range of 20 to 30 m, rigid frames comprising beams or ribs spaced at 3 to 4 m intervals connected by a slab is more economical. Fig. 12.1(b) shows the typical structural elements of a rigid frame bridge with tee beam ribs and slab.

Rigid frame bridges do not require separate abutments since the vertical columns retain the earth and serve as retaining walls thus eliminating the construction of abutments and their foundations. The use of double barrel type rigid frames eliminate the use of separate central pillars since the central column serves as a median between the up and down traffic lanes.

12.2 ADVANTAGES OF RIGID FRAME BRIDGES

Reinforced concrete rigid frame bridges have the following unique advantages in comparison with the other traditional types of bridges.[2,3]

1. Rigid frame bridge comprising a portal frame being a monolithic structure, eliminates the use of separate abutments. The vertical sides of the rigid frame serves as retaining walls to retain earth in road crossings of embankments.

2. Slab type rigid frame bridges can be easily cast *in-situ* since plain moving form work can be used for rapid construction work.

3. Rigid frame bridges can be advantageously adopted in fly over crossings where roads criss cross at different levels.

4. Rigid frame bridges have high stability against lateral forces like wind, earth quake and soil pressure.

5. Rigid frame bridges do not require any separate bearings since hinged supports can be provided with structural advantages.

6. Rigid frame bridges are aesthetically superior and consume less quantity of materials compared to R.C.C. beam and slab bridges due to variable depth of the members.

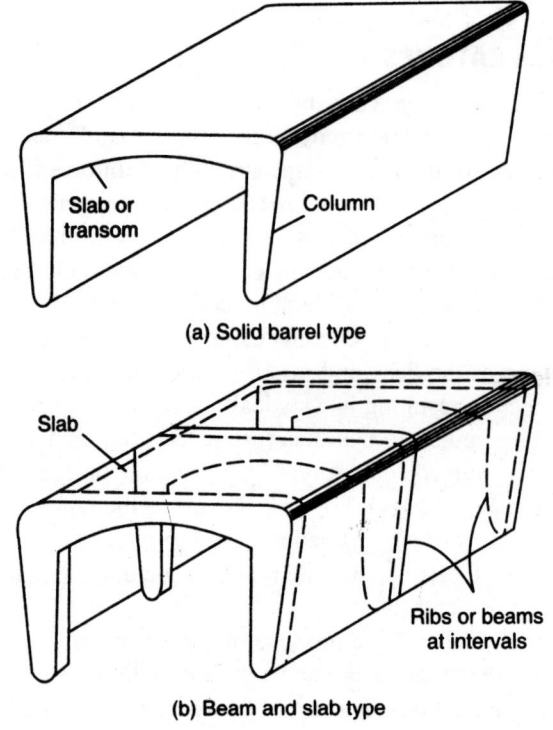

(a) Solid barrel type

(b) Beam and slab type

Fig. 12.1 Types of Rigid Frame Bridges.

12.3 DESIGN CRITERIA OF RIGID FRAME BRIDGES

1. Selection of Preliminary Sections of Structural Elements

The selection of dimensions at salient points in the frame is based on empirical relations developed on the basis of past practical experience.

The typical dimensions at the crown and junction of horizontal and vertical members can be expressed in terms of the clear span. Figure 12.2 shows the rigid frame bridge with the salient dimensions expressed in terms of the clear span.

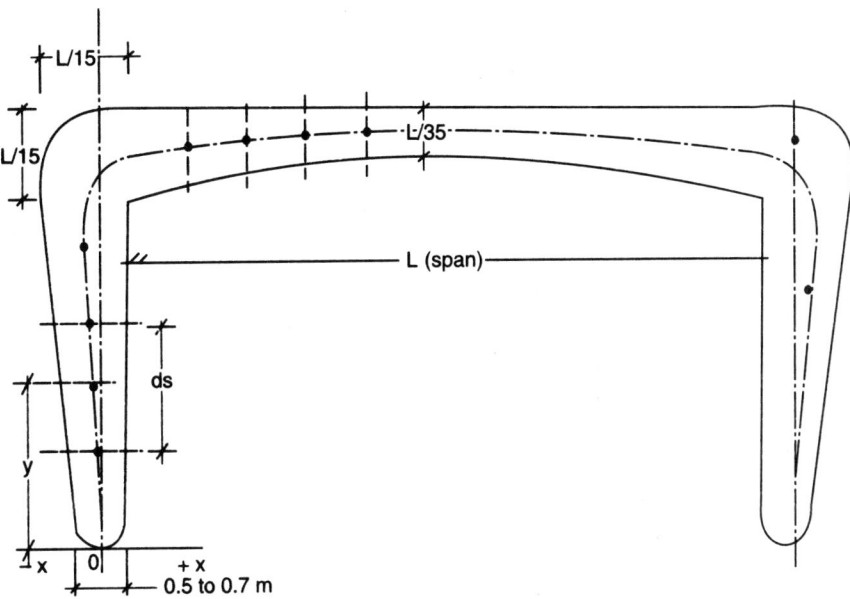

Fig. 12.2 Dimensions of Rigid Frame Bridge.

2. Computation of Frame Constants

The rigid frame is divided into a number of discrete elements with the origin at the left hand column base as shown in Fig. 12.2. The centroid of each element from the origin along the y-axis is computed along with the moment of inertia of each of the elements. The frame constant $\Sigma\ (y^2\ ds/I)$ is computed as the summation for all the elements, where

y = distance of the centroid of elements from origin

ds = Length of elements

I = second moment of area of each element

The rib shortening effect is computed as (L/A_c) where

L = Length of transom or horizontal member

A_c = Cross sectional area of crown section

The value of $\Sigma\ (M_F \cdot y \cdot ds/I)$ is computed for the entire frame,

where M_F = bending moment due to unit load at the centroids of the various elements.

3.　Influence Line Diagram for Horizontal Thrust

The influence line ordinates for the horizontal thrust at the hinged support is computed by the relation,

$$H = \left\{ \frac{\Sigma\,(M_F \cdot y \cdot ds\,/\,I)}{[(Y^2 \cdot ds\,/\,I) + (L\,/\,A_c)]} \right\}$$

The influence line diagram for horizontal thrust will be useful in computing the horizontal thrust developed due to the dead and live loads.

4.　Earth Pressure Computations[4]

The earth pressure developed due to the earth filling against the sides of the columns is computed by assuming a suitable distribution of earth pressure which depends upon the density of soil, height of filling and the angle of internal friction.

5.　Horizontal Thrust due to Temperature Changes[5]

The change of temperature is likely to influence the span length which in turn induces the horizontal thrust similar to a two hinged arch. The magnitude of the horizontal thrust induced is given by the relation,

$$H = [(\alpha\,t\,.\,L\,.\,E)/\!\int(y^2\,.\,ds/I)]$$

where　　　　L = span length

α = coefficient of linear expansion of concrete generally having a value of 6.5×10^{-6}

H = horizontal thrust

t = temperature change in degrees

E = Modulus of elasticity of concrete computed by the empirical relation $5700\sqrt{f_{ck}}$

f_{ck} = characteristic compressive strength of concrete.

6.　Dead load Computations

The dead loads of the various elements are computed assuming the density of concrete for unit length of deck for computations of bending moments due to dead load. The dead loads are assumed to act at the centroids of each element.

7.　Influence Lines for Bending Moment at Various Salient Sections[6]

Using the influence line ordinates of horizontal thrust and the reactions developed at the hinged supports, influence line ordinates are computed for salient sections like the crown, mid height of column and the junction of column and beam elements.

8. Dead Load and Live Load Bending Moments

Using the influence lines for bending moments, the dead load bending moments and maximum live load bending moments due to the moving I.R.C. loads are computed.

9. Design of Sections

The reinforcements in the various salient sections are designed for the maximum bending moment are thrust using interaction charts of SP-16, design aids for reinforced concrete.

10. Design of Hinged Footing[7]

The reinforcements for the hinged footing is designed for the maximum reaction developed at the supports assuming the safe bearing capacity of the soil at site. Suitable curved mating surfaces are provided at the junction of the column and footing to facilitate rotation at the column ends.

The use of these design principles is illustrated by the following example of a slab type rigid frame with hinged supports.

12.4 DESIGN EXAMPLE OF REINFORCED CONCRETE RIGID FRAME BRIDGE

Design a reinforced concrete rigid frame bridge to suit the following data:

Clear span = 15 m
Width of carriage way = 7.5 m
Width of kerbs on either side = 600 mm
Height of rigid frame soffit at centre of span = 6 m
Type of Loads: IRC Class AA tracked vehicle
Temperature variations at site = 30°C
Coefficient of linear expansion of concrete = (6.5×10^{-6})
Characteristic compressive strength of concrete = f_{ck} = 20 N/mm^2
Characteristic tensile strength of concrete = f_{ctd} = 1.9 N/mm^2
Characteristic tensile strength of HYSD reinforcements = f_y = 415 N/mm^2
End conditions of columns: Hinged supports
Modulus of elasticity of concrete = E_c = 29 kN/mm^2
Modulus of elasticity of steel = E_s = 200 kN/mm^2

1. Selection of Sectional Dimensions of Transom and Columns

Preliminary dimensions of the rigid frame structural elements are assumed based on empirical relations for the thickness at crown and at junction of transom and columns of the rigid frame, expressed as a function of the span.

Clear span = L = 15 m
Depth at crown (centre of span) section = $(L/35)$ = $[(15 \times 10^3)/35]$ = 428.5 mm

Adopt an overall depth at crown = 450 mm

Depth at junction of transom and column = $(L/15)$ = $[(15 \times 10^3)/15]$ = 1000 mm

The dimensions adopted for the rigid frame critical sections are shown in Fig. 12.3

The section is analyzed for frame constants and influence lines for horizontal thrust and bending moments at salient points, like crown, junction of transom and columns and for typical points in the vertical member are obtained by a detailed analytical procedure.

The area of steel can be neglected for computations of the second moment of area for the various sections. Horizontal thrust due to earth pressure is calculated and the moments due to the lateral earth pressure are determined. The maximum bending moment and thrust at various salient sections are determined due to live loads for designing the reinforcements.

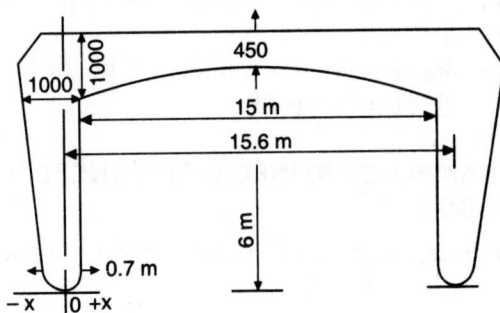

Fig. 12.3 Rigid Frame Bridge.

2. Rigid Frame Constants

For purposes of analysis, the frame is divided into suitable number of parts with the x and y axis as shown in Fig. 12.4.

The transom is divided into segments of 1.5 m each and the column is divided into three parts of 1.8 m each. The frame constant $(y^2 ds/I)$ is computed as shown in Table 12.1.

$$\Sigma y^2 (ds/I) = 11693$$

For half arch ... 11693
× 2

For full arch ... 23386

Rid shortening effect = (L/A_c) = (15/0.45) = 33.3 = 34

$$[y^2 ds/I) + (L/A_c)] = 23420$$

3. Determination of $\Sigma M_F \cdot (y \cdot ds/I)$

The computation of $M_F \cdot (y \cdot ds/I)$ for different position of the unit load at points 5, 6, 7, 8 and 9 different sections 5 – 5′ to 9 – 9′ is shown in Table 12.2.

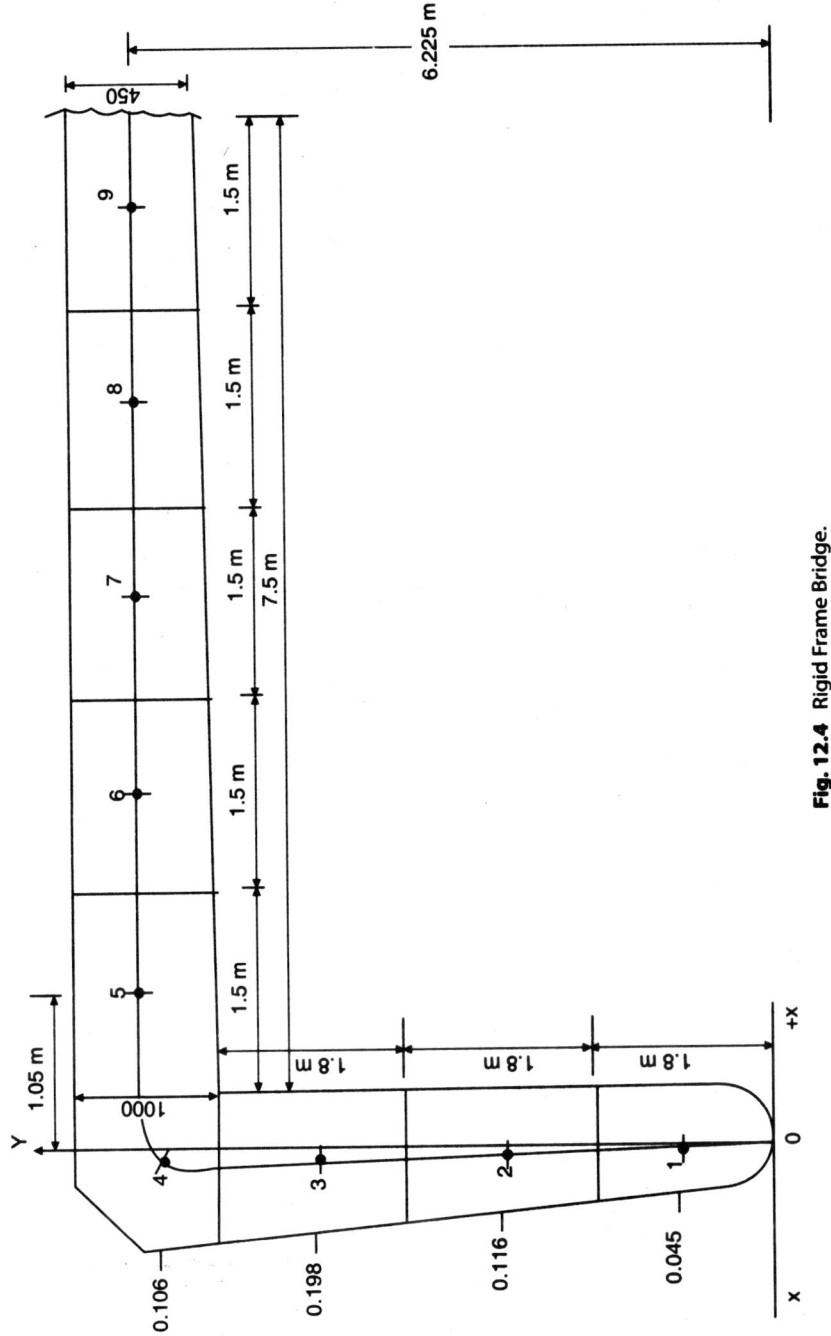

Fig. 12.4 Rigid Frame Bridge.

Table 12.1 Frame Constants

Section	Size (m)	x (m)	y (m)	y^2	$I = bd^3/12$ (m^4)	ds (m)	yds/I	y^2ds/I
1–1′	0.7 × 1	−0.045	0.90	0.81	0.028	1.8	57.8	52.0
2–2′	0.84 × 1	−0.116	2.70	7.29	0.049	1.8	99.1	267.5
3–3′	1 × 1	−0.198	4.50	20.25	0.083	1.8	97.5	438.7
4–4′	1.22 × 1	−0.106	5.80	33.64	0.151	1.06	40.7	236.0
5–5′	1 × 1	1.05	6.02	36.24	0.083	1.5	108.7	654.3
6–6′	0.88 × 1	2.55	6.06	36.72	0.056	1.5	162.3	983.5
7–7′	0.76 × 1	4.05	6.11	37.33	0.036	1.5	254.5	1554.9
8–8′	0.64 × 1	5.60	6.15	37.82	0.021	1.5	439.2	2701.0
9–9′	0.52 × 1	7.10	6.20	38.44	0.012	1.5	775.0	4805.0
								11693.9

Table 12.2 Determination of $M_F \cdot y \, (ds/I)$

Section	$\left(\dfrac{y \cdot ds}{I}\right)$	Load at				
		5 a = 1.05 m	6 a = 2.55 m	7 a = 4.05 m	8 a = 5.55 m	9 a = 7.05 m
5–5′	108.7	$M_F = 0.9786$ 106.37	0.8778 95.40	0.7700 83.60	0.6700 73.50	0.5750 62.55
6–6′	162.3	$M_F = 0.8760$ 142.1	2.13 345.9	1.88 305.1	1.64 266.5	1.39 226.7
7–7′	245.5	$M_F = 0.7700$ 189.0	1.8800 461.5	2.9900 735.7	2.6000 640.3	2.2100 544.8
8–8′	439.2	$M_F = 0.6700$ 294.2	1.6300 715.8	2.6000 1141.9	3.5700 1569.7	3.0400 1335.7
9–9′	775.0	$M_F = 0.5700$ 441.7	1.3900 1077.2	2.2100 1712.7	3.0400 2358.0	3.8600 2991.5
$\sum\left(\dfrac{M_F \cdot y \cdot ds}{I}\right)$		1173.37	2695.8	3979	4908	5161

4. Influence Line Ordinates for Horizontal Thrust

The influence line ordinate for horizontal thrust is given by

$$H = [(M_F \cdot y \cdot ds/I)/(y^2 \cdot ds/I) + (L/A_c)]$$

Hence the influence line ordinates at the various points 5, 6, 7, 8 and 9 are tabulated as shown in Table 12.3. The influence line for horizontal thrust is shown in Fig. 12.5.

5. Earth Pressure Calculations

The determination of horizontal thrust due to earth pressure requires detailed computations. The earth pressures acting on the column portion of the frame are as shown in Fig. 12.6.

Table 12.3 Influence Line Ordinates for Horizontal Thrust

5	6	7	8	9
$\left(\dfrac{1173.37}{23420}\right)$	$\left(\dfrac{2695.8}{23420}\right)$	$\left(\dfrac{3979}{23420}\right)$	$\left(\dfrac{4908}{23420}\right)$	$\left(\dfrac{5161}{23420}\right)$
= 0.05	= 0.115	= 0.169	= 0.209	= 0.22

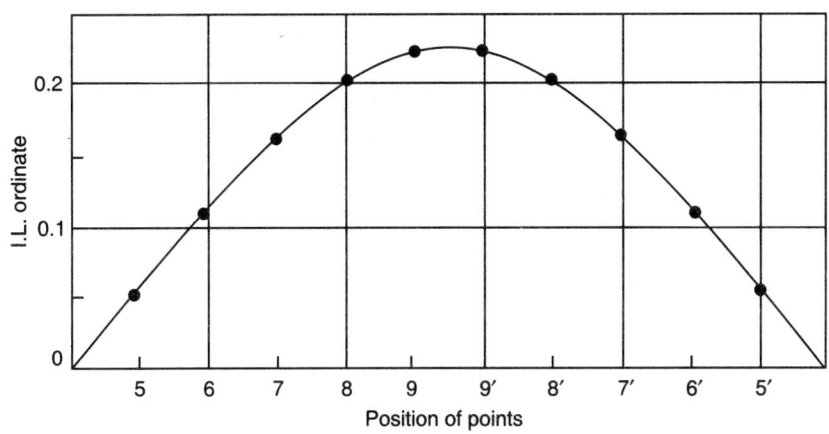

Fig. 12.5 Influence line for Horizontal Thrust.

$$p = [wh(1 - \sin \phi)/(1 + \sin \phi)]$$

Assuming $\phi = 36° - 53'$

$w = 16 \text{ kN/m}^3, p = [16h(1 - 0.6)/(1 + 0.6)] = 4h$

$p_1 = (4 \times 1) = 4 \text{ kN/m}^2$

$p_2 = (4 \times 2.8) = 11.2$

$p_3 = (4 \times 4.6) = 18.4$

$p_4 = (4 \times 6.4) = 25.6$

$P_1 = (1/2 \times 4 \times 1) = 2 \text{ kN}$

$P_2 = (4 + 11.2)/2 \ (1.8) = 13.68 \text{ kN}$

$P_3 = (11.2 + 18.4)/2 \ (1.8) = 26.64 \text{ kN}$

$P_4 = (18.4 + 25.6)/2 \ (1.8) = 39.60 \text{ kN}$

The reaction at the hinged support A of the rigid frame due to earth pressures is computed as shown in Fig. 12.7.

The horizontal thrust '*H*' due to earth pressure only and the moments at various sections due to earth pressure from left only, right only and from both sides is computed as shown in Table 12.4.

6.   Horizontal Thrust due to Temperature Changes

For spans greater than 30 m, the temperature effect will be considerable

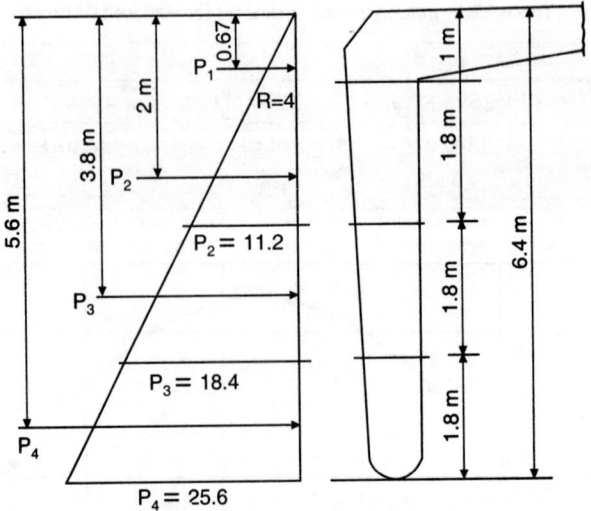

Fig. 12.6 Earth Pressure Calculations.

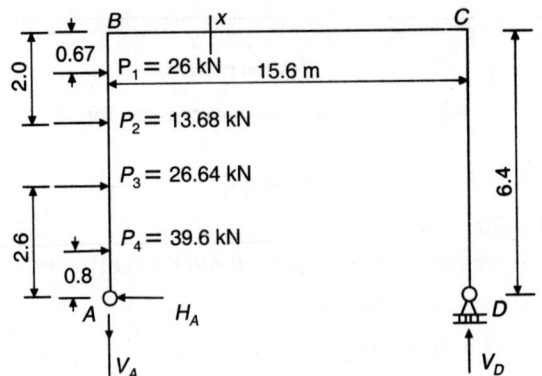

$M_X = H_A \cdot Y - M'_f - V_A \cdot x$

where M'_f = Moment due to earth pressures
   P_1, P_2, P_3 and P_4 only
Taking Moments about D
$V_A \cdot 15 = (39.6 \times 0.8) + (26.64 \times 2.6) + (13.68 \times 4.4) + (2 \times 5.73)$
$\therefore V_A = 11.50$ kN

Fig. 12.7 Determination of Reaction VA.

Change in span length $= (\alpha t L) = \int (M.yds)/EI$

$$= \int (M_F - H.y)(y.ds)/EI$$

$\therefore$ $\alpha t L = \int (M_F.y.ds/EI) - H \int (y^2.ds/EI)$

Table 12.4 Determination of H (Earth Pressure Only)

1	2	3	4	5	6	7	8	9	10	11	12	13	14	15	16
Section	x	y	Earth Pressure P(kN)	$p.y$	$H_A.y$ $= (81.92.y)$	M_F	$V_A.x$ $(11.50.x)$	$V_D(L-x)$	Total MF $(H_A.y - V_A.x - M_F)$	$y.ds/I$	$\dfrac{M_F.y.ds}{I}$	$H.y$ $= 15.y$	$M_L =$ $(M_F - H_y)$ E.P. Left only	M_R E.P. Right only	$M_L + M_R$ E.P. Both sides
1	−0.045	0.90	39.60	35.64	73.72	0.1×39.60 = 3.96	−0.517	—	70.27	57.8	4061	13.5	56.77	−14.01	42.76
2	−0.116	2.70	26.64	71.92	221.18	77.90	−1.334	—	144.61	99.1	14330	40.5	104.11	−41.83	62.28
3	−0.198	4.50	13.68	61.56	368.64	198.48	−2.277	—	172.43	97.5	16811	67.5	104.93	−69.77	35.16
4	−0.106	5.80	2.00	11.60	475.13	302.53	−1.219	—	173.81	40.7	7074	87.0	86.81	−88.21	−1.40
5	1.05	6.02	81.92	493.15	493.15	320.55	12.07	—	160.53	108.7	17449	90.3	70.23	−78.20	−7.97
6	2.55	6.06	= H_A		496.43	323.82	29.32	—	143.29	162.3	23255	90.9	52.39	−61.50	−9.11
7	4.05	6.11			500.53	324.65	46.57	—	129.31	254.5	32909	91.6	37.71	−45.00	−7.29
8	5.60	6.15			503.80	325.45	64.40	—	113.95	439.2	50046	92.2	21.75	−27.80	−6.05
9	7.10	6.20			507.90	326.70	81.65	—	99.55	775.0	77151	93.0	6.55	−11.35	−4.80
9'		↓ Reflect						81.65	81.65	775.0	63278	93.0	−11.35	6.55	−6.05
8'								64.40	64.40	439.2	28284	92.2	−27.80	21.75	−6.05
7'			Reflect					46.57	46.57	254.5	11852	91.6	−45.00	37.71	−7.29
6'								29.32	29.32	162.3	4758	90.9	−61.50	52.39	−9.11
5'								12.07	12.07	108.7	1312	90.3	−78.20	70.23	−7.97
4'								−1.219	−1.219	40.7	−50	87.0	−8.21	86.81	−1.41
3'								−2.277	−2.277	97.5	−222	67.5	−69.77	104.93	35.36
2'								−1.334	−1.334	99.1	−132	40.5	−41.83	104.11	62.28
1'								−0.517	−0.517	57.8	−30	13.5	−14.01	56.77	42.76
									$\Sigma M_F \cdot y \cdot ds/I = 352136$						

$H = [\Sigma M_F \cdot y \cdot Tds/I]/[(\Sigma y^2 ds/I + L/A_c)] = (352136/23420) = 15$ kN
(Earth Pressure only).

$$= 0 - H.\int (y^2 ds)/EI$$

Due to temperature variation only

$$H = (\alpha t L E)/\int (y^2./ds/I)$$

If $\alpha = 0.0000065$

$t = \pm 30°$ $L = 15$ m

$E = 5700\sqrt{f_{ck}}$

$= 5700\sqrt{20} = 25491\,\text{N/mm}^2 = 25.491 \times 10^6\,\text{kN·m}^2$

∴ $H = (0.0000065 \times 30 \times 15 \times 25.491 \times 10^6)/23420 = 3.18$ kN

The moments at various sections due to temperature changes are compiled in Table 12.5.

Table 12.5 Moment at Various Sections Due to Temperature Changes

Section	y	$H.y = 3.18\, y$ kN·m	Remarks
1	0.90	2.862	
2	2.70	8.586	
3	4.50	14.300	
4	5.80	18.440	
5	6.02	19.140	
6	6.06	19.270	
7	6.11	19.420	
8	6.15	19.550	
9	6.20	19.710	

7.   Dead Load Computations

The dead loads of the various sections 1–1′ to 9–9′ are calculated as shown in Table 12.6. The total load for half the frame is 264.00 kN.

Table 12.6 Dead Load Calculations

Section	Dimensions	Weight (kN)
1–1′	0.700 × 1.8 × 24	30.24
2–2′	0.840 × 1.8 × 24	36.28
3–3′	1.000 × 1.8 × 24	43.20
4–4′	1.22 × 1.0 × 24	29.28
5–5′	1 × 1 × 24	24.00
6–6′	0.88 × 1.5 × 24	31.68
7–7′	0.76 × 1.5 × 24	27.36
8–8′	0.64 × 1.5 × 24	23.04
9–9′	0.52 × 1.5 × 24	18.72
	Total for half frame	264.00

8. Influence Lines for Bending Moment

The reaction at A or D is computed using the Fig. 12.8.

The influence line for bending moment at crown is computed as shown in Table 12.7.

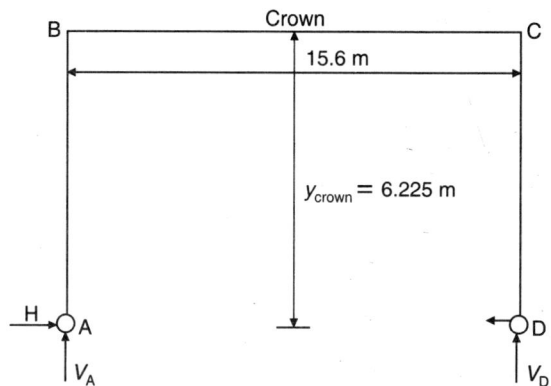

Fig. 12.8 Computation of Vertical Reactions.

The influence line for bending moment at crown is shown in Fig. 12.9. In a similar way the influence line ordinates for bending moment at points 5 and 2 are calculated as shown in Tables 12.8 and 12.9. The influence lines for bending moment at point 5 and 2 are shown in Figs. 12.10 and 12.11.

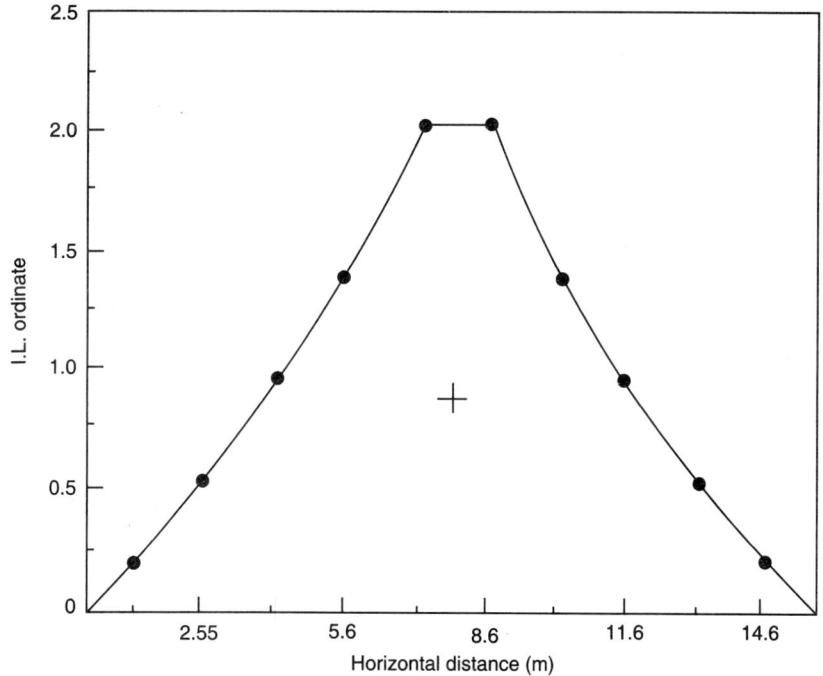

Fig. 12.9 Influence Line for Bending Moment at Crown.

Table 12.7 Effect of Live Loads—Influence Line for B.M. at Crown

Particulars	Load at									
	5	6	7	8	9	9'	8'	7'	6'	5'
Reaction at A or D	$\left(\dfrac{1.05}{15.6}\right)$ $= 0.067$	$\left(\dfrac{2.55}{15.6}\right)$ $= 0.163$	$\left(\dfrac{4.05}{15.6}\right)$ $= 0.26$	$\left(\dfrac{5.60}{15.6}\right)$ $= 0.356$	$\left(\dfrac{7.10}{15.6}\right)$ $= 0.45$	0.45	0.356	0.26	0.163	−0.067
$M_F = (R_A$ or $R_D) \times 7.5$	0.502	1.222	1.95	2.67	3.375	3.375	2.67	1.95	1.222	0.502
$H \cdot y$ (H from Table 11.3) $= 6.225\,H$	(6.225) $\times (0.05)$ $= 0.31$	(6.225) $\times (0.115)$ $= 0.715$	(6.225) $\times (0.169)$ $= 1.05$	(6.225) $\times (0.209)$ $= 1.30$	(6.225) $\times (6.225)$ $= 1.36$	1.36	1.30	1.05	0.715	0.31
$M = (M_F - H \cdot y)$	0.192	0.51	0.90	1.37	2.01	2.01	1.37	0.90	0.51	0.192

Table 12.8 Influence Line for B.M. at Point 5

Details	Load at									
	5	6	7	8	9	9'	8'	7'	6'	5'
Reaction at A (V_A)	0.933	0.837	0.740	0.644	0.550	0.450	0.356	0.260	0.163	0.067
$M_F = V_A$ (1.05)	0.979	0.878	0.777	0.676	0.577	0.472	0.373	0.273	0.171	0.070
$H \cdot y = 6.02\,H$	(6.02) $\times (0.05)$ $= 0.301$	(6.02) $\times (0.115)$ $= 0.692$	(6.02) $\times (0.164)$ $= 0.987$	(6.02) $\times (0.209)$ $= 1.258$	(6.02) $\times (0.22)$ $= 1.324$	1.324	1.258	0.987	0.692	0.301
$M = (M_F - H \cdot y)$	0.678	0.186	−0.21	−0.582	−0.747	−0.852	−0.885	−0.714	−0.521	−0.231

Table 12.9 Influence Line for B.M. at Point 2

Details	Load at										
	5	6	7	8	9	9'	8'	7'	6'	5'
Reaction at A (V_A)	0.933	0.837	0.740	0.644	0.550	0.450	0.356	0.260	0.163	0.067
$M_F = V_A$ (−0.116)	−0.108	−0.097	−0.085	−0.074	−0.063	−0.052	−0.041	−0.030	−0.0189	−0.007
$H \cdot y = 2.70\,H$	(2.70) × (0.05) = 0.135	(2.70) × (0.115) = 0.310	(2.70) × (0.164) = 0.442	(2.70) × (0.209) = 0.564	(2.70) × (0.22) = 0.594	0.594	0.564	0.442	0.310	0.135
$M = (M_F − H \cdot y)$	−0.243	−0.407	−0.527	−0.638	−0.657	−0.646	−0.605	−0.472	−0.329	−0.142

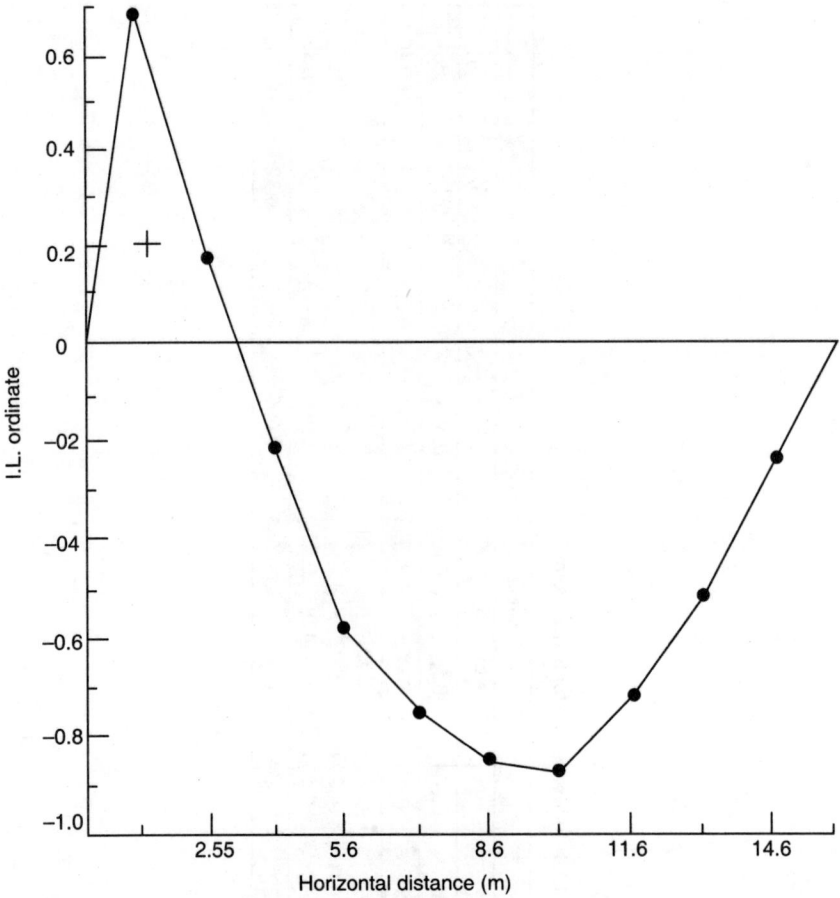

Fig. 12.10 Influence Line for Bending Moment at Point 5.

9. Dead Load Bending Moments

The dead load bending moments developed at crown, point 5 and point 2 and the corresponding horizontal thrust at these points are computed as shown in Table 12.10.

10. Live Load Bending Moments

Width of road = 7.5 m
Width of Kerbs = 600 mm wide
Total road width = 8.7 m
Centre line length of bridge = 15.6 m

$$(B/L) = (8.7/15.6) = 0.557$$

From Table 3.1, $K = 1.85$

Effective width of dispersion for one wheel is given by

$$b_e = K \times (1 - x/L) + b_w$$

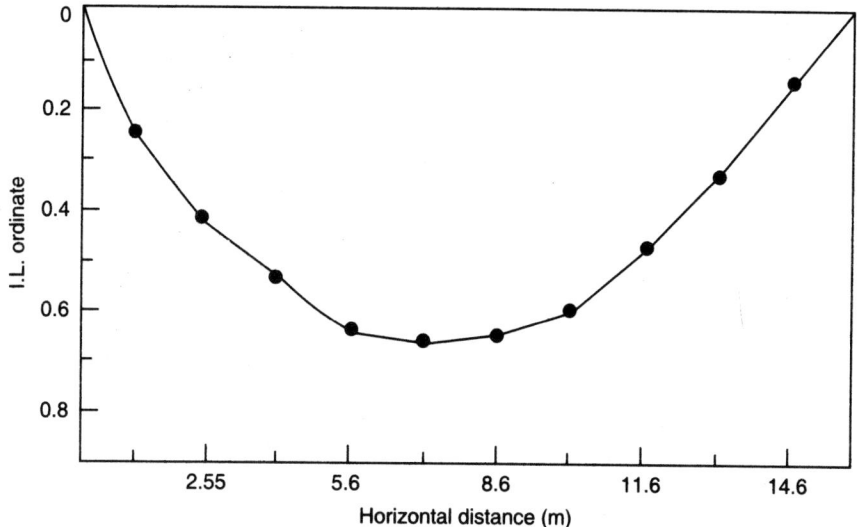

Fig. 12.11 Influence Line for Bending Moment at 2.

Referring to Fig. 12.12

We have $\quad x = 7.8$ m

$\qquad L = 15.6$ m

$\qquad B = 8.7$ m

$\qquad b_w = (0.85 + 2 \times 0.10) = 1.05$

$\qquad b_e = 1.85 \times 7.8 \,(1 - 7.8/15.6) + 1.05 = 8.265$ m

The tracked vehicle is placed close to the kerb with the required minimum clearance as shown in Fig. 12.13.

Net effective width of dispersion = 8.407 m

Total load of two tracks with impact = $(700 \times 1.10) = 770$ kN

Average intensity of load = $770/(8.407 \times 4.7) = 20$ kN/m

(a) Live Load B.M. at Crown

The live load bending moment at crown is calculated using the influence diagram as shown in Fig. 12.14.

Live load maximum B.M. at crown = (Area of influence line diagram under the load) × (intensity of load)

$\qquad = (8.43 \times 20) = 168.6$ kN·m

(b) Corresponding Thrust for Maximum B.M. at Crown computed by referring to Fig. 12.15.

The corresponding thrust for maximum B.M. at crown is given by Maximum Thrust

$\qquad = [0.5 \,(0.20 + 0.22) \times 4.7 \times 20] = 19.74$ kN

Table 12.10 Dead Load Bending Moments at Various Sections

Section	Dead Load (kN)	I.L.O. For Thrust (kN)	Thrust (kN)	Crown I.L.O.	Crown Moment (kN·m)	Point 5 I.L.O.	Point 5 Moment (kN·m)	Point 2 I.L.O.	Point 2 Moment (kN·m)
5	24.00	0.05	1.200	0.192	4.60	(0.678 – 0.231) = 4.447	10.73	(0.243 – 0.142) = – 0.385	–9.24
6	31.68	0.115	3.643	0.51	16.15	(0.186 – 0.521) = – 0.335	–10.61	(–0.407 – 0.329) = – 0.736	–23.3
7	27.36	0.169	4.623	0.90	24.62	(–0.21 – 0.714) = –0.924	–25.28	(–0.527 – 0.472) = –0.999	–27.33
8	23.04	0.209	4.815	1.37	31.56	(–0.582 – 0.885) = –1.467	–33.79	(–0.638 – 0.605) = –1.243	
9	18.72	0.22	4.118	2.01	36.72	(–0.747 – 0.852) = –1.599	–29.93	(–0.657 – 0.646) = –1.303	–24.39

For Half Frame = (18.4 × 2) = (113.65 × 2)

For Full Frame = 36.8 kN = 227.3 kN·m

For Full Frame = – 88.88 kN·m

For Full Frame = –112.90 kN·m

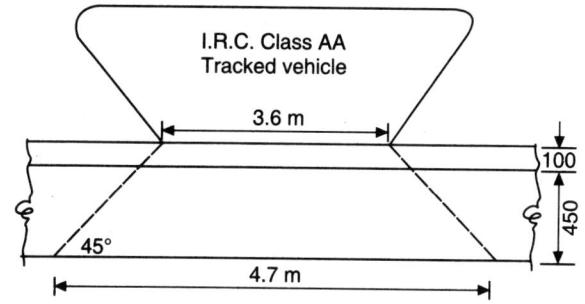

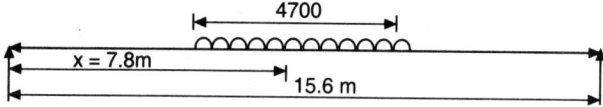

Fig. 12.12 I.R.C. Class AA Tracked Vehicle.

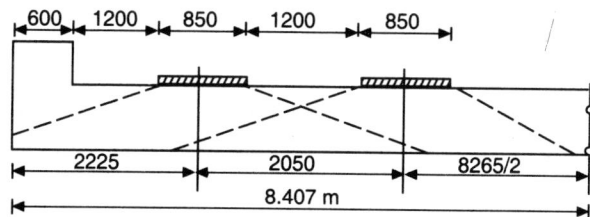

Fig. 12.13 Effective Width of Dispersion.

(c) Live Load B.M. for Point 5

Referring to Fig. 12.16
Maximum live load B.M. at point 5 (Negative)

$$= [0.5 (0.72 + 0.885) \times 4.7 \times 20] = 75.44 \text{ kN·m}$$

(d) Corresponding Thrust for Maximum B.M. at point 5 Referring to Fig. 12.17
Maximum Thrust = $[0.5 (0.22 + 0.17) \times 4.7 \times 20] = 18.33$ kN

(e) Live Load B.M. for Point 2

Referring to Fig. 12.18
Maximum B.M. = $[0.5 (0.60 + 0.657) \times 4.7 \times 20] = 59$ kN·m

(f) Corresponding Thrust for Maximum B.M. at Point 2 is Computed by Referring to Fig. 12.19.

Maximum Thrust = $[0.33 (0.19 + 0.22 + 0.21) \times 4.7 \times 20] = 19.43$ kN

11. Design Service Load Moments and Thrusts

The maximum service load design moments and the corresponding thrusts at various sections are compiled in Table 12.11. In this example, the ultimate moments and thrusts are calculated by applying a uniform load factor of 1.5 which will result in a conservative design. However, the reader can attempt an economical design by using separate load factors of 1.35 for dead load and 1.5 for live loads as specified in IRC: 6-2014[9].

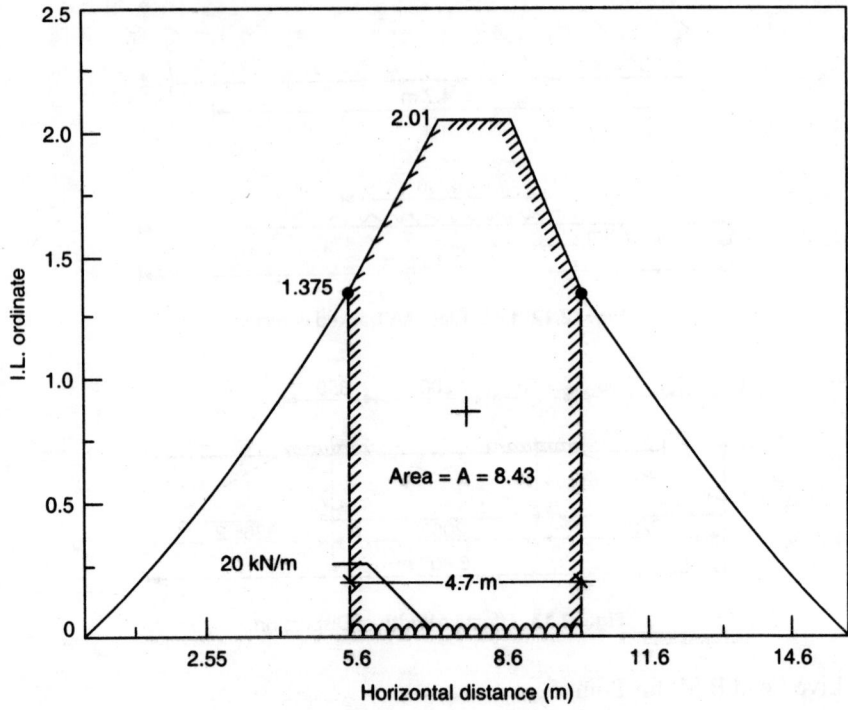

Fig. 12.14 Live Load Bending Moment at Crown.

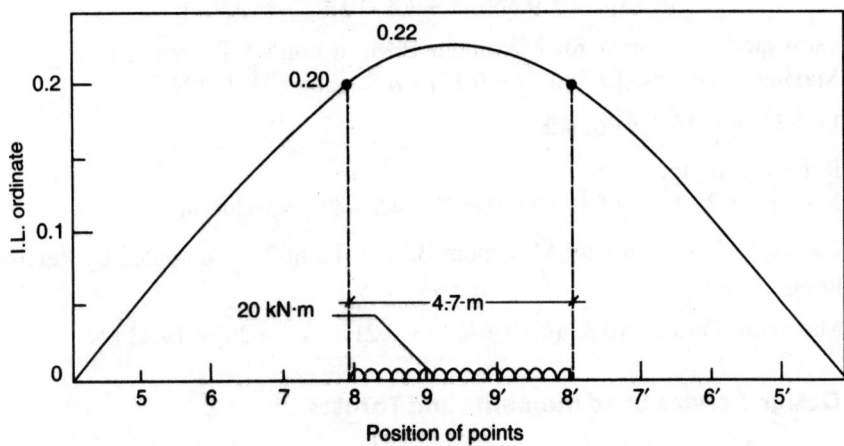

Fig. 12.15 Corresponding Thrust for Maximum Bending Moment in Crown.

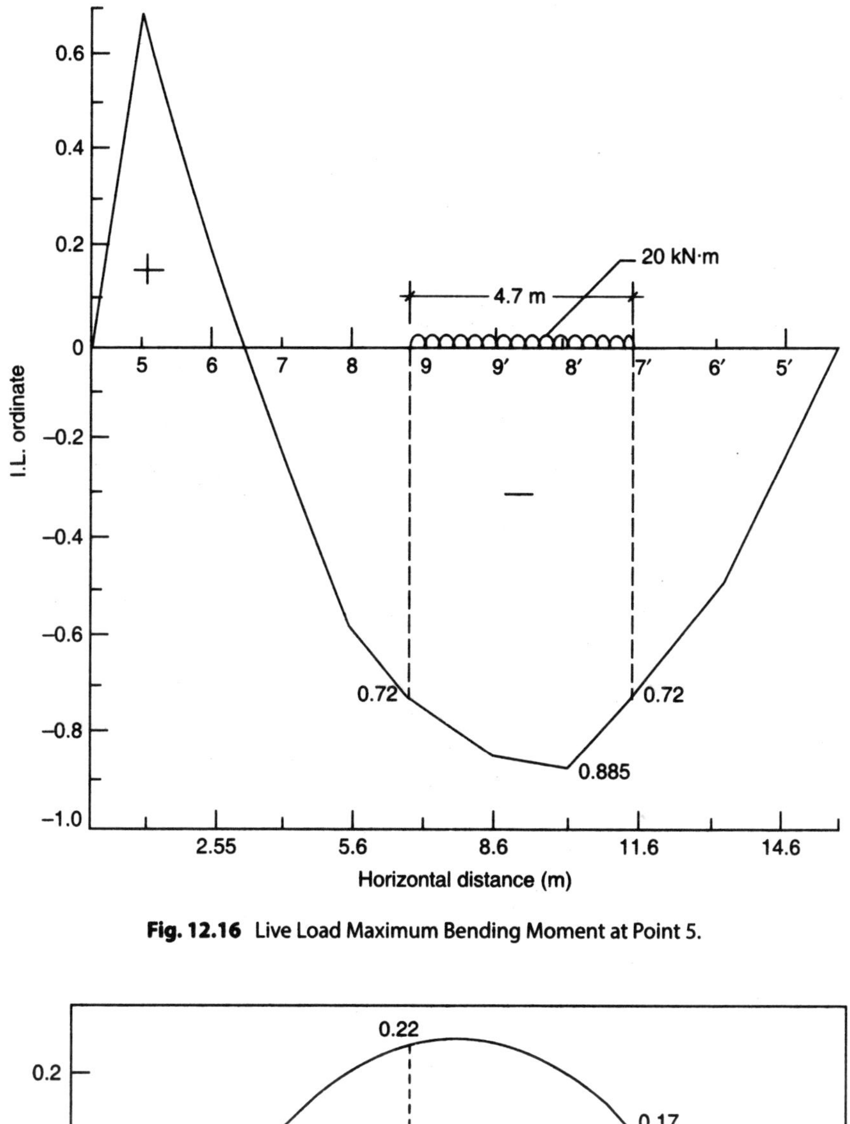

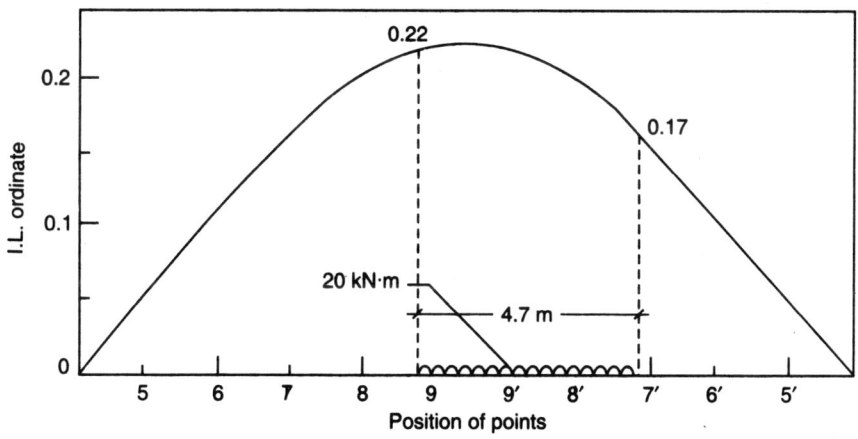

Fig. 12.16 Live Load Maximum Bending Moment at Point 5.

Fig. 12.17 Corresponding Thrust for Maximum Bending Moment at Point 5.

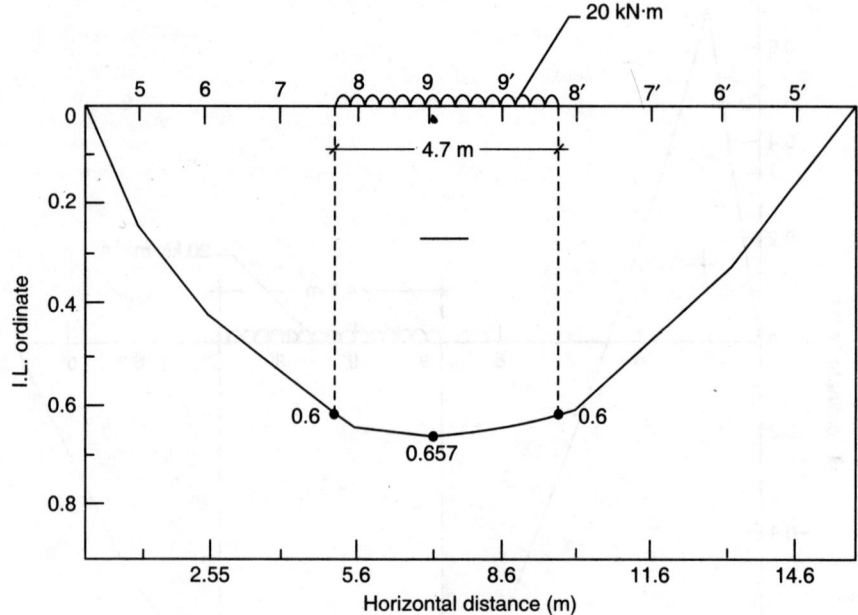

Fig. 12.18 Live Load Maximum Bending Moment at Point 2.

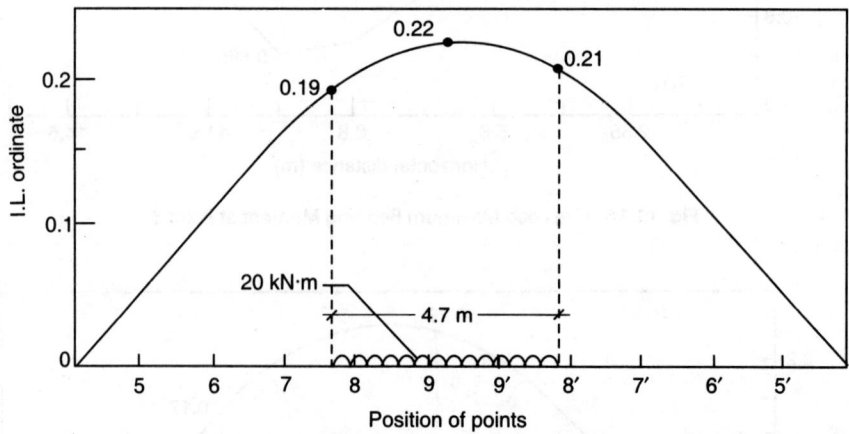

Fig. 12.19 Corresponding Thrust for Maximum Bending Moment at Point 2.

12. Design of Sections

(a)   Crown Section

Overall depth of section = D = 450 mm
Cover = d' = 50 mm
Design moment = M = 422.16 kN·m
Design thrust = P = 38.36 kN

Table 12.11 Abstract of Moments and Thrusts

Particulars of Load	Crown		Point 5		Point 2	
of Load	B.M. (kN·m)	Thrust (kN)	B.M. (kN·m)	Thrust (kN)	B.M. (kN·m)	Thrust (kN)
Dead Loads	227.3	36.8	−88.88	36.8	−112.9	36.8
Live Loads	168.6	19.74	−76.44	18.33	−59.0	19.43
Earth Pressure −ve or +ve or both whichever produces worst effect	E.P. from left only 6.55	−15	E.P. form 1 right only −78.20	−15	−41.83	−15
Temperature action taken so as to produce worst moments for the section	19.71	−3.18	−19.14	+3.18	−8.586	+3.18
Total	422.16	38.36	−261.66	43.31	−222.316	44.41

Ultimate moment = M_u = (1.5 × 422.16) = 633.24 kN·m
Ultimate thrust = P_u = (1.5 × 38.36) = 57.54 kN
Using M-20 grade concrete, f_{ck} = 20 N/mm^2

$(P_u/f_{ck} \cdot b \cdot D)$ = [(57.54 × 10^3)/(20 × 1000 × 450)] = 0.0063
$(M_u/f_{ck} \cdot b \cdot D^2)$ = [(633.24 × 10^6)/(20 × 1000 × 450^2)] = 0.156

Using Chart-32 of SP-16 (Design Aids to IS: 456),
for (d'/D) = 0.10, the value of (p/f_{ck}) = 0.10
A_s = $(pbD/100)$ = [(0.10 × 20 × 1000 × 450)/100] = 9000 mm^2
Use 25 mm diameter bars at 100 mm centres both at top and bottom.

(b)   Section 5

Overall depth of section = D = 1000 mm
Cover = d' = 50 mm
Design Moment = M = −261.66 kN·m
Design thrust = P = 43.31 kN
Ultimate moment = M_u = (1.5 × 261.66) = 392.5 kN·m
Ultimate thrust = P_u = (1.5 × 43.31) = 65 kN
Using M-2/3 grade concrete, f_{ck} = 20 N/mm^2

$(P_u/f_{ck} \cdot b \cdot D)$ = [(65 × 10^3)/(20 × 1000 × 1000)] = 0.0032
$(M_u/f_{ck} \cdot b \cdot D^2)$ = [(392.5 × 10^6)/(20 × 1000 × 1000^2)] = 0.0196

Using Chart-31 of SP-16[8] (Design Aids to IS: 456),
for (d'/D) = 0.05, the value of (p/f_{ck}) = 0.01
A_s = $(pbD/100)$ = [(0.01 × 20 × 1000 × 1000)/100] = 2000 mm^2
Minimum reinforcement = [(0.8/100) × 1000 × 1000] = 8000 mm^2
Use 25 mm diameter bars at 100 mm centres at top and bottom.

(c)   Section 2

Overall depth of section = D = 840 mm
Cover = d' = 50 mm
Design Moment = M = –222.316 kN·m
Design thrust = P = 44.41 kN
Ultimate moment = M_u = (1.5 × 222.316) = 334 kN·m
Ultimate thrust = P_u = (1.5 × 44.41) = 67 kN
Using M-20 grade concrete, f_{ck} = 20 N/mm²

$$(P_u/f_{ck} \cdot b \cdot D) = [(67 \times 10^3)/(20 \times 1000 \times 840)] = 0.004$$
$$(M_u/f_{ck} \cdot b \cdot D^2) = [(334 \times 10^6)/(20 \times 1000 \times 840^2)] = 0.204$$

Using Chart-31 of SP-16 (design Aids to IS: 456),
for (d'/D) = 0.05, the value of (p/f_{ck}) = 0.02

$$A_s = (pbD/100) = [(0.02 \times 20 \times 1000 \times 840)/100] = 3360 \text{ mm}^2$$

Minimum reinforcement = $[(0.8/100) \times 1000 \times 840]$ = 6720 mm²
Use 25 mm diameter bars at 100 mm centres on both faces.
Distribution bars = $[(0.12/100) \times 1000 \times 840]$ = 1080 mm²
Provide 16 mm diameter bars at 300 mm centres.

13.   Design of Footing

(a)   Calculation of Maximum Reactions

The load position for maximum shear is shown in Fig. 12.20.

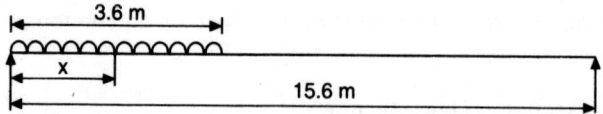

3.6 m

x

15.6 m

Fig. 12.20 Position of Load for Maximum Shear.

Effective width of dispersion
$$b_e = K \times (1 - x/L) + b_w$$
$$x = 1.8 \text{ m}$$
$$L = 15.6 \text{ m}$$
$$B = 8.7 \text{ m}$$
$$b_w = 1.05 \text{ m}$$

For (B/L) = (8.7/15.6) = 0.557
From Table 3.1, K = 1.85

$$b_e = 1.85 \times 1.8 \ (1 - 1.8/15.6) + 1.05 = 4 \text{ m}$$

Effective width of dispersion for 2 wheels (Referring to Fig. 12.13)
$$= [2225 + 2050 + (4000/2)] = 6275 \text{ mm} = 6.275 \text{ m}$$

Intensity of load = 700/(6.275 × 3.6) = 31 kN/m²
Reaction at support = (31 × 3.6 × 13.8)/15.6 = 99 kN

(b)   Footing Area and Reinforcements

Live load reaction at support   =   99 kN
Dead load reaction at support   =   264 kN
Total load   =   363 kN

Assuming safe bearing capacity of soil at site as 250 kN/m²,
Area of footing required = (363/250) = 1.452 m²
Adopt a footing 1.5 m wide throughout the length of 8.7 m
Referring to Fig. 12.21,

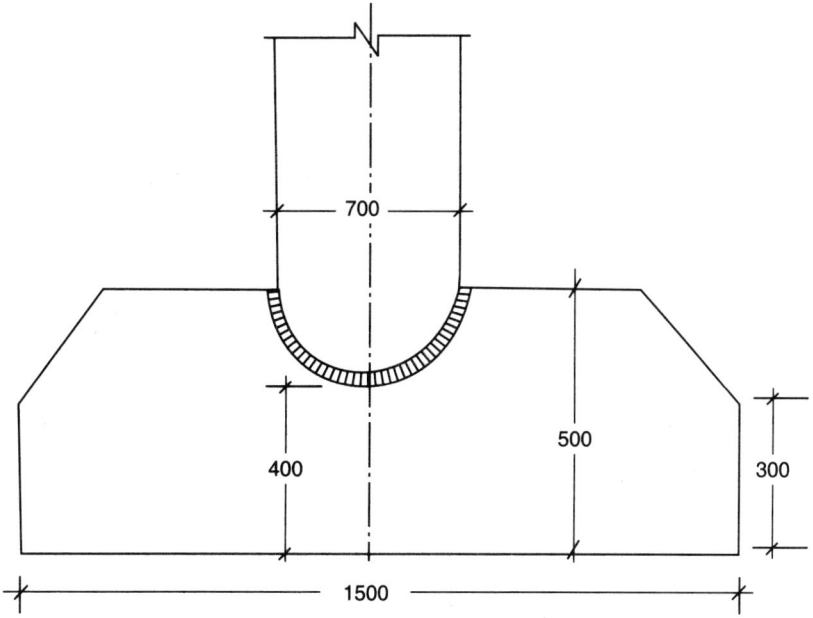

Fig. 12.21 Details of Footing.

Intensity of soil pressure = 363/(1.5 × 1) = 242 kN/m²
Maximum service load bending moment at the centre of footing is

$$M = (242 \times 0.75 \times 0.5 \times 0.75) = 68 \text{ kN·m/m}$$

Design Ultimate load bending moment = M_u = (1.5 × 68) = 102 kN·m
The effective depth required using M-20 Grade concrete and Fe-415 HYSD bars is
given by

$$d = \sqrt{\frac{M_u}{0.138 \times f_{ck} \times b}} = \sqrt{\frac{102 \times 10^6}{0.138 \times 20 \times 1000}} = 192.2 \text{ mm}$$

From shear considerations, the depth required will be larger, hence adopt an
Effective depth = d = 350 mm and overall depth = 400 mm at centre of footing.

The area of reinforcement required is interpolated from Table 2 of SP: 16 design charts
By computing the parameter $(M_u/bd^2) = [(102 \times 10^6)/(1000 \times 350^2)] = 0.83$
Corresponding to this value, read out the percentage reinforcement $= p_t = 0.242$

$$A_{st} = \left(\frac{p_t.b.d}{100}\right) = \left(\frac{0.83 \times 1000 \times 350}{100}\right) = 2905 \text{ mm}^2$$

Provide 25 mm diameter bars at 150 mm spacing (A_{st} provided = 3272 mm^2)
Area of Distribution bars = $(0.0013 \times 1000 \times 350) = 455$ mm^2
Provide 10 mm diameter bars at 150 mm centres (A_{st} provided = 524 mm^2)
The ultimate shear resistance of the section is checked by using the relation,

$$V_{Rd.c} = [0.12K(80\,\rho_1 f_{ck})^{0.33}]b_w.d$$

where
$$K = 1 + \sqrt{\frac{200}{d}} \le 2.00 = \left[1 + \sqrt{\frac{200}{350}}\right] = 1.75$$

and
$$\rho_1 = \left(\frac{A_{sl}}{b_w.d}\right) \le 0.02$$

$$= \left(\frac{3272}{1000 \times 350}\right) = 0.0093$$

$$V_{Rd.c} = [0.12 \times 1.75(80 \times 0.0093 \times 20)^{0.33}](1000 \times 350) \text{ N}$$
$$= (176 \times 1000) \text{ N}$$
$$= 176 \text{ kN}$$

The required Ultimate shear resistance at a distance '*d*' from the centre is computed as

$$V_u = (1.5 \times 242 \times 0.40 \times 1) = 145.2 \text{ kN} < V_{Rd.c}$$
$$\text{(Hence safe against shear)}$$

The reinforcement details in the rigid frame and footing are shown in Figs. 12.22 and
12.23 Conforming to SP: 34-1987[10].

12.5   DESIGN EXAMPLE OF PRESTRESSED CONCRETE RIGID FRAME BRIDGE

Design a prestressed concrete rigid frame bridge with hinged supports for a national
High way crossing to suit the following data:

1.   Data

Type of super structure: Two pinned portal frame
Span (Length of transom) = 15 m
Height of column legs = 6 m
Width of road way = 7.5 m
Foot path: one metre on either side
Thickness of wearing coat = 80 mm

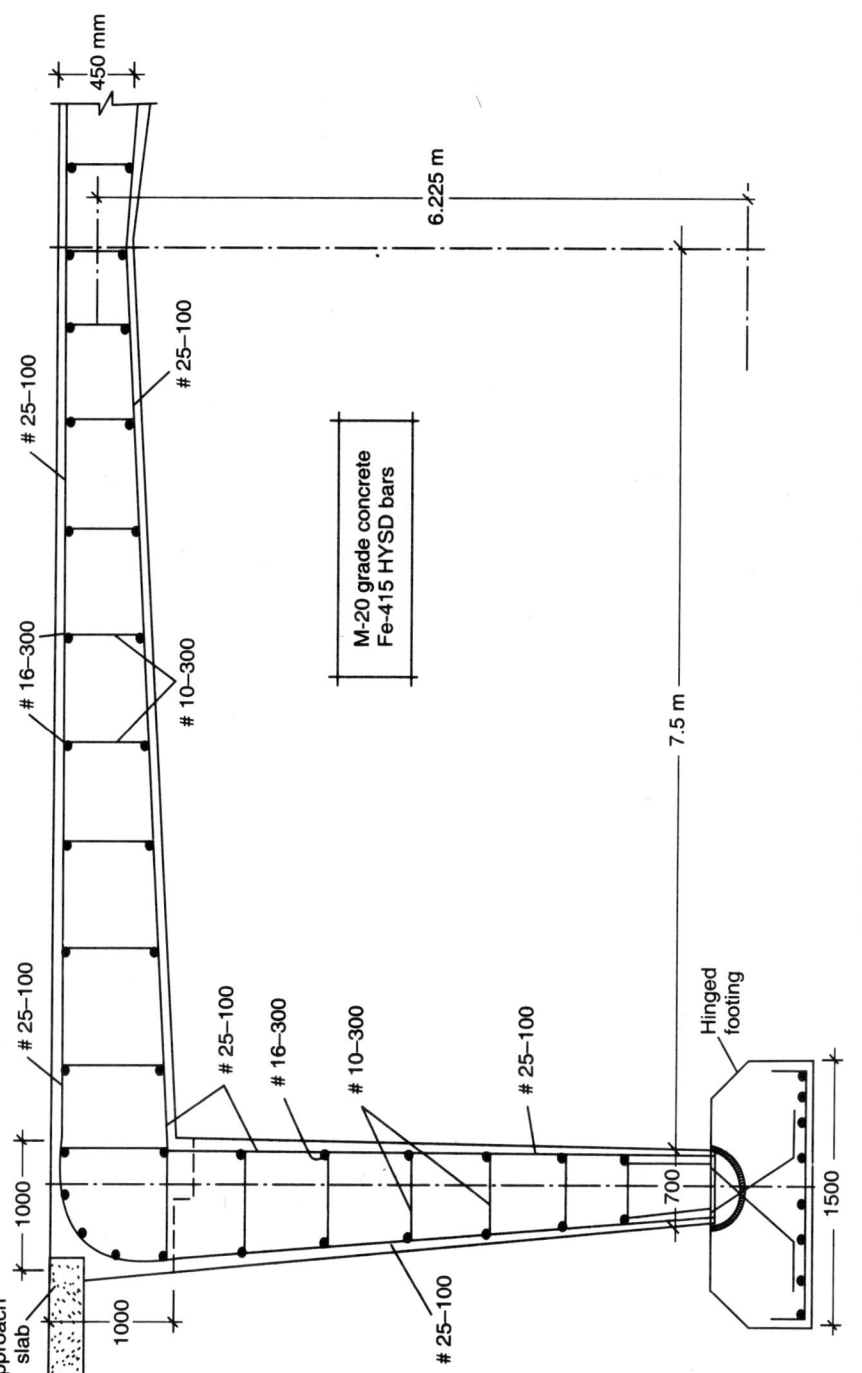

Fig. 12.22 Longitudinal Section of Rigid Frame.

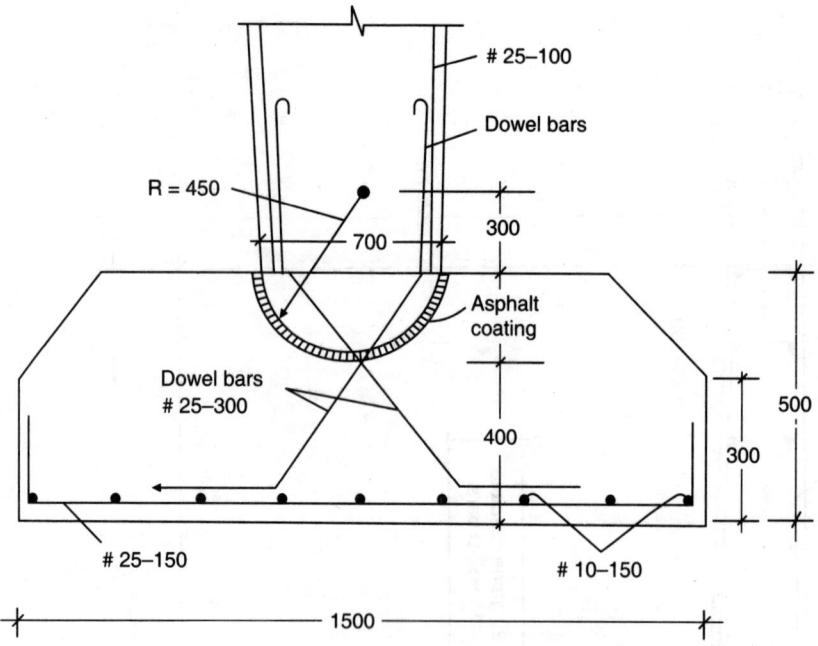

Fig. 12.23 Reinforcement Details in Hinged Footing.

Live load: IRC Class AA Tracked vehicle
Grade of concrete: M-45 for portal frame and M-20 of footing
Loss ratio = 0.85
Safe bearing capacity of soil at site = 250 kN/m²

Adopt 7K-15 type Freyssinet high tensile steel strands conforming to IS: 6006-1983 for prestressing the members of portal frame.
Fe-415 HYSD bars for supplementary reinforcements and footing foundation.
Design the rigid frame bridge as Class 1 type structure conforming to the specifications of Indian Roads congress codes, IRC: 6-2014, IRC: 112-2011 and IS: 1343-2012.

2. Maximum Permissible Stresses in Concrete and Steel

According to the given data, we have for M-45 grade concrete
Compressive strength of concrete $= f_{ck} = 45$ N/mm²
Tensile strength of concrete $= f_{ctd} = 3.3$ N/mm²
Adopt permissible stress in concrete at transfer $= f_{ct} = 15$ N/mm²
Permissible tensile stress (Class 1 type structure) $= f_{tt} = f_{tw} = 0$
Ultimate tensile strength of 7K-15 type Freyssinet H.T strands with $f_p = 1862$ N/mm²
Modulus of elasticity of concrete $= E_c = 34$ kN/mm²
For M-20 grade concrete and Fe-415 HYSD bars adopt the following parameters.
$f_{ck} = 20$ N/mm² and $f_y = 415$ N/mm²

3. Dimensions of Portal Frame and Cross Section of Deck

Rigid rectangular portal frame comprising of transom and column members is proposed to have a uniform cross-section.

Span of transom = L = 15 m
Thickness of transom = $(L/30)$ = $[(15 \times 1000)/30]$ = 500 m

Adopt overall depth of 600 mm for transom and column members.
The dimensions of the portal frame and the cross-section of the deck is shown in Figs. 12.24 and 12.25.

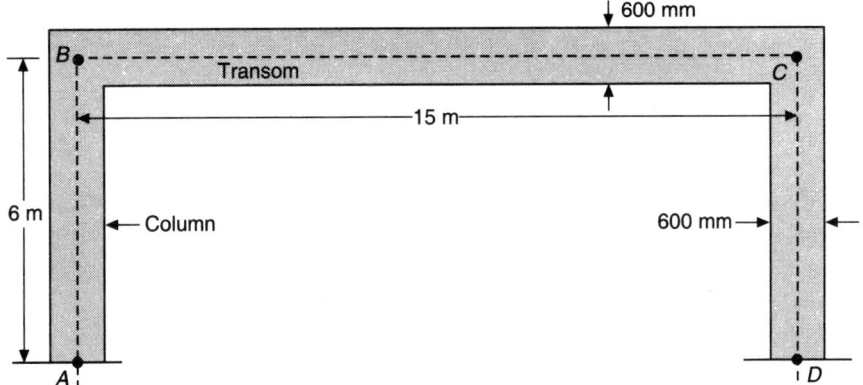

Fig. 12.24 Dimensions of Portal Frame.

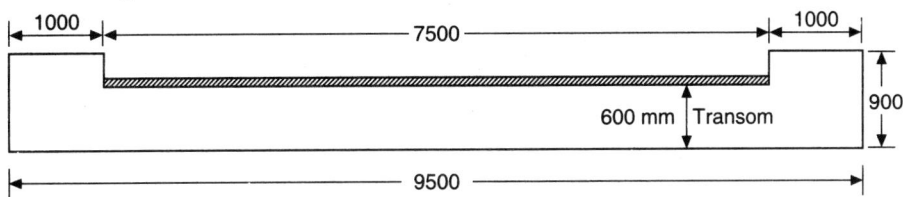

Fig. 12.25 Cross-section of Bridge Deck.

4. Dead Loads

Self weight of transom = (0.6×24) = 14.40 kN/m^2
Self weight of wearing coat = (0.08×22) = 1.76
Weight of parapet, railing, etc. (Lumpsum) = 1.84
Total dead load = 18.00 kN/m^2

5. Dead Load Bending Moments

The frame ABCD shown in Fig. 12.26 is analysed for dead load bending moments by any of the wall established methods like moment distribution of column analogy computing the stiffness coefficients of the transom and column members and distribution coefficients at the joints. The resulting bending moment diagram is shown in Fig. 12.27. The critical moments for design are the maximum positive moment at centre of transom and the negative moment developed at the junction of column and transom.

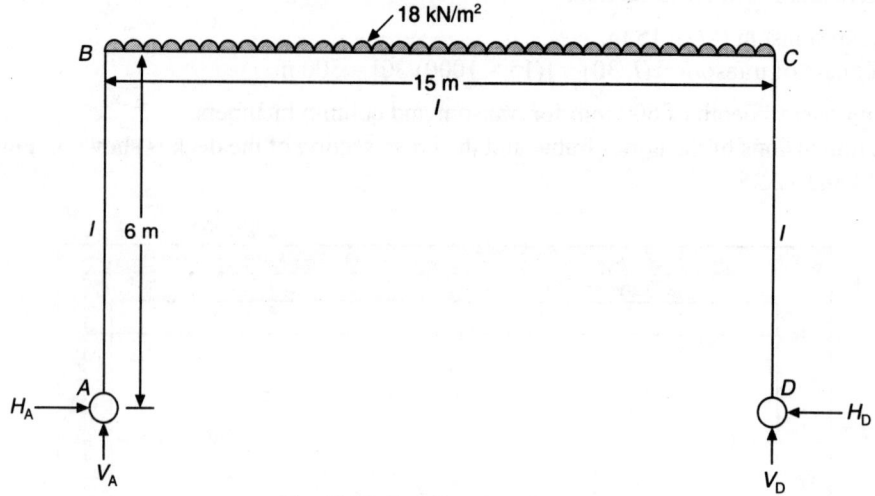

Fig. 12.26 Dead Loads on Frame.

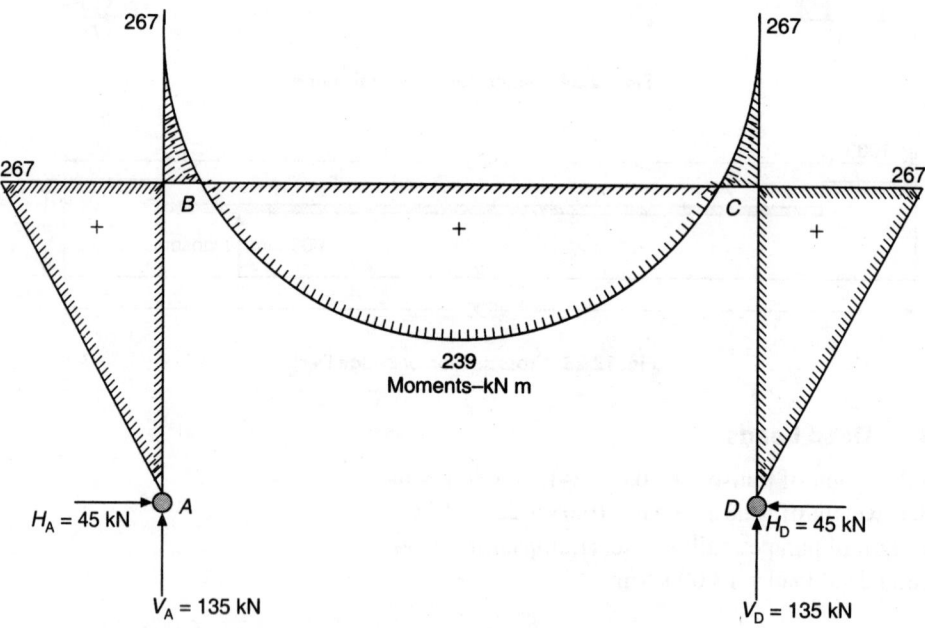

Fig. 12.27 Dead Load Bending Moments.

6. Live Load Distribution on Deck

The deck is subjected to IRC Class AA tracked vehicle loading comprising an army tank of 700 kN spread over two tracks.

Impact Factor = 10 per cent for a span of 15 m
The tracked vehicle is placed symmetrically at centre of span on the deck.
Effective length of load = [3.6 + 2(0.6 + 0.08)] = 4.96 m
Effective width of slab perpendicular to the span is expressed as

$$b_e = Kx\left(1 - \frac{x}{L}\right) + b_w$$

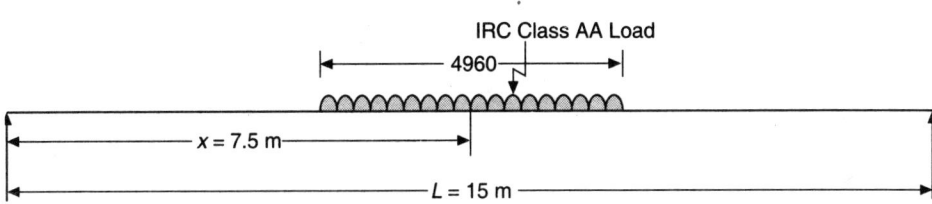

Fig. 12.28 Position of Load for Maximum Bending Moment.

Referring to Fig. 12.28

$$x = 7.5 \text{ m}, L = 15 \text{ m}, B = 9.5 \text{ m}$$

Ratio $(B/L) = (9.5/15) = 0.63$

$$b_w = [0.85 + (2 \times 0.08)] = 1.01 \text{ m}$$

From Table 4.6 for $(B/L) = 0.63$ and continuous slab read out the value of $K = 1.88$

∴ $$b_e = (1.88 \times 7.5)\left[1 - \frac{7.5}{15}\right] + 1.01 = 8.06 \text{ m}$$

The IRC Class AA tracked vehicle is placed close to the kerb with required minimum clearance as shown in Fig. 12.29.

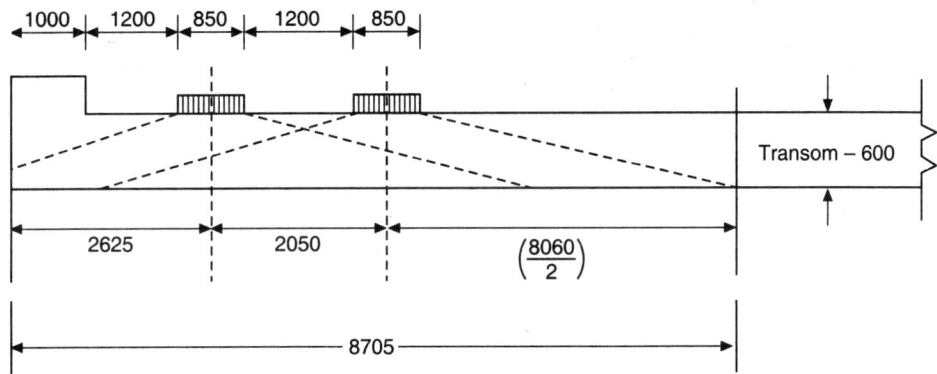

Fig. 12.29 Effective Width of Dispersion for IRC Class AA Tracked Vehicle.

Not effective width of dispersion = 8.705 m
Total load of two tracks with impact = (700 × 1.10) = 770 kN

$$\text{Average intensity of load} = \left[\frac{770}{(4.96 \times 8.705)}\right] = 17.8 \text{ kN/m}^2$$

7. Live Load Bending Moments in Portal Frame

The portal frame is analysed for maximum positive and negative bending moments when the live load is in the centre of transom. The corresponding live load bending moment diagram for the frame is shown in Fig. 12.30.

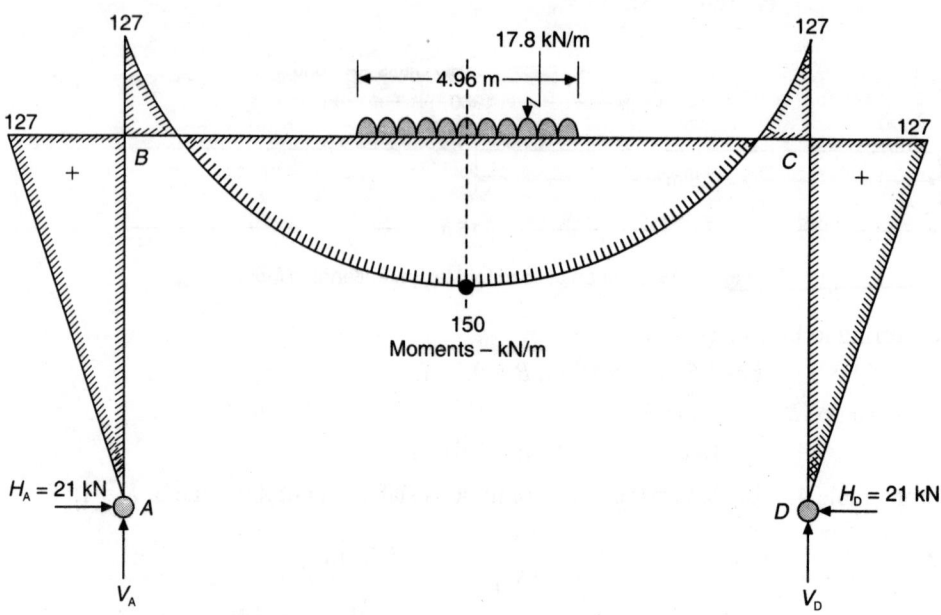

Fig. 12.30 Live Load Moments in Frame.

8. Design Bending Moments in Frame

Maximum positive moment at mid point of transom = $(150 + 239) = 389$ kN·m
Maximum negative moment at knee = $(127 + 267) = 394$ kN·m
Maximum horizontal thrust at hinge = $(21 + 45) = 66$ kN
Maximum vertical reaction in legs = $(45 + 135) = 180$ kN

9. Dimensions of Critical Section of Transom

For transom, maximum positive design moment at centre is $M_L = 389$ kN·m
and $(N_1 - N_2) = -66$ kN

Assuming 1 m width of deck and using Eq. 8.8, we have

$$M_L = (1/6) f_c b h^2 + (1/6) (N_1 - N_2)h$$

In this example, $f_c = 15$ N/mm²

$$M_L = (389 \times 10^6)\ \text{N·mm}$$

h = overall depth of the section

$b = 1000$ m

Substituting these values in the moment equation, we have

$$(389 \times 10^6) = [(1/6) \times 15 \times 1000 \times h^2] + [(1/6)(-66 \times 10^3) h]$$

Solving, $\quad h = 401$ mm

Similarly for the column members, $M_L = 394$ kN·m and $(N_2 - N_1) = -180$ kN

$$(394 \times 10^6) = [(1/6) \times 15 \times 1000 \times h^2] + [(1/6)(-180 \times 10^3) h]$$

Solving, $\quad h = 409$ mm

The section adopted for transom and column has an overall depth of 600 mm. Hence the section assumed is adequate to safely resist the design moments.

10. Design of Prestressing Force

The prestressing force required in the transom and legs is computed using the following Equations[7]:

For transom,

$$N_p = [(1/2) f_c \, b \, h - N_g - N_1]$$

where $\quad N_g = 45$ kN and $N_1 = 21$ kN

Hence $\quad N_p = \left(\dfrac{1}{2} \times \dfrac{15}{1000} \times 1000 \times 600 \right) - 45 - 21$

$$= 4434 \text{ kN}$$

For legs,

$$N_p = [(1/2) f_c \, b \, h - N_g - N_1]$$

where $\quad N_g = 135$ kN and $N_1 = 45$ kN

$$N_p = \left(\dfrac{1}{2} \times \dfrac{15}{1000} \times 1000 \times 600 \right) - 45 - 135$$

$$= 4320 \text{ kN}$$

11. Determination of the Cable Zone

The permissible tendon zone for each member is determined for various load conditions. The core limits for a beam of rectangular section of overall depth 'h' is ($\pm h/6$), but due to the presence of applied loading thrusts, they have to be modified by multiplying them by the factor (N/N_p).

The factor (N/N_p) for the various members is computed as detailed below:

(a) Transom (Under dead load only)

$$\frac{N}{N_p} = \left[\frac{4434 + 45}{4434} \right] = 1.01$$

(b) Transom (Under live load + dead load)

$$\frac{N}{N_p} = \left[\frac{4434 + 45 + 21}{4434} \right] = 1.014$$

(c) Top of Legs (Under dead load only)

$$\frac{N}{N_p} = \left[\frac{4320 + 135}{4320}\right] = 1.031$$

(d) Top of Legs (Under dead load + live load)

$$\frac{N}{N_p} = \left[\frac{4320 + 135 + 45}{4320}\right] = 1.041$$

The low values of the ratio of (N/N_p) indicate that the effect of axial thrust is negligible in transom and legs.

12. Limiting Zone for Prestressing Force

The limiting zone can be obtained as the ratio of the moments to the thrust and are evaluated for transom and legs as detailed below:

1. **Leg Foot Zone Limits:** Since the column foot is hinged at supports, there are no moments and hence the core limit is computed as

$$\frac{h}{6} = \frac{600}{6} = 100 \text{ mm}$$

The distance measured from the centroidal axis are as follows:

(a) Upper limit (towards outside of frame) = $(-100 \times 1.031) = -103$ mm

(b) Lower limit (inside the frame) = $(100 \times 1.041) = 104$ mm

2. **Leg Top Zone Limits**

(a) Upper (outer limit)

$$\left[\frac{M_2 + M_g}{N_p} - \frac{h}{6} \cdot \frac{N}{N_p}\right] = \left[\frac{-267 \times 10^3}{4320} - (100 \times 1.03)\right] = -165 \text{ mm}$$

(b) Lower (inner limit)

$$\left[\frac{M_1 + M_g}{N_p} + \frac{h}{6} \cdot \frac{N}{N_p}\right] = \left[\frac{-394 \times 10^3}{4320} + (100 \times 1.04)\right] = 13 \text{ mm}$$

3. **Transom Mid Span Zone Limits**

(a) Upper limit

$$\left[\frac{M_2 + M_g}{N_p} - \frac{h}{6} \cdot \frac{N}{N_p}\right] = \left[\frac{389 \times 10^3}{4434} - (100 \times 1.014)\right] = -13.6 \text{ mm}$$

(b) Lower limit

$$\left[\frac{M_1 + M_g}{N_p} + \frac{h}{6} \cdot \frac{N}{N_p}\right] = \left[\frac{239 \times 10^3}{4434} + (100 \times 1.014)\right] = 155.3 \text{ mm}$$

4. Transom End Zone Limits

(a) Upper limit

$$\left[\frac{M_2 + M_g}{N_p} - \frac{h}{6} \cdot \frac{N}{N_p}\right] = \left[-\frac{267 \times 10^3}{4434} - \left(100 \times 1.014\right)\right] = -161.6 \, \text{mm}$$

(b) Lower limit

$$\left[\frac{M_1 + M_g}{N_p} + \frac{h}{6} \cdot \frac{N}{N_p}\right] = \left[-\frac{394 \times 10^3}{4434} + \left(100 \times 1.014\right)\right] = 12.6 \, \text{mm}$$

13. Determination of Bending Concordant Profile

The bending concordant profile lying within the cable zone is determined by considering the moments developed for the two cases of loading. The eccentricity of the cable can be determined by the relation,

$$e = \left[\frac{M_g}{P} + \frac{M_1 + M_2}{2P}\right]$$

At transom mid span,

$$e = \left[\frac{239 \times 10^3}{4434} + \frac{150 \times 10^3}{2 \times 4434}\right] = 70.80 \, \text{mm}$$

At transom end,

$$e = \left[-\frac{267 \times 10^3}{4434} - \frac{127 \times 10^3}{2 \times 4434}\right] = -74.52 \, \text{mm}$$

Leg top zone,

$$e = \left[-\frac{267 \times 10^3}{4320} - \frac{127 \times 10^3}{2 \times 4320}\right] = -76.50 \, \text{mm}$$

The cable profile conforming to these values of eccentricity is straight in the column portion and parabolic in the transom. The limiting zone and bending concordant profile of the cable is shown in the Fig. 12.31

14. Design of Prestressing Force in Cables

(a) Transom

Prestressing force = 4434 kN/m
Using Freyssinet 7K-15 type seven wire strands in 65 mm diameter cable ducts conforming to IRC: 6006–1983[8],

Force in each cable = $(7 \times 0.8 \times 260.7) = 1459$ kN

$$\text{Number of cables/metre} = \left(\frac{4434}{1459}\right) = 3.04$$

$$\text{Spacing of cables} = \left(\frac{1000}{3.04}\right) = 329 \text{ mm}$$

Adopt cables at 300 mm centres

(b) Column

Prestressing force = 4320 kN/m

Adopt the same type of cables and spacing since the prestressing force is more or less similar in magnitude.

15 Check the Stresses at Service Loads

The stress developed at the extreme fibres of the critical sections of the transom and columns are checked for various loading conditions.

(a) Properties of cross-section

$$b = 1000 \text{ mm}$$

$$h = 600 \text{ mm}$$

$$A = (1000 \times 600) = (6 \times 10^5) \text{ mm}^2$$

$$Z_b = Z_t = Z = \left(\frac{bh^2}{6}\right) = \left(\frac{1000 \times 600^2}{6}\right) = (60 \times 10^6) \text{ mm}^2$$

Loss ratio $= h = 0.85$

(b) Stress in Transom (N/mm²)

P (N)	*A* (mm²)	*(P/A)* (N/mm²)	*Centre of Span* e = 70.8 mm	*Supports* e = –74.5 mm
(4434×10^3) $(\eta P/A) = 6.28$ N/mm²	(6×10^5)	7.35	$(Pe/Z) = 5.23$ $(\eta Pe/Z) = 4.44$ $M_g = (239 \times 10^6)$ $M_L = (150 \times 10^6)$ $(M_g/Z) = 3.98$ $(M_L/Z) = 2.50$	$(Pe/Z) = 5.50$ $(\eta Pe/Z) = 4.67$ $M_g = (-267 \times 10^6)$ $M_L = (-127 \times 10^6)$ $(M_g/Z) = 4.45$ $(M_L/Z) = 2.11$

Resultant Stresses (Transom) Case-1 (Prestress + Dead Load) (+ Compression and – Tension)

Stress at	*Centre of Span*	*Supports*
Soffit	$f_b = [(P/A) + (Pe/Z) - (M_g/Z)]$ $= [7.39 + 5.23 - 3.98]$ $= 8.64 \text{ N/mm}^2$	$f_b = [(P/A) - (Pe/Z) + (M_g/Z)]$ $= [7.39 - 5.50 + 4.45]$ $= 6.34 \text{ N/mm}^2$
Top	$f_t = [(P/A) - (Pe/Z) + (M_g/Z)]$ $= [7.39 - 5.23 + 3.98]$ $= 6.14 \text{ N/mm}^2$	$f_t = [(P/A) + (Pe/Z) - (M_g/Z)]$ $= [7.39 + 5.50 - 4.45]$ $= 8.44 \text{ N/mm}^2$

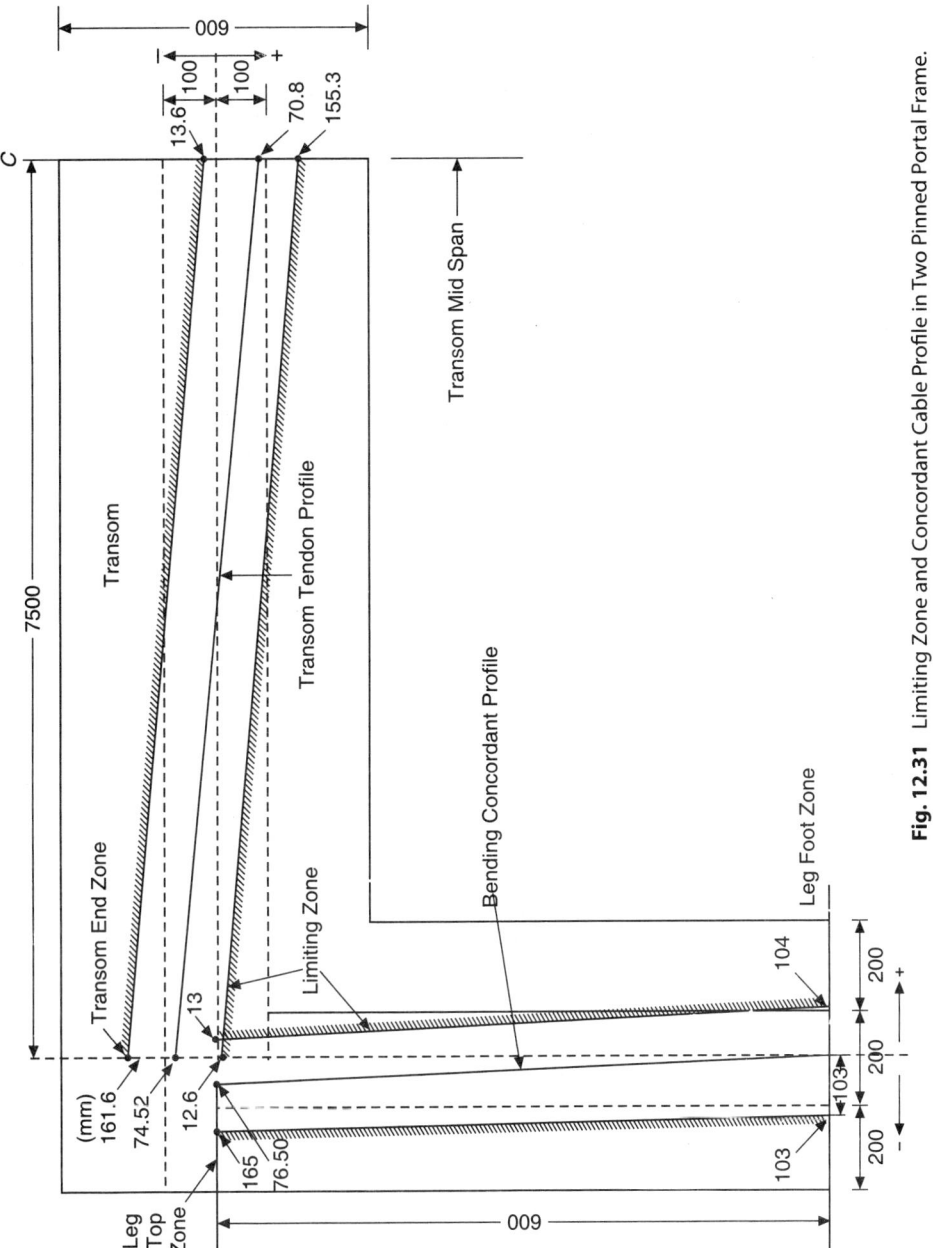

Fig. 12.31 Limiting Zone and Concordant Cable Profile in Two Pinned Portal Frame.

Case-2 (Effective Prestress + Dead Load + Live Load) (+ Compression and – Tension)

Stress at	Centre of Span	Supports
Soffit	$f_b = \left(\dfrac{\eta P}{A} + \dfrac{\eta Pe}{Z} - \dfrac{M_g}{Z} - \dfrac{M_L}{Z} \right)$ $= 6.28 + 4.44 - 3.98 - 2.5$ $= 4.24 \text{ N/mm}^2$	$f_b = \left(\dfrac{\eta P}{A} - \dfrac{\eta Pe}{Z} + \dfrac{M_g}{Z} + \dfrac{M_L}{Z} \right)$ $= (6.28 - 4.67 + 4.45 + 2.11)$ $= 8.17 \text{ N/mm}^2$
Top	$f_t = \left(\dfrac{\eta P}{A} - \dfrac{\eta Pe}{Z} + \dfrac{M_g}{Z} + \dfrac{M_L}{Z} \right)$ $= (6.28 - 4.44 + 3.98 + 2.5)$ $= 8.32 \text{ N/mm}^2$	$f_t = \left(\dfrac{\eta P}{A} + \dfrac{\eta Pe}{Z} - \dfrac{M_g}{Z} - \dfrac{M_L}{Z} \right)$ $= (6.28 + 4.67 - 4.5 - 2.11)$ $= 4.39 \text{ N/mm}/^2$

Resultant stresses (Column top)

$$P = (4320 \times 10^3) \text{ N} \qquad Z = (60 \times 10^6) \text{ m}^3$$
$$A = (6 \times 10^5) \text{ mm}^2 \qquad M_g = (267 \times 10^6) \text{ N·mm}$$
$$e = -76.50 \text{ mm} \qquad M_L = (127 \times 10^6) \text{ N·mm}$$

$$\left(\frac{\eta P}{A} \right) = 6.12 \text{ N/mm}^2 \qquad \left(\frac{Pe}{Z} \right) = 5.50 \text{ N/mm}^2$$

$$\left(\frac{\eta Pe}{Z} \right) = 4.68 \text{ N/mm}^2 \qquad \left(\frac{M_g}{Z} \right) = 4.45 \text{ N/mm}^2$$

$$\left(\frac{M_L}{Z} \right) = 2.11 \text{ N/mm}^2$$

Resultant Stresses (N/mm²)

Location	Prestress + Dead Load	Prestress + Dead Load + Live Load
Outside	$f_o = \left(\dfrac{P}{A} + \dfrac{Pe}{Z} - \dfrac{M_g}{Z} \right)$ $= (7.20 + 5.50 - 4.45)$ $= 8.25 \text{ N/mm}^2$	$f_o = \left(\dfrac{\eta P}{A} + \dfrac{\eta Pe}{Z} - \dfrac{M_g}{Z} - \dfrac{M_L}{Z} \right)$ $= (6.12 - 4.68 + 4.45 + 2.11)$ $= 4.24 \text{ N/mm}^2$
Inside	$f_i = \left(\dfrac{P}{A} + \dfrac{Pe}{Z} - \dfrac{M_g}{Z} \right)$ $= (7.20 - 5.50 + 4.45)$ $= 6.15 \text{ N/mm}^2$	$f_i = \left(\dfrac{\eta P}{A} - \dfrac{\eta Pe}{Z} + \dfrac{M_g}{Z} + \dfrac{M_L}{Z} \right)$ $= (6.12 - 4.68 + 4.45 - 2.11)$ $= 8 \text{ N/mm}/^2$

All the stresses in concrete under different loading conditions are well within the safe permissible limits.

16. Check for Ultimate Flexural Strength

(a) Centre of Transom

Consider one metre width of section

Overall depth $(h) = 600$ mm

Prestressing steel comprises of 7K-15 type seven wire strands in 65 mm cables

Provided at 300 mm at an eccentricity (e) of 70.8 mm

Effective depth $= d = (300 + 70.8) = 370.8$ mm

A_p (each cable) $= (7 \times 140) = 980$ mm^2

Number of cables per metre $= (1000/300) = 3.33$

A_p/metre $= (3.33 \times 980) = 3263$ mm^2

$\qquad f_p = 1862$ N/mm^2

Dead load moment $= G = 239$ kN·m

Live load moment $= Q = 150$ kN·m

According to IRC: 6-2014

Required $M_u = (1.35 \, M_g + 1.5 \, M_q)$
$$= [(1.35 \times 239) + (1.5 \times 150)]$$
$$= 547.65 \text{ kN·m}$$

Compute the ratio $\left(\dfrac{A_p f_p}{b.d.f_{ck}} \right) = \left(\dfrac{3263 \times 1862}{1000 \times 370.8 \times 45} \right) = 0.364$

Refer Table 11 of IS: 1343-2012 and interpolate the values of the ratios given below:

$$\left(\frac{f_{pb}}{0.87 f_p} \right) = 0.79 \text{ and } \left(\frac{x_u}{d} \right) = 0.61$$

$$f_{pb} = (0.79 \times 0.87 \times 1862) = 1279.7 \text{ N/mm}^2$$
$$x_u = (0.61 \times 370.8) = 226.18 \text{ mm}$$

$\therefore \qquad M_u = (f_{pb} \, A_p)[d - 0.42x_u]$
$$= (1279.7 \times 3263) \, [370.8 - 0.42 \times 226.18]$$
$$= (1151 \times 10^6) \text{ N.mm}$$
$$= 1151 \text{ kN·m} > 547.65 \text{ kN·m (Hence Safe)}$$

(b)   Transom Support Section

Copy up to Required M_u and resume new material as

Required $M_u = (1.35 \, M_g + 1.5 \, M_q)$
$$= [(1.35 \times 267) + (1.5 \times 127)]$$
$$= 550.95 \text{ kN·m}$$

Compute the ratio $\left(\dfrac{A_p f_p}{b.d.f_{ck}} \right) = \left(\dfrac{3263 \times 1862}{1000 \times 374.5 \times 45} \right) = 0.360$

Refer Table 11 of IS: 1343-2012 and interpolate the values of the ratios given below:

$$\left(\frac{f_{pb}}{0.87 f_p} \right) = 0.79 \text{ and } \left(\frac{x_u}{d} \right) = 0.61$$

$$f_{pb} = (0.79 \times 0.87 \times 1862) = 1279.7 \text{ N/mm}^2$$
$$x_u = (0.61 \times 374.5) = 228.44 \text{ mm}$$

$\therefore$

$$M_u = (f_{pb} A_p)[d - 0.42 x_u]$$
$$= (1279.7 \times 3263)[374.5 - 0.42 \times 228.44]$$
$$= (1163 \times 10^6) \text{ N·mm}$$
$$= 1163 \text{ kN·m} > 550.95 \text{ kN·m (Hence Safe)}$$

17. Check for Ultimate Shear Strength

The transom ends the shear force is maximum due to dead and live loads

Required design shear strength $= V_u = (1.35 V_g + 1.5 V_q)$
$$= [(1.35 \times 135) + (1.5 \times 44)]$$
$$= 248 \text{ kN}$$

According to IRC: 112-2011, the shear resistance of prestressed members without shear reinforcements is limited by the tensile strength of concrete and is expressed by the relation,

$$V_{Rd.c} = \left(\frac{Ib_w}{s}\right)\sqrt{(f_{ctd})^2 + k_1 \sigma_{cp} f_{ctd}}$$

$f_{ctd} = 3.3 \text{ N/mm}^2$, For post tensioned members, $k_1 = 1$

$\sigma_{cp} = (P/A_c) = [(4320 \times 10^3)/(1000 \times 600)] = 7.2 \text{ N/mm}^2$

$$I = \left(\frac{bD^3}{12}\right) = \left(\frac{1000 \times 600^3}{12}\right) = (18 \times 10^9) \text{ mm}^4$$

$$S = \left(\frac{1000 \times 300 \times 300}{2}\right) = (45 \times 10^7) \text{ mm}^3$$

$$V_{Rd.c} = \left(\frac{18 \times 10^9 \times 1000}{45 \times 10^7}\right)\sqrt{(3.3)^2 + 1 \times 7.2 \times 3.3}$$

$$= (267 \times 10^3) \text{ N}$$
$$= 267 \text{ kN} > 248 \text{ kN (Hence safe)}$$

18. Design of Hinged Footing

(a) Maximum Live Load Reaction

The live load position for maximum support reaction is shown in Fig 12.32 Effective width of dispersion is computed as

$$b_e = K \times (1 - x/L) + b_w$$

where
$x = 1.8 \text{ m}$
$L = 15 \text{ m}$
$B = 9.5 \text{ m}$

$$b_w = 1.01 \text{ m}$$

For $(B/L) = (9.5/15) = 0.63$, Read out from Table 4.6 the value of K as 1.88

$$b_e = [(1.88 \times 1.8) \{1 - (1.8/15)\} + 1.01]$$
$$= 3.98 \text{ m}$$

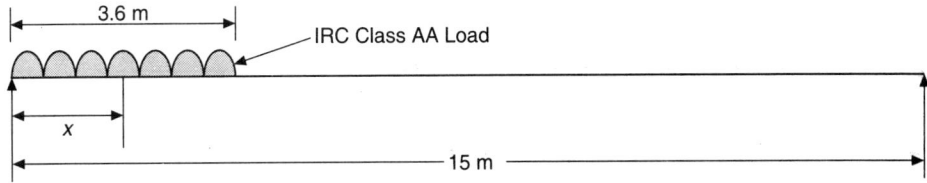

Fig. 12.32 Position of Load for Maximum Shear.

For 2 wheels (Refer Fig. 12.29)

Effective width of dispersion = $[(2625 + 2050 + (3980/2)] = 6665$ mm = 6.665 m

Intensity of live load $= \left(\dfrac{700 \times 1.1}{6.665 \times 3.6} \right) = 32 \text{ kN/m}^2$

Live load reaction at support $= \left[\dfrac{32 \times 3.6 \times 13.2}{15} \right] = 101 \text{ kN}$

(b) Dead Load Reaction

Reaction due to dead load of frame/metre = $0.5(6 + 6 +15) (0.6 \times 1 \times 24) = 194$ kN

Total load on footing = $(101 + 194) = 295$ kN

Self weight of footing (Lump sum) = 30 kN

Total load .... = 325 kN

Assuming safe bearing capacity of soil at site as 250 kN/m²

Area of footing required $= \left[\dfrac{325}{250} \right] = 1.3 \text{ m}^2$

Adopt a footing of 1.5 m wide throughout the length of 9.5 m,

Referring to Fig. 12.33,

Intensity of soil pressure $= \left(\dfrac{325}{1.5 \times 1.0} \right) = 217 \text{ kN/m}^2 < 250 \text{ kN/m}^2$

Maximum ultimate load bending moment is calculated as given below:

Dead Load.................. = 194 kN

Self weight of footing = 30 kN

Total dead load........... = 224 kN

Live load Reaction..... = 101 kN

Soil pressure due to dead load = $[224/(1.5 \times 1)] = 149.33$ kN/m²

Soil pressure due to live load = $[101/(1.5 \times 1)] = 67.33$ kN/m²

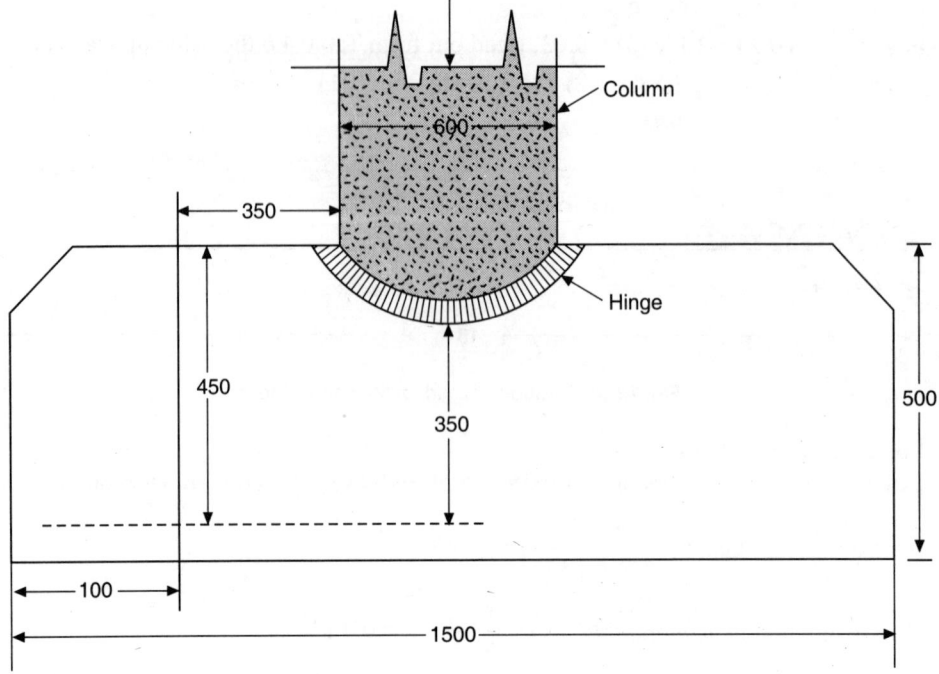

Fig. 12.33 Hinged Footing for Column of Portal Frame.

Maximum B.M at centre of footing due to dead load is
$$M_g = (149.33 \times 0.75 \times 0.5 \times 0.75) = 42 \text{ kN·m}$$
Maximum B.M at centre of footing due to live load is
$$M_q = (67.33 \times 0.75 \times 0.5 \times 0.75) = 19 \text{ kN·m}$$
The ultimate B.M is computed by applying appropriate load factors as
$$\begin{aligned} M_u &= (1.35\ M_g + 1.5\ M_q) \\ &= [(1.35 \times 42) + (1.5 \times 19)] \\ &= 85.2 \text{ kN·m} \end{aligned}$$
Using M-20 grade concrete and Fe-415 HYSD reinforcements, the depth of footing is

$$d = \sqrt{\frac{M_u}{0.138 \times f_{ck} \times b}} = \sqrt{\frac{85.2 \times 10^6}{0.138 \times 20 \times 1000}} = 175.6 \text{ mm}$$

Since shear considerations require larger depth, provide an effective depth of 350 mm and overall depth of 400 mm.

Refer Design Table 2 of SP-16 and interpolate the percentage reinforcement required corresponding to the parameter $(M_u/bd^2) = [(85.2 \times 10^6)/(1000 \times 350^2)] = 0.69$
$$\begin{aligned} p_t &= 0.2 = (100\ A_{st}/bd) \\ A_{st} &= [(0.2 \times 1000 \times 350)/100] = 700 \text{ mm}^2/\text{m} \end{aligned}$$
Use 16 mm diameter bars at a spacing 200 mm ($A_{st} = 1005 \text{ mm}^2$)

(c)   Check for Ultimate Shear Strength

Ultimate dead load shear at the face of column junction = $(0.45 \times 149.33 \times 1) = 67.2$ kN
Ultimate live load shear at the face of column junction = $(0.45 \times 67.33 \times 1) = 30.2$ kN
Total Ultimate load shear force = $V_u = (67.2 + 30.2) = 97.4$ kN

The ultimate shear strength of the reinforced concrete footing is checked by using the equation 4.7.

$$V_{Rd.c} = [0.12K\,(80\rho_1 f_{ck})^{0.33}]\,b_w.d$$

where $\qquad K = 1 + \sqrt{\dfrac{200}{350}} = 1.75$ but limited to 2.00

and $\qquad \rho_1 = \left(\dfrac{A_{sl}}{b_w.d}\right) \le 0.02$

$$= \left(\frac{1005}{1000 \times 350}\right) = 0.0028$$

$$\begin{aligned}
V_{Rd.c} &= [0.12 \times 1.75\,(80 \times 0.0028 \times 20)^{0.33}](1000 \times 350)\ \text{N}\\
&= (121.2 \times 1000)\ \text{N}\\
&= 121.2\ \text{kN} > 97.4\ \text{kN (Hence safe)}
\end{aligned}$$

19. Supplementary Reinforcements

Using M-20 grade concrete and Fe-415 HYSD reinforcements, the minimum reinforcements according to IRC: 112-2011 is given by

$$A_{s,min} = 0.0013\,bd = (0.0013 \times 1000 \times 350) = 455\ \text{mm}^2/\text{m}$$

Provide 12 mm diameter distribution bars at a spacing of 200 mm ($A_{st} = 565$ mm^2)
The details of reinforcements provided in the rigid frame are shown in Fig. 12.34.

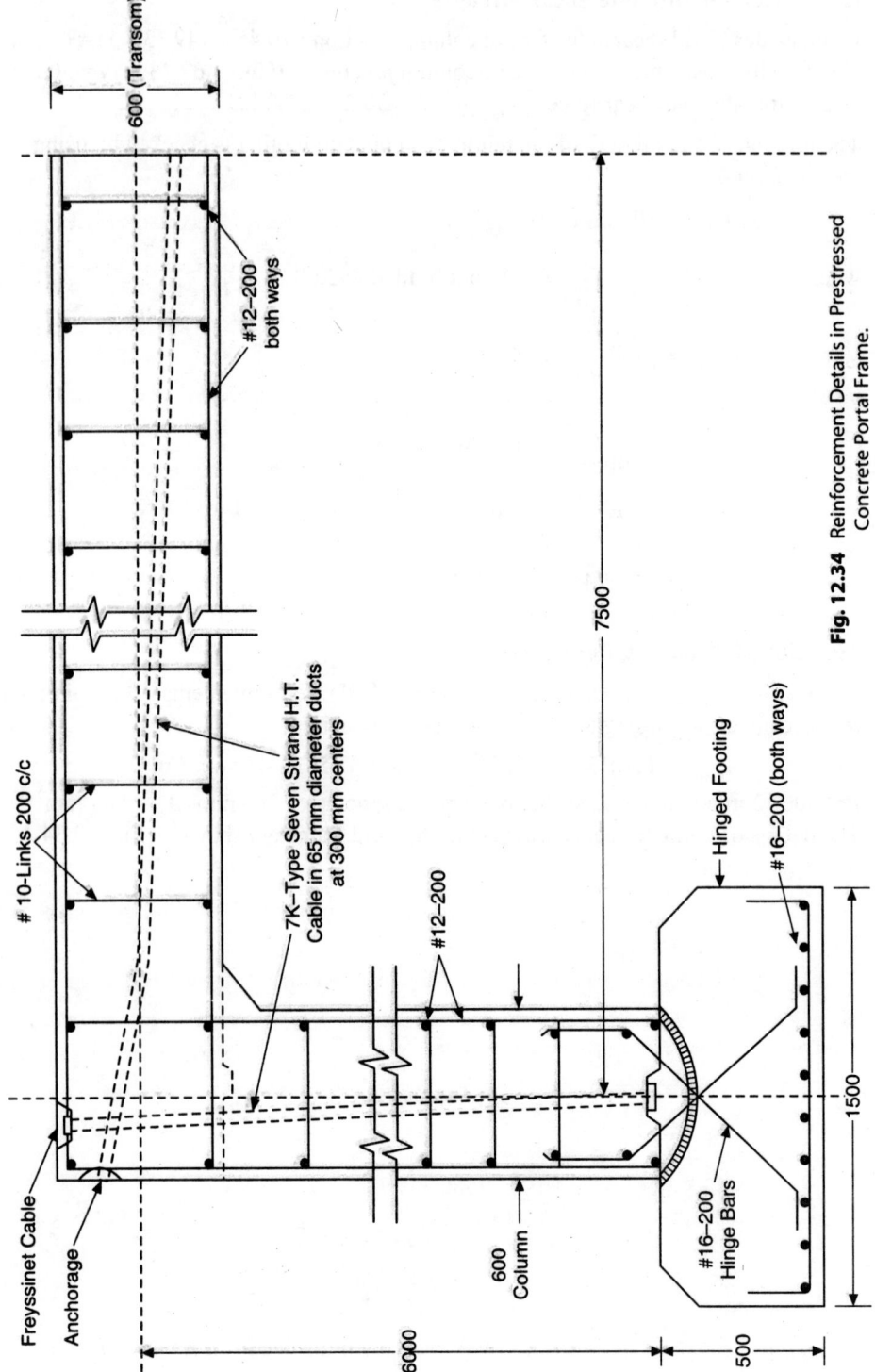

Fig. 12.34 Reinforcement Details in Prestressed Concrete Portal Frame.

EXAMPLES FOR PRACTICE

1. A reinforced concrete figid frame bridge is to be designed to suit the following data:

   Clear span = 20 m
   Road width = 7.5 between kerbs
   Footpath = 1.5 m on each side, Height = 7.5 m
   End conditions = Hinged
   Loading: I.R.C. Class AA or Class A whichever produces the worst effect.
   Temperature change = 40°C
   Coefficient of linear expansion of concrete
   $$= 6.5 \times 10^{-6}$$
   Modulus of elasticky of concrete $E_c = 25.5$ kN/mm^2
   Concrete grade: M-20, tor steel of Fe-415 grade is available for use.

   Design the rigid frame and sketch the details of reinforcements.

2. A reinforced concrete rigid frame bridge is required for the crossing of a National Highway to suit the following particulars:

   Clear span = 25 m
   Road width = 15 m between kerbs
   Footpath = 1.5 m on each side, Height = 7.5 m
   End conditions = Hinged
   Loading: I.K.C. Class AA or Class whichever produces the worst effect
   Temperature change = 35°C
   Coefficient of linear expansion of concrete
   $$= 6.5 \times 10^{-6}$$
   Modulus of elasticity of concrete $E_c = 25.5$ kN/mm^2
   Concrete grade M-20, steel quality—tor steel Fe-415 grade
   Safe bearing capacity of soil = 100 kN/m^2

   Design the rigid frame and a suitable footing for the frame and sketch the details of reinforcements.

3. Design a reinforced concrete rigid frame bridge for a National Highway to suit the following data:

   Clear span = 18 m
   Road width between kerbs = 7.5 m
   Kerbs 600 mm wide on either side
   End conditions = Hinged footings
   Loading: I.R.C. Class 70R tracked vehicle
   Temperature change at bridge site = 30°C
   Coefficient of linear expansion of concrete = 6.5×10^{-6}
   Concrete M-25 Grade and Fe-415 HYSD bars
   Safe bearing capacity of soil at site = 150 kN/m^2

   Design the rigid frame bridge deck and sketch the details of reinforcements.

4. Design an R.C.C Rigid Frame bridge in a highway network to suit the following data:

Clear span = 15 m
Height = 8 m at centre of span
Road width = 7.5 m
Foot Paths: 1 m wide on either side with kerbs of 600 mm
End conditions: Hinged Footing
Loading: IRC Class A A tracked vehicle
Temperature change = 40 degrees Centigrade
Coefficient of linear expansion of concrete = 6.5×10^{-6}
Concrete Grade: M-30
Type of reinforcement: Fe-415 HYSD bars
Safe bearing capacity of soil at site = 200 kN/m^2

Design the rigid frame bridge deck and sketch the details of reinforcements in the frame.
The design should conform to the specifications of relevant IRC codes.

5. Design a prestressed concrete two pinned rigid frame to suit the following dada:

Clear span = 20 m
Width of carriage way = 15 m between kerbs
Clear Height to the soffit of transom = 6 m
Loading: IRC Class AA tracked vehicle
Thickness of wearing coat = 80 mm
Safe bearing capacity of soil at site = 200 kN/m^2
Adopt 7K-15 type Freyssinet cables with high tensile steel strands conforming to IS: 6006 for prestressing the transom and columns.
Grade of concrete; M-60 for rigid frame and M-30 for the hinged footing

Design the bridge as Class 1 type structure conforming to the specifications of IRC: 6-2014, IRC: 112-2011 and IS: 1343-2012.

6. Design a two pinned portal frame R.C.C. foot bridge having a uniform cross section for the transom and columns using the following data:

Clear span = 20 m
Height of column legs = 6 m
Width of pedestrian walk way = 2 m
Live load on bridge = 3 kN/m^2
Grade of concrete: M-30 grade
Type of steel: Fe-415 HYSD bars
Safe bearing capacity of soil at site = 150 kN/m^2
Design the pedestrian bridge conforming to the specifications of IRC: 6-2014, IRC: 112-2011 and IS: 456-2000.

REFERENCES

1. Raina, V.K., Concrete Bridge Practice, Analysis, Design & Economics, Tata McGraw Hill Publishing Co. Ltd., New Delhi, 1991, pp. 196–200.
2. Victor, J.D., Essentials of Bridge Engineering (Fifth Edition), Oxford & IBH Publishing Co. Pvt. Ltd., New Delhi, 2001, pp. 164–166.
3. Chettoe, C.S and Adams, H.C., Reinforced Concrete Bridge Design, Chapman & Hall London, 1952, pp. 1–416.
4. Dunham, C.W., Advanced Reinforced Concrete, McGraw Hill Book Co. Ltd., New York, 1964.
5. Taylor, F.W, Thompson, S.E, and Smulski, E., Reinforced Concrete Bridges, John Wiley & Sons, New York, 1955.
6. Park, R and Paulay, T., Reinforced Concrete Structures, John Wiley & Sons, New York, 1975.
7. Krishna Raju, N., Prestressed Concrete Bridges, CBS, New Delhi, 2009, pp. 165–186.
8. SP: 16-1980, Design Aids for Reinforced Concrete to IS: 456, Bureau of Indian Standards, New Delhi, 1980, pp. 1–232.
9. IRC: 6-2014, Standard Specifications and Code of Practice for Road Bridges, Section II, Loads and Stresses (Revised Edition), Indian Roads Congress, New Delhi, 2014, pp. 1–84.
10. SP: 34-1987, Hand Book on Concrete Reinforcement and Detailing, Bureau of Indian Standards, New Delhi, pp. 1–81.

REVIEW QUESTIONS

1. Briefly explain the various structural elements of a rigid frame structure mentioning their practical applications in bridges.
2. What are the salient advantages of rigid frame bridges in comparison with the traditional types of reinforced and prestressed concrete bridges.
3. What are the span ranges for which the rigid frame bridges are generally adopted? Distinguish between the solid barrel and beam and slab types of rigid frame bridges.
4. Briefly explain the significance of frame constants with reference to rigid frame bridges and the method of determining the constants.
5. Explain briefly the necessity of using influence lines for determining the maximum bending moments at various critical cross sections of the rigid frame bridge.
6. Mention the various factors influencing the horizontal thrust in a rigid frame bridge and outline the method of calculating the design thrust associated with the design moment.
7. Briefly discuss the various combinations of load effects to be considered in determining the maximum design moments at critical sections of a rigid frame bridge.

8. What is the effect of variation in temperature on the horizontal thrust developed in a rigid frame bridge? Mention the various parameters to be considered in the computation of reactions due to temperature changes.

9. What are the advantages of using hinged footings for rigid frame bridges? In what way this type is preferable to the other types of foundations?

10. Briefly outline the method of designing a hinged footing? How do you check the flexural and shear strength of footings?

OBJECTIVE TYPE QUESTIONS

1. Reinforced and Prestressed concrete Rigid frame bridges can be classified as
   a) Determinate structures
   b) Indeterminate structures
   c) Complex structures

2. Rigid frame bridge structure resist the lateral earth pressure by
   a) Abutments
   b) The vertical column elements
   c) Wing walls

3. Reinforced concrete rigid frame bridges are generally used in the span range of
   a) 5 to 10 m
   b) 10 to 30 m
   c) 30 to 40 m

4. In a rigid frame bridge, the maximum positive bending moments due to dead and live loads develop at
   a) The junction of transom and columns
   b) The centre of transom
   c) The base of the column

5. The maximum horizontal thrust in a hinged portal frame develops when the IRC Class AA tracked vehicle is located at
   a) The quarter span point
   b) The supports
   c) The centre of span

6. The load factors specified in the IRC: 112-2011 code for dead loads is
   a) 1.50
   b) 1.25
   c) 1.35

7. The hinged footing at the base of the column in a rigid frame bridge should be designed for
   a) Bending moment and thrust
   b) Bending moment and shear
   c) Vertical reaction

8. In a reinforced concrete portal frame rigid bridge, the thickness at the centre of horizontal transom member, expressed as a ratio of the span is generally
   a) L/15
   b) L/20
   c) L/35
9. In a prestressed concrete rigid portal frame bridge, the magnitude of steel reinforcement required is least because of the use of
   a) High strength concrete
   b) Prestressing of members
   c) Larger cross sections of members
10. In a prestressed concrete rigid frame bridge designed as a Class 1 type structure
   a) Limited tensile stresses are allowed
   b) Tensile stresses are not permitted
   c) Shear stresses are allowed

Steel Trussed Bridges

13.1 GENERAL FEATURES

Steel trussed bridges[1] are generally economical in the span range of 100 to 200 m. These type of bridges are generally preferred for long span railway bridges. Trussed bridges are economical since the members are subjected to direct forces and the open web construction facilitates the use of larger depths with a reduction in the self-weight. The construction of steel trussed bridge is faster due to lightness of members and fabrication of joints at site. However the maintenance costs of a steel bridge is higher since the members have to be painted periodically to prevent their deterioration due to rusting. In addition steel bridges should be provided with rocker and roller bearings which need periodical inspection and maintenance.

13.2 TYPES OF TRUSSES

The most common type of steel truss used for bridges is the Warren truss.[2] This type can be advantageously used both for through and deck type bridges. The configuration of the Warren trusses used in different forms as reported by Victor are shown in Fig. 13.1(a) to (g). When the depth of the bay is twice its length, K-bracing is adopted as shown in Fig. 13.1(e). In the case of long spans the top chords can be gradually curved towards the supports as shown in Fig. 13.1(f) and (g).

A steel trussed bridge deck consists of the flooring, stringer beams, cross girders, supported by the main truss system. The flooring is either made up of steel plates or reinforced concrete slab which is generally adopted due to its rigidity.

13.3 DESIGN FEATURES

The deck slab is designed as a two-way slab to support the I.R.C. loads using Pigeaud's curvesr[3]. The stringer beams are generally rolled steel joists and they are designed to resist the dead and live loads on the beams. The cross girders are designed as plate girders to resist the load transmitted by the stringer beams.

The main truss members are designed on the assumption that the members are subjected to axial forces only with the loads applied at the nodal points. The depth of the truss for a high-way bridge is generally in the range of 1/6 to 1/20 span and for railway bridges, it ranges from 1/5 to 1/10 span. The panel length is selected such that the slope of the diagonals is not less than 45 degrees with the horizontal.

The design of a trussed bridge deck is illustrated in the following example as reported by Krishna Murthy[4].

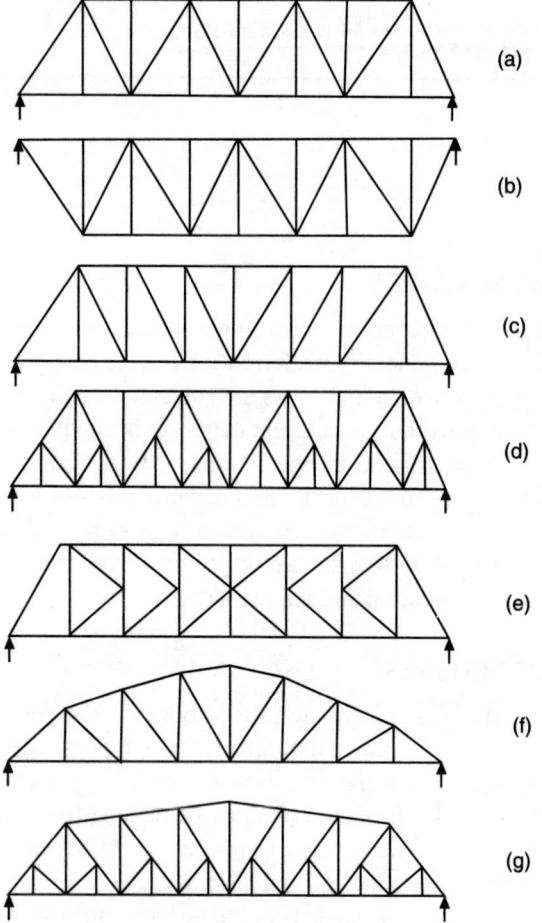

(a)

(b)

(c)

(d)

(e)

(f)

(g)

Fig. 13.1 Typical Bridge Trusses.

13.4 DESIGN EXAMPLE

Design a steel trussed bridge to suit the following data:

1. Data

Effective span = 30 m
Roadway: 7.5 m (two lane)

Kerbs: 600 mm
Loading: I.R.C. Class AA tracked vehicle
Materials: M-25 Grade concrete and Fe-415 HYSD bars for deck slab. Rolled steel
sections with an yield stress of 236 N/mm^2 conforming to IS: 226[5] and IRC: 24[6] are
available for use.

2. Arrangement of Members

For a span of 30 m, it is proposed to provide a Warren truss with 6 panels of 5 m each.
Cross girders are provided at 5 m intervals joining the nodal points. The stringers are
spaced at 1.875 m centres. The configuration of the Warren truss and the arrangement
of cross girders and stringers are shown in Fig. 13.2.

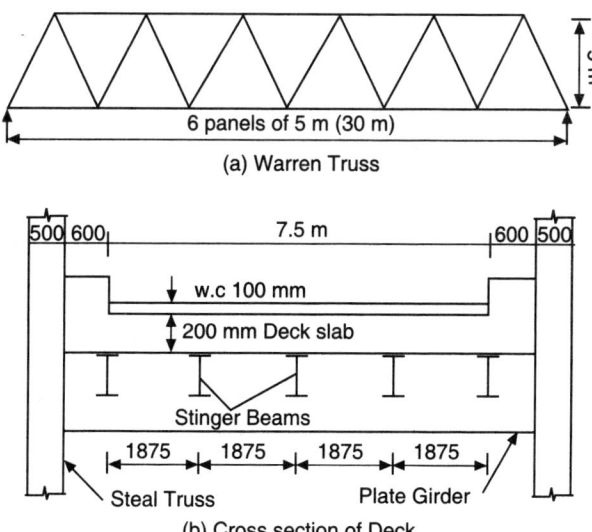

(a) Warren Truss

(b) Cross section of Deck

Fig. 13.2 Details of Deck and Truss.

3. Loads

Self-weight of deck slab = $(0.2 \times 24) = 4.8$ kN/m^2
Weight of wearing coat = $(0.1 \times 22) = 2.2$
Total dead load = 7.0 kN/m^2
Live load is I.R.C. class AA tracked vehicle with two tracks of 3.6 m long and 0.85 m
wide carrying a load of 350 kN. Impact factor is 25 percent for spans less than 9 m.

4. Design of Deck Slab

Class AA tracked vehicle, one wheel placed at the centre of panel as shown in Fig.
13.3.

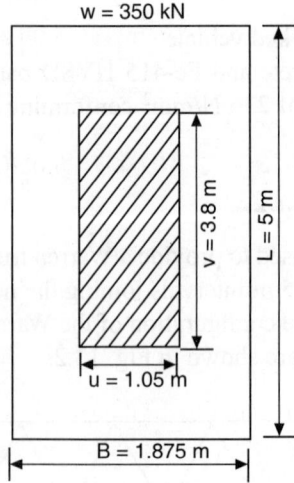

Fig. 13.3 I.R.C. Class AA Tracked Vehicle Wheel Load.

$$u = (0.85 + 2 \times 0.1) = 1.05 \text{ m}$$
$$v = (3.6 + 2 \times 0.1) = 3.80 \text{ m}$$
$$L = 5 \text{ m} \qquad B = 1.875 \text{ m}$$
$$(u/B) = (1.05/1.875) = 0.56$$
$$(v/L) = (3.80/5.0) = 0.76$$
$$K = (B/L) = (1.875/5) = 0.375$$

Referring to Pigeaud's curves with $K = 0.4$ (Refer Fig. 4.6)

$$m_1 = 8.5 \times 10^{-2} \qquad m_2 = 0.9 \times 10^{-2}$$
$$M_B = W(m_1 + 0.15\, m_2)$$
$$= 350\,(0.085 + 0.15 \times 0.009)$$
$$= 30.1 \text{ kN·m}$$

As the slab is continuous, applying impact and continuity factors

$$M_B = (1.25 \times 0.8 \times 30.1) = 30.1 \text{ kN·m}$$
$$M_L = W(m_2 + 0.15\, m_1)$$
$$= 350\,(0.009 + 0.15 \times 0.085)$$
$$= 3.57 \text{ kN·m}$$

Design $M_L = (1.25 \times 0.8 \times 3.57) = 3.57 \text{ kN·m}$

(a) Dead Load bending Moment

Total load on panel = $(5 \times 1.875 \times 7.0) = 65.6 \text{ kN}$

Referring to Pigeaud's curves (Refer Fig. 4.13)

$$(u/B) = 1 \qquad\qquad\qquad (v/L) = 1$$
$$K = (B/L) = 0.375 \text{ (Nearly 0.4)} \quad 1/K = 2.5$$

$$m_1 = 0.045 \qquad\qquad m_2 = 0.004$$
$$M_B = 65.6\,(0.045 + 0.15 \times 0.004) = 3.00 \text{ kN·m}$$

Taking continuity into effect

$$M_B = (0.8 \times 3.00) = 2.4 \text{ kN·m}$$
$$M_L = 65.6\,(0.004 + 0.15 \times 0.045) = 0.656 \text{ kN·m}$$

Taking continuity into effect

$$M_L = (0.8 \times 0.656) = 0.53 \text{ kN·m}$$

5. Design of Slab Section

The total design ultimate load moments are

Short span moment = $M_{Bu} = [1.35\,M_d + 1.5\,M_L]$
$$= [(1.35 \times 2.4) + (1.5 \times 30.1)]$$
$$= 48.50 \text{ kN·m/m}$$

Long span Ultimate moment $M_{Lu} = [(1.35 \times 0.53) + (1.5 \times 3.57)]$
$$= 6.06 \text{ kN·m/m}$$

Effective depth of slab $= d = \sqrt{\dfrac{M_u}{0.138 f_{ck}.b.}} = \sqrt{\dfrac{48.50 \times 10^6}{0.138 \times 25 \times 1000.}} = 118 \text{ mm}$

From practical considerations, adopt effective depth, $d = 200$ mm and overall depth of 250 mm

Using 12 mm diameter bars

Effective depth provided = 200 mm

$$\left(\frac{M_u}{b.d^2}\right) = \left(\frac{48.5 \times 10^6}{1000 \times 200^2}\right) = 1.21, \text{ Using M-25 grade concrete and Fe-415 HYSD bars}$$

Read out the percentage of reinforcement required from Table 3 of SP: 16 Design Aids

$$p_t = 0.36 = \left(\frac{100 A_{st}}{b.d}\right)$$

Solving $\quad A_{st} = \left(\dfrac{0.36 \times 1000 \times 200}{100}\right) = 720 \text{ mm}^2$

For short span, Provide 12 mm diameter bars at 120 mm centers (A_{st} provided = 942 mm²)

For long span, provide 10 mm diameter bars at 150 mm centers.

6. Design of Stringer Beams

Dead load due to self weight of slab and wearing coat = (7×1.875)
$$= 13.125 \text{ kN/m}$$

Self-weight of stringer (assumed) = 1 kN/m

Total dead load on stringer = (13, 125 + 1) = 14.125 kN/m

Maximum B.M. due to dead load = (14.125 × 5²)/8 = 44.14 kN·m

Maximum shear force due to dead load = (14.125 × 5)/2 = 35.3 kN

The stringer is subjected to maximum bending moment when one of the tracks is directly on it as shown in Fig. 13.4(a).

Maximum B.M. due to live load = [(0.5 × 350 × 2.5) – (0.5 × 350 × 0.25 × (3.6)]

$$= 280 \text{ kN·m}$$

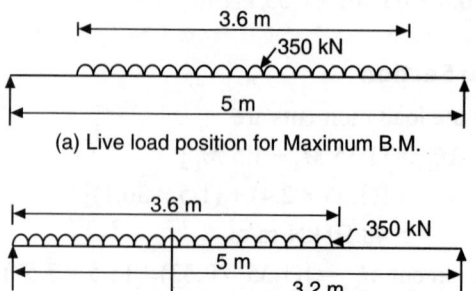

(a) Live load position for Maximum B.M.

(b) Live load position for Maximum shear

Fig. 13.4 Live Load on Stringer Beams.

Referring to Fig. 13.4(b)

Maximum S.F. due to live load = (350 × 3.2)/5 = 224 kN

B.M. including impact at 25% = (1.25 × 280) = 350 kN·m

Design B.M. = (350 + 44.14) = 395 kN·m

Design S.F. = (1.25 × 224 + 35.3) = 316 kN

Section Modulus $Z = (M/\sigma_b) = (395 \times 10^6)/150 = 2.63 \times 10^6 \text{ mm}^3$

Use ISWB – 550 ($Z = 2.72 \times 10^6 \text{ mm}^3$)

Shear stress = $(316 \times 10^3)/(10.5 \times 550) = 54.7 \text{ N/mm}^2 < 85 \text{ N/mm}^2$ (Safe)

7. Design of Cross Girders

(a) Bending Moments and Shear Forces

Span of the cross girder = (7.5 + 2 × 0.6) = 8.7 m

Impact factor (refer Fig. 2.19) for steel bridges (for a span of 9 m) = 40%

Dead Load due to slab and W.C........................= (7 × 5) = 35,00 kN

Dead load due to stringer beams = (1.125 × 5) = 5.625 kN

Add load due to connectors............................= 0.375

Total load ....= 6.00 kN

Self weight of cross girder = (0.2L + 1) = (0.2 × 8.7 + 1) = 3 kN/m

Total uniformly distributed load = (35 + 3) = 38 kN/m

Referring to Fig. 13.5(a), the dead load bending moments and shear forces are computed as

$$M_g = [(180.3 \times 4.35) - (38 \times 4.35 \times 0.5 \times 4.35) - (6 \times 3.75) - (6 \times 1.875)]$$

$$= 391 \text{ kN·m}$$
$$V_g = 180.3 \text{ kN}$$

Referring to Fig. 13.5(b), the maximum live load bending moment including impact occurs when the two tracks are spaced symmetrically from the centre of cross girder.

$$M_q = 1.4[(0.5 \times 350 \times 8.7) - (0.5 \times 350 \times 2.05)]$$
$$= 1630 \text{ kN·m}$$

Maximum shear force due to live loads occur when one of the edges of the track is 1.2 m from the kerb.

Maximum shear force including impact is computed as

$$V_q = 1.4[(350 \times 6.475)/8.7] + 1.4[(350 \times 4.425)/8.7]$$
$$= 613.8 \text{ kN}$$

Total design B.M = (391 + 1630) = 2021 kN
Total design S.F = (180.3 + 613.8) = 794 kN

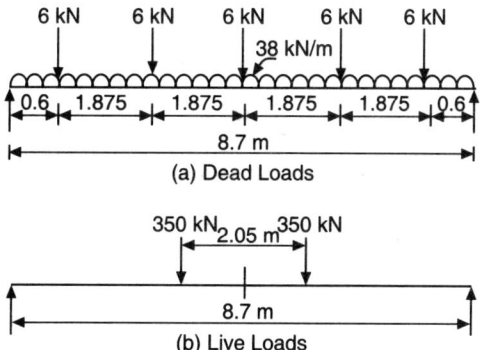

Fig. 13.5 Position of Dead and Live Loads for Maximum Bending Moment.

(b) Trial Section of Plate Girder[7]

Economical depth of plate girder $= d = \left(\dfrac{Mk}{f_y}\right)^{0.33}$

where $k = (d/t_w) = 180$ for thin webs and $f_y = 250 \text{ N/mm}^2$

substituting the values, $d = \left(\dfrac{2021 \times 10^6 \times 180}{250}\right)^{0.33} = 1130 \text{ mm}$

Web depth based on shear considerations, assuming 10 mm thick plate and spacing of stiffeners as equal to the depth of web plate [Refer Table IRC: 24 and interpolate τ for the ratio of $(d/t_w) = 100$] is computed as,

$$d = \left[\frac{V}{(\tau \times 12)}\right] = \left[\frac{794 \times 10^3}{(100 \times 10)}\right] = 794 \text{ mm}$$

Adopt a web plate of size 1000 mm depth and thickness of 10 mm

(c) Flange Plates

Approximate area of flange plates is evaluated as,

$$A_f = \left[\left(\frac{M}{\sigma_b.d}\right)\right] \text{ where, } \sigma_b = \text{permissible bending stress in girders as per IRC: 24-2011}$$

And $\sigma_b = 0.62\,f_y = (0.62 \times 250) = 155 \text{ N/mm}^2$

$$A_f = \left[\left(\frac{2021 \times 10^6}{155 \times 1000}\right)\right] = 13038 \text{ mm}^3$$

Adopt flange width = 500 mm
Thickness of plate required = (A_f/B) = [13038/500] = 26 mm
Adopt flange plates of size 500 mm by 25 mm
The section selected is shown in Fig. 13.6

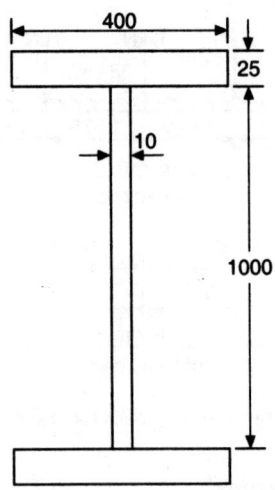

Fig. 13.6 Section of Plate Girder.

(d) Check for Maximum Bending Stresses

The second moment of area of the cross section of the girder about axis XX and YY are computed to determine the permissible stresses.

$$I_{xx} = \left[\left(500 \times \frac{1050^3}{12}\right) - (490 \times 1000^3/12)\right]$$

$$= (71 \times 10^8) \text{ mm}^4$$

$$I_{yy} = \left[\frac{(2 \times 25 \times 500^3)}{12} + (1000 \times 10^3)/12\right]$$

$$= (521 \times 10^6) \text{ mm}^4$$

$$A = [(2 \times 25 \times 500) + (1000 \times 10)] = 35000 \text{ mm}^2$$

Minimum radius of gyration $= r_y = \sqrt{(I_{yy}/A)} = \sqrt{\left[\dfrac{521 \times 10^6}{35000}\right]} = 122$

According to IRC: 24-2011, the permissible bending stress (σ_{bc}) is interpolated from Table 9.1 (Table 8.2 of IRC: 24-2001) corresponding to the values of f_{cb} and f_y, and the equation,

$$f_{cb} = k_1(X + k_2 Y)(c_2/c_1) \text{ MPa}$$

where

$$X = Y\sqrt{L + \left(\frac{1}{20}\right)\left[(LT)/(r_y.D)\right]^2} \text{ MPa}$$

and

$$Y = [(2650000)/(L/r_y)^2]$$

f_{cb} = elastic critical stress

c_1 and c_2 are lesser and greater distances from neutral axis to extreme fibers

D = overall depth of girder

T = mean thickness of the compression flange

L = effective length of compression flange (Between cross bracings at 6 m intervals)

r_y = radius of gyration

k_1 = coefficient to allow for reduction in thickness or breadth of flanges between points of effective lateral restraint

k_2 = coefficient to allow for unequal flanges

In the present case, c_1, c_2, k_1 and k_2 have values of 1.0

The simplified equation for the value of elastic critical stress is given by

$$f_{cb} = (X + Y) \text{ MPa}$$

However, the code provides for a simpler method for evaluating the value of X and Y which involves the computation two parameters such as $(D/T) = (1050/25) = 42$ and $(L/r_y) = (5000/122) = 40.9$

Using these values, X and Y can be interpolated from Table 9.2 (Table 8.5 of IRC: 24-2001 code) for known values of the ratio (D/T) and (L/r_y).

Using this method, the value of $X = 1630$ and $Y = 1500$ and

$$f_{cb} = (1630 + 1500) = 3130 \text{ N/mm}^2$$

Using the values of f_{cb} and $f_y = 250$ N/mm^2, the permissible compressive stress σ_{bc} can be interpolated from Table 9.1 as $\sigma_{bc} = 161$ N/mm^2.

Actual maximum bending stress is given by

$$\sigma_{bc} = \sigma_{bt} = \left(\frac{My}{I}\right) = \left(\frac{2021 \times 10^6 \times 525}{71 \times 10^8}\right) = 149 \text{ N/mm}^2 < 161 \text{ N/mm}^2 \text{ (Hence safe)}$$

(e) Check for Maximum Shear Stress

Shear stress depends upon the ratio of $(d/t) = (1000/10) = 100$

Using stiffener spacing $= c = 0.33$ to $1.5 d$

Adopting, $c = 1000$ mm, $(d/t) = 100$ and spacing of stiffeners at 'd'

Referring to Table 9.3 (Table 8.6 of IRC: 24-2001), interpolate the average permissible shear stress as $\tau_{va} = 100$ N/mm^2.

$$\text{Actual average shear stress} = \left(\frac{V}{d.t_w}\right) = \left(\frac{794 \times 10^3}{1000 \times 10}\right) = 79.4 \,\text{N/mm}^2 < 100 \,\text{N/mm}^2$$
$$\text{(Hence safe)}$$

(f) Connections Between Flanges and Web[8]

Maximum shear force at the junction of web and flange is given by

$$\tau = \left(\frac{VAy}{I}\right)$$

Where
$$V = 794 \text{ kN}$$
$$A = (500 \times 25) = 12500 \text{ mm}^2$$
$$y = 512.5 \text{ mm}$$
$$I = 71 \times 10^8 \text{ mm}^4$$
$$\tau = \left(\frac{VAy}{I}\right) = \left(\frac{794 \times 10^3 \times 12500 \times 512.5}{71 \times 10^8}\right) = 716 \text{ N/mm}^2$$

Refer Table 9.4 and read out the size of weld required as 6 mm with a strength of 793 N/mm. Provide 6 mm continuous fillet welds on either side.

(g) Intermediate Stiffeners[9]

Since the ratio of $(d/t) = (1000/10) = 100 > 85$, vertical stiffeners are required

Adopting the spacing of stiffeners $= c = 1000$ mm

Greater unsupported panel dimension of the web $= 1000$ mm $< 270 \, t_w < (270 \times 10) = 2700$ mm

The Intermediate stiffeners are designed to have a minimum moment of inertia specified as

$$I = \left[\frac{1.5 d^3 t^3}{c^2}\right] = \left[\frac{1.5 \times 1000^3 \times 10^3}{1000^2}\right] = (150 \times 10^4) \text{ mm}^4$$

Using 10 mm thick plate, outstand of stiffener should be not greater than

$$12t = (12 \times 10) = 120 \text{ mm}$$

Adopt a plate 10 mm by 80 mm, having,

$$I = (10 \times 80^3)/3 = (170 \times 10^4) > (150 \times 10^4) \text{ mm}^4$$

(h) Connections of Vertical Stiffener to Web

Shear on welds connecting stiffener to web $= (126t^2/h)$

Where $\qquad t =$ web thickness (mm) $= 10$ mm

$\qquad\qquad h =$ outstand of stiffener (mm) $= 80$ mm

Shear on welds $= [(126 \times 10^2)/80] = 157.5$ N/mm

Size of weld $= s = [157.5/(0.7 \times 158)] = 1.42$ mm

Effective length of weld should be not less than $10t = (10 \times 10) = 100$ mm

Provide 100 mm long, 5 mm fillet welds alternately on either side.

8. Design of Steel Truss[10]

A warren truss with 6 panels of 5 m each is used

Span of the truss $= 30$ m

Height of Truss $= (1/6)$ span $= (30/6) = 5$ m

(a) Loads

Dead loads due to deck slab, wearing coat, stringer beams and cross girders acting at each node $= 181$ kN

Self-weight of truss $= (0.15 L + 5.5) = 10$ kN/m

$\qquad\qquad\qquad\qquad = (0.15 \times 30 + 5.5) = 10$ kN/m

Self-weight at each node point $= (5 \times 10) = 50$ kN

Total dead load $= (181 + 50) = 231$ kN

Live loads: I.R.C. Class AA loading. Maximum B.M. is produced when the class AA vehicle is closest to main girder.

Maximum load transferred when one track is at 1.625 m from the edge of the kerb as shown in Fig. 13.7.

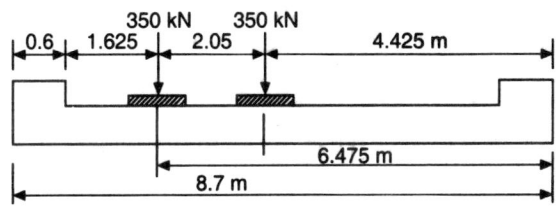

Fig. 13.7 Load Position for Truss Design.

Maximum load transferred when one track is at 1.625 m from the edge of the kerb

$\qquad\qquad = [(350 \times 6.475)/8.7] + [(350 \times 4.425)/8.7] = 439$ kN

Impact factor $= 10\%$

Live Load including impact $= (439 \times 1.1) = 483$ kN

$\therefore$ Average u.d.l $= (483/3.6) = 135$ kN·m

Forces in the members of the truss is determined when a rolling load of length 3.6 m, shorter than span rolls on the bridge.

9. Forces in Truss Members

Influence lines are drawn for forces in the various members of the truss as shown in Fig. 13.8 and 13.9.

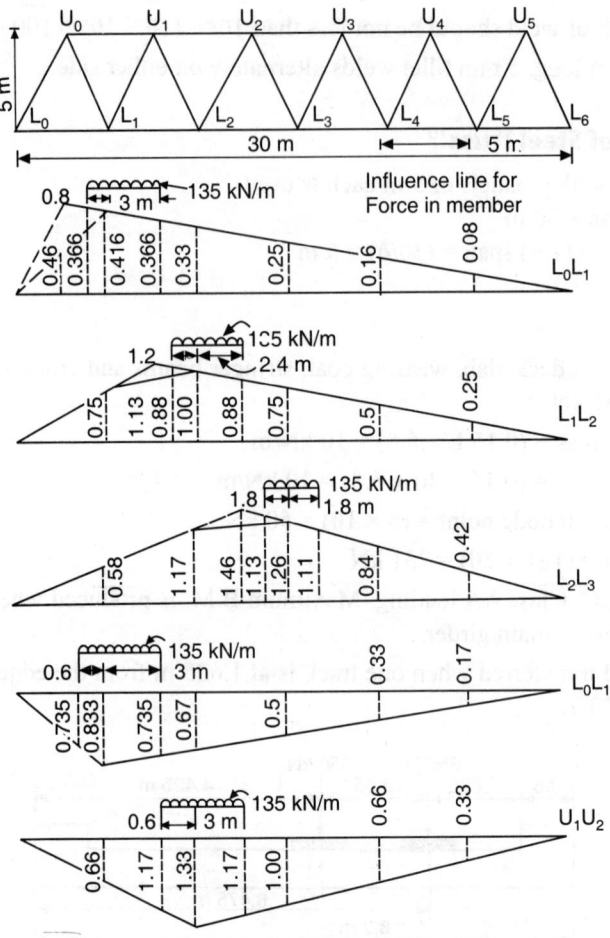

Fig. 13.8 Influence Lines for Forces in Truss Members.

1. Member L_0L_1

Taking moments about U_0, influence line ordinate at U_0

$$= (\text{B.M. about } U_0)/(\text{Perpendicular distance})$$
$$= (2.5 \times 27.5)/(30 \times 5) = 0.46$$

Influence line ordinate at $L_1 = (0.46 \times 25)/(27.5)$
$$= 0.416$$

Using the influence line diagram, force due to dead loads

$$= 231 \, (0.416 + 0.33 + 0.25 + 0.16 + 0.08)$$
$$= 286 \text{ kN (Tension)}$$

Force due to live loads $= [0.5 \, (0.416 + 0.366) \, 3.6 \times 135]$

$$= 190 \text{ kN (Tension)}$$

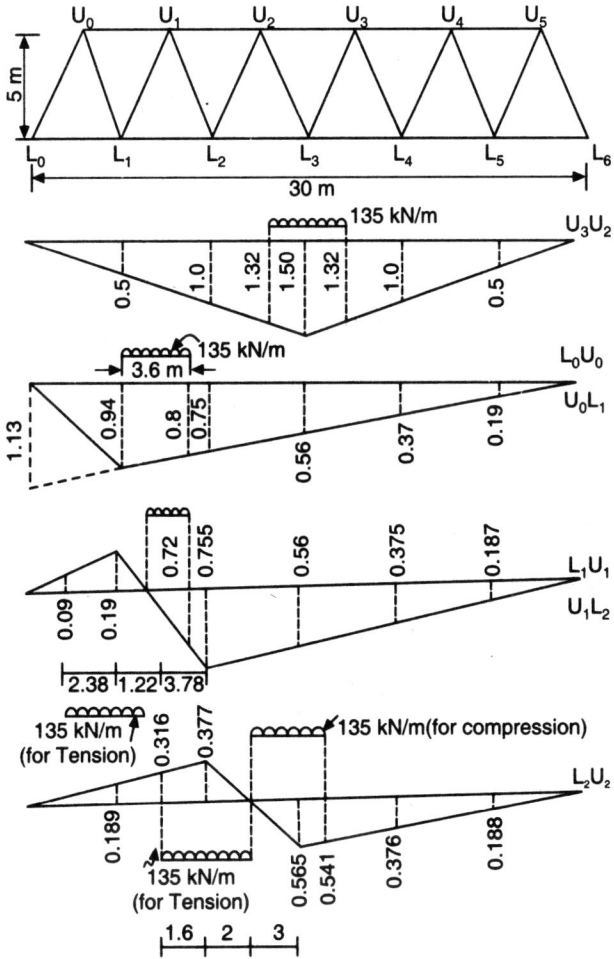

Fig. 13.9 Influence Lines for Forces in Truss Members.

2. Member L_1L_2

Taking moments about U_1

I.L. ordinate at $U_1 = (7.5 \times 22.5)/(30 \times 5) = 1.13$

I.L. ordinate at $L_1 = (1.13 \times 5)/7.5 = 0.75$

I.L. ordinate at $L_2 = (1.13 \times 20)/22.5 = 1.0$

Force due to dead load = 231 (0.75 + 1 + 0.75 + 0.5 + 0.25) = 751 kN

Force due to live load = [0.5 (0.88 + 0.94) 3.6 × 135] = 442 kN (Tension)

3. Member L_2L_3

Taking moments about U_2

I.L. ordinate at U_2 = (12.5 × 17.5)7(30 × 5) = 1.46

I.L. ordinate at L_2 = (1.46 × 10)712.5 = 1.17

I.L. ordinate at L_3 = (1.46 × 15)717.5 = 1.26

Force due to dead loads = 231 (1.17 + 1.26 + 0.58 + 0.84 + 0.42)

$$= 989 \text{ kN (Tension)}$$

Force due to live load = 0.5 (1.13 + 1.11) 3.6 × 135 = 544 kN (Tension)

4. Member U_0U_1

Taking moments about L_1

I.L. ordinate at L_1 = (5 × 25)/(30 × 5) = 0.833

Force due to dead loads

$$= 231 (0.833 + 0.67 + 0.5 + 0.33 + 0.17) = 578 \text{ kN (Compression)}$$

Force due to live loads = 0.5 (0.735 + 0.883) (3.6 × 135)

$$= 381 \text{ kN (Compression)}$$

5. Member U_1U_2

Taking moments about L_2

I.L. ordinate for force at L_2 = (10 × 20)/(30 × 5) = 1.33

Force due to dead loads = 231 (0.66 + 1.33 + 1.0 + 0.66 + 0.33)

$$= 920 \text{ kN (Compression)}$$

Force due to live loads = 0.5 (1.17 + 1.33) (3.6 × 135) = 608 kN (Compression)

6. Member U_2U_3

Taking moments about L_3

I.L. ordinate for force at L_3 = (15 × 15)/(30 × 5) = 1.5

Force due to dead loads = 231 (0.5 + 1 + 1.5 + 1 + 0.5)

$$= 1040 \text{ kN (Compression)}$$

Force due to live loads = 0.5 (1.5 + 1.32) 3.6 × 135 = 686 kN (Compression)

7. Member L_0U_0

Force in L_0U_0 is governed by the shear force in panel L_0L_1

I.L. ordinate at L_1 = [(30 – 5)/30] sec θ = (25/30) × 1.13 = 0.94

Force due to dead loads

$$= 231 (0.94 + 0.75 + 0.56 + 0.37 + 0.19) = 650 \text{ kN (Compression)}$$

Force due to live loads

$$= 0.5 (0.94 + 0.8) 3.6 × 135 = 423. \text{kN (Compression)}$$

8. Member U_0L_1

Influence line is the same as that for member L_0U_0

Force due to dead loads = 650 kN (Tension)

Force due to live loads = 423 kN (Tension)

9. Member U_1L_1

Force in U_1L_1 is governed by shear force is panel L_1L_2

I.L. ordinate at $L_1 = (5/30) \times 1.13 = 0.19$

I.L. ordinate at $L_2 = (20/30) \times 1.13 = 0.75$

Force due to dead loads

$$= 231 \, (-0.19 + 0.75 + 0.56 + 0.375 + 0.187) = 389 \text{ kN (Compression)}$$

Force due to live loads:

Maximum compressive force = $(0.5 \times 0.72 \times 3.6 \times 135) = 175$ kN

Maximum tensile force = $(0.5 \times 0.19 \times 1.22 \times 135)$
$$+ \, 0.5 \, (0.09 + 0.19) \, (2.38 \times 135) = 61 \text{ kN}$$

10. Member U_1L_2

Force in member U_1L_2 is the same as in U_1L_1 but with different sign.

Force due to dead loads = 389 kN (Tension)

Maximum tensile force = 175 kN

Maximum compressive force = 61 kN

11. Member U_2L_2

Force in member is governed by shear force in panel L_2L_3

I.L. ordinate at $L_2 = 1.13 \times (10/30) = 0.377$

I.L. ordinate at $L_3 = 1.13 \times (15/20) = 0.565$

Force due to dead load = $-231 \, (0.377 + 0.189) + 231 \, (0.565 + 0.376 + 0.188)$
$$= (-131 + 261) = 130 \text{ kN (Compression)}$$

Force due to live load:

Maximum compressive force = $(1/2 \times 3 \times 0.565 \times 135) + [0.565 + 0.542)/2]$
$$\times \, (0.6 \times 135) = 160 \text{ kN}$$

Maximum tensile force = $(0.5 \times 0.377 \times 2 \times 135) + [(0.316 + 0.377)/2]$
$$\times \, (1.6 \times 135) = 126 \text{ kN}$$

12. Member U_2L_3

Force in member U_2L_3 is the same as in U_2L_2 but with different sign.

Force due to dead loads = 135 kN (Tension)

Maximum compressive force = 126 kN

Maximum tensile force = 160 kN

10. Design Forces

The design forces in the various truss members are compiled in Table 13.1.

Table 13.1 Design Forces in Truss Members

Member	Due to L.L. (kN)		Due to D.L. (kN)		Combined Load (kN)		Design Force (kN)
	Compression	Tension	Compression	Tension	Max	Min	
L_0L_1	—	190	—	286	−476	−286	−476
L_1L_2	—	442	—	751	−1193	−751	−1193
L_2L_3	—	544	—	986	−1530	−986	−1530
U_0U_1	381	—	578	—	+959	+578	+959
U_1U_2	608	—	920	—	+1528	+920	+1528
U_2U_3	686	—	1040	—	+1726	+1040	+1726
U_0L_0	423	—	650	—	+1073	+650	+1073
U_0L_1	—	423	—	650	−1073	−650	+1073
U_1L_1	175	61	389	—	+564	−61	−61
							+564
U_1L_2	61	175	—	389	−564	+61	−564
							+61
							−564
U_2L_2	160	126	130	—	+290	−126	+290
							−126
U_2L_3	126	160	—	130	−290	+126	−290
							+126

11. Design of Members

(a) Member L_2L_3

Design Force in Member L_2L_3 = 1530 kN (Tension)
Try a section with

2 plates of 350 × 12 mm with a cross sectional area ......= 8400 mm^2
4 angles 75 × 75 × 10mm with a cross sectional area.....= 4600 mm^2
Total are provided ....= 13200 mm^2

Safe load on section = (13200 ×150)/1000 = 1980 kN
8 mm filet welds are used to connect the members

(b) Member U_2U_3

Design force in member U_2U_3 = 1726 kN (Compression)
Try a section with

1 top flange plate (500 × 8) with an area ....= 4000 mm^2
2 ISMC 350 = (2 × 5366)..= 10732 mm^2
Total area..= 14732 mm^2

$$r_{xx} = 0.4\,d = (0.4 \times 350) = 140 \text{ mm}$$
$$r_{yy} = 0.34\,b = (0.34 \times 500) = 170 \text{ mm}$$

Effective length $= L = (0.85 \times 5000) = 4250$ mm

Slenderness ratio $= \lambda = (L/r) = (4250/140) = 30.2$

Refer Table 9.5 of text or Table 11.1 of IRC: 24-2001 and interpolate he permissible compressive stress as

$$\sigma_{sc} = 145 \text{ N/mm}^2$$

Safe load on member $= (14732 \times 145)/1000 = 2136$ kN > 1726 kN (Hence safe)

Use 75 mm by 12 mm flat lacing to connect the channels.

(c) Member U_0L_1 (Diagonal Tension Member)

Force in member $= 1073$ kN (Tension)

The width of the member should be such that it can be accommodated in the bottom and top chord members.

Adopt a width of 280 mm

Use 1 plate (280×15) of area.........................4200 mm^2

4 angles of ($75 \times 75 \times 10$)..............................5608

Total cross sectional area ....9808 mm^2

Using welded connections and permissible tensile stress of 150 N/mm^2

Safe permissible load $= (9808 \times 150)/1000 = 1471$ kN

(d) Member U_1L_1 (Diagonal Member)

Force in member $= 564$ kN (Compression) and 61 kN (Tension)

Length of the member $= 5.6$ m

Width of the member $= 280$ mm

Adopt a section with I plate (280×10) ....$= 2800$ mm^2

4 angles ($100 \times 100 \times 10$) ....$= 7600$

Total area..$= 10400$ mm^2

Radius of gyration, $r_{xx} = 0.39\,d = (0.39 \times 280) = 109$ mm
$$r_{yy} = 0.21\,b = (0.21 \times 210) = 44 \text{ mm}$$

Slenderness ratio $= \lambda = (L/r) = (085 \times 5600/44) = 108$

Refer Table 11.1 of IRC: 24-2001 and interpolate he permissible compressive stress as

$$\sigma_{sc} = 72 \text{ N/mm}^2$$

Safe load on member $= (10400 \times 72)/1000 = 748$ kN > 564 kN (Hence safe)

The detailed drawings of the truss consisting of the cross sectional details at the typical junctions of the members are compiled in Figs. 13.10 and 13.11.

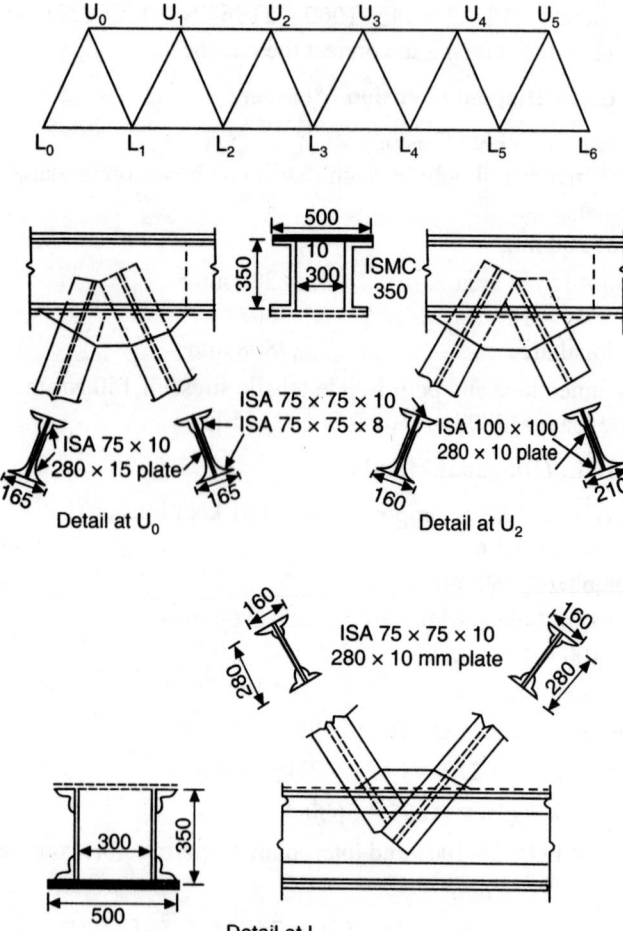

Fig. 13.10 Details of Truss Members.

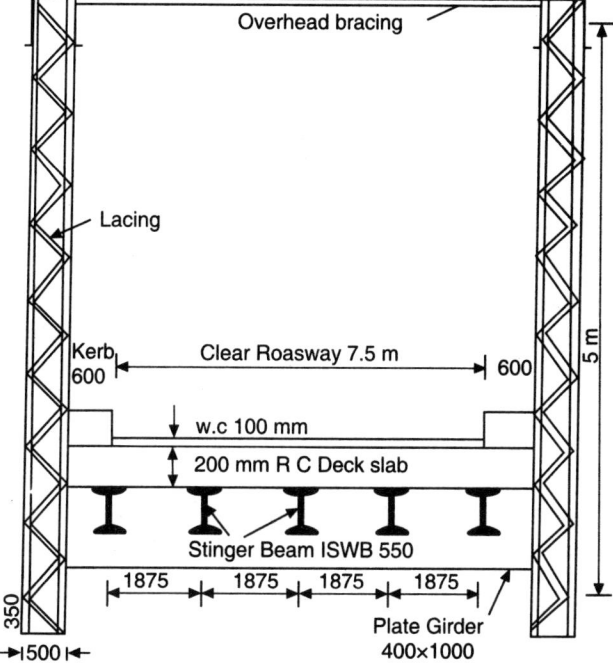

Fig. 13.11 Cross Section of Steel Trussed Bridge.

EXAMPLES FOR PRACTICE

1. A curved chord steel truss is proposed for a highway bridge as shown in Fig. 13.12.

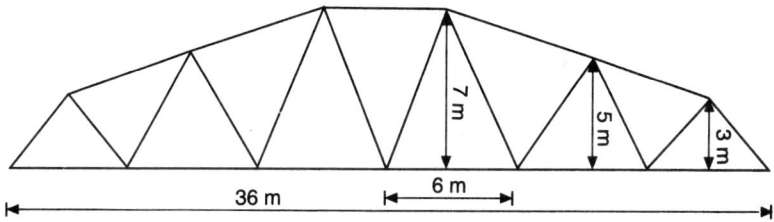

Fig. 13.12 Curved Chord Steel Truss.

The span of the through type bridge is 36 m made up of 6 panels of 6 m each. Cross girders are placed at 6 m intervals.

Spacings of stringer beams = 1.875 m, Loading: I.R.C. class AA (tracked vehicle). Road width = 7.5 m (two lane carriageway), Kerbs = 600 mm on each side.

Design the reinforced concrete deck slab, stringer beams, cross girder and typical members of the curved chord steel truss.

2. A through type steel trussed bridge of N-type is required for a highway crossing. The span of the bridge is 48 m. The spacing of the cross girders is 6 m and the spacing of the stringer beams is 2.25 m.
   Road width = 7.5 m
   Footpaths = 1.25 m on either side
   Loading: I.R.C. class AA, Materials: M-20 grade confrete and Fe-415 grade tor steel. Adopt rolled steel sections for the truss members. Design the R.C.C. deck slab, stringer beams, cross girder and the typical members of the N-truss.

3. A Warren truss is proposed for a deck type steel trussed bridge to cover a span of 30 m. The cross girders are spaced at 5 m intervals. Road width = 7.5 m Kerbs = 600 mm on each side. Spacing of stringer beams = 1.875 m.
   Loading: I.R.C. Class AA. Materials: M-20 grade concrete, Fe-415 tor steel. Adopt rolled steel sections for the truss members. Design the deck slab and truss members.

4. A warren type truss with 5 panels of 5 m each is required for a through type highway bridge of span 25 m. The cross girders are spaced at 5 m intervals. The bridge deck has to carry a two lane 7.5 m road way comprising of reinforced concrete deck slab supported on stringer beams spaced 2 m apart. Width of foot paths on either side being 1 m. The deck has to be designed for I.R.C. Class AA tracked vehicle loading. Adopting M-20 grade concrete and Fe-415 grade HYSD fears design the deck slab. Also design the stringer beams, cross girders and the warren truss members using rolled steel sections and plates with an yield stress of 236 N/mm² and conforming to IS:226 and the permissible stresses should conform to the provisions of IRC:24 code. Sketch the cross section of the bridge deck showing the details of R.C.C. deck, stringer beams, cross girder and the truss. Design and sketch the typical joints of the truss.

5. A warren type truss with intermediate verticals having panels of 6 m is required for a deck type railway bridge to support a broad gauge track. The cross girders are spaced at 6 m and the effective span of the bridge is 48 m. design the various structural elements of the deck type bridge conforming to IS: 226 and IRC: 24-1981 Codes. Sketch the longitudinal and cross sections of the bridge showing the details of Stringer beams, cross girders and the main truss members. Also design and sketch the typical joints of the truss.

REFERENCES

1. Johnson Victor, D., Essentials of Bridge Engineering (Fifth Edition), Oxford & IBH Publishing Co. Pvt. Ltd., New Delhi, 2001, pp. 230–240.
2. Krishnamachar, B.S. and Ajith Simha, Design of Steel Structures, Tata McGraw Hill Publishing Co. Ltd., New Delhi, 1978.
3. Rowe, R.E., Concrete Bridge Design, C.R. Books Ltd., London, 1962, 336 pp.
4. Krishna Murthy, D., Structural Design & Drawing, Vol. III (Steel Structures), C.B.S. Publishers & Distributors, New Delhi, 1985, pp. 1–182.

5. IS: 226–1975, Indian Standard Specifications for Structural Steel (Standard Quality), Bureau of Indian Standards, New Delhi, 1962, pp. 1–12.

6. IRC: 24-2001, Standard Specifications and Code of Practice for Road bridges, Section-V, Steel Road Bridges, Indian Roads Congress, New Delhi, 2001, pp. 157.

7. Vawter, I and Clark, J.G., Elementary Theory and Design of Flexural Members, John Wiley & Sons, New York, 1955, pp. 1–360.

8. Kazimi, S.M.A and Jindal, Design of Steel Structures (Second Edition), Prentice Hall of India, New Delhi, 1987, pp. 39–185.

9. IS: 800–1984, Indian Standard Code of Practice for general Construction in Steel (Second Revision), Bureau of Indian Standards, New Delhi, 1985, pp. 84–85.

10. Arya, A.S and Ajmani, J.L., Design of Steel Structures, Nemchand Bros, Roorke, (Fifth Edition), 1996, pp. 1–901.

REVIEW QUESTIONS

1. Specify the situations under which you would select the steel trussed option in highway bridges.
2. What are the advantages of using steel trussed bridges in comparison with the reinforced and prestressed concrete bridges for national high way crossings?
3. Mention the common types of trusses used for long span national high way crossings with sketches.
4. Sketch the typical cross section of a steel trussed bridge for a high way traffic showing the various structural elements in the bridge.
5. What methods do you follow for analyzing the forces in the various truss members due to the IRC loads on the high way?
6. Briefly mention the method of designing the reinforced concrete deck slab supported on stringer beams in a trussed bridge for IRC loads.
7. Discuss briefly the function of stringer beams and cross girders in a steel trussed bridge.
8. What are the salient design considerations to be followed while designing the cross girders in a steel trussed bridge?
9. What type of steel members are preferred for the top and bottom chord members of a long steel trussed bridge? Explain with sketches.
10. Discuss briefly the advantages and disadvantages of steel trussed bridges for long span National high way crossings.

OBJECTIVE TYPE QUESTIONS

1. Steel trussed bridges are economical for long span high way crossings mainly because the members are subjected to
   a) Bending and compression
   b) Combined bending and thrust
   c) Direct compression or tension

2. In a long span steel trussed bridge, the reinforced concrete deck slab is supported on
   a) Cross girders
   b) Stringer beams
   c) Bottom chord truss members
3. The reinforced concrete deck slab in a long span steel trussed bridge should be designed as a slab supported on
   a) Two sides only
   b) All the four sides
   c) Three sides
4. The maximum bending moment in a stringer beam of a long span steel trussed bridge occurs when the IRC Class AA tracked load is positioned at
   a) The supports
   b) The centre of span
   c) The quarter span
5. Cross girders of a long span steel trussed bridge are located at the
   a) Centre of the bottom chord members
   b) Junctions of the bottom chord members
   c) Junctions of the top chord members
6. The analysis of forces developed in the truss members due to the moving high way loads is generally done by using
   a) Slope deflection method
   b) Moment distribution method
   c) Influence lines
7. The bottom chord members of a long span steel trussed bridge is generally a
   a) Single rolled steel section
   b) Plate girder
   c) Built up section
8. The diagonal members of a steel warren truss used for a long span steel bridge should be designed for
   a) Tension only
   b) Compression only
   c) Tension and compression
9. The reinforced concrete deck slab of a long span steel trussed bridge used for a national high way crossing should be designed using
   a) Elastic method
   b) Ultimate load method
   c) Limit state method
10. A comparative analysis of the maintenance costs incurred in various types of high way bridges indicates the least value for
    a) Reinforced concrete bridges
    b) Composite bridges
    c) Steel bridges

Balanced
Cantilever Bridges

14.1 GENERAL FEATURES

Balanced cantilever bridges offer the unique advantage of continuity in construction ideally suited for medium to long spans. The super structure basically comprises of simply supported, suspended, cantilever and end spans combined together so that the entire structure, has the advantages of a continuous bridge with the simplicity of a determinate structure for purposes of analysis. According to Taylor et al.[1], balanced cantilever type of bridges are among the several new innovative types of bridges which were planned and constructed after the second World War in Germany. The following advantages of balanced cantilever type of designs over simply supported girder type bridges are reported by Victor[2].

1. Economy in the use of concrete, steel and form work

2. Reactions at the piers are vertical and central permitting slender piers

3. The balanced cantilever design requires only one bearing at every pier while the simply supported design needs two bearings. Hence the width of the pier can be smaller.

4. Fewer expansion bearings are required for the entire structure, resulting in lower initial cost and maintenance.

The only disadvantage of this type is that it requires skilled planning on the part of the designer and a more elaborate detailing of the reinforcements. Also the variation of bending moments is less favourable than in continuous beams.

14.2 ARRANGEMENT OF SPANS AND SUPPORTS

Raina[3] has categorized the various typical balanced cantilever type bridges with hinges or articulations at different locations as shown in Fig. 14.1(a) to (d). The connection between the suspended span and the edge of the cantilever is termed as articulation. The bearings at articulations are normally comprised of the roller-rocker arrangement, sliding plates or elastomeric pad bearings. Generally the hinges are positioned in the vicinity of low and zero bending moment under dead load. By convenient location of the hinges, the distribution of dead load bending moments can be made almost identical to that in continuous decks of the same shape and similar loading conditions. Balanced cantilever bridges constructed in Germany[2] have resulted in central depth to main span ratios as low as I in 35 with spans up to 45 m. Since this type of structure is statically determinate with vertical reactions at the supports, relatively light piers can be used. Normally the depth of girder at the support will be about 2 to 4 times that at the mid span and the ratio of main span to the depth at mid span is generally in the range of 15 and 35.

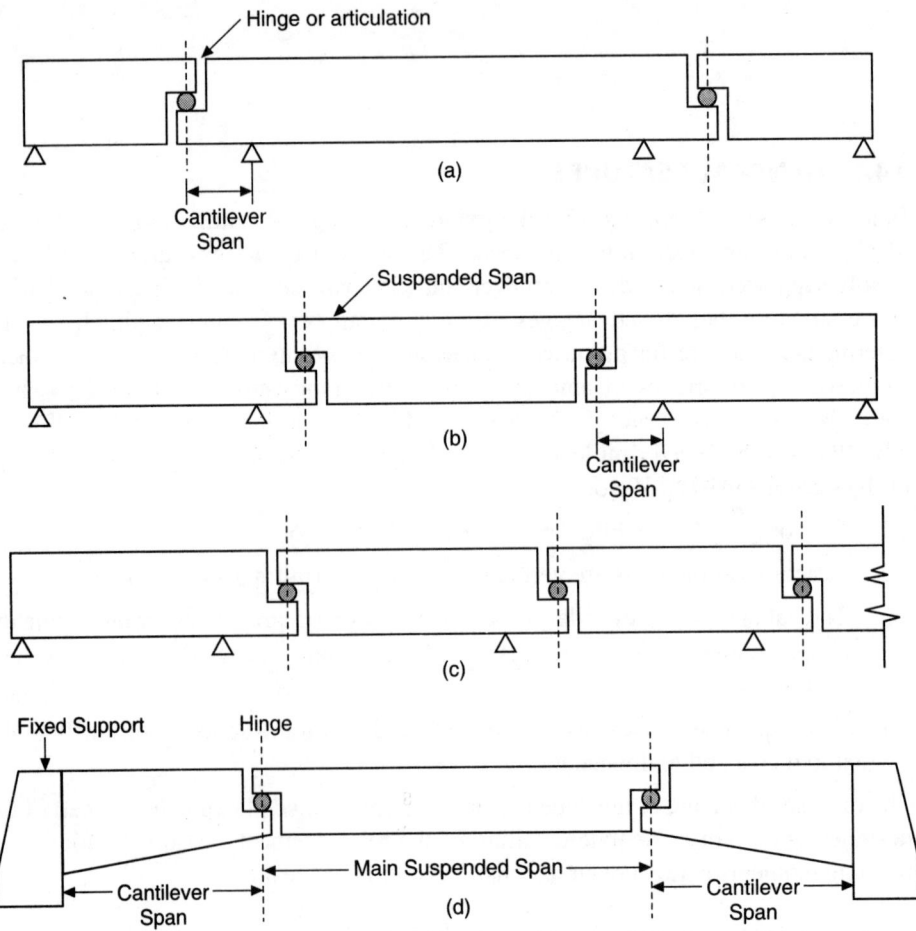

Fig. 14.1 Types of Balanced Cantilever Bridges with Position of Hinges.

Typical arrangement of simply supported and cantilever spans of balanced cantilever bridge types is shown in Fig. 14.2. For economy, the main criterion is that the maximum moments in the beams are the least. For this criterion[3], the dimensions of x and y should be adjusted accordingly. Under normal conditions the ratio

$$\left(\frac{x}{y}\right) = 4 \text{ to } 5$$

This reduces the bending moment to about 50% of the moment for a simply supported beam of span $(x + 2y)$. The simplest arrangement of balanced cantilever bridge is of minimum three spans as shown in Fig. 14.3. The central span can be up to 30 m. If the length of the bridge exceeds, then a 5 span arrangement is selected, with a simply supported span at the centre.

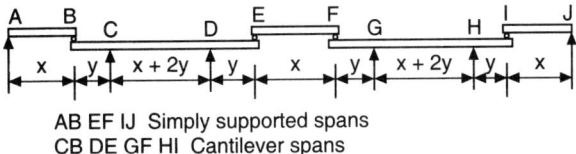

AB EF IJ Simply supported spans
CB DE GF HI Cantilever spans

Fig. 14.2 Typical Balanced Cantilever Bridge.

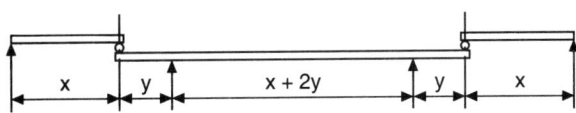

Fig. 14.3 Double Cantilever Bridge.

Normally the deck consists of main girders of tee beam and slab and the spacing's between the beams being arranged depending upon the lanes of traffic.

14.3 DESIGN FEATURES

In the case of balanced cantilever bridges, the depth of the girders are usually of variable cross section with maximum depth at the supports to resist the negative bending moments and with minimum depth at the centre of span and end supports. The width of the carriage way is determined based on the lanes of traffic on the high way and the cross sectional details are finalized by selecting the number of main girders spaced at 1.5 to 2.5 m.

Analysis of moments and shear forces in the girders of variable depth along the span is generally simplified by using influence lines for all critical sections such as the supports and centre of spans. The maximum bending moments at critical sections are obtained by arranging the IRC loads in conjunction with the relevant influence lines. The design ultimate load moments and shear forces are evaluated at the critical sections by applying the appropriate load factors to the service load values as prescribed in IRC: 6-2014[4]. The critical sections are designed by using the limit state method conforming to the specifications of the Indian Roads Congress code IRC: 112-2011.

14.4 SHEAR VARIATION IN VARIABLE DEPTH GIRDERS

In the case of beams of varying depth, the net shear force at the section is calculated by the relation given IS: 456-2000[5] as

$$V_{u.net} = \left(V_u \mp \frac{M_u}{d} \tan \beta \right)$$

Where V_u = shear force at the support

         M_u = bending moment at the section

         d = effective depth at the section

         β = angle between the top and bottom edges of the beam

The negative sign in the formula applies when the bending moment M_u increases numerically in the same direction as the effective depth d increases and the positive sign when the moment decreases numerically in this direction. Fig. 14.4 shows both the cases of computing the net shear force in a beam of variable depth.

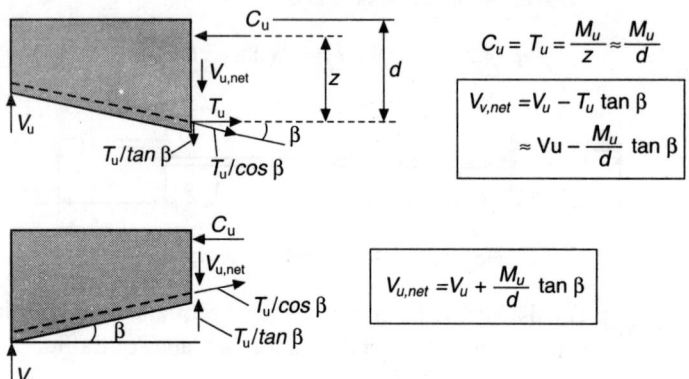

Fig. 14.4 Design Shear Force in Beams of Variable Depth.

14.5 ARTICULATION

The junction of cantilever and simply supported span is referred to as articulation[6] or a hinge which permits rotation and translation avoiding the development of moments. At articulations heavy loads are transmitted resulting in bending moments and tension requiring detailed analysis of forces developed near the articulation. Fig. 14.5 shows the various critical sections 1, 2, 3 & 4 etc inclined at angles θ_1, θ_2, θ_3, & θ_4 to the vertical and the moments, normal thrust and shear developed at these sections are analysed for designing suitable reinforcements at the articulations. The beams are supported on steel or concrete rocker or neoprene pad bearings which permit limited rotations and translations.

14.6 DESIGN EXAMPLE

Design a double cantilever bridge to suit the following data:

Total length of the bridge = 77 m

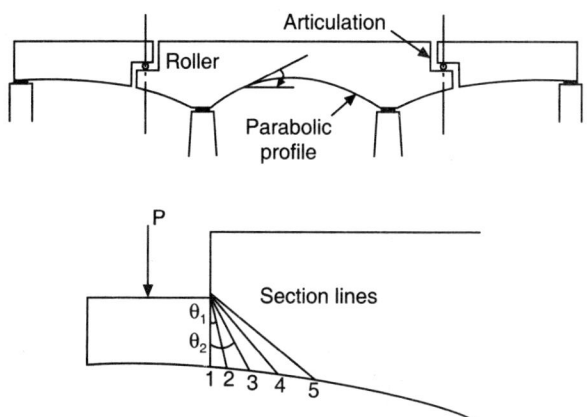

Fig. 14.5 Analysis of Critical Section at Articulation.

Road width = 7.5 m between kerbs
Footpaths = 1.8 m on either side
Spacing of the Tee beams = 1.8 m
Loading: I.R.C. class AA tracked vehicle
Materials: M-25 grade concrete, Fe-415 grade HYSD bars
Design the salient structural elements of the bridge and sketch the details of reinforcements.

1.   Arrangements of Spans

Total length of the bridge deck = 77 m
Central span = 27 m End span = 25 m

$$(x/y) = 4 \text{ and } (x + y) = 25$$
$$y = 5 \text{ m and } x = 20 \text{ m}$$

The general arrangement of spans is shown in Fig. 14.6.
Spacings of beams = 1.8 m
The width of beam = 450 mm
Footpaths = 1.8 m on either side

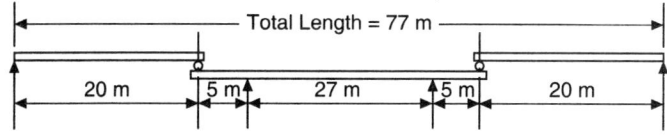

Fig. 14.6 Arrangement of Spans in Double Cantilever Bridge.

The cross-section of the bridge deck showing the general arrangement of Tee beams and foot path is shown in Fig. 14.7.

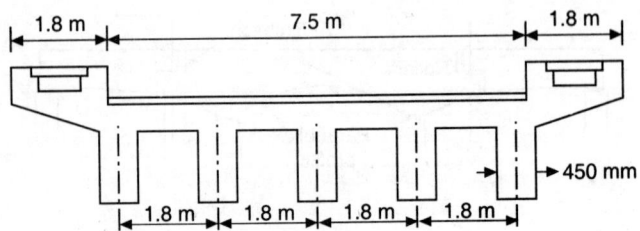

Fig. 14.7 General Arrangement of Tee Beams.

2. Design of Deck Slab[6]

Thickness of deck slab assumed as 200 mm
Thickness of the wearing coat = 80 mm
Width of main girder = 450 mm
Spacing between main girders = 1.8 m
The slab is supported and continuous over Tee beams.
Dead weight of slab = (0.2×24) = 4.80 kN/m^2
Dead weight of W.C. = (0.08×22) = 1.76
Total dead load   = 6.56 kN/m^2
Since the slab is continuous
Maximum D.L.B.M. = $(0.8 \times 6.56 \times 1.8^2)/8 = 2.12$ kN·m

The live load B.M. will be maximum when the I.R.C. Class AA tracked vehicle wheel
is placed at the centre of span as shown in Fig. 14.8.

Effective width of slab, perpendicular to span is expressed as

$$b_e = K \times (1 - x/L) + b_w$$

In this case

$$x = 0.9 \text{ m}$$
$$B = 27 \text{ m} \qquad (B/L) = 15$$
$$L = 1.8 \text{ m}$$
$$b_e = 3.6 + (2 \times 0.08) = 3.76 \text{ m}$$

From Table 4.6 for $(B/L) = 15$
Continuous slab, value of $K = 2.60$

∴ $b_e = 2.6 \times 0.9 \,[1 - (0.9/1.8)] + 3.76 = 4.93$ m

Average intensity of load = $350/(1.01 \times 4.93) = 70.3$ kN/m^2

$M_{max} = [0.5 \,(70.3 \times 1.01 \times 0.9) - 0.5 \,(70.3 \times 1.01) \,(0.25 \times 1.01)] = 23$ kN·m

Live load moment = $(1.25 \times 0.8 \times 23) = 23$ kN·m

Total service load moment = $(M_g + M_q) = (2.12 + 23) = 25.12$ kN·m

Design ultimate moment is computed by applying load factors as

$$M_u = [1.35 \, M_g + 1.5 \, M_q) = [(1.35 \times 2.12) + (1.5 \times 23)] = 37.4 \text{ kN·m}$$

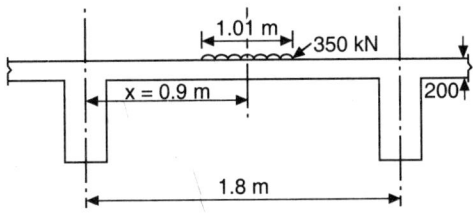

(a) Position of Load for Maximum B.M.

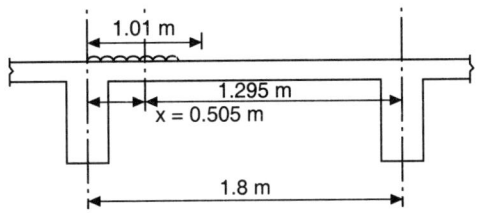

(b) Position of Load for Maximum Shear

Fig. 14.8 Live Load Position on Slab for Maximum B.M. and S.F.

Effective depth of slab required $= d = \sqrt{\dfrac{M_u}{0.138 f_{ck}.b.}} = \sqrt{\dfrac{37.4 \times 10^6}{0.138 \times 25 \times 1000.}} = 104$ mm

Adopt effective depth $= d = 150$ mm and overall depth $= 200$ mm

Using 16 mm diameter bars and effective cover of 50 mm
Effective depth provided $= 150$ mm

$\left(\dfrac{M_u}{b.d^2}\right) = \left(\dfrac{37.4 \times 10^6}{1000 \times 150^2}\right) = 1.66$, using M-25 grade concrete and Fe-415 HYSD bars

Read out the percentage of reinforcement required from Table 3 of SP: 16[7] Design Aids

$$p_t = 0.500 = \left(\dfrac{100 A_{st}}{b.d}\right)$$

Solving $\quad A_{st} = \left(\dfrac{0.500 \times 1000 \times 150}{100}\right) = 750$ mm^2

From shear considerations larger quantity of steel is required.
For short span, Provide 20 mm diameter bars at 150 mm centers (A_{st} provided $= 2094$ mm^2)
For long span, provide 10 mm diameter bars at 150 mm centers.

Maximum shear force in the slab occurs when the load position is as shown in Fig. 14.8(b). For this position of load,

$\quad x = 0.505$ m, $\quad L = 1.8$ m, $\quad B = 27$ m, $\quad (B/L) = 15$, $\quad b_w = 3.76$ m, $\quad K = 2.60$

From Table 4.6

$$b_e = 2.6 \times 0.505 \left(1 - \frac{0.505}{1.8}\right) + 3.76 = 4.7\,\text{m}$$

Average intensity of load = $[(350 \times 1.25)/(1.01 \times 4.70)] = 92.16\,\text{kN/m}^2$
Maximum live load shear = $[(92.16 \times 1.01 \times 1.295)/1.8] = 67\,\text{kN}$
Dead load shear = $(6.56 \times 1.8)/2 = 5.9\,\text{kN}$
Total Design service load shear = $V = (67 + 5.9) = 72.9\,\text{kN}$
Design ultimate load shear = $[(1.35 \times 5.9) + (1.5 \times 67)] = 108.5\,\text{kN}$

The Ultimate Shear Strength of the section is calculated as per specifications of IRC: 112-2011[8] in reinforced concrete members without shear reinforcements, given as

$$V_{\text{Rd.c}} = [0.12K\,(80\rho_1 f_{ck})^{0.33}]b_w.d$$

where

$$K = 1 + \sqrt{\frac{200}{d}} \le 2.00 = \left[1 + \sqrt{\frac{200}{200}}\right] = 2.00 \text{ (taken as 2.0)}$$

and

$$\rho_1 = \left(\frac{A_{sl}}{b_w.d}\right) \le 0.02$$

$$= \left(\frac{2094}{1000 \times 150}\right) = 0.009$$

$$\begin{aligned} V_{\text{Rd.c}} &= [0.12 \times 2.00\,(80 \times 0.01396 \times 25)^{0.33}](1000 \times 150)\,\text{N} \\ &= (109.8 \times 1000)\,\text{N} \\ &= 109.8\,\text{kN} > 108.5\,\text{kN (Hence safe)} \end{aligned}$$

3. Preliminary Design of Longitudinal Girder Sections

Preliminary design of the critical sections is carried out by using the elastic method to determine the depth of the girders. The self weight of the girder strips are determined and used to calculate the dead load moments using the influence lines.

(a) Preliminary Design

Assuming an approximate section for portion A, B, C as shown in Fig. 14.9.

(i) Loads on AB

Weight of slab $(0.2 \times 1.8 \times 24)$	= 8.64 kN
Weight of W.C. $(0.08 \times 1.8 \times 22)$	= 3.17 kN
Self-weight of girder $(1.5 \times 0.45 \times 24)$	= 16.20 kN
Weight of fillets $(150 \times 150 \text{ size})$:	
$(2 \times 0.5 \times 0.15 \times 0.15 \times 24)$	= 0.54 kN
Total load/m	= 28.55 kN
Dead load shear at $A = (28.55 \times 20)/2$	= 285.5 kN
Live load shear at A (Approximate value)	= 350.0 kN
Total shear	= 635.5 kN

Permitting a shear stress $\tau_v = 1$ N/mm^2
(with shear reinforcements)

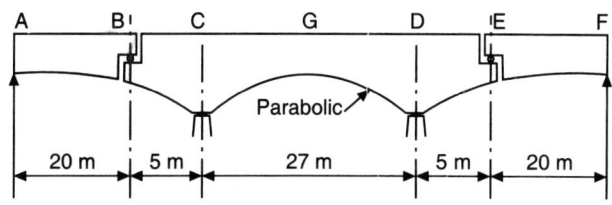

Section at A.C.G. are arrived at by rough
design. The section at A is designed for shear
and section C.G. for B.M.

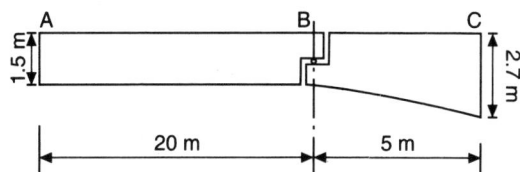

Fig. 14.9 Preliminary Design.

Effective depth $d = (V/b.\tau_v)$
$$= (635.5 \times 10^3)/(450 \times 1) = 1412 \text{ mm}$$

Adopt an overall depth of 1500 mm at A

(ii) Cantilever B.C.

Referring to Fig. 14.10(a)
Reaction at B due to dead weight of girder AB
$$= 285.5 \text{ kN}$$

B.M. at C due to reaction at $B = (285.5 \times 5)$
$$= 1427.5 \text{ kN·m}$$

B.M. due to self weight of cantilever at C
$$= (1.5 \times 0.45 \times 5 \times 5/2) + (1/2 \times 1.2 \times 5 \times 0.45 \times 5/3)24$$
$$= 256.3 \text{ kN·m}$$

B.M. due to weight of slab, W.C. and fillets over cantilever B.C.
$$= 0.5 (8.64 + 3.17 + 0.54) 5^2 = 154.4 \text{ kN·m}$$

Total dead load B.M. at $C = 2123.7$ kN·m
Referring to Fig. 14.10(b)
Live load B.M. at C is given by

$$0.5 (4.3 + 5) (3 + 0.6) (350/3.6) = 1627.5 \text{ kN·m}$$

Total design B.M. at $C = 3751.2$ kN·m
Using M-25 Grade Concrete and Fe-415 HYSD bars, $M = 1.10 \, bd^2$

$$\therefore \qquad d = \sqrt{(3751.2 \times 10^6)/(1.10 \times 450)} = 2752 \text{ mm}$$

Since section C will be designed as a doubly reinforced section an overall depth of 2.7 m can be adopted.

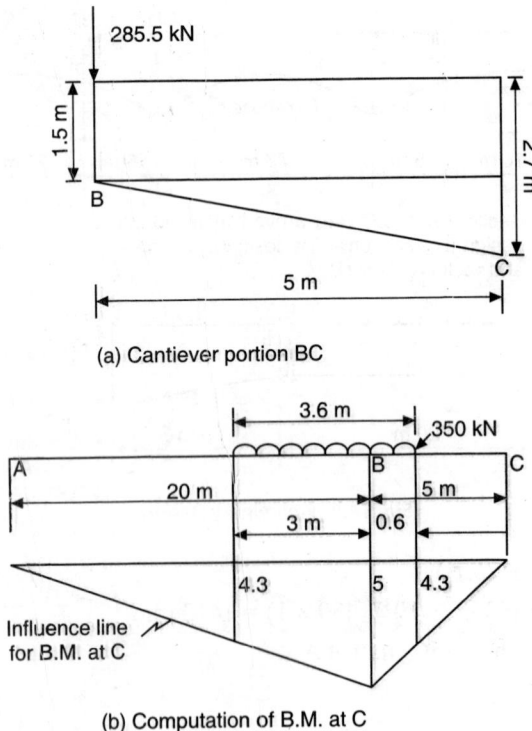

(a) Cantilever portion BC

(b) Computation of B.M. at C

Fig. 14.10 Preliminary Design of Section at C.

(iii) **Section for Mid Span at G**

The preliminary section assumed at centre of span G is shown in Fig. 14.11.
Loads on Span CGD
Dead load due to deck, W.C., and fillets

$$= (8.64 + 3.17 + 0.54) = 12.35 \text{ kN/m}$$

Self-weight of girder

$$= 0.5 \,(2.7 + 2.1) \,(0.45 \times 24) = 25.92$$

Total load $= 38.27$ kN/m

+ ve moment due to D.L. $= (1/2 \times 27 \times 6.75 \times 38.27) = 3487$ kN·m
– ve moment due to D.L. $= (1/2 \times 5 \times 2.5 \times 2 \times 38.27)$
$$= -478 \text{ kN·m}$$

– ve moment due to cantilever loads
$$= (285.5 \times 2.5 \times 2) = 1427 \text{ kN·m}$$

Net + ve B.M. due to dead load at $G = 1582$ kN·m
Maximum + ve B.M. due to live load at G

$$= 2 \times 1/2 \, (5.8 + 6.75) \, 1.8 \, (350/3.6) = 2196 \text{ kN·m}$$

Total design B.M. at $G = 3778$ kN·m
for rectangular beam action at centre

Approximate depth $d = \sqrt{(3778 \times 10^6)/(1.10 \times 450)} = 2762$ mm

Since Tee beam action will be prevalent, an over all depth of 2100 mm will be sufficient.

(iv) The sections chosen at different parts such as cantilevers, mid span and supports is detailed below:

Referring to Fig. 14.12.
Equation of curve AC' (origin at C') is

$$y = (4r/L^2) \times (L - x)$$
$$= (4) \, (1.2) \times (50 - x)/50^2$$
$$= 0.00192 \times (50 - x)$$

The ordinates of the girder at sections 1, 2, 3, 4, 5, 6, 7, 8 are tabulated in Table 14.1.

Table 14.1 Ordinates of Girder AC.

Point	1	2	3	4	5	6	7	8
x (m)	1	2.33	3.66	5.0	10.0	15.00	20.0	25
Depth of girder (m)	2.6	2.5	2.4	2.3	1.95	1.70	1.56	1.5

Equation of curve for central span CD is given by

$$y = [(4 \times 0.6) \times (27 - x)]/27^2 = 0.0032 \times (27 - x)$$

The ordinates of the girder for half the span CG is shown in Table 14.2.

Table 14.2 Ordinates of Girder CG.

Point	1'	2'	3'	4'	5'
x (m)	2.70	5.40	8.10	10.8	13.50
Depth of girder (m)	2.50	2.30	2.20	2.18	2.10

The self weight of the various girder strips shown in Fig. 14.2, is compiled in Table 14.3.

(b) *The bending moments and shear forces at various sections are computed using the influence line diagrams and tables as detailed below:*

Refer Tables 14.4 to 14.19, and Figs. 14.13 to 14.28.

4. Influence Lines[9] for Bending Moments and Shear Forces

The influence lines for bending moment and shear force at various cross sections computed in Tables 14.4 to 14.19 are shown Figs. 14.13 to 14.28.

Table 14.3 Self-weight of Girder Strips

R.H. Strip	L.H. Ordinate (m)	Mean Ordinate (m)	Depth of Rib Ordinate (m)	Breadth (Mean 0.2 m) (m)	Length of Rib (m)	Weight of Strip (m)	Weight of Rib W(kN)	Total deck Slab W_1(kN)	Dist of Weight $(W + W_1)$ kN	C.G. Strip from C (m)
C–1	2.70	2.61	2.65	2.45	0.45	1.00	26.46	16.35	42.81	0.5
1–2	2.61	2.50	2.55	2.35	0.45	1.33	33.75	21.74	55.49	1.665
2–3	2.50	2.40	2.45	2.25	0.45	1.33	32.31	21.74	54.05	3.000
3–4	2.40	2.30	2.35	2.15	0.45	1.33	30.88	21.74	52.62	4.330
4–5	2.30	1.95	2.12	1.92	0.45	5.00	103.68	81.75	185.43	7.50
5–6	1.95	1.70	1.82	1.62	0.45	5.00	87.48	81.75	169.23	12.50
6–7	1.70	1.56	1.63	1.43	0.45	5.00	77.22	81.75	158.97	17.50
7–8	1.56	1.50	1.53	1.33	0.45	5.00	71.82	81.75	153.57	22.50
C–1'	2.50	2.70	2.60	2.40	0.45	2.70	70.00	44.14	114.14	1.35
1'–2'	2.50	2.30	2.40	2.20	0.45	2.70	64.15	44.14	108.29	4.05
2'–3'	2.30	2.20	2.25	2.05	0.45	2.70	59.77	44.14	103.90	6.75
3'–4'	2.20	2.18	2.19	1.99	0.45	2.70	58.02	44.14	102.16	9.45
4'–5'	2.18	2.10	2.14	1.94	0.45	2.70	56.57	44.14	100.70	12.15

Table 14.4 Bending Moment for Point—1'

	(Refer Fig. 14.11)				
Load-Point	Maximum Ordinate	x or x'	Ordinate at Point	Load (kN)	B.M. (kN.m)
+ ve B.M.					
L.H.S. of L.L		2.5	2.1 & 2.45		
R.H.S. of L.L	2.45	20.7	2.1 & 2.45	385	876
C–1'	—	1.35	1.22	114.14	139
1'–2'	—	22.95	2.30	108.29	250
2'–3'	—	20.25	2.04	103.90	212
3'–4'	—	17.55	1.76	102.16	181
4'–5'	—	14.85	1.49	100.70	150
5'–6'	—	12.15	1.22	100.70	124
6'–7'	—	9.45	0.94	102.16	97
7'–8'	—	6.75	0.67	103.90	70
8'–9'	—	4.05	0.40	108.29	44
9'–D	—	1.35	0.13	114.14	15
Total + ve B.M. due to D.L. alone					1282
–ve B.M.					
Reaction at B	4.60	—	(4.6 + 0.51)	347	1773
Reaction at E	0.51		= 5.11		
$(3-4)_L + (3-4)_R$	—	4.33	4.33	52.62	228
$(2-3)_L + (2-3)_R$	—	3.00	3.00	54.05	162
$(1-2)_L + (1-2)_R$	—	1.67	1.67	55.49	93
$(C-1)_L + (C-1)_R$	—	0.50	0.50	42.81	22
Total – ve B.M. due to D.L. alone					2278
–ve B.M.					
L.H.S. of L.L.	4.6	16.97	4.6	385	1598
			and		
R.H.S. of L.L.	4.6	4.25	3.7		

Table 14.5 Bending Moment for Point—2′

Load-Point	Maximum Ordinate	x or x′	Ordinate at Point	Load (kN)	B.M. (kN.m)
+ ve B.M.		4.5	3.65		
L.H.S. of L.L	4.32		and	385	1525
R.H.S. of L.L		18.73	4.32		
C–1′	—	1.35	1.09	114.14	125
1′–2′	—	4.05	3.27	108.29	354
2′–3′	—	20.25	4.09	103.90	425
3′–4′	—	17.55	3.50	102.16	358
4′–5′	—	14.85	3.00	100.70	302
5′–6′	—	12.15	2.45	100.70	247
6′–7′	—	9.45	1.90	102.16	194
7′–8′	—	6.75	1.36	103.90	141
8′–9′	—	4.05	0.81	108.29	88
9′–D	—	1.35	0.27	114.14	31
Total + ve B.M. due to D.L. alone					2265
–ve B.M.					
Reaction at B	4.15	—	(4.15 + 1.03)	347	1797
Reaction at E	1.03		= 5.18		
$(3-4)_L + (3-4)_R$	—	4.33	4.33	52.62	228
$(2-3)_L + (2-3)_R$	—	3.00	3.00	54.05	162
$(1-2)_L + (1-2)_R$	—	1.67	1.67	55.49	93
$(C-1)_L + (C-1)_R$	—	0.50	0.50	42.81	21
Total – ve B.M. due to D.L. alone					2301
– ve B.M.		16.8	3.5		
L.H.S. of L.L.	4.15		and	385	1473
R.H.S. of L.L.		4.5	4.15		

(Refer Fig. 14.12)

Table 14.6 Bending Moment for Point—3'

Load-Point	Maximum Ordinate	x or x'	Ordinate at Point	Load (kN)	B.M. (kN.m)
(Refer Fig. 14.13)					
+ ve B.M.		4.5	3.65		
L.H.S. of L.L	4.32		and	385	1525
R.H.S. of L.L		18.73	4.32		
C–1'	—	1.35	0.95	114.14	109
1'–2'	—	4.05	2.84	108.29	308
2'–3'	—	6.75	4.75	103.90	494
3'–4'	—	17.55	5.28	1.02.16	540
4'–5'	—	14.85	4.48	100.70	452
5'–6'	—	12.15	3.66	100.70	369
6'–7'	—	9.45	2.84	102.16	291
7'–3'	—	6.75	1.95	103.90	203
8'–9'	—	4.05	1.17	108.29	127
9'–D	—	1.35	0.39	114.14	45
Total + ve B.N. due to D.L. alone					2938
–ve B.M.					
Reaction at B	3.70	—	(3.7 + 1.60)	347	1839
Reaction at E	1.60		= 5.3		
$(3-4)_L + (3-4)_R$	—	4.33	4.33	52.62	228
$(2-3)_L + (2-3)_R$	—	3.00	3.00	54.05	162
$(1-2)_L + (1-2)_R$	—	1.67	1.67	55.49	93
$(C-1)_L + (C-1)_R$	—	0.50	0.50	42.81	21
Total – ve B.N. due to D.L. alone					2938
–ve B.M.		8.1	3.70		
L.H.S. of L.L	3.70		and	385	1328
R.H.S. of L.L.		4.5	3.20		

Table 14.7 Bending Moment for Point—4′

Load-Point	Maximum Ordinate	x or x′	Ordinate at Point	Load (kN)	B.M. (kN.m)
(Refer Fig. 14.14)					
+ ve B.M.		9.00	6.5		
L.H.S. of L.L	6.5		and	385	1525
R.H.S. of L.L		14.24	5.6		
C–1′	—	1.35	0.81	114.14	93
1′–2′	—	4.05	2.45	108.29	266
2′–3′	—	6.75	4.09	103.90	425
3′–4′	—	9.45	5.70	102.16	585
4′–5′	—	14.85	6.00	100.70	604
5′–6′	—	12.15	4.90	100.70	494
6′–7′	—	9.45	3.80	102.16	390
7′–8′	—	6.75	2.72	103.90	283
8′–9′	—	4.05	1.63	108.29	176
9′–D	—	1.35	0.54	114.14	62
Total + ve B.N. due to D.L. alone					3378
–ve B.M.					
Reaction at B	3.1	—	(3.1 + 2.1)	347	1804
Reaction at E	2.1		= 5.2		
$(3-4)_L + (3-4)_R$	—	4.33	4.33	52.62	228
$(2-3)_L + (2-3)_R$	—	3.00	3.00	54.05	162
$(1-2)_L + (1-2)_R$	—	1.67	1.67	55.49	93
$(C-1)_L + (C-1)_R$	—	0.50	0.50	42.81	21
Total – ve B.N. due to D.L. alone					3378
–ve B.M.			2.5		
L.H.S. of L.L.	3.1	—	and	385	1078
R.H.S. of L.L.			3.1		

Table 14.8 Bending Moment for Point—5′

Load-Point	Maximum Ordinate	x or x'	Ordinate at Point	Load (kN)	B.M. (kN.m)
+ ve B.M. L.H.S. of L.L R.H.S. of L.L	6.75	13.50	5.80 and 6.75	385	2416
C–1′	—	1.35	0.68	114.14	78
1′–2′	—	4.05	2.04	108.29	222
2′–3′	—	6.75	3.40	103.90	354
3′–4′	—	9.45	4.70	102.16	488
4′–5′	—	12.15	6.13	100.70	618
5′–6′	—	12.15	6.13	100.70	618
6′–7′	—	9.45	4.70	102.16	488
7′–8′	—	6.75	3.40	103.90	354
8′–9′	—	4.05	2.04	108.29	222
9′–D	—	1.35	0.68	114.14	78
Total + ve B.N. due to D.L. alone				3520	
–ve B.M. Reaction at B	2.57	—	(2.57 + 2.57) = 5.14	347	1784
Reaction at E	2.57				
$(3-4)_L + (3-4)_R$	—	4.33	4.33	52.62	228
$(2-3)_L + (2-3)_R$	—	3.00	3.00	54.05	162
$(1-2)_L + (1-2)_R$	—	1.67	1.67	55.49	93
$(C-1)_L + (C-1)_R$	—	0.50	0.50	42.81	21
Total – ve B.N. due to D.L. alone				2288	
–ve B.M. L.H.S. of L.L. R.H.S. of L.L.	2.57		2.1 and 2.57	385	899

(Refer Fig. 14.15)

Table 14.9 Bending Moment Support—C

Load-Point	Maximum Ordinate	x or x'	Ordinate at Point	Load (kN)	B.M. (kN.m)
– ve B.M. Reaction at B	5	—	5	347	1735
$(3-4)_L$	—	4.33	4.33	52.62	228
$(2-3)_L$	—	3.00	3.00	54.05	162
$(1-2)_L$	—	1.67	1.67	55.49	93
$(C-1)_L$	—	0.50	0.50	42.81	22
Total – ve B.N. due to D.L. alone				2240	
–ve B.M. L.H.S. of L.L. R.H.S. of L.L.	5	—	4.3 and 5.0	385	1790
Total – ve B.N. due to D.L. alone					**1790**

(Refer Fig. 14.16)

Table 14.10 Bending Moments for Points—1, 2 and 3

(Refer Fig. 14.17)					
(a) B.M. for Point – 1					
Load-Point	Maximum Ordinate	*x* or *x'*	Ordinate at Point	Load (kN)	B.M. (kN.m)
– ve B.M. Reaction at B	4	—	4	347	1388
1 – 2	—	1.67	0.6	55.49	33
2 – 3	—	3.00	2.0	54.05	108
3 – 4	—	4.33	3.25	52.62	171
Total B.M. (– ve) due to D.L. alone					1700
– ve B.M. L.H.S. of L.L. R.H.S. of L.L.	—	—	4 and 0.33	385	836
(b) B.M. for Point – 2					
– ve B.M. Reaction at B	2.66	—	2.66	347	923
2 – 3	—	3.00	0.80	54.05	43
3 – 4	—	4.33	2.00	52.62	105
Total – ve B.M. due to D.L. alone					1071
– ve B.M. L.H.S. of L.L. R.H.S. of L.L.	—	—	2.6 and 2.2	385	924
(c) B.M. for Point – 3					
– ve B.M. Reaction at B	1.33	—	1.33	347	462
3 – 4	—	4.33	0.70	52.62	37
Total (– ve) B.M. due to D.L. alone					499
–ve B.M. L.H.S. of L.L. R.H.S. of L.L.	—	—	1.33 and 1.1	385	468

Table 14.11 Shear Correction Table (refer Figure 14.18 and 14.19)

	Properties of Section			Corrected I.L. ordinate at point				Corrected I.L.O. at B				Corrected I.L.O. at E			
Section	Depth (m)	Effective Depth h (m)	tan α	Ordinary Shear Ordinate	Moment M at point when unit load placed at point	(M/h) tan α	Corrected Shear Ordinate	Ordinary Shear Ordinate	Moment at point when unit load placed at B	(M/h) tan α	Correct Shear Ordinate	Ordinary Shear Ordinate	Moment at point when unit load placed at E	(M/h) tan α	Corrected Shear Ordinate
B	2.3	2.15	−0.076	−1.0	—	—	−1.0	−1.0	—	—	—	—	—	—	—
3	2.4	2.25	−0.080	−1.0	—	—	−1.0	−1.0	−1.33	+0.047	−0.953	—	—	—	—
2	2.5	2.35	−0.085	−1.0	—	—	−1.0	−1.0	−2.67	+0.096	−0.903	—	—	—	—
1	2.61	2.46	−0.090	−1.0	—	—	−1.0	−1.0	−4.00	+0.146	−0.853	—	—	—	—
C_L	2.70	2.55	−0.093	−1.0	—	—	−1.0	−1.0	−5.00	+0.182	−0.817	—	—	—	—
C_R	2.70	2.55	+0.089	+1.0	—	—	+1.0	+0.19	−5.00	−0.174	+0.016	−0.19	—	—	−0.19
1'	2.50	2.35	+0.072	+0.9	2.43	+0.074	+0.974	+0.19	−4.51	−0.138	+0.052	−0.19	0.5	−0.015	−0.205
2'	2.30	2.15	+0.054	+0.8	4.32	+0.108	+0.908	+0.19	−4.00	−0.100	+0.09	−0.19	1.0	−0.025	−0.215
3'	2.20	2.05	+0.036	+0.7	5.67	+0.099	+0.799	+0.19	−3.51	−0.062	+0.128	−0.19	1.5	−0.026	−0.216
4'	2.18	2.03	+0.008	+0.6	6.48	+0.025	+0.625	+0.19	−3.00	−0.012	+0.178	−0.19	2.0	−0.008	−0.198
5'	2.10	1.95	+0.000	+0.5	6.75	—	+0.500	+0.19	−2.50	—	0.190	−0.19	2.5	—	−0.190

Table 14.12 Shear at Point—C_R

(Refer Fig. 14.20)						
Load	Point	Maximum Ordinate	x or x' (m)	Ordinate at Point	Load (kN)	Shear (kN)
+ ve L.L.		1.0	25.12	0.93	385	358
+ ve D.L.	C – 1'	—	25.65	0.95	114.14	109
	1 – 2'	—	22.95	0.85	108.29	92
	2' – 3'	—	20.25	0.75	103.90	78
	3' – 4'	—	17.55	0.65	102.16	67
	4' – 5'	—	14.85	0.55	100.70	56
	5' – 6'	—	12.15	0.45	100.70	45
	6' – 7'	—	9.45	0.35	102.16	36
	7' – 8'	—	6.75	0.25	103.90	26
	8' – 9'	—	4.05	0.15	108.29	16
	9' – D	—	1.35	0.05	114.14	6
+ ve Reaction at B		+0.016	5.00	+0.016	347	6
$(3 – 4)_L$		—	4.33	+0.014	52.62	0.74
$(2 – 3)_L$		—	3.00	+0.009	54.05	0.48
$(1 – 2)_L$		—	1.67	+0.005	55.49	0.28
$(C – 1)_L$		—	0.50	+0.001	42.81	0.04
Total + ve shear due to D.L. alone						539
Reaction at E		–0.19	—	–0.19	347	–66
$(3 – 4)_R$		—	4.33	–0.16	52.62	–8.4
$(2 – 3)_R$		—	3.00	–0.11	54.05	–5.9
$(1 – 2)_R$		—	1.67	–0.06	55.49	–3.5
$(C – D)_R$		—	0.50	–0.019	42.81	–0.8
Total + ve shear due to D.L. alone						–84
Design shear for CR						813

Table 14.13 Shear at Point—1'

Load	Point	Maximum Ordinate	x or x' (m)	Ordinate at Point	Load (kN)	Shear (kN)
+ ve L.L.		0.974	22.42	0.898	385	346
+ ve D.L.	1' – 2'	—	22.95	0.920	108.29	100
	2' – 3'	—	20.25	0.810	103.90	84
	3' – 4'	—	17.55	0.703	102.16	72
	4' – 5'	—	14.85	0.595	100.70	60
	5' – 6'	—	12.15	0.486	100.70	49
	6' – 7'	—	9.45	0.378	102.16	39
	7' – 8'	—	6.75	0.270	103.90	28
	8' – 9'	—	4.05	0.162	108.29	18
	9' – D	—	1.35	0.054	114.14	6
Reaction B		0.052	5.00	0.052	347.0	18
Total + ve shear due to D.L. alone						474
– ve Reaction at E		0.205	—	–0.205	347	–71
C – 1'		0.026	1.35	–0.013	114.14	–1.48
$(3 - 4)_L + (3 - 4)_R$		—	4.33	–0.135	52.62	–7.10
$(2 - 3)_L + (2 - 3)_R$		—	3.00	–0.093	54.05	–5.00
$(1 - 2)_L + (1 - 2)_R$		—	1.67	–0.049	55.49	–2.71
$(C - 1)_L + (C - 1)_R$		—	0.50	–0.0139	42.81	–0.59
Total + ve shear due to D.L. alone						–84
Design shear for Point—1'						733

Table 14.14 Shear at Point—2′

(Refer Fig. 14.22)						
Load	Point	Maximum Ordinate	x or x′ (m)	Ordinate at Point	Load (kN)	Shear (kN)
+ ve L.L.		0.908	19.72	0.83	385	320
+ ve D.L.	2′ – 3′	—	20.25	0.852	103.90	89
	3′ – 4′	—	17.55	0.74	102.16	76
	4′ – 5′	—	14.85	0.625	100.70	63
	5′ – 6′	—	12.15	0.511	100.70	52
	6′ – 7′	—	9.45	0.398	102.16	41
	7′ – 8′	—	6.75	0.284	103.90	30
	8′ – 9′	—	4.05	0.170	108.29	18
	9′ – D	—	1.35	0.054	114.14	7
Reaction at B			5.00	0.090	347	31
Total + ve shear due to D.L. alone						407
– ve Reaction at E		0.215	—	–0.205	347	–74.60
C – 1′		—	1.35	–0.023	114.14	–2.62
1′ – 2′		—	4.05	–0.069	108.29	–7.47
$(3-4)_L + (3-4)_R$		—	4.33	–0.109	52.62	–5.73
$(2-3)_L + (2-3)_R$		—	3.00	–0.074	54.05	–4.00
$(1-2)_L + (1-2)_R$		—	1.67	–0.039	55.49	–2.16
$(C-1)_L + (C-1)_R$		—	0.50	–0.0112	42.81	–0.48
Total – ve shear due to D.L.						–97
Design shear for Point—2′						733

Table 14.15 Shear at Point—3′

Load	Point	Maximum Ordinate	x or x′ (m)	Ordinate at Point	Load (kN)	Shear (kN)
+ ve L.L.		0.799	17.02	0.72	385	277
+ D.L.	3′ – 4′	—	17.55	0.742	102.16	76
	4′ – 5′	—	14.85	0.628	100.70	63
	5′ – 6′	—	12.15	0.514	100.70	52
	6′ – 7′	—	9.45	0.400	102.16	41
	7′ – 8′	—	6.75	0.286	103.90	30
	8′ – 9′	—	4.05	0.171	108.29	19
	9′ – D	—	1.35	0.054	114.14	7
Reaction at B		—	5.00	0.128	347	45
Total + ve shear due to D.L. alone						333
– ve Reaction at E		0.216	—	–0.216	347	–75.00
C – 1′		—	1.35	–0.033	114.14	–3.76
1′ – 2′		—	4.05	–0.100	108.29	–10.82
2′ – 3′		—	6.75	–0.168	103.90	–17.40
$(3-4)_L + (3-4)_R$		—	4.33	–0.077	52.62	–4.05
$(2-3)_L + (2-3)_R$		—	3.00	–0.52	54.05	–2.81
$(1-2)_L + (1-2)_R$		—	1.67	–1.55	55.49	–1.55
$(C-1)_L + (C-1)_R$		—	0.50	–0.0078	42.81	–0.33
Total – ve shear due to D.L. alone						–115
Design shear for Point—3′						495

(Refer Fig. 14.23)

Table 14.16 Shear at Point—4′

Load	Point	Maximum Ordinate	x or x′ (m)	Ordinate at Point	Load (kN)	Shear (kN)
		(Refer Fig. 14.24)				
+ ve L.L.		0.625	14.32	0.553	385	213
+ ve D.L.	4′ – 5′	—	14.85	0.575	100.70	58
	5′ – 6′	—	12.15	0.470	100.70	48
	6′ – 7′	—	9.45	0.36	102.16	37
	7′ – 8′	—	6.75	0.26	103.90	27
	8′ – 9′	—	4.05	0.15	108.29	17
	9′ – D′	—	1.35	0.052	114.14	6
Reaction at B			5.00	0.178	347	62
Total + ve shear due to D.L. alone						255
–ve Reaction at E		0.198	—	–0.198	347	–69
C – 1′		—	1.35	–0.046	114.14	–5.25
1′ – 2′		—	4.05	–0.140	108.29	–15.16
2′ – 3′		—	6.75	–0.234	103.90	–24.30
3′ – 4′		—	9.45	–0.330	102.16	–33.71
$(3 - 4)_L + (3 - 4)_R$		—	4.33	–0.017	52.62	–0.89
$(2 - 3)_L + (2 - 3)_R$		—	3.00	–0.012	54.05	–0.64
$(1 - 2)_L + (1 - 2)_R$		—	1.67	–0.006	55.49	–0.33
$(C - 1)_L + (C - 1)_R$		—	0.50	–0.0018	42.81	–0.0
Total – ve shear due to D.L. alone						–149
Design shear for Point—4′						319

Table 14.17 Shear at C_L, 1, 2, 3 and 4

Load-Point	Maximum Ordinate	x or x' (m)	Ordinate at Point	Load (kN)	Shear (kN)
(Refer Fig. 14.25)					
(a) Shear at C_L					
L.L.	1.0	23.12	0.927	385	357
C – 1	1.0	24.50	0.982	42.81	42
1 – 2	1.0	23.33	0.936	55.49	52
2 – 3	1.0	22.00	0.884	54.05	48
3 – 4	1.0	20.67	0.827	52.62	44
Reaction at B	0.817	20.00	0.817	347	284
Total – ve shear at C_L					827
(b) Shear at – 1					
L.L.	1.0	22.12	0.923	385	355
1 – 2	—	23.33	0.970	55.49	54
2 – 3	—	22.00	0.914	54.05	49
3 – 4	—	20.67	0.856	52.62	45
Reaction at B	0.853	20.00	0.853	347	296
Total – ve shear at Point—1					799
(c) Shear at – 2					
L.L.	1.0	20.78	0.921	385	355
2 – 3	—	22.00	0.970	54.05	52
3 – 4	—	20.67	0.910	52.62	48
Reaction at B	—	20.00	0.903	347	313
Total – ve shear at Point—2					**768**
(d) Shear at – 3					
L.L.	1.0	19.45	0.914	385	352
3 – 4	—	20.67	0.967	52.62	51
Reaction at B	—	20.00	0.953	347	331
Total – ve shear at Point—3					734
(e) Shear at – 4					
L.L.	—	18.12	0.907	385	349
Reaction at B	—	20.00	1.00	347	347
Total – ve shear at Point—4 (ARTICULATION)					696

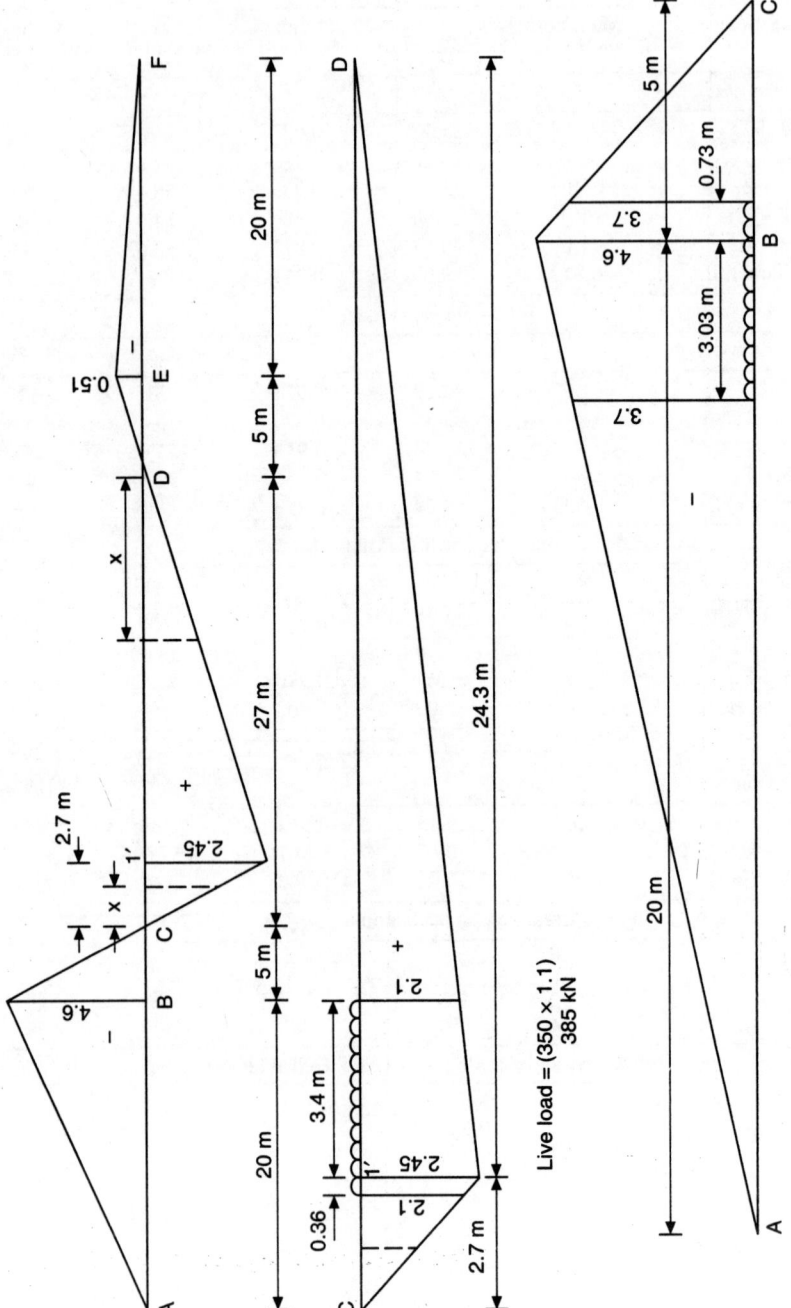

Live load = (350 × 1.1)
385 kN

Fig. 14.13 Influence Line for Bending Moment at Point 1'.

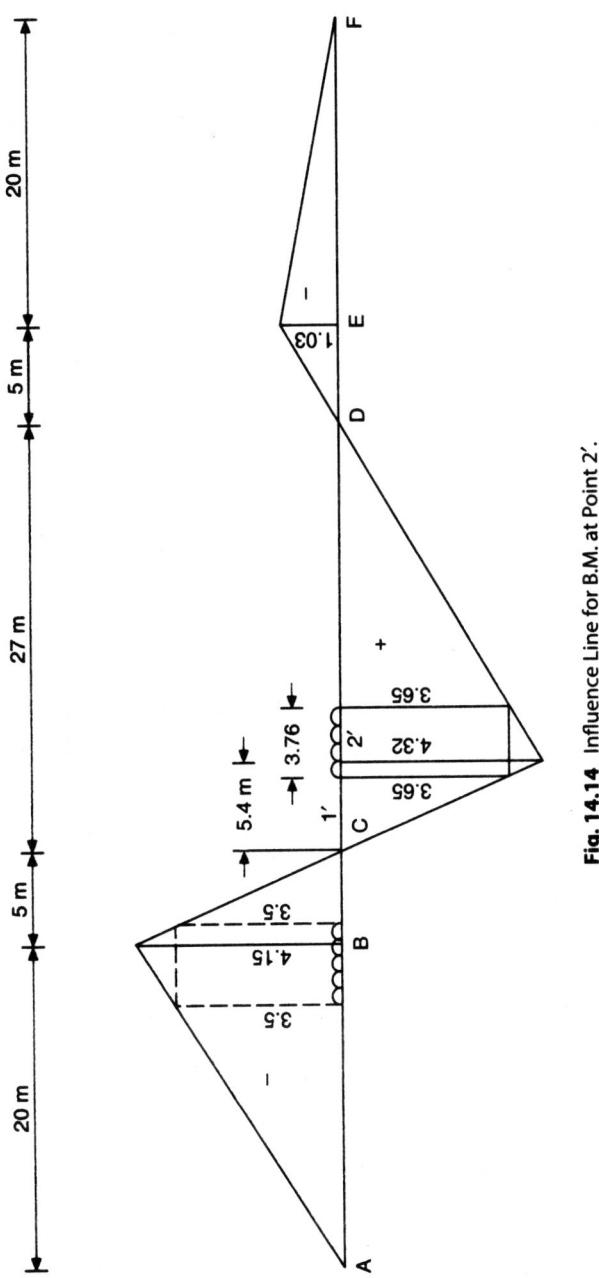

Fig. 14.14 Influence Line for B.M. at Point 2'.

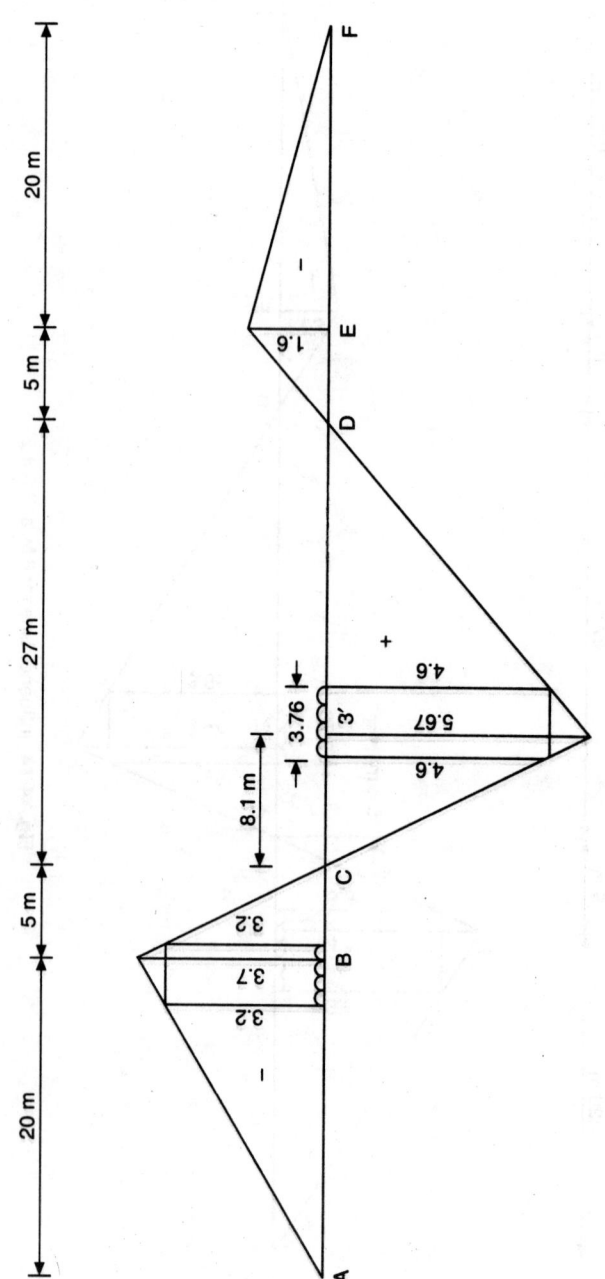

Fig. 14.15 Influence Line for B.M. at Point 3'.

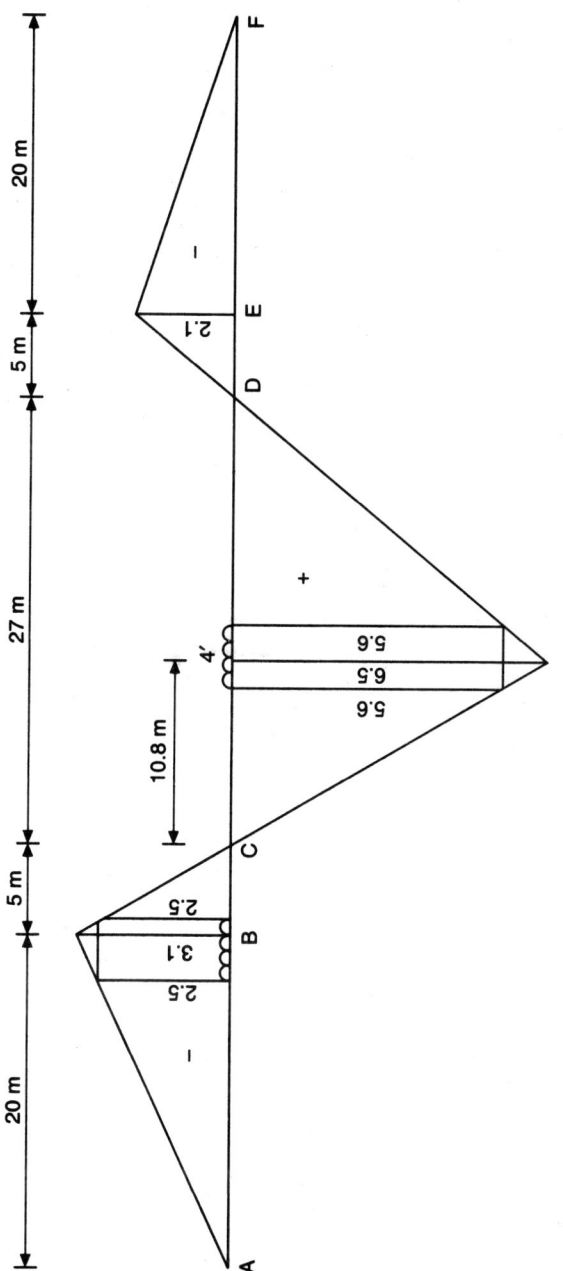

Fig. 14.16 Influence Line for B.M. at Point 4′.

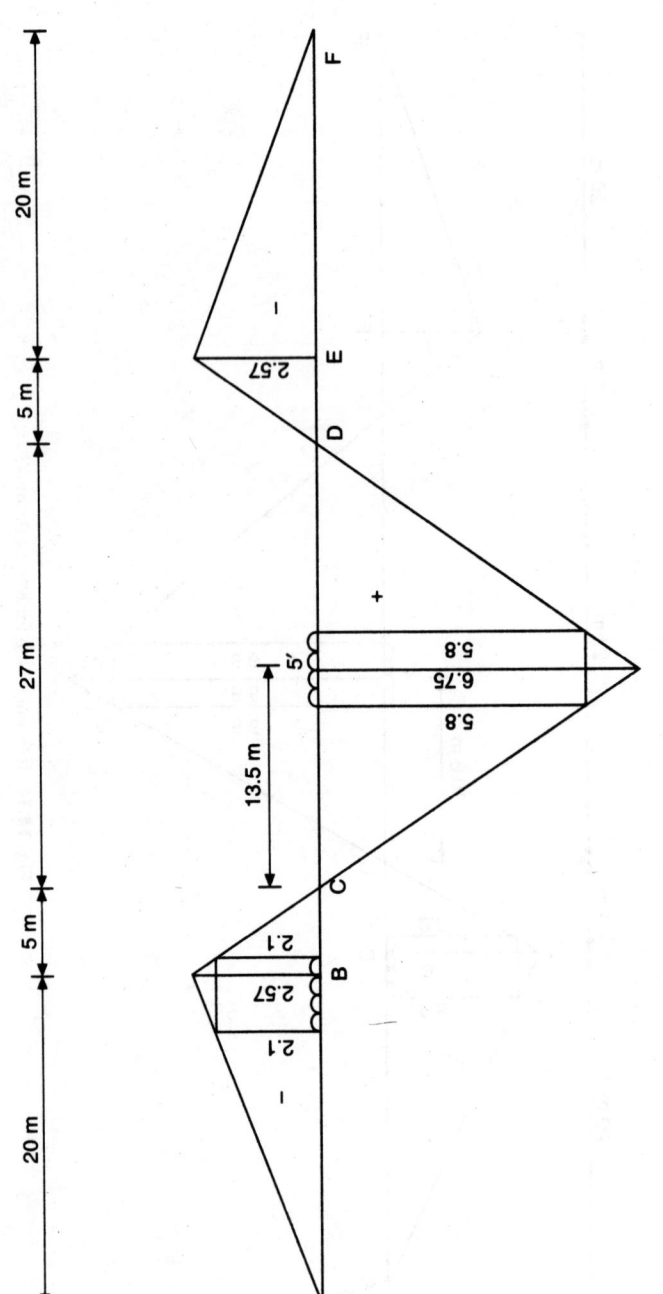

Fig. 14.17 Influence Line for B.M. at Point 5′.

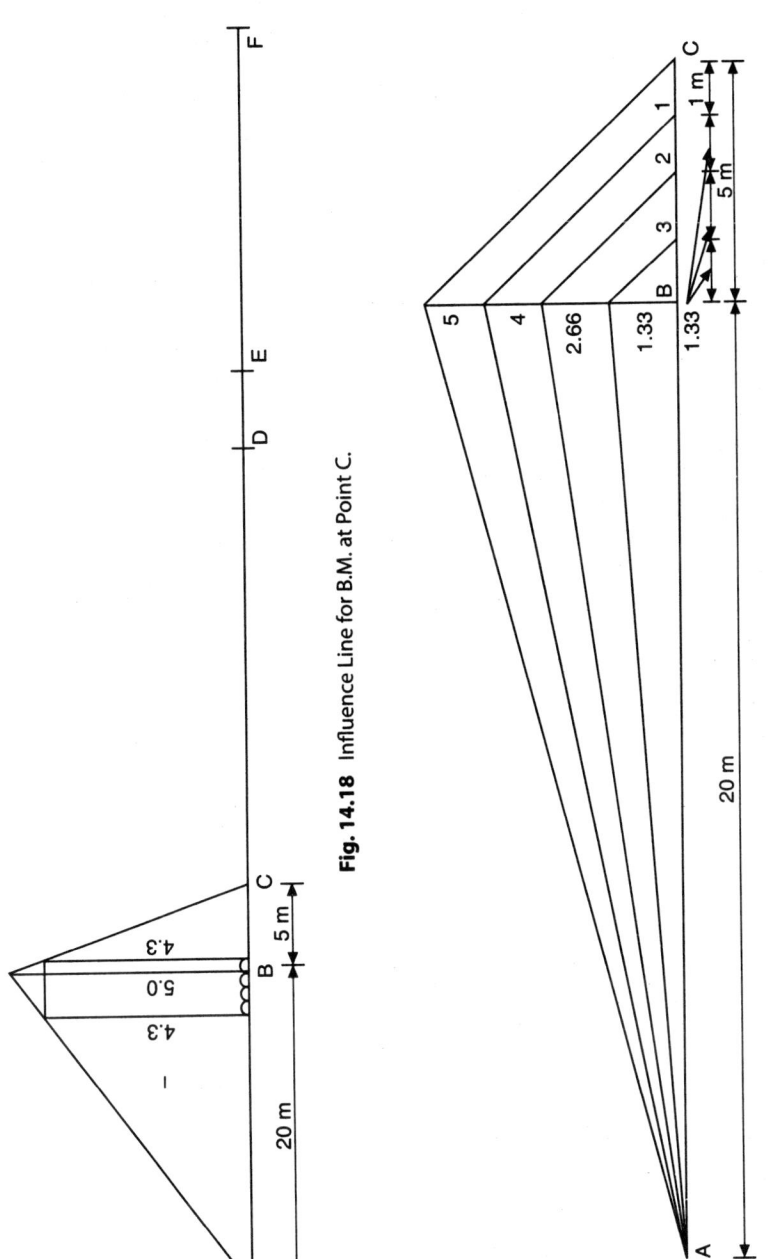

Fig. 14.18 Influence Line for B.M. at Point C.

Fig. 14.19 Influence Line for B.M. at Point 1, 2 and 3.

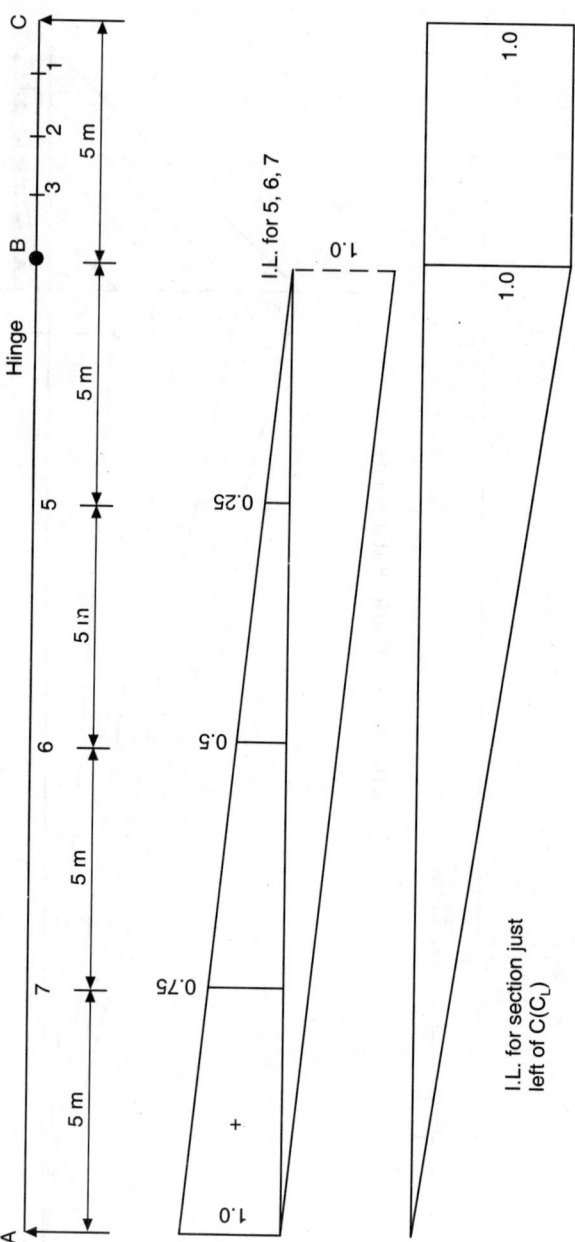

Fig. 14.20 Shear Influence Line (Span AC).

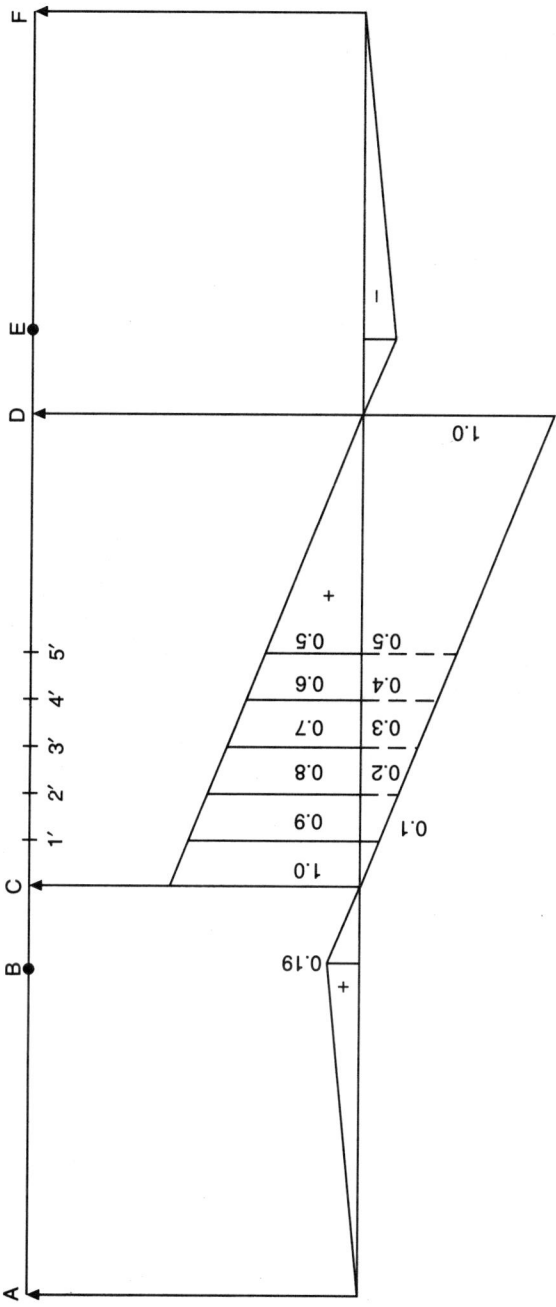

Fig. 14.21 Influence Line for Point Just Right of $C(C_R)$.

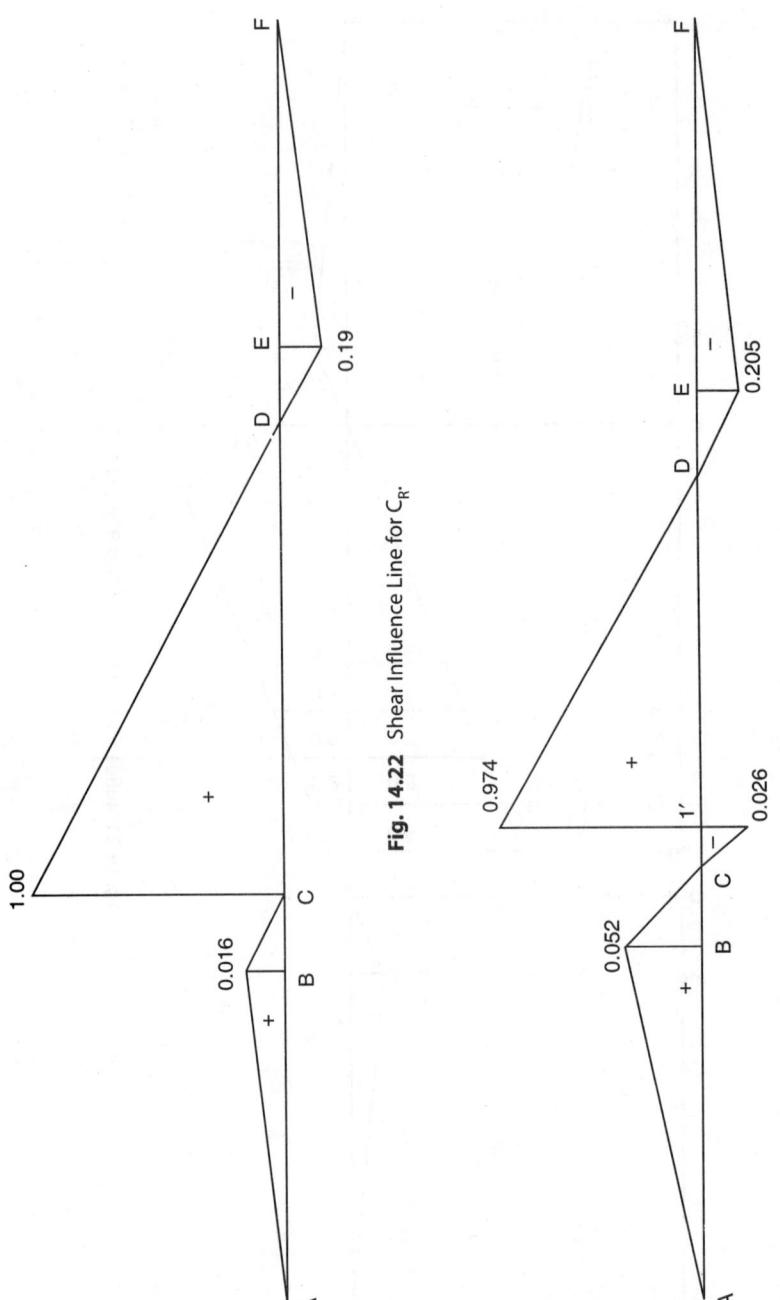

Fig. 14.22 Shear Influence Line for $C_{R'}$.

Fig. 14.23 Shear Influence Line for Point 1′.

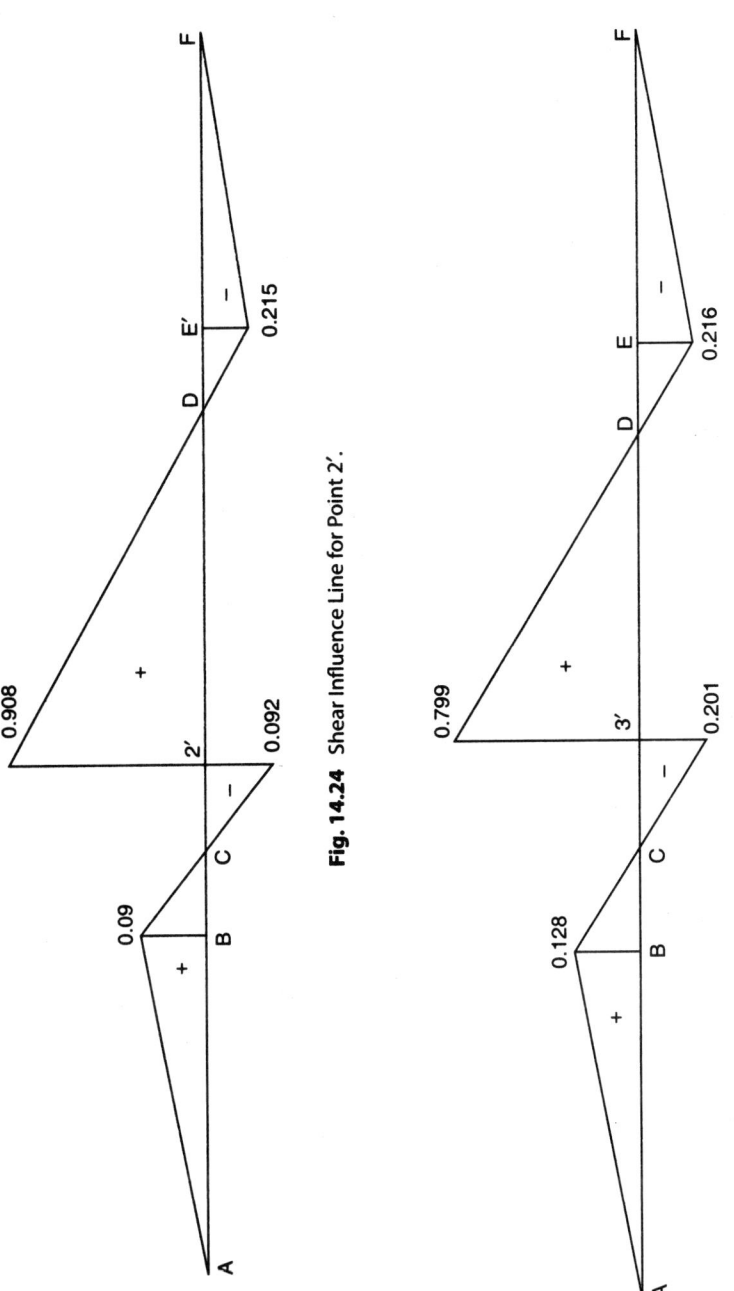

Fig. 14.24 Shear Influence Line for Point 2′.

Fig. 14.25 Shear Influence Line for Point 3′.

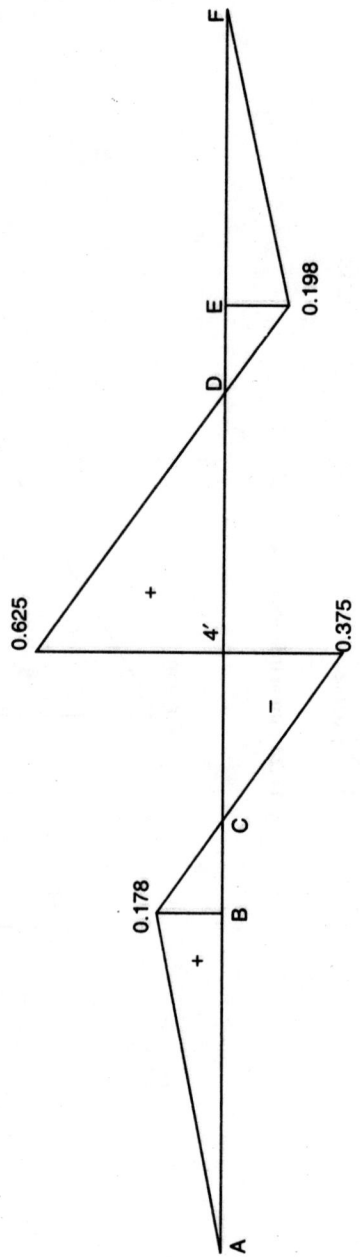

Fig. 14.26 Shear Influence Line for Point 4'.

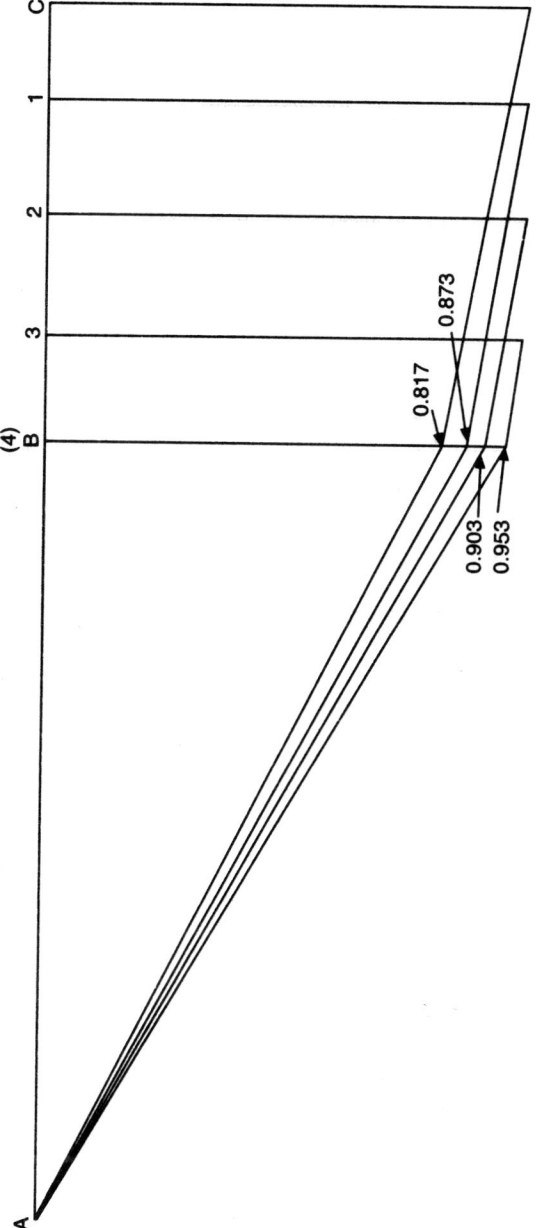

Fig. 14.27 Shear Influence Line for Point C_L, 1, 2, 3 and 4.

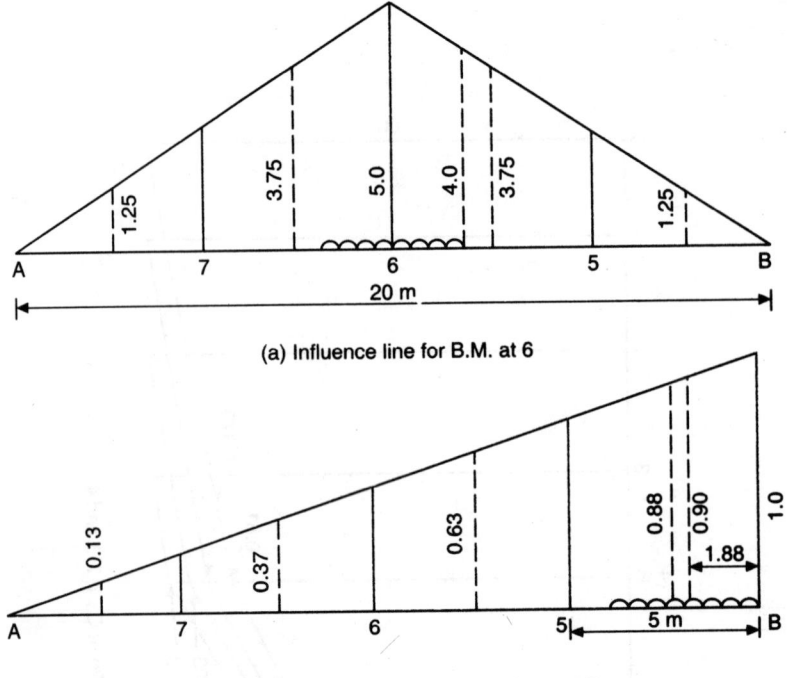

(a) Influence line for B.M. at 6

(b) Influence line for shear force at B

Fig. 14.28 Influence Lines for Bending Moment at 6 and Shear Force at B.

5. Abstract of Service Load Bending Moments and Shear Forces at Various Sections

The design moments and shear forces calculated from Tables are computed in Table 14.18 and 14.19.

6. Design Ultimate Moments and shear Forces at Critical Sections

a) Support Section-C

$M_u = [(1.35 \, M_g + 1.5 \, M_q)] = [(1.35 \times 2240) + (1.5 \times 1790)] = -5709 \text{ kN·m}$

$V_u = [(1.35 \, V_g + 1.5 \, V_q)] = [(1.35 \times 455 + 1.5 \times 358)] = 1151 \text{ kN}$

b) Centre of Middle Span-G

$M_u = [(1.35 \, M_g + 1.5 \, M_q)] = [(1.35 \times 1232) + (1.5 \times 2416)] = 5287 \text{ kN·m}$

c) Centre of Simply Supported Span-6

$M_u = [(1.35 \, M_g + 1.5 \, M_q)] = [(1.35 \times 1655) + (1.5 \times 1733)] = 4834 \text{ kN·m}$

Table 14.18 Abstract of Design Bending Moments and Shear Forces

Section	Dead Load (kN.m)		Moment	Live Load Moment		Design (kN.m)	Moment	Design shear (kN)
	+ ve	− ve	Net	+ ve	− ve	+ ve	− ve	—
Central – Span								
C_R	—	—	—	—	—	—	—	813
1′	1282	2278	−996	876	−1598	—	−2594	733
2′	2265	2301	−36	1525	−1473	+1489	−1509	630
3′	2938	2343	+595	1977	−1328	+2572	−733	495
4′	3378	2308	+1070	2329	−1078	+3399	−8	319
5′	3520	2288	+1232	2416	−899	+3648	—	—
Cantilever – BC								
C	—	2240	−2240	—	−1790	—	−4030	—
1	—	1700	−1700	—	−836	—	−2536	799
2	—	1071	−1071	—	−924	—	−1995	768
3	—	499	−499	—	−468	—	967	734
4	—	—	—	—	—	—	—	696
C_L	—	—	—	—	—	—	—	827

Table 14.19 Design Moments and Shears for Simply Supported Span AB

Load-Point	Maximum Ordinate	x or x'	Ordinate at Point	Load (kN)	Moment (kN.m)
(Refer Fig. 14.26)					
Bending Moment at Centre of AB (Point – 6)					
B – 5	5.00	—	1.25	185.43	232
5 – 6	—	—	3.75	169.23	635
6 – 7	—	—	3.75	158.97	596
7 – 8	—	—	1.25	153.57	192
Live Load	5.00	—	5 & 4	385	1733
Total B.M. at Point – A					3388
Shear Force at Support – B					Shear (kN)
B – 5	—	—	0.88	185.43	163
5 – 6	—	—	0.63	169.22	107
6 – 7	—	—	0.37	158.97	59
7 – 8	—	—	0.13	153.57	20
Total Dead Load Shear					349
L.L.	—	—	0.903	385	347
Total Design Shear at B					696

d) Support Section at A

$$V_u = [(1.35\ V_g + 1.5\ V_q)] = [(1.35 \times 320) + (1.5 \times 347)] = 952\ \text{kN}$$

e) Articulation Section-B

$$V_u = [(1.35\ V_g + 1.5\ V_q)] = [(1.35 \times 349) + (1.5 \times 347)] = 992\ \text{kN}$$

7. Design of Reinforcements at Critical Sections

Using M-25 Grade concrete and Fe-415 HYSD reinforcements

$$f_{ck} = 25\ \text{N/mm}^2\ \text{and}\ f_y = 415\ \text{N/mm}^2$$

Width of Girder = b = 450 mm, Overall depth provided = D = 2700 mm

a) Support Section-C

$$M_u = -5709\ \text{kN·m}$$
$$V_u = 1151\ \text{kN}$$

Effective depth of beam required = $d = \sqrt{\dfrac{M_u}{0.138 f_{ck}.b.}} = \sqrt{\dfrac{5709 \times 10^6}{0.138 \times 25 \times 450}} = 1917$ mm

Providing an effective depth = d = 2550 mm

Compute the parameter ratio $(M_u/b.d^2) = [(5709 \times 10^6)/(450 \times 2550^2)] = 1.95$ N/mm^2

Refer SP: 16 Design Table 3 and read out the reinforcement percentage corresponding to

$$f_{ck} = 25\ \text{N/mm}^2\ \text{and}\ f_y = 415\ \text{N/mm}^2\ \text{as}$$
$$p_t = (100\ A_{st}/b.d) = 0.601$$
∴ $$A_{st} = [(0.601 \times 450 \times 2550)/100] = 6897\ \text{mm}^2$$

Use 8 bars of 36 mm diameter (A_{st} = 8143 mm)

Nominal shear stress = $\tau_v = \left[\dfrac{V_u}{b_w.d}\right] = \left[\dfrac{1151 \times 10^3}{450 \times 2550}\right] = 1.00 < \tau_{c\,max} = 3.1$ N/mm^2

For p_t = 0.601, interpolate τ_c from table 19 of IS: 456-2000 as τ_c = 0.54 N/mm^2

Design shear strength of concrete = $V_c = (\tau_c.b.d) = (0.54 \times 450 \times 2550) = (619 \times 10^3)$ N

Since $V_u > V_c$, shear reinforcements are required

Balance shear force = (1151 – 619) = 532 kN

Using 10 mm diameter 4-legged vertical stirrups, the spacing is obtained as

$$S_v = \left(\frac{0.87 f_y A_{sv} d}{V_{us}}\right) = \left(\frac{0.87 \times 415 \times 4 \times 79 \times 2550}{532 \times 1000}\right) = 546\ \text{mm}$$

Provide the vertical stirrups at a spacing of 300 mm near supports gradually increased to 500 mm towards the centre of span.

b) Centre of Middle Span-G

$M_u = 5287$ kN·m

The moment being almost similar in magnitude to that at support, provide the same reinforcement of 8 bars 0f 36 mm diameter on the tension side, with nominal stirrups of 10 mm 4-legged at a spacing of 500 mm.

c) Centre of Simply Supported Span-6

$M_u = 4834$ kN·m, $b_w = 450$ mm, $b_f = 1800$ mm, $D = 1700$ mm, $d = 1550$ mm, $D_f = 200$ mm

Section designed as tee beam with slab thickness of 200 mm, overall depth of 1700 mm, width of beam as 450 mm.

The ultimate moment capacity of the flange section alone is computed assuming $(x_u = 200$ mm)

$$M_{uf} = [0.36 f_{ck} \times b_f \times D_f (d - 0.42\, x_u)]$$
$$= [0.36 \times 25 \times 1800 \times 200\, (1550 - 0.42 \times 200]/10^6$$
$$= 4750 \text{ kN·m}$$

The area of tensile steel to resist this moment is given by

$$A_{st} = \left(\frac{M_{uf}}{0.87 f_y \left(d - 0.42 \times D_f\right)} \right) = \left(\frac{4750 \times 10^6}{0.87 \times 415 \left(1550 - 0.42 \times 200\right)} \right) = 8974 \text{ mm}^2$$

Balance moment = $(4834 - 4750) = 84$ kN·m. The area of tensile steel to resist this moment is small, increase the area of steel by 10 percent so that the total steel required at this section is

$$A_{st} = (8974 + 896) = 9870 \text{ mm}^2$$

Provide 10 bars of 36 mm diameter ($A_{st} = 10178$ mm²)

d) Articulation Section-B (Section-4)

The articulation has to be designed for the following forces:

1. The separating force

2. Cantilever action

3. Combined bending and direct stresses

4. Shear force

The direct and bending stresses on oblique planes are found out from which the worst section can be located. The shear force normal thrust and moment at the section can then be calculated. Depth of girder at articulation = 2.3 m.

A roller of 23 cm diameter is provided at the articulation.

Articulation reaction = 696 kN

Oblique planes inclined at 0, 13°, 23°, 32°, 36° and 45′ to the vertical are considered and the forces on these sections are tabulated as shown in Table 14.20 and Fig. 14.29.

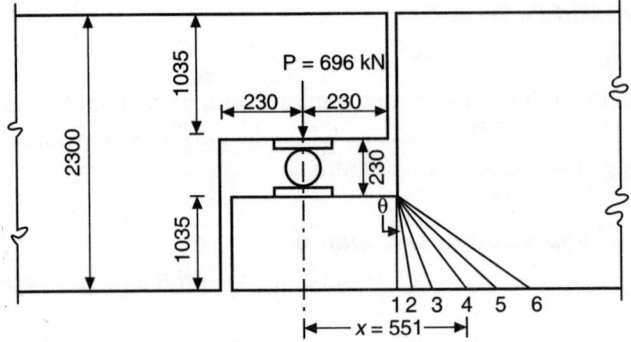

Fig. 14.29 Articulation at Section B.

Service load bending moment = 383×10^6 N.mm
Direct service load tension = 368×10^3 N
Depth of the girder = 1150 mm
Width of girder = 450 mm
Fig. 14.30 shows the forces acting at the articulation section-4

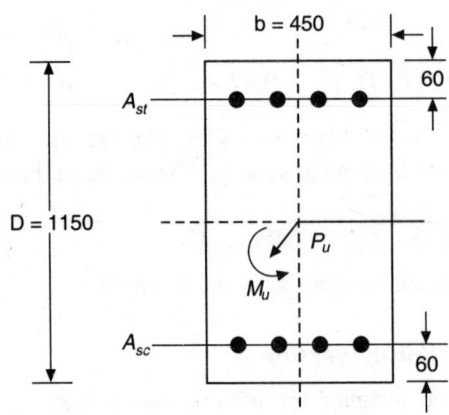

Fig. 14.30 Forces Acting at Articulation Section.

The service dead & L.L shear acting on the Critical section are computed using Table 14.17.

Shear due to service loads at section-4 of articulation (Critical section at $\theta = 32°$)
Dead load shear = V_g = 347 kN
Live load shear = V_q = 349 kN
Horizontal Distance of centroid of section from load = x = 551 mm = 0.551 m
Dead load moment = $M_g = (V_g x) = (347 \times 0.551) = 191.19$ kN·m
Live load moment = $M_q = (V_q x) = (349 \times 0.551) = 192.29$ kN·m
Direct tension due to dead load = T_g = 184 kN

Table 14.20 Determination of Worst Section at Articulation

Section	Breadth b(mm)	Effective depth h(mm)	Area F = bh (mm²)	bZ = 7F/8 (approx) Z = Lever arm	Section Modulus W = bh²/6	x = Distance (mm)	θ = Inclination of plane to vertical in degrees	sin θ
1	450	935	4.2×10^5	3.67×10^5	6.5×10^7	230	0	0
2	450	955	4.3×10^5	3.76×10^5	6.8×10^7	346	13	0.2250
3	450	1016	4.57×10^5	3.99×10^5	7.7×10^7	447	23	0.3907
4	450	1110	4.99×10^5	4.36×10^5	9.2×10^7	551	32	0.5299
5	450	1169	5.26×10^5	4.6×10^5	10.2×10^7	602	36	0.5878
6	450	1351	6.07×10^5	5.31×10^5	13.6×10^7	741	45	0.7071

cos θ	M = P·x (N·mm)	N = P·sin θ (N)	Q = P·cos θ (N)	A N/mm²	B N/mm²	C	D	E N/mm²
1	1.6×10^8	—	6.96×10^5	—	2.46	2.46	2.46	1.89
0.9744	2.4×10^8	1.56×10^5	6.78×10^5	0.36	3.50	3.86	3.14	1.80
0.9205	3.1×10^8	2.71×10^5	6.4×10^5	0.59	4.00	4.59	3.14	1.60
0.8480	3.83×10^8	3.68×10^5	5.9×10^5	0.73	4.16	4.89*	3.43	1.35
0.8090	4.18×10^8	4.09×10^5	5.63×10^5	0.77	4.09	4.86	3.32	1.22
0.7071	5.15×10^8	4.92×10^5	4.92×10^5	0.81	3.70	4.51	2.89	0.92

*Maximum value

A = Direct Tensile stress = (N/F) N/mm²

B = Bending Stress = (M/W) N/mm²

C = Total Tensile stress at top = [(N/F) + (M/W)]

D = Total compressive stress at bottom = [(M/W) – (N/F)]

E = Shear stress = (Q/bz) N/mm²

P = 696 RN

Note: The worst Section is plane-section '4' inclined at 32° to vertical where tensile stress is maximum.

Direct tension due to live load = T_q = 349 sin 32° = 185 kN
For θ = 32°, Sin θ = 0.5299 and Cos θ = 0.8480

$M_u = [(1.35\ M_g + 1.5\ M_q)] = [(1.35 \times 191.19) + (1.5 \times 192.29)] = 546.5$ kN·m

$T_u = [(1.35\ T_g + 1.5\ T_q)] = [(1.35 \times 184) + (1.5 \times 185)] = 526$ kN

The section-4 at articulation subjected to bending moment and tension is designed using SP-16 design hand book Chart-68.

Overall depth of section = D = 1150 mm, $P_u = T_u$ = 526 kN and M_u = 546.5 kN·m
Width of section = b = 450 mm

Cover to tension steel = d' = 60 mm and effective depth = d = (1150 – 60) = 1090 mm

Compute the parameters $(P_u/f_{ck}bD)$ and $(M_u/f_{ck}bD^2)$ for referring to Chart-68 of SP-16

$$\left(\frac{P_u}{f_{ck}bD}\right) = \left(\frac{526 \times 10^3}{25 \times 450 \times 1150}\right) = 0.04$$

$$\left(\frac{M_u}{f_{ck}bD^2}\right) = \left(\frac{546.5 \times 10^6}{25 \times 450 \times 1150^2}\right) = 0.036$$

Interpolate the value of the ratio (p/f_{ck}) = 0.04 where A_{st} = $(pbD/100)$

$$A_{st} = \left(\frac{pbD}{100}\right) = \left(\frac{0.04 \times 25 \times 450 \times 1150}{100}\right) = 5175 \text{ mm}^2$$

The area of steel is arranged equally on the compression and tension side.

Provide 4 bars of 36 mm diameter at the tension and compression face of the articulation.

Area of tensile steel provided = A_{st} = 4071 mm²

Maximum service load shear force at the section = $(V_g + V_q)$ = (347 + 349) = 696 kN

Design ultimate load shear = $V_u = [(1.35 \times 347) + (1.5 \times 349)] = 992$ kN

Nominal shear stress = $\tau_v = \left[\dfrac{V_u}{b_w.d}\right] = \left[\dfrac{992 \times 10^3}{450 \times 1090}\right] = 2.02 < \tau_{c\,max} = 3.1 \text{ N/mm}^2$

P_t = (100 A_{st}/bd) = [(100 × 4071)/(450 × 1090)] = 0.82

For p_t = 0.82, interpolate τ_c from table 19 of IS: 456-2000 as τ_c = 0.60 N/mm²

Design shear strength of concrete = $V_c = \tau_c.b.d) = (0.60 \times 450 \times 1090) = (294 \times 10^3)$ N

Since $V_u > V_c$, shear reinforcements are required

Balance shear force = V_{us} = (992 – 294) = 698 kN

Using 12 mm diameter 4-legged vertical stirrups, the spacing is obtained as

$$S_v = \left(\frac{0.87 f_y A_{sv} d}{V_{us}}\right) = \left(\frac{0.87 \times 415 \times 4 \times 113 \times 1090}{698 \times 1000}\right) = 254 \text{ mm}$$

Provide the vertical stirrups at a spacing of 100 mm near the articulations and gradually increased to 200 and 300 mm towards the centre of span.

7.   Reinforcement Details

The details of reinforcements in the deck slab and tee beams conforming to the specifications laid down in SP: 34-1987[10] are shown in Fig. 14.31.

The longitudinal section of the central half of the main span CD with the details of reinforcements is shown in Fig. 14.32.

The longitudinal section of the simply supported span AB with reinforcement details are shown in Fig. 14.33. The details of reinforcements at the articulation junction are shown in Fig. 14.34.

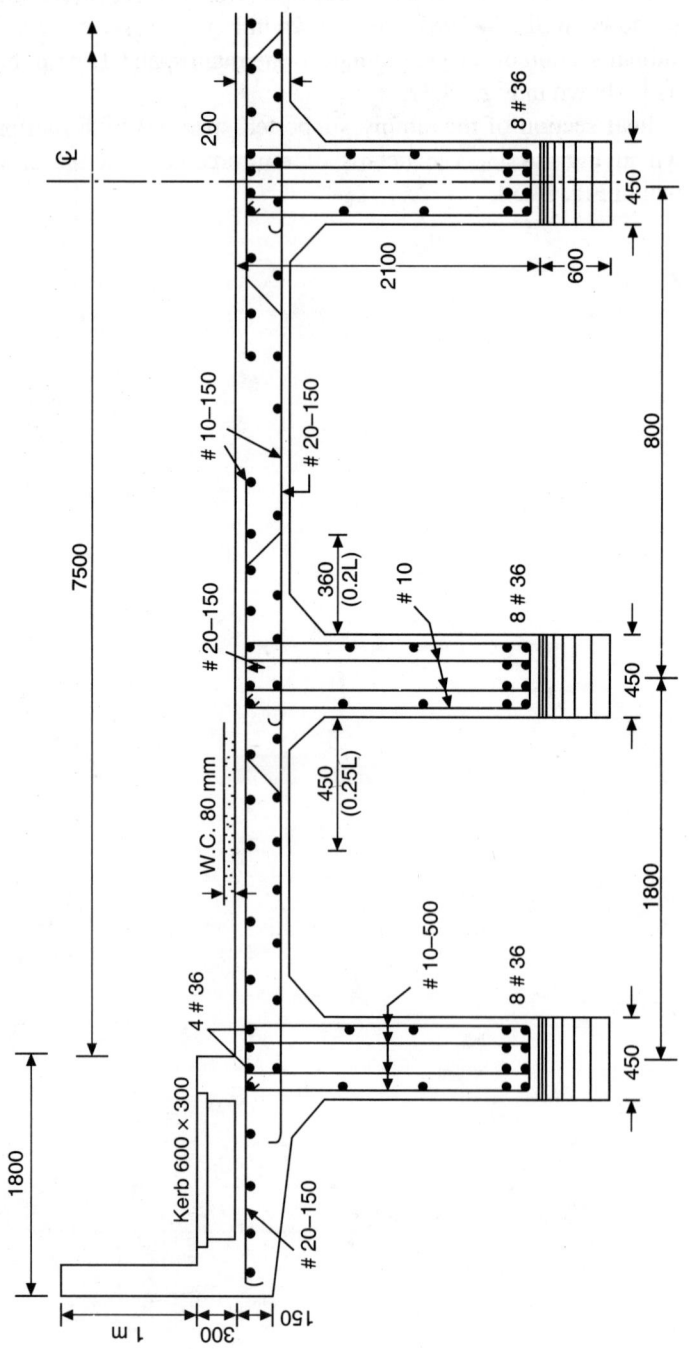

Fig. 14.31 Cross-section of Deck at Mid Span G.

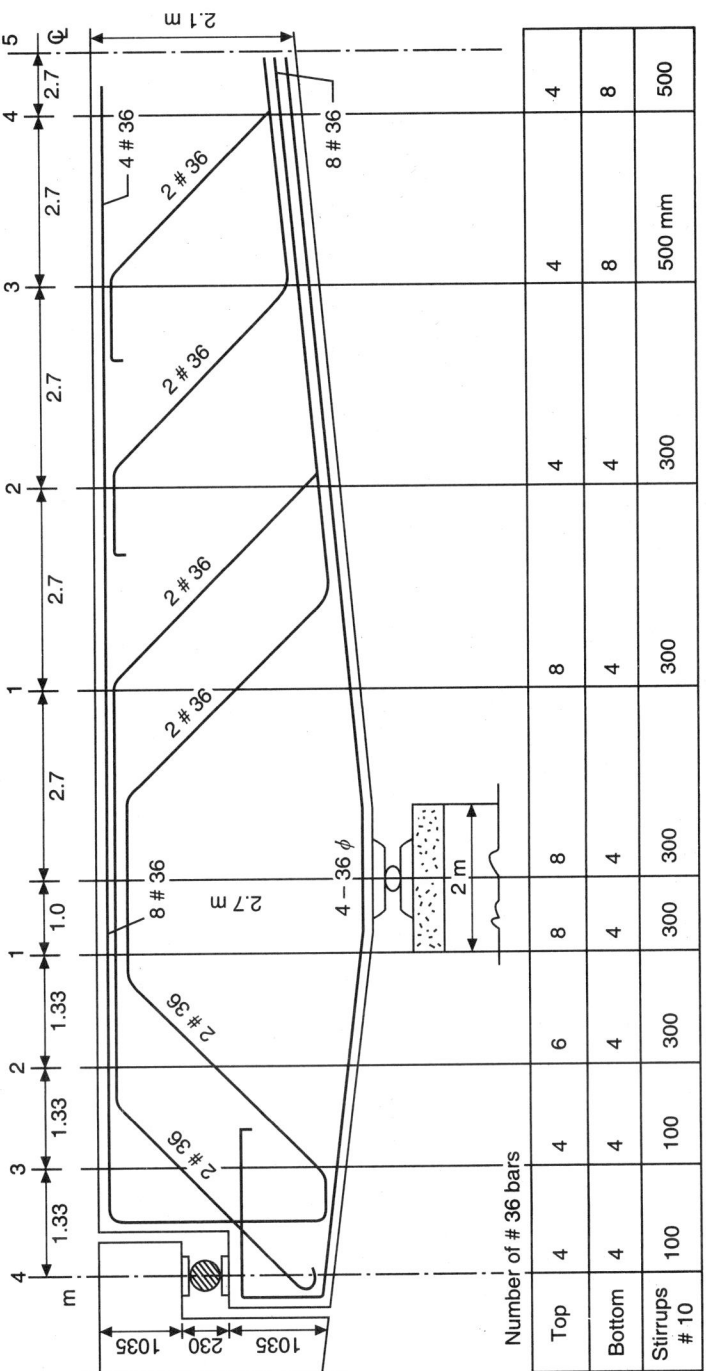

Fig. 14.32 Longitudinal Section of Central Span.

Number of # 36 bars															
Top	4	4	6		8	8	8		4		4		4		4
Bottom	4	4	4		4	4	4		4		4		8		8
Stirrups # 10	100	100	300		300	300	300		300		300		500 mm		500

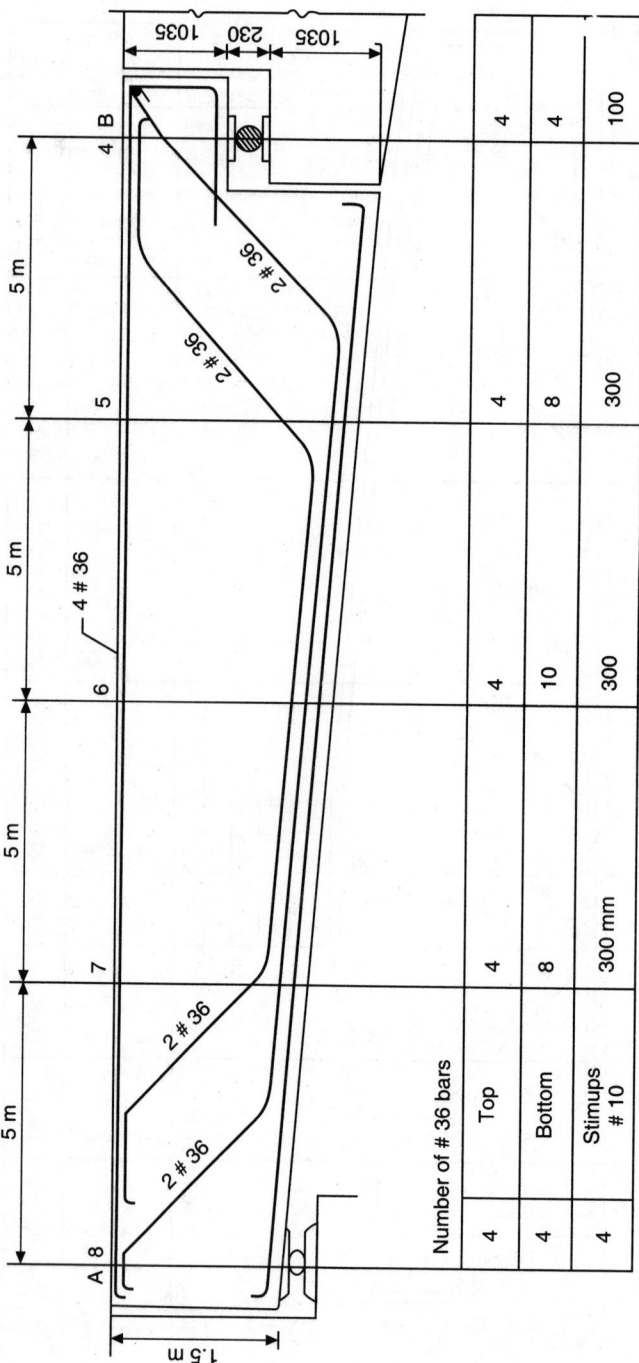

Fig. 14.33 Longitudinal Section of Simply Supported Beam AB.

Number of # 36 bars						
Top	4	4	4	4	4	4
Bottom	4	8	10	8	8	4
Stirrups # 10	300 mm	300	300	300	300	100

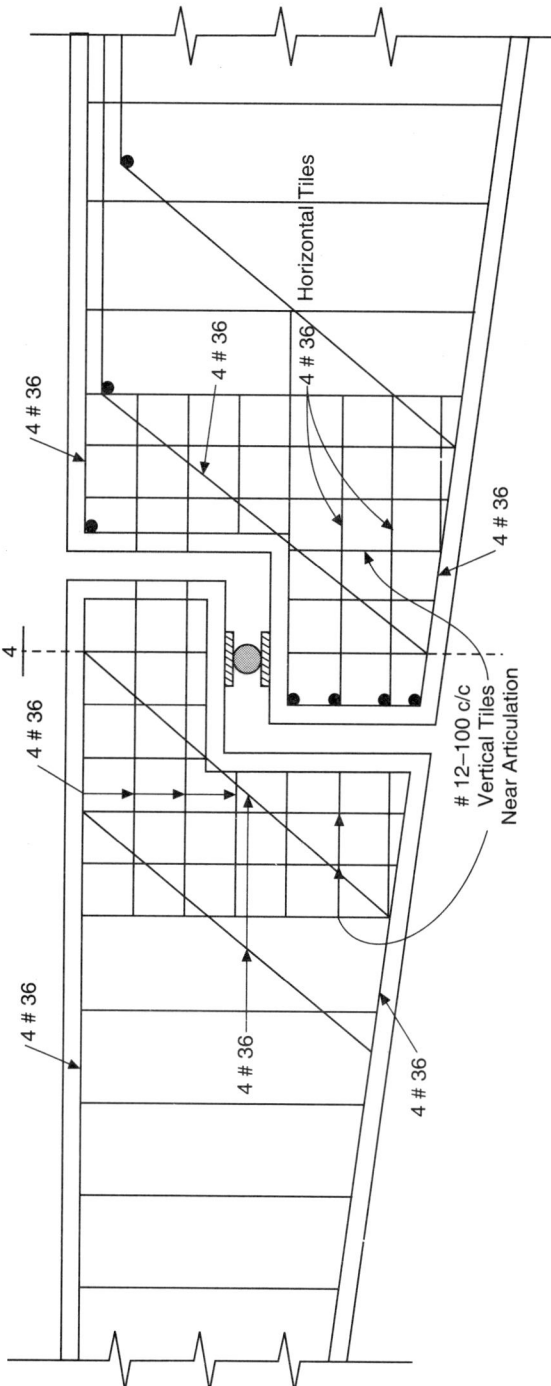

Fig. 14.34 Reinforcement Details at Articulation (Section-4).

EXAMPLES FOR PRACTICE

1. Design a double cantilever bridge to suit the following data:

   Total length of the bridge = 52 m
   Road width = 7.5 m (Two lane)
   Footpaths = 1.5 m on either side
   Spacings of Tee beams = 2 m
   Loading: I.R.C. class AA tracked vehicle
   Materials: M-20 grade concrete, Fe-415 grade tor steel
   Design the deck slab and one of the main girders and sketch the details of reinforcements.

2. A double cantilever bridge is to be designed for a road crossing in a mining area. The following data is available.

   Total length of the bridge = 95 m
   Road width = 7.5 m (Two lane)
   Footpaths = 1.75 m on either side
   Spacings of Tee beams = 1.8 m
   Loading I.R.C. class A or AA whichever gives the worst effect.
   Materials: M-20 grade concrete, Fe-415 grade tor steel.

   Design the deck slab and main girder and sketch the details of reinforcements.

3. Design a R.C.C. double cantilever bridge deck for a road crossing to suit the following data:

   Total length of the bridge = 100 m
   Four lane road 15 m wide with foot paths 1.5 m wide.
   Spacings of Tee beams = 2 m
   Loading: I.R.C. Class AA tracked vehicle
   Materials: M-25 Grade concrete & Fe-415 HYSD bars

   Design the deck slab and main girders and sketch the details of reinforcements.

4. Design a reinforced concrete balanced cantilever bridge for a national high way crossing to suit the following data:

   Total length of the bridge = 60 m
   Width of carriage way (Six lanes) with a dividing median 1.5 m at centre
   Foot paths = 1.5 m on either side
   Spacings of tee beams = 1.5 m
   Loading: IRC. Class A
   Materials: M-30 Grade concrete and Fe-500 HYSD reinforcements

   Design the salient structural elements of the bridge conforming to IRC: 112-2011, IRC: 6-2014 and IS: 456-2000 codes and sketch the details of reinforcements.

REFERENCES

1. Taylor, F.W, Thompson, S.E, and Smulski, E., Reinforced Concrete Bridges, John Wiley & Sons, New York, 1955, pp. 1-456.
2. Johnson Victor, D., Essentials of Bridge Engineering (Fifth Edition), Oxford & IBH Publishing Co, Ltd, New Delhi, 2002, pp.161-162.
3. Raina, V.K., Concrete Bridge Practice, Analysis, Design and Economics, Tata McGraw-Hill Publishing Co, Ltd, New Delhi, 1991, pp, 184-200.
4. IRC: 6-2014, Standard Specifications and Code of Practice for Road Bridges, Section II, Loads and Stresses (Revised Edition), Indian Roads Congress, New Delhi, 2014, pp. 1-84.
5. IS: 456-2000, Indian Standard Code of Practice for Plain and Reinforced Concrete, (Fourth Revision), Bureau of Indian Standards, New Delhi, 2000, p. 72.
6. Reynolds, C.E and Steedman, J., Reinforced Concrete Designers Hand Book, Concrete Publications Ltd, London, 1974.
7. SP: 16-1980, Design Aids for reinforced Concrete to IS: 456-1978, Bureau of Indian Standards, New Delhi, 1980, pp. 1-232.
8. IRC: 112-2011, Code of Practice for Concrete Road bridges, Indian Roads Congress, New Delhi, 2011, p. 181.
9. Krishna Raju, N and Gururaja, D.R., Advanced Mechanics of Solids & Structures, Narosa Publishing House, New Delhi, 1997, pp. 417-443.
10. SP: 34-1987, Hand Book of Concrete Reinforcement and Detailing, Bureau of Indian Standards, New Delhi, 1987.

REVIEW QUESTIONS

1. What are Balanced cantilever type bridges? Under what specific situations you would prefer them for National high way crossings?
2. Explain with sketches a typical balanced cantilever bridge showing the various structural elements of the bridge.
3. What is the specific advantage you would find in the balanced cantilever type bridges with particular reference to the analysis of various force components developed in the bridge deck due to dead and live loads.
4. Discuss briefly the advantages and disadvantages of using uniform cross sections for balanced cantilever bridge decks.
5. List the various steps involved in designing a typical double cantilever bridge for a major National high way crossing.
6. What method you would follow to determine the various design forces in a balanced cantilever bridge due to dead and live loads?
7. What is articulation in a balanced cantilever bridge? How do you analyze the forces at this section and design the reinforcements?

8. Explain the method of designing the critical section subjected to the maximum moments and shear forces in a reinforced concrete double cantilever bridge deck.
9. What type of design you would suggest for long span balanced cantilever bridge decks to reduce the depth of the girders?
10. Explain with sketches the method of detailing reinforcements in a typical longitudinal and cross section of a reinforced concrete double cantilever bridge deck.

OBJECTIVE TYPE QUESTIONS

1. In the case of selecting the type of bridge deck to be constructed on soils amenable for settlement, the most suitable type is
   a) Rigid frame bridge
   b) Continuous type bridge
   c) Balanced cantilever bridge
2. Balanced cantilever bridges are categorized as
   a) Indeterminate structures
   b) Complex structures
   c) Determinate structures
3. In comparison with various types of bridge decks, in balanced cantilever bridges, the moments due to loads are
   a) Increased
   b) Decreased
   c) Not altered
4. In the case of balanced cantilever bridges used for long spans, it is economical to use
   a) Constant cross section for girders
   b) Variable cross section for girders
   c) Minimum cross section at interior supports
5. The ratio of simply supported to cantilever spans in a balanced cantilever bridge deck is generally around
   a) 5 to 3
   b) 4 to 5
   c) 2 to 5
6. In the case of reinforced concrete balanced cantilever bridge decks subjected to dead and live loads, the maximum shear force develops at
   a) The middle of interior spans
   b) The middle of simply supported spans
   c) The interior supports
7. Reinforced concrete balanced cantilever bridges should be designed to conform to the specifications prescribed in the code
   a) IS: 456-2000
   b) IRC: 112-2011
   c) IS: 1343- 2012

8. The maximum negative moment in a double cantilever bridge deck subjected to dead and live loads, develop at
   a) The end supports
   b) Middle of simply sup[ported spans
   c) Interior supports
9. The design bending moments and shear forces in a balanced cantilever bridge deck are determined by using
   a) Moment distribution
   b) Influence line diagrams
   c) Slope deflection analysis
10. The articulation section in a balanced cantilever bridge deck should be designed to resist
    a) The moment
    b) The normal thrust
    c) Moment, shear and axial thrust

15

Continuous Girder Bridges

15.1 GENERAL FEATURES

Reinforced concrete continuous girder bridge decks supported on piers and abutments are ideally suited for long spans where good sub soil is available without any settlements. The bridge deck generally comprises of solid slab and tee beam or box girders continuous over several spans. Continuous solid slab bridge decks[1] are economical for shorter spans while tee beam and slab type is preferred for longer spans in the range of 10 to 35 m. For spans exceeding 40 m, single or multi-celled box girder types are more economical. The bending moments and shear forces at various critical sections are determined by constructing influence lines for these forces at the required sections.

In the case of continuous beams, the negative bending moments developed at the supports due to dead and live loads renders the top slab of a Tee beam deck ineffective. Also the magnitude of bending moments at the support is larger than at the centre of spans. In view of this, the girders are usually of variable cross section with the depth gradually increasing from the centre to the supports. In addition to this, the support section is strengthened with compression reinforcement together with the provision of thickened webs and a cross beam. In the case of continuous slab decks, the thickness of the slab at support sections is increased approximately 1.3 to 1.8 times the minimum thickness at mid span and the length of haunches will be about 0.2 to 0.25 times that of the span.

In continuous girder bridges extending over several spans, it is important to provide for movements of the super structure due to temperature changes. To facilitate the movements suitable bearings are invariably provided at the supports. Generally all but one of the bearings should be of the expansion type to take care of the longitudinal movements of the bridge due to temperature effects. Continuous bridges are adopted as units of three, four or five spans. The three span continuous bridge is the most common type generally adopted for high way bridges.

15.2 ADVANTAGES OF CONTINUOUS BRIDGES

In comparison with simply supported bridges, the bending moments developed in continuous bridges are considerably less and consequently smaller sections can be adopted resulting in economy of steel and concrete. The ultimate moment capacity of continuous bridge deck is greater than that of simply supported decks due to the phenomenon of redistribution of moments in continuous structures. Since longer spans can be adopted, the number of piers required are less. Continuous beams[2] require less number of expansion joints and bearings, thus resulting in lesser initial and maintenance costs. Generally the continuous girders are of variable cross-section so that the moment of inertia of the girder section is proportional to the bending moments developed at the section.

The main disadvantages of continuous girder designs are that uneven settlements may lead to the failure of the structure. The detailing and placing of reinforcements require extra care and skilled workmanship. Continuous bridge deck being a statically indeterminate structure, the analysis is more complicated than the simply supported beams.

15.3 SELECTION OF SPANS AND PROFILE OF GIRDERS

In continuous span bridges, the exterior spans are made shorter than the interior spans. This will result in the reduction of moments in the end spans. Generally the end spans are made about 16 to 20 percent smaller than the intermediate spans. The bending moments at the interior supports will in general be larger and hence the depths of girder at intermediate supports should be larger than at centre of spans. The girder depth at supports is generally 1.5 to 2.5 times the minimum thickness at mid span. A parabolic profile is generally adopted with maximum depth at intermediate supports and minimum depth at mid spans as shown in Fig. 15.1.

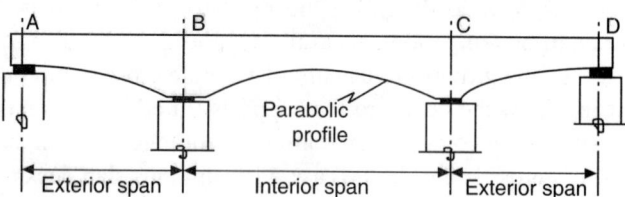

Fig. 15.1 Typical Profile of Three Span Continuous Bridge.

15.4 ANALYSIS OF CONTINUOUS GIRDER BRIDGES

The given IRC high way loads are positioned on the respective influence lines drawn for bending moments and shear forces at critical sections to evaluate the live load moments[3]. The dead load moments are computed using the influence lines[4] and the dead loads of the member divided into a number of parts along the span. The service load moments comprising the dead and live loads are determine for all the critical sections like the support and mid span sections. The design ultimate loads are evaluated by applying appropriate load factors specified in IRC codes for service load moments and shear forces.

15.5 INFLUENCE LINES FOR GIRDERS OF VARIABLE CROSS-SECTION

The procedure for drawing the influence line for bending moment and shear at a given section of the girder is outlined as follows:

(a) For the continuous girder ABCD shown in Fig. 15.1, the carry over factors C_{AB}, C_{BA}, C_{BC} and C_{CB} etc. are read out from the curves shown in Fig. 15.2.

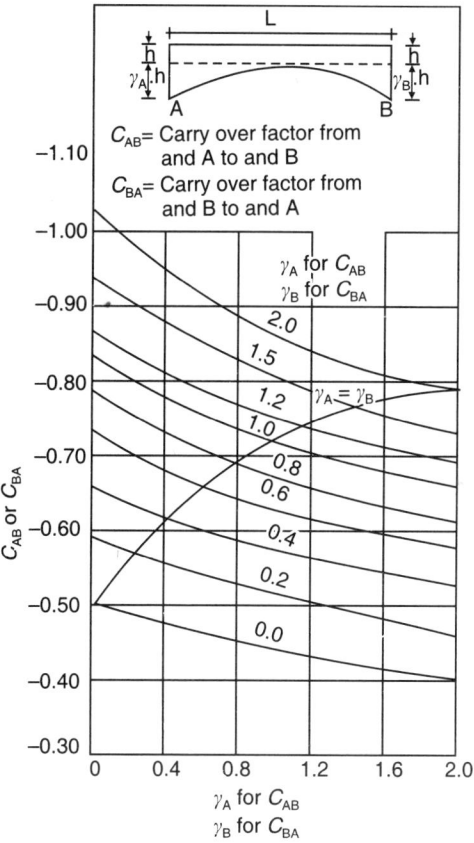

Fig. 15.2 Carry Over Factors.

(b) The distribution factors D_{BA}, D_{BC}, D_{CB} etc. are computed using the relationship

$$D = (K/\Sigma K) = (k.I_c/L)/\Sigma(k.I_c)E)/L$$

where k = Stiffness coefficient of joint end of members (K_{AB}, K_{BA}) obtained from the curves shown in Fig. 15.3.

   L = Span length of members

   I_c = Moment of inertia at centre of the members

E = Modulus of elasticity of concrete

K = Stiffness of the member

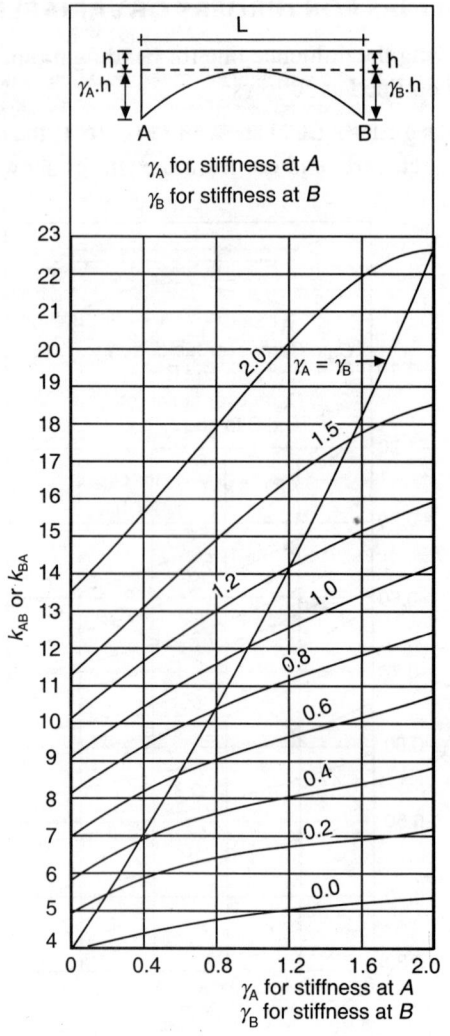

Fig. 15.3 Stiffness Coefficients.

(c) The fixed end moments $\overline{M}_{AB}$ and $\overline{M}_{BA}$ for different positions of the load in the span is read out from Figs. 15.4 to 15.12.

Using the following notations, the final moments at supports are computed in terms of fixed end moments.

$$\overline{M}_{AB}, \overline{M}_{BA}, \overline{M}_{BC}, \text{ etc.} = \text{Fixed end moments}$$

$$C_{AB}, C_{BA}, C_{BC}, \text{ etc.} = \text{Carry over factors}$$

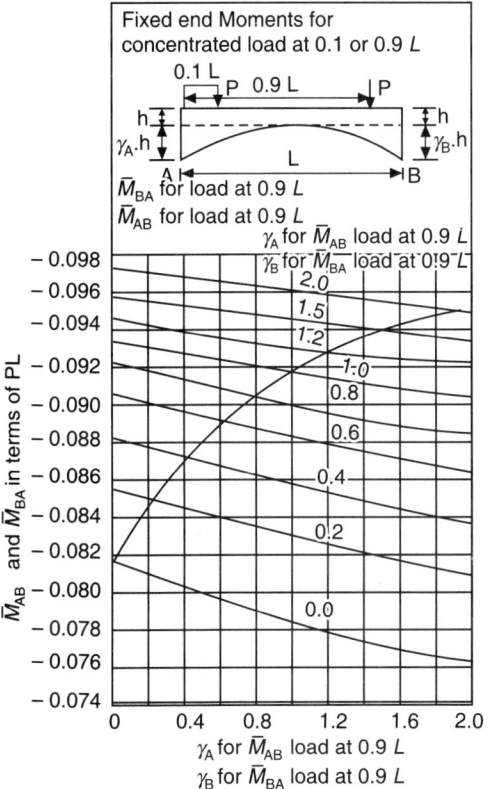

Fig. 15.4 Fixed End Moments.

D_{AB}, D_{BA}, D_{BC}, etc. = Distribution factors

D_{AB}, D_{BA}, D_{BC}, etc. = Distribution factors

$$M_1 = \overline{M}_{BA} - C_{AB} \cdot \overline{M}_{AB}$$
$$M_2 = \overline{M}_{BC} - C_{CB} \cdot \overline{M}_{CB}$$
$$M_3 = \overline{M}_{CD} - C_{DC} \cdot \overline{M}_{DC}$$
$$U = C_{BC} \cdot C_{CB} \cdot D_{BC} \cdot D_{CB}$$
$$V = C_{BC} \cdot D_{BC} \cdot D_{CD}$$
$$W = C_{CB} \cdot D_{CB} \cdot D_{BA}$$

Using these notations, the support moments M_B and M_C in a three span continuous beam for loads in different spans are given by the relations:

(1) Load in Span AB

$$M_B = [(1 - D_{BA} - U)/(1 - U)] M_1$$
$$M_C = [V/(1 - U)] M_1$$

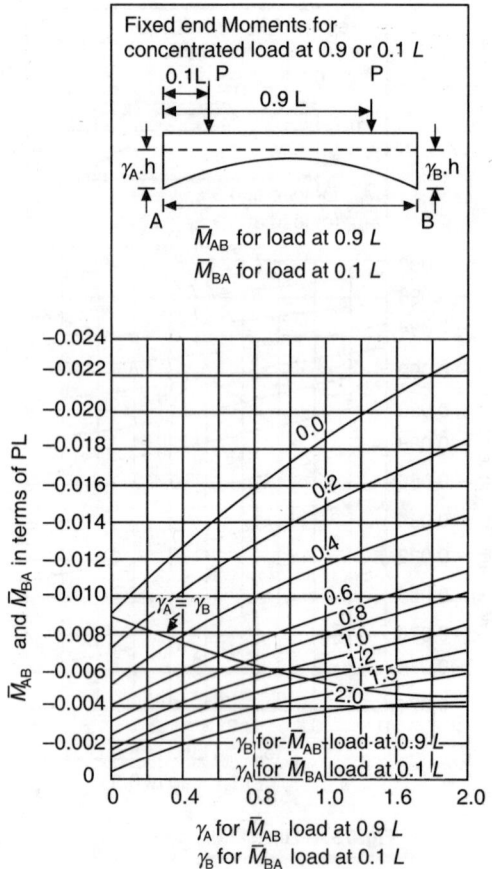

Fig. 15.5 Fixed End Moments.

(2) Load in Span BC

$$M_B = [(D_{BA} \cdot \overline{M}_{BC} - W \cdot \overline{M}_{CB})/(1 - U)]$$

$$M_C = [(D_{CD} \cdot \overline{M}_{CB} - V \cdot \overline{M}_{BC})/(1 - U)]$$

(3) Load in Span CD

$$M_B = [W/(1 - U)]M_3$$

$$M_C = [(1 - D_{CD} - U)/(1 - U)]M_3$$

(d) Influence lines for support moment M_B for different position of unit load is obtained by placing the load at successive points along each span. The influence line for support moment is used to derive the influence line for moment at any other section within the span by the method of superposition. For a continuous beam ABCD of three spans shown in Fig. 15.13, the influence line for the support moment M_B is first drawn. The influence line for bending moment at p distant 'xL' from A can be obtained by superposition.

$$M_p = M_s + (M_B/L)(xL)$$

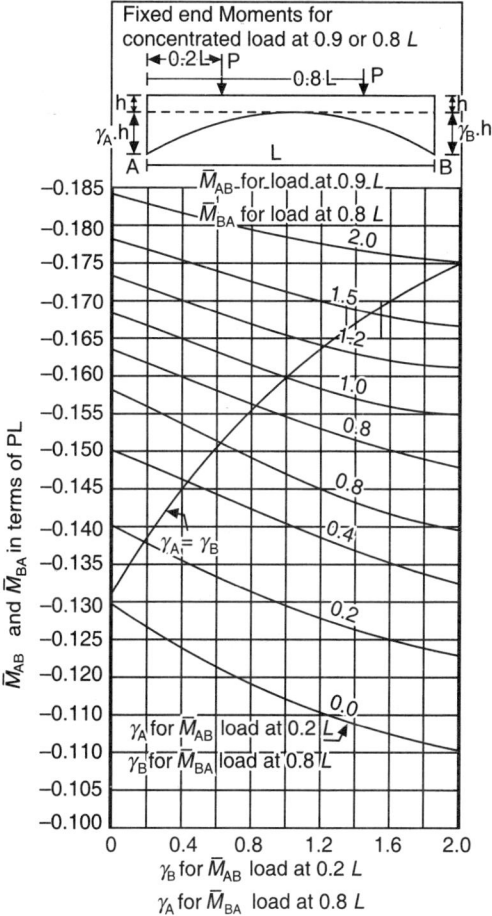

Fig. 15.6 Fixed End Moments.

Where M_s = simple bending moment at the section due to unit load located any where in span AB. If the load crosses the span AB, the bending moment at the section P is obtained as, $M_P = (M_B/L)(xL)$.

In this manner, a table can be prepared for calculating the influence line ordinates at any desired section.

(e) Influence lines for shear at a given section is derived by superposing the influence line ordinates at B and C and the simple beam under shear component.

Thus if M_B = Influence line ordinate at support B

M_C = Influence line ordinate at support C

μ = Shear ordinate

for a given position of load along the span, the influence line ordinate for stiear is obtained as, $[\mu + (M_C - M_B)/L]$

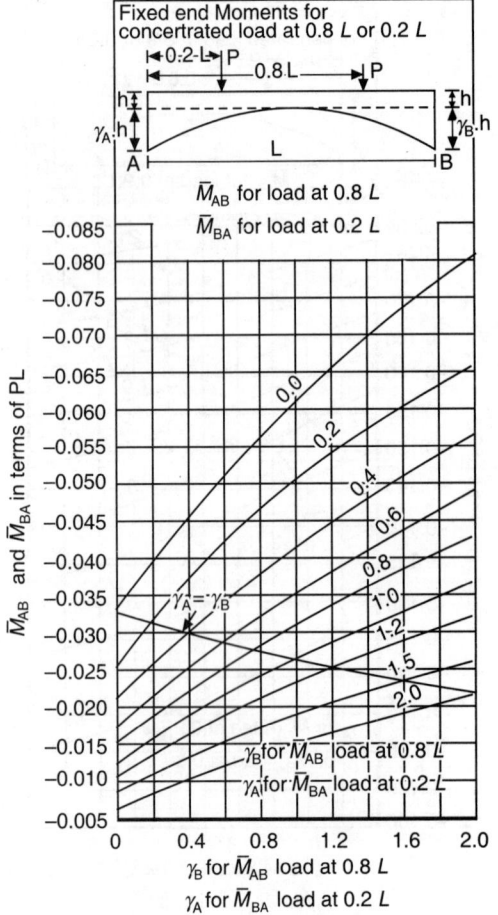

Fig. 15.7 Fixed End Moments.

15.6 COMPUTATION OF DESIGN MOMENTS AND SHEARS

The given I.R.C. loads are positioned on the influence lines such that maximum live load moment is obtained at each of the sections. The dead load moments are computed using the influence lines and the dead loads of the member divided into a number of parts between the sections. The live load and dead load moments and shear force are combined to obtain the design moments and shears at each of the sections.

15.7 DESIGN OF CRITICAL SECTIONS OF BRIDGE DECK

According to the latest IRC codes (IRC: 6-2014[5] and IRC: 112-2011[6]), the reinforced concrete sections of a bridge deck should be designed by using the principles of limit state design. The designed sections should conform to the limit state of ultimate strength and serviceability. Using the given grade of concrete and type of steel reinforcements, the empirically assumed depths at critical sections are checked for flexural and shear

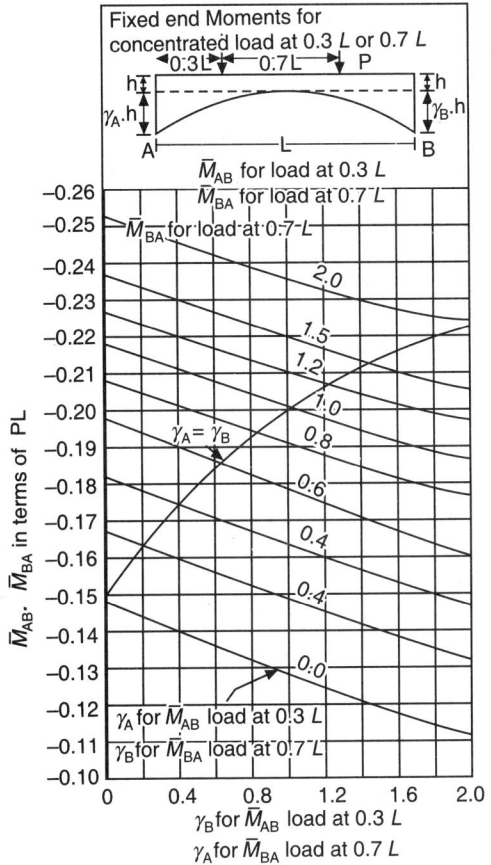

Fig. 15.8 Fixed End Moments.

strength. For long span girders, the reinforcements required being larger, the bars are arranged in 2 to 3 rows on the tension side. The sections are checked for ultimate shear strength and sufficient shear reinforcements are designed comprising of two or four legged stirrups.

The design of a three span continuous girder bridge for a National high way crossing is illustrated by the following example.

15.8 DESIGN EXAMPLE OF A REINFORCED CONCRETE CONTINUOUS GIRDER BRIDGE

A three span reinforced concrete continuous bridge with girders of variable cross section is required for the crossing of a National High way. Design the decks slab and main girders of the bridge to suit the following data:

Total length of the bridge = 70 m
Central span = 30 m

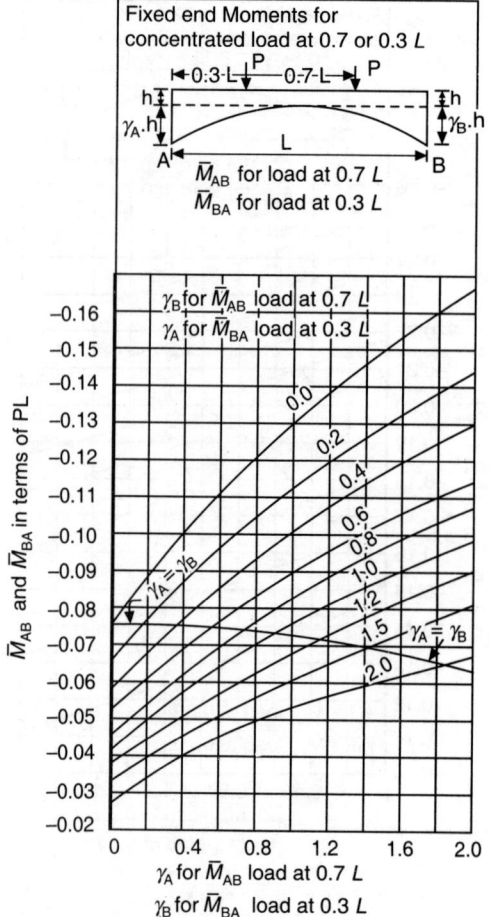

Fig. 15.9 Fixed Moments.

End spans = 20 m
Width of carriage way = 7.5 m
Loading: IRC Class AA tracked vehicle
Kerbs: 600 mm on either side
Spacing's of main girders = 2.9 m
Spacing's of cross girders = 4 m
Materials: Concrete of M-20 Grade, Fe-415 HYSD bars

Design the bridge deck and draw typical sections showing the details of reinforcements in the decks slab and girders. The design should conform to the specifications of IRC: 6-2014 and IRC: 112-2011.

1. Selection of Dimensions of Girders

The longitudinal elevation showing the main girders of the three span AB, BC and

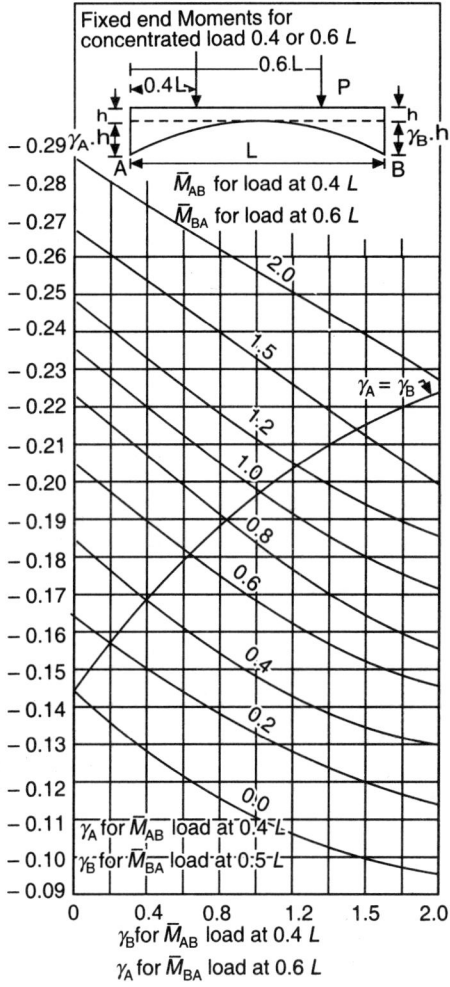

Fig. 15.10 Fixed End Moments.

CD is shown in Fig. 15.14. Assuming a parabolic profile for girders, depth of girder at $A = 1/20$ span

$\therefore \qquad h = (1/20 \times 20) = 1 \text{ m}$

Depth of girder at $B = (h + 2h) = 3h = 3$ m

$\therefore \qquad r_B = 2$

The cross-section of the deck showing the dimensions of the various structural components is shown in Fig. 15.15.

Thickness of deck slab = 250 mm
Width of girders = 500 mm
Spacings of girders = 2.9 m

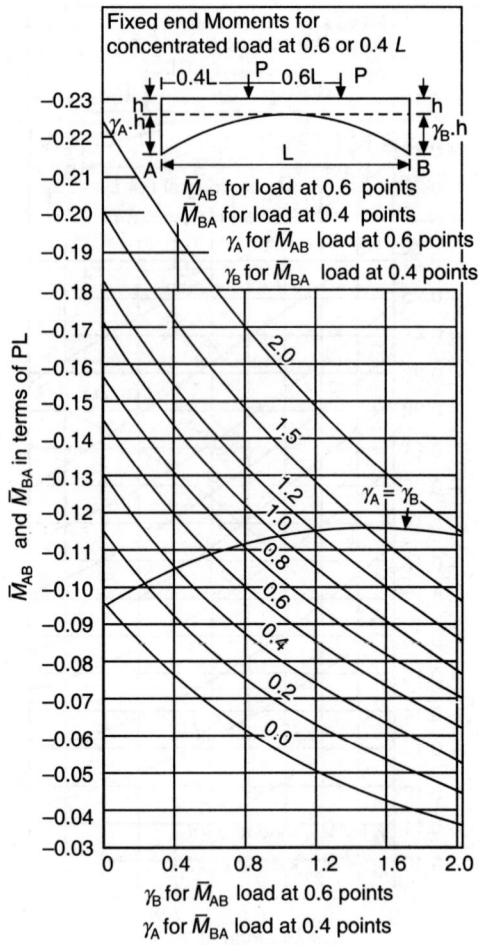

Fig. 15.11 Fixed End Moments.

Thickness of wearing coat = 80 mm

Kerbs 600 mm wide by 300 mm deep are provided on either side.

Cross girders are provided at 4 m intervals in end spans and 6 m intervals in central span.

Width of cross girder = 300 m

2. Design of Deck Slab

(a) Live Load Bending Moments

Live load is class AA tracked vehicle. One wheel is placed at the centre of panel as shown in Fig. 15.16.

$$u = (0.85 + 2 \times 0.08) = 1.01 \text{ m}$$

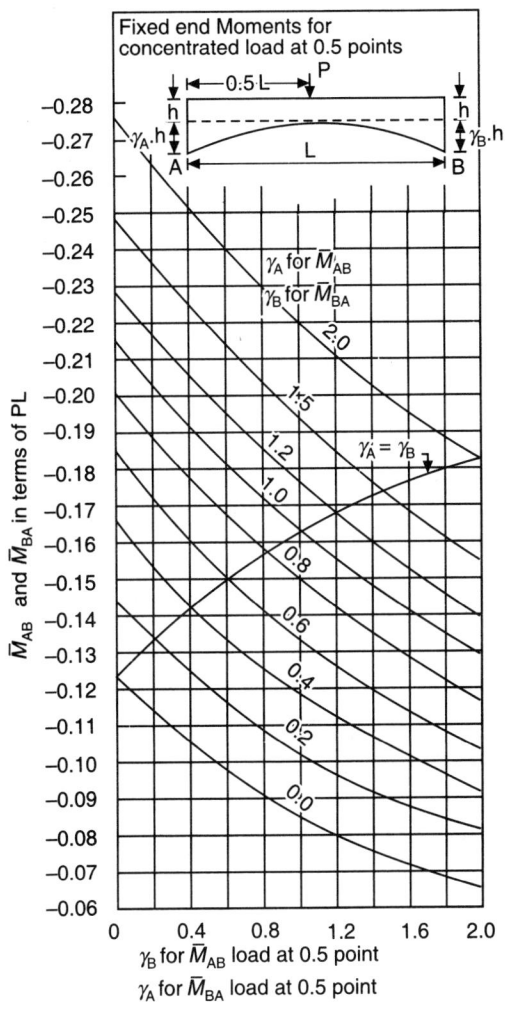

Fig. 15.12 Fixed End Moments.

$$v = (3.60 + 2 \times 0.08) = 3.76 \text{ m}$$
$$(u/B) = (1.01/2.7) = 0.348$$
$$(v/_L) = (3.76/4.0) = 0.94$$
$$K = (B/L) = (2.9/4.0) = 0.725$$

Referring to Pigeaud's curves (Refer Fig. 4.9)

$$m_1 = 0.09 \quad m_2 = 0.035$$
$$M_B = W(m_1 + 0.15\, m_2)$$
$$= 350\,(0.09 + 0.15 \times 0.035)$$
$$= 33.32 \text{ kN·m}$$

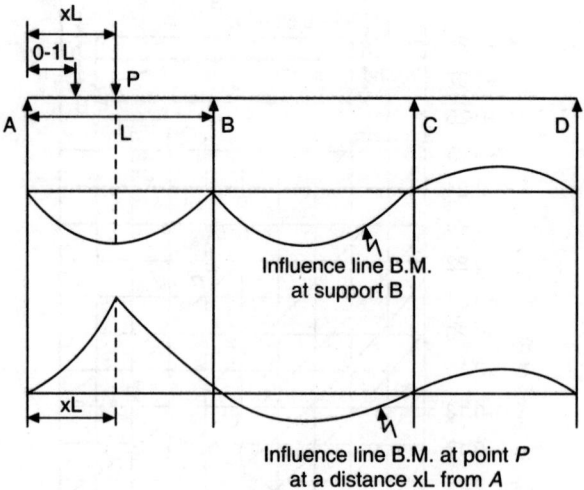

Fig. 15.13 Influence Line for Bending Moment.

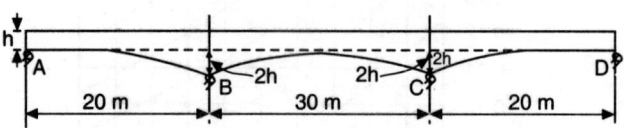

Fig. 15.14 Longitudinal Elevation of Main Girder.

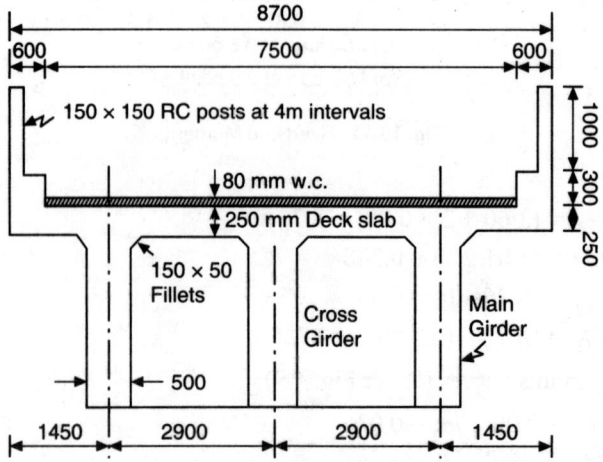

Fig. 15.15 Cross Section of Bridge Deck.

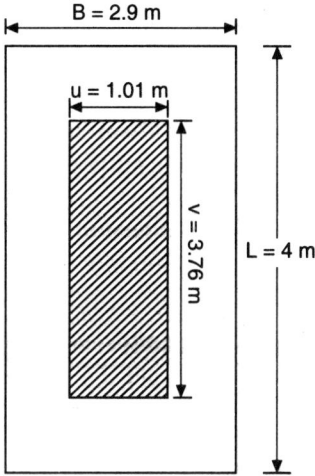

Fig. 15.16 I.R.C. Class AA Wheel Load on Slab Deck.

Design B.M. including impact and continuity factor is given by M (short span)
$$= (1.25 \times 0.8 \times 33.32) = 33.32 \text{ kN·m}.$$
Similarly, M (long span) $= 350 (0.035 + 0.15 \times 0.09) = 16.975$ kN·m

(b) Dead Load Bending Moments

Dead weight of slab $= (1 \times 1 \times 0.25 \times 24)$..................$= 6.00$ kN/m^2
Dead weight of W.C $= (0.08 \times 22)$..............................$= 1.76$ kN/m^2
Total dead load of deck...$= 7.76$ kN/m

Total load on one panel $= (40 \times 2.9 \times 7.76) = 90$ kN
   $(u/B) = (v/L) = 1$ as panel is loaded with uniformly distributed load
   $K = (B/L) = (2.9/4.0) = 0.725, (UK) = 1.379$
From Pigeaud's curves (Refer Fig. 3.9)
$$m_1 = 0.048 \text{ and } m_2 = 0.025$$
$$M_B = 90 (0.048 + 0.15 \times 0.025) = 4.65 \text{ kN·m}$$
Taking continuity into effect
$$M_B = (0.8 \times 4.65) = 3.726 \text{ kN·m}$$
$$M_L = 90 (0.025 + 0.15 \times 0.048) = 2.88 \text{ kN·m}$$
Taking continuity into account
$$M_L = (0.8 \times 2.88) = 2.304 \text{ kN·m}$$

(c) Design Service Load Bending Moments

Short span: $M_g = 3.726$ kN·m and $M_q = 33.32$ kN·m
Long span: $M_g = 2.304$ kN·m and $M_q = 16.975$ kN·m

(d) Design Ultimate Load Bending Moment

Design Ultimate load moments are derived by applying suitable load factors to the service load moments as per the principles of limit state design[7,8] and IRC. Code specifications.

Short span: $M_u = [1.35 \, M_g + 1.5 \, M_q) = [(1.35 \times 3.726) + (1.5 \times 33.32)] = 45.0$ kN·m

Long Span: $M_u = [1.35 \, M_g + 1.5 \, M_q) = [(1.35 \times 2.304) + (1.5 \times 16.975)] = 28.5$ kN·m

(e) Design of Section

Effective depth of slab required $= d = \sqrt{\dfrac{M_u}{0.138 f_{ck}.b.}} = \sqrt{\dfrac{45 \times 10^6}{0.138 \times 20 \times 1000.}} = 127.6$ mm

Adopt effective depth $= d = 200$ mm and overall depth $= 250$ mm

For short span, compute the parameter

$$\left(\frac{M_u}{b.d^2}\right) = \left(\frac{45 \times 10^6}{1000 \times 200^2}\right) = 1.12,$$ using M-20 grade concrete and Fe-415 HYSD bars

Read out the percentage of reinforcement required from Table 2 of SP: 16[11], Design Aids

$$p_t = 0.335 = \left(\frac{100 A_{st}}{b.d}\right)$$

Solving $\qquad A_{st} = \left(\dfrac{0.335 \times 1000 \times 200}{100}\right) = 670 \, \text{mm}^2$

For short span, Provide 12 mm diameter bars at 150 mm centers (A_{st} provided $= 754 \, \text{mm}^2$)

For long span, provide 10 mm diameter bars at 150 mm centers

The reinforcements in the deck slab are shown in Fig. 15.17.

3. Stiffness and Distribution Factors

The stiffness and distribution factors depend upon the dimensions, length and support conditions of the members using the principles of structural analysis[9,10]

Referring to Fig. 15.14, the value of r_B and $r_C = 2.0$, $r_A = r_D = 0$. For these values from Fig. 15.3, the stiffness coefficients are obtained as,

$$K_{BA} = 13.5 \qquad K_{BC} = 22.8$$

From Fig. 15.2, the carry over factors are,

$$C_{AB} = -1.01, \quad C_{BC} = -0.78 \quad C_{CD} = -0.40$$
$$C_{BA} = -0.40, \quad C_{CB} = -0.78 \quad C_{DC} = -1.01$$

The end A is simply supported. Hence the stiffness factor K_{BA} is modified by using the equation,

Modified value of $K'_{BA} = (1 - C_{BA} \, C_{AB}) \, K_{BA}$
$$= (1 - 1.4 \times 1.01) \, 13.5 = 8.046$$

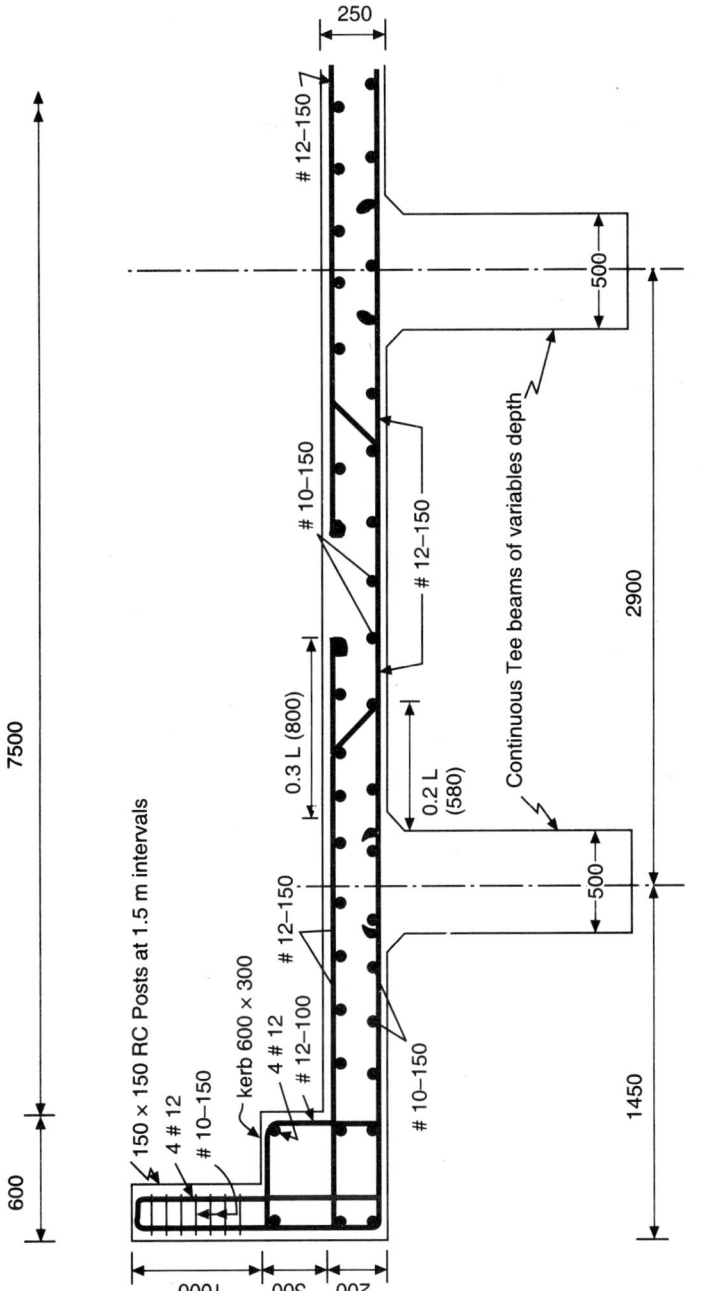

Fig. 15.17 Cross-section of Deck Slab.

The distribution factors are calculated using the following equation

$$D_{BA} = (K/\Sigma K) = [(kI_CE)/L]/[\Sigma(kI_CE)/L]$$
$$= (8.046/20)/[(8.046/20)] + [(22.8/30)]$$
$$= 0.3468 = D_{CD}$$
$$D_{BC} = (1 - 0.3468) = 0.6532 = D_{CB}$$

4. Moments at Supports B and C

The final moments at support B and C in terms of fixed end moments are calculated as follows:

(a) Load in Span AB

$$M_B = [(1 - D_{BA} - U)/(1 - U)]M_1$$
where
$$U = C_{BC} \cdot C_{CB} \cdot D_{BC} \cdot D_{CB}$$
$$= (-0.78)(-0.78)(0.6532)(0.6532)$$
$$= 0.2595$$
$$M_B = [(1 - 0.3468 - 0.2595)/(1 - 0.2595)]M_1$$
$$= 0.5316\ M_1$$
$$M_C = (V/1 - U)\ M_1$$
$$V = C_{BC} \cdot D_{BC} \cdot D_{CD}$$
$$= (-0.78 \times 0.6532 \times 0.3468) = -0.1766$$
$$M_C = [(-0.1766)/(1 - 0.2595)]\ M_1 = -0.2384\ M_1$$

(b) Load in Span BC

$$M_B = [(D_{BA} \cdot \overline{M}_{BC} - W \cdot \overline{M}_{CB})/(1 - U)]$$
$$W = C_{CB} \cdot D_{CB} \cdot D_{BA}$$
$$= (-0.78 \times 0.6532 \times 0.3468) = -0.1766$$
$$M_B = [(0.3468\ \overline{M}_{BC} + 0.1766\ \overline{M}_{CB})/(1 - 0.2595)]$$
$$= (0.4683\ \overline{M}_{BC} + 0.2384\ \overline{M}_{CB})$$
$$M_C = (D_{CD} \cdot \overline{M}_{CB} - V \cdot \overline{M}_{BC})/(1 - U)$$
$$= (0.3468\ \overline{M}_{CB} + 0.1766\ \overline{M}_{BC})/(1 - 0.2595)$$
$$= (0.4683\ \overline{M}_{CB} - 0.2384\ \overline{M}_{BC})$$

(c) Load in Span CD

$$M_B = [(W/1 - U)]\ M_3 = [(-0.1766)/(1 - 0.2595)]\ M_3$$
$$= -0.2384\ M_3$$
$$M_C = [(1 - D_{CD} - U)/(1 - U)]\ M_3$$
$$= [(1 - 0.3468 - 0.2595)/(1 - 0.2595)]\ M_3$$
$$= 0.5316\ M_3$$

Also
$$M_1 = \overline{M}_{BA} - C_{AB} \cdot \overline{M}_{AB}$$
$$M_3 = \overline{M}_{CD} - C_{DC} \cdot \overline{M}_{DC}$$

5. Influence Line Coefficients for Moments at Support B

The influence line coefficients for bending moment at support B is calculated for incremental positions of load in spans, AB, BC and CD respectively as compiled in Table 15.1.

In this table the coefficients are in terms of the length 'L'. The span lengths of. the design problem, $L_1 = 20$ m, $L_2 = 30$ m, $L_3 = 20$ m. The coefficients for M are multiplied by the respective span lengths depending upon the position of the load on span and influence line coefficients are derived.

The influence line ordinates for bending moment at support B is derived by multiplying the respective lengths of the spans, L_1, L_2 and L_3 depending upon the load position from 0.1 to 3.0. The influence line ordinates are compiled in Table 15.2 and the influence line plotted on the span is shown in Fig. 15.18.

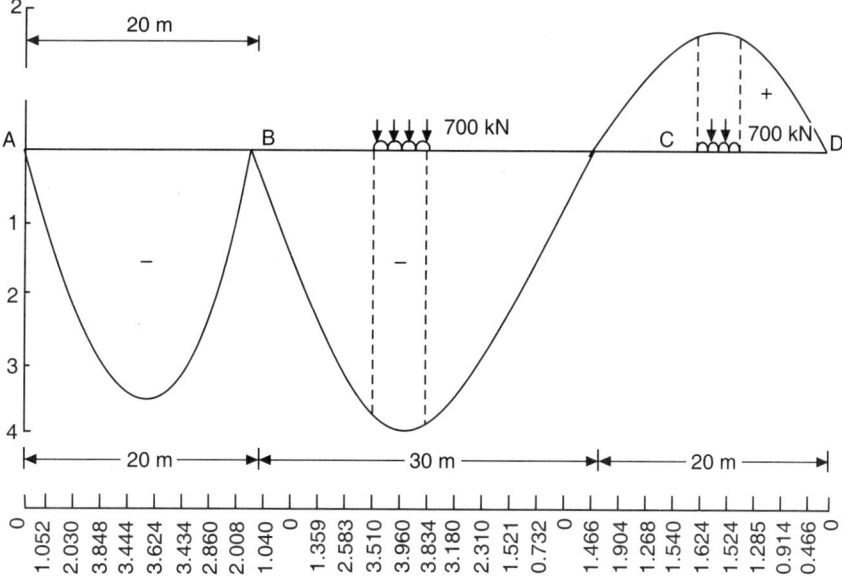

Fig. 15.18 Influence line for Bending Moment at Section of Support B.

6. Influence Line Ordinates at Various Sections

Similarly the influence line ordinates are derived for bending moment at sections 0.2 L, 0.4 L, 0.5 L, 0.6 L, 0.8 L, 1.2 L, 1.4 L, and 1.5 L and they are compiled in Tables 15.3 to 15.10. The influence line ordinates for these various sections are shown in Figs. 15.19 to 15.26.

Table 15.1 Coefficients for influence line for Moment at Support—B

A — 1 kN at αL_1 — B — 1 kN at αL_2 — C — 1 kN at αL_3 — D

Spans: L_1, L_2, L_3

Load Position	$\overline{M}_{AB}$	$\overline{M}_{BA}$	M_1	$\overline{M}_{BC}$	$\overline{M}_{CB}$	$\overline{M}_{CD}$	$\overline{M}_{DC}$	M_3	Load in Span AB M_B	Load in Span BC M_B	Load in Span CD M_B
0.1	−0.0755L	−0.0230L	−0.0990L	−0.0950L	−0.0040L	−0.0970L	−0.0010L	−0.0980L	−0.0526L_1	−0.0453L_2	+0.0233L_3
0.2	−0.1100L	−0.0800L	−0.1910L	−0.1740L	−0.0200L	−0.1840L	−0.0060L	−0.1900L	−0.1015L_1	−0.0861L_2	+0.0452L_3
0.3	−0.1120L	−0.1550L	−0.2680L	−0.2220L	−0.0550L	−0.2500L	−0.0160L	−0.2660L	−0.1424L_1	−0.1170L_2	+0.0634L_3
0.4	−0.0950L	−0.2250L	−0.3240L	−0.2250L	−0.1120L	−0.2805L	−0.0380L	−0.3230L	−0.1722L_1	−0.1320L_2	+0.0770L_3
0.5	−0.0660L	−0.2750L	−0.3410L	−0.1810L	−0.1810L	−0.2750L	−0.0660L	−0.3410L	−0.1812L_1	−0.1278L_2	+0.0812L_3
0.6	−0.0380L	−0.2850L	−0.3230L	−0.1120L	−0.2250L	−0.2250L	−0.0950L	−0.3200L	−0.1717L_1	−0.1060L_2	+0.0762L_3
0.7	−0.0180L	−0.2510L	−0.2690L	−0.0550L	−0.2220L	−0.1550L	−0.1150L	−0.2700L	−0.1430L_1	−0.0770L_2	+0.0643L_3
0.8	−0.0052L	−0.1840L	−0.1890L	−0.0200L	−0.1740L	−0.0820L	−0.1100L	−0.1920L	−0.1004L_1	−0.0507L_2	+0.0457L_3
0.9	−0.0010L	−0.0970L	−0.0980L	−0.0040L	−0.0950L	−0.0230L	−0.0755L	−0.0980L	−0.0520L_1	−0.0244L_2	+0.0233L_3

Table 15.2 Influence Line Ordinates for B.M. at Support B

Load Position	Influence Line Coefficient	I.L.O
0.1	$-0.0526\ L_1$	-1.052
0.2	$-0.1015\ L_1$	-2.030
0.3	$-0.1424\ L_1$	-2.848
0.4	$-0.1722\ L_1$	-3.444
0.5	$-0.1812\ L_1$	-3.624
0.6	$-0.1717\ L_1$	-3.434
0.7	$-0.1430\ L_1$	-2.860
0.8	$-0.1004\ L_1$	-2.008
0.9	$-0.0520\ L_1$	-1.040
1.0	0.0	0.0
1.1	$-0.0453\ L_2$	-1.359
1.2	$-0.0861\ L_2$	-2.583
1.3	$-0.1170\ L_2$	-3.510
1.4	$-0.1320\ L_2$	-3.960
1.5	$-0.1278\ L_2$	-3.834
1.6	$-0.1060\ L_2$	-3.180
1.7	$-0.0770\ L_2$	-2.310
1.8	$-0.0507\ L_2$	-1.521
1.9	$-0.0244\ L_2$	-0.732
2.0	0.0	0
2.1	$+0.0233\ L_3$	$+0.466$
2.2	$+0.0452\ L_3$	$+0.904$
2.3	$+0.0634\ L_3$	$+1.268$
2.4	$+0.0770\ L_3$	$+1.540$
2.5	$+0.0812\ L_3$	$+1.624$
2.6	$+0.0762\ L_3$	$+1.524$
2.7	$+0.0643\ L_3$	$+1.286$
2.8	$+0.0457\ L_3$	$+0.914$
2.9	$+0.0233\ L_3$	$+0.466$
3.0	0	0

Similarly influence line ordinates for shear force at support sections A and B compiled in Tables 15.11 and 15.12.

The influence for shear at supports are shown in Figs. 15.27 and 15.28.

7. Dead Load Bending Moments

(a) Self-weight of Deck Slab, Wearing Coat and Kerbs

Total dead load of deck slab and wearing coat = 7.76 kN/m²
Load due to kerb, R.C. posts etc. (Lump sum) = 0.24
Total load = 8 kN/m²

Table 15.3 Influence Line Ordinates for B.M. at Section 0.2 L
(4 m from Support A).

Load Position	Coefficient μ	Coefficient M_B	$(\mu + M_B)$	I.L.O.
0.1	0.08	−0.0105	+0.0695	+1.3900
0.2	0.16	−0.0203	+0.1397	+2.7940
0.3	0.14	−0.0284	+0.1116	+2.2320
0.4	0.12	−0.0344	+0.0856	+1.7120
0.5	0.10	−0.0362	+0.0638	+1.2760
0.6	0.08	−0.0343	+0.0457	+0.9140
0.7	0.06	−0.0286	+0.0314	+0.6280
0.8	0.04	−0.0200	+0.0200	+0.4000
0.9	0.02	−0.0104	+0.0096	+0.1920
1.0	0	0	0	0
1.1	—	−0.0090	−0.0090	−0.2700
1.2	—	−0.0172	−0.0172	−0.5160
1.3	—	−0.0234	−0.0234	−0.7020
1.4	—	−0.0264	−0.0264	−0.7920
1.5	—	−0.0255	−0.0255	−0.7650
1.6	—	−0.0212	−0.0212	−0.6360
1.7	—	−0.0154	−0.0154	−0.4620
1.8	—	−0.0101	−0.0101	0.3030
1.9	—	−0.0048	−0.0048	−0.1440
2.0	—	0	0	0
2.1	—	+0.0046	+0.0046	+0.0920
2.2	—	+0.0090	+0.0090	+0.1800
2.3	—	+0.0126	+0.0126	+0.2520
2.4	—	+0.0154	+0.0154	+0.3080
2.5	—	+0.0162	+0.0162	+0.3240
2.6	—	+0.0152	+0.0152	+0.3040
2.7	—	+0.0128	+0.0128	+0.2560
2.8	—	+0.0091	+0.0091	+0.1820
2.9	—	+0.0046	+0.0046	+0.0920
3.0	—	0	0	0

Loads transmitted to girders at 0.1 L sections are as follows:

In span AB = (8 × 2.9 × 2) = 47 kN
In span BC = (8 × 2.9 × 3) = 70 kN
Load transmitted at A = (8 × 2.9 × 1) = 23.5 kN
Load transmitted at B = (8 × 2.9 × 2.5) = 58.0 kN

(b) Self-weight of Main Girders

The self-weight of girders acting at various sections from 0.1 L to 1.5 L is compiled in Table 15.13. The main girders are of varying depth and of constant width of 500 mm. The depth of main girder at various sections is shown in Fig 15.29.

Table 15.4 Influence Line Ordinates for B.M. at Section 0.4 L

(8 m from Support A)

Load Position	Coefficient μ	Coefficient M_B	$(\mu + M_B)$	I.L.O.
0.1	0.06	−0.0210	+0.0390	+0.7800
0.2	0.12	−0.0406	+0.0794	+1.5880
0.3	0.18	−0.0569	+0.1231	+2.4620
0.4	0.24	−0.0688	+0.1712	+3.4240
0.5	0.20	−0.0724	+0.1276	+2.5520
0.6	0.16	−0.0686	+0.0914	+1.8280
0.7	0.12	−0.0572	+0.0628	+1.2560
0.8	0.08	−0.0401	+0.0399	+0.7980
0.9	0.04	−0.0208	+0.0192	+0.3840
1.0	0	0	0	0
1.1	—	−0.0181	−0.0181	−0.5430
1.2	—	−0.0344	−0.0344	−1.0320
1.3	—	−0.0468	−0.0468	−1.4040
1.4	—	−0.0528	−0.0528	−1.5840
1.5	—	−0.0511	−0.0511	−1.5330
1.6	—	−0.0424	−0.0424	−1.2720
1.7	—	−0.0308	−0.0308	−0.9240
1.8	—	−0.0202	−0.0202	−0.6060
1.9	—	−0.0097	−0.0097	−0.2910
2.0	—	0	0	0
2.1	—	+0.0093	+0.0093	+0.1860
2.2	—	+0.0180	+0.0180	+0.3600
2.3	—	+0.0253	+0.0253	+0.5060
2.4	—	+0.0308	+0.0308	+0.6160
2.5	—	+0.0324	+0.0324	+0.6480
2.6	—	+0.0304	+0.0304	+0.6080
2.7	—	+0.0257	+0.0257	+0.5140
2.8	—	+0.0182	+0.0182	+0.3640
2.9	—	+0.0093	+0.0093	+0.1860
3.0	—	0	0	0

(c) Self-weight of Cross Girders

Cross girders have the same depth as that of main girders and they are spaced at 4 m intervals in end spans and 6 m intervals in central span. Width of cross girder = 300 mm. The self-weight of cross girder acting at various sections is compiled in Table 15.14.

(d) Dead Load bending Moments at Various Sections

The total dead load acting at various sections is shown in Table 15.15.

(i) Section 0.2 L (4 m from Support A)

Referring to Table 15.14 and influence line shown in Fig. 15.19 we compute the bending moment as

Table 15.5 Influence Line Ordinates for B.M. at Section 0.5 *L*
(10 m from Support A)

Load Position	Coefficient μ	Coefficient M_B	$(\mu + M_B)$	I.L.O.
0.1	0.05	−0.0263	+0.0237	+0.474
0.2	0.10	−0.0507	+0.0493	+0.986
0.3	0.15	−0.0712	+0.0788	+1.576
0.4	0.20	−0.0861	+0.1139	+2.278
0.5	0.25	−0.0906	+0.1594	+3.188
0.6	0.20	−0.0858	+0.1142	+2.284
0.7	0.15	−0.0715	+0.0785	+1.570
0.8	0.10	−0.0502	+0.0498	+0.996
0.9	0.05	−0.0026	+0.0024	+0.048
1.0	0	0	0	0
1.1	—	−0.0226	−0.0226	−0.678
1.2	—	−0.0430	−0.0430	−1.290
1.3	—	−0.0585	−0.0585	−1.755
1.4	—	−0.0660	−0.0660	−1.980
1.5	—	−0.0639	−0.0639	−1.917
1.6	—	−0.0530	−0.0530	−1.590
1.7	—	−0.0385	−0.0385	−1.155
1.8	—	−0.0253	−0.0253	−0.759
1.9	—	−0.0122	−0.0122	−0.366
2.0	—	0	0	0
2.1	—	+0.0116	+0.0116	+0.232
2.2	—	+0.0226	+0.0226	+0.452
2.3	—	+0.0317	+0.0317	+0.634
2.4	—	+0.0385	+0.0385	+0.770
2.5	—	+0.0406	+0.0406	+0.812
2.6	—	+0.0381	+0.0381	+0.762
2.7	—	+0.0321	+0.0321	+0.642
2.8	—	+0.0229	+0.0229	+0.458
2.9	—	+0.0117	+0.0117	+0.234
3.0	—	0	0	0

$$= [72 \,(1.39 + 0.092) + 96 \,(2.794 + 0.182) + 78 \,(2.230 + 0.256)$$
$$+ \,110 \,(1.712 + 0.304) + 83 \,(2.276 + 0.324) + 123 \,(0.194 + 0.308)$$
$$+ \,93 \,(0.628 + 0.252) + 141 \,(0.400 + 0.18) + 107 \,(0.192 + 0.092)]$$
$$- \,[149 \,(0.270 + 0.144) + 173 \,(0.516 + 0.303) + 117 \,(0.702 + 0.462)$$
$$+ \,133 \,(0.792 + 0.636) + 142 \,(0.765)] = 647 \text{ kN·m}$$

(ii)   Section 0.4 L (8 m from Support A)

Referring to the influence line shown in Fig. 15.20, we compute the bending moment as

$$= [72 \,(0.780 + 0.186) + 96 \,(1.588 + 0.364) + 78 \,(2.462 + 0.514)$$
$$+ \,110 \,(3.424 + 0.608) + 83 \,(2.552 + 0.648) + 123 \,(1.828 + 0.616)$$
$$+ \,93 \,(1.256 + 0.506) + 141 \,(0.798 + 0.360) + 107 \,(0.384 + 0.186)]$$

Table 15.6 Influence Line Ordinates for B.M. at Section 0.6 *L*

(12 m from Support A)

Load Position	Coefficient μ	Coefficient M_B	$(\mu + M_B)$	I.L.O.
0.1	0.04	−0.0315	+0.0085	+0.170
0.2	0.08	−0.0609	+0.0191	+0.382
0.3	0.12	−0.0854	+0.0346	+0.692
0.4	0.16	−0.1033	+0.5067	+1.134
0.5	0.20	−0.1087	+0.0913	+1.826
0.6	0.24	−0.1030	+0.1370	+2.740
0.7	0.18	−0.0858	+0.0942	+1.884
0.8	0.12	−0.0602	+0.0598	+1.196
0.9	0.06	−0.0312	+0.0288	+0.576
1.0	0	0	0	0
1.1	—	−0.0271	−0.0271	−0.813
1.2	—	−0.0516	−0.0516	−1.548
1.3	—	−0.0702	−0.0702	−2.106
1.4	—	−0.0792	−0.0792	−2.376
1.5	—	−0.0766	−0.0766	−2.298
1.6	—	−0.0636	−0.0636	−1.908
1.7	—	−1.0462	−0.0462	−1.386
1.8	—	−0.0304	−0.0304	−0.912
1.9	—	−0.0146	−0.0146	−0.438
2.0	—	0	0	0
2.1	—	+0.0139	+0.0139	+0.278
2.2	—	+0.0271	+0.0271	+0.542
2.3	—	+0.0380	+0.0380	+0.760
2.4	—	+0.0462	+0.0462	+0.924
2.5	—	+0.0487	+0.0487	+0.974
2.6	—	+0.0457	+0.0457	+0.914
2.7	—	+0.0385	+0.0385	+0.770
2.8	—	+0.0274	+0.0274	+0.548
2.9	—	+0.0139	+0.0139	+0.278
3.0	—	0	0	0

$$- [149 (0.543 + 0.291) + 173 (1.032 + 0.606) + 117 (1.404 + 0.924)$$
$$+ 133 (1.584 + 1.272) + 142 (1.533)] = 660 \text{ kN·m}$$

(iii) Section 0.5 L (10 m from support A)

Referring to the influence line shown in Fig. 15.21, we compute the bending moment as

$$= [72 (0.474 + 0.234) + 96 (0.986 + 0.458) + 78 (1.576 + 0.642)$$
$$+ 110 (2.278 + 0.762) + 83 (3.188 + 0.812) + 123 (2.284 + 0.770)$$
$$+ 93 (1.57 + 0.634) + 141 (0.996 + 0.452) + 107 (0.480 + 0.232)]$$
$$- [149 (0.678 + 0.366) + 173 (1.29 + 0.759) + 117 (1.755 + 0.155)$$
$$+ 133 (1.980 + 1.590) + 142 (1.917)] = 293 \text{ kN·m}$$

Table 15.7 Influence Line Ordinates for B.M. at Section 0.8 L
(16 m from Support A)

Load Position	Coefficient μ	Coefficient M_B	$(\mu + M_B)$	I.L.O.
0.1	0.02	−0.0420	−0.0220	−0.440
0.2	0.04	−0.0812	−0.0412	−0.824
0.3	0.06	−0.1139	−0.0539	−1.078
0.4	0.08	−0.1377	−0.0577	−1.154
0.5	0.10	−1.1449	−0.0449	−0.898
0.6	0.12	−0.1373	−0.0173	−0.346
0.7	0.14	−0.1144	+0.0256	+0.512
0.8	0.16	−0.0803	+0.0797	+1.594
0.9	0.80	−0.0416	+0.0384	+0.768
1.0	0	0	0	0
1.1	—	−0.0362	−0.0362	−1.086
1.2	—	−0.0688	−0.0688	−2.064
1.3	—	−0.0936	−0.0936	−2.808
1.4	—	−0.1056	−0.1056	−3.168
1.5	—	−0.1022	−0.1022	−3.066
1.6	—	−0.0848	−0.0848	−2.544
1.7	—	−0.0616	−0.0616	−1.848
1.8	—	−0.0405	−0.0405	−1.215
1.9	—	−0.0195	−0.0195	−0.585
2.0	—	0	0	0
2.1	—	+0.0186	+0.0186	+0.372
2.2	—	+0.0361	+0.0361	+0.0722
2.3	—	+0.0507	+0.0507	+1.014
2.4	—	+0.0616	+0.0616	+1.232
2.5	—	+0.0649	+0.0649	+1.298
2.6	—	+0.0609	+0.0609	+1.218
2.7	—	+0.0514	+0.0514	+1.028
2.8	—	+0.0365	+0.0365	+0.730
2.9	—	+0.0186	+0.0186	+0.372
3.0	—	0	0	0

(iv) Section 0.6 L (12 m from Support A)

Referring to the influence line shown in Fig. 15.22 we compute the bending moment as

$$= [72\,(0.170 + 0.278) + 96\,(0.382 + 0.548) + 78\,(0.692 + 0.770)$$
$$+ 110\,(1.134 + 0.914) + 83\,(1.826 + 0.974) + 123\,(2.74 + 0.924)$$
$$+ 93\,(1.884 + 0.760) + 141\,(1.196 + 0.542) + 107\,(0.576 + 0.278)]$$
$$- [149\,(0.813 + 0.438) + 173\,(1.548 + 0.912) + 117\,(2.106 + 1.386)$$
$$+ 133\,(2.376 + 1.908) + 142\,(2.298)] = -190 \text{ kN·m}$$

(v) Section 0.8 L (16 m from Support A)

Referring to the influence line shown in Fig. 15.23 we compute the bending moment as

Table 15.8 Influence Line Ordinates for Bending Moment at Section 1.2 L
(6 m from Support B)

Load Position	Coefficient μ	Coefficient M_B	Coefficient M_C	$(\mu + M_B + M_C)$	I.L.O.
0.1	—	−0.0420	+0.0046	−0.0374	−0.748
0.2	—	−0.0812	+0.0091	−0.0721	−1.442
0.3	—	−0.1139	+0.0128	−0.1011	−2.022
0.4	—	−0.1377	+0.0152	−0.1225	−2.450
0.5	—	−0.1449	+0.0162	−0.1287	−2.574
0.6	—	−0.1373	+0.0154	−0.12'19	−2.438
0.7	—	−0.1144	+0.0126	−0.1018	−2.036
0.8	—	−0.0803	+0.0090	−0.0713	−1.426
0.9	—	0.0416	+0.0046	−0.0370	−0.7400
1.0	0	0	0	0	0
1.1	0.08	−0.0362	−0.0048	+0.0390	+1.170
1.2	0.16	−0.0688	−0.0101	+0.0811	+2.433
1.3	0.14	−0.0936	−0.0154	+0.0310	+0.930
1.4	0.12	−0.1056	−0.0212	−0.0068	−0.204
1.5	0.10	−0.1022	−0.0255	−0.0277	−0.831
1.6	0.08	−0.0848	−0.0264	−0.0312	−0.936
1.7	0.06	−0.0616	−0.0234	−0.0250	−0.750
1.8	0.04	−0.0405	−0.0172	−0.0177	−0.53 1
1.9	0.02	−0.0195	−0.0090	−0.0085	−0.255
2.0	0	0	0	0	0
2.1	—	+0.0186	−0.0104	+0.0082	+0.164
2.2	—	+0.0361	−0.0200	+0.0161	+0.322
2.3	—	+0.0507	−0.0286	+0.0221	+0.442
2.4	—	+0.0616	−0.0343	+0.0273	+0.546
2.5	—	+0.0649	−0.0362	+0.0287	+0.574
2.6	—	+0.0609	−0.0344	+0.0265	+0.530
2.7	—	+0.0514	−0.0284	+0.0230	+0.460
2.8	—	+0.0365	−0.0203	+0.01620	+0.324
2.9	—	+0.0186	−0.0105	+0.0081	+0.162
3.0	—	0	0	0	0

$$= [(93 \times 0.512) + (141 \times 1.594) + (107 \times 0.768) + (72 \times 0.372)$$
$$+ (96 \times 0.73) + (78 \times 1.08) + (110 \times 1.218) + (83 \times 1.298)$$
$$+ (123 \times 1.232) + (93 \times 1.014) + (141 \times \smile.722) + (107 \times 0.372)]$$
$$- [(72 \times 0.444) + (96 \times 0.824) + (78 \times 1.078) + (110 \times 1.154)$$
$$+ (83 \times 0.898) + (123 \times 0.346) + 149 (1.086 + 0.585)$$
$$+ 173 (2.064 + 1.215) + 117 (2.808 + 1.848) + 133 (3.168 + 2.544)$$
$$+ (142 \times 3.066)] = -1835 \text{ kN·m}$$

(vi)   Section 1.0 L (Support Section B)

Referring to the influence line in Fig. 15.18 we compute the bending moment as

$$= [72 (0.466 - 1.052) + 96 (0.914 - 2.030) + 78 (1.286 - 2.848)$$
$$+ 110 (1.524 - 3.444) + 83 (1.624 - 3.624) + 123 (1.540 - 3.434)$$

Table 15.9 Influence Line Ordinates for Bending Moment at Section 1.4 *L*
(12 m from Support B)

Load Position	Coefficient μ	Coefficient M_B	Coefficient M_C	$(\mu + M_B + M_C)$	I.L.O.
0.1	—	−0.0315	+0.0093	−0.0222	0.444
0.2	—	−0.0609	+0.0182	−0.0427	−0.854
0.3	—	−0.0854	+0.0257	−0.0597	−1.194
0.4	—	−0.1033	+0.0304	−0.0729	−1.458
0.5	—	−0.1087	+0.0324	−0.0763	−1.526
0.6	—	−0.1030	+0.0308	−0.0722	−1.444
0.7	—	−0.0858	+0.0253	−0.0605	−1.210
0.8	—	−0.0602	+0.0180	−0.0422	−0.844
0.9	—	−0.0312	+0.0093	−0.0219	−0.438
1.0	0	0	0	0	0
1.1	0.06	−0.0271	−0.0097	+0.0232	+0.696
1.2	0.12	−0.0516	−0.0202	+0.0482	+1.446
1.3	0.18	−0.0702	−0.0308	+0.0790	+2.370
1.4	0.24	−0.0792	−0.0424	+0.1184	+3.552
1.5	0.20	−0.0766	−0.0511	+0.0723	+2.169
1.6	0.16	−0.0636	−0.0528	+0.0436	+1.308
1.7	0.12	−0.0462	−0.0468	+0.0270	+0.810
1.8	0.08	−0.0304	−0.0344	+0.0152	+0.456
1.9	0.04	−0.0146	−0.0181	+0.0073	+0.219
2.0	0	0	0	0	0
2.1	—	+0.0139	−0.0208	−0.0069	−0.138
2.2	—	+0.0271	−0.0401	−0.0130	−0.260
2.3	—	+0.0380	−0.0572	−0.0192	−0.384
2.4	—	+0.0462	−0.0686	−0.0224	−0.448
2.5	—	+0.0487	−0.0724	−0.0237	−0.474
2.6	—	+0.0457	−0.0688	−0.0231	−0.462
2.7	—	+0.0385	−0.0569	−0.0184	−0.368
2.8	—	+0.0274	−0.0406	−0.0132	−0.264
2.9	—	+0.0139	−0.0210	−0.0071	−0.142
3.0	—	0	0	0	0

$$+ 93 (1.268 - 2.860) + 141 (0.904 - 2.008) + 107 (0.466 - 1.040)]$$
$$- [(149 (1.359 + 0.732) + 173 (2.583 + 1.521) + 117 (3.510 + 2.31)$$
$$+ 133 (3.96 + 3.18) + 142 (3.834)] = - 4443 \text{ kN·m}$$

(vii) Section 1.2 L (6 m from Support B)

Referring to the influence line shown in Fig. 15.24 we compute the bending moment as

$$= [72 (0.162 - 0.748) + 96 (0.324 - 1.442) + 78 (0.46 - 2.022)$$
$$+ 110 (0.53 - 2.45) + 83 (0.574 - 2.574) + 123 (0.546 - 2.438)$$
$$+ 93 (0.442 - 2.036) + 141 (0.322 - 1.426) + 107 (0.164 - 0.74)$$
$$+ 149 (1.170 - 0.225) + 173 (2.433 - 0.531)$$
$$+ 117 (0.930 - 0.750) - (133 \times 0.204) - (142 \times 0.831)$$
$$- (133 \times 0.836)] = - 1030 \text{ kN·m}$$

Table 15.10 Influence Line Ordinates for Bending Moment at Section 1.5 L
(15 m from Support B)

Load Position	Coefficient μ	Coefficient M_B	Coefficient M_C	$(\mu + M_B + M_C)$	I.L.O.
0.1	—	−0.0263	+0.0117	−0.0146	−0.292
0.2	—	−0.0507	+0.0229	−0.0278	−0.556
0.3	—	−0.0712	+0.0321	−0.0391	−0.782
0.4	—	−0.0861	+0.0381	−0.0048	−0.960
0.5	—	−0.0906	+0.0406	−0.0554	−1.108
0.6	—	−0.0858	+0.0385	−0.0473	−0.946
0.7	—	−0.0715	+0.0317	−0.0398	−0.796
0.8	—	−0.0502	+0.0226	−0.0276	−0.552
0.9	—	−0.0026	+0.0116	−0.0144	−0.288
1.0	0	0	0	0	0
1.1	0.05	−0.0226	−0.0122	+0.0152	+0.456
1.2	0.10	−0.0430	−0.0253	+0.0317	+0.951
1.3	0.15	−0.0585	−0.0385	+0.0530	+1.590
1.4	0.20	−0.0660	−0.0530	+0.0081	+2.430
1.5	0.25	−0.0639	−0.0639	+0.1222	+3.660
1.6	0.20	−0.0530	−0.0660	+0.0081	+2.430
1.7	0.15	−0.0385	−0.0585	+0.0530	+1.590
1.8	0.10	−0.0253	+0.0430	+0.0317	+0.951
1.9	0.05	−0.0122	−0.0226	+0.0152	+0.456
2.0	0	0	0	0	0
2.1	—	+0.0116	−0.0260	−0.0144	−0.288
2.2	—	+0.0226	−0.0502	−0.0276	−0.552
2.3	—	+0.0317	−0.0715	−0.0398	−0.796
2.4	—	+0.0385	−0.0858	−0.0473	−0.946
2.5	—	+0.0406	−0.0906	−0.0050	−0.108
2.6	—	+0.0381	−0.0861	−0.0048	−0.960
2.7	—	+0.0321	−0.0712	−0.0391	−0.782
2.8	—	+0.0229	−0.0507	−0.0278	−0.556
2.9	—	+0.0117	−0.0263	−0.0146	−0.292
3.0	—	0	0	0	0

(viii) Section 1.4 L (12 m from Support B)

Referring to the influence line shown in Fig. 15.25 we compute the bending moment as

$$= [72 \, (-0.444 - 0.142) + 96 \, (-0.854 - 0.264) + 78 \, (-1.194 - 0.368)$$
$$+ 100 \, (-1.458 - 0.462) + 83 \, (-1.526 - 0.474)$$
$$+ 123 \, (-1.444 - 0.448) + 93 \, (-1.210 - 0.384)$$
$$+ 141 \, (-0.844 - 0.260) + 107 \, (-0.438 - 0.138)]$$
$$+ [149 \, (0.696 + 0.219) + 173 \, (1.446 + 0.456) + 117 \, (2.370 + 0.810)$$
$$+ 133 \, (3.552 + 1.308) + (142 \times 2.169)] = 545 \text{ kN·m}$$

(ix) Section 1.5 L (15 m from Support B)

Referring to the influence line shown in Fig. 15.26 we have the bending moment as

$$= [72 \,(- \,2 \times 0.292) + 96 \,(- \,2 \times 0.556) + 78 \,(- \,2 \times 0.782)$$
$$+ \,110 \,(- \,2 \times 0.960) + 83 \,(- \,2 \times 1.108) + 123 \,(- \,2 \times 0.946)$$
$$+ \,93 \,(- \,2 \times 0.796) + 141 \,(- \,2 \times 0.552) + 107 \,(- \,2 \times 0.288)]$$
$$+ \,[149 \,(2 \times 0.456) + 173 \,(2 \times 0.951) + 117 \,(2 \times 1.590)$$
$$+ \,133 \,(2 \times 2.430) + (142 \times 3.660)] = 1740 \text{ kN·m}$$

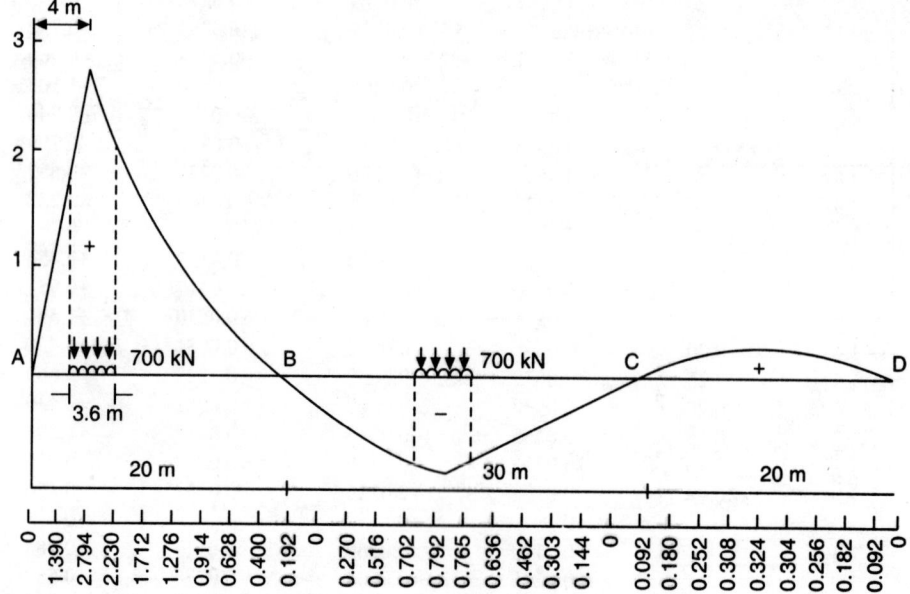

Fig. 15.19 Influence Line for Bending Moment at Section 0.2 *L*
(4 m from Support A).

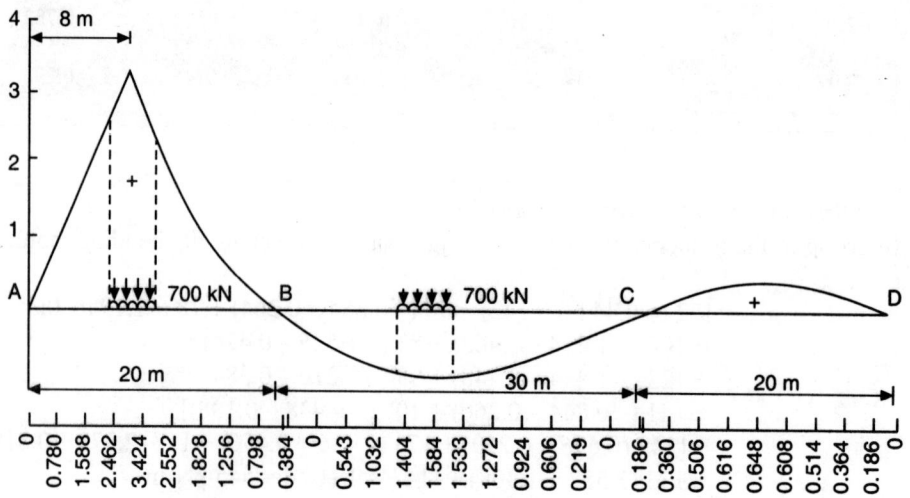

Fig. 15.20 Influence Line for Bending Moment at Section 0.4 *L*
(8 m from Support A).

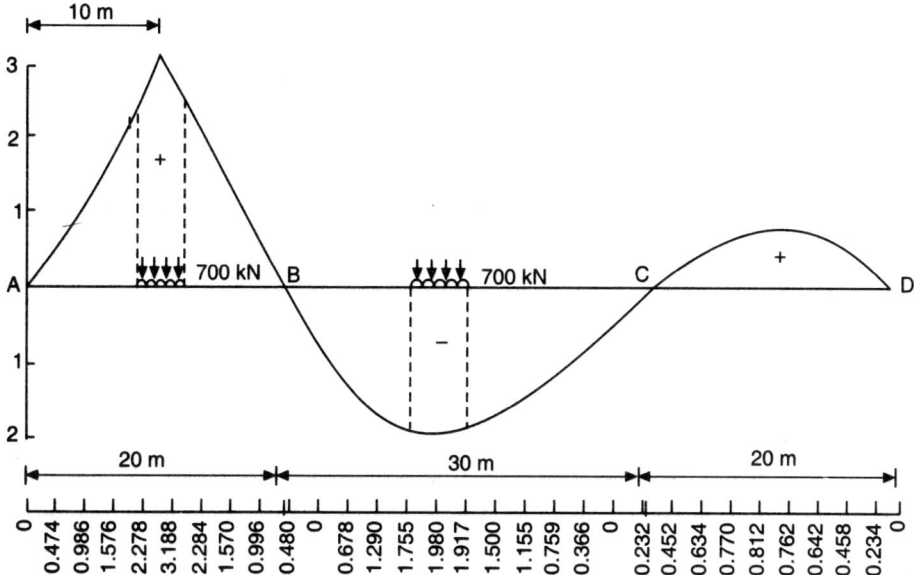

Fig. 15.21 Influence Line for Bending Moment at Section 0.5 L
(10 m from Support A).

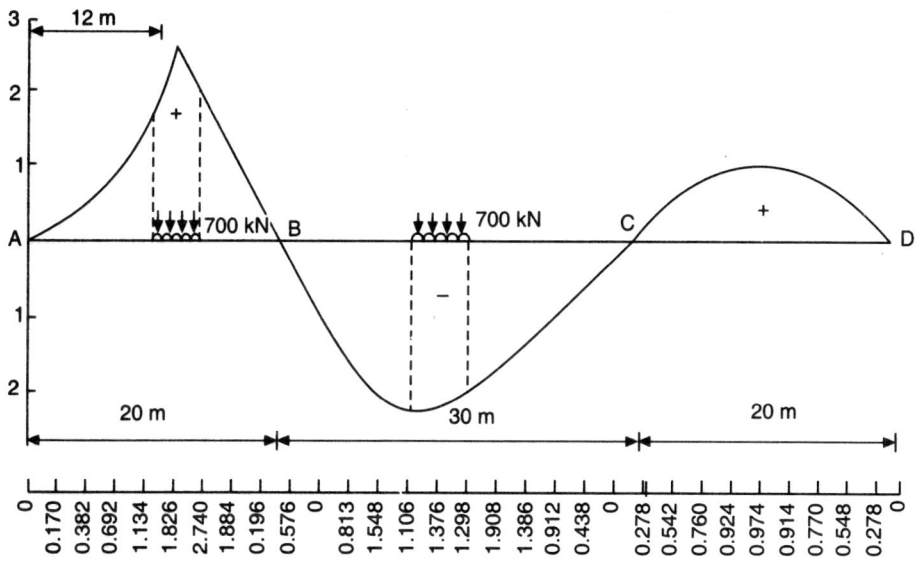

Fig. 15.22 Influence Line for Bending Moment at Section 0.6 L
(12 m from Support A).

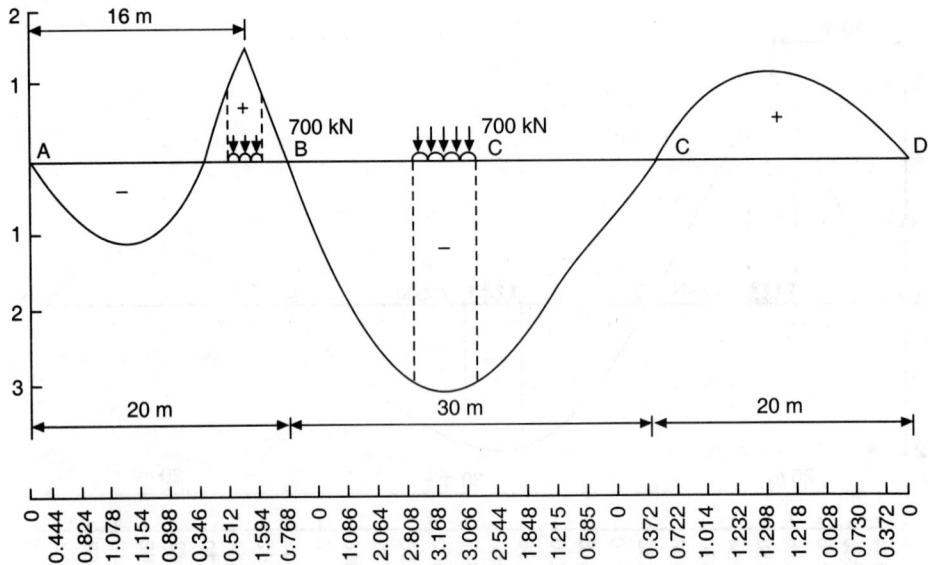

Fig. 15.23 Influence Line for Bending Moment at Section 0.8 L
(16 m from Support A).

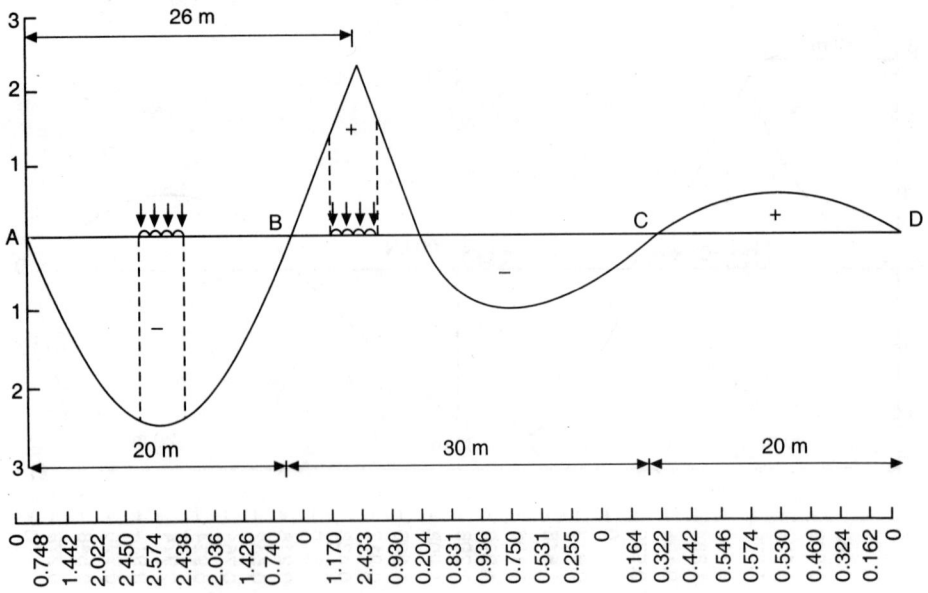

Fig. 15.24 Influence Line for Bending Moment at Section 1.2 L
(6 m from Support B).

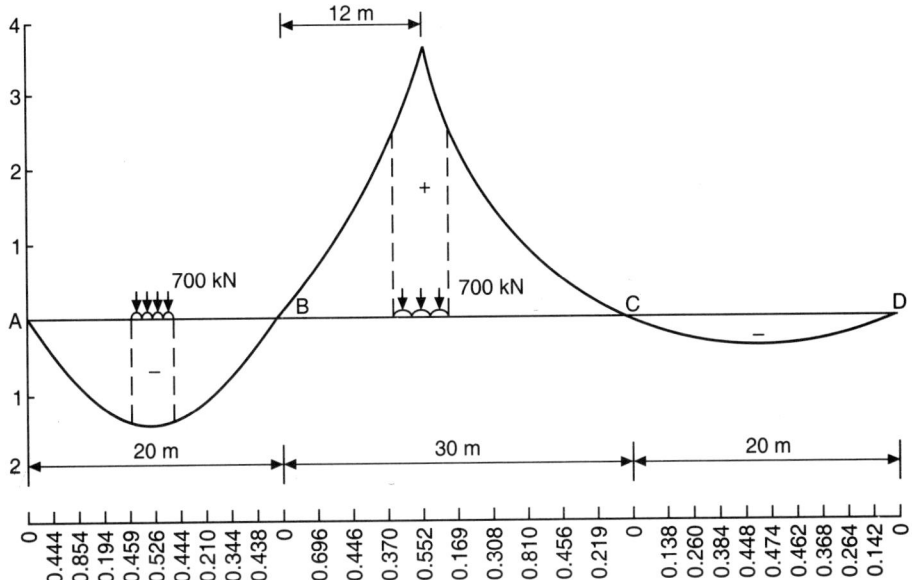

Fig. 15.25 Influence Line for Bending Moment at Section 1.4 L
(12 m from Support B).

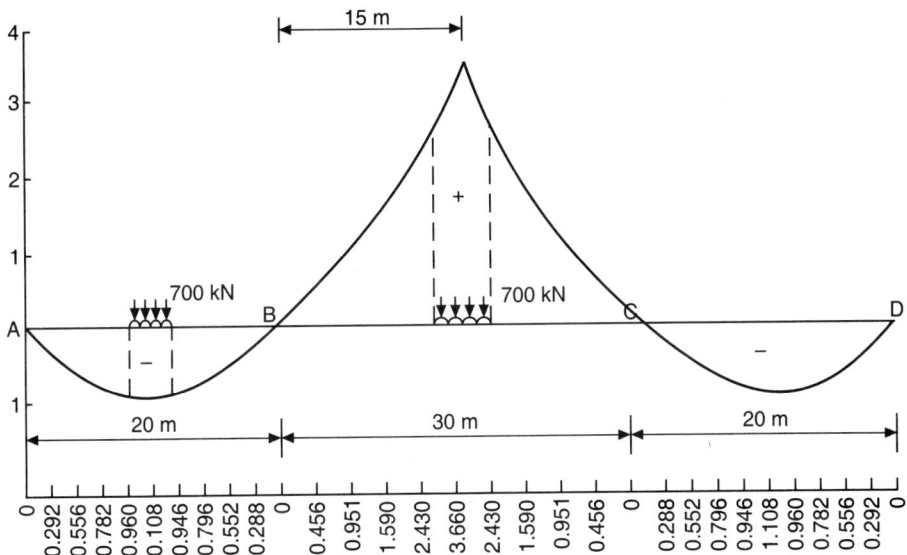

Fig. 15.26 Influence Line for Bending Moment at Section 1.5 L
(15 m from Support B).

Table 15.11 Influence Line Ordinate for Shear Force at Support A

Load Position	Coefficient μ	M_B	M_B/L_1	$(μ + M_B/L_1)$
A	1.0	0	0	+1.0000
0.1	0.9	−1.052	−0.0526	+0.8474
0.2	0.8	−2.030	−0.1015	+0.6985
0.3	0.7	−2.848	−0.1424	+0.5576
0.4	0.6	−3.444	−0.1722	+0.4278
0.5	0.5	−3.624	−0.1812	+0.3188
0.6	0.4	−3.434	−0.1717	+0.2283
0.7	0.3	−2.860	−0.1430	+0.1570
0.8	0.2	−2.008	−0.1004	+0.0996
0.9	0.1	−1.040	−0.052	+0.0480
1.0	0	0	0	0
1.1	—	−1.359	−0.0679	−0.0679
1.2	—	−2.583	−0.1291	−0.1291
1.3	—	−3.510	−0.1755	−0.1755
1.4	—	−3.960	−0.1980	−0.1980
1.5	—	−3.834	−0.1917	−0.1917
1.6	—	−3.180	−0.1590	−0.1590
1.7	—	−2.310	−0.1155	−0.1155
1.8	—	−1.521	−0.0760	−0.0760
1.9	—	−0.732	−0.0366	−0.0366
2.0	—	0	0	0
2.1	—	+0.446	+0.0233	+0.0233
2.2	—	+0.904	+0.0452	+0.0452
2.3	—	+1.268	+0.0634	+0.0634
2.4	—	+1.540	+0.0077	+0.0077
2.5	—	+1.624	+0.0081	+0.0081
2.6	—	+1.524	+0.0076	+0.0076
2.7	—	+1.286	+0.0064	+0.0064
2.8	—	+0.914	+0.0457	+0.0457
2.9	—	+0.466	+0.0233	+0.0233
3.0	—	0	0	0

8.   Live Load Bending Moments

(a)   Reaction Factors

The main girders are connected rigidly by cross girders and deck slab. Hence Courbon's theory for the load distribution is adopted. The I.R.C. class AA tracked vehicle is arranged for maximum eccentricity as shown in Fig. 15.30.

Reaction factor for outer girder is

$$R_A = (2W_1/3) [1 + (37 \times 2.9 \times 1.1)/(27 \times 2.9^2)] = 1.045\ W_1$$

Reaction factor for inner girder is

$$R_B = (2W_1/3) [1 + 0] = 0.667\ W_1$$

Table 15.12 Influence Line Ordinate for Shear Force at Support B
(Left of the Section)

Load Position	Coefficient μ	M_B	M_B/L_1	$(\mu + M_B/L_1)$
A	0	0	0	0
0.1	−0.1	−1.052	−0.0526	−0.1526
0.2	−0.2	−2.030	−0.1015	−0.3015
0.3	−0.3	−2.848	−0.1424	−0.4424
0.4	−0.4	−3.444	−0.1722	−0.5722
0.5	−0.5	−3.624	−0.1812	−0.6812
0.6	−0.6	−3.434	−0.1717	−0.7717
0.7	−0.7	−2.860	−0.1430	−0.8430
0.8	−0.8	−2.008	−0.1004	−0.9004
0.9	−0.9	−1.040	−0.0520	−0.9520
1.0	−1.0	0	0	0
1.1	—	−1.359	−0.0679	−0.0679
1.2	—	−2.583	−0.1291	−0.1291
1.3	—	−3.510	−0.1755	−0.1755
1.4	—	−3.960	−0.1980	−0.1980
1.5	—	−3.834	−0.1917	−0.1917
1.6	—	−3.180	−0.1590	−0.1590
1.7	—	−2.310	−0.1155	−0.1155
1.8	—	−1.521	−0.0760	−0.0760
1.9	—	−0.732	−0.0366	−0.0366
2.0	—	0	0	0
2.1	—	+0.466	+0.0233	+0.0233
2.2	—	+0.904	+0.0452	+0.0452
2.3	—	+1.268	+0.0634	+0.0634
2.4	—	+1.540	+0.0077	+0.0077
2.5	—	+1.624	+0.0081	+0.0081
2.6	—	+1.524	+0.0076	+0.0076
2.7	—	+1.286	+0.0064	+0.0064
2.8	—	+0.914	+0.0457	+0.0457
2.9	—	+0.466	+0.0233	+0.0233
3.0	—	0	0	0

If $\qquad W = $ (Axle load $= 700$ kN, $W_1 = 0.5\ W$)

∴ $\qquad R_A = (1.045 \times 0.5\ W) = 0.5225\ W$

$\qquad R_B = (0.667 \times 0.5\ W) = 0.3335\ W$

Impact factor for class AA tracked vehicle is 10% upto a span of 40 m. The influence lines shown in Figs. 15.18 to 15.26, are used to compute the maximum live load bending moments at the various sections. The load position giving the maximum positive and negative bending moment is shown on the influence lines drawn for various sections. Corresponding to the load position shown, the values of the bending moments obtained for I.R.C. class AA tracked vehicle is compiled in Table 15.16.

9. Design Bending Moments

The live load and dead load bending moments at various sections are compiled in Table 15.17.

The design bending moments are obtained by combining the dead load and live load moment for inner and outer girders.

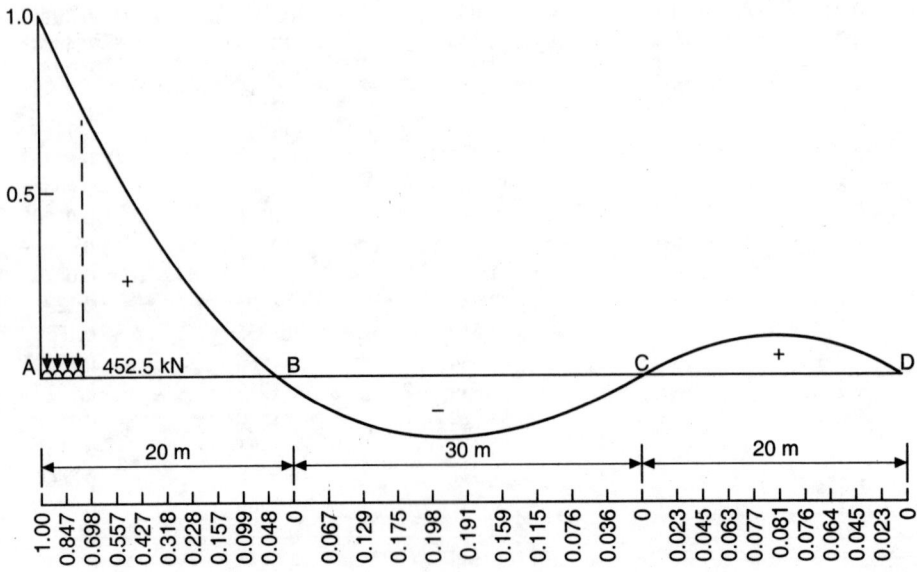

Fig. 15.27 Influence Line Ordinates for Shear Force at Support A.

Table 15.13 Self-weight of Main Girder at Various Sections

Section	Load Calculations	Load (kN)
A	(0.5 × 1 × 1 × 24)	12.0
0.1	(0.5 × 1.05 × 2 × 24)	25.2
0.2	(0.5 × 1.1 × 2 × 24)	26.4
0.3	(0.5 × 1.3 × 2 × 24)	31.2
0.4	(0.5 × 1.4 × 2 × 24)	33.6
0.5	(0.5 × 1.5 × 2 × 24)	36.0
0.6	(0.5 × 1.7 × 2 × 24)	40.8
0.7	(0.5 × 1.9 × 2 × 24)	45.6
0.8	(0.5 × 2.1 × 2 × 24)	50.4
0.9	(0.5 × 2.5 × 2 × 24)	60.0
1.0	(0.5 × 3.0 × 2.5 × 24)	90.0
1.1	(0.5 × 2.2 × 3 × 24)	79.2
1.2	(0.5 × 1.8 × 3 × 24)	64.8
1.3	(0.5 × 1.3 × 3 × 24)	46.8
1.4	(0.5 × 1.1 × 3 × 24)	39.6
1.5	(0.5 × 1 × 3 × 14)	72.0

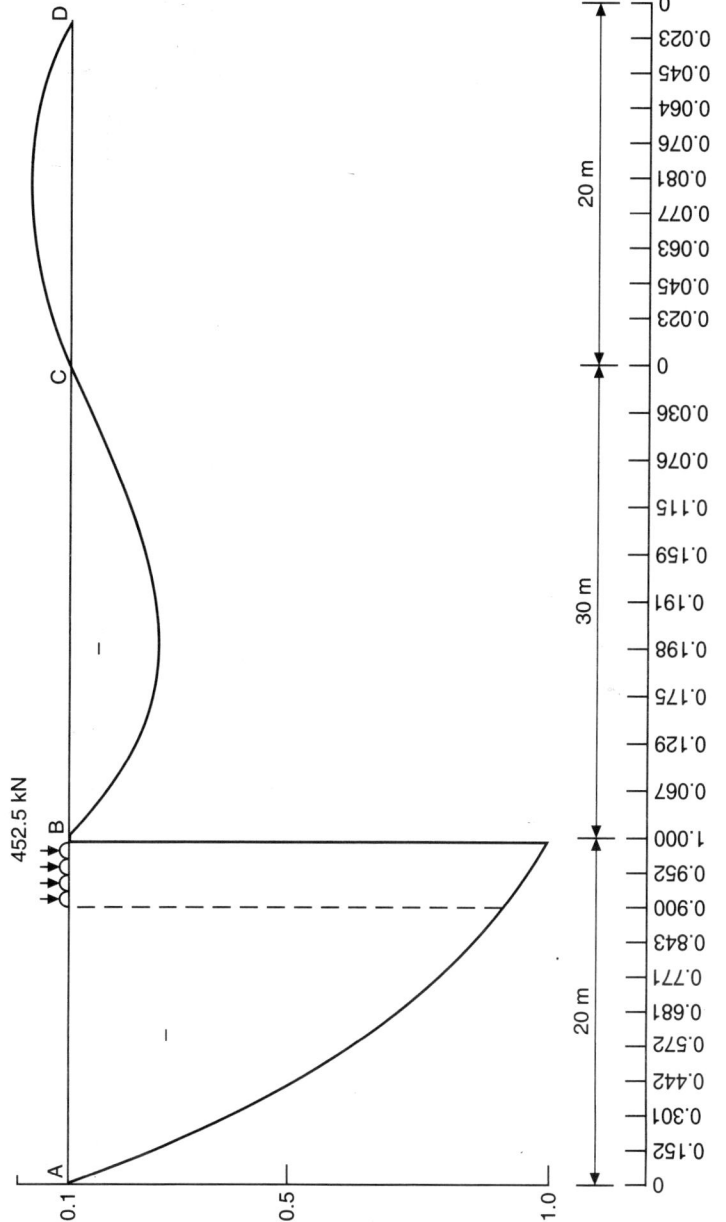

Fig. 15.28 Influence Line Ordinates for Shear Force at Support B (Left of the Section).

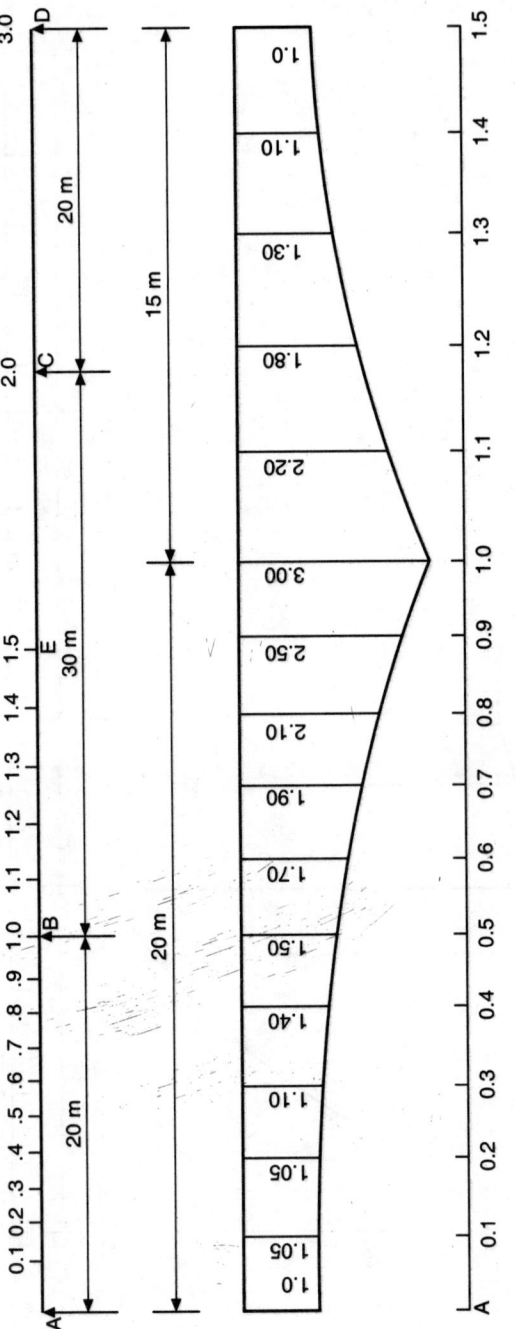

Fig. 15.29 Depth of Main Girder at Various Sections.

Table 15.14 Self-weight of Cross Girder at Various Sections

Section	Load Calculations	Load (kN)
A	$(0.5 \times 1 \times 1 \times 24)$	12.0
A	$(0.3 \times 1 \times 2.9 \times 24)$	20.9
0.2	$(0.3 \times 1.1 \times 2.9 \times 24)$	22.9
0.4	$(0.3 \times 1.4 \times 2.9 \times 24)$	29.2
0.6	$(0.3 \times 1.7 \times 2.9 \times 24)$	35.5
0.8	$(0.3 \times 2.1 \times 2.9 \times 24)$	43.8
1.0	$(0.3 \times 3 \times 2.9 \times 24)$	62.6
1.2	$(0.3 \times 1.8 \times 2.9 \times 24)$	37.6
1.4	$(0.3 \times 1.1 \times 2.9 \times 24)$	22.9

Table 15.15 Total Dead Load Acting at Various Sections

Section	Load due to Deck Slab	Load due to Main Girder	Load due to Cross Girder	Total Load (kN)
A	23.5	12.0	20.9	57
0.1	47.0	25.2	—	72
0.2	47.0	26.4	22.9	96
0.3	47.0	31.2	—	78
0.4	47.0	33.6	29.2	110
0.5	47.0	36.0	—	83
0.6	47.0	40.8	35.5	123
0.7	47.0	45.6	—	93
0.8	47.0	50.4	43.8	141
0.9	47.0	60.0	—	107
1.0	58.0	90.0	62.6	211
1.1	70.0	79.2	—	149
1.2	70.0	64.8	37.6	173
1.3	70.0	46.8	—	117
1.4	70.0	39.6	22.9	133
1.5	70.0	72.0	—	142

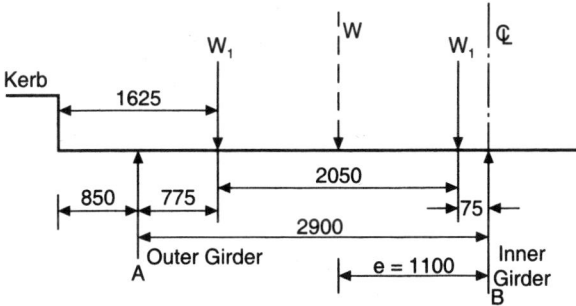

Fig. 15.30 Position for Wheel of Maximum Eccentricity.

Table 15.16 Maximum Live Load Bending Moments at Various Sections

Section	Computation of Live Load Bending Moment (kN·m)	Live Load B.M. (kN·m) Outer Girder (R.F. = 0.5225)	Inner Girder (R.F. = 0.3335)
0.2	+(1.1 × 700) [(2.794 + 2.25)/2] = 1941.9	1015	648
	−(1.1 × 700) [(0.792 + 0.765)/2] = 599.4	313	200
0.4	+(1.1 × 700) [(3.424 + 2.552)/2] = 2300	1202	768
	−(1.1 × 700) [(1.584 + 1.533)/2] = 1200	627	400
0.5	+(1.1 × 700) [(3.188 + 2.284)/2] = 2107	1101	703
	−(1.1 × 700) [(1.980 + 1.917)/2] = 1500	784	500
0.6	+(1.1 × 700) [(2.74 + 1.884)/2] = 1780	930	594
	−(1.1 × 700) [(2.376 + 2.298)/2] = 1800	940	600
0.8	+(1.1 × 700) [(1.594 + 0.768)/2] = 909	475	303
	−(1.1 × 700) [(3.168 + 3.066)/2] = 2400	1254	800
1.0 (B)	+(1.1 × 700) [(1.624 + 1.540)/2] = 1218	636	406
	−(1.1 × 700) [(3.960 + 3.834)/2] = 3000	1567	1000
1.2	+(1.1 × 700) [(2.433 + 1.170)/2] = 1387	725	463
	−(1.1 × 700) [(2.574 + 2.450)/2] = 1934	1010	645
1.4	+(1.1 × 700) [(3.552 + 2.370)/2] = 2280	1191	760
	−(1.1 × 700) [(0.526 + 1.444)/2] = 1143	597	381
1.5	+(1.1 × 700) [(3.660 + 2.430)/2] = 2345	1225	782
	−(1.1 × 700) [(1.108 + 0.968)/2] = 800	418	267

Table 15.17 Design Service Load Bending Moments

Section	Live Load B.M. (kN·m) Outer Girder +	−	Inner Girder +	−	Dead Load B.M. (kN·m) +	−	Design B.M. (kN·m) Outer Girder +	−	Inner Girder +	−
0.2	1015	313	648	200	647	—	1662	313	1295	200
0.4	1202	627	768	400	660	—	1862	627	1428	400
0.5	1101	784	703	500	293	—	1394	784	996	500
0.6	930	940	594	600	—	190	930	1130	594	790
0.8	475	1254	303	800	—	1835	475	3089	303	2635
1.0	636	1567	406	1000	—	4443	636	6010	406	5443
1.2	725	1010	463	645	—	1030	725	2040	463	1675
1.4	1191	597	760	381	545	—	1736	597	1305	381
1.5	1225	418	782	267	1740	—	2965	418	2522	267

10. Maximum Shear Force in Main Girders

The load position for maximum shear force in any girder is as shown in Fig. 15.31.
Load on girder A = 350 + 350 (0.85/2.90) = 452.5 kN
The total load of 452.5 kN will be acting over a length of 3.6 m. Using the influence lines shown in Fig. 15.27 and 15.28, the maximum live load shear force at support A and B are computed. Live Load shear force at support A

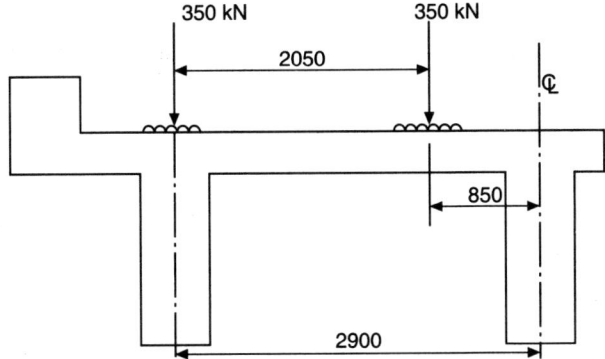

Fig. 15.31 Position of Live Load for Maximum Shear Force.

$1.1 \times 452.5 \, (1 + 0.7)/2 = 423$ kN

Live load shear force at support B.

$$= 1.1 \times 452.5 \, (1 + 0.9)/2 = -473 \text{ kN}$$

The dead load shear force is computed using the influence lines. Dead load shear force at A.

$$\begin{aligned}
&= [(57 \times 1) + 72 \, (0.847 + 0.023) + 96 \, (0.698 + 0.045) \\
&\quad + 78 \, (0.557 + 0.064) + 110 \, (0.427 + 0.076) + 83 \, (0.318 + 0.081) \\
&\quad + 123 \, (0.228 + 0.077) + 93 \, (0.157 + 0.063) + 141 \, (0.099 + 0.045) \\
&\quad + 107 \, (0.048 + 0.023)] - [(149 \, (0.067 + 0.036) \\
&\quad + 173 \, (0.129 + 0.076) + 117 \, (0.175 + 0.115) + 133 \, (0.198 + 0.159) \\
&\quad + 142 \, (0.191)] = 255 \text{ kN}
\end{aligned}$$

Dead Load Shear Force at B

$$\begin{aligned}
&= [72 \, (-0.152 + 0.023) + 96 \, (-0.301 + 0.045) + 78 \, (-0.442 + 0.064) \\
&\quad + 110 \, (-0.572 + 0.076) + 83 \, (-0.681 + 0.081) + 123 \, (-0.771 + 0.077) \\
&\quad + 93 \, (-0.843 + 0.063) + 141 \, (-0.900 + 0.045) \\
&\quad + 107 \, (-0.952 + 0.023)] - [149 \, (0.067 + 0.036) + 173 \, (0.129 + 0.076) \\
&\quad + 117 \, (0.175 + 0.115) + 133 \, (0.198 + 0.159) + 142 \, (0.191)] \\
&= -705 \text{ kN}
\end{aligned}$$

Design shear force at support A

$$= V_{A} = (\text{live load shear} + \text{dead load shear}) = (423 + 255) = 678 \text{ kN}$$

Design shear force at support B

$$= V_{B} = (\text{live load shear} + \text{dead load shear})$$
$$= (-423 - 705) = -1178 \text{ kN}$$

11. Design Ultimate Load Bending Moments and Shear Forces at Critical Sections

i) Bending Moments

1) At Section 0.5 (Mid span of AB)

a) Outer Girder

$$M_u = [(1.35 \, M_g + 1.5 \, M_q)] = [(1.35 \times 293) + (1.5 \times 1101)] = +2047 \text{ kN·m}$$

b) Inner Girder

$$M_u = [(1.35 \, M_g + 1.5 \, M_q)] = [(1.35 \times 293) + (1.5 \times 703)] = +1450 \text{ kN·m}$$

2) At Section 1 (Support-B)

a) Outer Girder

$$M_u = [(1.35 \, M_g + 1.5 \, M_q)] = [(1.35 \times 4443) + (1.5 \times 1567)] = -8348 \text{ kN·m}$$

b) Inner Girder

$$M_u = [(1.35 \, M_g + 1.5 \, M_q)] = [(1.35 \times 4443) + (1.5 \times 1000)] = -7498 \text{ kN·m}$$

3) At Section 1.5 (Mid span of BC)

a) Outer Girder

$$M_u = [(1.35 \, M_g + 1.5 \, M_q)] = [(1.35 \times 1740) + (1.5 \times 1225)] = +4186 \text{ kN·m}$$

b) Inner Girder

$$M_u = [(1.35 \, M_g + 1.5 \, M_q)] = [(1.35 \times 1740) + (1.5 \times 782)] = +3522 \text{ kN·m}$$

ii) Shear Forces

1) Shear Force at Support B

$$V_u = [(1.35 \, V_g + 1.5 \, V_q)] = [(1.35 \times 705) + (1.5 \times 423)] = 1586 \text{ kN}$$

2) Shear Force at Support A

$$M_u = [(1.35 \, M_g + 1.5 \, M_q)] = [(1.35 \times 255) + (1.5 \times 423)] = 1009 \text{ kN}$$

12. Design of Beam Sections

M-20 Grade Concrete and Fe-415 HYSD reinforcements

(1) At Section 0.5 (Mid span of AB)

Overall depth = D = 1500 mm
Effective depth = d = 1350 mm
Width of beam = b = 500 mm

$$M_u = +2047 \text{ kN·m}$$

Effective depth required

$$= d = \sqrt{\frac{M_u}{0.138 f_{ck} \cdot b}} = \sqrt{\frac{2047 \times 10^6}{0.138 \times 20 \times 500}} = 1217 \text{ mm} < 1350 \text{ mm}$$

Compute the parameter

$$\left(\frac{M_u}{b.d^2}\right) = \left(\frac{2047 \times 10^6}{500 \times 1350^2}\right) = 2.2, \text{ using M-20 grade concrete and Fe-415 HYSD bars}$$

Read out the percentage of reinforcement required from Table 2 of SP: 16[11], Design Aids

$$p_t = 0.717 = \left(\frac{100 A_{st}}{b.d}\right)$$

$$A_{st} = \left(\frac{0.717 \times 500 \times 1350}{100}\right) = 4840 \text{ mm}^2$$

Provide 12 bars of 25 mm diameter ($A_{st} = 5890$ mm^2)

(2)   At Section 1 (Support B)

Overall depth = $D = 3000$ mm
Effective depth = $d = 2850$ mm
Width of beam = $b = 500$ mm

$$M_u = -8348 \text{ kN·m}$$

Effective depth required

$$= d = \sqrt{\frac{M_u}{0.138 f_{ck}.b.}} = \sqrt{\frac{8348 \times 10^6}{0.138 \times 20 \times 500}} = 2459 \text{ mm} < 2850 \text{ mm}$$

Compute the parameter

$$\left(\frac{M_u}{b.d^2}\right) = \left(\frac{8348 \times 10^6}{500 \times 2850^2}\right) = 2.0, \text{ using M-20 grade concrete and Fe-415 HYSD bars}$$

Read out the percentage of reinforcement required from Table 2 of SP: 16[11], Design Aids

$$p_t = 0.640 = \left(\frac{100 A_{st}}{b.d}\right)$$

$$A_{st} = \left(\frac{0.640 \times 500 \times 2850}{100}\right) = 9120 \text{ mm}^2$$

Provide 12 bars of 32 mm diameter ($A_{st} = 9651$ mm^2)

Design Shear Force at B = $V_u = 1586$ kN

Nominal shear stress = $\tau_v = \left[\dfrac{V_u}{b_w.d}\right] = \left[\dfrac{1586 \times 10^3}{500 \times 2850}\right] = 1.11 < \tau_{c\,max} = 2.8$ N/mm^2

$$P_t = (100 A_{st}/bd) = [(100 \times 9651)/(500 \times 2850)] = 0.67$$

For $p_t = 0.67$, interpolate τ_c from Table 19 of IS: 456-2000[12] as $\tau_c = 0.52$ N/mm^2
Design shear strength of concrete = $V_c = (\tau_c.b.d) = (0.52 \times 500 \times 2850) = (741 \times 10^3)$ N
Since $V_u > V_c$, shear reinforcements are required

Balance shear force $= V_{us} = (1586 - 741) = 845$ kN

Using 10 mm diameter 4-legged vertical stirrups, the spacing is obtained as

$$S_v = \left(\frac{0.87 f_y A_{sv} d}{V_{us}} \right) = \left(\frac{0.87 \times 415 \times 4 \times 79 \times 2850}{845 \times 1000} \right) = 384 \text{ mm}$$

Provide the vertical stirrups at a spacing of 300 mm increased to 300 mm throughout the span.

(3) At Section 1.5 (Mid span of BC)

The section will be designed as tee beam with flange thickness of 250 mm

$M_u = 4186$ kN·m, $b_w = 500$ mm, $b_f = 2900$ mm, $D = 1000$ mm, $d = 850$ mm, $D_f = 250$ mm

The ultimate moment capacity of the flange section alone is computed assuming

$$(x_u = 250 \text{ mm})$$
$$M_{uf} = [0.36 f_{ck} \times b_f \times D_f (d - 0.42 x_u)]$$
$$= [0.36 \times 20 \times 2900 \times 250 (850 - 0.42 \times 250]/10^6$$
$$= 3888 \text{ kN·m}$$

The area of tensile steel to resist this moment is given by

$$A_{stf} = \left(\frac{M_{uf}}{0.87 f_y \left(d - 0.42 \times D_f \right)} \right) = \left(\frac{3888 \times 10^6}{0.87 \times 415 \left(850 - 0.42 \times 250 \right)} \right) = 14454 \text{ mm}^2$$

Balance moment $= (4186 - 3888) = 298$ kN·m

Steel required to resist this moment is computed as (assuming $x_u = D_f$)

$$M_u = 0.87 f_y A_{st} [d - 0.42 x_u]$$
$$(298 \times 10^6) = (0.87 \times 415 \times A_{st}) [850 - 0.42 \times 250]$$

Solving $A_{st} = 1107$ mm^2

Total area of tensile steel $= (14454 + 1107) = 15561$ mm^2

Provide 16 bars of 36 mm diameter ($A_{st} = 16286$ mm^2)

Nominal shear reinforcements of 10 mm 4-legged stirrups at a spacing of 300 mm are provided throughout the span.

13. Cross Girders

Cross girders 300 mm wide are provided at regular intervals of 4 m connecting the main girders. Four bars of 32 mm diameter at top and bottom with 10 mm diameter 4-legged stirrups are provided in the cross girders to improve the structural integrity of the bridge deck.

14. Details of Reinforcements in Bridge Deck

The reinforcement details in the main girders provided according to the specifications of SP: 34-1987[13] are shown in Figs. 15.32 to 15.34.

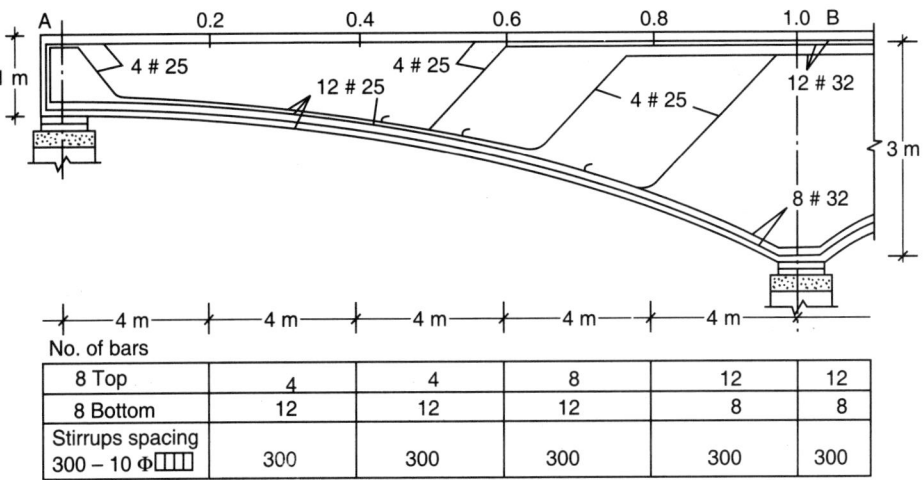

Fig. 15.32 Reinforcement Details in end Span AB.

No. of bars					
8 Top	4	4	8	12	12
8 Bottom	12	12	12	8	8
Stirrups spacing 300 – 10 Φ	300	300	300	300	300

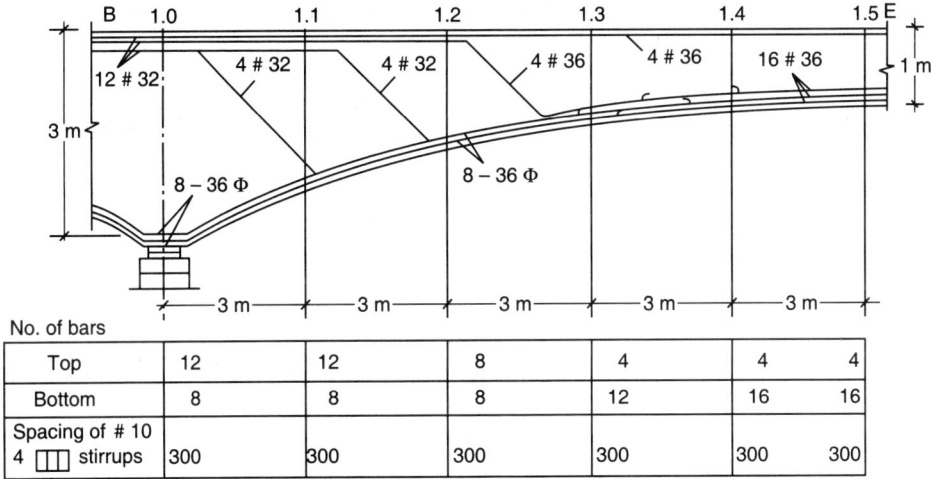

Fig. 15.33 Reinforcement Details in Central Span.

No. of bars						
Top	12	12	8	4	4	4
Bottom	8	8	8	12	16	16
Spacing of # 10 4 stirrups	300	300	300	300	300	300

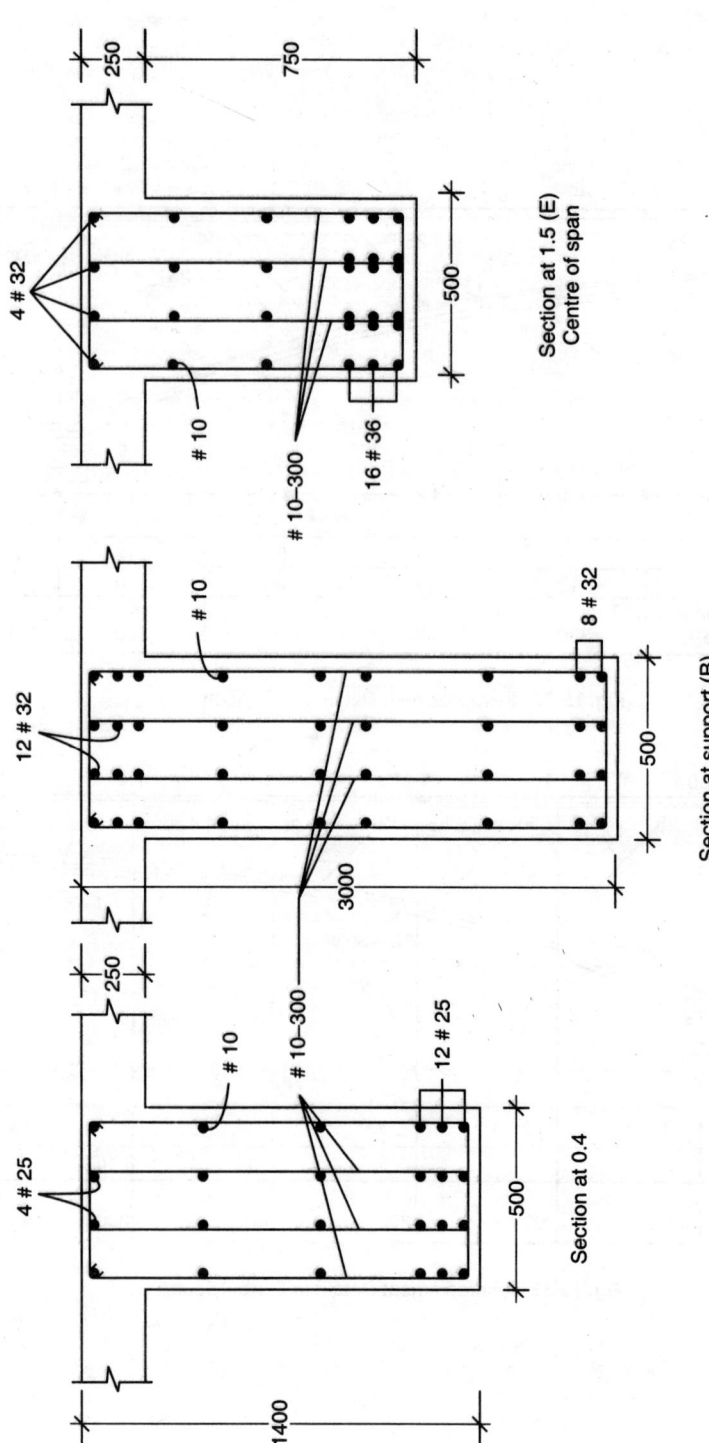

Fig. 15.34 Cross Section of Main Beams.

EXAMPLES FOR PRACTICE

1. A three span continuous reinforced concrete bridge with girders for variable cross-section is required for the crossing of a national highway bridge. Design the deck slab and main girders to suk-the following data:

   Total length of bridge = 60 m
   Span lengths, end span =18 m
   Central span = 24 m
   Width of carriage way = 7.5 m (two lane)
   Footpaths = 1.6 m on either side
   Kerbs = 600 m on either side
   Spacings of main girders = 3 m
   Spacings of cross girders = 4 m
   Concrete: M-20 grade, Steel: Fe-415 tor steel
   Loadings I.R.C. class A or AA which never produces the worst effect.

   Design the bridge deck and draw typical sections showing reinforcement details in the deck slab and girders.

2. A two span continuous reinforced concrete bridge with girders of variable cross-section is required for a bridge, crossing. Design the deck slab and main girders to suit the following data:

   Total length of bridge = 50 m
   Span lengths: two equal spans of 25 m each
   Width of carriage way = 7.5 m
   Kerbs = 600 mm on either side
   Footpaths = 1 m on either side
   Spacings of main girders = 3 m
   Spacings of cross girders = 5 m
   Concrete: M-20 grade, Steel: Fe-415 tor steel
   Loading: I.R.C. class A or AA whichever gives the worst effect.

   Design the bridge deck and draw typical sections showing reinforcements details in the deck slab and girders.

3. A four span continuous reinforced concrete bridge with girders of variable cross-section is used for a highway bridge crossing. Design the deck slab and main girders to suit the following data:

   Total length of the bridge = 84 mm
   Span lengths: End spans = 18 m, Central span = 24 m
   Width of carriage way = 7.5 m
   Footpaths = 1.0 m either side
   Kerbs = 600 mm on either side
   Spacings of main girders = 3 m
   Spacings of cross girders = 4 m
   Concrete: M-20 grade, Steel: Fe-415 tor steel
   Loading: I.R.C. class A or AA whichever produces the worst effect.

Design the bridge deck and draw typical sections showing reinforcements details in the deck slab and girders.

4. A three span continuous reinforced concrete bridge with girders of variable cross section is to be designed for the crossing of a national highway bridge, deck. Design the deck slab and main girders using the following data:

Total length of the bridge = 100 m

Span length: Central span = 40 m, end spans = 30

Width of carriage way = 7.5 m

Foot paths = 1 m on either side

Spacings of main girders = 2.5 m

Spacings of cross girders = 5 m

Materials: Concrete: M-25 Grade & Fe-415 Grade HYSD bars

Loading: I.R.C. Class A or AA whichever produces the worst effect.

Design the bridge deck and draw typical sections showing reinforcement details in the deck slab and girder.

REFERENCES

1. Krishna Raju, N., Advances in design and Construction of Concrete Bridges, Construction India Annual Publication, Bombay, 1992, pp. 50-53.

2. Johnson Victor, D., Essentials of Bridge Engineering (Fifth Edition), Oxford & IBH Publishing Co. Pvt. Ltd, New Delhi, 2001, pp. 162-164.

3. Reynolds, C.E. and Steedman,J., Reinforced Concrete Designer's Hand Book, Concrete Publications Ltd, London, 1974.

4. Aswani, M.G., Vazirani, V.N, and Ratwani, N.M., Design of Concrete Bridges, Khanna Publishers, New Delhi, 1975, pp. 1-396.

5. IRC: 6-2014, Standard Specifications and Code of Practice for Road Bridges, Section II, Loads and Stresses (Revised Edition), Indian Roads Congress, New Delhi, 2014, pp. 1-84.

6. IRC: 112-2011, Code of Practice for Concrete Road bridges, Indian Roads Congress, New Delhi, 2011, p. 181.

7. Rowe, R.E., Cranston, W.B., and Best, B.C., New Concepts in the Design of Concrete, Structural Engineer, Vol.43, 1965, pp.339-403.

8. Bate, S.C.C., Why Limit State Design, Concrete, London, March 1968, pp. 103-108.

9. Turner, M.J, Clough, R.W, Martin, H.C, and Topp, L.J., Stiffness and deflection Analysis of Complex Structures, Journal of Aeronautical Science, Vol. 23, No.9, Sept 1956, pp. 805-24.

10. Krishna Raju, N and Gururaja, D.R., Advanced Mechanics of Materials and Structures, Narosa Publishing House, New Delhi, 1997, pp. 299-416.

11. SP: 16-1980, Design Aids for reinforced Concrete to IS: 456-1978, Bureau of Indian Standards, New Delhi, 1980, pp. 1-232.

12. IS: 456-2000, Indian Standard Code of Practice for Plain and Reinforced Concrete, (Fourth Revision), Bureau of Indian Standards, New Delhi, 2000, p. 72.
13. SP: 34-1987, Hand Book of Concrete Reinforcement and Detailing, Bureau of Indian Standards, New Delhi, 1987.

REVIEW QUESTIONS

1. Briefly explain the necessity of using continuous girder bridges in preference to simply supported spans.
2. What is the main disadvantage of continuous girder bridges? Under what situations you would avoid the use of continuous bridges?
3. Why should you adopt variable depth for continuous girders instead of constant depth in long span bridges?
4. What methods do you recommend for the analysis of moments and shear forces in reinforced concrete Continuous girder bridges?
5. What are influence lines? Explain the advantages of using influence lines to determine the maximum moments and shear forces under live loads.
6. In the case of continuous bridge decks, locate the sections at which maximum positive and negative moments develop due to dead and live loads.
7. In a three span continuous girder bridge subjected to moving high way loads, specify the section at which maximum shear force develops.
8. Briefly explain the method of designing reinforcements in critical sections for flexure using the limit state method.
9. How do you design reinforcements for resisting the ultimate shear force at critical sections in a continuous girder?
10. Explain with typical sketches the method of reinforcing continuous girder bridge decks for flexure and shear.

OBJECTIVE TYPE QUESTIONS

1. Continuous girder bridges are more economical for use in national high way crossings in preference to the simply supported bridges due to
   a) Reduction in bending moments
   b) Ease of construction
   c) Simpler analysis and design
2. The type of multi span bridge which requires the least number of expansion joints and bearings is
   a) Simply supported slab bridge
   b) Simply supported girder bridge
   c) Continuous girder bridge

3. Continuous girder bridges are generally classified under the group
   a) Determinate structures
   b) Complex structures
   c) Indeterminate structures
4. The live load moment analysis in the deck slab of a continuous girder bridge is generally done by using
   a) Moment distribution
   b) Pigeaud's curves
   c) Matrix analysis
5. The design maximum moments and shear forces in the girders of a continuous multi span bridge deck are determined by using
   a) Slope deflection method
   b) Influence line diagrams
   c) Column analogy method
6. In the case of a three span reinforced concrete continuous girder bridge deck, maximum negative design moments develop at
   a) Centre of middle span
   b) End supports
   c) Interior supports
7. The maximum shear force due to dead and live loads in a multi span continuous girder bridge, develops at
   a) The centre of middlie span
   b) The end support
   c) The penultimate support
8. The maximum shear force due to dead and live loads in a multi girder continuous bridge develops in the
   a) outer girder
   b) Inner girder
   c) In between the girders
9. For constructing the Influence lines for bending moment and shear force at any section of a multi span variable depth girders, the prerequisite is to compute the
   a) Dead Loads of the girder
   b) Stiffness of the members
   c) Live load on the girders
10. According to the latest Indian Roads Congress codes, bridge structures should be designed by using the
    a) Working stress method
    b) Ultimate load method
    c) Limit state method

16

Bridge Bearings

16.1 GENERAL FEATURES

Bearings are structural contraptions provided at the top of piers and abutments to support the girders of the super structure. The main function of the bearings is to accommodate the movements of the super structure. The movements are induced due to the various reasons compiled by Lee[1], Victor[2] and Long[3] are listed below:

1. Translational movement of the super structure like expansion and contraction due to the variations in the temperature at the bridge site.

2. Rotational movements due to the vertical deflections of the super structure arising out of highway loads.

3. Vertical movements due to the sinking of supports.

4. Movements due to shrinkage of the super structure

5. Movements due to pre stressing or creep

Thermal movement of bridge decks depend upon the coefficient of linear expansion of the material and temperature range at bridge site. Typical coefficient of linear expansion of concrete is of the order of 6.5×10^{-6} per degree Centigrade. If the span of the bridge is 30 m and the temperature at site changes from a minimum of 10 degrees to a maximum of 40 degrees Centigrade, the change in length 'δL' of the bridge deck is computed as

$$\delta L = (\alpha.L.t) = 6.5 \times 10^{-6} \times 30 \times 10^3 \times 30) = 5.85 \text{ mm}$$

Hence suitable provision for the movement of the deck should be made by providing bearings at the supports. In addition to the horizontal movement, the bridge girders rotate near the supports. The magnitude of rotation at support depends upon the magnitude of deflection at centre of span and the span length.

If θ = rotation of the girder at the supports

                L = span length (30 m)

                e = Maximum deflection at centre of span (60 mm)

Assuming the deflected profile of the beam to be parabolic, the rotation at the supports of the girder is computed as

$$\theta = (4e/L) = (4 \times 60)/(30 \times 10^3) = 8 \times 10^{-3} \text{ radians}$$
$$= (8 \times 10^{-3} \times 180)/\pi = 0.46 \text{ Degrees}$$

For the purpose of preliminary estimates, Long has suggested that the maximum movement due to all causes expressed as a function of the span of the girders may be assumed as given below depending upon the type of the bridge deck:

(a) *In-situ* Reinforced Concrete .... 9×10^{-4}

(b) Precast Reinforced Concrete .... 7×10^{-4}

(c) *In-situ* Prestressed Concrete .... $16 - 10^{-4}$

(d) Precast Prestressed Concrete .... 11×10^{-4}

(e) Steel .... 9×10^{-4}

(f) Composite Steel and Concrete .... 8×10^{-4}

16.2 TYPES OF BEARINGS

Generally bearings are classified as

(a) Fixed bearings and

(b) Expansion bearings.

Fixed bearings permit rotations while, preventing expansion. Expansion bearings accommodate both horizontal movements and rotations. The type of bearing to be selected depends upon the type of super structure, type of supports and also the span length.

A simply supported span is generally provided with a fixed bearing at one end and an expansion bearing at the other support. For a two span continuous girder, a fixed bearing is provided at the central support and expansion bearings at the end supports. In the case of major bridges the cost of bearings are in the range of 10 to 15 percent of the total cost of the bridge.

For culverts with reinforced concrete slab of small spans, no special bearings are required. However a thick layer of Kraft paper is provided between the slab and the bed block.

In the case of girder bridges, the following types of bearings are invariably adopted.

1. Expansion type bearings

    (a) Sliding plate bearing

    (b) Sliding cum Rocker bearing

    (c) Steel Roller cum Rocker bearing

    (d) R.C. Rocker cum roller bearing

    (e) Elastomeric Bearing

2. Fixed type bearings

    (f) Steel Rocker bearing

    (g) R.C. hinge (rocker) bearing

(a) Sliding Plate Bearing

This is the simplest form of expansion bearing used for girder bridges of spans up to 20 m. When the contact surfaces are flat, teflon coating should be used to prevent the development of frictional resistance and to facilitate smooth movement due to expansion.

The current practice is to provide a curved shape to the top plate to reduce the contact area and frictional resistance. Typical details of a sliding plate bearing is shown in Fig. 16.1.

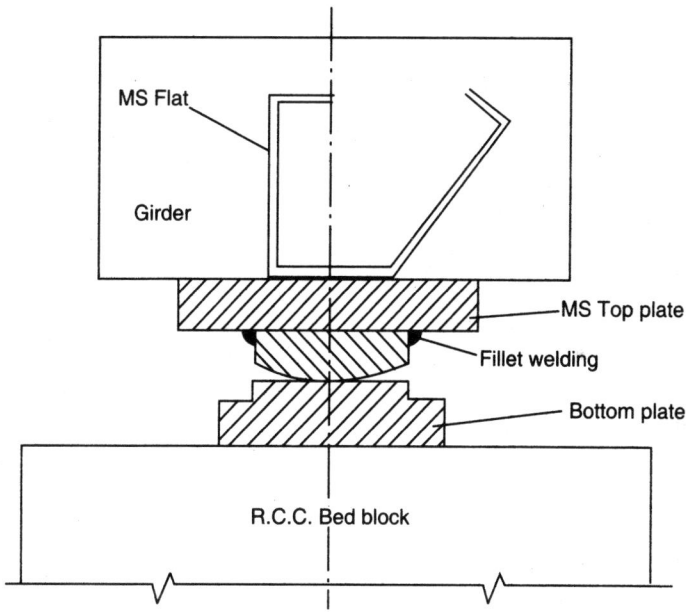

Fig. 16.1 Sliding Plate Bearing.

(b) Sliding Cum Rocker Bearing

In the case of bridges with curved alignment, the bearings provided should permit sliding movement and also rotation in different directions. Hence sliding cum rocker bearings shown in Fig. 16.2 are adopted for supporting curved decks. Essentially the bearing comprises a sliding and tilting plate with a pressure pad and base plate. The mating surfaces should be coated with teflon to reduce the frictional resistance.

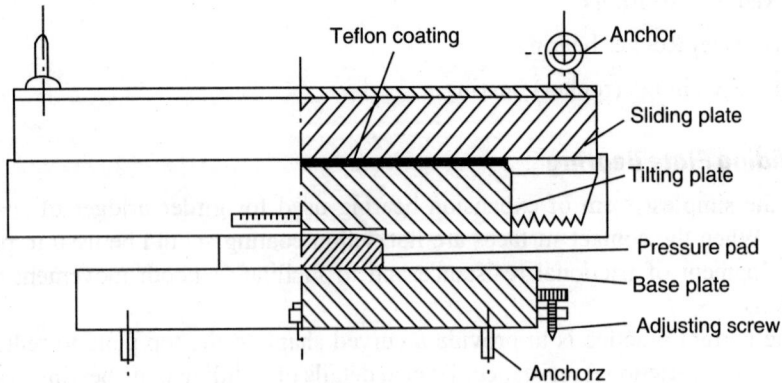

Fig. 16.2 Sliding Cum Rocker Bearing.

(c) Steel Roller Cum Rocker Bearing

Generally roller cum rocker bearings permit longitudinal movement by rollers and rotational movements by the rocker. For long span bearings cast steel roller bearings are generally used. Rollers of diameter 100 to 150 mm are generally preferred. Single or two large diameter rollers are preferred to a nest of small rollers. Figure 16.3 shows the typical details of a steel roller cum rocker bearing.

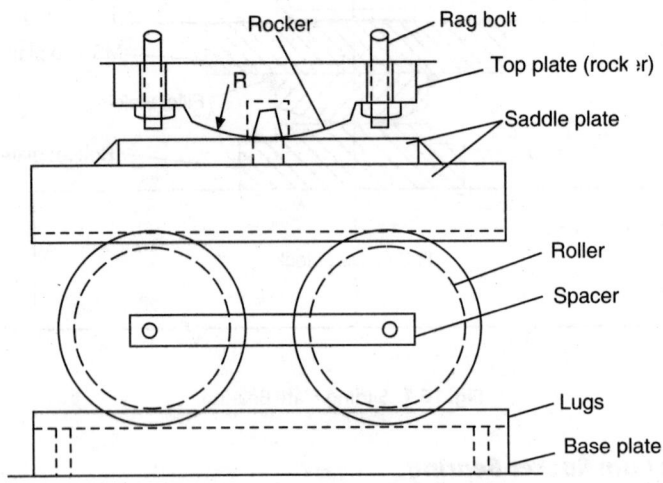

Fig. 16.3 Steel Roller Cum Rocker Bearing.

The Indian roads Congress Code IRC: 83-1999 (Part-1)[4] provides some guide lines for the selection of the diameter of rollers.

(d) R.C. Rocker Cum Roller Bearing

In many cases of bridge structures, reinforced concrete rocker bearings have been used to support the reinforced concrete decks[5,6]. They are designed as compression members with sufficient reinforcements to support the axial loads from bridge deck according to the specifications of the Indian standard code IS: 456-2000.

Reinforced concrete rocker bearings are less expensive in comparison with steel rocker bearings. A typical R.C. rocker cum roller bearing comprises of rigid reinforced concrete block with lead sheets provided at the top and bottom of the pedestal. The length of the lead sheet coincides with that of the girder width while its breadth should be sufficient to limit the stresses on the sheet within permissible limits. The R.C. pedestal is designed as a short column to support the reaction from the girder. Mild steel dowel rods are used to resist the shear due to longitudinal forces. The lead sheet permits the girder to rotate.

Both rotations and longitudinal movements of the girder are permitted by this type of bearing. For rotation, the girder compresses the lead sheet along the inside edge and the block tilts inside. For longitudinal movements, the lower lead sheet is compressed along the outside edge making the block tilt inside. For longitudinal movements, the lower lead sheet is compressed along the outside edge making the block tilt outside. If the height of the concrete rocker is more, a smaller angular compressive load is sufficient to accommodate the desired longitudinal expansion. Typical details of a reinforced concrete rocker cum roller bearing used in the Dimni bridge is shown in Fig. 16.4.

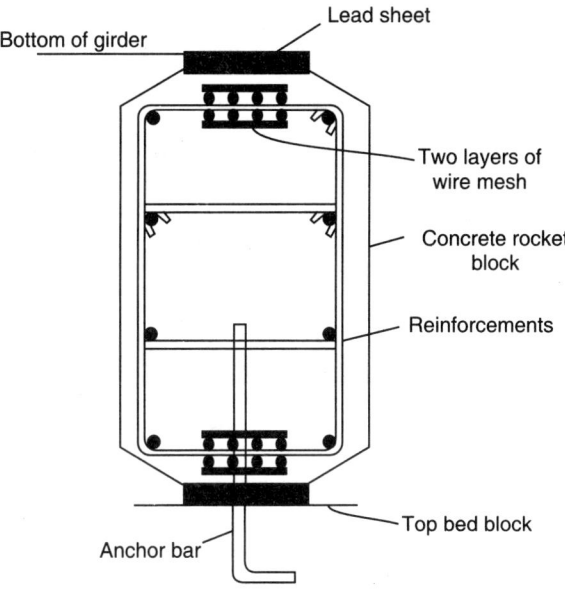

Fig. 16.4 R.C. Rocker Bearing.

(e) Elastometric Bearings

The present trend is to use elastomeric bearings in preference to metallic bearings which are expensive in initial cost and maintenance. Besides occupying a smaller space, elastomeric bearings are easy to maintain and also to replace when damaged.

Chloroprene rubber termed as neoprene is the most commonly used type of elastomer in bridge bearings. Typical details of an elastomeric pad and pot bearings are shown in Figs. 16.5 and 16.6. The design of elastomeric pad bearings should conform to the specifications prescribed in IRC: 83-1987 (Part-II). In the case of fixed type bearings the longitudinal movement is prevented by the rocker pin along the axis of the bearing.

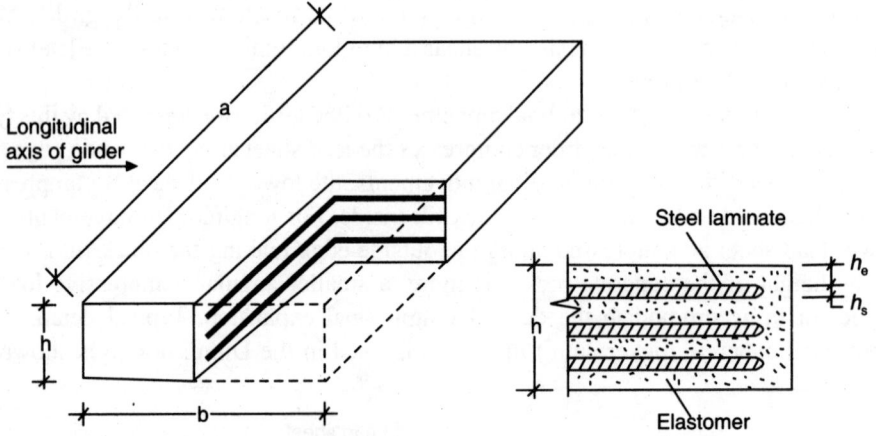

Fig. 16.5 Elastomeric Pad Bearing.

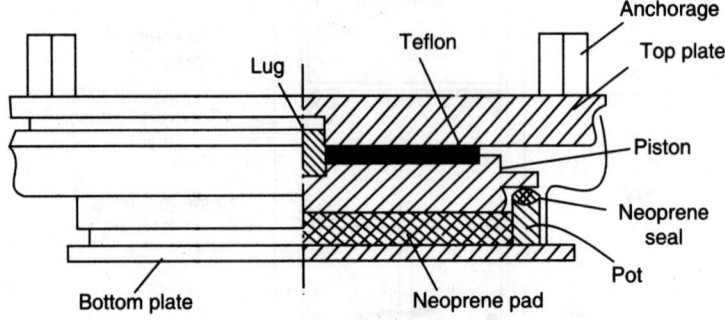

Fig. 16.6 Elastomeric Pot Bearing.

The following types are generally used:

(f) Steel Rocker Fixed bearing

Steel rocker bearings are generally used for longer spans exceeding 15 m. A typical steel rocker bearing comprises a top portion with a curved contact surface rocking over the bottom plate which has flat contact surface.

The rocker pin is designed to resist the horizontal shear. A typical steel rocker bearing to support a reaction of 1250 kN is shown in Fig. 16.7.

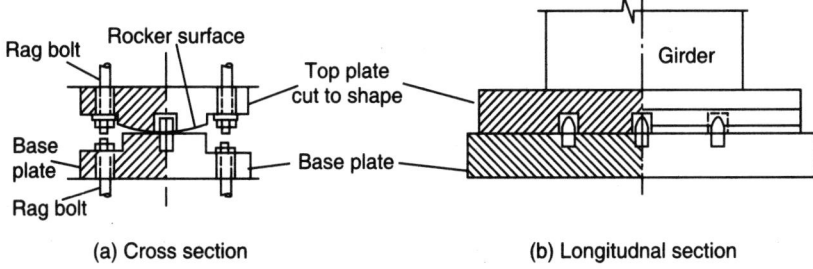

(a) Cross section   (b) Longitudnal section

Fig. 16.7 Steel Rocker Fixed Bearing.

(g) R.C. Rocker bearing

Concrete rocker fixed bearings are also referred to as linear concrete hinges and these are simple and cheap to produce but require proper detailing and care during construction. This type of bearing permits large rotations if constructed properly and accurately with correct design dimensions.

This type of hinge is usually built in with the super and sub structure and require no maintenance and has a long life. A typical linear concrete hinge capable of rocking (rotating) about its longitudinal axis is shown in Fig. 16.8.

A narrow throat and a very low height of the hinge (generally about 2 cm) with circular curved faces allround is provided to ensure a three dimensional confinement of concrete. This region of confined concrete in the throat can resist compressive stresses nearly 7 times that of the standard 28 day strength of concrete since the concrete in thc throat is in a state of triaxial compression. Cracking developed due to rotation heals itself due to creep under the high compressive stress. These type of hinges rarely require any reinforcements crossing the throat unless the horizontal shear exceeds about one-eighth of the co-existing vertical load and or transverse moment on the hinge. However reinforcements are provided above and below the throat in the form of hair pin loops.

Experimental investigations[7] have indicated that a throat plain section 150 mm by 700 mm under a working load of 4500 kN can tolerate rotations up to 0.012 radians under 37 million rotational loadings without showing any signs of cracking or distress.

16.3 DESIGN PRINCIPLES OF STEEL ROCKER AND ROLLER BEARINGS

The design principles of metallic bearings using different types of steels like mild and high tensile steels are given Indian standard code IRC: 83-1999[4]. The Permissible stresses in the various structural parts of the metallic bearings specified in the code are compiled in Table 16.1.

For cylindrical rollers on flat surfaces, the allowable working load per unit length of roller shall be

(a) For mild steel

    (i) Single and double rollers, 8 d_1 N per mm length,

    (ii) Three or more rollers, 5 d_1 N per mm length.

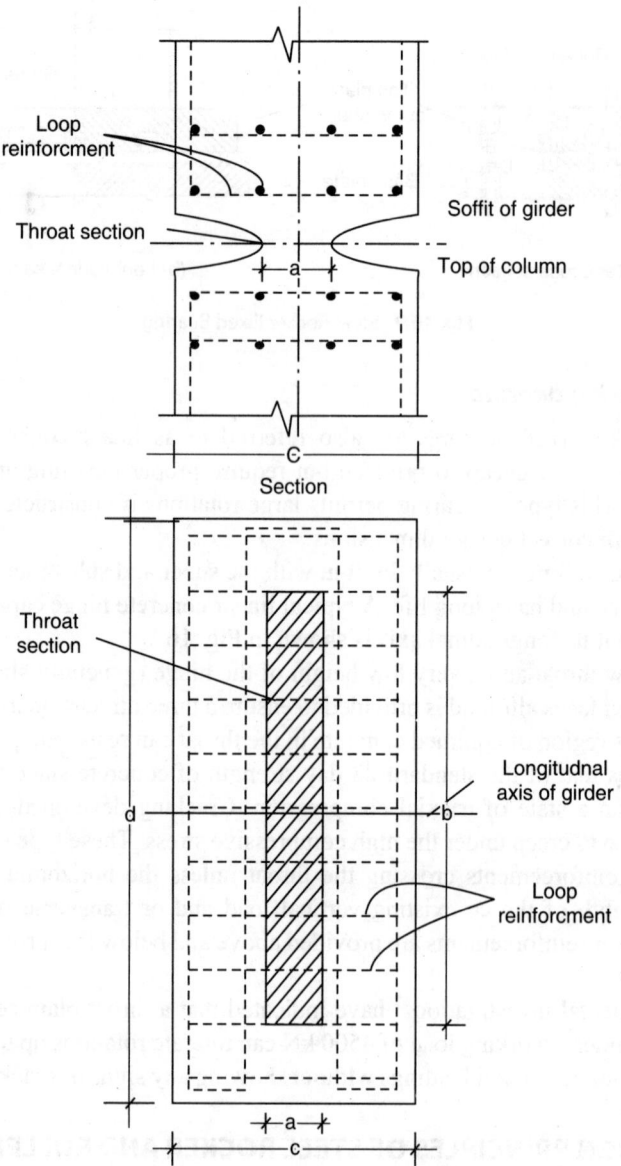

Fig. 16.8 R.C. Rocker (Hinge) Fixed Bearing.

(b) For High tensile steel

   (i) Single and Double rollers, 10 d_1 N per mm length

   (ii) Three or more rollers, 7 d_1 N per mm length,

   where d_1 = diameter of roller expressed in mm.

Table 16.1 Permissible Stresses in Steel (IRC: 83-1982)

No.	Description	Mild Steel (N/mm^2)	High Tensile Steel (N/mm^2)
1.	Part in bending (tensile or Compression) on effective sectional area for extreme fiber stress		
	(a) For plates, flats, rounds, square and similar sections	160	200
	(b) For pins	205	295
2.	Parts in Shear		
	(a) maximum shear stress on plates	105	140
	(b) Maximum shear stress for turned and fitted bolts and pins	100	$0.43 f_Y$ where f_Y is the yield stress
	(c) maximum shear stress in black bolts and rocker pins	85	$0.37 f_Y$
3.	Parts in Bearings		
	(a) On flat surface	185	240
	(b) Knuckle pin and black bolts	200	$0.87 f_Y$

The minimum diameter of the roller shall be not less than 75 mm. The ratio of the length of the roller to its diameter shall normally be not more than 6, but not more than 10 in any case. The gap between the rollers shall be not less than 50 mm in the case of multiple full rollers.

16.4 DESIGN EXAMPLE OF STEEL ROCKER BEARING

Design a steel rocker bearing for transmitting a vertical reaction of 1000 kN and a horizontal reaction of 100 kN at the support of a bridge girder, assuming the following permissible stresses according to IRC: 83–1999[4].

Permissible compressive stress in concrete bed block = 4 N/mm^2
Permissible bending stress in steel plate = 160 N/mm^2
Permissible bearing stress in steel plate = 185 N/mm^2
Permissible shear stress in steel = 105 N/mm^2
Sketch the typical details of the rocker bearing.

Details of design

1. Bed Plate

Area of bed plate = $(1000 \times 10^3)/4 = 25 \times 10^4$ mm^2
Provide a bed plate of overall size 400 × 650 × 40 mm
and top plate of overall size 400 × 600 × 40 mm

2. Rocker Diameter

Let R = Radius of rocker surface in contact with the flat surface of bottom plate.

Vertical design load per unit length $\not> \dfrac{170R^2\sigma_u^3}{E^2}$

where σ_u = Nominal ultimate tensile strength of material (250 N/mm²)

                R = radius of rocker spherical surface

                E = Modulus of elasticity of material (200 kN/mm²)

Design load per unit length = (1000 × 10³)/650 = 1538 N/mm

Hence   1538 = (170 × R² × 250³)/(200 × 10³)²

Solving   R = 152 mm

Provide a radius of 200 mm for the rocker surface.

3. Rocker Pin

Providing 2 rocker pins, the horizontal shear force to be resisted by each pin

        (100/2) = 50 kN

If d = diameter of rocker pin,

$$\left(\frac{\pi d^2}{4} \times 105\right) = (50 \times 10^3)$$

∴ d = 24.6 mm

Adopt a tapering pin with a top diameter of 25 mm and bottom diameter of 30 mm and height 55 mm.

4. Thickness of Base Plate

Maximum bending moment about central axis is base plate = (500 × 10³ × 100) = 5 × 10⁷ N·mm

If t = thickness of base plate required.

Section Modulus = $Z = (b.t^2/6) = (M/\sigma_t)$

∴ $t = \sqrt{\dfrac{6M}{b\sigma_t}} = \sqrt{\dfrac{6 \times 5 \times 10^7}{650 \times 150}} = 53.7$ mm

Provide an overall thickness of 72 mm for the central portion of the base plate

5. Check for Bearing Stress

Assuming 50 percent contact area between top and bottom plates,

$$\text{Bearing stress} = \left(\frac{1000 \times 10^3}{650 \times 100}\right)$$

$$= 153.8 \text{ N/mm}^2 < 185 \text{ N/mm}^2$$

Hence bearing stresses are within safe permissible limits. Figure 16.9 shows the dimensional details of the rocker bearing.

16.5 DESIGN OF STEEL ROCKER-ROLLER BEARING

Design a steel rocker-roller bearing to transmit a load of 1000 kN using the following data:

Allowable working load on single and double rollers = 8 N per rnm diameter per mm length.

Permissible comprfessive stress on concrete bed block = 4 N/mm^2

Permissible shear stress in steel = 105 N/mm^2

Permissible bending stress in steel = 160 N/mm^2

Permissible bearing stress in steel = 185 N/mm^2

Design Details

The design of top plate, bed plate, rocker pin are similar those presented in section 16.4.

Assuming two rollers of 200 mm diameter each,

permissible load = 8 N/mm diameter/mm length

If L = Total length of all rollers,

Load taken by rollers = 8 × L × diameter

$(1000 \times 10^3) = (8 L \times 200)$

Solving $L = 625$ mm

Use 2 rollers of 400 mm length having a diameter of 200 mm.

A typical cross section of the rocker-roller bearing is shown in Fig. 16.10.

16.6 DESIGN OF REINFORCED CONCRETE ROCKER BEARING

Design a reinforced concrete rocker bearing to transmit a support reaction of 600 kN. Adopt M–30 Grade concrete and Fe-415 grade HYSD bars. Permissible bearing stress in concrete is 8 N/mm^2. Sketch the details of reinforcements in the rocker bearing.

Design Details

Area of lead sheet over concrete bearing block = $(600 \times 10^3)/8 = 75000$ mm^2

Assuming a width of 200 mm,

Length of legd sheet = (75000/200) = 375 mm

Adopt a lead sheet at top and bottom of bearing of size 200 mm by 400 mm and concrete block of size 250 mm by 400 mm with a height of 450 mm.

Total Bursting Tension = (1/3) support reaction

$= (1/3) (600) = 200$ kN

Reinforcements are provided in the horizontal and vertical directions.

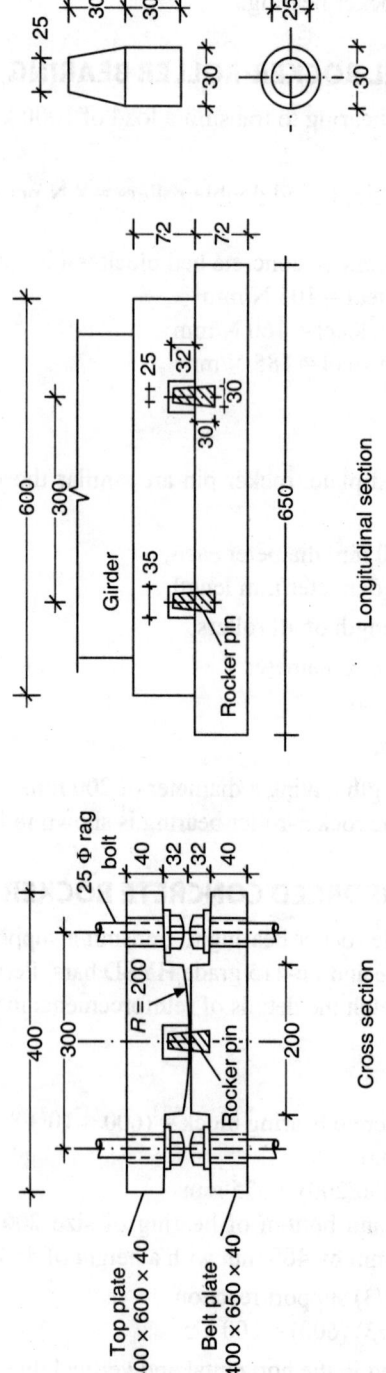

Fig. 16.9 Steel Rocker Bearing of 1000 kN Capacity.

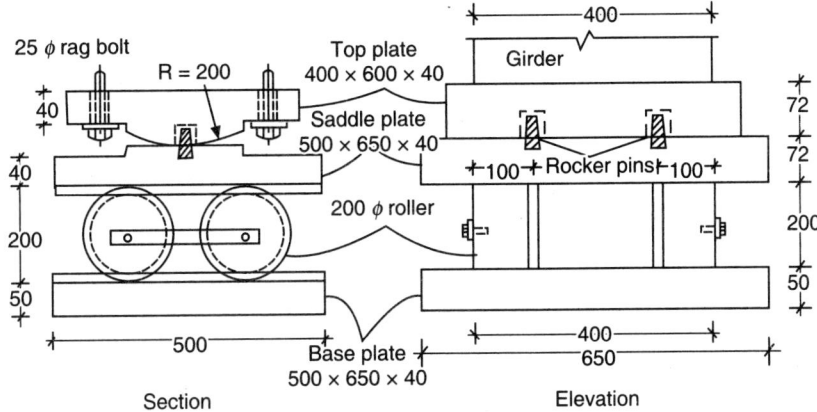

Fig. 16.10 Mild Steel Roller Cum Rocker Bearing of 1000 kN Capacity.

The rocker is designed as a column for axial loads. The load being small, minimum longitudinal reinforcement of 0.8 percent of the cross sectional area is provided with hoop reinforcement to resist bursting tension.

$$\text{Vertical reinforcement} = \left(\frac{0.8}{100} \times 250 \times 400 \right) = 800 \, \text{mm}^2$$

Using 8 mm diameter bars,
Number of bars = (800/50) = 16
Horizontal reinforcement in the form of hoops

$$= \left(\frac{200 \times 10^3}{230} \right) = 869.5 \, \text{mm}^2$$

using 8 mm diameter bars,
Number of hoops = 0.5 (869.5/50) = 8.69
Provide 9 hoops of 8 mm diameter in 3 layers and 3 rows and 20 vertical bars as shown in Fig. 16.11.

16.7 ELASTOMERIC PAD BEARING

(a) General Aspects

Elastomeric pad bearings comprise of synthetic rubber and neoprene pads which are elastic in nature to provide both translation and rotation. These type of bearings are widely used for bridges due to their economy and negligible maintenance costs. Neoprene pad bearings are compact, weather and fire resistant. Hence, now a days elastomeric bearings have more or less completely replaced the metallic bearings. Chloroprene is the raw material prescribed to be used by Indian Roads Congress Code IRC: 83-1987 (Part-II) for the manufacture of elastomeric pad bearings.

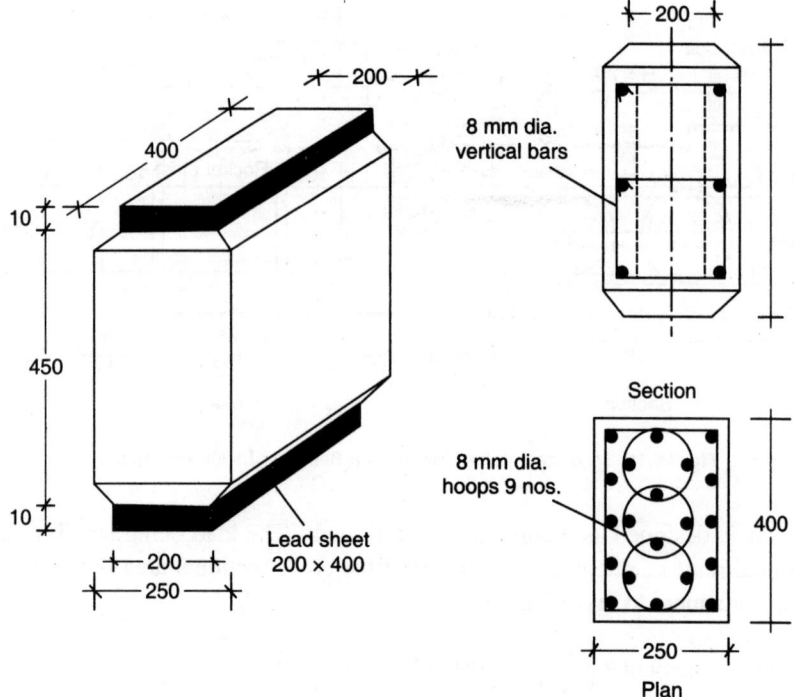

Fig. 16.11 R.C. Rocker Bearing.

The elastomer used for the bearing pads should have the following properties:

Hardness values should be in the ranger of (60 ∓ 5) IRHD (international rubber hardness scale).

This scale extends from 0 to 100. Hardness values for an eraser being 30 and for a car tyre is 60. The minimum ultimate tensile strain at failure should be not less than 400 percent. The shear mofulus of the elastomeric bearing should be not less than 0.8 N/mm², not greater than 1.20 N/mm².

(b) Design Procedure

The basic deformational characteristics of the elastomeric bearing under loads are shown in Fig. 16.12. The guide lines specified for the designer of elastomeric pad bearings as per IRC: 83 (Part-II)–1987 are as follows:

1. Standard plan dimensions compiled in Table 16.2 are to be preferred. However interpolation is permitted provided the design criteria is satisfied.

2. The bearing area of the pad should be such that the compressive stresses developed in concrete are within specified limits.

3. The design vertical load N_d should be within the limits of N_{max} and N_{min} specified in Table 16.2.

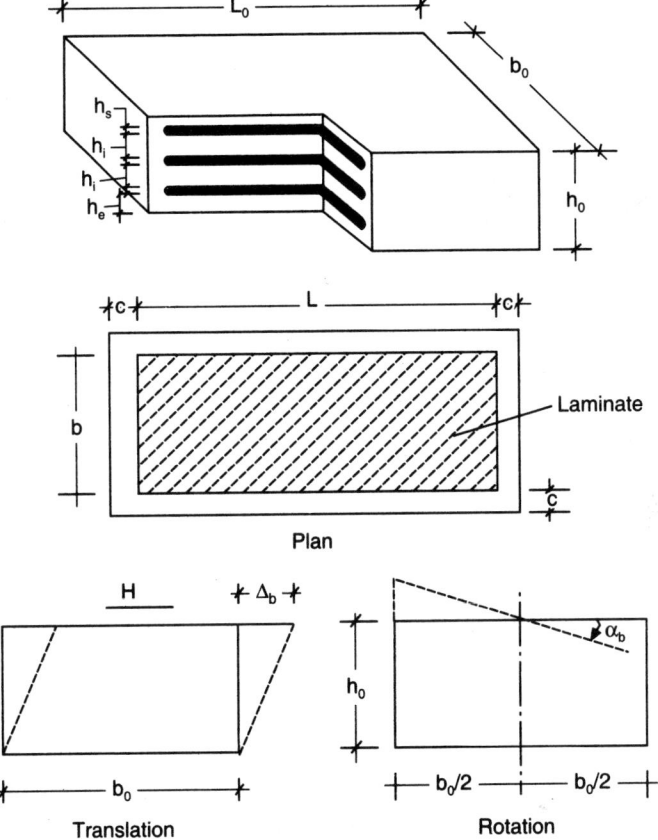

Fig. 16.12 Deformations of Elastomeric Bearings.

4. The ratio of overall length to breadth should be equal to or less than 2.

5. The overall thickness of the bearing should be in the range of 1/5 and 1/10 of the overall breadth.

6. The thickness of the internal layer of elastomer h_i, the thickness of the laminate h_s, and the elastomer cover at the top and bottom h_e should correspond to the following dimensions:

   h_i (mm)............ 8   10   12   16
   h_s (mm) 3   3   4   6
   h_e (mm) 4   5   6   6

7. The side cover of elastomer for the steel laminates is 6 mm.

8. The thickness of the elastomeric pad bearing should be adequate to limit the shear strain to horizontal load and movements due to creep, shrinkage and temperature to a value of less than 0.7. In the absence of more accurate analysis,

the shear strain Δbd due to creep, shrinkage and temperature can be computed assuming a total strain of 5×10^{-4} for common reinforced concrete bridge decks.

9. The ghape factor S should be greater than 6 and less than 12.

10. The number of elastomer layers provided shall satisfy the relation

$$\alpha_d = \beta.n.\alpha_{bi.max}$$

where α_d = angle of rotation which may be taken as $(400\,M_{max}\,L)/(EI)10^{-3}$
   n = number of elastomer layers

$$\beta = (\sigma_m)/(\sigma_{m.max})$$

where σ_m = average compressive stress
   $\sigma_{m.max} = 10\ \text{N/mm}^2$
   $\sigma_{bi.max} = (0.5\,\sigma_m\,h_i)/(b.s^2)$

11. Under critical, loading conditions, the following limit shall be satisfied to ensure adequate friction:

Shear strain $\leq 0.2 + 0.1\ \sigma_m$
   $10\ \text{N/mm}^2 \geq \sigma_m \geq 2\ \text{N/mm}^2$

12. The total shear stress due to normal and horizontal loads and rotation should be less than $5\ \text{N/mm}^2$.

Shear stress due to normal load $= \tau_c = (1.5\ \sigma_m)/S\ \text{N/mm}^2$
Shear stress due to horizontal load = shear strain $= \tau_r\ \text{N/mm}^2$
Shear stress due to rotation $= \tau_\alpha = 0.5\ (b/h_i)\ \alpha_{bi}\ \text{N/mm}^2$

13. Standard plan dimensions and design data specified in IRC: 83 is compiled in Table 16.2.

16.8 DESIGN EXAMPLE OF ELASTOMERIC PAD BEARING

Design an elastomeric pad bearing to support a Tee beam girder of a bridge using the following data:

Maximum dead load reaction per bearing = 300 kN
Maximum live load reaction per bearing = 700 kN
Longitudinal force due to friction per bearing = 45 kN
Effective span of the girder = 16 m
Estimated rotation at bearing of the girder due to dead and live loads = 0.002 radians
Concrete for Tee beam and bed block = M–20 Grade
Total estimated shear strain due to creep, shrinkage and temperature = 6×10^{-4}.

Design of Bearing

1. Selection of Bearing Pad Dimensions

Maximum vertical load on bearing $= N_{max} = (300 + 700) = 1000$ kN
Minimum vertical load $= N_{min} = 300$ kN

Table 16.2 Standard Plan Dimensions for Elastomeric Bearings (IRC: 83-Part-II-1987)

Size Index No	b_O (mm)	L_O (mm)	$A.10^{-4}$ (mm)^2	N_{max} (kN)	N_{min} (kN)	h_i (mm)	n_{max}	n_{min}	h_{max}	h_{min}	$\alpha_{bi.max} \times 10^{-3}$
1	160	250	3.5	350	70	8	3	1	32	16	8.0
2	160	320	4.6	460	90	8	3	1	32	16	7.0
3	200	320	5.8	580	120	8	4	2	40	24	4.0
4	200	400	7.3	730	150	8	4	2	40	24	3.5
5	250	400	9.2	920	180	10	4	2	50	30	3.0
						12	3	1	48	24	3.5
6	250	500	11.6	1160	230	10	4	2	50	30	3.0
						12	3	1	48	24	5.5
7	320	500	15.0	1500	300	10	5	2	60	30	2.0
						12	4	2	60	36	3.0
8	320	630	19.5	1900	380	10	5	2	60	30	1.5
						12	4	2	66	36	2.5
9	400	630	23.9	2400	480	12	6	3	84	48	1.5
10	400	800	30.6	3100	600	12	6	3	84	48	1.3

Note: (1) See Fig. 16.12 for notations.
(2) Marginal values in Nmax not exceeding 10 percent over the specified value may be permitted.
(3) Where two values of hi are given, the higher values may be adopted only when the lower value cannot cater for $\alpha_{bi.max}$ specified.

Referring to Table 16.2, select plan dimensions of bearing pad of size 320 mm by 500 mm

Loaded area $= A_2 = 15 \times 10^{-4}$ mm^2

According to IRC: 83,

Allowable contact pressure $= 0.25 f_c \sqrt{A_1/A_2}$

where A_1 = Concrete bed block area over pier

A_2 = Elastomeric pad area

The ratio (A_1/A_2) is limited to 2

Allowable contact pressure $= \sigma_c = (0.25 \times 20\sqrt{2}) = 7.07$ N/mm^2

Effective bearing area required $= (N_{max}/\sigma_c)$

$\qquad\qquad\qquad\qquad\qquad = (1000 \times 10^3)/7.07$

$\qquad\qquad\qquad\qquad\qquad = 14.14 \times 10^4$ mm$^2 < 15 \times 10^4$ mm^2

$\qquad\qquad\qquad\qquad\qquad$ (Hence safe)

Bearing stress $= \alpha_m = [(1000 \times 10^3)/(15 \times 10^4)] = 6.6$ N/mm^2

Referring to Table 16.2 and IRC: 83 Clause 916.2,

thickness of individual elastomer layers $= h_i = 10$ mm

Thickness of outer layer $= h_e = 5$ mm

Thickness of steel laminates $= h_s = 3$ mm

Side covering $= c = 6$ mm

Adopt 3 laminates with two internal layers.

Total thickness of elastomeric pad

$$h_o = (2h_e + 3h_s + 2h_i)$$
$$= [(2 \times 5) + (3 \times 3) + (2 \times 10)]$$
$$= 39 \text{ mm}$$

2. Shape factor $= \left[\dfrac{\text{Loaded surface area of an internal layer of elastomer (excluding side covers)}}{\text{Surface area free to bulge}} \right]$

$$= \left[\frac{(500-12)(320-12)}{(2 \times 10)(500+320)} \right]$$

$$= 9.16 > 6 < 12 \text{ (Hence safe)}.$$

Shear strain due to creep, shrinkage and temperature

per bearing $= (0.5 \times 6 \times 10^{-4}) = 3 \times 10^{-4}$

Shear strain due to translation per bearing

$$= \gamma_d = \left[\begin{array}{c} \text{Shear strain due to creep} \\ \text{Shrinkage and temperature} \end{array} \right] + \left[\begin{array}{c} \text{Shear strain due to} \\ \text{longitudinal force} \end{array} \right]$$

$$= \left[\frac{3 \times 10^{-4} \times 16 \times 10^3}{39} \right] + \left[\frac{45 \times 10^3}{15 \times 10^4} \right]$$

$$= (0.123 + 0.3)$$
$$= 0.423 < 0.7 \text{ (Hence safe)}$$

Assuming $\sigma_{m.max} = 10 \text{ N/mm}^2$

3. Maximum permissible angle of rotation of a single internal layer of elastomer corresponding to sm value of 10 N/mm² is given by

$$\alpha_{bi.max} = \left[\frac{0.5\sigma_m h_i}{b.s^2}\right] = \left[\frac{(0.5 \times 10 \times 10)}{308 \times 9.16^2}\right]$$

$$= 0.00193 \text{ radians}$$

Permissible rotation $= \alpha_d = \beta.n.\alpha_{bi\,max}$

where $\beta = 0.1\,\sigma_m = (0.1 \times 6.6) = 0.66 \text{ N/mm}^2$

and n = number of internal elastomeric layers = 2

$\alpha_d = (0.66 \times 2 \times 0.00193) = 0.0025 > 0.002 \text{ (actual), Hence safe.}$

4. Friction

Shear strain computed = 0.423

Under critical loading conditions,

Shear strain $\leq 0.2 + 0.1\,\sigma_m$

$\leq 0.2 + (0.1 \times 6.6)$

$\leq 0.86 > 0.423$, Hence safe.

Also $\sigma_m = 6.6 \text{ N/mm}^2$, satisfies the criterion that

$10 \text{ N/mm2} \geq \sigma_m \geq 2 \text{ N/mm}^2$

5. Total Shear Stress

Shear stress due to compression $= 1.5\,(\sigma_m/S)$

$= 1.5\,(6.6/9.16)$

$= 1.09 \text{ N/mm}^2$

Shear stress due to horizontal deformation

$= \tau_r = \gamma_d = 0.423 \text{ N/mm}^2$ as per computation due to translation

Shear stress due to rotation $= 0.5\,(b/h_i)^2\,\alpha_{bi}$

$= 0.5\,(308/10)^2 \times 0.00193$

$= 0.915 \text{ N/mm}^2$

Total shear stress $= (1.09 + 0.423 + 0.915)$

$= 2.428 \text{ N/mm}^2 < 5 \text{ N/mm}^2$, Hence safe.

Hence adopt an elastomeric pad bearing of overall dimensions 320 mm by 500 mm with a total thickness of 39 mm having two internal elastomeric layers of 10 mm thickness and three steel laminates of thickness 3 mm each having bottom and top covers of 5 mm. The designed elastomeric pad bearing is shown in Fig. 16.13.

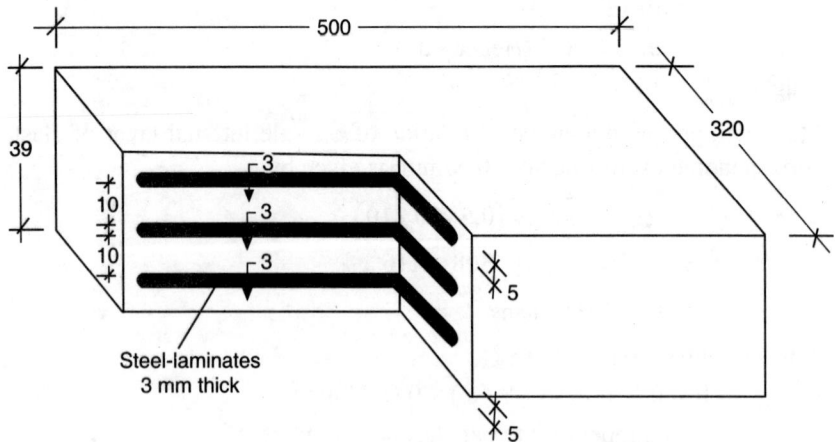

Fig. 16.13 Elastomeric Pad Bearing for Support Reaction of 1000 kN.

16.9 ELASTOMERIC POT BEARINGS

Elastomeric pot bearings are generally preferred for use in Long span bridge deck to transmit very heavy loads to the sub structure. These type of bearing permit large translations and rotations associated with heavy loads and aggressive temperature variations. Freyssinet spherical bearings (tetron type S-3 range) are available in the form of fixed and sliding bearings. The Indian Roads Congress code IRC: 83-2002 (Part-III)[8] covers the specifications of various types of Pot, Pot cum PTFE, Pin and Metallic guide bearings for use in Bridges. The bearings comprise the following materials:

(a) Tetron S-3 bases and rockers made of maintenance free alluminium alloy.

(b) Sliding plates are made of mild steel faced with high quality stainless steel.

(c) Sliding surfaces are lined with PTFE conforming to BS: 3784 specifications.

(d) Pins for side restraints are special spring steel with minimum yield strength of 1100 N/mm^2.

(e) All exposed steel surfaces are corrosion protected with a metallic zinc rich epoxy coating followed by chlorinated rubber paint.

(f) The average base contact stress of the bearings is of the order of 17.5 N/mm^2.

A typical tetron disc pot bearing of the fixed type is shown in Fig. 16.14. The principal dimensions of the bearings together with maximum sustainable vertical and horizontal loads are compiled in Table 16.3.

16.10 EXPANSION JOINTS FOR BRIDGE DECKS

According to Lee[1], construction and expansion joints are invariably provided in all concrete bridge decks at locations of structural discontinuity between two elements.

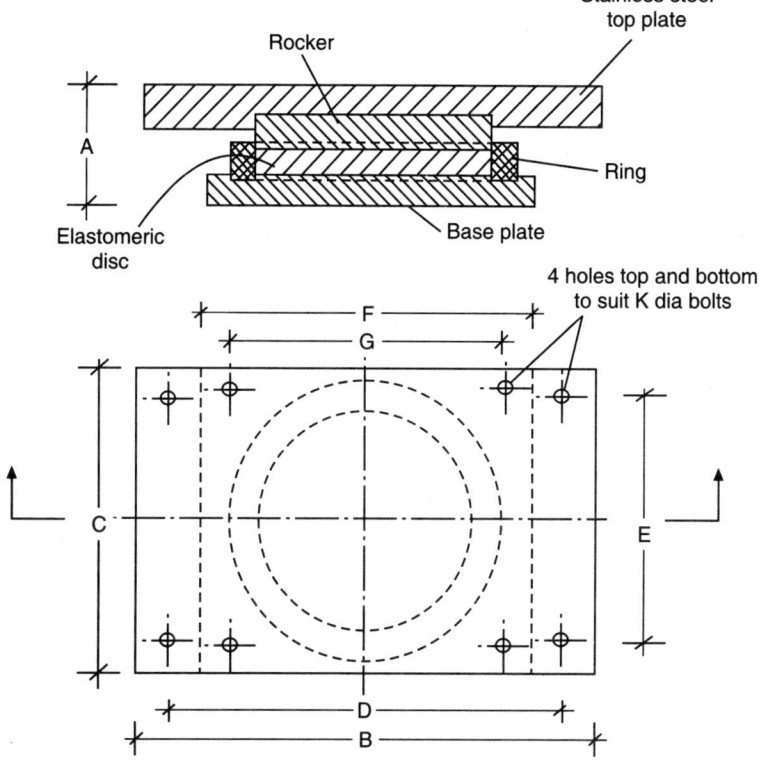

Fig. 16.14 Tetron D3T Fixed Type Elastomeric Pot Bearing.

Construction joints should be planned in advance and preferably they should be located at points of minimum shear and they should be nearly perpendicular to the principal lines of stress. Keys should be made by embedding water soaked beveled timbers in spft concrete and they should be removed after the concrete has set. When the work is resumed the surface of the concrete previously placed should be thoroughly cleaned of dirt, scum, laitance, loosely projecting aggregates and other soft material using stiff wire brushes. The surface should then be thoroughly soaked with clear water for two or three hours before further concreting using a thin layer.

Construction joints are generally either vertical or horizontal. In road way slabs, construction joints shall be formed vertical and in true alignment. Shear keys in construction joints should be constructed as shown in working site plans. In the case of box girder webs, these shear keys are normally shown on the plans to the full width. Before resuming the day's concrete work, all joints should be roughened and prepared for the next pour of concrete in accordance with the standard specifications. An expansion joint implicitly also refers to a contraction joint and hence it is more rational to designate it as a movement joint. These joints become necessary due to the following reasons:

Table 16.3 Details of Tetron D3T Fixed Type Pot Bearing

Bearing Type	Principal Dimensions (mm)								Maximum Vertical Load (kN)	Maximum Horizontal Load (kN)
	A	B	C	D	E	F	G	K		
D3T-50	58	235	170	195	120	170	130	M12	500	100
D3T-80	75	340	235	280	155	235	175	M20	800	150
D3T-100	80	355	250	295	170	250	190	M20	1000	150
D3T-125	83	375	270	315	190	270	210	M20	1250	190
D3T-160	92	395	290	335	210	290	230	M20	1600	220
D3T-200	97	460	335	385	240	335	260	M20	2000	250
D3T-250	97	485	360	410	270	360	285	M20	2500	280
D3T-325	116	575	410	475	280	410	310	M20	3250	300
D3T-400	127	615	450	515	330	450	350	M24	4000	360
D3T-500	132	680	515	580	390	515	410	M30	5000	500
D3T-650	141	770	570	645	420	570	440	M30	6500	600
D3T-800	156	835	635	710	490	635	510	M30	8000	650
D3T-1000	175	950	710	805	540	710	560	M30	10000	700
D3T-1250	179	1015	785	870	620	785	640	M30	12500	900
D3T-1600	203	1140	870	970	680	870	700	M30	16000	1000
D3T-2000	203	1260	985	1090	780	985	800	M36	20000	1300
D3T-2500	232	1425	1100	1220	875	1100	895	M42	25000	1600
D3T-3000	257	1550	1230	1350	1000	1230	1030	M42	30000	2000

1. Differential shrinkage of concrete

2. Creep or inelastic deformation of concrete

3. Thermal expansion and contraction of the super structure and in some cases even the sub structure

4. Elastic shortening of concrete due to prestress

5. Displacements of the structure under load or any other action

The joint should be designed to allow free translation, deflection and rotation of the structure at the edges without damage or inconvenience to the traffic. The expansion joint should be strong enough to with stand the knocking of wheels of vehicles passing over the bridge deck. The modern trend is to adopt elastomeric (Neoprene) compression seals for expansion joints in bridge decks. Different types of standard compression seals are shown in Fig. 16.15. These elastomeric seals are made of polychloroprene otherwise known as Neoprene. Single compression seal is used for small movements and newer type modular joint system using several seals are preferred to accommodate larger movements. Typical details of expansion joints for small and large movements are shown in Fig. 16.16(a) & (b) respectively.

There are many organizations which manufacture the proprietary expansion joints. According to Raina[6], the well known brands are: Maurer, Freyssinet International,

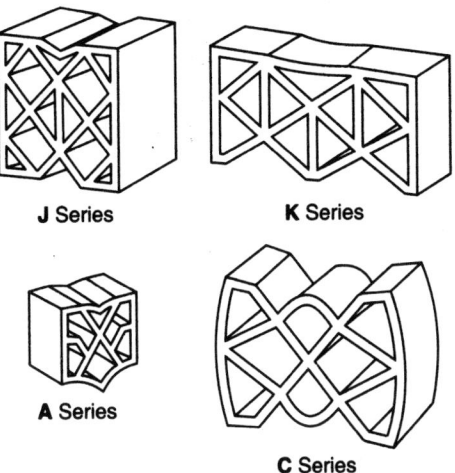

Fig. 16.15 Standard Compression Seals.

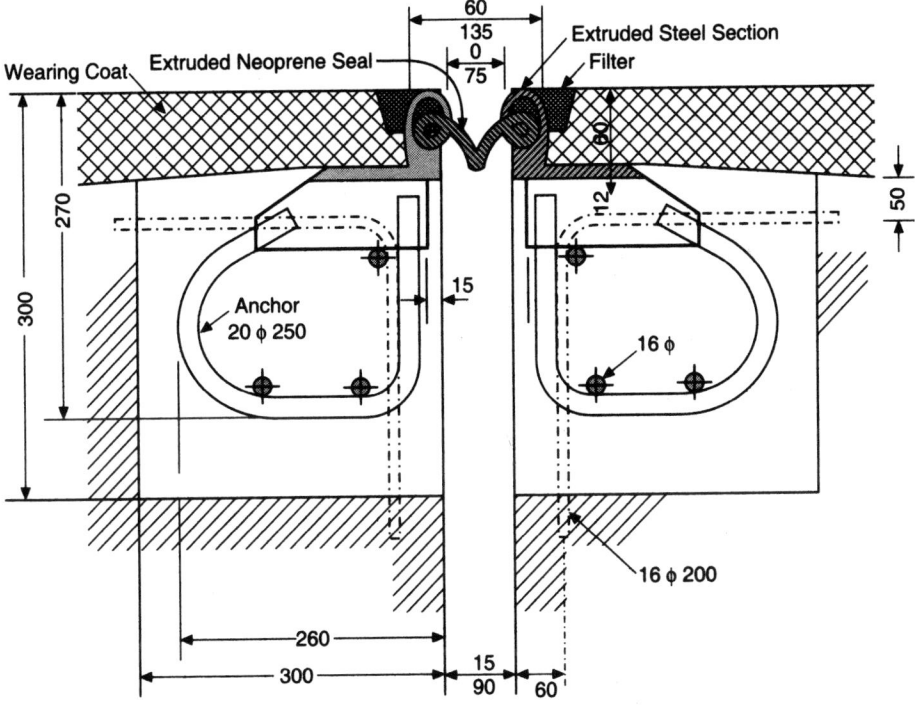

Fig. 16.16(a) Expansion Joint for Small Movements.

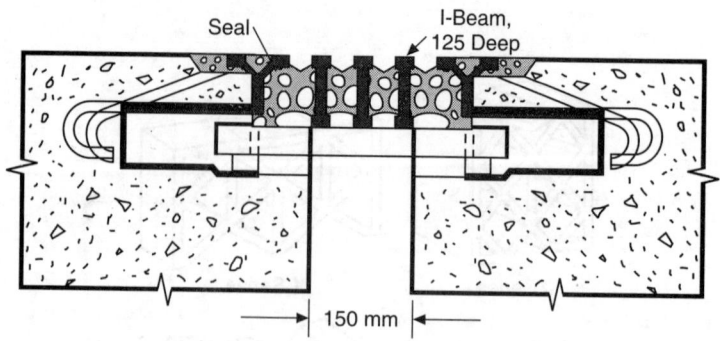

Seal

I-Beam, 125 Deep

150 mm

Fig. 16.16(b) Expansion Joint for Large Movements.

Thormajoint, Transflex, Waboflex, Gutehoffhungshutte (GMH) and Demag. In selecting an appropriate joint the Engineer must carefully weigh the manufacturer's claims on performance against the actually required functions of rotation and translation and the required directions of the incumbent movement, the fixing details, the initial costs and the maintenance problems. Strict adherence to quality controls and actual performance record should be the main criteria for impartial selection of a particular brand of joint seal.

For additional information on expansion and contraction joints, the reader may refer to other Manographs[9] and hand books SP: 34-1987[10].

EXAMPLES FOR PRACTICE

1. Design a steel rocker bearing for transmitting a vertical reacting of 800 kN and a horizontal reaction of 120 kN at the support of a girder bridge. Assume the following permissible stresses according to IRC: 83-1982.

   Permissible compressive stress on concrete bed block = 4 N/mm^2
   Permissible flexural stress in steel plate = 160 N/mm^2
   Permissible bearing stress on steel plate = 185 N/mm^2
   Permissible shear stress in steel = 105 N/mm^2

   Sketch the typical details of the rocker bearing.

2. Design a steel rocker and roller bearing to transmit a load of 1200 kN from a plate girder at the supports of a major bridge using the following data:

   Allowable working load on single and double rollers = 8 N/mm diameter per mm length.
   Permissible compressive stress on concrete bed block = 4 N/mm^2
   Permissible shear stress in steel = 105 N/mm^2
   Permissible bending stress in steel = 160 N/mm^2
   Permissible bearing stress in steel = 185 N/mm^2

   Sketch the typical details of the steel rocker-roller bearing.

3. Design a reinforced concrete rocker bearing to transmit a support reaction of 800 kN. Adopt M-30 grade concrete and Fe-415 grade HYSD bars. Permissible bearing stress in concrete = 8 N/mm². Sketch the details of reinforcements in the concrete rocker bearing.

4. Design an elastomeric pad bearing to support a Tee beam girder of a major bridge using-the following data:

Maximum dead load reaction per bearing = 330 kN
Maximum live load reaction per bearing = 500 kN
Longitudinal force due to friction for each bearing = 30 kN
Effective span of the girder = 20 m
Estimated rotation at bearing of the girder due to dead and live loads = 0.003 radians
Concrete used for Tee beam and bed block = M-20 Grade
Total estimated shear strain due to creep, shrinkage and temperature = 5×10^{-4} units

Design the elastomeric pad bearing to conform to the specifications of IRC: 83 (Part-II)–1987 and sketch the salient details of the bearing.

REFERENCES

1. Lee, D.J., Theory and Practice of bearings and Expansion Joints for Bridges, Cement and Concrete Association, London, 1971, pp. 1–65.
2. Victor, J.D., Essentials of Bridge Engineering (Fifth Edition), Oxford & IBH Publishing Co, Pvt, Ltd, New Delhi, 2001, pp. 318–349.
3. Long, J.E., Bearings in Structural Engineering, Newness-Butterworth Publications, London, 1974, pp. 1–162.
4. IRC: 83-1999, (Part-1), Standard Specification and Code of Practice for Road bridges, Section IX, Part-1, Metallic Bearings, Indian Roads Congress, New Delhi, 1999. pp. 1–27.
5. IS: 456-2000, Indian Standard Code of Practice for Plain and Reinforced Concrete, (Fourth Revision), Bureau of Indian Standards, New Delhi, 2000, pp. 1–100.
6. Reynolds, C.E and Steedman, J.C and Thvelfall, A.J., reinforced Concrete designer's Hand Book, Eleventh Edition, Psychology Press, London, 2008, pp. 1–401.
7. Raina, V.K., Concrete Bridges Practice, Analysis, Design & Economics, Tata McGraw Hill Publishing Co. Ltd., New Delhi, 1991, pp. 140–157.
8. IRC: 83-2002, (Part-III), Standard Specification and Code of Practice for Road bridges, Section IX, Part-III, Pot, Pot cum PTFE, Pin and Metallic Guide Bearings, Indian Roads Congress, New Delhi, 2002. pp. 1–70.
9. Krishna Raju, N., Reinforced Concrete Structural Elements, New Age International Publishers, New Delhi, 2016, pp. 403–518.

10. SP: 34-1987, Hand Book of Concrete Reinforcement and Detailing, Bureau of Indian Standards, New Delhi, 1987.

REVIEW QUESTIONS

1. Explain the necessity of using bearings in Bridge. What are the main functions of bearings used in bridge structures?
2. Briefly discuss the classification of bearings with examples of the different types and their specific use in bridges.
3. What are expansion bearings? Where do you use these type of bearings? Sketch a typical sliding plate bearing showing the various structural parts.
4. Explain with sketches a typical steel Rocker-Roller bearing used in bridges. Mention the structural functions of the various parts of the bearing.
5. What is a steel rocker fixed bearing? Explain the situation in which this type of bearing is adopted.
6. What are the design principles of steel rocker and roller bearings? Mention the IRC code provisions in designing such bearings.
7. What are the advantages of reinforced concrete rocker bearing? Sketch a typical concrete rocker bearing suitable for supporting the girders of major bridge.
8. Discuss briefly the design procedure of reinforced concrete rocker bearing. Sketch a typical reinforced concrete rocker bearing suitable for supporting the girders of a tee beam and slab bridge deck.
9. What are elastomeric pad and pot bearings? Bring out clearly the difference between the two types. What are the special advantages of these bearings in comparison with the traditional steel bearings?
10. Explain clearly the purpose of using expansion joints in bridge decks. What is the necessity of using the expansion joints?

OBJECTIVE TYPE QUESTIONS

1. Bearings are generally used in bridge decks and they are located at
   a) The deck of the bridge
   b) The centre of spans
   c) The soffit of decks over the supports
2. Due to shrinkage and creep of the bridge deck experiences
   a) Vertical movements
   b) Translations
   c) Rotations
3. The magnitude of movements in a concrete bridge deck can be estimated by using the material property of
   a) Compressive strength
   b) Modulus of elasticity
   c) Coefficient of Linear expansion

4. The deflections in the super structure of a bridge deck due to dead and live loads results in
   a) Translational movements
   b) Rotational movements
   c) Vertical displacements.

5. The cost of maintenance of bearings in bridges is the least with the use of
   a) Steel rocker roller bearings
   b) Reinforce concrete rocker bearings
   c) Elastomeric pad bearings

6. In the case of selecting the type of bearing for a major bridge located in the west coast of India, the ideal type is a
   a) Steel rocker-roller bearing
   b) Sliding plate bearing
   c) Reinforced concrete rocker bearing

7. In a reinforced concrete rocker bearing designed to support a heavy load from a bridge girder, the hoop reinforcements are designed based on the
   a) Grade of concrete
   b) Bearing stress
   c) Hoop tension

8. The shape factor of an elastomeric pad bearing designed to support a bridge girder should have value in the range of
   a) 4 to 8
   b) 6 to 12
   c) 12 to 16

9. The ratio of overall length to breadth of an elastomeric pad bearing should be equal to or less than
   a) 6
   b) 4
   c) 2

10. Expansion joints in bridge decks are normally provided at locations of
    a) Maximum bending moment
    b) Maximum shear force
    c) Structural discontinuity

17

Cable Stayed Bridges

17.1 GENERAL FEATURES

During the last two decades cable stayed bridges have found wide applications, especially in Western Europe and to a limited extent in other parts of the world. Cable stayed bridges were developed in Germany during the post war years, mainly to obtain optimum structural performace from material like steel which was in short supply during the post war years.

The successful application of cable stayed systems was realized only recently, with the introduction of high strength steels, orthotropic type decks, development of welding techniques and progress in structural analysis. The development and application of electronic computers opened up new and practically unlimited possibilities for the exact solution of these highly statically indeterminate systems and for precise statical analysis of their three-dimensional performance.

Basically the following important factors helped for the successful development of cable stayed bridges.

1. The development of orthotropic steel decks

2. Application of high strength steels, new methods of fabrication and erection.

3. The development of methods of structural analysis of highly statically indeterminate structures and application of electronic computers.

4. Experience with previously built bridges containing basic elements of cable stayed bridges.

5. The ability to analyse such structure through model studies.

The first modern cable stayed bridge being the stromstund bridge in Sweden designed by Dischinger and constructed in the year 1955.

In 1952 Leonhardt[1] designed the cable stayed bridge across the Rhine in Dusseldorf. After the first two cable stayed bridges of modern design had proved to be very stiff under traffic load, aesthetically appealing, economical and relatively simple to erect, the way was open for further wide and successful application. The new system became rapidly popular among German bridge engineers and about ten years later, in several other countries also. It is now increasingly applied by designers all round the world.

A list of world's prominent cable stayed bridges constructed in the various countries and their salient features are compiled in Table 17.1.

Table 17.1 Cable Stayed Bridges.

No.	Name of Bridge	Material	Span (m)	Tower Height (m)	Year
1	Saint - Nazaire (France)	Steel	404	67	1975
2	Luling (U.S.A)	Steel	376	75	1980
3	Duisburg (West-Germany)	Steel	350	50	1970
4	West Gate (Australia)	Steel	336	46	1974
5	Brazo-Largo (Argentina)	Steel	330	48	1976
6	Zarace (Argentina)	Steel	330	67	1976
7	Kholbrand (West-germany)	Steel	325	98	1974
8	Knie (West-Germany)	Steel	320	114	1969
9	Brotonne (France)	Concrete	320	65	1976
10	Wadikuf (Libiya)	Concrete	282	50	1972
11	Mesapatomia (Argentina)	Concrete	340	97	1972
12	Dames Point (U.S.A)	Concrete	396	92	1995
13	Rafael Urdanela (Venczula)	Concrete	235	68	1962
14	Polcevera (italy)	Concrete	208	45	1967
	Examples of Cable Stayed bridges in India				
1	Haradwar (India)	Concrete	130	20	1990
2	Jogighopa (India)	Concrete	5 × 286	50	1995
3	Akkar (India)	Concrete	152	56	1995
4	Vidyasagar Sethu (India)	Concrete & Steel	452	100	1995
	Longest Cable Stayed Bridges				
1	Normandie Bridge (France)	Steel & Concrete	624 Main Span	–	U.C (1994)
2	Messina Straits (Italy)	Steel	1800	–	1982

17.2 COMPONENTS OF CABLE STAYED BRIDGES[2]

The various structural components of a typical, cable stayed bridge are the following:

1. Towers or Pylons

2. Deck System

3. Cable System Supporting the Deck

Figure 17.1 shows a typical cable stayed bridge (North bridge) constructed in Dusseldorf, Germany.

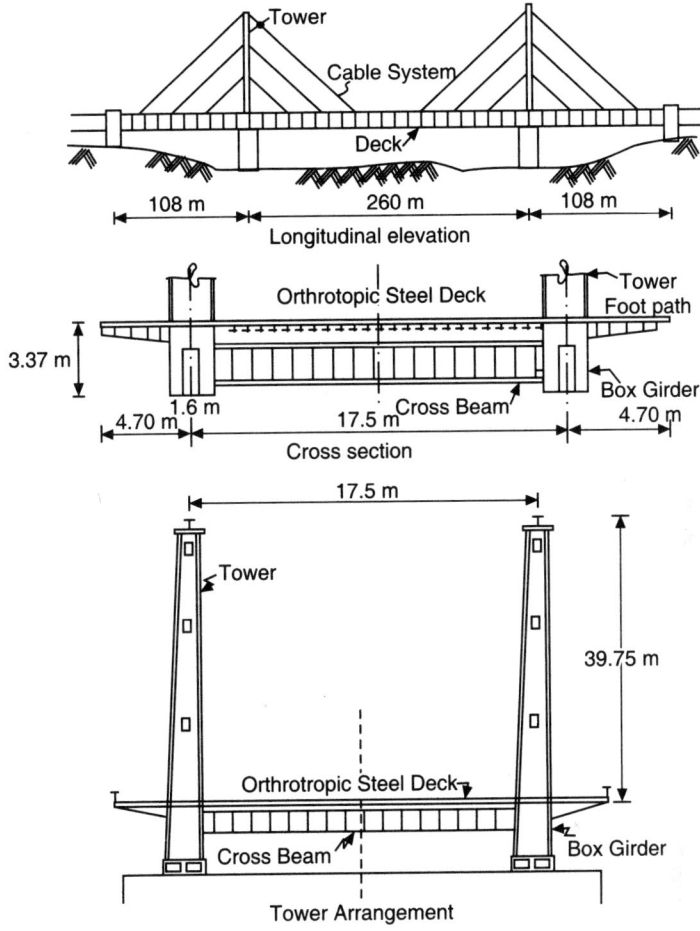

Fig. 17.1 Typical Cable Stayed Bridge.

The bridge consists of continuous suspended box girders. The cable in a harp configuration are in two planes supported by single towers 40.9 m high which are built into the deck structure. The tower saddle for the centre cable is fixed but the saddles for the upper and lower cables are supported on rocker type bearings.

The cross-section of the deck consists of two box shaped main girders 3.37 m high and 1.60 m wide. The orthotropic steel deck consists of a steel plate 14.3 mm thick stiffened by 203 × 102 × 11.1 mm angles spaced at 0.41 m and supported by cross beams spaced at 1.83 m intervals. In addition to a 14.3 wide roadway, 1.83 m bicycle paths and 2.14 m pedestrian walks of reinforced concrete run along both sides, making the total width of deck of 25.9 m.

17.3 TOWERS OR PYLONS[3]

The towers are the principal compression members transmitting the load to the foundations. Towers are of different types to accomodate different cable arrangements, bridge site conditions, design features aesthetics and economical considerations. Generally the arrangement of the cable stays determines the design of both the pylon and the deck.

The pylons can be arranged to support one axial layer of cable stays or two lateral layer of cable stays. Figures 17.2 and 17.3 shows the arrangement of pylons for single and double layer of cable stays.

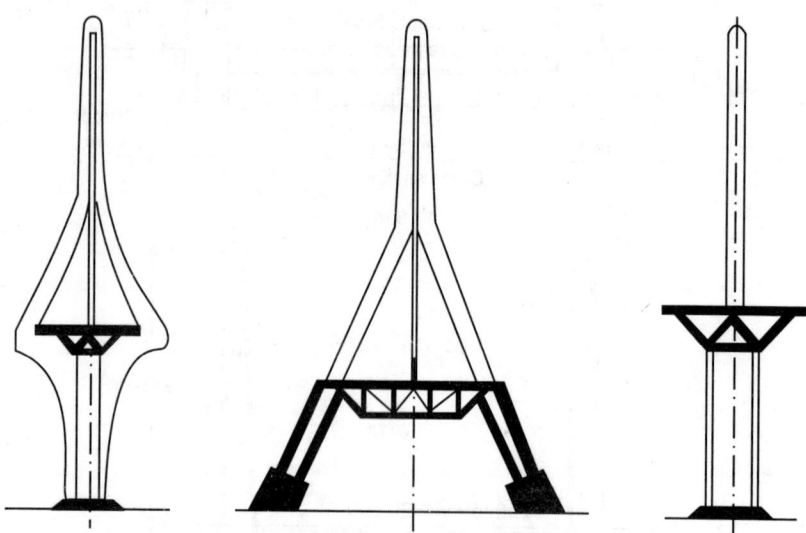

Fig. 17.2 Transverse Arrangement of Pylons (One axial layer of stays).

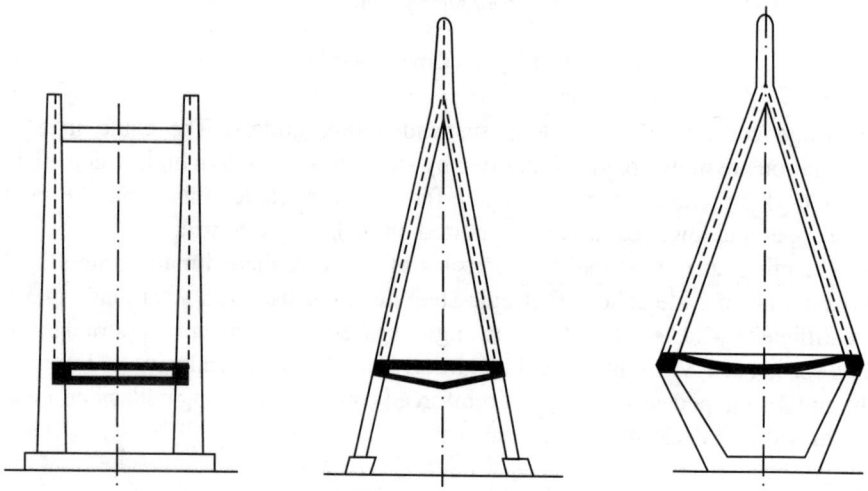

Fig. 17.3 Transverse Arrangement of Pylons (Two lateral layers of stays).

The transfer of forces from the cable stays to the pylon is achieved by three different arrangements:

(a) A saddle permitting the continuity of the cable stay are shown in Fig. 17.4.

(b) The cable stays fixed to the top of the tower may cross each other inside the pylon as shown in Fig. 17.5.

(c) A relay device incorporated into the top of the tower connecting the upper anchorages of associated cable stays as shown in Figs. 17.6 and 17.7.

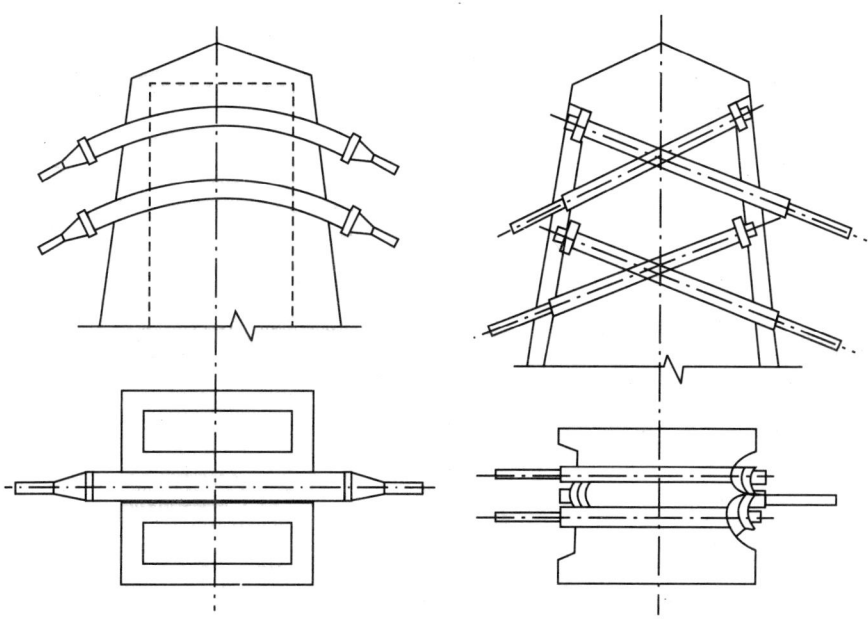

Fig. 17.4 Saddle.   **Fig. 17.5** Crossing of Stays Cables.

17.4 TYPES OF CABLE STAYS[4]

The cable stays are made up of high tensile steel of different types with an ultimate tensile strength in the range of 1500 to 2000 N/mm^2. The different types of cable stays used are:

(a) Twisted cable,

(b) Parallel wires,

(c) Parallel strands, and

(d) Locked coil cable. Typical cross sections of these types are shown in Fig. 17.8.

Locked coil cable system generally used in suspended bridges comprise of stranded cables with an outer layer of Z-shaped wires interlaced to form a locked unit and this is the first type used in the construction of cable stayed bridges. The stress variations

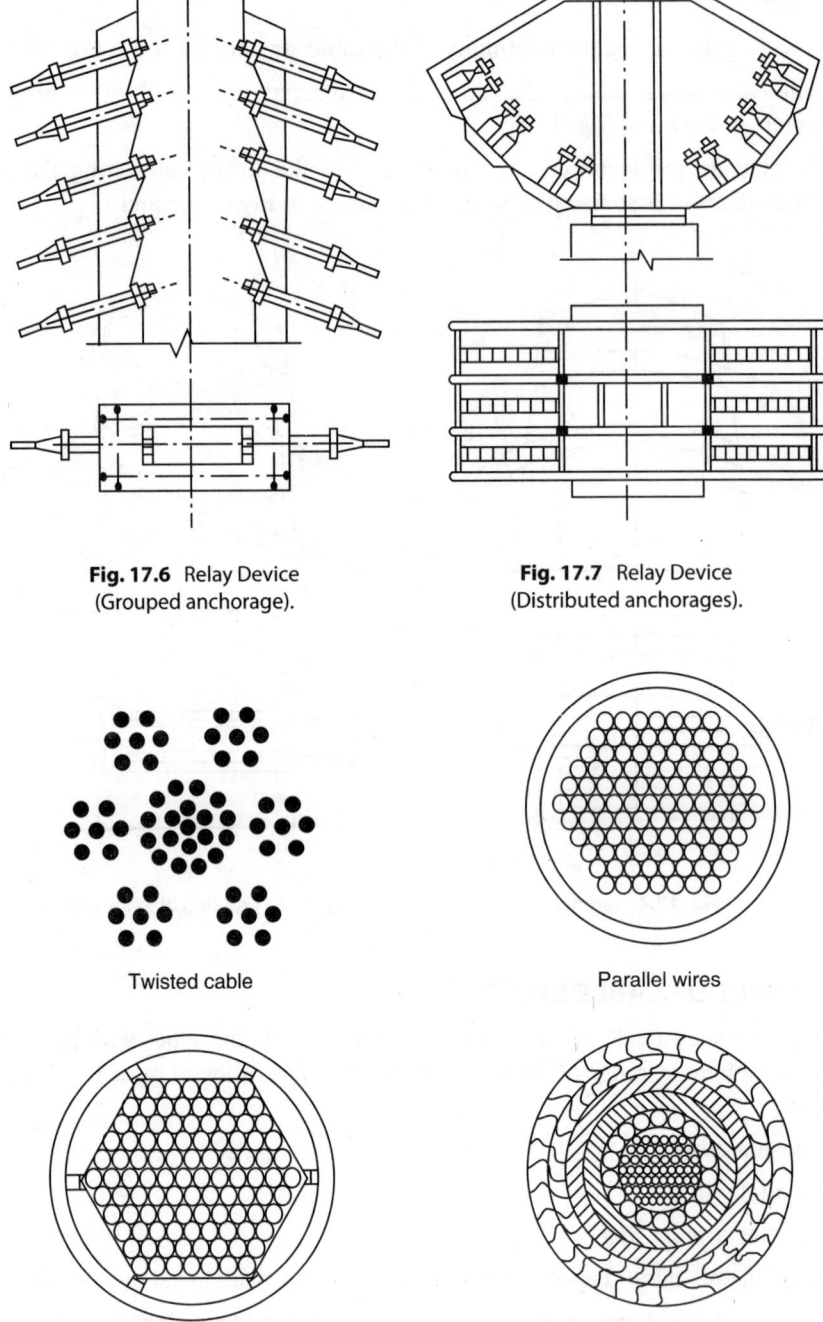

Fig. 17.6 Relay Device
(Grouped anchorage).

Fig. 17.7 Relay Device
(Distributed anchorages).

Twisted cable

Parallel wires

Parallel strands

Locked coil cable

Fig. 17.8 Different Types of Cable Stays.

being high, this type was not entirely satisfactory. Later the use of parallel wire cables developed from prestressing techniques. During the period from 1960 to 1980, seven wire strands were widely used for prestressing and the research worr undertaken on cable stays resulted in the use of strands with a very high yield point (1860 N/mm^2) and excellent fatigue resistance.

The recent types include the Freyssinet cable stay comprising a bundle of parallel strands of 15 mm diameter enclosed in a tube generally of polyethylene which allows the threading and grouting of the stay. A spiral of steel wire inside the polyehthylene ensures that proper grout cover is provided around the bundle of strands as shown in Fig. 17.9.

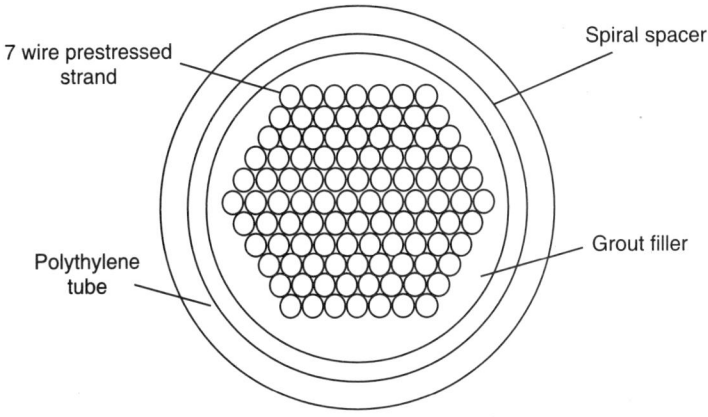

Fig. 17.9 Section of Freyssinet Cable Stay.

The B.B.R.-HIAM (High Amplitude) stay and anchorage developed in Switzerland consists of parallel wire bundles. The wires are provided with a variant of B.B.R buttonhead anchorage with special features to improve their fatigue strength.

The strand used for the stay cables must comply with the specifications prescribed in (a) B.S: 5896: 1980 (b) A.S.T.M-A 416-80, (c) EURONORM 138-79 or any other equivalent standard. The strands should withstand a stress range of 195 N/mm^2 with the upper stress limit of 0.8 times the breaking stress with a fatigue life of not less than 2 million cycles. The standard specifications of strands used for cable stays are compiled in Table 17.2. Table 17.3 shows the ultimate load characteristics of the various types of cable stays of different sizes.

The cable stay comprises of 3 types of zones viz.,

(a) The free length zone

(b) The transition zone

(c) The anchorage zone

A typical general layout of the Freyssinet cable stay is shown in Fig. 17.10.

Table 17.2 Standard Specifications for Strands for Cable Stays

Standard	A.S.T.M A 416-80	EURONORM 138-79	BS 5896:1980
Nominal Diameter (mm)	15.24	15.70	15.70
Nominal Tensile Strength (MPa)	1862	1770	1770
Nominal Steel Area (mm²)	140	150	150
Nominal Weight (kg/m)	1.102	1.180	1.180
Specified Characteristic Breaking Load (kN)	260.7	265.0	265.0
Specified Characteristic 0.1% proof load (kN)	–	225.0	225.0
Load at 1 % Elongation (kN)	221.5	233.0	233.0
Minimum Elongation at Maximum Load (%)	3.5	3.5	3.5
for L (mm)	610	500	500

Table 17.3 Ultimate Load Characteristics of Freyssinet Cable Stays

Type of Strand		Single Stand	19H15	27H15	37H15	48H15	61H15	75H15	91H15
15.24 mm	*A*	140	2660	3780	5180	6720	8540	10500	12740
Grade-270	*Af*$_{pu}$	260.7	4953	7039	9646	12514	15902	19552	23724
ASTM A-426	*g*	1.102	20.94	29.75	40.77	52.90	67.22	82.65	100.28
15.77	*A*	150	2050	4050	5550	7200	9150	11250	13650
Super Strand	*Af*$_{pu}$	265	5035	7155	9805	12720	16165	19875	24115
EuroNorm EU-138	*g*	1.19	22.42	31.86	43.66	56.64	71.98	88.50	107.28
15.2 mm	*A*	139	2641	3753	5143	6672	8479	10425	12649
Class III French	*Af*$_{pu}$	252.1	4790	6807	9328	12100	15378	18907	22941
Circular N° 73-175	*g*	1.091	20.73	29.46	40.37	52.37	66.55	81.82	107.28

A – Nominal Cross sectional area (mm²)
Af$_{pu}$ – Ultimate Tensile force (kN)
g – Nominal weight (kg/m)

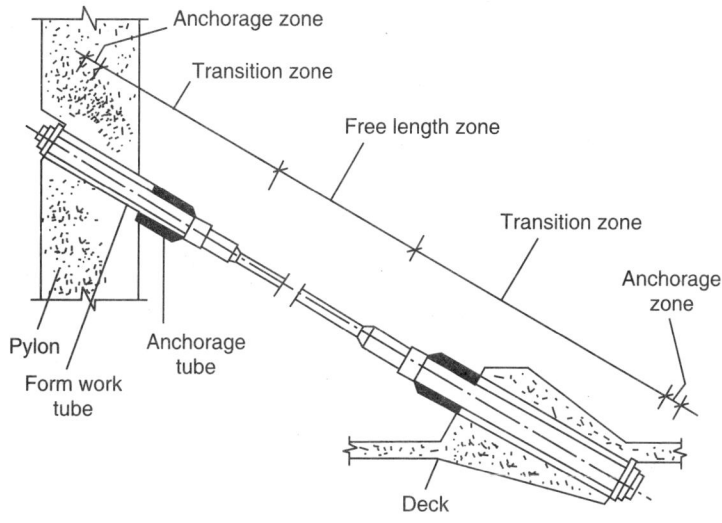

Fig. 17.10 General Layout of the Freyssinet Cable Stay.

17.5 LONGITUDINAL CABLE ARRANGEMENT

The arrangement of cables in the longitudinal direction depends upon several factors such as clear span, tower height, spacing of towers and level of approach roads. For shorter span lengths a single forstay and a backstay with a single pylon is sufficient to satisfy the loading requirements. A typical example of this type used in Severin bridge at Cologne, Germany is shown in Fig. 17.11.

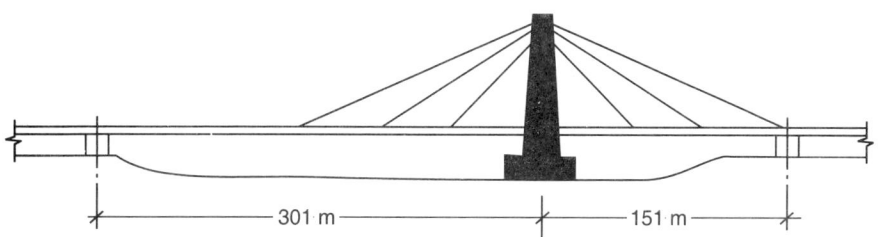

Fig. 17.11 Severin Bridge at Cologne, Germany.

Basically there are four types of cable configurations generally used' and they are classified as

(a) Fan type,

(b) Harp type,

(c) Mixed type and

(d) Star type.

Typical arrangements of these types are shown in Fig. 17.12. The choice of the system is influenced by span, type of loading number of roadway lanes, height of towers, economy and aesthetic considerations. The fan type is more aesthetic and as a rule the most economical for a pylon of slenderness ratio $(h/L) \leq 0.3$. For an equal tower height, the average inclination of the cable stays is lower. However the cable stays are longer and converge towards a single point at the top of the tower posing problems of anchoring arrangement and any subsequent stay replacement is difficult. To obviate this problem often, a mixed type solution is adopted by spreading the anchorages along the mast according to their dimensions closer to the top.

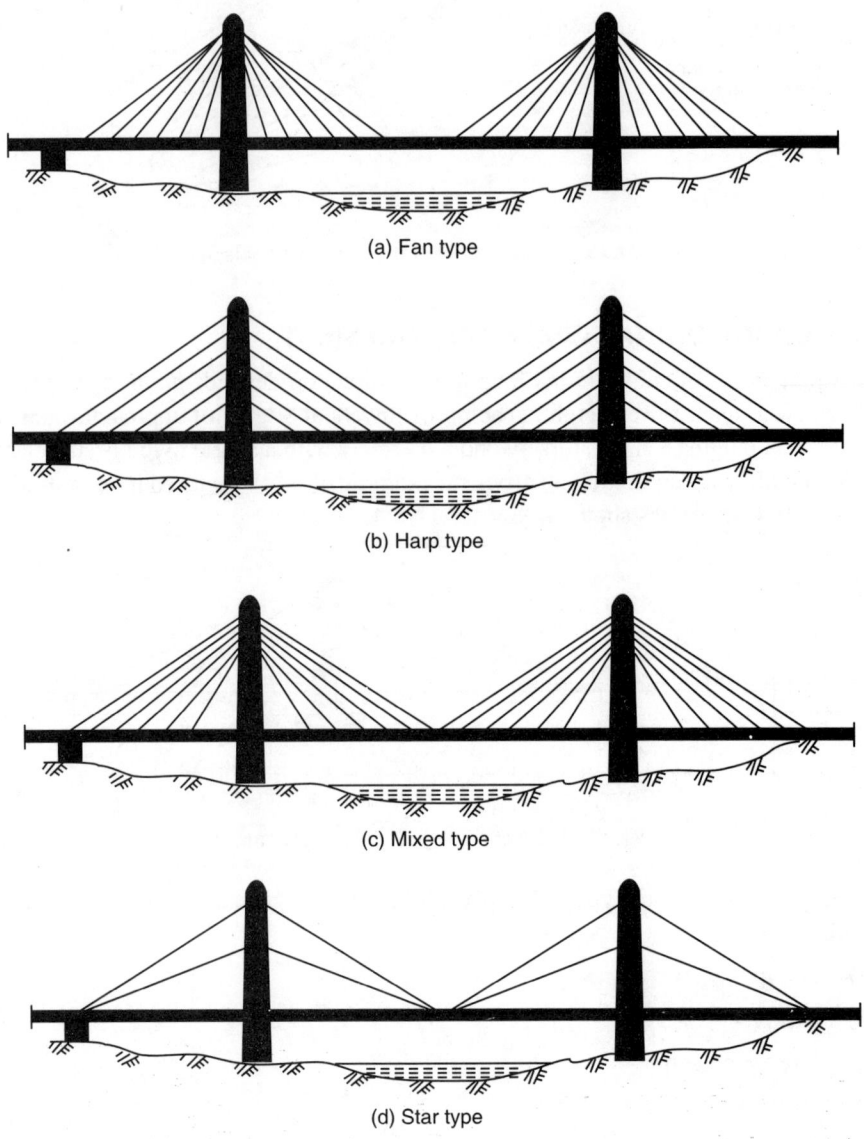

(a) Fan type

(b) Harp type

(c) Mixed type

(d) Star type

Fig. 17.12 Longitudinal Cable Arrangements.

The harp system is perferred in a double plane system as it minimises the intersection of cables when viewed from an oblique angle. The motorist may find the harp system more attractive. In the harp system the cable connections are distributed throughout the height of the tower and hence results in an efficient tower design in comparison with the fan type. The structural behaviour of the tower varies depending upon the type of cable system. The fan type increases buckling problems due to grater effective strut length and the harp type increases bending moments.

The mixed type represents a compromise between the extremes of the harp and the fan systems and it is useful when it becomes difficult to accommodate all cables at the top of the tower. The star system may be preferred due to its unique aesthetic appearance.

17.6 ADVANTAGES OF CABLE STAYED BRIDGES

Cable stayed bridge is an innovative structure and is preferred to conventional steel suspension bridges for long spans mainly due to the reduction in moments in the stiffening girders resulting in smaller section of the girders leading to economy in overall costs. Figure 17.13 shows the comparison of bending moment in a five span conventional continuous girder system with that of the cable stayed girder for the same span. The ratio of maximum bending moment in the cable stayed girder is nearly I/IOth of that of the conventional continuous girder system. In addition, the moments can be controlled to make them more uniformly distributed along the girder length resulting in efficient material utilisation even with a very low depth to span ratio of 1/90.

17.7 BASIC CONCEPTS OF STRUCTURAL ANALYSIS5

(a) Approximate Structural Analysis

A cable stayed bridge system is generally many times statically indeterminate. A linear analysis can be made by assuming a suitable statically determinate system. The deflections of the basic system under applied loads may be determined by applying the classical theory of structures by neglecting the deformations of the system when formulating the equilibrium conditions.

For a statically determined basic system, the resulting equations are linear in the loads and in the internal forces and linear superposition is valid for the internal forces caused by different loads or load groups. If Hooke's law is assumed to be valid, linear superposition applies also to the, displacements and therefore to the determination of the stresses of cable stayed bridge systems. The design process for a cable stayed bridge system with accepted geometrical layout may be divided into the following stages:

(a) A preliminary set of sectional properties is assumed for each member of the system.

(b) The sectional properties assumed in stage (a) are analysed, applying one of the statical methods of analysis. Stresses and displacements under the given loads on the system are determined and compared with the maximum unit stresses and maximum displacement span ratios allowed by the specifications.

(c) A new set of sectional properties is chosen to satisfy the requirements of the specifications.

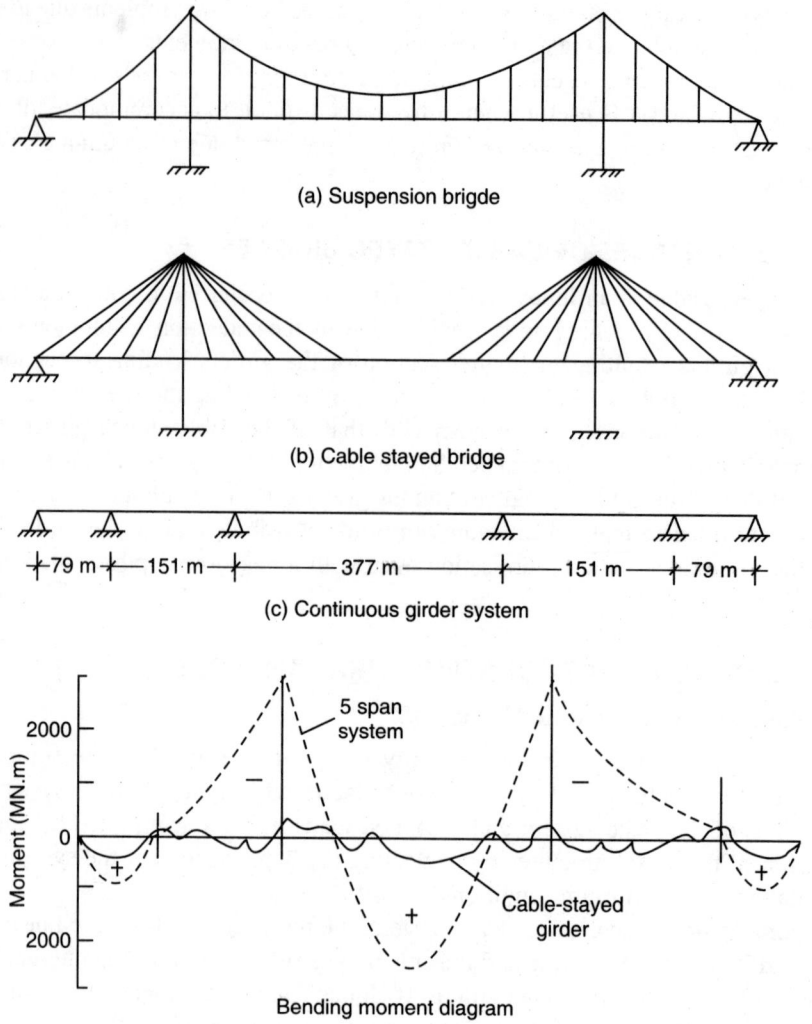

Fig. 17.13 Moment Variation in Cable Stayed and Five Span System.

The above stages are repeated until we obtain a specified relation between the sectional properties assumed in stage (a) and those obtained in stage (c).

Several general methods for the linear structural analysis of a cable stayed bridge system are available. These methods that do not require the use of an electronic computer are termed as 'Classical methods'.

(b)   Rigorous Methods of Structural Analysis[6]

A cable stayed bridge is a highly statically indeterminate structure in which the stiffening girder behaves as a continuous beam supported elastically at the points of cable attachments. Except in the case of a very simple cable-stayed bridge, a computer is necessary for the solution of this type of structure, its use being primarily in analysis rather than in design applications.

Computer programmes are necessary to generate the influence diagrams for cable forces, stiffening girder bending moments and shears and tower and pier reactions. The computer is also required for the rapid solution of various parametric efforts and loadings that have to be taken into account in achieving a reasonably efficient design. Probably the most important problems are the determination of the optimum section of the stiffening girder section and cable configuration and size.

In a simplified approach to the solution, the structure is assumed to be a linear elastic system which may be analysed using the standard stiffness or flexibility method. Several general computer programmes are available such as FRAN, STRESS, STRUDL, which use this approach.

The non-linear behaviour of cables, whose sag varies with changing axial load, presents problems in the solution of the bridge system more complex than those of a structure of linear behaviour. A convenient method of accounting for the non-linear behaviour of the cable stayed bridge system is to introduce the concept of a straight line chord member with a modified or ideal modulus of elasticity substituted for the actual cable member. The use of this concept allows the application of a plane frame computer programme properly adopted to account for the non-linearty by an iteration procedure. For rigorous analysis and design of cable stayed bridge the reader may refer to the excellent monographs and research papers of Troitsky[7], Smith[8], Tang Man-Chung[9], Schreier[10], compiled in the references.

17.8   APPROXIMATE STRUCTURAL ANALYSIS

Approximate structural analysis of cable stayed bridge system includes the computation of the following basic items:

(a) Optimum inclination of cables

(b) Height of tower and length of panels

(c) Cable forces

(d) Approximate weight of stiffening girders

(e) Self weight of cables

(f) Degree of redundancy

The design procedure for a cable stayed bridge system with a predetermined geometrical layout may be classified into the following stages:

(1) A preliminary set of sectional properties is assumed for each of the system.

(2) The sectional properties assumed are analysed applying one of the statical methods of analysis.

(3) Stresses and displacements under the given loads on the system are found and copipared with the maximum permissible displacement-span ratios allowed by the specifications.

(4) A new set of sectional properties are chosen to satisfy the requirements of the specifications.

The above steps are repeated until a satisfactory convergence is obtained between the sectional properties assumed in stage1 and those obtained in stage 4.

The classical method of structural analysis can be carried out by manual computations. However the disadvantage of this method is that a linear elastic behaviour of the bridge system is essential to simplify the computations.

(a) Determination of Optimum Inclination of Cables

The height of the tower significantly influences the stiffness of the bridge system. As the angle of inclination of the cable with respect to the stiffening girder increases, the stresses in the cables decrease, as does the required cross-section of the tower. However, as the height of the tower increases, the length of the cables and therefore their axial deformations also increase, as well as the amount of metal in the cables.

In order to find the optimum amount of material and inclination of the cables, the following simplified bridge system hinged at the locations of the cable connections to the stiffening girder is considered. The following notations are used in the computations.

A_n = Cross sectional area of cable (mm^2)

L_n = Length of the cable (m)

y = Specific weight of the cable material (kN·m^3)

W = Weight of the cable (kN)

α_n = The angle of inclination of the cable

$na = L_n \cos \alpha_n$ = Horizontal projection of cable length

n = The corresponding number of the panels

a = Length of each panel (m)

E = Modulus of elasticity of the cable material (kN·m^2)

F_n = Force in the cable (kN)

f = Permissible stress in the cable (kN/m^2)

P_n = Vertical component of the force in the cable (kN)

Referring to Fig. 17.14

The weight of the cable is computed as

$$W = A_n L_n y \tag{1}$$

and
$$A_n = (F_n/f), \text{ also } L_n = (na/\cos \alpha_n)$$
$$F_n = (P_n/\sin \alpha_n)$$
∴
$$A_n = (P_n/f \sin \alpha_n) \tag{2}$$

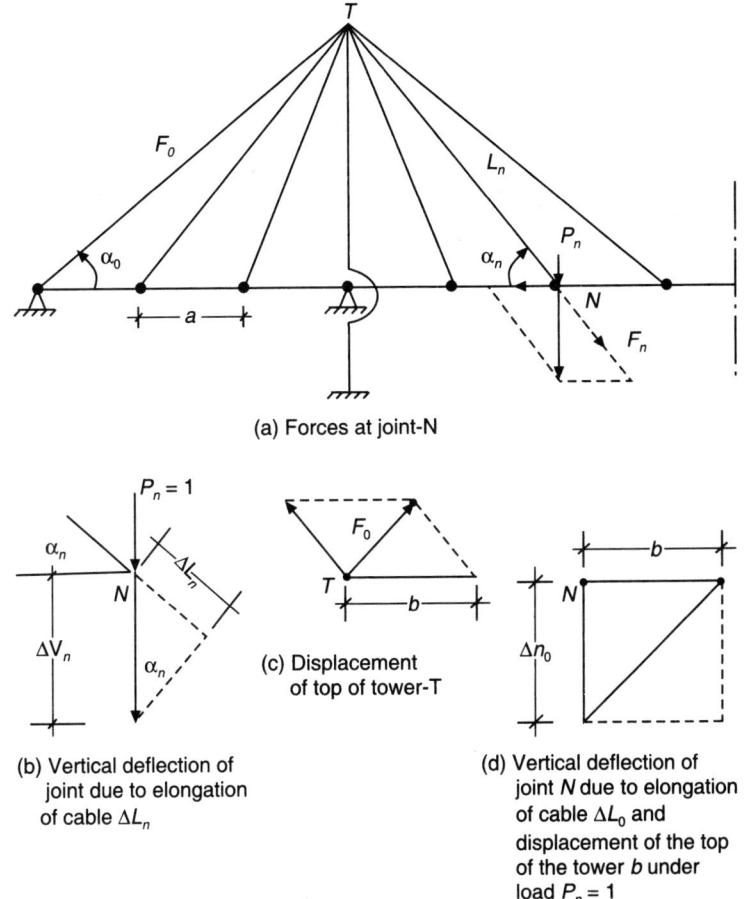

(a) Forces at joint-N

(b) Vertical deflection of
joint due to elongation
of cable ΔL_n

(c) Displacement
of top of tower-T

(d) Vertical deflection of
joint N due to elongation
of cable ΔL_0 and
displacement of the top
of the tower b under
load $P_n = 1$

Fig. 17.14 Anaylsis of Simplified Cable Stayed Bridge System.

By substituting Eq (l) in Eq (2), we have
$$W = (P_n \, nay)/(f \sin \alpha_n \cos \alpha_n)$$
Assuming $P_n = 1$ and designating $(nay/f) = C$, we have

$$W = \left[\frac{C}{\sin \alpha_n \cos \alpha_n} \right] \tag{3}$$

Hence the weight of the cable is a function of $1/(\sin \alpha_n \cos \alpha_n)$. The force in the cable P_n due to the Load $P_n = 1$ at the Joint N is expressed as
$$F_n = P_n / \sin \alpha_n \tag{4}$$
The corresponding elongation of the cable is

$$\Delta L_n = \left(\frac{P_n L_n}{EA_n \sin \alpha_n} \right) \tag{5}$$

The force in the upper cable F_o transferred by the tower from the cable F_n is

$$F_o = \left(\frac{F_n \cos \alpha_n}{\cos \alpha_n} \right) = \left(\frac{P_n \cot \alpha_n}{\cos \alpha_n} \right) \tag{6}$$

The corresponding elongation of the cable F_o is

$$\Delta L_n = \left(\frac{P_n L_o \cot \alpha_n}{EA_o \cos \alpha_n} \right) \tag{7}$$

and the corresponding displacement of the top of the tower is

$$b = (\Delta L_o / \cos \alpha_o) \tag{8}$$

The vertical deflection of the joint n due to the elongation ΔL_n of the cable is expressed as

$$\Delta_{vn} = \left(\frac{\Delta L_n}{\sin \alpha_n} \right) = \left(\frac{P_n L_n}{EA \sin^2 \alpha_n} \right) \tag{9}$$

The vertical deflection of the joint n due to the elongation L_o of the cable and displacement b of the top of the tower, due to the load P_n at the joint n is

$$\Delta_{no} = b \cot \alpha_n = \left(\frac{L_n \cot \alpha_n}{\cos \alpha_n} \right) + \left(\frac{P_n L_n \cot^2 \alpha_n}{EA_o \cos^2 \alpha_n} \right) \tag{10}$$

therefore the total vertical deflection of the joint n under the load P is given by

$$\Delta_{tot} = (\Delta_{vn} + \Delta_{no}) = \left(\frac{P_n L_n}{E A_n \sin^2 \alpha_n} \right) + \left(\frac{P_n L_o \cot^2 \alpha_n}{E A_n \cos^2 \alpha_n} \right) \tag{11}$$

After substituting in expression (11), the values

$$P_n L_n = F_n \, n \, a \tan \alpha_n$$

$$P_n L_n = F_n \, n_1 \, a \left(\frac{\sin \alpha_n}{\cos \alpha_n} \right)$$

the expression for total deflection is given by

$$\Delta_{tot} = \left(\frac{C_1}{\sin \alpha_n \cos \alpha_n} \right) + \left(\frac{C_2 \cos^2 \alpha_n}{\sin \alpha_n \cos^3 \alpha_n} \right) \tag{12}$$

where the constants

$$C_1 = \left(\frac{F_n \, na}{E \, A_n} \right) \text{ and } C_2 = \left(\frac{F_n \, n_1 \, a}{E \, A_o} \right)$$

If the angle and the number of panels are equal,

$$na = n_1 a \quad \text{also} \quad \alpha_o = \alpha_n = \alpha \quad \text{and} \quad C_1 = C_2$$

The final expression for the total deflection is given by

$$\Delta_{tot} = C_3 / (\sin \alpha \cos \alpha) \tag{13}$$

where $(C_1 + C_2) = 2C_1 = 2C_2 = C_3$

Comparison of the expression (3) and (13) indicates that the displacement of the joint in the stiffening girder, and therefore the bending moment, follows the same pattern as the change in the weight of the cable. The values of Δ_{tot} as function of the angle α_n is shown in Fig. 17.15.

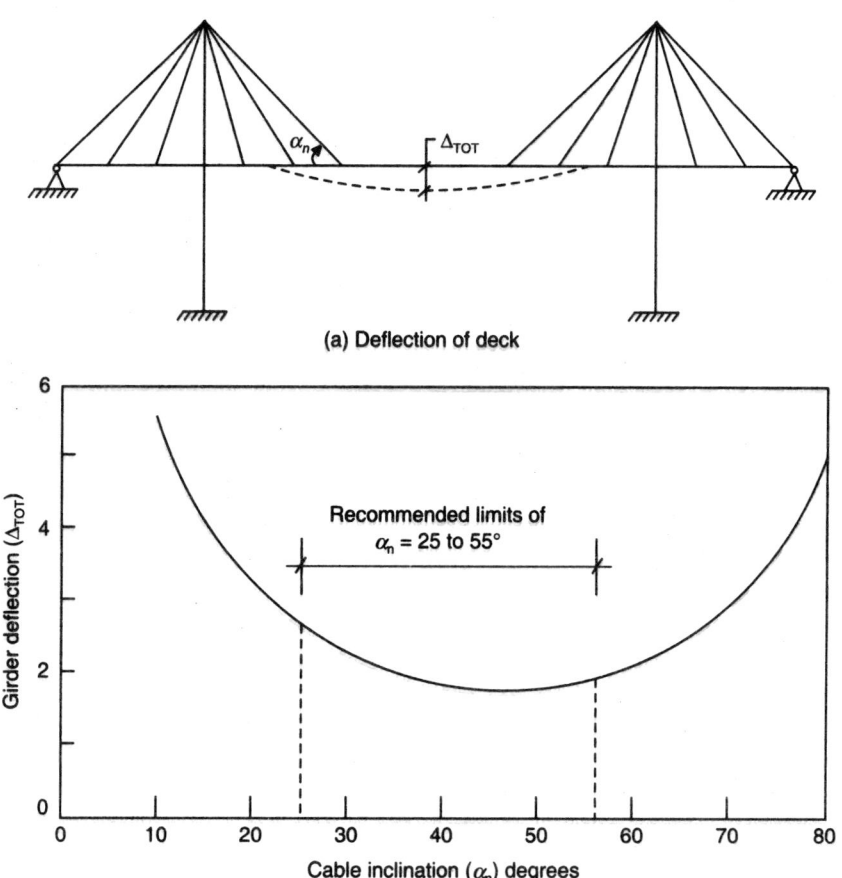

(a) Deflection of deck

Fig. 17.15 Relation between Cable Inclination and Deflection of Girder.

The relation between the cable inclination and deflection of joint indicates that the optimum angle of the cable inclination is 45° and may vary in the reasonable limits of 25 to 65°. The low values of the angle of inclination correspond to the external cables while the greater values correspond to the cable nearest to the tower.

(b) Determination of Height of Tower and Length of Panels

The height of the tower as a function of the panel length na, may be expressed as follows:

$$h = na \tan 25° = 0.465\ na \tag{14}$$

Using three cables on each side of the tower, $n = 3$

$$h = 0.465 \times 3a = 1.4a \tag{15}$$

and with four cables, we have

$$h = 0.465 \times 4a = 1.86a \tag{16}$$

The middle panel is usually longer than the remaining panels and may be taken as 1.3a. In that case, the ratio of the tower height to the length of the mid span considering a total of six panel is obtained as

$$\left(\frac{h}{L}\right) = \left(\frac{1.4a}{(6+1.3)a}\right) + \left(\frac{1}{5.2}\right)$$

Hence $\qquad h = (L/5.2)$ \hfill (17)

The number and length of the panels are basically determined by the bridge system and its structural characteristics.

It is possible to reduce the moment of inertia of the girder and for this purpose it is necessary to reduce the panel length. However, the reduction of the girder's depth is limited because of the connection of the cable to the girder. Technically it is certainly convenient to have the minimum number of cable connections to reduce the number of anchorages and for regulation of forces in the cables.

A comparison of the existing structures-indicate the following optimum values of the panel lengths:

(1) For central spans in the range of 137 to 150 m, panels of 20 m length are recommended.

(2) For the smaller central spans, the panels should be in the range of 15 to 17 m.

(3) For central spans longer than 170 m, panels should be 30 m in length.

The middle panel performs differently from the other panels since it is not compressed by the horizontal component of the cable forces and therefore it is possible to use comparatively a longer penel. The size of the middle panel substantially affects the distribution of the loadings between the remaining parts of the stiffening girder. With greater stiffness of the girder at the middle panel, the non loaded part contributes more to the increased carrying capacity of the loaded part.

The optimum size of the middle panel is determined under the assumption of full use of the material of the girder. Experience indicates that the length of the middle panel may be 20 to 30 percent longer than the other panels.

(c) Determination of Cable Forces

For the preliminary design it is possible for a cable system with five equal panels to use the empirical relation,

$$M_{max} = 0.007\ w\ L^2$$

and for one with seven panels

$$M_{max} = 0.006 \, w \, L^2$$

Where w = Total load (dead + Live) (kN/m)

L = Length of the panel (m)

The maximum bending moment in the stiffening girder can be estimated using these empirical equations. The cable stay forces depend on factors such as the length of span, number and size of panels and angles of inclination of cables, dead weight of deck and live loads. Referring to Fig. 17.16 the force in the cable is computed using the following notations:

P = Force in the cable

S = Spacing of cable

W = Total load per metre of deck

α = Angle of inclination of cable with horizontal

R = Vertiatal reaction at cable stay node = $S \cdot w$

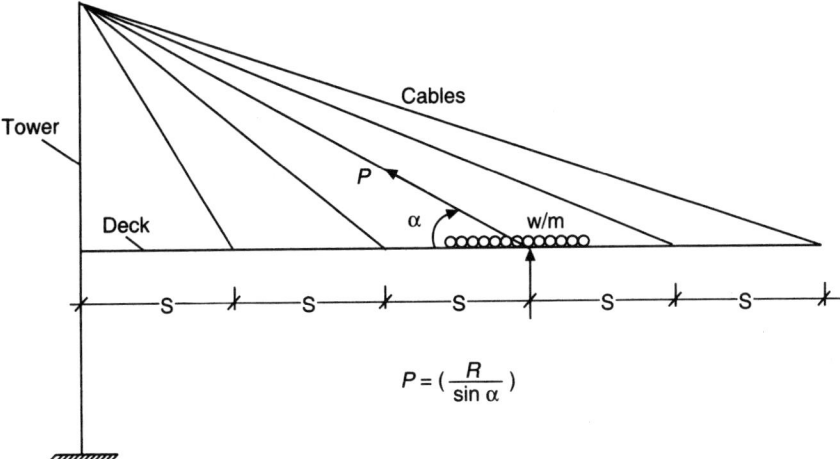

Fig. 17.16 Force in Cables.

The force in the cable stay is computed as

$$P = \left(\frac{R}{\sin \alpha} \right)$$

The number of wires or strands required in the cable can be designed for the designed cable force.

The following example illustrates the design of a typical cable for the Akkar bridge in Sikkim shown in Fig. 17.17.

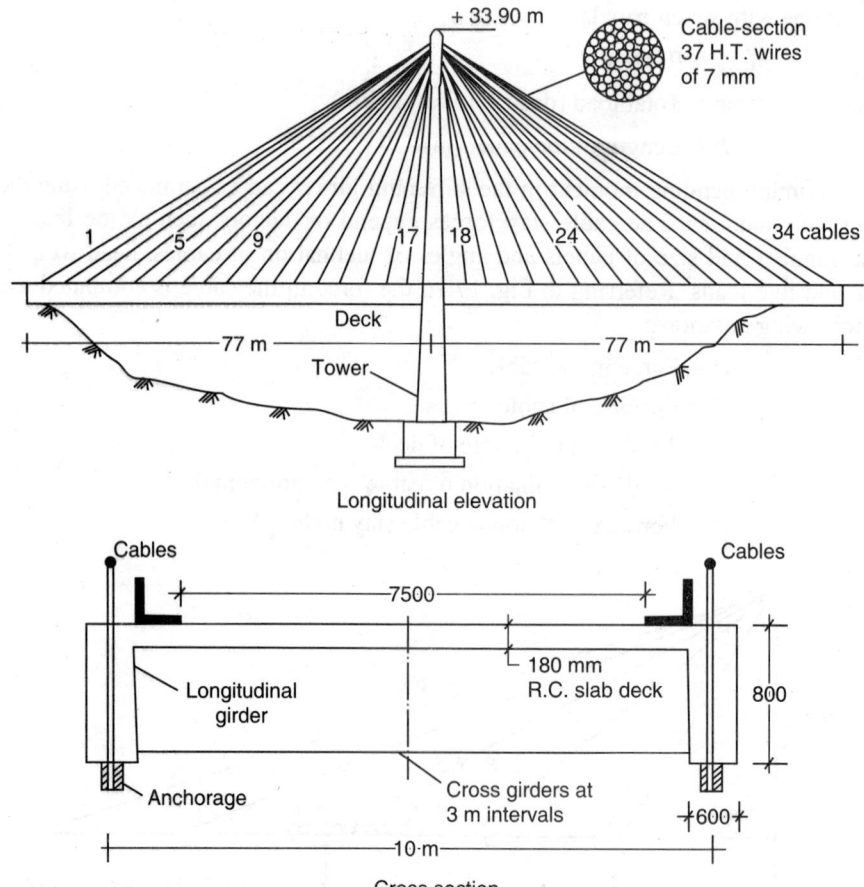

Fig. 17.17 Akkar Bridge (Sikkim).

The data of the Akkar bridge in Sikkim is as follows:

Total span length = 160 m
Width of Deck = 10 m
Width of Road way = Two lane 7.5 m
Number of cables = 34
Spacing of cables = 6 m
Thickness of R.C.C. Deck slab = 180 mm
Wearing coat = 70 mm
Longitudinal girders (Two Nos) = 600 mm by 800 mm
Cross girders at 3m intervals = 450 by 800 mm
Loading: I.R.C. Class AA tracked vehicle

design a typical cable assuming that the cables are arranged as shown in Fig. 17.17.

(i) Live Loads

Referring to Fig. 17.18 showing the critical loading position for maximum reaction due to I.R.C. Class AA vehicle loading,

$$\text{Reation } R_q = \left(\frac{700 \times 7.4}{10}\right) = 518 \text{ kN}$$

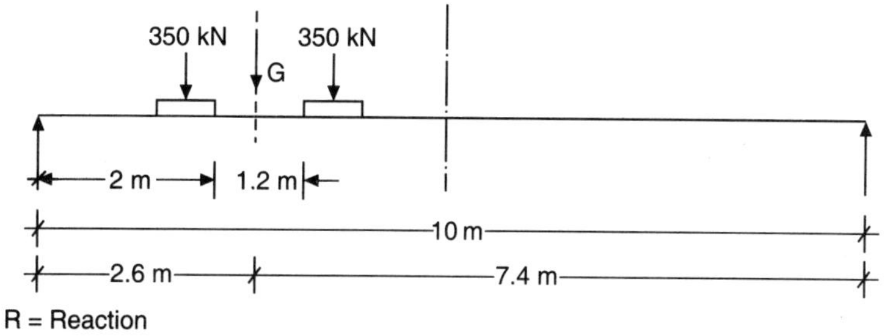

Fig. 17.18 Loading Position for Maximum Reaction.

(ii) Dead Loads

Weight of deck slab = $(0.25 \times 24) = 6 \text{ kN/m}^2$
and wearing coat
Weight of longitudinal girder = $(0.6 \times 0.8 \times 24) = 11.5 \text{ kN/m}$
Weight of cross girders = $(0.45 \times 0.8 \times 24) = 8.5 \text{ kN/m}$

(iii) Total Weight of Deck for 6 m Length and 10 m Wide

Weight of deck slab = $(6 \times 10 \times 6) = 360 \text{ kN}$
Weight of longitudinal girder = $(2 \times 11.5 \times 6) = 138 \text{ kN}$
Weight of cross girders = $(3 \times 10 \times 8.5) = 255 \text{ kN}$
Total Load = 750 kN
Force transmitted on each cable = $(0.5 \times 750) = 375 \text{ kN}$

(iv) Total Reaction

Due to Live loads = 518 kN
Due to dead load = 375 kN
Add for Foot paths
Parapet railing etc = 107 kN
Total reaction (R) = 1000 kN

(v) Design of Cable[11]

Referring to Fig. 17.19, the force in the cable is given by

$$P = \left(\frac{R}{\sin \alpha}\right) = \left(\frac{1000}{0.65}\right) = 1538 \text{ kN}$$

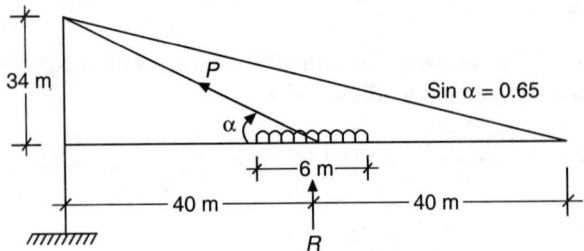

Fig. 17.19 Computation of Force in Cables.

Using 7 mm diameter hight tensile wires initially stressed to 1200 N/mm²,

Force in each wire = (38.5 × 1200)/1000 = 46 kN

∴ Number of wires = (1538/46) = 35

In Akkar bridge 37 high tensile wires of 7 mm diameter have been used in the cable having an ultimate strength of (37 × 38.5 × 1600)/1000 = 2280 kN.

(d) Determination of Approximate Self Weight of Girders

The problem of the determination of the approximate self weight of stiffening girders for the preliminary design is not yet properly developed. Technical literature provides very little information on this subject. However on the basis of the analysis performed in the previous section it is possible to estimate the approximate value of the self weight of the girder using the formulae for the maximum bending moments in the five and seven panel bridge systems.

The maximum bending moment developed in the stiffening girder including the effect of change of temperature may be expressed by the empirical relation,

$$M = 0.05 \, (\psi \, g + p + q) \, L^2$$

Where L = Span length

g = the theoretical weight of the stiffening girder per unit length of span

p = the uniformly distributed weight of deck per unit length of span

q = the uniformly distributed live load carried by a single girder per unit length

ψ = the construction coefficient of the stiffening girder.

If the permissible flexural stress is denoted as/, the'required section modulus of the girder is given by

$$Z = \left[\frac{0.05 \, (\psi \, g + p + q) L^2}{f} \right]$$

The section modulus of the stiffening girder may "be expressed by its cross sectional properties. If the girder consists of an *I* or box section, assuming equal areas of the top and bottom chords, it is possible to express the section modulus as

$$Z = \left[\frac{bh^2}{6} + \frac{2A_c (h/2)^2}{(h/2)} \right] = A_w (h/6) + A_c h$$

Assuming that the cross sectional areas of a single chord and web are equal, we have $A_c = A_w$ and therefore the total area can be expressed as $A = 3A_w$. Hence the section modulus is written as

$$Z = 1.17 A_w h = 1.17 (A/3)h$$

$$\therefore \qquad A = (2.5Z/h)$$

Substituting the value of the section modulus, we have

$$A = \left[\frac{0.0125 (\psi g + p + q) L^2}{f h} \right]$$

By multiplying the theoretical cross-sectional area of the girder by the specific weight of the material γ, the theoretical weight of the stiffening girder per unit length can be expressed as

$$g = \left[\frac{0.0125 (\psi g + p + q) L^2 \psi}{f h} \right]$$

and after transformation this becomes

$$g = \left[\frac{p + q}{(f h / 0.0125 L^2 \gamma) - \psi} \right]$$

Assuming the depth of the girder as one-hundredth of span ($h = L/100$) the theoretical weight is expressed as

$$g = \left[\frac{p + q}{(f / 1.25 L \gamma) - \psi} \right]$$

This relation has been developed without taking into consideration the axial force acting in the stiffening girder. However in the middle panel there is no axial force and at sections where large axial forces exist, the bending moments as a rule are relatively small. Therefore the relation developed for the weight of the girder may be used as the first approximation assuming the construction coefficient $\psi = 1.4$. The empirical relation corresponds to the five panel system. In the case of seven panels, the corresponding formula is given by

$$g = \left[\frac{p + q}{(f / 1.071 L \gamma) - \psi} \right]$$

A comparative analysis of the empirical relations for the self weight of stiffening girders indicate that for the larger spans it is more economical to divide the span into a greater number of panels. The empirical formulas may provide good estimates of the weight of the girders only for spans in the range of 270 to 400 m. Larger spans should be divided into nine, eleven, and even greater number of panels to obtain a relatively light stiffening girder.

(e) Determination of Weight of Cables

Using the empirical formulas developed for expressing the weight of the stiffening girder, it is possible to determine the approximate weight of the cable. The forces developed in the cables depend on the number of cables supporting the deck. Reduction in the number of cables increases the load on the cables. Hence it may be assumed that the weight of the cables depend to some extent on the number of cables.

Referring to Fig. 17.20 and using the following notations,

F_0 = Force developed in the backstay cable

F_1 = Force developed in the first cable of length L_1

F_2 = Force developed in the second cable of length L_2

Q_1 = Weight of First Cable

Q_2 = Weight of Second cable

g = Specific weight of cable material

α_1 and α_1 are angles subtended by first and second cable with the horizontal

L = Span length of the girder

The weight of the cables for a five panel bridge is determined by using the following empirical relations:

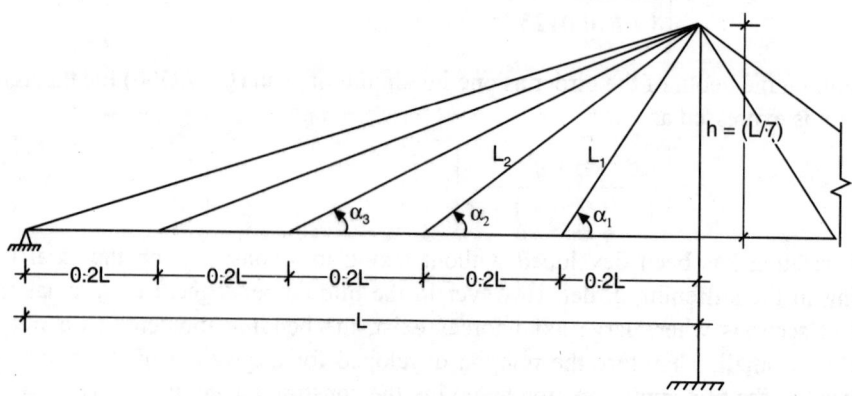

Fig. 17.20 Determination of Weight of Cables.

$$F_1 = \left[\frac{0.237(g+p+q)L}{\sin \alpha_1} \right]$$

$$F_2 = \left[\frac{0.174(g+p+q)L}{\sin \alpha_2} \right]$$

The lengths of the first and second cables are obtained as

$$L_1 = \left(\frac{0.2L}{\cos\alpha_1}\right) \text{ and } L_2 = \left(\frac{0.4L}{\cos\alpha_2}\right)$$

Assuming the allowable stress as f and the specific weight of cables as γ, the weights of the cables are evaluated as

$$Q_1 = \left(\frac{F_1 L_1}{f\cos\alpha_1}\right) = \frac{0.047(g+p+q)L^2\gamma}{\sin\alpha_1\cos\alpha_1 f}$$

$$Q_2 = \left[\frac{0.0696(g+p+q)L^2\gamma}{\sin\alpha_2\cos\alpha_2 f}\right]$$

Assuming the height of the tower as $h = (L/7) = 0.143 L$

$$\tan\alpha_1 = 0.713 \qquad \tan\alpha_2 = 0.357$$
$$\alpha_1 = 35° - 30' \qquad \alpha_1 = 19° - 40'$$
and
$$\sin\alpha_1\cos\alpha_1 = 0.463$$
$$\sin\alpha_2\cos\alpha_2 = 0.312$$

Therefore

$$Q_1 = \left[\frac{0.102(g+p+q)\gamma L^2}{f}\right]$$

$$Q_2 = \left[\frac{0.223(g+p+q)\gamma L^2}{f}\right]$$

After distributing the weight of the four cables uniformly along the span, the theoretical weight per unit length of the span is obtained as

$$g_c = \left(\frac{0.65(g+p+q)\gamma L}{f}\right)$$

The backstay cable force is

$$F_0 = \left[\frac{F_1\cos\alpha_1}{\cos\alpha_0} + \frac{F_2\cos\alpha_2}{\cos\alpha_0}\right]$$

$$= \left(\frac{0.237(g+p+q)L}{\tan\alpha_1\cos\alpha_0} + \frac{0.174(g+p+q)L^2}{\tan\alpha_2\cos\alpha_0}\right)$$

By assuming $\alpha_0 = 30°$, the cable force is

$$F_0 = 0.948(g+p+q)L$$

Then considering the length of the two back stay cable stays as equal to 08L, their weight can be expressed as,

$$Q_o = \frac{F_o}{f}(0.8L) = \left[\frac{0.78(g+p+q)L^2}{f} \right]$$

and the weight per unit length is

$$g_o = \left[\frac{0.78(g+p+q)L}{f} \right]$$

Hence the total weight of all cable stays is expressed as

$$g_{tot} = (g_c + g_o) = \left[\frac{1.43(g+p+q)L}{f} \right]$$

The number of cables significantly influences the anchorage system and consequently the reinforcement in the stiffening girder to transfer moment, shear and axial forces. A relatively deep girder is required to span large distances between the stay cable attachments. A large number of cable stays supporting a continuous elastic medium simplifies the anchorage and distribution of forces to the girder and permits the use of shallower depth girders. Although more stays are used, the additional cost is more than offset by simpler connection details for the smaller cable and lesser force in cable stays. The erection work is also simplified since the deck structure can be constructed by cantilever method from stay connection point without any auxiliary means.

(f) Degree of Redundancy

The degree of statical indeterminacy of the cable stayed bridge system is determined by the formula

$$I = C + 2S - H - 3$$

where

C = Total number of cables

S = Total number of stiffening girder supports

H = Number of moveable connections or hinges considering even the moveable supports of the cables on towers.

For example in the Akkar Bridge in Sikkim shown in Fig. 17.17 the degree of redundancy is computed using the following data:

Number of cables = $C = 34$

Number of supports = $S = 3$

Number of moveables connections = $H = 3$

Hence the degree of redundancy is computed as

$$I = C + 2S - H - 3$$
$$= 34 + (2 \times 3) - 3 - 3$$
$$= 34$$

Hence 34 equations have to be formulated by selecting suitable redundants and the equations are solved using a digital computer.

17.9 STRUCTURAL ANCHORAGES

The axial force in the stiffening girder depends upon the method of anchoring the cables and the provision of expansion joints and their location in the structure. Basically three different types of anchored systems are considered such as

(1) Self anchored system

(2) Fully anchored system

(3) Partially anchored system

1. Self Anchored System

Referring to Fig. 17.21(a) in the self anchored system, there is no restraint at the supports to the horizontal components of the cable force. In this case the axial force distribution in the girder will vary from zero at the centre of main span to maximum compression near the towers.

2. Fully Anchored System

In this type, no provision is made for movement at the supports but expansion joints are provided at the towers. The axial force distribution changes as shown in Fig. 17.21(b) with zero force at towers increasing to a maximum value at the centre of span.

3. Partially Anchored System

In the partially anchored system, the axial forces are considerably reduced using a combination of the above two systems by providing horizontal restraint at the abutments with no expansion joints or expansion joints provided only in the end spans as shown in Fig. 17.21(c).

The anchorages must be designed to allow adjustment of length and replacing a cable damaged by an accident without interrupting the traffic. They must be further designed to prevent bending stresses in the wires or strands at the socket due to change of sag or due to slight oscillations. The anchorage should further comprise dampers to prevent resonance oscillations of the cables caused by wind eddies. At the tower head, cables running over a saddle like in a suspension bridge should be avoided because their replacement would be rather difficult. It is preferable to anchor the cable at each side individually and design suitable devices to carry the horizontal component of the cable forces through the tower. At the top of the towers, the box must be wide enough to allow access and handling of the jacks.

17.10 DYNAMIC BEHAVIOUR AND AERODYNAMIC STABILITY

When the first cable stayed bridge, the Stromsund bridge, was built in Sweden in 1955, the problem of aerodynamic stability in bridge design did receive considerable study. However, that study did not lead to explicit design rules and formulas. The aerodynamic phase of the problem is a real challenge to the bridge engineers. Collapse of the Tacoma Narrows bridge in 1940 prompted bridge engineers to do research on aero dynamic stability. The problem of aero dynamic stability is more important in the case lighter bridges covering long spans such as the suspension and the cable stayed bridge systems.

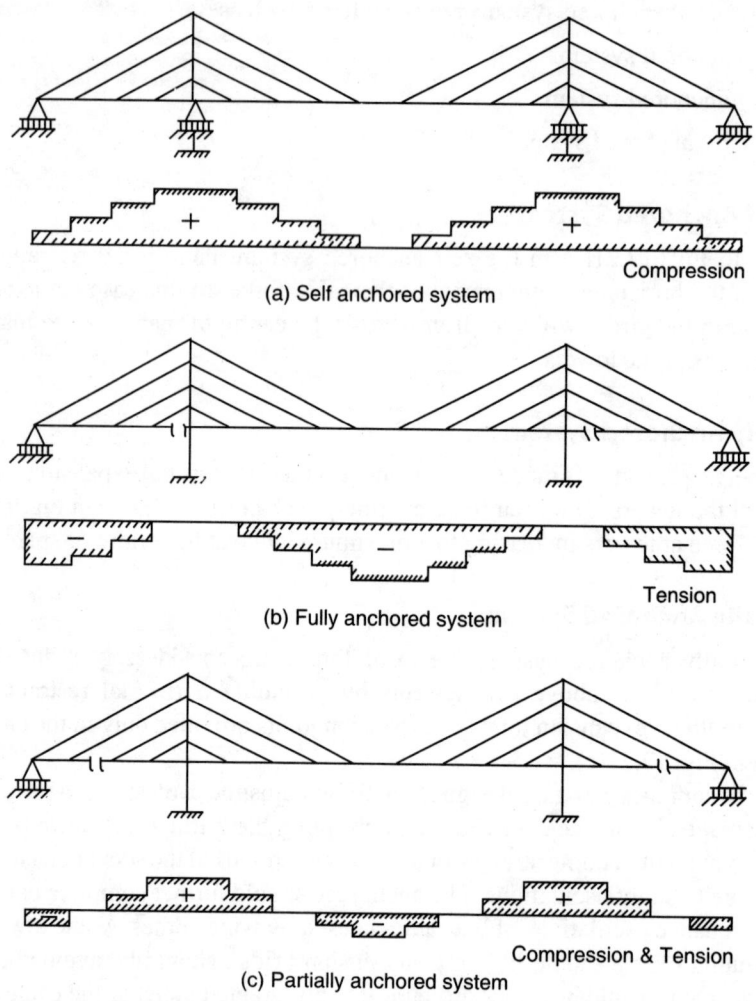

(a) Self anchored system

(b) Fully anchored system

(c) Partially anchored system

Fig. 17.21 Structural Anchorage Systems.

According to Leonhardt, the cable stayed bridge with concrete decks and highly stressed cables has very favourable dynamic behaviour. The deflection under live loads are extremely small because the effective depth of the large cantilever truss formed by the cables is much larger than for beam girders. The main advantage of the multicable system being that the increase of amplitude due to resonance oscillation is prevented by system damping caused by the interference of the multicable system. Measurements made at the Tjorn bridge in Sweden indicated that the damping increases with increasing amplitudes and the logarithmic decrement reaches a value well above 0.10. This damping is very favourable for the stability of the structure under wind loads. Current theories which calculate critical wind speeds do not adequately represent the actual behaviour and have only limited validity. The same is true for wind tunnel tests with sectional models in which only a constant damping factor is applied. More tests should be made of actual bridges to improve our theories based on observed facts.

Based on the present knowledge, Leonard! has suggested the following geometrical relations for obtaining wind stability in the case of concrete bridge decks supported with cables in two planes along the edges.

Referring to Fig. 17.22 and using the following notations,

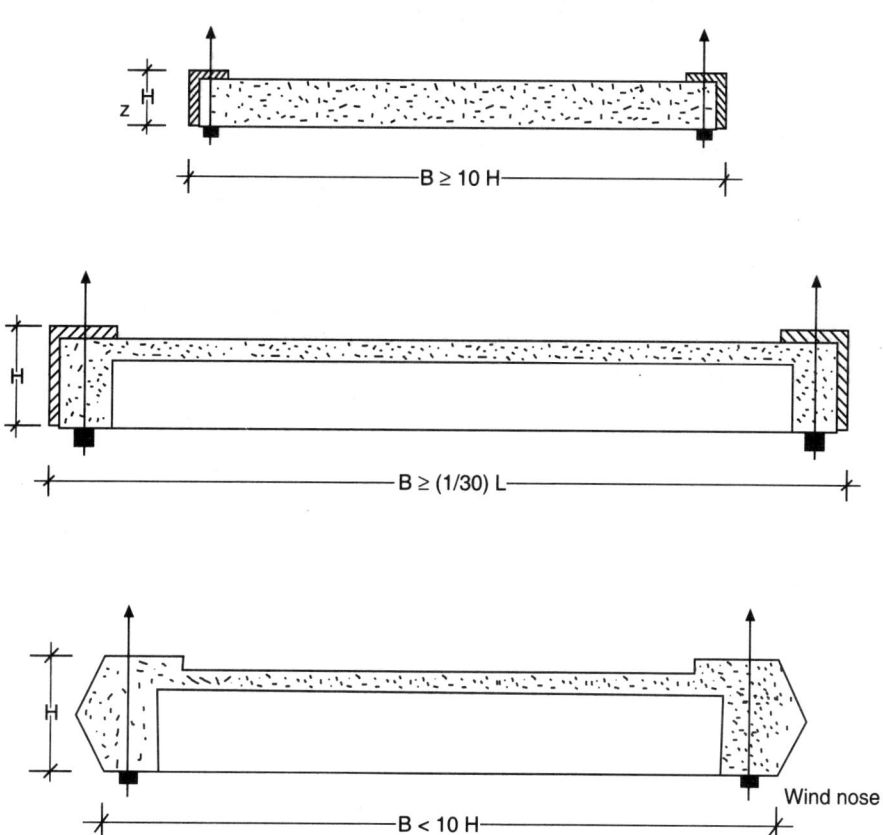

Fig. 17.22 Geometrical Relations for Obtaining Wind Stability with Cables at Beam Edges.

L = Span of the bridge deck
B = Width of bridge deck
H = Depth of Stiffening girder

the bridge will be safe against wind loads if the following geometrical relations are satisfied:

(a) $B \geq 10H$

(b) For $B < IOH$, a wind nose should be provided

(c) $B \geq (1/30)$ L, which indicates that the width of the bridge should not be too small in relation to the main span. If thrs ratio gets smaller, then

A-shaped towers and wind shaping of the cross section must be used. The A-shaped tower provides a triangular shape of the cable planes and the deck, which increases the torsional rigidity. Bridges supported with cables in one plane along the centre line have negligible damping for torsional oscillations.

17.11 CONSTRUCTION METHODS

In the case of cable stayed bridge decks, the method of construction significantly affects the nature of stresses developed in the super structure and hence suitable precautions have to be taken during the construction procedure.

The prominent methods generally used for the erection of cable stayed bridges are:

(1) The Staging Method

This method is adopted when low clearance is required below the deck and supporting form work does not interfere with traffic. This method facilitates rapid construction by maintaining correct geometry of the structure with relatively low cost. This method has been used in the construction of the Rhine river bridge at Maxau, Japan.

(2) The Push-out Method

In this method, large precast sections of the bridge deck are pushed out over the piers on rollers or sliding teflon bearings. The deck units are pushed out from both abutments towards the centre or from one abutment all the way to the other abutment. Assembling the components in an erection bay and progressively pushing the components out into span simplifies construction and reduces costs. This technique was used in the construction of Julicher-Strasse bridge, Germany and Paris Massena bridge in France.

(3) The Cantilever Method

The cantilever method of erection is the latest and the most popular method of construction of cable stayed bridge decks. The erection proceeds from either side of the pylon in the form of two free cantilevers which balance each other. The units may also be supported directly by the stay cables depending upon the stay spacing and size of each precast unit. For bridges with box girders, prefabricated segments can be used with match cast joints but using a paste in the joint which compensates for differential shortening during the curing and hardening period.

Bridges with a deck of composite beams allow the simplest and quickest erection. The grid of steel cross girders and light steel edge girders, including the cable anchors, are installed with light derricks and then the .prefabricated concrete slabs are placed, leaving gaps for overlapping reinforcement and shear connectors which are closed by cast *in-situ* place concrete. This method was adopted for the construction of the Annacis bridge in Vancouver, British Columbia.

The main advantage of the cantilever method of construction is that the traffic below the bridge is not hindered during erection. This technique has been used in the construction of Pascokennewick intercity bridge, U.S.A., Kniebrucke bridge. Germany, Seven bridge, Germany and the Stromsund bridge, Sweden.

During the last decade, construction methods have shown major progress towards simplification and reducing erection equipment. The construction procedure must however be well planned using sequential computations for the alignment, forces, exact lengths and angles considering temperature and creep influences, which depend on seasonal, climatic and even daily conditions. The collective experience and knowledge gained during the past thirty years will help in evolving innovative applications and methods in the construction of cable stayed bridges in the future in various countries throughout the world to serve the needs of human society.

17.12 ECONOMIC STUDIES[12]

Detailed analysis by Fritz Leonhardt indicates that Cable stayed bridges are structurally efficient and cost effective for low, medium and long spans, ranging from 40 m to 1800 m. Pedestrian bridges with only 40 m span comprising a prestressed concrete deck with a depth of 250 to 300 mm supported by a few cable stays have been successfully built in Germany. Highway bridges can be built of prestressed concrete with spans up to 700m and rail road bridges up to about 400 m. If composite action between the steel girders and a concrete deck slab is utilized, then spans of about 1000 and 600 m for highway and railway bridges respectively can be safely used. For the crossing of the Messina Straits in Italy, Leonhardt has designed and constructed an all steel cable stayed bridge with a main span of 1800 m for six lanes of road traffic and two railway tracks without encountering any structural or construction difficulties.

Recent studies have indicated that cable stayed bridges are much more economical, stiffer and aerodynamically safer when compared with a suspension bridge. Figure 17.23 shows the comparison of the quantity of cable steel for suspension and cable stayed bridges in the span range of 500 to 1800 m. For a bridge with 1800 m main span and 38 m width, a suspension bridge requires 46000 t of steel, whereas a cable stayed bridge needs only 19200 t of steel, indicating more than 50 percent savings in the quantity of steel required in the cable stayed bridge.

A comparative analysis indicates that a suspension bridge requires a stiffening girder with a flexural stiffness which must be about ten times larger than that required for a cable stayed bridge covering the same span. The suspension bridge needs additional heavy anchor blocks which can be prohibitively costly if the navigation clearance is high and foundation conditions are poor. The total cost of such a suspension bridge can easily be 20 to 30 percent more than the cost of a cable stayed bridge.

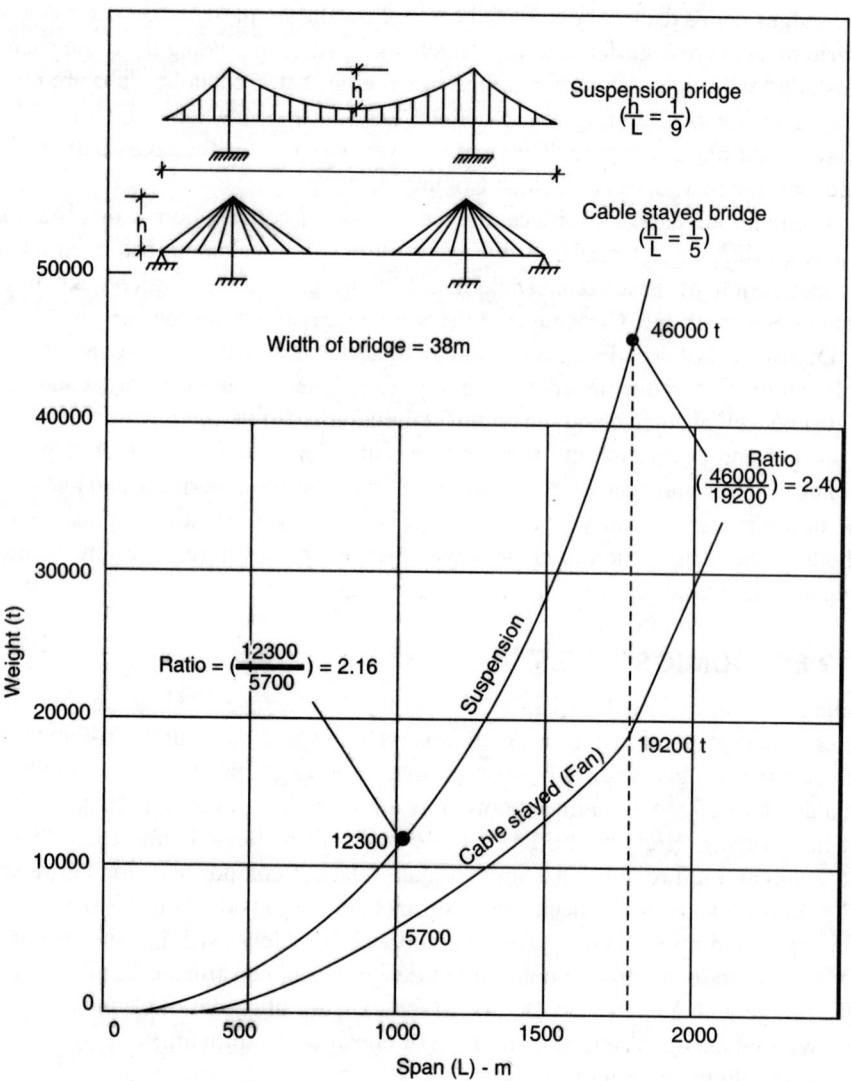

Fig. 17.23 Comparison of Quantity of Cable Steel for Supension and Cable Stayed Bridges.

The second Hooghly bridge[13] (Vidyasagar sethu) at Kolkata is an excellent example of a cable stayed bridge comprising a main span of 457.2 m and two side spans of 182.8 m each. The deck is made up of a concrete slab 230 mm thick, two outer steel I-girders 28.1 m apart and a central I-girder. The deck is suspended by cable stays comprising parallel wire cables of BBR-HIAM type with their own anchorage system. The bridge provides for two 3-lane carriage ways, 12.3 each and 2.5 m footways. The cable stayed bridge costing 600 million rupees was found to be cost effective in comparison with other types.

Economic studies of bridges constructed in Russia clearly indicate the cost effectiveness of cable stayed bridges. In this study, five types of concrete and four types of steel bridges were included. The concrete bridges are cable stayed, arch cantilever arch rigid frame, suspension and continuous box girder types. The steel bridges are cable stayed, conventional suspension, arch and continuous type. Economic evaluation included the cost of piers, super structure and erection work. The analysis indicates that the volume of concrete per unit area of bridge deck was the least in the case of cable stayed bridges in the span range of 60 to 300 m. Cost studies of the various types of bridges in U.S.A., also indicate that the cost per unit area of bridge deck is the least for cable stayed bridges in the span range of 60 to 400 m.

The survey indicates that cable stayed bridges of prestressed concrete and steel have wide application in the future in countries throughout the world to serve the needs of human society.

EXAMPLES FOR PRACTICE

1. A suitable high tensile strand cable is to be selected for use in a typical cable stayed bridge. Using the following data, select a suitable cable from the Freyssinet system (K-range) (IS: 6006-1983) to suit the following data:

   Total span length of bridge deck = 200 m
   Width of bridge deck = 13.5 m
   Bridge deck supported between two R.C.C. Girders at the edges to which the cables are anchored.
   The bridge deck supported by two sets of fan type cables passing over a central tower and anchored to the edge girders at regular intervals.
   Height of tower above deck = 50 m
   The inclination of a typical H.T. cable to the horizontal = 45°
   The total reaction transmitted to the cable at deck level comprising of live and dead loads is estimated as 1030 kN.

   Calculate the force in the cable and select a suitable Freyssinet system cable suitable for use in the cable stayed bridge.

2. A cable stayed bridge is planned for a river crossing in west coast of India. Design a typical high tensile cable supporting the deck using the data given below:

   Total span length of the bridge = 200 m
   Width of bridge deck between longitudinal girders = 12 m
   A central tower supports the cables at top at a height of 40 m above deck
   Spacing of cables = 5 m
   Thickness of R.C.C. deck slab = 200 mm
   Thickness of Wearing coat = 80 mm
   Longitudinal girders (Two Nos) of size 600 mm by 1000 mm
   Cross girders at 4 m intervals of size 400mm by 800 mm
   Loading: IRC Class AA tracked vehicle

   Design the size of a typical High tensile cable assuming fan type cable arrangement supporting the deck and anchored to the edge girders.

REFERENCES

1. Leonhardt, F. and Zellner, W., Cable Stayed Bridges, International Association for Bridge & Structural Engineering Surveys, S-13/80, IABSE Periodical, 2/1980, May 1980.
2. Podolony, Walter and Scalzi, J.B., Construction and Design of Cable Stayed Bridges, John Wiley & Sons, New York, 1976.
3. Leonhardt, F., Latest Developments in Cable Stayed Bridges for Long Spans, Bygningsstatiske Meddelesor, Copenhagen, 1974
4. Raina, V.K., Concrete Bridge Practice, Analysis, Design and Economics, Tata McGraw Hill Publishing Co. Ltd., New Delhi, 1991, p. 525.
5. Marshall, W.T and Nelson, H.M., Structures, Pitman Publications, London, 1970, pp. 1–442.
6. Krishna Raju, N and Gururaja, D.R., Advanced Mechanics of Solids & Structures, Narosa Publishing House, New Delhi, 1997, pp. 1–450.
7. Troitsky, M.S., Cable Stayed Bridges, Theory & Design, Crossby Lockwood Staples, London. 1977, pp. 1–383.
8. Smith, B.S., The Single Plane Cable Stayed Bridge, A Method of Analysis Suitable For Computer use, Proceedings of the Institution of Civil Engineers, London, Vol.37, July 1967, pp. 183–194.
9. Tang Mang-Chung, Analysis of Cable Stayed Girder Bridges, Journal of the Structural Division, Proceedings of the American Society of Civil Engineers, ST-5, May 1970, pp. 1481–1496.
10. Schreier, G., North Bridge at Dusseldorf, Analysis, Design, Fabrication & Erection of the bridge Spanning the River, Acier Sthal, Steel, No. 9,1958, pp. 369–385.
11. Krishna Raju, N., Prestressed Concrete (IV Edition) Tata McGraw Hill Publishing Co. Ltd., New Delhi, 2007, p. 686.
12. Leonhardt, F., New Trends in design and Construction of Long Span bridges and Preliminary Publication of Eighth Congress of International Association for Bridge and Structural Engineering, New York, 1968.
13. Victor, J.D., Essentials of Bridge Engineering (Fifth Edition), Oxford & IBH Publishing Co. Pvt. Ltd., New Delhi, 2001, pp. 414–415.

REVIEW QUESTIONS

1. What are the advantages of cable stayed bridges in comparison with the traditional suspension bridges?
2. Explain the various structural components of a cable stayed bridge mentioning their functions.
3. What are towers or Pylons with reference to a cable stayed bridge system? What is the structural function of these towers?

4. Explain with sketches the different types of cable stays mentioning their ultimate load characteristics.

5. What are the main types of longitudinal cable arrangement? Discuss briefly their advantages and disadvantages.

6. Explain briefly the variation of moments in cable stayed, suspension and continuous girder bridge systems of similar span range.

7. List the various basic steps involved in the approximate structural analysis of a cable stayed bridge.

8. Discuss briefly the influence of cable inclination on the deflection of the bridge deck. Explain the method of determining the height og tower and the length of panels in a cable stayed bridge

9. What are the prominent methods used for the construction of cable stayed bridges?

10. Bring out the advantages of a cable stayed bridge in comparison with a suspension bridge by a comparative analysis of the quantity of cable steel required by these systems in relation to the variation in span.

OBJECTIVE TYPE QUESTIONS

1. Cable stayed bridges are generally preferred for bridges in the span range of
   a) 8 to 10 m
   b) 15 to 20 m
   c) Above 100 m

2. In the case of cable stayed bridges, the loads on the deck are transferred to the towers through
   a) Cross girders
   b) Main girders
   c) Stay cables

3. The total dead and live loads of the cable stayed bridge deck is ultimately transferred to the foundations by the
   a) Main girders
   b) Pylons
   c) Stay cables

4. The optimum utilization of material with overall economy is possible by adopting
   a) Suspension bridges
   b) Cable stayed bridges
   c) Continuous girder bridges

5. The recommended limits of inclination of the cables to the horizontal deck in a cable stayed bridge deck should be in the range of
   a) 15 to 25°
   b) 25 to 65°
   c) 45 to 85°

6. The stay cables supporting the deck are generally anchored to the
   a) Cross girders
   b) Deck slab
   c) Main girders

7. A comprehensive survey of the existing cable stayed bridges, indicate that the optimum values of the panel lengths in a cable stayed bridge with the central span in the range of 130 to 150 m should be
   a) 10 m
   b) 15 m
   c) 20 m

8. Analysis of the weight of stay cables in a cable stayed bridge indicates that the total weight of all cables is Inversely proportional to the
   a) span
   b) dead and live loads
   c) permissible stress in the cable

9. The degree of statical indeterminacy of a cable stayed bridge depends upon the
   a) Span length
   b) Height of towers
   c) Total Number of cables

10. An economic survey of the cost/unit area of various types of bridge decks in the span range of 60 to 400 m, indicate that the cost is least by adopting
    a) Suspension bridges
    b) Rigid frame bridges
    c) Cable stayed bridges

Piers and Abutments

18.1 GENERAL FEATURES

The substructure[1] of a bridge comprises the piers and abutments which are located below the level of the bearings and rest above the foundations. Figure 18.1 shows the longitudinal elevation of a typical bridge with the superstructure supported by the substructure, which, in turn, rests on the foundations located below the ground level. The piers and abutments generally built up of brick or stone masonry or concrete are supported on foundation like spread footings, piles, wells or caissions[2]. The superstructure comprising the slab or beams transmit the reactions to the piers and abutments through the bridge bearings and bed blocks.

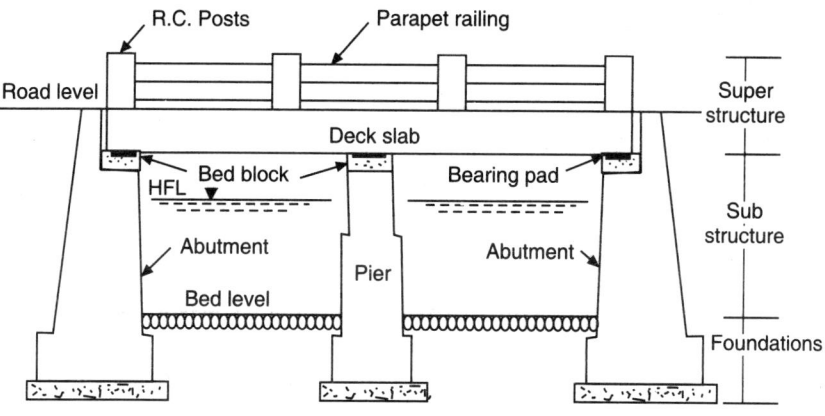

Fig. 18.1 Elevation of a Typical Bridge Structure.

18.2 BED BLOCK

A reinforced concrete bed block resting over the top of the piers and abutments is generally provided to evenly distribute the dead and live loads on to the pier and abutments.

Figure 18.2 shows the cross-section and plan of a typical reinforced concrete bed block used in major bridges[3]. The bed block is generally cast with M-15 grade concrete and reinforced with steel bars of area equal to 0.3 per cent of the cross-sectional dimensions and distributed as mesh reinforcement near the top and bottom surfaces of the bed block.

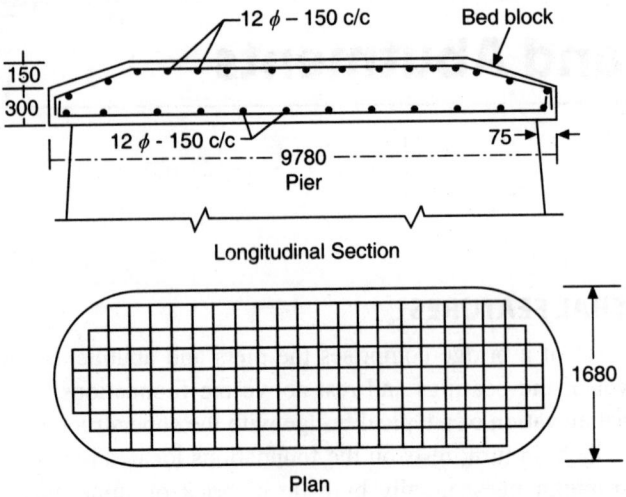

Fig. 18.2 Reinforcement Details in Bedblock over Piers.

18.3 MATERIALS FOR PIERS AND ABUTMENTS

The following types of materials are generally used for the construction of piers and abutments:

(a) Mass concrete of M-10 grade corresponding to mix proportions of 1 : 3 : 6 with 40 mm maximum size aggregates.

(b) Reinforced concrete of M-15 grade corresponding to mix proportions of 1 : 2 : 4.

(c) Coursed Rubble masonry in cement mortar of proportions 1 : 4.

(d) Brick masonry in cement mortar of proportions 1 : 6.

(e) Prestressed concrete for piers particularly in viaducts with tall piers. Concrete of M-30 to M-40 grade is the minimum requirement for prestressed concrete piers.

The maximum permissible compressive and tensile stresses in the various types of materials in the substructure are compiled in Table 18.1 as reported by Victor[4].

18.4 TYPES OF PIERS

Depending on the type, size and dimensions of the superstructure, the following types of piers are in general use:

Table 18.1 Maximum Permissible Stress in Various Materials Used for Substructure

Sl. No.	Material	Maximum permissible compressive stress (N/mm²)	Maximum permissible tensile stress (N/mm²)
1.	Mass Concrete (1 : 3 : 6 mix by volume)	2.00	0.25
2.	Reinforced concrete (M-15 grade)	4.00	0.50
3.	Coursed Rubble masonry in cement mortar 1 : 4	1.50	0.30
4.	Brick masonry in cement mortar 1 : 6	1.00	0.02
5.	Prestressed concrete (M-40 grade)	14.00	Not to exceed 3N/mm² in Type-2 members

(a) Solid Type Pier

The solid type pier is generally built using brick or stone masonry or concrete. This type with cut ease water is widely used for river bridges. Figure 18.3(a) shows the cross-sectional plan of a typical solid concrete pier.

(b) Trestle Type Pier

The trestle type pier comprises of a number of reinforced concrete columns with a connecting cap at the top as shown in Fig. 18.3(b). The trestle type of pier finds wide applicability in the case of flyovers and elevated roadways generally used for crossings in city roads.

(c) Hammer-head Type Pier

Hammer-head type of pier consists of a massive single pier with cantilever caps on opposite sides resembling the head of a hammer. This type of pier is generally suitable for elevated roadways and when used in river bridges, there is minimum restriction of waterway. A typical hammer-head type pier is shown in Fig. 18.3(c).

(d) Cellular Type Pier

For the construction of massive piers carrying multilane traffic, it is economical to use cellular type reinforced concrete piers which results in savings of cbncrete. However cellular piers require costly shuttering and additional labour for placing of reinforcements. For tall piers, slip forming work can be adopted for rapid construction. A typical cellular type pier is shown in Fig. 18.3(d).

(e) Framed Type Piers

R.C. framed type piers are aesthetically superior and rigid due to the monolithic joints between the vertical, or inclined and horizontal members as shown in Fig. 18.3(e) and (f). This type of piers are ideally suited to reduce the span length of main girders on either side of the centre line of the pier resulting in savings in the cost of superstructure. However this type of construction requires two expansion joints at close intervals with increase of maintenance costs.

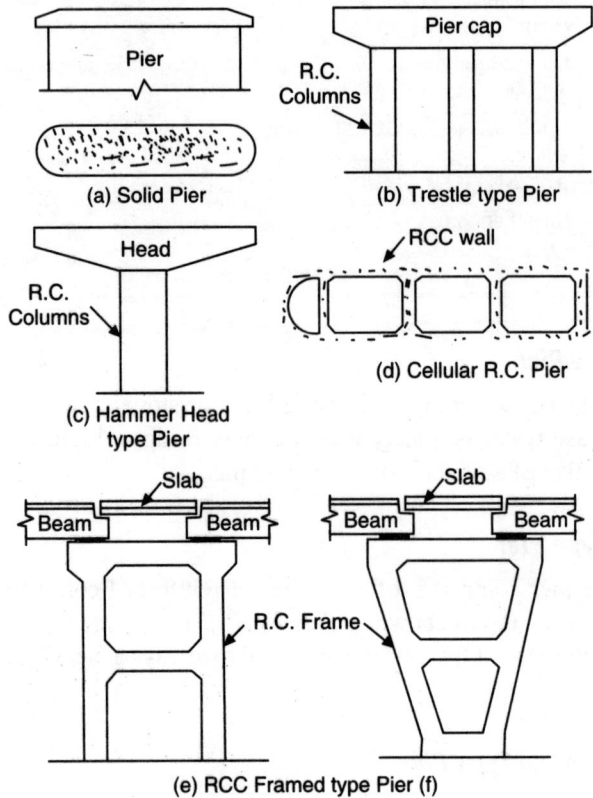

Fig. 18.3 Typical Types of Piers.

18.5 FORCES ACTING ON PIERS[5]

The various forces to be considered in the design of piers are as follows:

1. Dead load of superstructure and pier.
2. Live loads of vehicles moving on the bridge.
3. Effect of eccentric live loads.
4. Impact effect for different classes of loads.
5. Effect of buoyancy on the submerged part of the pier.

6. Effect of wind loads acting on the moving vehicles and the superstructure.

7. Forces due to water current.

8. Forces due to wave action.

9. Longitudinal forces due to tractive effort of vehicles.

10. Longitudinal forces due to braking of vehicles.

11. Longitudinal forces due to resistance in bearings.

12. Effect of earthquake forces.

13. Forces due to collision for piers in navigable rivers.

The stability analysis for the piers is generally made by considering some of the critical forces which will have significant effect on the stresses developed in the piers.

18.6 DESIGN OF PIER

The salient dimensions of pier like the height, pier width and batter are determined as follows:

(a) Height

The top level of pier is fixed 1 to 1.5 m above the high flood level, depending upon the depth of water on the upstream side. Sufficient gap between the high flood level and top of pier is essential to protect the bearings from flooding.

(b) Pier Width

The top width of pier should be sufficient to accommodate the two bearings. It is usually kept at a minimum of 600 mm more than the outer to outer dimension of the bearing places.

(c) Pier Batter

Generally the sides are provided with a batter of 1 in 12 to 1 in 24. Short piers have vertical sides. The increased bottom width is required to restrict the stresses developed under loads within safe permissible values.

(d) Cut and Ease Waters

The pier ends are shaped for streamlining the passage of water. Normally the cut and ease waters are either shaped circular or triangular as shown in Fig. 18.4.

18.7 STABILITY ANALYSIS OF PIERS

A pier shown in Fig. 18.5 supports the deck forming a major highway. The various forces acting on the pier are listed below:

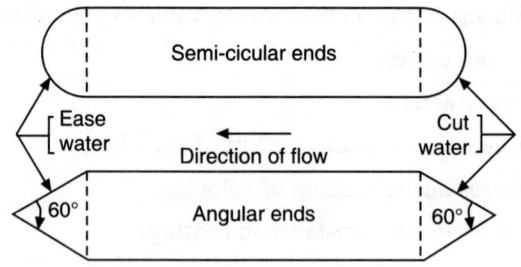

Fig. 18.4 Pier with Cut and Ease Water.

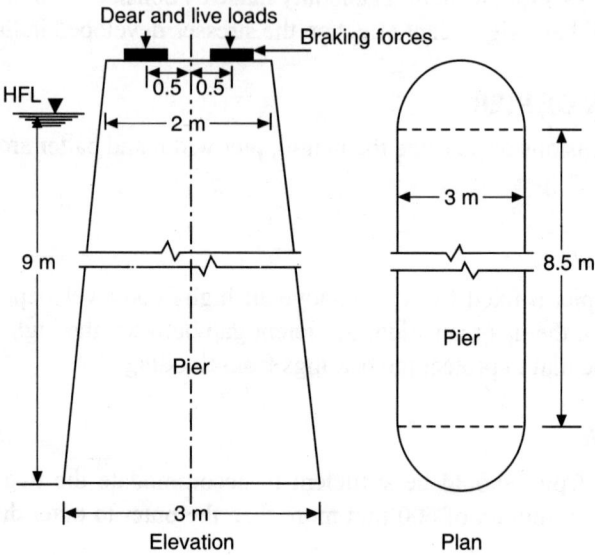

Fig. 18.5 Forces Acting on Pier.

Dead loads from each span   = 2000 kN
Reaction due to live load on one span = 1000 kN
Braking forces   = 140 kN
Wind pressure on pier   = 2.4 kN·m^2
Material of pier   = 1 : 3 : 6 cement concrete
Density of concrete   = 24 kN/m^3

Calculate the stress developed at the base of the pier due to the following cases:

(1) Dead load and self-weight of pier

(2) Effect of buoyancy

(3) Due to eccentricity of live load

(4) Due to longitudinal braking forces

(5) Due to wind pressure.

Estimate the maximum and minimum stresses developed at the base of pier due to the critical combinations of the various loads.

Design Computations

(1) *Stress due to dead loads and self-weight of pier*

Referring to Fig. 18.5.

Dead load from superstructure = (2 × 2000) = 4000 kN

Self-weight of pier = 8.5 × 0.5 (2 + 3) 10 × 24 = 5100 kN

Total Direct load = (400 + 5100) = 9100 kN

Compressive stress at base of pier $= \left(\dfrac{9100}{8.5 \times 3}\right) = 356.8 \text{ kN/m}^2$

(2) *Effect of buoyancy*

Width of pier at H.F.L. = 19 m

Submerged Volume of pier = 8.5 × 0.5 (1.9 + 3.0) 9 = 187.4 m^3

Reduction in weight of pier due to buoyancy = (187.4 × 10) = 1874 kN

Tensile stress at base due to buoyancy $= -\left(\dfrac{1874}{8.5 \times 3}\right)$

$$= -73.5 \text{ kN/m}^2$$

(3) *Stresses due to eccentricity of live load*

Reaction due to Live load from one span is 1000 kN acting at an eccentricity of $e = 0.5$ m

Moment about base = $M = (1000 \times 0.5) = 500$ kN·m

Section modules $= Z = \left(\dfrac{8.5 \times 3}{6}\right)$

$$= 12.75 \text{ m}^3$$

Stresses developed at base of pier due to the eccentricity of the live load;

$$= \left(\frac{1000}{8.5 \times 3}\right) \pm \left(\frac{500}{12.75}\right)$$

$$= 39.2 \pm 39.2$$

$$\sigma_{max} = 78.4 \text{ kN/m}^2$$

$$\sigma_{min} = 0$$

(4) *Stresses due to Longitudinal braking forces*

Braking force at bearing level = 140 kN

Moment about base of pier = (140 × 10)

$$M = 1400 \text{ kN·m}$$

$$\text{Stresses at base} = \pm\left(\frac{M}{Z}\right) = \pm\left(\frac{1400}{12.75}\right)$$

$$= \pm 109.8 \text{ kN/m}^2$$

(5) *Stresses due to wind pressure*

Total wind pressure on pier

$$= (\text{Area}) \times (\text{wind intensity})$$

$$= \left(\frac{2+3}{2}\right)10 \times 2.4$$

$$= 60 \text{ kN}$$

Assuming the wind to act mid-height of the pier,

Moment about base of pier $= (60 \times 5) = 300$ kN·m

$$\text{Modulus of section at base} = Z = \left(\frac{3 \times 8.5^2}{6}\right)$$

$$= 36\ 1.25 \text{ m}^3$$

Stress developed at base due to wind loads

$$= \pm\left(\frac{M}{Z}\right) = \pm\left(\frac{300}{361.25}\right) = \pm 0.83 \text{ kN/m}^2$$

The maximum and minimum stress developed are computed by combining the stresses due to the various load combinations as shown in Table 18.2.

The material of the pier being 1 : 3 : 6 cement concrete, the maximum permissible compressive stress in concrete (Refer Table 18.1) is 2 N/mm² or 2000 kN/m².

Hence the stresses developed at the base of the pier are within safe permissible limits.

Table 18.2 Maximum and Minimum Stresses at Base of Pier

Sl. No.	Type of Load	Stress (kN/m²)	
		When Dry	During Floods
1.	Dead load and self-weight	356.8	356.8
2.	Buoyancy	—	−73.5
3.	Eccentric live load	78.4	78.4
4.	Braking Forces	+109.8	+109.8
5.	Wind pressure	+0.83	+0.83
	Maximum stress	545.83	472.33
	Minimum stress	324.57	251.07

18.8 GENERAL FEATURES OF ABUTMENTS

Abutments are end supports to the superstructure of a bridge and they retain earth on their back side which serves as an approach to the bridge. In the case of river bridges, the abutment also protects the embankment from scour of the stream. Abutments are generally built using solid stone, brick masonry or concrete. An abutment comprises three distinct structural components.

(a) The breast wall which directly supports the dead and live loads of the superstructure and retains the earthfilling on the rear side.

(b) The wing walls which act as extensions of the breast wall, retains the earthfill without resisting any loads from the superstructure.

(c) The back wall is a small retaining wall located just behind the bridge seat and it prevents the earthfill from flowing into the bridge seat and bearings.

The typical details of an abutment with its structural components are shown in Fig. 18.6.

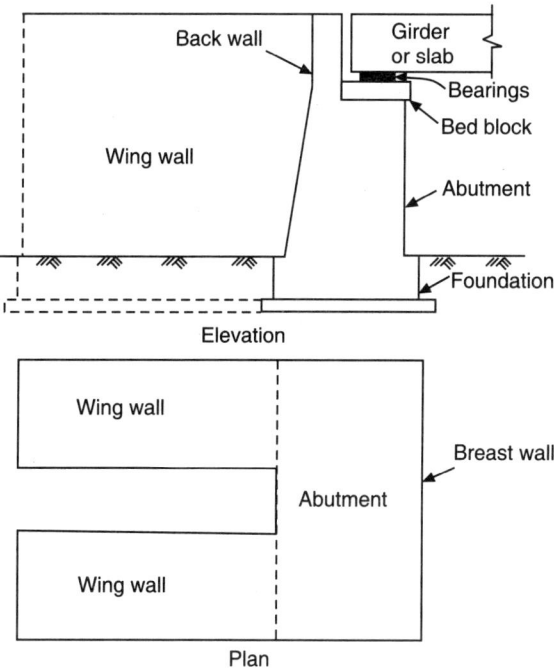

Fig. 18.6 Structural Elements of Abutment.

18.9   DESIGN OF ABUTMENT

(a) *Height:* The height of abutment is kept equal to that of the piers.

(b) *Abutment batter:* The water face is kept vertical or a small batter of 1 in 24 to 1 in 12 is given. The earth face is provided with a batter of 1 in 3 to 1 in 6 or it may be stepped down.

(c) *Abutment width:* The top width should provide enough space for bridge bearings and bottom width is dimensioned as 0.4 to 0.5 times the height of the abutment.

(d) *Length of abutment:* The length of the abutment must be at least equal to the width of the bridge.

(e) *Abutment cap:* The bed block over the abutment is similar to the pier cap with a thickness of 450 to 600 mm.

18.10 FORCES ACTING ON ABUTMENTS[6]

The various forces to be considered in the design of abutments are as follows:

1. Dead Load due to superstructure
2. Live load on the superstructure
3. Self-weight of the abutment
4. Longitudinal forces due to tractive effort and braking
5. Forces due to temperature variation
6. Earth pressure due to backfill

The abutment should be designed to resist all these forces. Typical forces acting on an abutment and the resulting stress distribution at the base of the abutment is shown in Fig. 18.7.

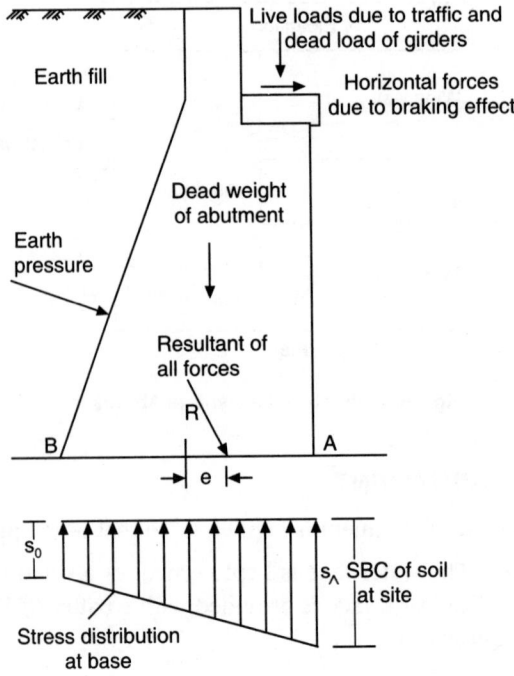

Fig. 18.7 Typical Forces Acting on Abutment.

18.11 STABILITY ANALYSIS OF ABUTMENTS

Figure 18.8 shows the section of a stone masonry abutment used for a highway bridge together with the forces acting per unit length of abutment.

Safe bearing capacity of soil = 150 kN/m^2
Coefficient of friction between masonry and soil = 0.5
Density of stone masonry = 25 kN/m^3

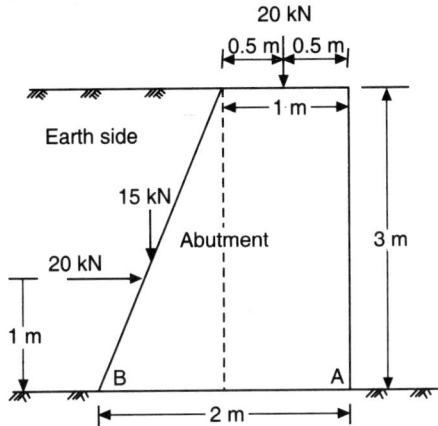

Fig. 18.8 Stone Masonry Abutment.

Compute the stresses developed at the base and check for the stability of the abutment.
Analysis of forces
Total vertical forces = W

$$W = W_1 + W_2 + W_3 + W_4$$

where $\quad W_1$ = Weight of rectangular portion of abutment
$$= (1 \times 3 \times 25)$$
$$= 75 \text{ kN}$$
$\quad W_2$ = Weight of triangular portion of abutment
$$= (0.5 \times 1 \times 3 \times 25)$$
$$= 37.5 \text{ kN}$$
$\quad W_3$ = Live and dead loads = 20 kN
$\quad W_4$ = Vertical load due to earth = 15 kN

Total vertical forces are

$$W = (75 + 37.5 + 20 + 15) = 147.5 \text{ kN}$$

Considering the moments of all forces about the toe A, we have

$$M = (95 \times 0.5) + (37.5 \times 1.33) + (15.1.67) - (20 \times 1)$$
$$= 102.3 \text{ kN·m}$$

Position of resultant R from A (Refer Fig. 18.9)

$$Z = \left(\frac{M}{W}\right) = \left(\frac{102.3}{147.5}\right) = 0.69 \text{ m}$$

$$\text{Eccentricity} = e = \left[\frac{b}{2} - z\right]$$

$$= [1 - 0.69] = 0.31 \text{ m}$$

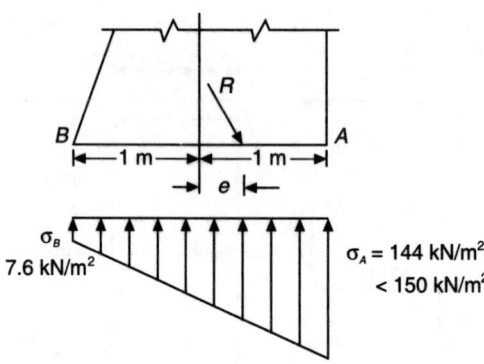

Fig. 18.9 Stress Distribution at Base of Abutment.

But $(b/6) = (2/6) = 0.33$ m
Since $e < (b/6)$, the stresses are compressive at both heel and toe.
Maximum and

$$\text{Minimum stress} = \frac{W}{b}\left[1 \pm \frac{6e}{b}\right]$$

$$= \frac{147.5}{2 \times 1}\left[1 \pm \frac{6 \times 0.31}{2}\right]$$

$$= 73.5 \,[1 \pm 0.93]$$

$$\sigma_A = 73.75 \,(1 + 0.93) = 144 \text{ kN/m}^2$$

$$\sigma_B = 73.75 \,(1 - 0.93) = 7.6 \text{ kN/m}^2$$

The maximum stress σ_A is less than the safe bearing capacity of the soil. Hence the stresses are within safe permissible limits.

Check for safety against sliding

Total vertical forces $W = 147.5$ kN

$$mW = (0.5 \times 147.5) = 73.75 \text{ kN}$$

$$\text{Factor of safety} = \left(\frac{73.75}{20}\right) = 3.68 > 2.00$$

Hence the abutment has sufficient factor of safety against sliding.

18.12 TYPES OF WING WALLS

There are two types of wing walls depending on the type of embankment and approaches to the bridge. The main function of the wing wall is to retain the earthfill without resisting any loads from the superstructure.

(a) Return or Box Type Wing Wall

When the approaches to the bridge are in cutting or small embankments, return type wing Wall is generally provided. The length of the return type wing wall depends upon the slope of the embankment and side slopes of the streams. Figure 18.10 shows the elevation and plan of a typical return type wing wall.

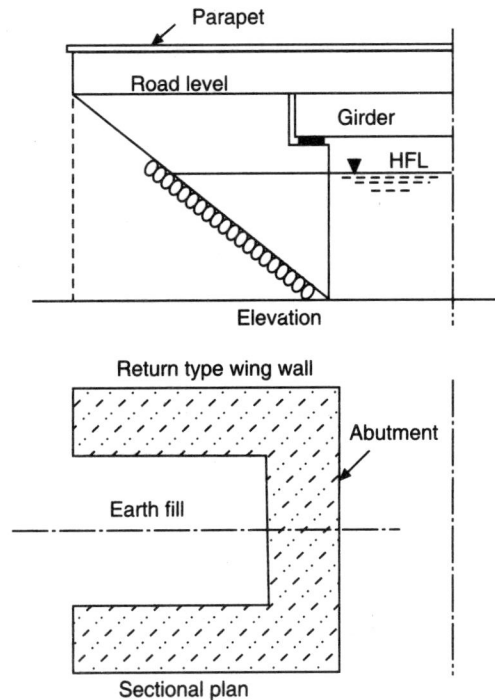

Fig. 18.10 Return Type Wing Wall.

(b) Splayed Type Wing Wall

In the case of bridges with heavy approach embankments, it becomes necessary to provide splayed type wing walls as shown in Fig. 18.11. The thickness of the splayed type wing wall is maximum at the junction of the abutment and gradually reduces to a minimum of 300 to 500 mm towards the bottom of the embankment. The splayed wing walls prevent the flow of soil towards the ventway under the bridge.

Wing walls are designed as retaining walls similar to those of abutments. The wing wall dimensions are generally fixed as follows:

Thickness at top = 0.5 m
Thickness at bottom = 0.45 to 0.5 h
where h = height of wing wall
Face batter = 1 in 12
Back batter = 1 in 6

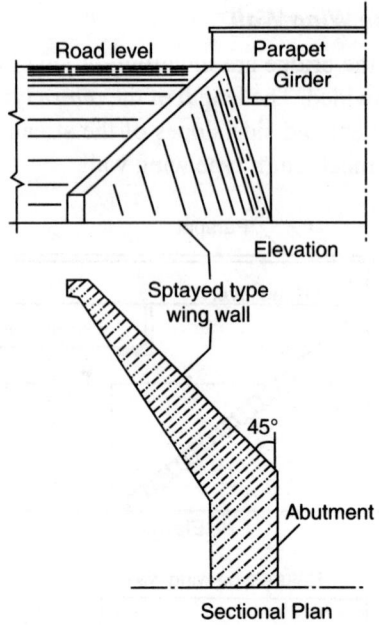

Fig. 18.11 Splayed Type Wing Wall.

18.13 APPROACHES

Approaches are usually provided at both ends of the bridge. As per I.R.C. specifications, the approaches should have a minimum straight length of 15 m on either side of the bridge. In the case of heavy embankments, the slope should be gradual so that the visibility of the vehicles approaching from the opposite side is not affected. Preferably the approaches should be in line with the longitudinal centre line of the bridge and in no case they should be curved at the entrance and exit of the bridge structure. Figure 18.12 shows the typical approaches to a major bridge on a National highway.

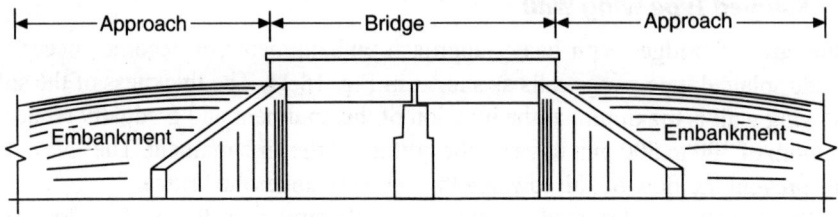

Fig. 18.12 Bridge with Approaches.

EXAMPLES FOR PRACTICE

1. A concrete pier of a major bridge is rectangular in section with dimensions of 3 m wide by 10 m in length and a height of 10 m above ground with hemispherical cut and ease waters the top section is 2m by 10 m. The dead load from the super structure is 1800 kN from each span acting at a distance of 0.5 m from the centre line of the pier. The live load reaction from one span is 1000 kN. Wind pressure on pier is 2 kN/m². The maximum permissible compressive stress on concrete pier is 2000 kN/m². Calculate the stresses developed at the base of the pier due to a) dead load and self weight b) Effect of buyoncy c) Eccentricity of live load d) wind pressure.

2. The concrete abutment of a major bridge has the following dimensions:

   Width at top = 1.5 m
   Width at foundation level = 3 m
   Height of abutment = 4 m
   The water face of the abutment is vertical and the earth side is sloping
   Live loads acting at the centre of top are 25 kN
   Safe bearing capacity of the soil = 200 kN/m²
   Density of soil at site = 16 kN/m³
   Angle of internal friction = 30°

   Compute the stresses developed at the base and check for the stability of the abutment

3. Design the pier of a major bridge using the following data:

   Dimensions of pier; Height of pier = 9 m, Width at top = 2.0 m, Width at bottom = 3.0 m
   Length of pier = 8.5 m
   Maximum water level = 8 m above base of pier
   Girder bearings located at 500 mm from centre of pier on either side
   Dead load from each span = 2300 kN
   Reaction due to live load on one span = 1000 kN
   Maximum mean velocity of current = 3 m/sec
   Material for pier; M-20 Grade concrete
   Live load: IRC Class AA or Class A whichever produces the worst effect

   Check the adequacy of the dimensions provided by computing the stresses in the pier.

REFERENCE

1. Anderson, P., Substructure Analysis and design, Ronald Press New York, 1956.
2. IRC: 78-1983, Standard Specifications and Code of Practice for Road bridges (Section VII), Foundations and Substructures, Indian Roads Congress, New Delhi, 1994, pp. 1–72.

3. Raiker, R.N., Outstanding Concrete Structures of India (Transportation Structures), Compiled by Maharashtra India Chapter of American Concrete Institute, Presented at A.C.I Fall convention at New York, 1984, pp. 105–134.

4. Johnson Victor, D., Essentials of Bridge Engineering (Fifth Edition), Oxford & IBH Publishing Co. Pvt. Ltd., New Delhi, 2001, pp. 277–294.

5. Xanthakos, P.P., Bridge Substructure and Foundation Design, Prentice Hall, PTR, New Jersey, 1995, pp. 1–844.

6. Dunham, C.W., Foundations of Structures, McGraw Hill Book Co. Ltd., New York, International Student Edition, 1962, pp. 1–722.

REVIEW QUESTIONS

1. Briefly explain the different types of piers used in High way bridge structures. What are the normal type of materials used for constructing piers?
2. List the various forces acting on piers and abutments.
3. How do you fix the various dimensions of piers? Mention the function of cut waters and ease waters used in piers.
4. What are the various forces to be considered in the design of piers? How do you evaluate these forces?
5. Describe the method of checking the stability of a pier. What is the effect of buoyancy during floods on the stability of the pier?
6. How do you calculate the braking forces due to live loads? What is the effect of these forces on the stability of a pier?
7. Briefly explain the necessity of using abutments. What materials are generally used for the construction of abutments?
8. List the various forces to be considered in the design of abutments of a bridge.
9. Explain the method of checking the stability of abutments subjected to dead, live and earth pressure from backfill.
10. Explain with sketches the different types of wing walls used mentioning the suitability of each type.

OBJECTIVE TYPE QUESTIONS

1. The minimum clearance from the edge of the bearing plate to the pier edge should be not less than
   a) 600 mm
   b) 100 mm
   c) 300 mm
2. Due to the effect of buoyancy during floods, the dead weight of the pier
   a) Increases
   b) Remains the same
   c) Decreases

3. For stability of the pier subjected to various types of loads, it is safer to ensure that the eccentricity of the resultant force acting at the base of the pier of width '*b*'should not exceed
   a) $b/2$
   b) $b/6$
   c) $b/12$

4. The maximum permissible compressive stress allowed on a reinforced concrete pier adopting M-15 grade concrete is limited to a value of
   a) 2 N/mm^2
   b) 6 N/mm^2
   c) 4 N/mm^2

5. In the case of massive concrete piers required for supporting decks of multilane traffic, savings in concrete is possible by selecting
   a) Solid piers
   b) Hammer head pier
   c) Cellular pier

6. The longitudinal braking forces due to the passage of live loads acting at the level of the bearings located at the top of pier results in
   a) Compressive stresses
   b) Shear stresses
   c) Bending stresses

7. The wind pressure acting on a pier develops minimum flexural stress on the
   a) Leeward side
   b) Mid portion of pier
   c) Windward side

8. The factor of safety against sliding in the case of abutments subjected to various forces should be greater than
   a) 4
   b) 2
   c) 3

9. In the case of approaches at the ends of a bridge are in embankments, it is preferable to use
   a) Return type wing walls
   b) Box type wing walls
   c) Splayed type wing walls

10. The necessity of checking the stability due to the action of forces developed due to earth arises only in the case of
    a) Piers
    b) Deck slab
    c) Abutments

19
Bridge Foundations

19.1 GENERAL ASPECTS

The design of bridge foundations is an integral part of the overall design of a bridge and for major-bridges, the cost of foundations depends upon the type of soil at site, span of the deck, type of foundation and it significantly influences the overall cost of the bridge. The design of bridge foundations in general should conform to the standard specifications and code of practice for bridges prescribed in IRC bridge code section-7 (IRC: 78-1983)[1].

The design of foundation for a bridge depends upon the various factors like

(a) The maximum scour depth

(b) The minimum grip length required

(c) The soil pressure at base and

(d) The stresses in the structures comprising the foundation.

19.2 TYPES OF FOUNDATIONS[2]

The foundations generally used for bridge piers and abutments are classified as:

(a) Shallow foundations and

(b) Deep foundations.

Shallow foundations involving open excavation are suitable for small span bridges constructed on hard gravelly soil and rocky strata. Footings and raft foundations are typical examples of shallow foundations. Shallow foundations transfer the load from the super and sub structure to the ground by bearing at the bottom of foundations and the maximum compressive stresses developed at the base due to dead and live loads should not exceed the permissible bearing pressure prescribed for the soil at site.

For major bridges, deep foundations[3] are generally adopted and they are further classified as:

(a) Pile foundations

(b) Well foundations or open caissons

(c) Pneumatic Caissons

19.3 PILE FOUNDATIONS[4]

Piles can be made up of (i) timber, (ii) steel, (iii) reinforced cement concrete and (iv) prestressed concrete.

Timber piles are used only for temporary structures such as jettys and not for bridge foundations. Nowadays, R.C.C. piles are the most common types used in major bridge foundations.

Concrete piles are further classified as a) Cast *in-situ* and b) precast piles.

In the case of cast-*in-situ* piles, a steel shell is driven first to the required depth and concreting is done after placing the reinforcement cage in the hole. If the shell is left in place, than it is called a shell pile. If the shell is removed, it is referred to as shell less pile. The world's longest shell pile[5] is provided at Walt Disney world in Florida, U.S.A. The length of this pile is 114 m. During 1950 to 1960, cast *in-situ* piles were commonly adopted since the technique of precasting was not well developed. With the introduction of better quality cement and precasting techniques, nowadays precast piles are invariably preferred in place of cast *in-situ* piles since bridge structures generally involve foundations under water or in soils with a high water table.

Precast piles[6] can be made with a high degree of quality control regarding dimensions and strength and hence have superior structural properties in comparison with cast *in-situ* piles. Precast piles c an be cast to various shapes such as: (i) circular, (ii) square, (iii) rectangular and (iv) octogonal. Generally square and circular section piles are preferred to other shapes.

During the last three decades, prestressed concrete piling has been extensively used as versatile foundation solution for major national high way bridge structures. The main advantage of prestressed concrete piles over the traditional reinforced concrete piles are:

1. High load moment carrying capacity

2. Standardization in design for mass production

3. Excellent durability under adverse environmental conditions

4. Ease of handling, transporting and driving

5. Crack free characteristics under handling and driving

6. Overall economy in production and installation

Prestressed concrete piles have been used have been widely used for the foundations of a number of bridges constructed in India under the Golden Quadrilateral scheme.

Typical details of reinforcements in a precast concrete pile are shown in Fig. 19.1. A bridge sub structure may require several piles to support the loads. In such cases, groups of piles are used and they are combined by a pile cap which supports the sub structure. Fig. 19.2 shows the typical shapes of pile caps. Piles are designed as compression members conforming to the National codes. Typical reinforcement details in pile caps are shown in Fig. 19.3.

The design of pile foundation for the piers of a major bridge transmitting heavy loads is illustrated by the following example.

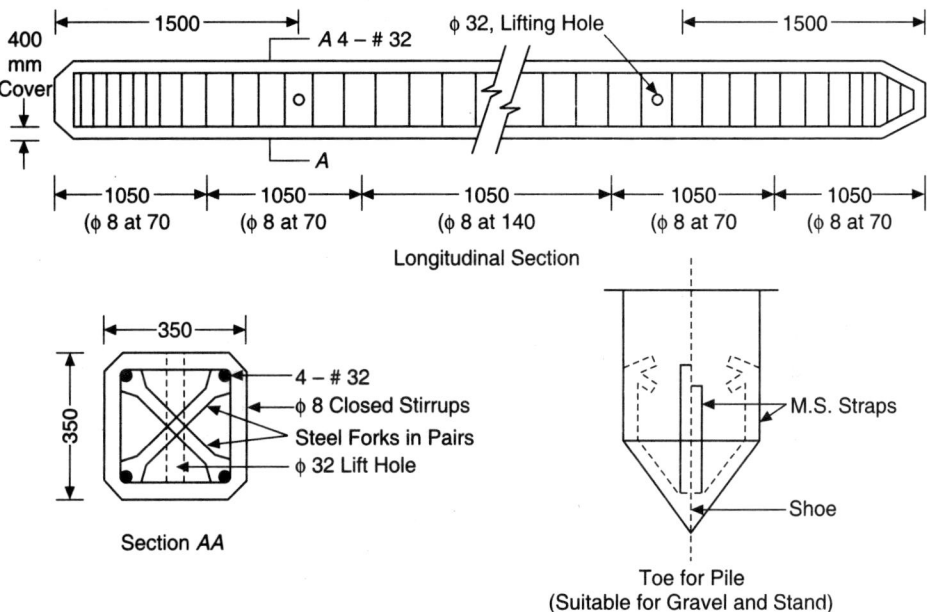

Fig. 19.1 Typical Details of Reinforcements in a Precast Concrete Pile.

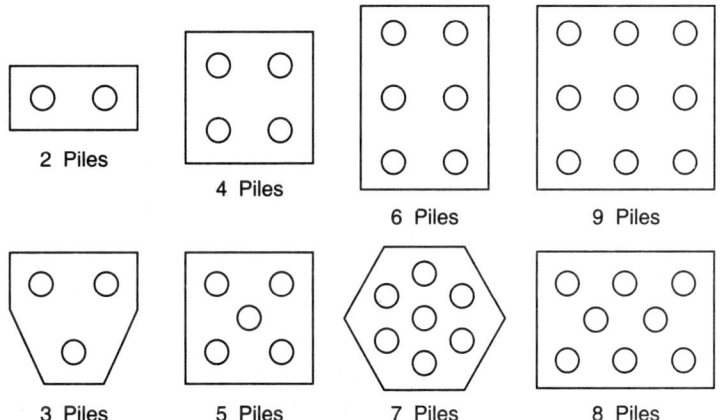

Fig. 19.2 Typical Shapes of Pile Caps.

The design of pile foundations for the piers of a major bridge transmitting heavy loads is illustrated by the following example:

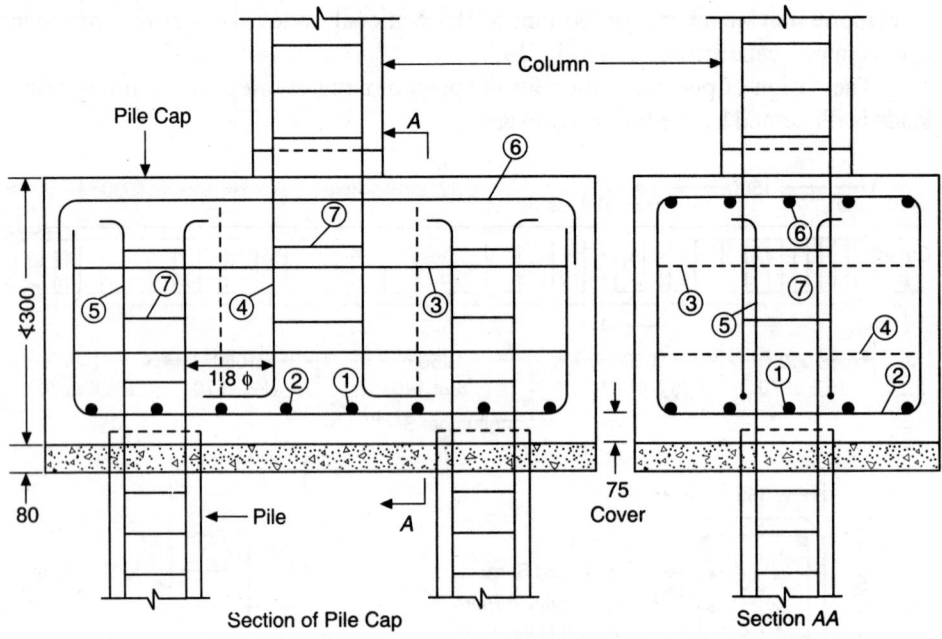

Bar Mark:
1, 2 – Main reinforcements   5 – Main bars in piles
   3 – Horizontal ties   6 – Top reinforcement in pile cap
   4 – Column reinforcement starter bars   7 – Ties for column and pile bars

Fig. 19.3 Typical Reinforcement Details in Pile Caps.

Design Example

The pier of a major fly over bridge transmits a load of 8400 kN at the foundation level. Design the number of precast R.C.C. piles and a suitable pile cap using the following data:

Width of pier = 1 m,   Length of pier = 9 m
Size of piles = 300 mm by 300 mm,   Spacings of piles = 1.5 m
Materials: M-20 Grade concrete and Fe-415 HYSD bars.
Hard strata is available at a depth of 6m below the ground level at bridge site.

1.   Arrangement of Piles and Piles Cap

Fourteen piles are arranged at a spacing of 1.5m as shown in Fig. 19.4.
Load on each pile = (8400/14) = 600 kN

2.   Pile Reinforcements

(a) Longitudinal Reinforcement

Length of pile above ground level = 0.6 m

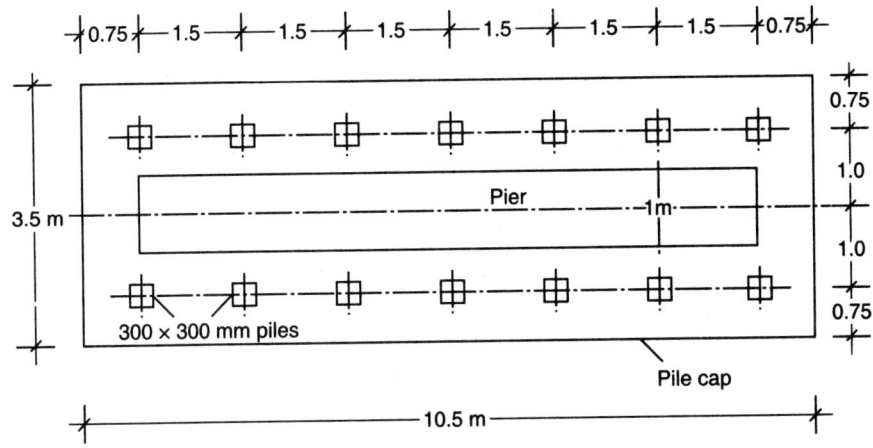

Fig. 19.4 Arrangement of Piles and Pile Cap.

Total length of pile = (6 + 0.6) = 6.6 m
Size of pile = 300 mm by 300 mm
L = 6.6 m and B = 0.3 m
Ratio (L/B) = (6.6/0.3) = 22 > 12
Hence the pile is designed as a long column

$$\text{Reduction Coefficient} = \left(1.25 - \frac{L}{48B}\right)$$

$$= \left(1.25 - \frac{22}{48}\right) = 0.792$$

Safe permissible stress in concrete = σ_{cc} = (0.792 × 5)
$\qquad\qquad\qquad\qquad\qquad\qquad\qquad\quad$ = 3.96 N/mm^2

Safe permissible stress in steel = (0.792 × 190) = 150 N/mm^2
Load carrying capacity of the pile is expressed as (IS: 456-2000)

$$P = [\sigma_{cc} A_{cc} + \sigma_{sc} A_{sc}]$$
$$(600 \times 10^3) = 3.96 [(300 \times 300) - A_{sc}] + 150 A_{sc}$$

Solving A_{sc} = 1643 mm^2

According to IRC: 78-1983, the longitudinal reinforcement $A_{sc} \not< 1.25$ percent of gross cross section for piles with a length less than 30 times the least width.

Hence $A_{sc} \not< (0.0125 \times 300 \times 300) \not< 1125$ mm^2

Adopt 8 bars of 20 mm diameter providing an area of A_{sc} = 2212 mm^2 with a clear cover of 40 mm.

(b) Lateral Reinforcement

In the body of the pile, the lateral reinforcement should be not less than 0.2 percent of the gross volume.

Using 8 mm diameter ties.

Volume of tie = 50 [4 (300 – 80)] = 44000 mm^3

If p = pitch of the ties

Volume of pile per pitch length = (300 × 300) p mm^3

$$= 90000 \text{ p mm}^3$$

Hence equating, we have

$$44000 = (0.002 × 90000 \text{ p})$$

$\therefore$ $\quad\quad\quad\quad$ p = 244 mm

Maximum permissible pitch = (0.5 × 300) = 150 mm

Hence provide 8 mm diameter ties at 150 mm centres in theonain body of the pile.

(c) Lateral Reinforcement Near Pile Head

Lateral reinforcement is of particular importance in resisting driving stresses near pile head provided for a length of $3B$ = (3 × 300) = 900 mm.

$\quad$ Spiral reinforcement is provided near pile head using 8 mm diameter helical ties (A_s = 50 mm^2).

Volume of spiral per mm length $= \dfrac{0.6}{100}\Big[(300 \times 300 \times 1)\Big] = 540 \text{ mm}^3$

If $\quad\quad\quad\quad$ p = pitch of the spiral

$$p = \left(\frac{\text{Circumference of spiral} \times A_s}{540}\right)$$

Providing a clear cover of 40 mm to the main longitudinal reinforcement of 20 mm diameter bars and using 8 mm diameter spiral ties inside the main reinforcements,

Diameter of spiral = [300 – (2 × 40) – 2(20) – 8] = 172 mm

$\therefore$ $\quad\quad\quad\quad$ $p = \left(\dfrac{\pi \times 172 \times 50}{50}\right) = 50 \text{ mm}$

Adopt 8 mm diameter spirals at a pitch of 50 mm for a length of 900 mm at the top of pile.

(d) Lateral Reinforcement Near Pile Ends

Lateral reinforcement of 0.6 percent of gross volume is provided in the form of ties for a distance of 3 times the least lateral dimension both at top and bottom of the pile.

Volume of the ties = 0.6 percent of gross volume for a

$\quad\quad\quad\quad\quad\quad\quad$ length of (3 × 300) = 900 mm

Using 8 mm diameter ties,

Volume of each tie = 50 [4 (300 – 80)] = 44000 mm^3

If P = pitch of the ties

Volume of piles per pitch length = (300 × 300 × p) = 90000 p

$$\therefore \quad 44000 = \left(\frac{0.6}{100} \times 90000 p\right)$$

Solving $\qquad p = 81.48$ mm

Adopt 8 mm diameter ties at 80 mm centres for a length of 900 mm from the ends of the pile both at top and bottom.

The longitudinal and cross sections of the pile with reinforcement details is shown in Fig. 19.5.

3. Pile Cap

The arrangement of piles with the pile cap is shown in Fig. 19.4. Referring to Figure 19.6 the maximum bending moment in the pile cap is computed as,

$$M_{zz} = (0.5 \ W \times 1 - 0.5 \ W \times 0.25) = 0.375 \ W \ \text{kN·m}$$

where $\qquad W = (2 \times 600) = 1200$ kN

$\therefore \qquad M_{zz} = (0.375 \times 1200) = 450$ kN·m

The effective depth required is given by

$$d = \sqrt{\frac{450 \times 10^6}{0.874 \times 1500}} = 585 \text{ mm}$$

Adopt $\qquad d = 600$ mm and overall depth $= 650$ mm

$$A_{st} = \left(\frac{450 \times 10^6}{230 \times 0.9 \times 600}\right) = 3623 \text{ mm}^2 \text{ per } 1.5 \text{ m width}$$

Using 25 mm diameter bars, spacing is given by

$$s = \left(\frac{1500 \times 491}{3623}\right) = 203 \text{ mm}$$

Adopt 25 mm diameter bars at 200 mm centres.

Distribution reinforcement $= 0.12$ percent of gross area

$$= (0.0012 \times 650 \times 1000)$$
$$= 780 \text{ mm}^2/\text{m}$$

Adopt 16 mm diameter bars at 250 mm centres as distribution steel.

Maximum shear force $= V = 600$ kN

Shear stress $= \tau_v = \left(\dfrac{v}{bd}\right) = \left(\dfrac{600 \times 10^3}{1500 \times 600}\right) = 0.40$

From Table 17 of IS:456-1978, $\tau_c = 0.28$ N/mm^2

$$V_{us} = (V - \tau_c \ b \ d)$$

$$= 600 - \left[\frac{0.28 \times 1500 \times 600}{1000}\right] = 348 \text{ kN}$$

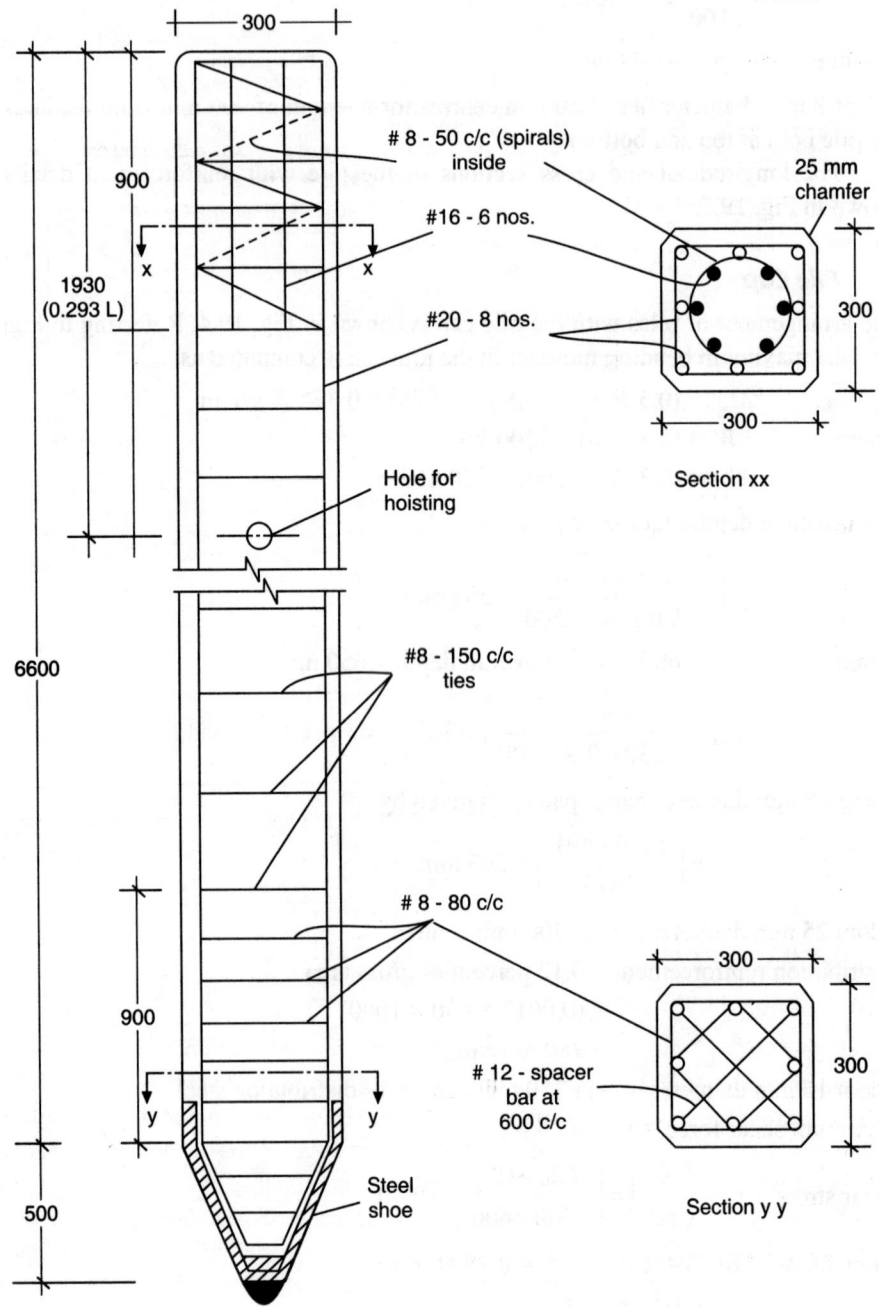

Fig. 19.5 Reinforcement Details in Precast Pile.

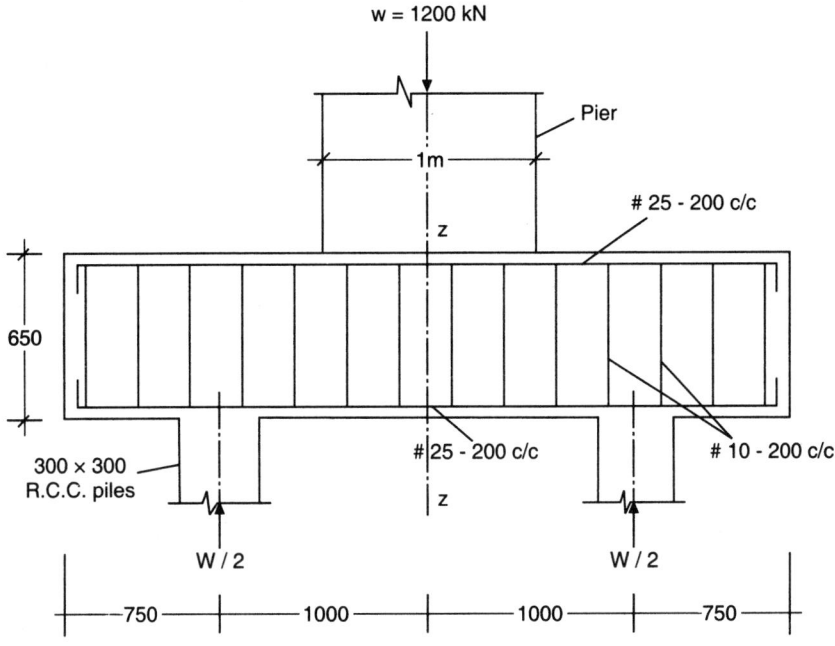

Fig. 19.6 Reinforcement Details in Pile Cap.

Using 10 mm diameter stirrups (8 legged), spacing is given by

$$s_v = \left[\frac{8 \times 79 \times 230 \times 600}{348 \times 10^3} \right] = 250 \text{ mm}$$

Adopt 10 mm diameter stirrups at 200 mm centers in a width of 1500 mm. The reinforcement details in the pile cap is shown in Fig. 19.3.

19.4 WELL FOUNDATIONS[7]

(a) General Aspects

Well foundations also referred to as open caissons are the most common type of foundations generally adopted for major bridges crossing rivers in India, where the soil strata comprises of sand or stiff clay. The foundation comprises of single large diameter well or a group of smaller wells of circular shape or double-D, square or rectangular shapes. The circular shaped well is generally preferred due to its simplicity for construction and sinking operations. In the case of foundations for large span cantilever, suspension or cable stayed bridges, larger rectangular wells with multiple dredge holes of square shape are commonly used. The typical shapes of wells used are shown in Fig. 19.7.

The size and shape of well for a particular foundation depends upon the various factors such as (a) size of pier (b) depth of foundation (c) nature of soil at site and the

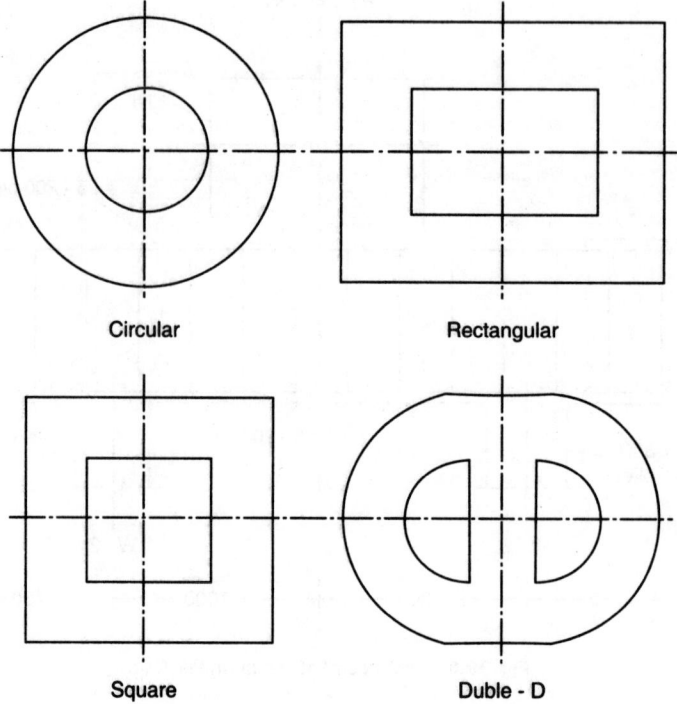

Circular

Rectangular

Square

Duble - D

Fig. 19.7 Shapes of Well Foundations.

possibility of pneumatic sinking. The dredge holes should be large enough to permit easy dredging. According to IRC: 78-1983, the minimum dimension of a dredge hole should be not less than 2 m.

(b) Components of Well Foundation[8]

A typical well foundation comprises of the following structural components:

(1) Steining

(2) Well Curb

(3) Bottom and Top Plugs

(4) Well Cap

(5) Sand filling

(6) Cutting Edge

Figure 19.8 shows the elevation of a typical well foundation with its various structural components.

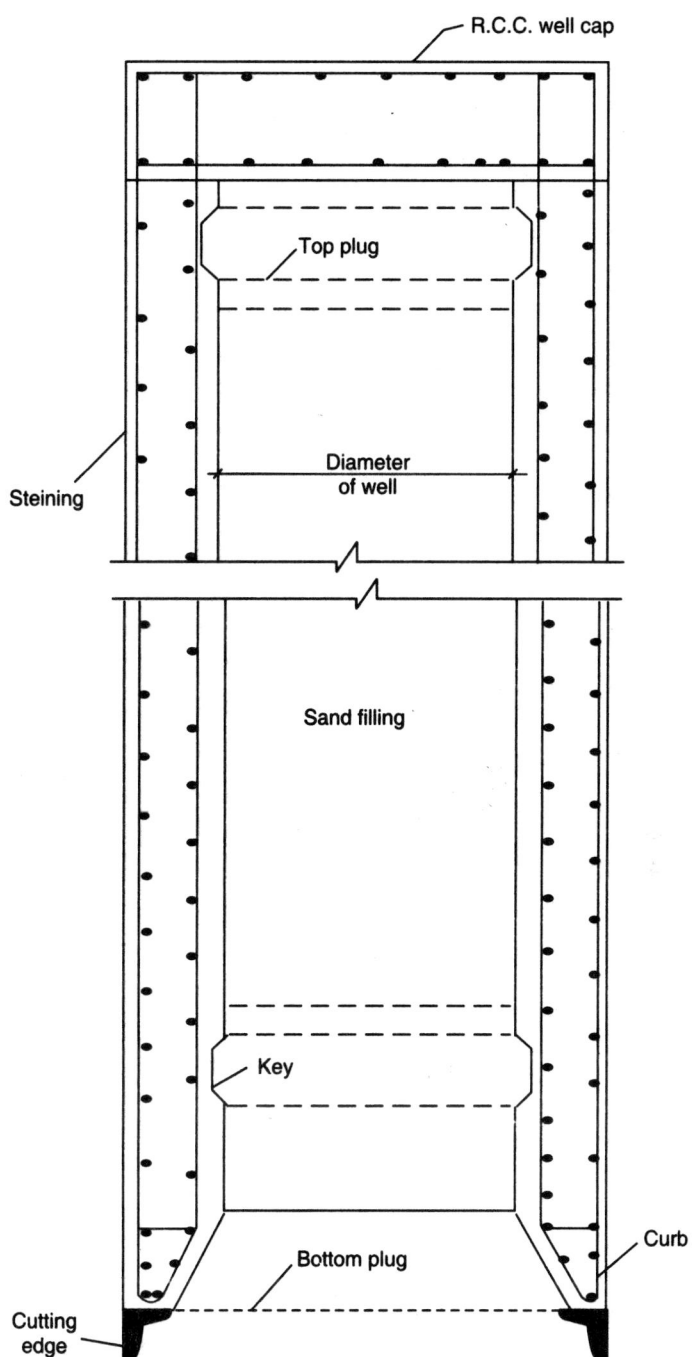

Fig. 19.8 Typical Well Foundation.

(c) Design Features[9]

(1) Well Steining

The minimum thickness of well steining should be not less than 500 mm and satisfy the following empirical relation,

$$h = K \, d_e \sqrt{L}$$

where $\quad$ h = Minimum thickness of steining (m)

$\quad d_e$ = External diameter of circular well (m)

$\quad L$ = Depth of well below low water level or ground level whichever is higher (m)

$\quad K$ = a constant depending upon the type of soil and material of well.

The value of K as outlined in IRC: 78-1983 for different types of wells are compiled in Table 19.1.

$\quad$ The stresses developed in the well steining at any level under various conditions of loading during sinking and service should be within safe permissible limits. For plain concrete wells, vertical reinforcement in the steining should be not less than 0.12 percent of the gross cross sectional area while for R.C.C. wells, the minimum vertical reinforcement distributed on both faces should be not less than 0.2 percent of the cross sectional area. The transverse reinforcement in the form of hoops should be not less than 0.04 percent of the volume per unit length of the steining.

Table 19.1 Values of Constant K (IRC: 78-1983)

Type of Shape of Well	Material of Well	Predominant Type of Soil	
		Sandy Soil	Clayee Strata
Single Circular or Dumb-bell type	Cement Concrete	0.030	0.033
–Do–	Brick Masonry	0.047	0.052
Twin D-type Wells	Cement Concrete	0.039	0.043
–Do–	Brick Masonry	0.062	0.068

(2) Well Curb

The well curb should be strong enough to be able to transmit superimposed loads from the steining to the bottom plug. The curb is generally of reinforced concrete having a wedge shape with the grade of concrete not leaner than M-20.

$\quad$ Minimum reinforcement in curb = 72 kg/m³ excluding bond rods.

(3) Bottom and Top Plugs

The bottom plug of concrete is provided with its top not lower than 300 mm in the centre above the top of the curb. The bottom of the plug is generally curved and it should be below the level of the cutting edge.

The concrete mix in the bottom plug should have a minimum cement content of 330 kg/m³ and a slump of about 150 mm to permit easy flow of concrete through tremie to fill up all cavities.

If grouting is used the cement mortar mix should be not leaner than 1 : 2 and the rate of pumping should be controlled so that the grout fills all the interstices up to the top of the plug.

The top plug comprising of 1 : 3 : 6 cement concrete is generally provided for a thickness of 500 mm below the well cap.

(4) Well Cap

Reinforced concrete well cap connecting multiple wells should be provided with its bottom preferably laid as low as possible taking into account the low water level and the longitudinal bars from the well steining are anchored into the well cap. The depth of the well cap should be designed to withstand the forces transmitted from the pier or abutment.

(5) Sand Filling

The well is filled with sand or excavated material free of organic matter if considered necessary between the bottom and top plugs.

(6) Cutting Edge

A cutting edge comprising of mild steel angle firmly anchored to the curb *i* generally provided to facilitate sinking of the well through the soil strata. The quantity of steel in the cutting edge should preferably be not less than 40 kg/rr.

The design of a typical well foundation is illustrated by the following example.

(d) Design Example

Design a well foundation for the pier of a major highway bridge to suit *t* following data:

Internal diameter of well = 2.5 m
Type of soil strata: Clayee ($K = 0.033$)
Depth of well = 25 m below bed level
Materials: M-20 Grade concrete
                Fe-4 15 Grade HYSD bars

(1) Thickness of Steining

Minimum thickness of steining is given by

$$h = K d_e \sqrt{L} = K \left(d_i + 2h \right) \sqrt{L}$$

where d_i = internal diameter of well = 2.5 m

h = Thickness of steining

L = Length of well = 25 m

K = Constant for Clayee soil = 0.033

∴ $h = 0.033(2.5 + 2h)\sqrt{25}$

Solving $h = 0.558$ m = 558 mm

Adopt a steining of 600 mm

(2) Reinforcement in Steining

For R.C.C. wells, Minimum longitudinal reinforcement

A_{sc} = 0.2 percent of gross cross sectional area

$$= \frac{0.2}{100}\left[\frac{\pi}{4}(3.7^2 - 2.5^2)\right] \times 10^6$$

= 11680 mm² or 5840 mm² for each face

Use 16 mm diameter bars at 300 mm centres on both faces.

Hoop reinforcement $\not< 0.04$ percent of volume/unit length

$$\not< \frac{0.04}{100}\left[\frac{\pi}{4}(3.7^2 - 2.5^2)\right]1$$

$\not< 0.002336$ m³/m

$\not< 0.002336 \times 7200$ kg/m

$\not< 16.8$ kg/m

Using 10 mm diameter bars

Average circumference of the hoop = $(\pi \times 3.1)$ = 9.734 m

Weight of one hoop = (0.62×0.374) = 6 kg

Number of hoops per metre $= \left(\dfrac{16.8}{6}\right) = 2.8$

∴ Spacings of hoops $= \left(\dfrac{1000}{2.8}\right) = 357$ mm

Use 10 mm diameter hoops at 600 mm centres on both faces.

(3) Well Curb

Minimum reinforcement = 72 kg/m²

Providing a curb, 1000 mm deep with a bottom and inside width of 150 mm.

Volume of concrete in curb $= \dfrac{\pi}{4}\left[(3.7^2 - 2.5^2)\right] - [0.5 \times 0.45 \times 0.85 \times \pi \times 2.65]$

= 4.22 m³

Total quantity of steel in Curb = (72 × 4.22) = 304 kg

Using hoops of average diameter = 3.1 m

Weight of one hoop of 20 mm diameter = (π × 3.1 × 2.47) = 24 kg

Weight of one hoop of 16 mm diameter = (π × 3.1 × 1.58) = 15.38 kg

Weight of one tie of 8 mm diameter = (3 × 0.39) = 1.17 kg

Adopting a spacing of 300 mm for ties

$$\text{Number of ties required} = \left(\frac{\pi \times 3.1}{0.3} \right) = 33$$

Weight of ties = (33 × 1.17) = 39 kg

Weight of 8 hoops of 20 mm diameter = (8 × 24) = 192 kg

Weight of 6 hoops of 16 mm diameter = (6 × 15.38) = 92 kg

Total quantity of steel provided in the Curb = (192 + 39 + 92)

$$= 323 \text{ kg} > 304 \text{ kg}$$

The details of reinforcements in the well and curbs is shown in Fig. 19.9.

19.5 CAISSON FOUNDATIONS[10]

Pneumatic caisson foundations consisting of steel cutting base and concrete pier is adopted for deep foundations in watery situations. Pneumatic caisson consisting of a working chamber and an air lock is used. The curb portion constitutes the working chamber and this is made up of steel sections. Air locks made of steel are attached to the working chambers.

The pressure inside the air lock can go up to 3.5 atmospheres above normal (limit of human endurance). This pressure is reached at a depth of 40 m. Hence it is not convenient to go beyond 40 m depths with pneumatic caisson and the cost will be prohibitive for larger depths. Two air locks are generally used, one for men and the other for removing the excavated muck. When the final foundation level is reached, the working chamber is filled with concrete and the air pressure is maintained until the concrete has attained the desired strength.

The sequence of operations in sinking caissons is shown in Fig. 19.10.

The sectional elevation showing the typical details of a pneumatic caisson used for the foundations of a major bridge in India is shown in Fig. 19.11. The working chamber must be sufficiently strong to support the weight of the pier and it should also be air tight. Two air locks made of steel are provided for transportation of men and excavated muck.

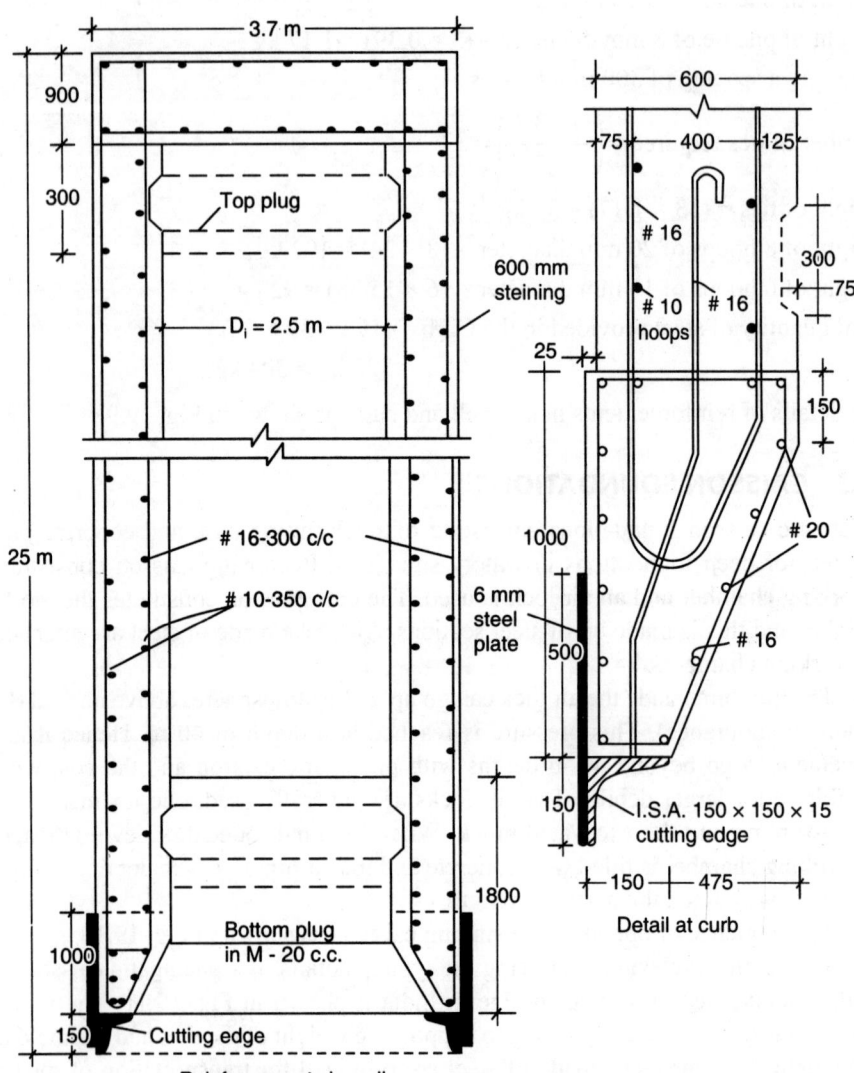

Fig. 19.9 Typical Well Foundation (Reinforcement Details).

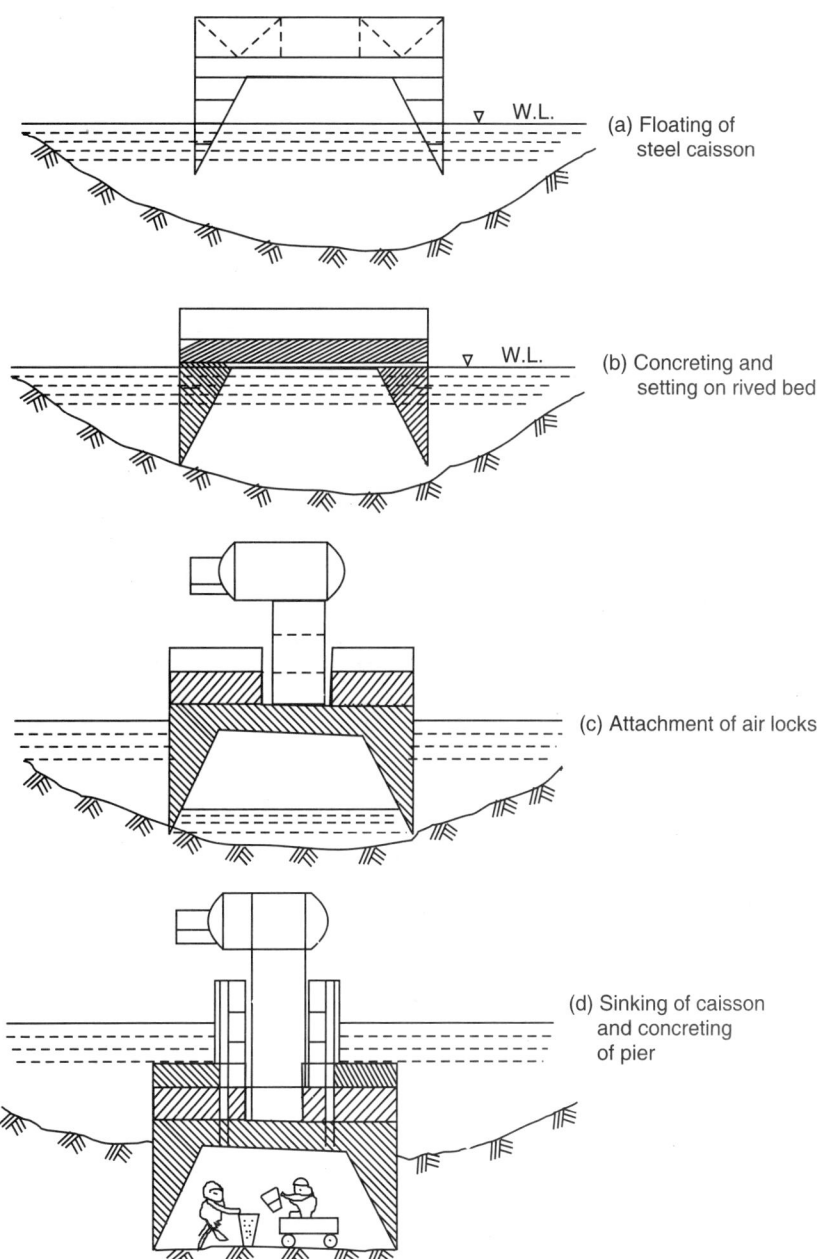

(a) Floating of steel caisson

(b) Concreting and setting on rived bed

(c) Attachment of air locks

(d) Sinking of caisson and concreting of pier

Fig. 19.10 Shrinking of Pneumatic Caisson with Steel Curb.

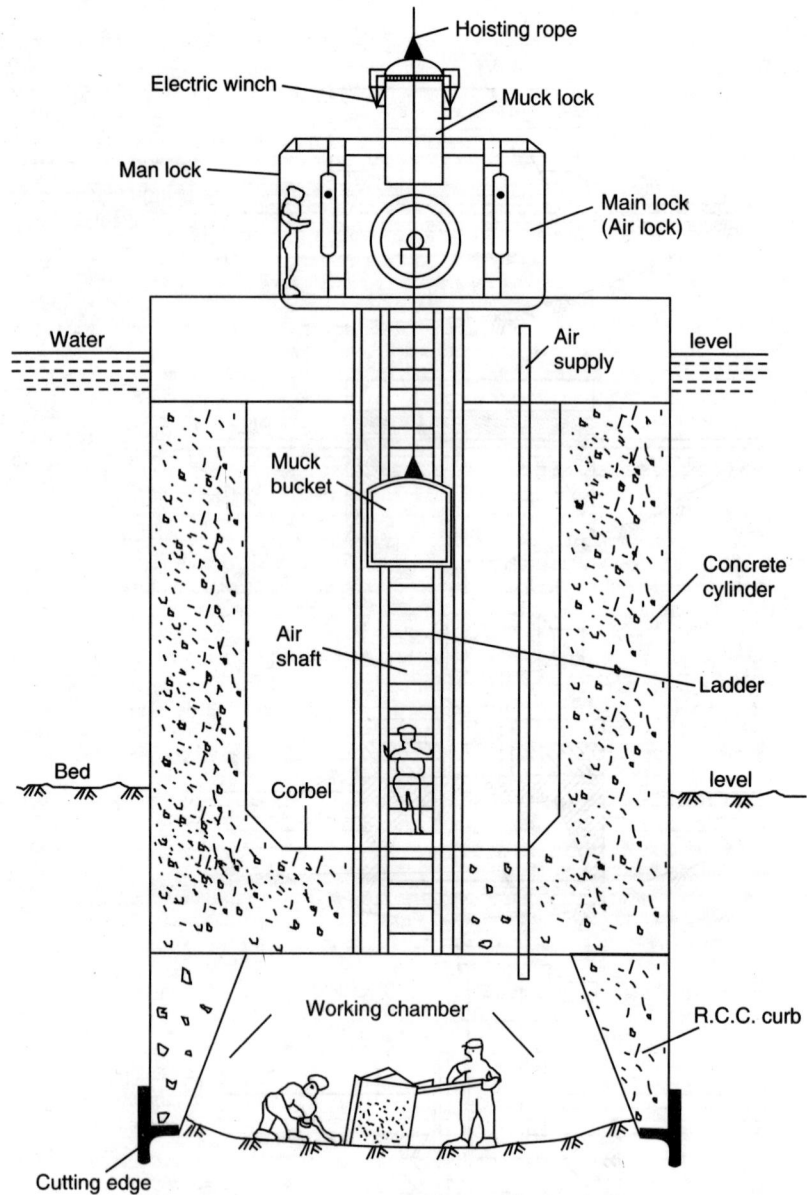

Fig. 19.11 Pneumatic Caisson with R.C. Curb.

EXAMPLES FOR PRACTICE

1. The pier of a bridge transmits a total load of 7200 kN at the foundation level. Precast R.C.C. piles 300 mm by 300 mm are proposed for the foundation. Width of pier = 0.9 m and length of pier = 8.75 m. Spacings of piles = 1.75 m.

Materials used are M-20 Grade concrete and Fe-415 Grade HYSD bars. Hard strata is available at a depth of 5 m below ground level. Design a typical pile and a suitable pile cap and sketch the details of reinforcements.

2. It is proposed to use 400 mm by 400 mm precast R.C.C. piles for the foundation of the pier of a major bridge. It is estimated that the load on each pile to be 800 kN. Design suitable reinforcements in the pile if the total length of the pile is estimated as 8 m using M-20 grade concrete and Fe-415 grade HYSD bars.

3. Design a suitable R.C.C. well foundation for the pier of a major highway bridge using the following data:

   Internal diameter of well = 3 m
   Type of soil strata (sandy) with constant $K = 0.030$
   Depth of well = 20 m below bed level
   Materials: M-20 Grade concrete
              Fe-415 Grade HYSD bars

4. A reinforced concrete well foundation is proposed for the pier of a bridge. Design suitable reinforcements in the well steining using the following data:

   Internal diameter of well = 2 m
   Type of soil strata (Clayee) with constant $K = 0.033$
   Depth of well = 24 m
   Materials: M-20 Grade concrete
              Fe-415 Grade HYSD bars

   Sketch the details of reinforcements in the well and curb.

REFERENCES

1. IRC: 78-1983, Standard Specifications and Code of Practice for Road Bridges (Section VII), Foundations and Sub Structure, Indian Roads Congress, New Delhi, 1994, pp. 1–72.
2. Peck, R.B, Hanson, W.E and Thornburn, T.H., Foundation Engineering, Asia Publishing House, Bombay, First Edition, 1959, pp. 1–410.
3. Shu-T'ien Li and Tony Chen-yeh Liu, Prestressed Concrete Piling-Contemporary Design Practice and Recommendations, Journal of the American Concrete Institute, Proceedings, Vol. 67, No.3, March 1970, pp. 201–220.
4. Tambe, S.R, Bongirwar, P.L and Kanhere, D.K., Critical Appraisal of Mumbai Fly Over Project, Journal of Indian Roads Congress, New Delhi, Vol. 60-2, October 1999, pp. 235–263.
5. World's Longest Pile? Foundation Facts, Raymond International Inc., Vol. VII, No. l, 1971, p. 20.
6. Bruce, R.N. and Hebert, D.C., Splicing of Precast Prestressed Concrete Piles, Parts 1 and 2, Journal of Prestressed Concrete Institute, Vol. 19, No.5, Sept/Oct 1974, pp. 70–97 and No. 6, Nov/Dec 1974, pp. 40–66.

7. Saxena, R.K., Well Foundations for Road Bridges, Journal of Indian Roads Congress Vol. 34, No. 2, Nov. 1971, pp. 391–435.
8. Johnson Victor, D., Essentials of Bridge Engineering (Fifth Edition), Oxford & IBH Publishing Co. Pvt. Ltd., New Delhi, 2001, pp. 307–311.
9. Lee, D.H., An Introduction to Deep Foundations and Sheet Piling, Concrete Publications Ltd., London, 1961, pp. 1–204.
10. Ponnuswamy, S., Bridge Engineering, Tata McGraw Hill Publishing Co. Ltd., New Delhi, 1986, pp. 1–544.

REVIEW QUESTIONS

1. What are the major classifications of bridge foundations? Under what situations you would adopt shallow and deep foundations?
2. Briefly explain the different types of pile foundations adopted for bridges.
3. What are the advantages of precast piles over cast-*in-situ* piles?
4. Explain with sketches the typical reinforcement details in a precast pile and pile cap.
5. Under what situations you would resort to well foundations for bridge structures?
6. Briefly explain the various structural components of a typical well foundation, specifying the function of each of these components.
7. Explain with typical sketches the shapes of wells and reinforcement details in the staining, well cap and bottom plug of a well foundation.
8. Write a brief note on the method of sinking wells for the piers of a major bridge.
9. What are Caisson foundations? Under what situations you would use them for bridge foundations?
10. Explain with sketches the reinforcement details of a typical caisson and the method of sinking a pneumatic caisson for a major bridge foundation.

OBJECTIVE TYPE QUESTIONS

1. For bridges located in river beds with clayey soil conditions, the type of foundation to be selected is
   a) Spread footings
   b) Caisson foundations
   c) Pile foundations
2. In the case of pile foundations, better quality control and economy is possible by using
   a) cast-*in-situ* piles
   b) Precast piles
   c) Steel piles

3. In the case of reinforced concrete piles, the longitudinal reinforcement expressed as a percentage of the cross section should be not less than
   a) 1.8 %
   b) 2.5%
   c) 1.25%
4. To withstand driving stresses in a precast pile, the pile heads should be provided with spiral reinforcement having a volume per mm length not less than
   a) 0.12%
   b) 0.6%
   c) 0.3%
5. Pile caps should be designed to resist
   a) Compression
   b) Combined flexure and shear
   c) Flexure
6. In the case of well foundations, sinking of the well is facilitated by using
   a) Square wells
   b) Rectangular wells
   c) Circular wells
7. In the steining of a reinforced concrete well, the minimum vertical reinforcement distributed on both faces, expressed as a percentage of the cross sectional area should be not less than
   a) 0.12%
   b) 0.15%
   c) 0.2%
8. In the curb of a reinforced concrete well foundation, the minimum reinforcement to be used should be not less than
   a) 50 kg/m^3
   b) 60 kg/m^3
   c) 72 kg/m^3
9. Pneumatic caisson foundations are invariably preferred for bridges in the case of
   a) Hard gravelly soil
   b) Sandy soils
   c) Deep watery situations
10. For sinking operation of caisson foundations in watery situations, the common device used is
   a) Dewatering pumps
   b) Air locks
   c) Drilling

20

Curved Bridge Decks

20.1 INTRODUCTION

Flyovers generally used in metropolitan city crossings necessitate the use of curved bridge decks. In the case of curved bridges[1], the support reactions are not in the same plane as that of dead and live loads. Consequently the dead and live load moments are accompanied by torque for equilibrium of the structural system.

In the case of bridge structures (built up of multicelled box girders) of moderate curvature, simple beam theory[2] involving the simple bending theory and torsion of thin walled members is generally sufficient. The longitudinal bending moments are computed as for a straight beam and the expression for the incremental torque is integrated along the beam from-points of known zero torque to give the final torque at any section.

20.2 METHODS OF ANALYSIS

(1) Simple Beam Theory

The theory is based on simple bending and torsion of thin walled members. The incremental torque at any section is computed as a function of the longitudinal beam for straight beam and applied torque along the span and the radius of curvature of the bridge deck as shown in Fig. 20.1.

Warping and distortion[3] shown in Fig. 20.2 are ignored in this method. Local bending moments in the deck slab are determined from the influence surfaces and the transverse moments around the box girder by moment distribution for the element as shown in Fig. 20.3.

(2) Folded Plate Analysis[4]

In this method, the structure is represented as an assembly of plates joined rigidly along the longitudinal edges. Individual plate stiffness matrices for the displacements at various points of the edges are derived from classical plate theory and the overall stiffness matrix for edge loading of the structure assembled. All points along the longitudinal edges have four degrees of freedom as shown in Fig. 20.4(a). Wherever loads are applied away from the edges of a plate, the fixed end moments and forces are determined as in Fig. 20.4(b). These are applied to the structure in the directions shown in Fig. 20.4(c).

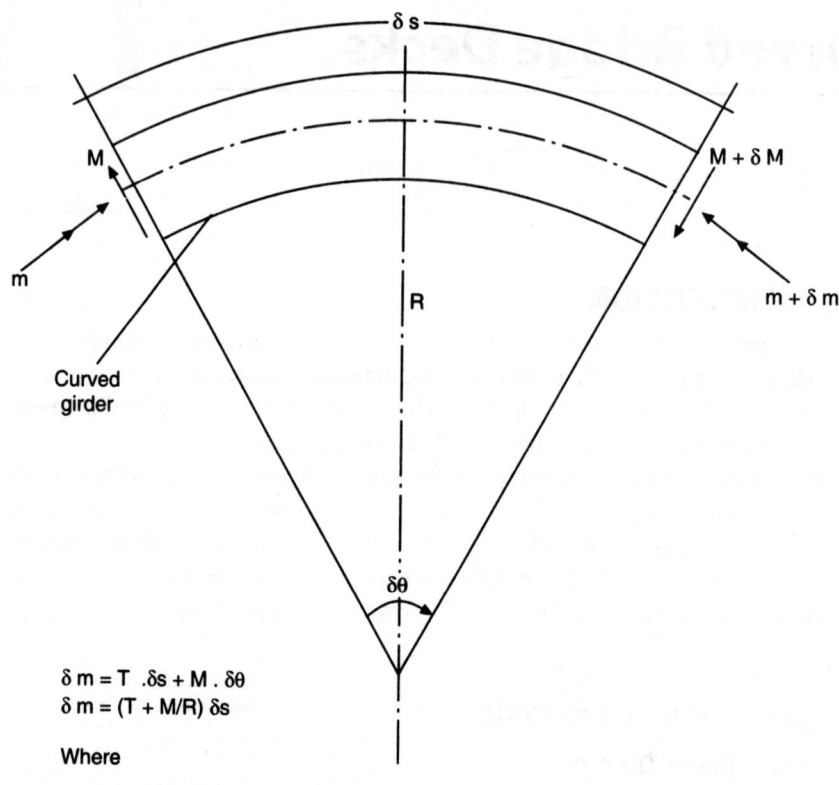

$$\delta m = T \,.\delta s + M \,.\, \delta\theta$$
$$\delta m = (T + M/R) \,\delta s$$

Where

M = Longitudina B.M. for straight beam
T = Applied torque along span
R = Radius of curvature
m = Induced torque at any section

Fig. 20.1 Analysis of Curved Girders.

The solution in all the cases is obtained using Fourier series and the final results are obtained by superposition of the results for each harmonic loading. This method is ideally suited for straight or curved, simply supported structures of constant cross section. Continuous structures are analysed by first considering the structure as simply supported and then eliminating displacements at the interiors supports as a second stage in the analysis. Due to the development of advanced methods like finite strip and elements, this method is rarely used in practice.

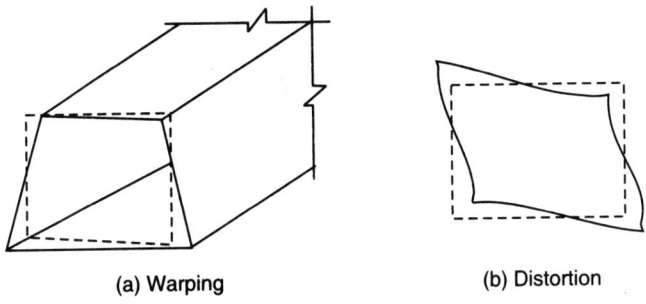

(a) Warping   (b) Distortion

Fig. 20.2 Warping and Distortion (Curved girders).

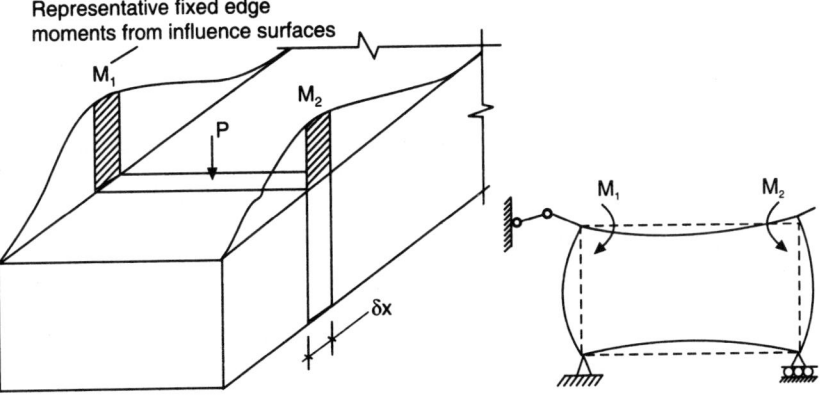

Fig. 20.3 Moment Distribution for Element dx.

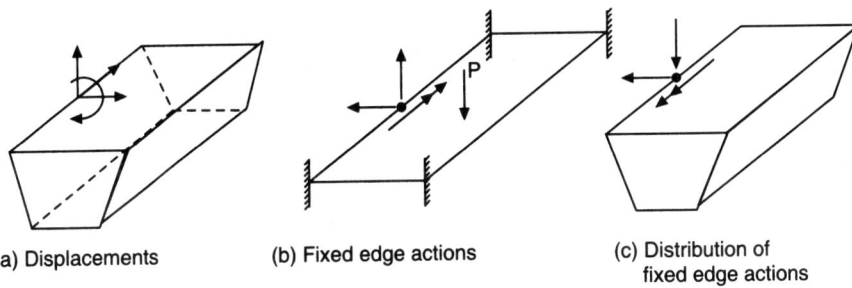

(a) Displacements   (b) Fixed edge actions   (c) Distribution of
fixed edge actions

Fig. 20.4 Folded Plate Anaysis.

(3) *Finite Strip Method*[5]

The finite strip method is a simplified version of the folded plate method in which the loaded surfaces are subdivided into a number of plates simulating finite strips as shown in Fig. 20.5. The degrees of freedom chosen are the same as that for the folded plate method and the solution is obtained by superposition of loading harmonics.

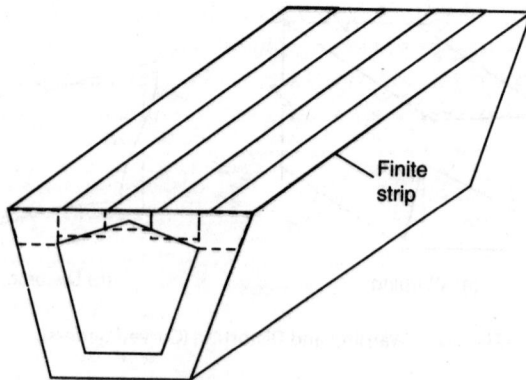

Fig. 20.5 Finite Strip Method of Analysis.

(4) Finite Element Method

In this method the entire structure is divided into small elements[6] and the stiffness of the structure is assembled from the membrane and the olate bending stiffnesses of each element. This method is generally applicable for all types of structures. Its accuracy depends upon the nature and number of elements- and the division of structure into elements as shown in Fig. 20.6.

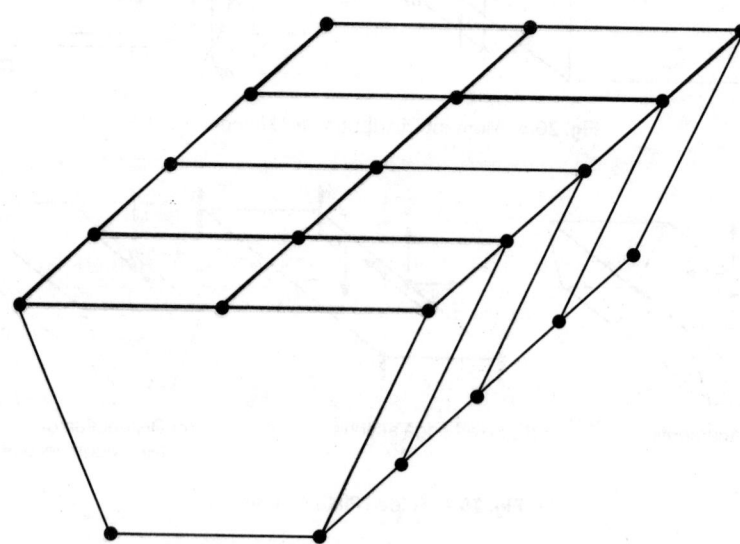

Fig. 20.6 Finite Element Method of Analysis.

Exhaustive research has been done to determine the most suitable types of elements for box girder bridges. Good agreement between theory and experiments have been demonstrated.

The disadvantage of this method is the large computer storage required and the time consuming input synthesis and output analysis.

The finite element method[7] is the most general method generally used for complex type of curved bridge deck analysis. Simplified commercial programmes for box girder structure are already available. Comparative analysis of the various methods indicate that for single or multiple cell boxes, straight in plan, simple beam theory can be used while for curved in plan decks that are wide and shallow in depth, finite element method is preferable.

The approximate methods outlined in 1 and 2 have been used successfully for the design of many major bridge decks in the past and they can be advantageously used for the analysis of simple bridge decks while the finite element method is the most general and versatile method suitable for the analysis of complex bridge decks.

REFERENCES

1. Raina, V.K., Concrete Bridge Practice, Analysis, Design and Economics, Tata McGraw Hill Publishing Co. Ltd., New Delhi, 1991, pp. 494–497.
2. Spyropoulos, P.J., Circularly Curved Beams Transversely Loaded, Journal of the American Concrete Institute, Vol. 60, No.10, October 1963, pp. 1457–1469.
3. Scordelis, A.C et al., Load Distribution in Concrete Box Girder Bridges, American Concrete Institute Special Publication SP: 23 on Concrete bridge Design, Detroit, 1969.
4. Chuk-h, E and Dudnik., Concrete Box Girder Bridges as Folded Plates, American Concrete Institute Special Publication SP: 23 on Concrete Bridge Design, Detroit, 1969.
5. Cheung, Y.K., Analysis of Box Girder bridges by the Finite Strip Method, American Concrete Institute Special Publication SP: 26 on Concrete Bridge Design, Detroit, 1971.
6. Sisodiya, R.G et al., New Finite Elements with Application to Box Girder Bridges, Proceedings of the Institution of Civil Engineers, Paper No: 7479S, London. 1972.
7. Lim, P.K.T et al., Finite Element Analysis of Curved Box Girder Bridges, Proceedings of the Conference on development in Bridge Design & Construction, Cardiff, 1971.

Dynamic Response of Bridge Decks

21

21.1 INTRODUCTION

Moving loads on bridge decks[1] causes the super structure comprising beams and slabs to deflect from its equilibrium position relatively quickly. The mass and inherent elasticity of the structure tends to restore the bridge deck to its equilibrium position thus causing a series of vibrations due to the motion of vehicles on the bridge deck. The effect of bridge deck vibrations result in

(a) structural damage if not properly designed for vibration effects,

(b) causes unpleasant physiological and psychological reactions on humans, and

(c) develops additional stresses of transient nature which are in addition the static effects.

The normal practice generally followed in several national codes to safeguard the bridge deck from the destructive effects of dynamic loads is to provide for impact factors for live loads which amplify the design static loads by a certain percentage. Consequently the bridge deck is rendered more rigid so that the dynamic effects are safely resisted with increased mass and elasticity of the structure.

Certain thumb rules incorporated in the codes include limiting the span/depth ratios of the deck and also limiting the deflection/span ratio. These provisions are not based on the data of frequency and amplitude of vibrations of the structure and hence cannot be taken as guarantee against the occurrence of undue vibrations even under normal loads.

21.2 FACTORS INFLUENCING BRIDGE VIBRATIONS

Bridge vibrations[2] are influenced by the following factors:

(1) Natural frequencies of the vehicle system and the suspension system

(2) Speed of passing vehicles

(3) Ratio of vehicle to bridge deck weight

(4) Flexural rigidity and natural frequency of vibration of the bridge deck

(5) Type of bridge deck and approaches

(6) Improper functioning of expansion joints

(7) Frequency of live load application due to the passage of multiple axles

(8) Motion induced in bridge before application of live loads (particularly important for continuous spans)

(9) The damping characteristics of the bridge and the vehicle

21.3 DYNAMIC RESPONSE OF BRIDGE DECKS

The dynamic response of a bridge deck to a moving load depends on mass stiffness, damping properties of the bridge and dynamic properties of the moving loads resulting in vibrations either at the natural frequency or at the frequency of the applied excited force.

The normal range of fundamental frequency of bridges varies between 1 and 20 cycles per second. This may coincide with the range of frequencies of moving vehicles resulting in the possibility of resonance leading to the failure of the bridge deck.

21.4 PRACTICAL APPROACH FOR VIBRATION ANALYSIS

A prerequisite for Vibration analysis of bridge decks is a proper understanding of the principles of dynamics of structures presented by Clough[3], Euro codes[4], Chopra[5] and Ashish Gupta[6]. A simpler procedure has been evolved by Lenzen, relating the natural frequency and the vibration amplitude shown in Fig. 21.1. This concept generally referred to as Lenzen's criteria is reported by Raina[2] is very helpful in identifying the degree of perception of vibrations in the structure by humans.

The following procedure is ideally suited for the designer for a logical approach for the vibration analysis of bridge decks:

(1) Estimate the maximum span deflection δ(mm) under a single 200 kN hypothetical point load at centre of span and cross section using the fully composite (average) flexural rigidity of the deck section of full width (EI)

(2) Estimate the fundamental natural frequency N_f of the deck span from the relation,

$$N_f = \frac{2}{L^2} \sqrt{\frac{EIg}{w_d}} \text{ cycles per second}$$

where

L = span (m)

EI = Flexural rigidity of the full width section of the bridge deck expressed in kN·m^2 units.

g = Acceleration due to gravity = 9.81 m/sec^2

w_d = Dead weight of deck including finishes expressed in kN/m of bridge span.

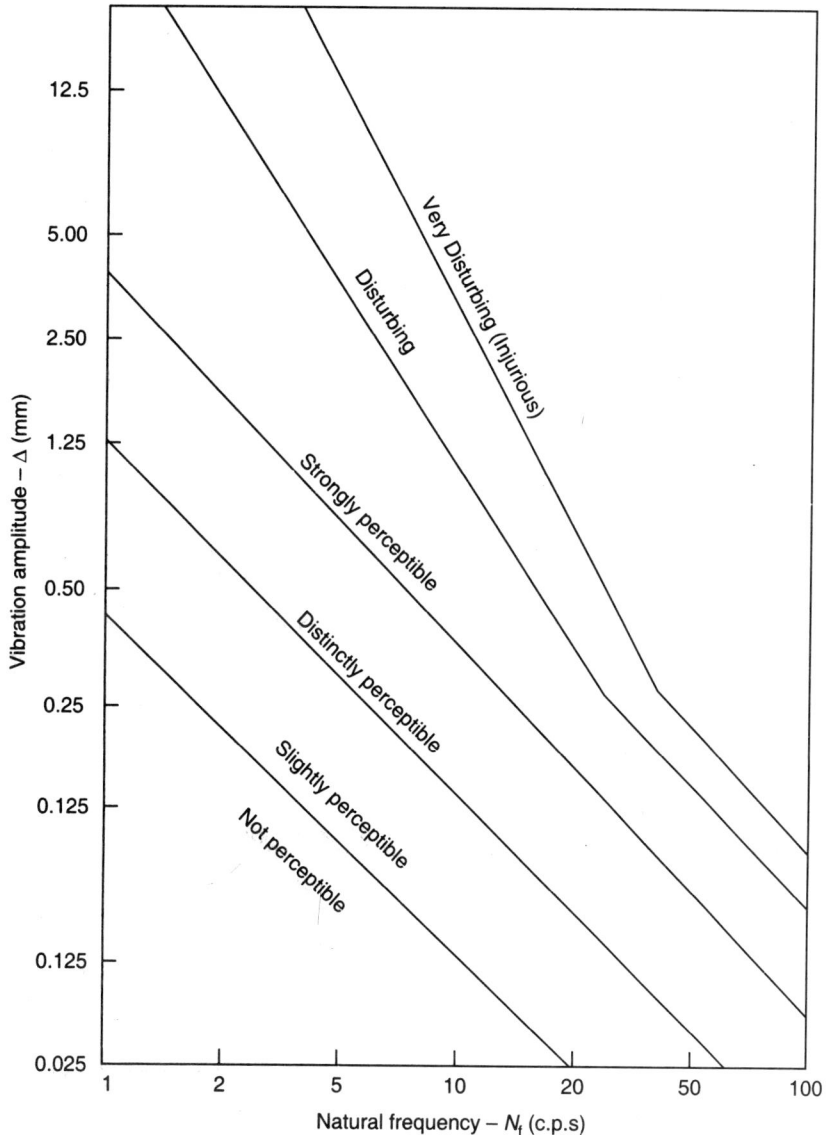

Fig. 21.1 Lenzen's Criteria

(3) Take $\quad \Delta = 0.40, \quad$ if $N_f > 4$ cycles per second

$\qquad \Delta = 0.75, \quad$ if $N_f < 4$ cycles per second

where $\quad \Delta$ = maximum amplitude of vibration (mm) and deflection δ expressed in mm.

(4) Estimate the maximum acceleration A from the equation

$$A = 40 \, \Delta \, N_f^2 \text{ mm/sec}^2$$

(5) Ensure that the product

$$A\Delta \not> 3226 \text{ mm}^2/\text{sec}^2$$

and that the vibration characteristic from Lenzen's criteria (from known value of N_f and Δ) is within desirable limits.

21.5 NUMERICAL EXAMPLES

1. A reinforced concrete slab deck for a culvert of effective span 6.5 m is 500 mm thick with two Lane road way 7.5 m wide and foot paths 1 m wide. Height of kerb = 300 mm, wearing coat is 100 mm thick with asphalt concrete. If M-20 Grade concrete is used in the deck slab, compute the natural frequency of the slab deck and check for the safety of the deck against failure due to dynamic effects.

   (1) Data: Effective span = 6.5 m

   M-20 Grade Concrete

   $$E = 5700\sqrt{f_{ck}} = 5700\sqrt{20} = 25490 \text{ N/mm}^2$$
   $$= 25.49 \times 10^6 \text{ kN·m}^2$$

   $$I = \left(\frac{9.5 \times 0.5^3}{12}\right) = 0.0989 \text{ m}^4$$

   Flexural rigidity = EI = (25.49 × 10⁶) (0.0989) = 2.52 × 10⁻⁶ kN·m²

   (2) Maximum Deflection:

   $$\delta = \left(\frac{WL^3}{48EI}\right) = \left(\frac{200 \times 6.5^3}{48 \times 2.52 \times 10^6}\right) = 454 \times 10^{-6} \text{ kN·m}^2$$

   (3) Natural Frequency of Vibration:

   $$N_f = \frac{2}{L^2}\sqrt{\frac{EIg}{w_d}} \text{ cycles per second}$$

Self weight = w_d		
Weight of deck slab = (9.5 × 0.5 × 24)	=	114 kN/m
Weight of foot paths = (2 × 0.3 × 0.1 × 24)	=	14.4 kN/m
Weight of wearing coat = (0.1 × 7.5 × 22)	=	16.5 kN/m
Weight of Parapet, railing etc. (Lumpsum)	=	5.1 kN/m
Total dead weight = w_d	=	150 kN/m

   $$N_f = \frac{2}{6.5^2}\sqrt{\frac{2.25 \times 10^6 \times 9.81}{150}}$$

   $$= 19.2 \text{ cycles per second}$$

Since $N_f > 4$ cycles per second

$$\Delta = 0.40\ \delta = (0.40 \times 0.454) = 0.1816\ \text{mm}$$
$$A = 40\ \Delta N_f^2 = (40 \times 0.1816 \times 19.2^2) = 2678\ \text{mm/sec}^2$$
$$A = (2678 \times 0.1816) = 486 < 3226\ \text{mm}^2/\text{sec}^2\ \text{(Hence safe)}$$

Compared with Lenzen's Criteria the vibration characteristic lies between the zone of distinctly to strongly perceptible.

(2) A reinforced concrete Tee beam and slab deck of effective span 16 m has a cross section with 3 longitudinal girders as shown in Fig. 21.2. M-20 Grade concrete is used for casting the slab and girders. Estimate the natural frequency of vibration of the Tee beam and slab deck and check for the safety of the deck against failure due to dynamic effects.

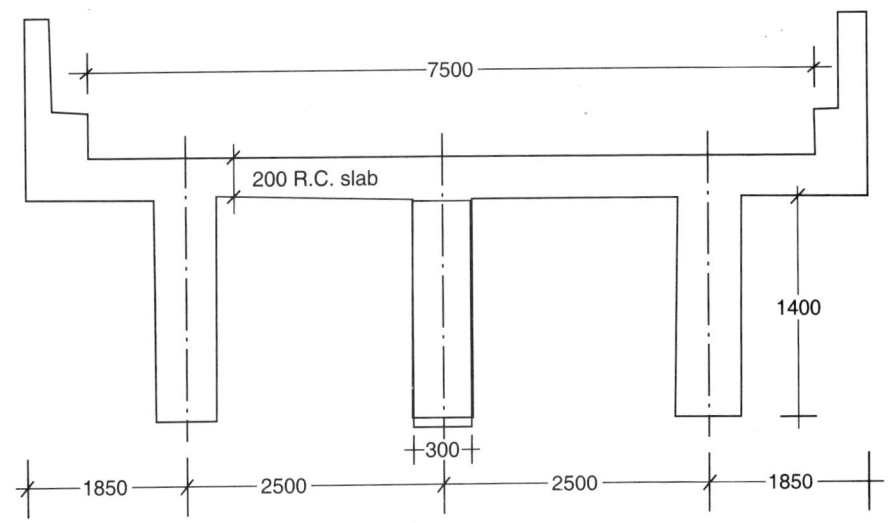

Fig. 21.2 Cross Section of Tee beam and Slab Deck.

(a) Data: Effective span = 16 m

M-20 Grade Concrete

$$E = 5700\sqrt{f_{ck}} = 5700\sqrt{20} = 25490\ \text{N/mm}^2$$
$$= 25.49 \times 10^6\ \text{kN/m}^2$$

(b) Sectional Properties: The cross section of Tee beam is shown in Fig. 21.3

The second moment of area = $I = 0.2162\ \text{m}^4$

For 3 girders, Effective $I = (3 \times 0.2162) = 0.6486\ \text{m}^4$

Flexural Rigidity[7] = $EI = (25.49 \times 10^6)\ (0.6486)$
$$= 16.53 \times 10^6\ \text{kN·m}^2$$

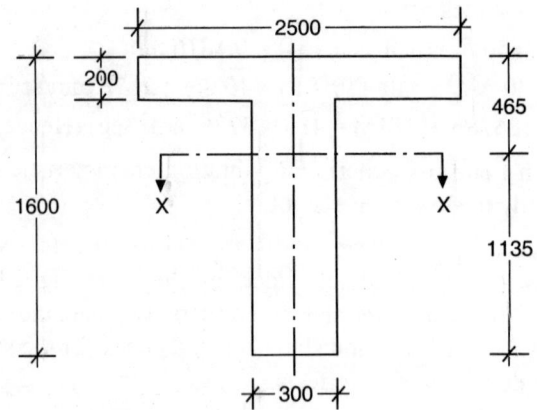

Fig. 21.3 Cross Section of Tee beam.

(c) Maximum Deflection:

Maximum deflection at centre of span under a hypothetical concentrated load of 200 kN is computed as

$$\delta = \left(\frac{200 \times 16^3}{48 \times 16.53 \times 10^6} \right) = 0.001032 \text{ m} = 1.032 \text{ mm}$$

(d) Dead Load of Deck:

Self weight of slab =	$(0.2 \times 1 \times 24 \times 8.7)$	= 41.76 kN/m
Weight of wearing coat =	$(0.08 \times 8.7 \times 22)$	= 15.31 kN/m
Weight of 3 girders =	$(3 \times 0.3 \times 1.4 \times 24)$	= 30.00 kN/m
Weight of kerb, parapet railing etc. (L.S.)		= 12.93 kN/m
Total dead load = w_d		= 100.00 kN/m

(e) Natural Frequency of Vibration:

$$N_f = \frac{2}{L^2} \sqrt{\frac{EIg}{w_d}}$$

$$= \frac{2}{16^2} \sqrt{\frac{16.53 \times 10^6 \times 9.81}{100}} = 9.94 \text{ cycles per second}$$

(f) Vibration Amplitude:

Since $N_f > 4$ cycles per second

$$\Delta = 0.40 \ \delta = (0.40 \times 1.032) = 0.4128 \text{ mm}$$

(g) Check for Dynamic Response:

$$A = 40 \ \Delta N_f^2 \text{ mm}^2/\text{sec}^2$$
$$= (40 \times 0.4128 \times 9.94^2)$$
$$= 1631 \text{ mm}^2/\text{sec}^2 < 3226 \text{ mm}^2/\text{sec}^2 \text{ (Hence safe)}$$

(h) Comparison with Lenzen's Criteria:

Compared with Lenzen's Criteria (Fig. 21.1) the vibration characteristic lies between the zone of distinctly to strongly perceptible.

21.6 CODAL PROVISIONS FOR DYNAMIC EFFECTS

The various codal provisions which correlate indirectly with the dynamic response of bridge decks are those pertaining to

(1) Impact factors

(2) Limitation of the ratio of deflection to span

(3) Restriction in the span/depth ratios.

These provisions have been evolved with time; based on a quantitative analysis of the bridge behaviour when subjected to live loads i.e. after accounting for the increase in the static values due to the dynamic effect.

I.R.C. codes specify different impact factors for steel and concrete bridge decks depending upon the type of loading.

For IRC-Class A or B loading, the impact factor is given by

$$I = \left(\frac{A}{B + L} \right)$$

where

I = Impact factor fraction

A = Constant having a value of 4.5 for R.C. bridges and 9 for steel bridges

B = Constant having a value of 6 for R.C. bridges and 13.5 for steel bridges

L = Span of the bridge deck in meters

For span less than 3 m, Impact factor = 0.5 for R.C. bridges

= 0.545 for steel bridges

For IRC-Class AA or 70R[8] loading, the impact factors are as follows:

(1) For spans less than 9 m

(a) Tracked vehicle: 25 per cent for spans up to 5m linearly reduced to 10 percent for span of 9 m.

(b) Wheeled vehicle: 25 percent

(2) For spans greater than 9 m

(a) Tracked vehicle: R.C. bridges, 10 percent up to span of 40 m and in accordance with Fig. 2.19 for spans exceeding 40 m

(b) Wheeled vehicle: As detailed in Fig. 2.19.

EXAMPLES FOR PRACTICE

1. A reinforced concrete slab deck for a culvert of effective span 5 m is 500 mm thick with a two lane road way 7.5 m wide with 600 mm kerbs on either side. Wearing coat is 80 mm thick. If M-20 Grade concrete is used in the deck slab, compute the natural frequency of the slab deck and check for the safety of the deck against failure due to dynamic effects.

2. A four lane R.C.C. slab deck with a total width of 17 m, spans of 6 m and the thickness of the slab is 500 mm. Thickness of wearing coat is 80 mm. If M-25 Grade concrete is used in the deck slab, estimate the natural frequency of vibration of slab deck and check for the safety of deck against dynamic failure.

3. A reinforced concrete Tee beam and slab deck of effective span 12 m comprises of three longitudinal girders spaced 2.5 m apart.

   Thickness of slab = 250 mm, Thickness of wearing coat = 100 mm
   Overall depth of girders = 1400 mm, Width of Roadway = 7.5 m

   Kerbs 600 mm wide on either side. M-20 Grade concrete is used for casting the slab and girders. Estimate the natural frequency of vibration of the Tee beam and slab deck and check for safety of deck against failure due to dynamic effects.

4. A.R.C.C. Tee beam slab deck has the following details:

   Effective span = 16 m. Road width = 7.5 m, Width of kerb = 600 mm
   Number of longitudinal girders = 4, Spacings of girders = 2 m
   Thickness of slab = 200 mm
   Overall depth of girders = 1600 mm

   The girders are reinforced with 12 bars of 36 diameter at an effective depth of 1450 mm. M-20 grade concrete and Fe-415 grade HYSD bars are used. Compute the natural frequency of vibration of the Tee beam and slab deck by considering the effective second moment of area of cross section including the effect of reinforcements in the girders. Also check for the safety of the deck using Lenzen's criteria.

REFERENCES

1. Johnson Victor, D., Essentials of Bridge Engineering (Fifth Edition), Oxford & IBH Publishing Co, Pvt, Ltd, New Delhi, 2001, pp. 1–431.
2. Raina, V.K., Concrete Bridge Practice, Analysis, Design and Economics, Tata McGraw-Hill Publishing Co, Ltd, New Delhi, 1991, 531–536.
3. Clough, R.W and Penzien, J., Dynamics of Structures, McGraw-Hill Publishers, New York, 1975, pp. 1-635.
4. Bridge Design and Euro Codes, Workshop on Bridge Design to Euro Codes, Vienna, October 2010, pp. 1–283.
5. Chopra, A.K., Dynamics of Structures, Prentice Hall International, 2015, pp. 1-984.

6. Ashish Gupta, Aman Deep Singh Ahuja., Dynamic Analysis of Railway Bridges Under High Speed Trains, Universal Journal of mechanical Engineering, 2(6), 2014, pp. 199–204.
7. Krishna Raju, N., Prestressed Concrete (Fifth Edition), McGraw-Hill Education, New Delhi, 2012, pp. 151–189.
8. IRC: 6-2014, Standard Specifications and Code of Practice for Road Bridges, Section-II' Loads and Stresses (Revised Edition), Indian Roads Congress, New Delhi, 2014, pp. 1-84.

REVIEW QUESTIONS

1. What are the effects of vibrations in bridge decks? What is the simplest method of resisting dynamic effects in bridge decks?
2. Mention the simple method incorporated in the codes to safe guard bridges from the destructive effects of vibrations.
3. What are the various factors influencing bridge vibrations?
4. Briefly explain the dynamic response of bridge decks to moving loads. What is the normal range of fundamental frequency of bridges?
5. What practical approach you would follow for the vibration analysis of a bridge deck?
6. Explain briefly the Lenzen's Criteria and its use in checking reinforced concrete bridge decks for safety against failure due to vibrations.
7. What are the various factors influencing the natural frequency of vibration of a tee beam and slab bridge deck?
8. How do you check for the dynamic response of a tee beam slab deck?
9. What provisions of the Indian standard codes correlate indirectly with the dynamic response of bridge decks?
10. In what way the flexural rigidity of the cross section of a bridge deck influences its dynamic response?

OBJECTIVE TYPE QUESTIONS

1. Dynamic analysis of bridge decks is absolutely necessary in the case of bridges located in
   a) Normal Atmospheric conditions
   b) Freezing atmospheric conditions
   c) Coastal aggressive windy zones
2. Dynamic effects are safely resisted in bridges designed to have
   a) More mass and rigidity
   b) Optimum weight and flexural rigidity
   c) Pinned joints

3. In the case of bridges having a high ratio of vehicle to bridge deck weight, damage due to vibrations are
   a) Not likely
   b) Very rare
   c) Most likely

4. The fundamental frequency of the bridge deck is inversely proportional to the
   a) Flexural rigidity of the bridge deck
   b) Square of the span length
   c) Acceleration due to gravity

5. Failure due to dynamic effects are generally rare in
   a) Concrete tee beam and slab bridge decks
   b) Steel suspension bridges
   c) Reinforced concrete rigid frame bridges

6. According to Lenzen's criteria, in a bridge deck, the dynamic effects are strongly perceptible if the vibration amplitude exceeds
   a) 0.125 mm
   b) 0.500 mm
   c) 4.000 mm

7. Bridge structures can be considered as safe if they are designed according to the
   a) The working stress method
   b) Limit state method
   c) Ultimate load method

8. The normal range of fundamental frequency of most of the bridges varies between
   a) 1 to 5 cycles/sec
   b) 20 to 30 cycles/sec
   c) 1 to 20 cycles /sec

9. To safe guard the bridge structure from dynamic effects, the IRC code has introduced the provision pertaining to
   a) Stresses in materials
   b) Impact factors
   c) Minimum reinforcements

10. The indirect codal method of safeguarding a bridge structure from dynamic effects is to restrict the instantaneous deflection of the deck due to live loads only to a value of
    a) Span/300
    b) Span/800
    c) Span/400

Planning and Economical Aspects of Bridges

22

22.1 INTRODUCTION

The logical steps to be followed before starting the construction of any bridge structure are invariably the process of comprehensive planning, detailed analysis and critical appraisal of alternative designs. From ancient times, construction of a structure has always been one of the most fascinating challenges to man ingenuity. Planning involves several known and unknown features such as management of materials and labour, mobilization of suitable cost effective techniques, treacherous foundation problems, adverse environmental conditions, scheduling of the construction process to a time bound frame, constant interaction with the design engineer, architect, site engineer, construction workers and the ability to take sound and daring decisions at times of crisis.

Developing countries like India to some extent lack the economic and social infrastructure which is always important in the process of planning. Many of the Public sector projects in India are planned according to the whims and fancies of the ruling politicians and bureaucrats, without a careful evaluation of the local needs of the population and ground realities. According to Raina[1], "Bridge Engineering is not just solving theoretical problems nor is it a matter of blind adherence to graphs, design charts and formulas. It is more meaningful to have an approximate solution to an exact problem rather than an exact solution to an approximate problem. Practical engineers must be more conceptual than mere perceptual, more creative than mere analytical and more visual than mere mathematical. Bridge engineers should have wide experience involving several types of bridge structural configurations rather than isolated narrow specialization. Expertise and original skills are attained from relentless study, research, understanding and actual practice rather than mere theoretical knowledge. Good and sound judgment are attained from wide practical experience and often experience comes from bad judgments".

22.2 ECONOMICAL SPAN RANGES AND CONFIGURATIONS FOR BRIDGE DECKS

The evolution of the super structure forming the deck portion of a bridge has seen revolutionary changes with the development and use of new materials and forms to keep pace with the ever increasing demand for covering longer spans. Materials like timber and iron have been replaced by new materials like high strength and high performance concrete, reinforced concrete, prestressed concrete, high yield strength deformed reinforcements, high tensile steel wires, bars, strands cables and polymeric materials.

The development of the revolutionary material "Prestressed Concrete" by Eugene Freyssinet[2] facilitated the rapid development and construction of innumerable number of bridges in Europe destroyed in World War-II. French, German and American engineers contributed immensely through research and practice for the widespread use of this material for medium and long span bridges.

The structural forms of various types of bridges and their respective economical span ranges are compiled in Table 22.1

Table 22.1 Economical Spans for Various Types of Bridges.

Span Range	Type of Bridge
4 m – 7 m	Reinforced Concrete Slab Culvert
7 m – 12 m	Prestressed Concrete Slab
15 m – 25 m	Reinforced Concrete Tee beam and Slab
	Reinforced Concrete Rigid Frame
25 m – 60 m	Prestressed Concrete Beam and Slab Steel Plate Girder
50 m – 100 m	Balanced Cantilever type, Reinforced or Prestressed Concrete
75 m – 150 m	Prestressed Concrete Box Girders using Cantilevered Construction, Steel Truss
150 m – 500 m	Cable Stayed bridges with Reinforced or Prestressed Concrete decks
500 m – 1800 m	Steel Suspension Bridges or Cable Stayed Bridges with Steel decks

22.3 COMPARATIVE COST ANALYSIS OF DIFFERENT TYPES OF BRIDGE DECKS

A comparative analysis of several types of designs using the available materials together with the cost of construction utilizing the locally available labour will lead to an economical and suitable design for a particular set of conditions. However, in general, the quantities of concrete and steel reinforcement used per unit area of bridge deck can be considered as indicative of the degree of economic cost although these figures are not the only factors which govern the overall cost of the bridge. The significant factors influencing the overall cost of the bridge are compiled in the following list:

1. The total width of bridge deck

2. The length of individual spans and overall length of the bridge

3. The number of longitudinal and cross girders

4. The type and class of live loads supported by the deck

5. The depth and type of foundations, excavations etc

6. The cost of form and false work

7. The cost of materials and labour

8. The type of construction such as precast or cast *in-situ*

9. The method of erection of precast segments

10. The time constraint for completion of the bridge project

Raina[1] has compiled the quantities of concrete and steel reinforcement required per unit area of different types of bridge decks with spans ranging from 35 to 140 m, based on comprehensive analysis and extensive practical experience. The collected data is shown in Fig. 22.1. For spans less than 35 m, simply supported spans are the cheapest form of construction. For spans exceeding 60 m, simply supported spans are prohibitively costly.

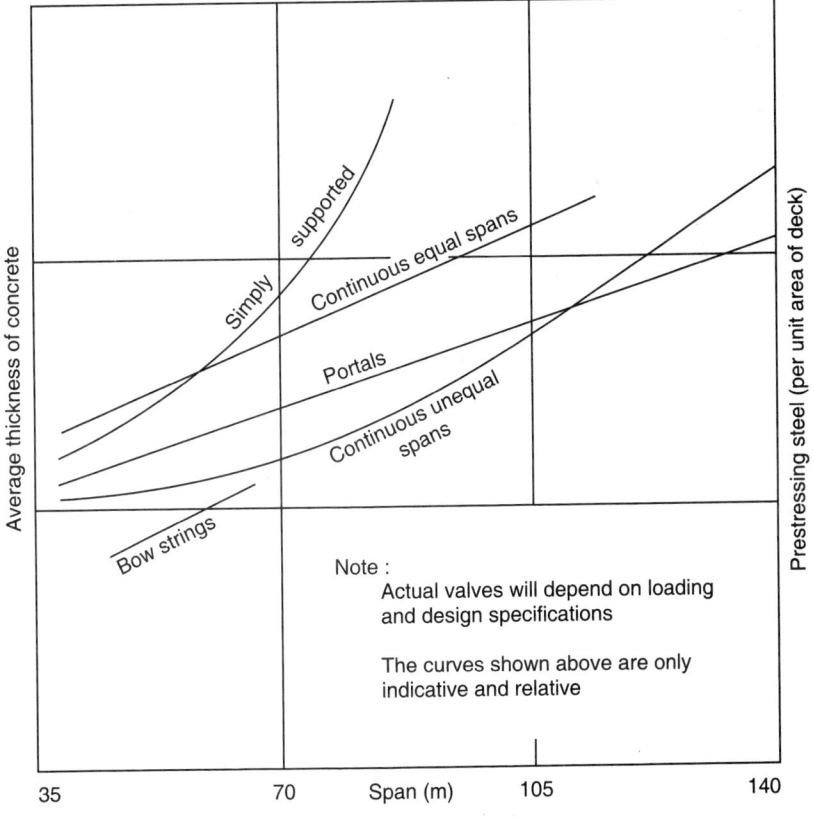

Fig. 22.1 Approximate Quantities of Concrete and Steel in Various Types of Bridge Decks.

Continuous bridge decks with unequal spans are more economical than equal spans in the span range of 50 to 100 m. Rigid Portal Frame design also uses the same quantity of materials as those of continuous girders with unequal spans. For spans less than 90 m, Portal frames are slightly more expensive than continuous beams of unequal spans. However for very long spans especially in crossing deep ravines, reinforced concrete arched bridges are more economical. Concrete arches are seldom prestressed since concrete is strong in compression and can easily resist the thrusts developed an arch. Bow string girders are economical, considering the material quantities for spans up to 50 m. The dimensions of the Tee beam of the bow string girder bridge can be considerably reduced by axial prestressing resulting in overall economy. Although the Bow string girder bridge is aesthetically superior to other types of girder bridges, the cost of form work being significantly higher, at present this type is rarely preferred.

The cross sections of various types of prestressed concrete bridge deck configurations are shown in Fig. 22.2. Cast *in-situ* post tensioned voided slabs are preferred for spans up to 40 m, either in simple or continuous spans. Span/depth ratios as high as 40 have been used resulting in high torsional resistance rendering it highly suitable for curved alignment with the bridge deck supported by single central columns for arterial highways. Precast Tee, I and Box shaped girders with cast *in-situ* slab are suitable for spans up to 50 m. The span/depth ratios vary from 18 to 20 for simply supported spans to 25 to 30 for continuous spans. Raina[3] has presented graphically the variation in the quantities of concrete and high tensile steel with average thickness of concrete deck as a function of span for different types of structural configurations used in Fig. 22.3. Generally the box sections are the most economical types requiring the least concrete content and prestressing steel compared to a voided slab for a given depth of bridge deck. A comparative analysis by Sarkar et al[4] has clearly established that prestressed concrete box girder decks on an average needed only 55 percent of the depth of composite I section decks when used in urban flyovers and autobahns.

A critical survey of the various types of bridges built during the last three decades clearly indicates the economical advantages of cable stayed bridges especially for long spans. Prof. Leonhardt[5] designed the first cable stayed bridge across the river Rhine in Dusseldorf in 1952. During the last two decades, hundreds of cable stayed bridges have been built in different countries with spans ranging from 100 to 624 m using concrete and steel and the longest cable stayed bridge with a span of 1800 m is the Messina Straights bridge with a steel deck located in Italy. The reader may refer to the introductory article by Subba Rao[6,7] and Krishna Raju[8,9] for a critical survey of concrete bridges in India.

22.4 LOW COST PEDESTRIAN SUSPENSION BRIDGES

An excellent example of appropriate technology suited to rural India is the small suspension bridge planned and constructed using locally available materials for pedestrian crossings of small rivers in the Western Ghats of south and north Kanara districts. Girish Bharadwaj[10] has built more than 50 low cost suspension bridges to help villagers to cross the rivers using locally available materials like steel ropes, rods, wooden and concrete planks. In many cases, existing trees on either side of the stream were used as pylons for the suspension bridge. In cases where trees are not available, concrete towers were cast at site over which steel ropes are drawn and anchored at the ends with steel rods suspended and supporting the deck made up of wooden or precast concrete planks. Girish Bharadwaj's first suspension bridge was 55 m long built in 1989 at Kushalnagar, Kodagu, Karnataka state to cross the Cauvery River.

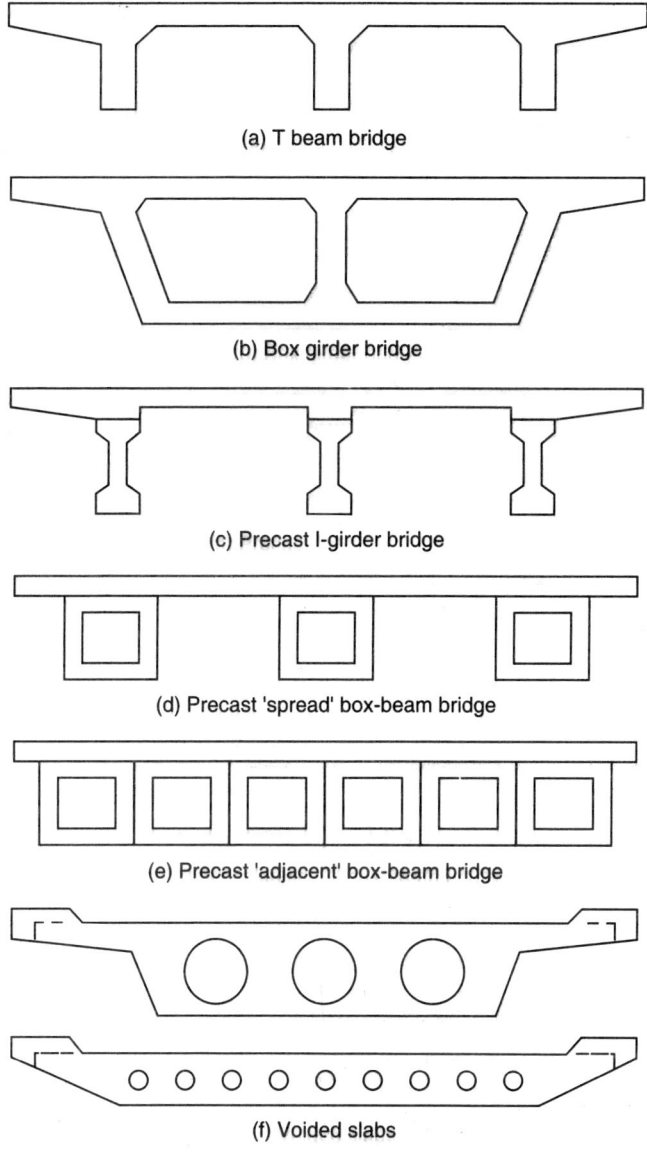

(a) T beam bridge

(b) Box girder bridge

(c) Precast I-girder bridge

(d) Precast 'spread' box-beam bridge

(e) Precast 'adjacent' box-beam bridge

(f) Voided slabs

Fig. 22.2 Typical Cross Section of Prestressed Concrete Bridge Decks.

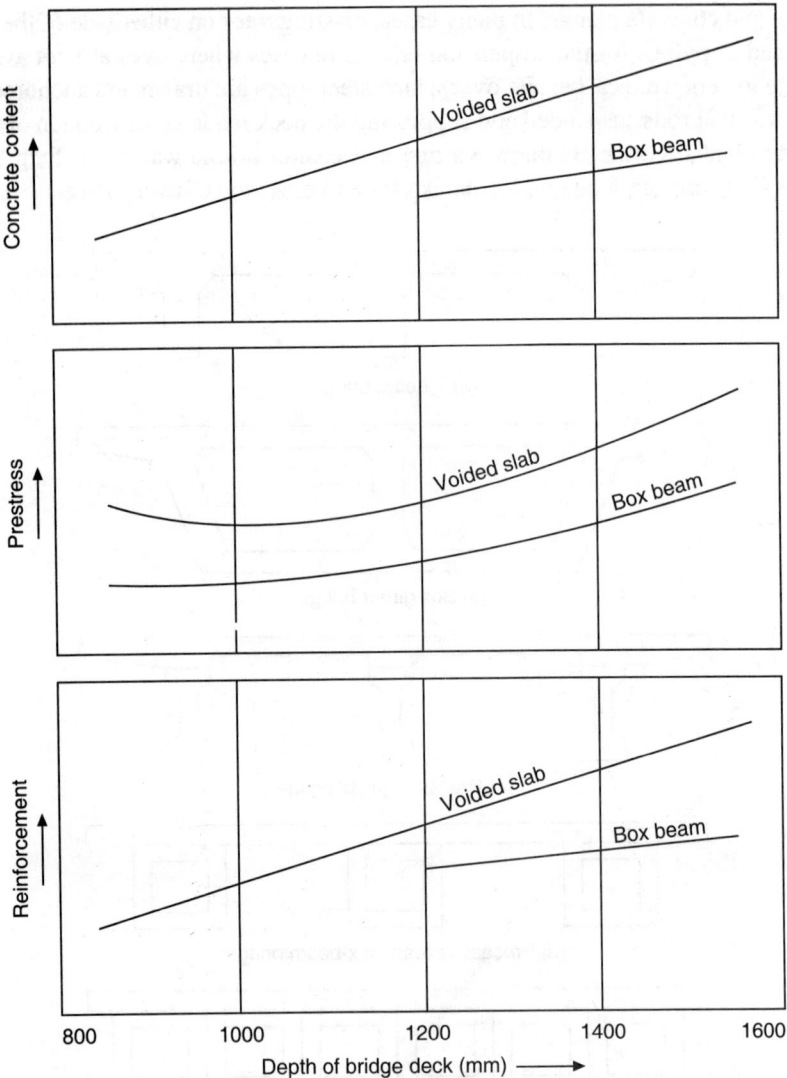

Fig. 22.3 Variation of Concrete Content, Prestress and Reinforcement with Depth of Bridge Deck.

The 87 m long suspension bridge to cross the Payasvini River was built by Girish Bharadwaj using two industrial cables hung from reinforced cement concrete pylons anchored to the ground. Steel rods formed the suspenders to support the pedestrian deck made up of wooden planks. The total cost of the bridge surprisingly was just one lakh of rupees which was raised by the Panchayat of the local community. In a suspension bridge near Nisargadhama in Kodagu district, he has used huge trees on the banks of the river to serve as pillars in place of costly reinforced concrete pylons with considerable reduction in the cost of the bridge. In a bridge near Shisila village in Belthngady, he has used second hand steel wire ropes from an abondened ship and employed the local villagers to construct the R.C.C pylons.

Out of the 52 suspension bridges built by Girish Bharadwaj, 31 are located in Karnataka State, 20 in Kerala State and one in Andhra Pradesh. The 288 m long suspension bridge built across the river Ghataprabha near Hukkeri, Belgaum district in Karnataka state is the longest and the 20 m bridge built near Sullya across Kootelu river is the smallest bridge. Fig. 22.4 shows a typical suspension bridge for pedestrian crossing constructed by him. According to Girish, "Trifles make perfection" but "Perfection is not a trifle". Wise words from some one who has lived his life perfecting his craft of low cost bridge construction for the village folk in the western Ghats and practicing appropriate technology suited to underdeveloped and developing countries with scarce resources.

Fig. 22.4 Low Cost Suspension Bridge for Pedestrian Crossing.

REFERENCES

1. Raina, V.K., Concrete Bridge Practice, Analysis, Design and Economics, Tata McGraw Hill Publishing Co. Ltd., New Delhi, 1991, pp. 1–680.
2. Freyssinet, E., The Birth of Prestressing, Cement & Concrete Association Translation No. CJ-59, London, 1956, pp. 1–44.
3. Raina, V.K., Concrete Bridge Practice, Construction, Maintenance & Rehabilitation, Tata McGraw Hill Publishing Co. Ltd., New Delhi, 1988, Second reprint 1993, pp. 87–111.
4. Sarkar, S, Kapla, M.S, Prasada Rao, A.S and Chhauda, J.N., Hand Book of Prestressed Concrete Bridges, Structural Engineering Research Centre, Roorke, U.P, 1969, pp. 226–230.
5. Leonhardt, F., Cable Stayed Bridges with Prestressed Concrete, Prestressed Concrete Institute Journal, September-October 1987 (Special Report), pp. 52–79.
6. Subba Rao, T.N., Application of Prestressed Concrete in India, Outstanding Concrete Structures of India, Compiled by the Maharashtra India Chapter of American Concrete Institute's Fall Convention at New York, 1984, pp. 1–5.
7. Subba Rao, T.N., Long Span Prestressed Concrete Bridges in India, Seminar on Problems of Prestressing, Preliminary Publication, Madras, 1970, pp. 1–113 to 1–130.
8. Krishna Raju, N., Advances in Design & Construction of Concrete Bridges, Construction India Manual, Bombay, Vol. 134, pp. 50–53.
9. Krishna Raju, N., Developments in Design & Construction of Concrete Bridges, Civil Engineering and Construction, Special Publication on Bridges, Vol. 6, No. 7 July 1993, pp. 12–15.
10. Bhanutez, N., The Bridges Built by Girish Bharadwaj that Saves Lives, The Week Magazine, March 28, 2004, pp. 16–17.

Construction Management of Bridge Decks

23

23.1 INTRODUCTION

Construction of a major bridge structure comprises multifarious functions like planning of labour, materials, machinery, finance and constant supervision to complete the bridge project within a specified target period utilizing optimally the available resources. The modern approach in construction management involves several diverse functionaries like structural designers, estimators, construction engineers, field supervisors, financial managers, corporate secretaries and tax planners working under professional managers.

According to Raina[1], construction management functions include the following major activities:

1. Tendering and winning the contract for the bridge project
2. Contract negotiations
3. Developing liaison with clients
4. Mobilizing financial resources for the project
5. Work planning
6. Supervision of work
7. Controlling and monitoring the progress of the project
8. Maintenance of good labour relations
9. Maintenance of proper accounts
10. Engineering and completion of the project

Major project controlling and monitoring techniques like Critical path method (CPM) and Project Evaluation and Review Techniques (PERT) are widely used. With advent of high speed computers, data processing, preparation of working drawings, work scheduling, materials and labour management, controlling the various activities of the project and updating tasks have been simplified and they can be efficiently handled with less paper work. In addition to these traditional management tools, optimization techniques[2,3] involving linear and non-linear programming have also contributed to optimal solutions with considerable savings in overall cost of the bridge project.

23.2 CONSTRUCTION MATERIALS FOR BRIDGE DECKS

The primary materials widely used for the construction of modern bridges comprise of High performance concrete, High strength concrete, structural steel like HYSD bars and high tensile wires, bars, strands and cables for prestressed concrete bridge decks. For steel bridges which are nor generally preferred to concrete due to prohibitive maintenance costs and rapid deterioration due to adverse environmental conditions, rolled standard steel sections conforming to the National Codes are easily available from factory outlets.

High performance concrete grade for reinforced concrete bridges according to the IRC: 112–2011 code[4] is specified as M-30 and M-35 for moderate and severe exposure conditions respectively when used with prestressed concrete or those with length exceeding 60 m. For culverts and other minor constructions, the grade of concrete can be reduced to M-20 and M-25 for moderate and severe exposure conditions. For prestressed concrete members the minimum grade of concrete specified according to the IS: 1343-2012[5] code is M-35 and M-40 for moderate and severe exposure conditions. The code also specifies that only 'Design Mix Concrete' should be used to ensure the desired strength and durability. The reader may refer to a separate monograph by the author[6] for detailed information regarding the design of concrete mixes for specified strength, workability and durability using the locally available cements, fine and coarse aggregates.

Weigh batching and machine mixing are essential to produce high performance concrete of uniform quality. Several types of admixtures are available to improve the workability of concrete without increasing the water/cement ratio which adversely affects the strength of concrete. Bridge deck concrete deposited in slabs and beams should be compacted by mechanical vibration using vibrators with frequencies in the range of at least 3200 to 3600 cycles per minute which ensures good compaction.

23.3 RHEODYNAMIC CONCRETE

Recent advances in the field of Polymer technology and Nano Science has made inroads in the field of concrete technology. Rheodynamic concrete generally referred to as self compacting concrete (SCC) is able to flow under its own weight and completely fill the form work, even in the presence of dense reinforcements without the need for any vibration whilst maintaining homogeneity and resulting in concrete of high early strength and durability. Degussa-MBT Construction Chemicals (India)[7] have developed innovative type of admixtures using Nano polymers which can be used to bring together functional groups aimed at targeted performances in concrete. Based on Nano science, they have developed a range of Nano polymers with the following applications:

1. **Zero Energy System:** A system of polymers with longer side chains and shorter main chains to facilitate high early strength in concrete without steam curing

and with specific applications in precast reinforced and prestressed concrete unit manufacturing industry.

2. **Glenium Sky:** A custom made Nano polymer which facilitates long haul concrete mix stability with development of high early strength coupled with high durability of hardened concrete suitable for prestressed concrete flyovers.

3. **Rheo Fit:** A Nano polymer range which meets wider expectations such as aesthetics, economics, durability and performance of manufactured concrete products.

The application of Nano Technology in the production of Nano Polymers has revolutionized the concrete industry. Self compacting concrete is commonly used in Europe, Japan, North America, Singapore and Taiwan. In India, this technique has gained popularity with some of the major projects such as Delhi Metro Rail Corporation, Nuclear Power Corporation and Indian Space Research Organization.

23.4 LONG SPAN BRIDGE CONSTRUCTION TECHNIQUES

New construction techniques rapidly developed in America, Russia, Europe and even India during the last several decades has facilitated the construction of long span cellular box girder bridges with spans in the range 100 to 300 m. Morandi's Bridge across the lake of Maricabo with a span of 235 m was constructed by the help of stay cables. In 1970, the longest span in Japan exceeded 300 m for cable stayed bridges. Cantilever construction techniques developed by Ulrich Finisterwalder[8] revolutionized the prestressed concrete bridge construction by eliminating ground supporting form work for long span bridges. The span range for economical application of box beam bridges built by cantilever construction technique is believed to lie normally in the range of 50-200 m. However Hamana Bridge[9] in Japan has pushed this upper limit further with a span of 240 m and Urado Bay Bridge in Japan has a span of 270 m built by using the technique of cantilever construction.

23.4.1 Cantilever Construction

The universally adopted cantilever construction for long span bridges can be used for cast *in-situ* work or precast structural elements. Cantilever construction is a novel method of progressively constructing a cantilever in segments and stitching them to the previously completed segments by prestressing. The cantilever segmental construction generally starts from a pier, extending on both sides. This method eliminates the use of expensive form work and scaffolding especially for bridges in deep valleys and rivers with large depth of water resulting in faster rate of work progress coupled with overall economy. Basically the cantilever construction technique is classified under the following two groups:

(1) Cast *in-situ* Construction

(2) Construction using Precast Segments

(1) Cast *in-situ* Construction

In the cast *in-situ* method, the piers of the bridge are first constructed and the bridge deck is cast *in-situ* with units of 2.5 to 3 m length cantilevering symmetrically on either side of the pier. The form work for cast *in-situ* construction is supported by steel frame work attached to the completed part of the bridge deck and the form work is moved progressively from one completed section to the next part. The *in-situ* construction is done by a pair of traveling gantries, each weighing about 40 t for casting 2.5 to 3 m segments of a two lane bridge deck. The gantry systems proceed in a systematic manner from section to section on either side of the pier after the prestressing of the segment last cast. The gantry system also supports a suspended scaffolding for constructional convenience and workers safety.

The cast *in-situ* construction method generally involves a working cycle extending over 10 days with the following cycle of operations:

1. Shifting of travelling gantry system .... 1 day
2. Completion of entire shuttering, placing of reinforcements and prestress-ing cables for the segment .... 3 days
3. Casting concrete in the segment .... 1 day
4. Curing the segment .... 4 days
5. Prestressing operations and Miscellaneous works .... 1 day
   Total cycle time for completion of segmental unit .... 10 days

The working cycle can be reduced to 6 days from ten by resorting to more organized and mechanical infrastructure. Fig. 23.1 shows the cast *in-situ* cantilever method of construction of Zuari Bridge[1] in Karnataka, India having 122 m spans.

Fig. 23.1 Cast *in-situ* Cantilever Method of Construction of Zuari Bridge (122 m Spans) in Karnataka, India.

(2) Construction Using Precast Segments

When long span bridges have to be constructed using the cantilever construction, it is more advantageous and economical to adopt precast segmental construction in preference to the cast *in-situ* method. In this method, the bridge segments comprising single cell or multi cell box girders are match cast in a casting yard in the vicinity of the bridge site by using special forms and the cured units are transported to the work site. The precast segments are placed in position by means of a mobile launching girder or when access under the bridge is possible with barges or trucks, using a crane or a mobile hoist located at the extremity of the cantilevers.

The primary advantage of precast segmental construction is that the units can be cast on ground near the bridge site well in advance and the quality of units are superior to those in the cast *in-situ* method.

The fully cured segments can be transported to the site by means of heavy duty trucks and lifted by cranes to join them to the previously erected units by using temporary stressing cables. The rate of progress achieved in this method is faster compared to the cast *in-situ* method. In both the methods, a typical cross section would be a box girder of constant or variable depth to suit the longitudinal profile of the bridge deck.

Fig. 23.2 shows the bridge between Oleron Island and the continent in France built by the cantilever system of construction using precast segments. The bridge is made up of 26 main spans of 79 m each by using the single cell precast box girders. Mahatma Gandhi Sethu Bridge at Patna in Bihar was built by using single cell precast box girders using the segmental method of construction. In this case the precast box girder units were transported on a barge and hoisted by a crane to assemble the bridge deck as shown in Fig. 23.3.

Fig. 23.2 Precast Segmental Cantilever Construction of Bridge between Oleron Island and the Continent (France).

Fig. 23.3 Transportation and Hoisting of Single Cell Box Girders of Mahatma Gandhi Sethu at Patna.

23.4.2 The Staging Method

In the case of long viaduct structures with relatively shorter spans, the staging method of span by span construction of the bridge deck is particularly advantageous and economical. The superstructure is executed in one direction span by span by means of a supporting traveler. The form carrier in effect provides a type of factory operation transplanted to the job site. The traveller may be supported on the piers or from the edge of the previously constructed at the joint location and the forward pier. Fig. 23.4 shows a typical arrangement of the staging method. Many long bridges have been built in Germany, France and other countries using this technique. Long Key Bridge and Seven Miles bridge with spans of 36 and 40 m respectively were built in Florida Keys using this method. The staging method of span by span construction seems to have a great potential for trestle type structures in terms of speed of construction and economy.

23.4.3 Progressive Placement Construction Method

In the progressive placement method of construction, the precast elements are placed from one end of the structure to the other in successive cantilevers on the same side of the various piers, rather than by balanced cantilevers on each side of the pier. At present this method has been found to be practicable and economical in the span range of 30 to 90 m.

The main feature of this method comprises a moveable temporary stay arrangement to limit the cantilever stresses during construction to a reasonable level. The erection procedure is illustrated in Fig. 23.5. Precast segments are transported over the completed portion of the deck to the tip of the cantilever span under construction, where they are positioned by a swivel crane that proceeds from one segment to the next. The segments are held in position by temporary external ties and by two stays passing through a tower located over the preceding pier. The stays are anchored to the top flange of the box girder segments so that the tension in the stays can be adjusted by light jacks. This method was first used in France on several structures and later adopted in U.S.A for the construction of the Linn Cove Viaduct in North Carolina.

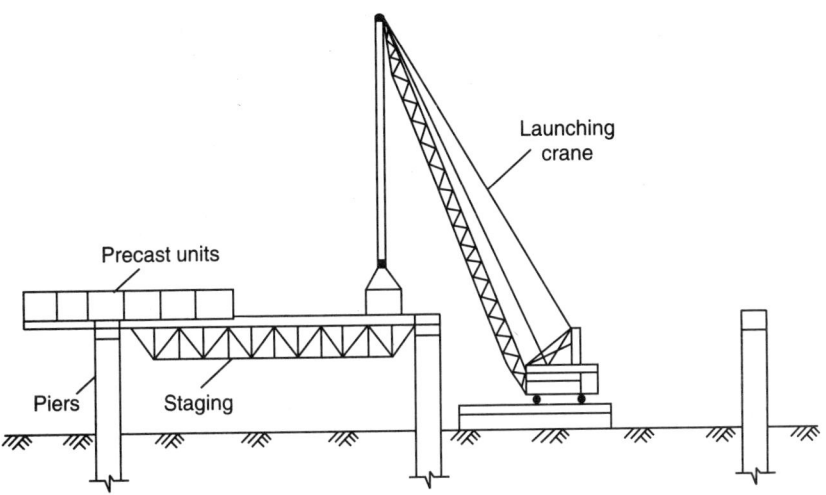

Fig. 23.4 Staging Method of Span by Span Assembly of Precast Segments.

23.4.4 Incremental Launching or Push out Technique

This method of construction was first developed by Willi Baur and Dr. Fritz Leonhardt[8] and used for the construction of Caroni Bridge in Venezuela in 1962. The precast segments of the bridge deck are cast near the site in lengths of 10 to 30 m in stationery forms located behind the abutments. Each unit is cast directly against the previous unit. After sufficient concrete strength is achieved, the new unit is post tensioned to the previous one by post tensioning. The assembly of units is pushed forward in a step wise manner to permit casting of the succeeding segments. Fig. 23.6 shows the incremental launching sequence of the push out technique in building the bridge deck.

Normally a work cycle of one week is required to cast and launch a segment regardless of its length. To allow the deck to move forward, special low friction sliding bearings are provided at the top of various piers with proper lateral guides. The main problem is to ensure the safety of stresses in the super structure under its own self weight during all stages of launching at various critical sections. To achieve this criterion, the first stage prestress is applied concentrically to the whole cross section and in successive increments over the entire length of the super structure.

A fabricated structural steel launching nose is attached to the lead segment to reduce the large negative bending moments developed in the front portion (particularly just before the super structure reaches a new pier). If the spans are large, they can be sub divided by temporary piers to control the magnitude of bending moments within safe limits. According to Raina[1] this method has been used for the construction of the bridge over River Danube near Worth, Germany having a ma .imum span length of 168 m. The incremental launching technique was also used for the fist time in U.S.A for the construction of Wabash River Bridge at Covington, Indiana.

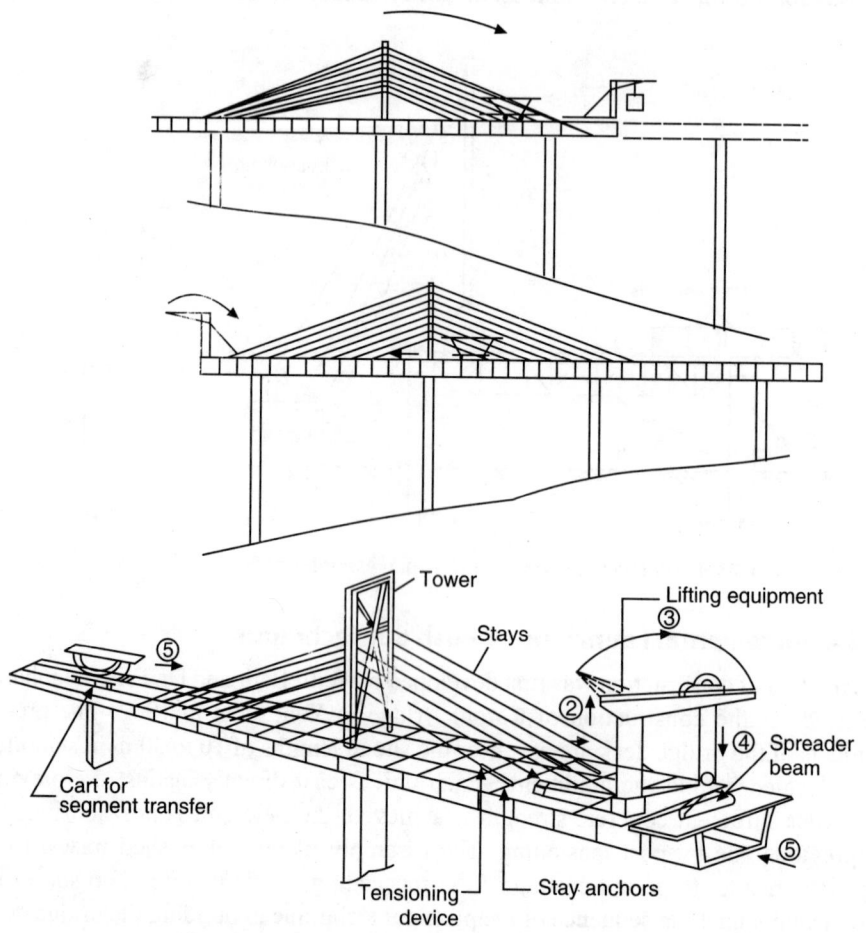

Fig. 23.5 Progressive Placement Construction Method.

Significant developments in the field of construction materials, techniques and management during the last several decades have showr major progress towards simplification and reduction of erection equipment coupled w... faster construction. The construction procedure must however be well planned using sequential computations for the alignment, forces exact lengths and angles considering temperature and creep influences which depend upon seasonal climatic and daily environmental conditions. The collective experience and knowledge gained through research and practice during last five decades will help in evolving innovative applications and management techniques in the domain of bridge construction.

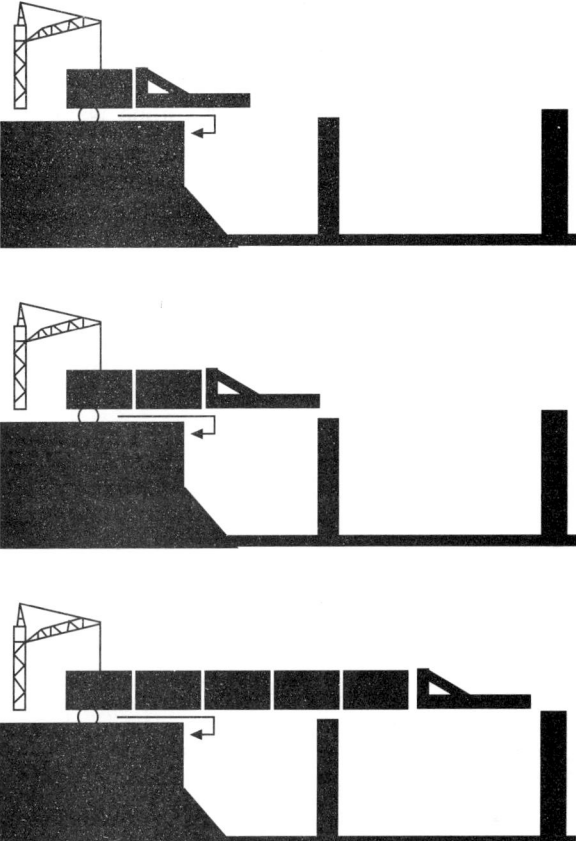

Fig. 23.6 Incremental Launching or Push-out Method of Construction.

REFERENCES

1. Raina, V.K., Concrete Bridge Practice, Construction, Maintenance and Rehabilitation Tata McGraw-Hill Publishing Co. Ltd., New Delhi, 1988, pp. 381–396.
2. Gallaghar, R.H and Zienkiewicz, O.C., Optimum Structural Design, Theory and Applications, John Wiley & Sons, New York, 1973, pp. 7–17.
3. Ramaswamy, G.S and Raman, N.V., Optimum Design of Prestressed Concrete Sections for minimum Cost by Non-Linear programming, F.I.P. VI Congress, Prague, June 1970, pp. 1–7.
4. IRC: 112-2011, Code of Practice for Concrete Road Bridges, Vol. 1, 2014, New Delhi. pp. 1–139.
5. IS: 1343-2012, Code of Practice for Prestressed Concrete (Second-Revision), New Delhi, 2012, pp. 1–54.

6. Krishna Raju, N., Design of Concrete Mixes, (Fourth Edition), CBS Publishers & Distributors, New Delhi, 2002, pp. 1–316.
7. Degussa Construction Chemicals (India) Pvt. Ltd., Product Promotional Pamphlet, C-68, MIDC, Thane, Balapur Road, Yurbhe, Nava Mumbai, July 2006.
8. Leonhardt, F., New Trends in Design and Construction of Long Span Bridges and Viaducts (Skew, flat slabs, torsion box), Preliminary Publication of Eighth Congress of International Association for Bridge and Structural Engineering (IABSE), New York, 1968.

Maintenance and Rehabilitation of Bridge Decks

24

24.1 INTRODUCTION

Bridge deck maintenance management system is an essential component to ensure long term conservation of bridge deck to serve its intended purpose involving several vital functions like periodic inspection, repairs and rehabilitation. Effective maintenance will ensure that the bridge deck will function satisfactorily at the various limit states of strength and serviceability[1,2]. Good maintenance practice requires periodical surveillance, identification of local damage, deterioration and loss of durability of the deck due to environ-mental and other load effects. Maintaining highway bridge decks and keeping them in a sound and fit condition ensuring safe and uninterrupted traffic flow, is the primary function of a bridge maintenance engineer.

The life of a bridge deck depends entirely on preventive maintenance with well programmed repairs and rehabilitation. It is a well established fact that the total number of bridges built is much more in comparison with the new bridges under construction, but the amount of energy and resources spent on preventive maintenance and or rehabilitation of distressed bridges is negligible in comparison with that spent on new bridges. Hence increasing number of bridges are becoming unserviceable and if let unserviced, they have to be pulled down. The cost of dismantling the damaged bridge may in some cases be several times that required for periodical maintenance and or rehabilitation. Hence it is more prudent to inspect the bridge at regular intervals for detection of any signs of deterioration and initiate rehabilitation measures to restore the bridge structure to a state of full serviceability.

24.2 PRIMARY FUNCTIONS OF BRIDGE MAINTENANCE AND REHABILITATION

The primary objective of bridge maintenance is to ensure the integrity of the structure during its intended life span so that it functions without any disruptions. The maintenance management system must also provide guide lines and methodologies to enable the local Engineers to reach rational cost effective measures and timely decisions regarding maintenance and rehabilitation of distressed bridge decks.

According to Gokhale and Rohra[3], periodical inspections, repairs and rehabilitation measures constitute the primary aspect of good and effective maintenance. This aspect is of paramount importance due to the rapid increase in the number of reinforced and prestressed concrete bridges built during the last few decades of the 20th century as a result of easy availability of good quality cement, steel and epoxy compounds and various other building materials along with innovative methods of construction of bridges.

Many of the concrete bridges built on the West Coast National Highway (NH-17) have shown signs of distress due to severe environmental conditions typically prevailing in coastal regions. These bridges built during 1950's are more than 50 years old and they have not been maintained due to the absence of any systematic periodical surveillance and credible maintenance organization. Rehabilitation of bridge structures may become essential due to several reasons. Some common causes are design and or constructional deficiency, environmental effects, overloading of the bridge deck either due to unanticipated loading or due to accidents and user made changes in the structure during its service life.

In general, bridge repair and rehabilitation is unique for the particular structure. Hence the use of common techniques for rehabilitation of various bridges is limited. Several new cementatious materials, epoxy resins and compounds have been developed during the last decades which are highly effective in protecting the basic structure from the destructive effects of severe environmental exposure conditions. It is pertinent to note that maintenance engineers must study the basic designs, history of construction, changes in loading patterns on the bridge deck and environmental changes etc. before embarking on repairs. Detailed analysis of all these parameters will enable the engineer to work out a strategy for long lasting rehabilitation measures for the distressed bridge deck.

Recent developments in the domain of instrumentation have resulted in various types of instruments which could monitor the *in-situ* strength of concrete, microcracking in concrete and rusting in steel reinforcements. Also methods have been codified[4] to evaluate the *in-situ* strength of slabs to sustain the designed loads by actual load testing of the slab panel and monitoring the deflections developed at the soffit of the slab. Structural slabs and beams exhibiting local distress can be repaired by external bonding of steel plates to the soffit by epoxy adhesives. Honey combing, cracks and cavities in concrete can be repaired by the process of guniting, pressure grouting and shotcreting procedures.

24.3 CATEGORIES OF BRIDGE INSPECTION

Effective maintenance system requires periodical inspection of bridges to locate any signs of distress in the early stages which can be repaired with minimum costs. Generally all types of remedial and preventive maintenance measures including minor repairs and replacement of bridge components should be planned at periodical intervals without disrupting the traffic on the bridge deck. The data collected from the inspection reports should be scientifically evaluated from time to time to asses the need for remedial measures. According to Raina[5] the following categories of inspection are normally undertaken for bridge structures.

1. **Routine Inspection:** These are general inspections undertaken frequently by bridge maintenance engineers possessing practical knowledge of high way structures. This type of inspection does not need any expertise in design, detailing and constructional aspects. Routine inspection is necessary to identify minor deficiencies which could lead to accidents or maintenance problems at a future date. This type of inspections are normally undertaken at monthly or bimonthly intervals.

2. **Detailed Inspection:** This type is further categorized under General and Major groups depending upon the frequency and intensity of inspection respectively.

   (a) **General Inspection:** The general type of inspection is planned annually covering all the structural components of the bridge. The procedure includes visual inspection assisted by standard instrumental aids and invariably followed by a detailed written report.

   (b) **Major Inspection:** Major inspection is comprehensive involving detailed examination of all structural components. The procedure involves installation of access facilities like inspection platforms to examine the soffits of deck slab and girders, articulation locations and bearings at supports. Depending upon the importance of the structure, this type of inspection is conducted at 2 to 3 year intervals or may be at smaller intervals for important bridges specially exposed to aggressive environmental conditions (e.g. Bridges located in coastal areas, marine locations and abnormal wind zones).

3. **Special Inspection:** Special inspections are necessary in extraordinary situations such as earth quakes, high intensity/abnormal loadings, floods etc. These inspections are exhaustive comprising testing of structural elements (e.g. Non destructive testing using ultrasonic pulse velocity techniques[6] to detect internal micro cracks and excessive deflections using dial gauges etc). The results of tests are examined in the light of structural analysis and codified specifications. Experienced engineers are entrusted with this type of inspections.

Special inspections are necessary for bridge bearings and joints during extremes of temperature and the soffits of deck slab and girders should be inspected for cracks under abnormal loading conditions. Exposure to aggressive environment may result in cracking and spalling of concrete in pretensioned girders with thin webs. The inspection team should be lead by a qualified and experienced bridge engineer who is familiar with the design and constructional aspects of the bridge structure to be inspected so that the observations are properly and accurately assessed resulting in a meaningful technical report containing details of distress/deficiency and recommendations for relevant repairs and rehabilitation measures for the bridge deck.

24.4 INSPECTION INSTRUMENTATION

During the last few decades, several types of testing equipment have been developed for use in structures under distress. Modern testing equipment for measurement of *in-situ* strength, surface and internal cracks in concrete are widely used. Also electronic gadgets have been developed to measure the thickness of concrete cover to reinforcements and strains developed in concrete due to loading on the bridge deck. The testing equipments widely used by the specialized inspection team are listed below:

1. Electronic strain gauges for measurement of strains in concrete and steel

2. Ultrasonic pulse velocity apparatus for detection of cracks in concrete

3. Rebound hammer (Schmidt hammer) for *in-situ* measurement of compressive strength or grade of concrete

4. Snooper-crawler and adjustable ladders

5. Magnetic detector for measuring thickness of concrete cover and for locating reinforcement bars

6. Vibration measuring equipment

7. Hydraulic jacks, pressure transducers or load cells for measurement of forces

8. Electrical resistance meter (for rust pockets)

9. Pachometer to locate and measure the size of steel bars embedded in concrete

10. Optical microscope with light source to measure the width of cracks on the surface of concrete

11. Mechanical extensometer or Demec Gauge with stainless steel targets for measuring surface strains on concrete under loads

12. Dial gauges for measurement of deflections at soffits of deck slab and beams when the bridge is loaded.

The gadgets mentioned above are very useful in evaluating the *in-situ* strength of concrete in the bridge deck by non destructive testing techniques. The author[7] has used these instruments for testing the integrity of reinforced concrete slabs, beams and columns at Darus Salam hostel complex located in Bangalore and also for testing the roof slab of explosion bunker[8] at M/S. Astra Indian Detonators Ltd., Bangalore. Pachometer is used to measure the diameter of steel reinforcements embedded in concrete along with the cover to an accuracy of ± 3 mm.

For Inspection of the soffits of bridge decks greater than 10.7 m in height, a mechanical contraption widely referred to as Barin's Snooper vehicle is generally used. The snooper is mounted on a heavy duty truck with a swiveling platform to facilitate the inspection of soffit of the bridge deck using the hydraulically operated platform. Various types of snooper vehicles have been developed to suit the rugged terrain and watery situations. Fig. 24.1 shows the typical schematic diagram of a snooper system. Fig. 24.2 shows the inspection of the under side of the bridge deck using the snooper platform.

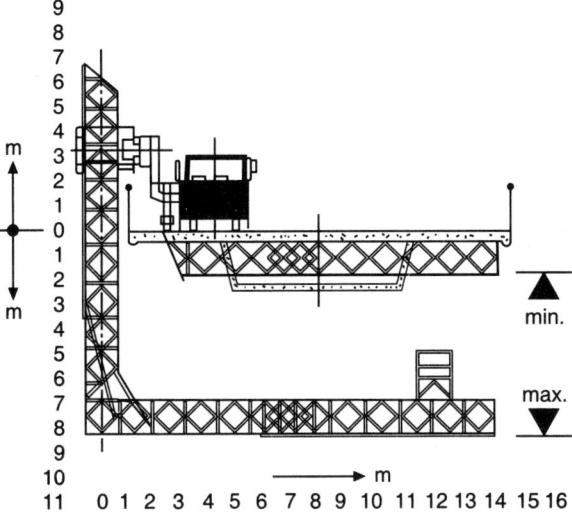

Fig. 24.1 Barin's Snooper System for Inspection of Bridges.

Fig. 24.2 Snooper Platform to Inspect the Soffit of Bridge Deck.

24.5 GENERAL ASPECTS OF REPAIRS AND REHABILITATION OF BRIDGE DECKS

Detailed inspection reports specify the nature and magnitude of distress or damage to the bridge deck. The reports are first analyzed in the light of previous history of the bridge before deciding upon the type of repair and rehabilitation which depends upon the degree of damage suffered by the bridge structure. Generally the degree of damage is classified under the following three major groups:

1. **Minor Damage:** Surface cracks developed on the bridge deck due to shrinkage of concrete as a consequence of temperature changes and minor spalling of concrete at edges of structural concrete elements are grouped under this category. The cracked surface is repaired by superficial patching by using epoxy grout or

guniting using shotcrete applied by a pneumatic gun. The damaged and spalled concrete is first removed by hand tools and the surface is thoroughly cleaned before the application of epoxy grout. All cracks should be sealed by the epoxy grout applied under pressure injection. Fig. 24.3 shows the typical epoxy resin sealing of cracks in the soffit of decks slab and webs of longitudinal girders of a highway bridge deck.

Fig. 24.3 Epoxy Sealing of Cracks in Deck Slab and Main Girder of a Bridge Deck.

2. **Moderate Damage:** Extensive spalling and cracking of concrete due to multifarious reasons in a bridge deck can be grouped under moderate damage. In such cases repairs and rehabilitation is done by removing the unsound or loose concrete by providing temporary supports to the girder to relieve dead load stresses. Fig. 24.4 shows the typical details of repairs to the spalled concrete of bridge deck girders. Expansion bolts or grout rebars are drilled into the sound concrete from the soffit and wire mesh is placed to the sides and welded to the existing bars. Gunite or shotcrete is applied under pressure to restore the girder to its predamaged condition. In the case of webs of prestressed concrete girders damaged by shear cracks near supports due to improper detailing or compaction of concrete, holes are drilled diagonally and rebars are placed and grouted to arrest the shear cracks as shown in Fig. 24.5.

3. **Severe Damage:** Bridge decks damaged severely require a detailed structural analysis and a design check based on the conditions of damage and the best engineering assumptions and judgement. A comprehensive review of design calculations and detailed examination of the damage will help in selecting a cost effective and appropriate restoration technique for the damaged structural elements of the bridge. Fig. 24.6 shows the repair technique adopted to restore

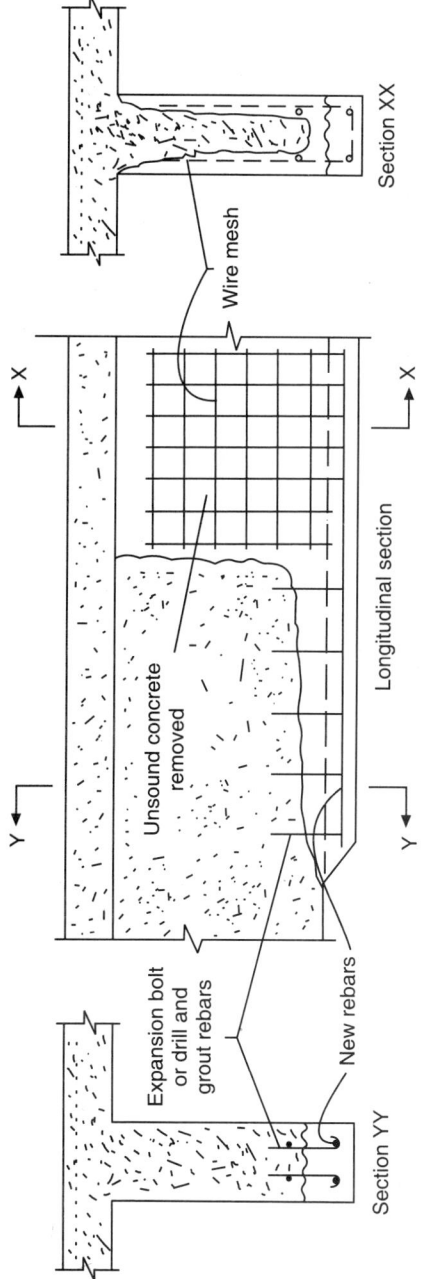

Fig. 24.4 Repair-Rehabilitation of Damaged/spalled Concrete.

the girder damaged by extensive shear cracks followed by spalling of concrete. In the jacketing stirrups technique, the deck slab concrete is carefully removed to permit positive anchoring of additional vertical stirrups placed around the existing beam. After priming the exposed surfaces, epoxy mortaring or shotcreting or guniting is done providing a new concrete jacket over the old girder.

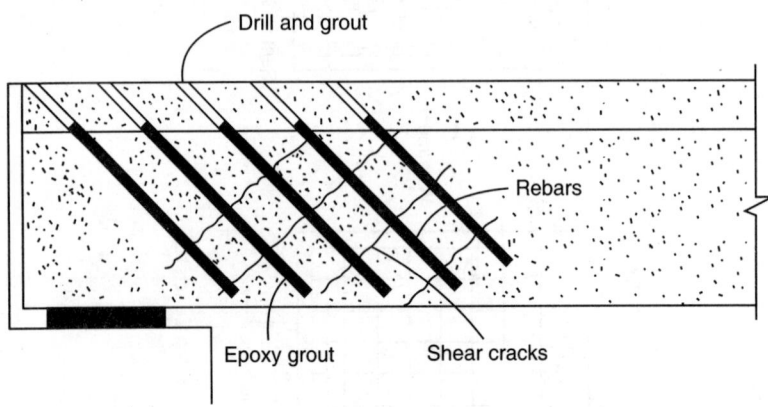

Fig. 24.5 Repairs of Shear Cracks by Stitching Rods and Epoxy Grovting.

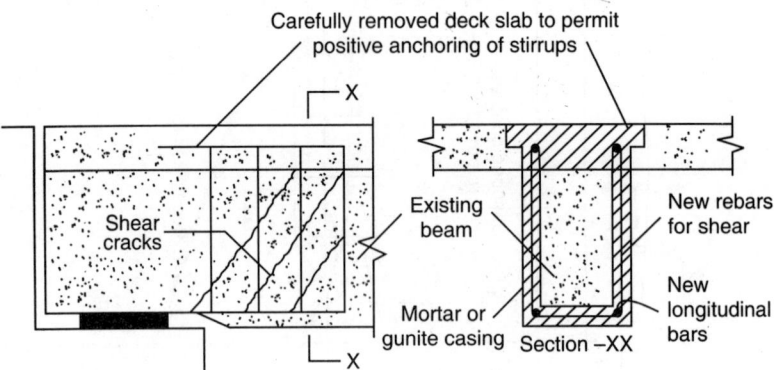

Fig. 24.6 Repairs of Beam with Shear Cracks by Jacketing Stirrups.

In coastal regions where bridge girders are subjected to severe exposure conditions, extensive spalling of concrete exposing the reinforcements is generally observed. In such cases the unsound and loose concrete around the girder is removed and repairing against the loss of concrete section is done by jacketing the girder using a steel box fixed to the girder as shown in Fig. 24.7. The gap between the girder and the steel box is filled by epoxy concrete grout ensuring the functional capacity of the girder.

24.6 REHABILITATION OF ARTICULATION JOINT

In the case of balanced cantilever bridge decks, the cantilevers projecting on either

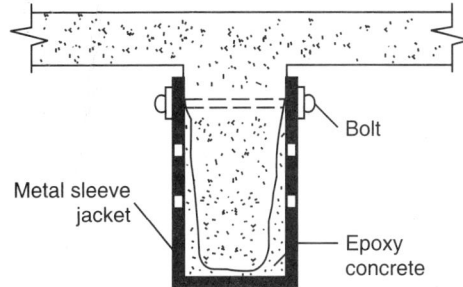

Fig. 24.7 Repairs of Concrete Girder by Metal Sleeve Jacket.

side of the central span rest on the end spans at halving joint generally referred to as articulation. Due to improper detailing of reinforcements or due to severe exposure conditions, distress may develop at the articulation joint. Repair and rehabilitation of articulation can be effectively done by using post tensioning techniques. In this method external anchor bars with plates and nuts are employed to arrest the fracture distress at a poorly detailed and overloaded articulation joint. Fig. 24.8(a) shows the anchor rods embeded to the top of the deck and stressed against the plates fixed to the bottom of the girders. The close up view of rehabilitation of the halving joint shown in Fig. 24.8(b) clearly indicates the principle of post tensioning to repair the distressed articulation joint in balanced cantilever bridges.

Fig. 24.8(a) Repairs to Articulation Joint using External Bars with Plates and Nuts.

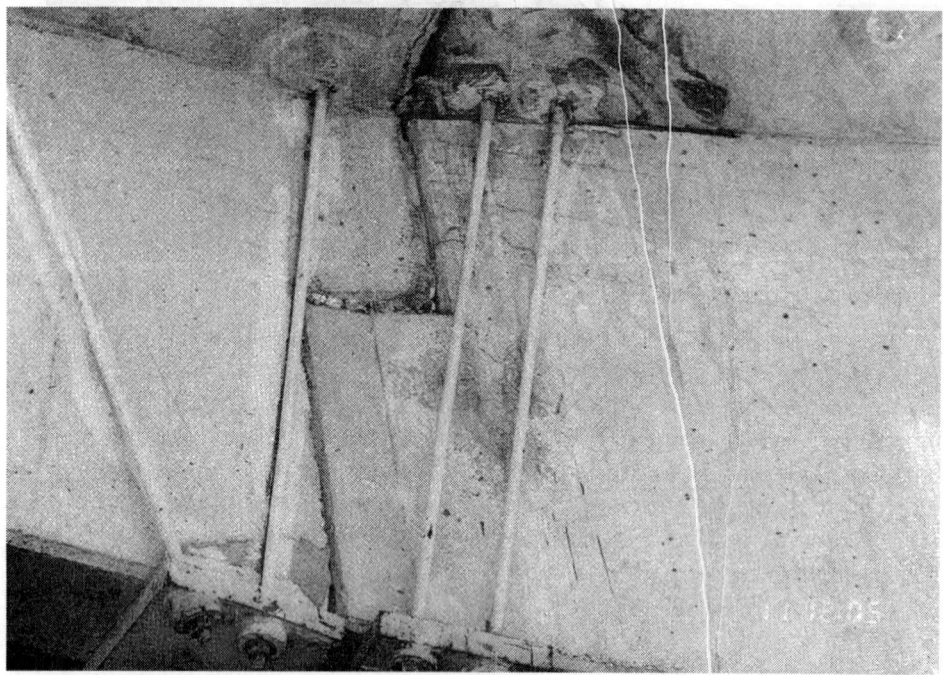

Fig. 24.8(b) Close up view of Repaired Articulation Joint.

24.7   REPAIRS OF GIRDERS DAMAGED BY COLLISION

Structural elements like pretensioned prestressed concrete girders and poles may get damaged during transportation or due to collision. If the damage is localized and not extensive, it can be repaired and rehabilitated. Fig. 24.9 shows the method of repairing the bottom flange of a pretensioned beam partially damaged at the sides. The damaged portion is thoroughly cleaned of all loose material and dowel bars or expansion bolts are introduced into the drilled holes to a depth not less than 75 mm and then covered using mortar or non shrink grout. If the damage extends over a larger depth covering the sloping sides, a welded steel wire mesh is embedded near the surface before applying the mortar or non shrink grout.

24.8   RESTORATION OF STRENGTH OF PRESTRESSED CONCRETE GIRDERS

Prestressed concrete girders are likely to get damaged due to severe exposure conditions especially in the coastal regions where humidity and temperature are high. When unbonded tendons are used, the high tensile strands may get damaged due to rusting resulting in loss of prestress in the girder. In such cases the effective cross sectional area of high tensile steel gradually decreases due to rusting leading to sudden explosive failure of the girder due to fracture of steel in tension[9]. Several methods have been used by Raina[10] to restore the strength of damaged prestressed concrete girders.

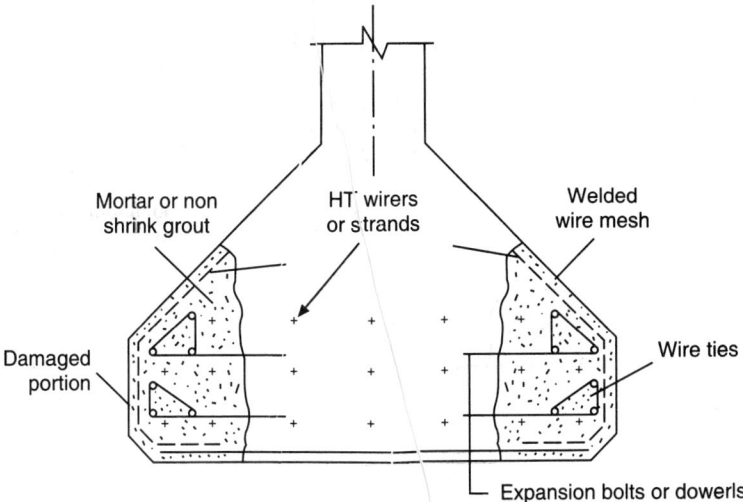

Fig. 24.9 Repairs to Damaged Pretensioned Girder.

Method-1: The principle of post tensioning is used to restore the damaged prestressed concrete I-girders. In this method post tensioning rods in conjunction with jacking (concrete) corbels located outside the damaged areas are used for restoration work. The calculated preload is applied and the damaged concrete is repaired. After the new concrete attains the desired strength, the preload is removed. Fig. 24.10 shows the details of locating the high strength rods on the roughened sloping bottom flange of the girder. After the jacking corbels are erected, the post tensioning of the rods is done as per design computations. Suitable spiral and link reinforcements are used in the jacking corbels to strengthen the concrete.

Method-2: In this method the reinforced concrete corbels are constructed and the damaged concrete is repaired by applying the required preload. Fig. 24.11 illustrates the method of adding external reinforced concrete to restore the strength of damaged beam. 16 mm diameter steel dowels are used to anchor the corbels to the bottom flange.

Method-3: Additional metal sleeve jacket is used to restore the damaged portion of the girder. This method does not restore the loss of prestress except that a partial prestress may be gained by preloading. The preload should be applied prior to the repair of the damaged concrete and removed after completion of repairs. Then a metal sleeve jacket should be installed around and beyond the damaged areas extending over a minimum length of 1 m. Finally the gap between the metal sleeve and the beam is filled with epoxy grout by pressure injection as shown in Fig. 24.12.

24.9 RESTORATION OF CONCRETE BRIDGE DECKS BY EXTERNALLY BONDED STEEL PLATES

The first reported attempts to strengthen and restore concrete beams by externally bonded steel plates were attempted by France around 1964-65. Practical applications

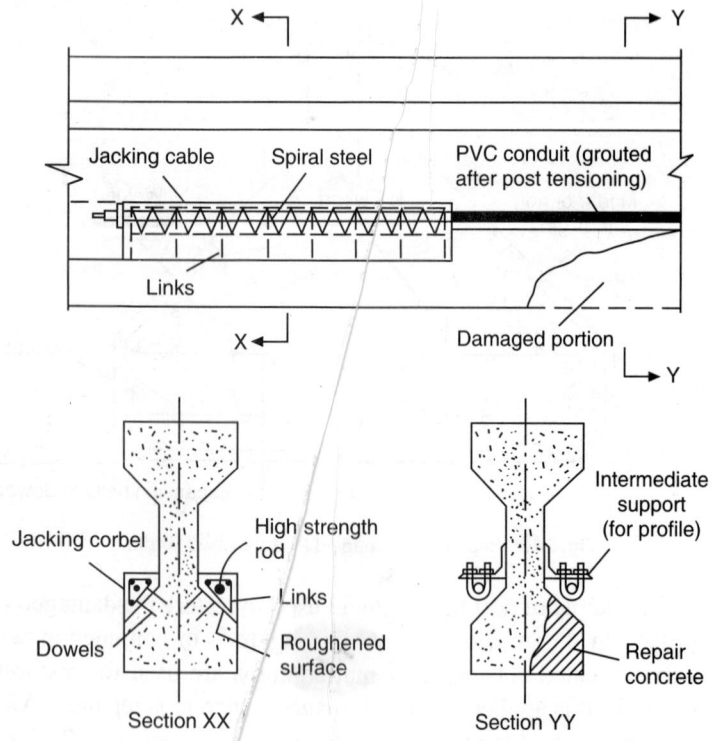

Fig. 24.10 Restoration of Girder by Post Tensioning.

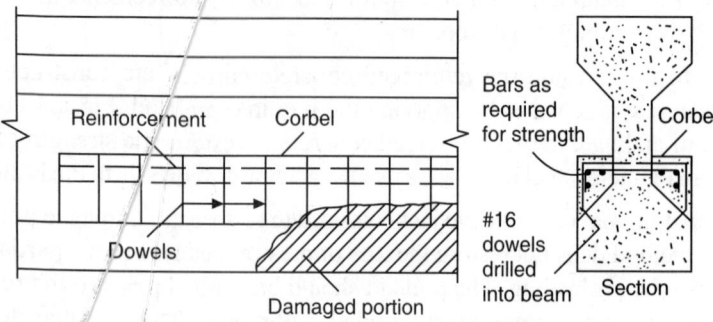

Fig. 24.11 Restoration of Girder by Adding External Reinforced Concrete.

date back to 1966-67 in France and South Africa followed by Japan and Russia. This method has been used extensively in Switzerland in both buildings and bridges for the last three decades. Detailed experimental investigations conducted at Transport and Road research Laboratory in United Kingdom by Irwin[11] and MacDonald[12] have conclusively established the efficacy of strengthening concrete beams by externally bonded steel plates.

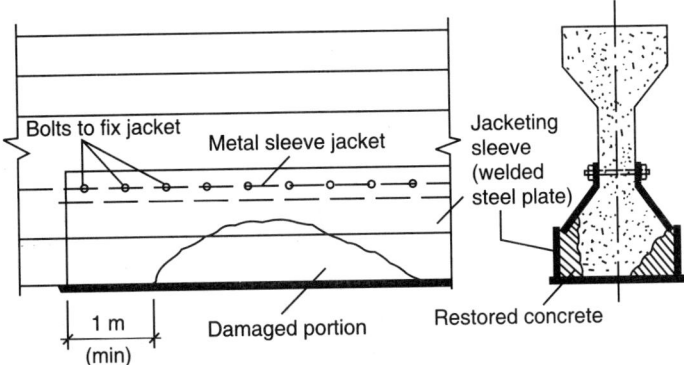

Fig. 24.12 Restoration of Girder by Addition of Metal Sleeve Jacket.

Experiments conducted by Krishna Raju and Nadgir[13] have shown that reinforced concrete beams when epoxy bonded with steel plates on the tension face exhibited significant increases of up to 30 percent in the ultimate flexural strength in comparison with non plated beams.

Two fundamental methods of strengthening/rehabilitating of flexural members with epoxy resin adhesives are:

(1) The total tensile reinforcement in the cross section is increased by epoxy bonding of thin steel plates on the tension face of the beam to enhance the flexural and shear strength.

(2) The overall depth of the flexural element is increased by adding a new layer of concrete on top of an existing cross section and bonding the old and new elements with modern epoxy resin adhesives.

In the case of bridge decks, it is not possible to add new concrete on top of deck and hence the first method of bonding steel plates to the soffit of slabs or girders is adopted. Generally any grade of structural steel is suitable for bonded reinforcing plates. Plate gauges below 3 mm are not suitable, because sand blasting can deform them. Steel plates between 6-16 mm thick have been used in some strengthening works in Switzerland and England. The adhesive joint is generally between 1-3 mm thick. Fig. 24.13 shows the strengthening operation by epoxy bonding of steel plates to the soffit of a girder. Rehabilitation of a bridge deck slab by fixing of steel plates to the underside is shown in Fig. 24. 14. Shear strengthening of beams in distress near supports is also possible by epoxy bonding of steel plates near supports as shown in Fig. 24.15.

24.10 CASE STUDIES OF REPAIRS AND REHABILITATION OF BRIDGE DECKS

After the Second World War, The construction activity received a boost in the form of infra structure development and in the war ravaged Germany alone, hundreds of reinforced and prestressed concrete bridges were built followed by the various countries in Europe, North America and even India. Many of the bridges built during

Fig. 24.13 Strengthening of Concrete Girders by Epoxy Bonded Steel Plates.

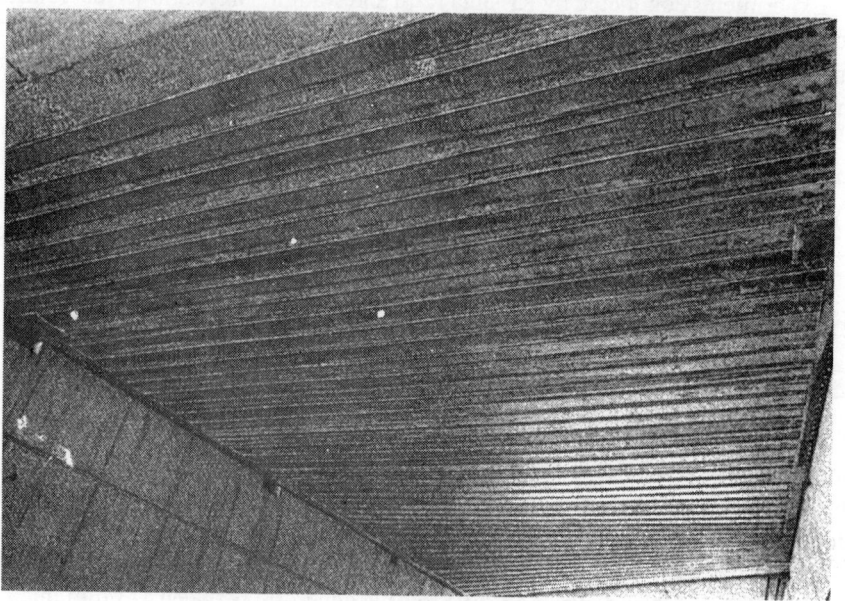

Fig. 24.14 Rehabilitation of Bridge Deck Slab by Epoxy Bonding of Steel Plates.

the later half of 20th century are showing signs of distress and deterioration due to various loading and environmental factors. Revolutionary developments in the field of cement, steel and chemical materials technology, have paved the way for effective repair and rehabilitation of bridges under distress. A brief survey of the various bridge structures repaired, strengthened and rehabilitated is compiled below:

1.   Katepura Bridge, Maharashtra[3]

The Katepura Bridge located in the state of Maharashtra was built using simply supported prestressed concrete girders in conjunction with reinforced concrete deck slab. The bridge extends over 4 spans, each of 37.8 m. The prestressed concrete girders were cast in place over temporary staging and side shifted to position after completion of post tensioning operations. Each girder has 16 high tensile cables and these were stressed in two stages. During the construction period, when the stage prestressing was being carried out for a girder, the girder cracked with sound exhibiting cracks near the end block as shown in Fig. 24.16. Detailed investigations revealed that the concrete in the end block portion of the girder was not homogeneous due to inadequate compaction.

Fig. 24.15 Shear Strengthening of Beams by Epoxy Bonding of Steel Plates.

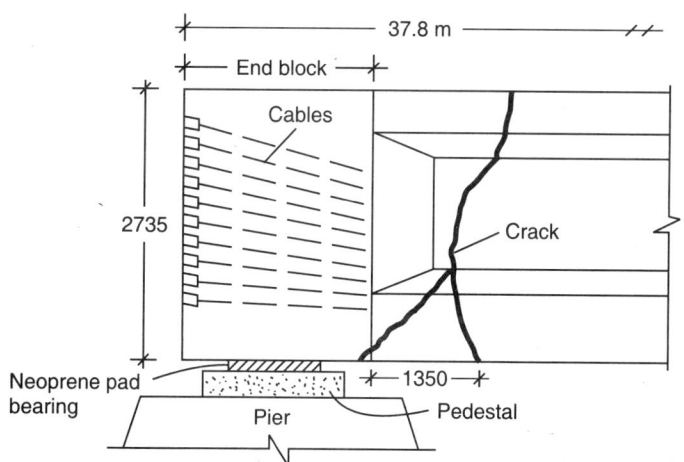

Fig. 24.16 Restoration of End Block of Prestressed Concrete Girder (Katepura Bridge Maharashtra).

Repair and restoration of the end block of the girder was done by completely dismantling the concrete in the end block after destressing of the cables. New reinforcement was welded with the existing reinforcement and the end block was rebuilt using new concrete with a vertical joint, taking extra care for proper compaction of concrete. After the concrete attained the desired compressive strength, prestressing operations were carried out in stages. The girder was finally load tested and found to be satisfactory at the serviceability limit states.

2. Gizenen Bridge, Switzerland[5]

The concrete in the deck slab and girders of Gizenen Bridge built in the year 1911 had to be strengthened to withstand planned future loading. After investigations it was decided that the concrete girders should be strengthened by epoxy bonding of steel plates. The damaged portions of the deck slab were repaired using epoxy resin mortar. A new cross beam was introduced at the joint located at the centre of span. The bonding work was started immediately after sand blasting the surfaces of the concrete and steel plates. 15 mm thick steel plates with a width of 200 mm were epoxy bonded to the main girders.

For transverse girders, 150 mm wide by 10 mm thick steel plates were used. The efficacy of the repair and rehabilitation was confirmed by load tests conducted by the Federal Material Testing Institute.

3. Quinton Bridges in United Kingdom[14]

The Qunton Bridges project comprises of four bridges located on the M-5 motorway at the Quinton Interchange west of Birmingham. The bridges have spans of 16.5, 27 m and the deck is made up of voided slabs 90-105 cm thick. During routine inspection, cracks were observed in the soffit of end and central sections. Design review indicated deficient tensile reinforcements at certain locations of the deck slab. The two rehabilitation methods examined for restoration of the bridge deck are:

(a) Epoxy bonding of steel plates

(b) Installation of prestressing elements

A comprehensive analysis indicated the option of strengthening the deck by epoxy bonding of steel plates to be more effective although the technology was new in the year 1975. Accordingly the end sections were strengthened by bonding 6 mm thick plates to the soffit of the voided slabs. A double layer of 6 mm thick plates was used at the middle of central spans. Along the sides of the central span where distress was more, three layers of 12 mm thick steel plates measuring about 3 m long and 250 mm wide were fastened to the soffits of the slab with screw plugs spaced at intervals of 450 or 900 mm. Before epoxy bonding of the steel plates, the soffit of the deck slab was pretreated to remove the unevenness and local undulations. Post restoration load tests indicated that the reinforced concrete bridge deck was flexurally stiffer indicating lower deflections after rehabilitation.

4.   Chambal Bridge in Uttar Pradesh[3]

This balanced cantilever bridge built across the river Chambal is located near Etwah connecting Uttar Pradesh and Madhya Pradesh. The bridge is 592 m long with a cellular box deck 11.1 m wide projecting on either side of the pier. The suspended span comprises of two prestressed concrete girders with reinforced concrete deck slab of span 40.6 m as shown in Fig. 24.17. Cast steel rocker and roller bearings have been provided at articulations for supporting the suspended span. Soon after the bridge was constructed and opened for traffic in 1975, it developed distress due to improper placing of roller bearings. The suspended span between two intermediate piers shifted towards down stream side at roller end by about 20 mm. Subsequently heavy loads were transported over the bridge by the Department of Atomic Energy. The deviation towards the down stream gradually increased with passage of time to about 110 mm.

Fig. 24.17 Prestressed Concrete Girder Supported on Rocker-Roller Bearings (Chambal Bridge).

Investigations revealed that the bearings were not at right angles to the axis of the bridge and the level of down stream side bearing was lower by 35 mm as compared to the elevation of upstream bearing. Hence due to transverse inclination of the bearings towards down stream, the span had a tendency to move in the transverse direction. The distress developed at the rocker-roller bearings is shown in Fig. 24.18.

Fig. 24.18 Distress at Rocker-Roller Bearings (Chambal Bridge).

The rehabilitation scheme comprised of providing an access from the deck to the roller end articulation to inspect and replace the bearings using a working platform suspended from the bridge deck. Lifting and rotating the suspended span was done by placing a heavy steel truss over hammer head and roller end articulation. Freyssi flat jacks and PTFE/stainless steel sliding arrangements were introduced under the trusses near articulation. The span was lifted by operating the flat jacks. The old roller bearings were replaced with new ones which were properly aligned and leveled with epoxy mortar. The traffic over the bridge was diverted only for 3 days during the period of rehabilitation work which was carried out by Uttar Pradesh State Bridge Corporation.

5.   Gautami Bridge in Andhra Pradesh[3]

Gautami Bridge located on the National Highway No. 5 between Chennai and Kolkata in Rajahmundry District of Andhra Pradesh is 2.337 km long and the super structure is made up of 5 nos of simply supported prestressed concrete girders of 47.8 m span with cross beams and deck slab as shown in Fig. 24.19. In the year 1978, durng monsoon floods, Foundation Well No. 2 tilted towards Well No. 1 on account of excessive scouring causing a tilt of about 300 mm at the top of pier P2. This resulted in toppling of all 10 roller bearings provided on pier P2. The restoration work included new cladding concrete for the pier and bonding it with old concrete by providing anchor bars in the holes drilled in the concrete. The new bearings were planned to rest partly on old concrete and partly on new concrete. To avoid concentration of stress due to differential shrinkage and strength, the old and new concretes were tied in upper portion by transverse prestressing cables located in the holes drilled in old concrete of the pier as shown in Fig. 24.20.

Fig. 24.19 Gautami Bridge (Andhra Pradesh).

Fig. 24.20 Transverse Prestressing Cables at the Top of Pier.

Neoprene pad bearings were placed over the pedestals and the span was lowered by using hydraulic jacks as shown in Fig. 24.21. Restoration work for both spans was completed in a short period of 7 days. After the completion of the repair work, the two affected spans were load tested for structural integrity. The performance of the bridge deck under service loads was found to be safe to support the normal traffic.

Fig. 24.21 Placing of Neoprene Pads over Pedestals using Hydraulic Jack.

6. Obra Singrauli Bridge No. 93[3]

Obra Singrauli Bridge located on the Eastern Railway of India is made up of two prestressed concrete girders stressed with Freyssinet system with a reinforced concrete deck slab. The bridge has 4 spans of 18.3 and one span of 24.4 m. The distress in the form of large number of cracks developed at the junction of girders and deck slab on both internal and external faces after 15 years in service. Also longitudinal cracks were observed in the bottom flange of the girders and vertical cracks at the junction of diaphragm and main girders. Some of the cracks were as large as 3 mm. Detailed investigations revealed the following main causes for development of cracks.

(a) Vertical stirrup reinforcements and the shear connectors at the junction of the top flange of the girder and deck slab were insufficient and did not comply with the revised Indian Standards.

(b) The bridge deck being on a railway line was subjected to repeated vibrations which could have contributed to distress in the form of cracks.

(c) Steel rocker and roller bearings were not properly maintained. The excessive resistance to their movement might have contributed to additional stresses.

The restoration of the bridge was done by strengthening the deck by pumping low viscosity epoxy resin to seal the various cracks developed in the deck. After sealing the cracks the longitudinal girders were prestressed vertically at 1.2 m intervals using 12.5 mm strands with anchorages provided at the top of the deck in conjunction with I-section girder and a steel saddle at the soffit as shown in Fig. 24.22.

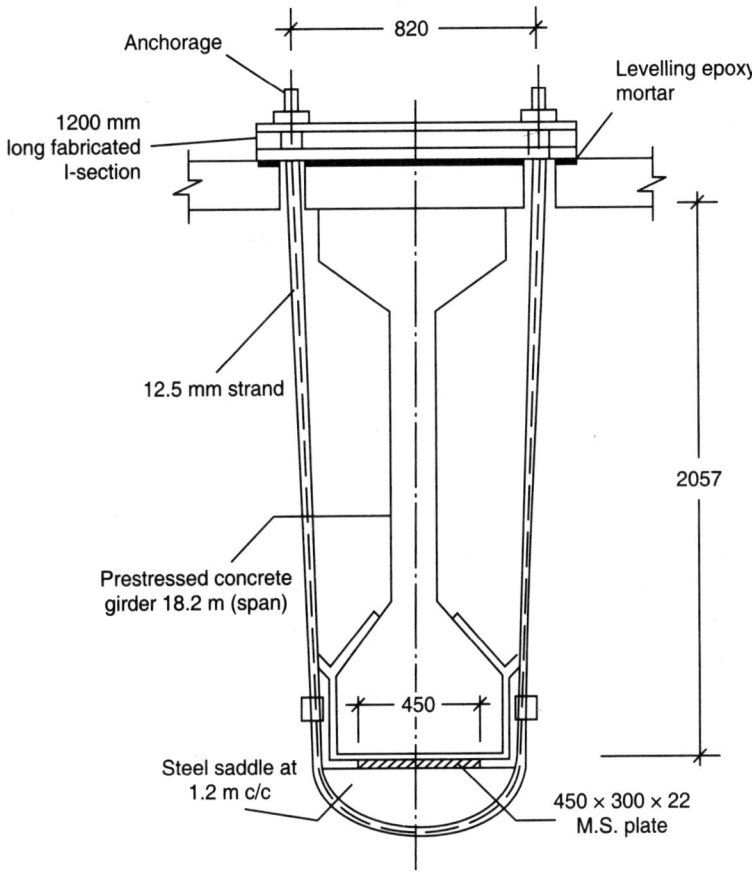

Fig. 24.22 Restoration of PSC Girder (Obra Sing Rauli Bridge No. 93).

7. Swanley Bridges in United Kingdom[14]

Swanley highway Bridges form part of the M-20 and M-25 motorway intersection. The super structure comprises of a continuous concrete slab supported on inclined piers. Shortly after the bridge was opened for traffic, cracks were observed on the soffit of deck slab at end sections. A detailed design review and inspection revealed that the steel reinforcements at the cracked locations were insufficient to resist the tensile stresses developed under service loads. Hence the deck slab was strengthened by epoxy bonding of steel plates 6 mm thick, 250 mm wide and 3 to 6 m long to the soffit of the slab. The plates were bonded in three layers in each strip extending over 12 m. A total of 15 strips were distributed over the entire width of the bridge deck.

The strengthening of the slab over the piers where tension develops at the top due to negative moments was done by epoxy bonding of four steel plates over a length of 12 m. Altogether 449 plates were epoxy bonded within a period of 20 days including pretreatment of concrete and plates. Sommerard14 has reported in detail the rehabilitation and testing of the repaired bridge structure under dynamic loading.

8. High way Bridges in France and Japan[5]

Bresson[15] has reported the rehabilitation of an inclined bridge on the French A6 motorway in France. The bridge over the 'Chemin departmental 126' was strengthened for flexure and shear by epoxy bonding of steel plates in the soffit and near the supports. Repair of a bridge built at the beginning of the 20th century and distressed due to severe weathering has been repaired using additional reinforcements in the form of plates. L'Hermite[16] et al., have reported the rehabilitation of a French bridge located on 'Route Natonale 186' crossing the saint Denis Canal.

During the early period of the 1970's, the road building authorities of Tokyo undertook a project involving the repair and rehabilitation of elevated motorways in the entire metropolitan zone. Today. Nearly 10 km of a total of about 100 km of elevated motorway has been strengthened using epoxy bonded reinforcements.

REFERENCES

1. Rowe, R.E, Cranston, W.B and Best, B.C., New Concepts in the Design of Structural Concrete, Structural Engineer, London, Vol. 43, 1965, pp. 399–403.
2. Krishna Raju, N., Limit State Design for Structural Concrete, proceedings of the Institution of Engineers (India), Vol. 51, Jan. 1971, pp. 138–143.
3. Gokhale, P.S and Rohra, M.R., Restoration of Distressed Concrete Structures, Case Studies, Concrete Structures of India, Compiled by Maharashtra India Chapter of the American Concrete Institute and Presented at A.C.I's Fall Convention at New York, 1984, First report, Jan. 1985, pp. 149–167.
4. IS: 456-2000, Indian Standard Code of Practice for Plain & Reinforced Concrete (Fourth Revision), BIS, New Delhi, 2000, pp. 30–31.
5. Raina, V.K., Concrete Bridge Practice, Construction, Maintenance & Rehabilitation, Tata McGraw Hill Publishing Co. Ltd., New Delhi, 1988, pp. 287–326.
6. Krishna Raju, N., Report of Rebound Hammer and Ultrasonic Pulse Velocity Tests on Prestressed Concrete Lattice Girders in Ramkumar Mills, Rajajinagar, Bangalore, Civil Engineering Department, M.S. Ramaiah Institute of Technology, Bangalore, Oct. 1992, pp. 1–13.
7. Krishna Raju, N., Technical Report on Rebound (Schimdt) Hammer tests on R.C.C. Beams, Columns and Slabs in Darus Salam Hostel and Office Complex at Site No.332, Queens Road, Bangalore, Civil Engineering Department, M.S. Ramaiah Institute of Technology, Bangalore, 1986, pp. 1–20.

8. Krishna Raju, N., Technical Report on Ultrasonic Pulse Velocity Tests on Roof Slab of Explosion Bunker, M/S. Astra Indian Detonators Limited, Bangalore, Civil Engineering department, M.S. Ramaiah Institiute of Technology, Bangalore, 1986, pp. 1–15.

9. Krishna Raju, N., Prestressed Concrete (Fifth Edition), M/S. Tata McGraw Hill Publishing Co. Ltd., New Delhi, 2012, pp. 190–231.

10. Raina, V.K., Concrete Bridge Inspection, Repairs, Strengthening, Testing, Load Capacity Evaluation, Tata McGraw Hill Publishing Co. Ltd., New Delhi, 1994, pp.10–63.

11. Irwin, C.A.K., The Strengthening of Concrete Beams by Bonded Steel Plates, Transport and Research Laboratory, Department of Environment, Supplementary Report-160UC, Crowthorne, Berkshire, U.K, 1975.

12. MacDonald, M.D., The Flexural Behaviour of Concrete Beams with Bonded External reinforcement, Transport and Research Laboratory, Department of Environment, Supplementary Report-415, Crowthorne, Berkshire, U.K, 1978.

13. Krishna Raju, N and Nadgir, N.S., Limit State Behaviour of Reinforced Concrete Beams Strengthened by Epoxy Bonded Steel Plates, Indian Concrete Journal, Vol. 65, No. 3, March 1991, pp. 124–129.

14. Sommerard, T., Swanley's Steel Plate Patch up, New Civil Engineer, London, No. 247, 16th June 1977, pp. 18–19.

15. Bressen. J., Reinforcement par collage d'armatures du passage inferieur du CD 126 Sous l'Autoroute du SUD, Annals de l'Instutut Technique du Batiment et des Travaux Publics, Supplement No. 297, September 1972, Concrete and Reinforcement Series No. 122.

16. L'hermite, R., Devars Du Mayne, R., Le Collage Structural et le reinforcement par Resines des Structures de la Construction. Annales de l'Institut Technique de Batiment et des Travaux Publics, No. 349, April 1977, general Construction Series No. 62, 13 papers are summarized under the title 'Le reinforcement du Beton par Resines' and 'les Applications du Reinforcement du Beton'.

Prominent Bridges of the World

25

25.1 INTRODUCTION

According to Thomas B. Macaulay[1], the most significant inventions after the alphabet and the printing press, are the bridges which abridge distances between places contributing the maximum benefit for the development of civilization and rapid dissemination of knowledge. Bridges have always figured prominently in human history. Historical battles and large scale migration would not have been possible without bridges. Great cities have sprung up at a bridgehead on the banks of rivers. The famous bridges in London, Paris, New York, Oxford and Cambridge enhance the vitality of the cities and promote the cultural, social and economic prosperity of the surrounding areas. The mobility of an army depends upon the availability of bridges and great battles have been fought for cities and their bridges.

The historical evolution of bridges from 4000 B.C to the beginning of 21st century is presented in great detail in Section 1.1. Starting from crude timber bridges, engineers have used stone, iron, steel, reinforced concrete and prestressed concrete for building bridges of ever increasing span. During the 19th century long span bridges were built using steel section decks supported by curved suspension cables between massive towers. In the present day long spans are covered using prestressed concrete cellular box girders or the deck is supported by cable stays. The present day cable stayed bridges have proved to be the most economical solution combined with economy of material, superior aesthetics, functional integrity and durability coupled with ease of maintenance.

25.2 WORLD'S LONG SPAN BRIDGES

In the nineteenth century, bridges having spans greater than 100 m were considered as long spans. However the word long span is only relative and depends on several factors like, the type of material available, developments in construction technology and expertise available at the time of construction. What is considered as long span during the early 19th century will be classified as short or medium span in the 21st century. Long span bridges are generally classified under the following three main groups depending upon the type of material used in the structural components of the bridge.

1. Steel Bridges

(a) Suspension Bridges

(b) Arched Bridges

(c) Trussed Bridges

(d) Plate Girder Bridges

(e) Cable Stayed Bridges

2. Concrete Bridges

(a) Prestressed Concrete Bridges

3. Steel/Concrete Hybrid Bridges

(a) Composite Bridge Decks

(b) Steel Cable Stayed Bridges with Concrete Deck

The earliest long span bridges were built using steel during the eighteenth century, soon after the development of steel by Bessemer, Sieman and Martin in 1856. Roebling's suspension bridge of 490 m span, marked the beginning of long span steel bridges all over the world. Rapid advances in the development of theoretical analysis of the load response of the structural system of suspension, cantilever and cable stayed bridges in the early part of the 20th century resulted in the construction of many elegant long span steel bridges. The Brooklyn Bridge with a main span of 486 m built in New York in 1883 and the Forth bridge having a main span of 521 m in Scotland, U.K are the earliest examples of suspension bridges built in the eighteenth century.

25.3 LONGEST SPAN SUSPENSION BRIDGES

Many long span steel suspension bridges with main spans exceeding 1000 m have been constructed in various advanced countries of the world during the 19th and 20th centuries. Lessons learnt by the failure of Tacoma Narrows suspension bridge due to aero dynamic instability in USA gave an impetus for detailed investigations of the behavior of slender structures like suspension bridges under environmental forces like severe winds. As a result of extensive theoretical and modeling studies, the structural engineers have realized the importance stabilizing the deck to resist extreme wind forces.

The present day suspension bridges are designed by considering the wind effects which are critical in long span slender structures like suspension bridges. Table 25.1 shows the list of longest span suspension bridges built/planned in different countries of the world. China has the maximum number of suspension bridges with as many as ten of the top twenty long span bridges with spans exceeding 1000 m.

Table 25.1 List of Longest Span Suspension Bridges in the World

No.	Bridge	Span [m]	Location	Country	Year	Notes
1	Akashi-Kaikyo	1991	Kobe-Naruto	Japan	1998	
2	Yangsigang	1700	Wuhan	China	2019	
3	Xihoumen	1650	Zhoushan	China	2009	
4	Great Belt East	1624	Korsor	Denmark	1998	
5	Izmit	1550	Gebze	Turkey	2017	
6	Yi-Sunsin	1545	Myodo-Gwangyang	South Korea	2013	
7	Runyang South	1490	Jiangsu	China	2005	
8	Nanjing-4	1418	Nanjing	China	2012	
9	Humber	1410	Kingston-upon-Hull	UK	1981	
10	Yavuz Sultan Selim	1408	Istanbul	Turkey	2016	
11	Jiangyin	1385	Jiangsu	China	1999	
12	Tsing Ma	1377	Hong Kong	China	1997	
13	Hardanger	1310	Vallavik-Bu	Norway	2013	
14	Verrazano-Narrows	1298	New York, NY	USA	1964	
15	Golden Gate	1280	San Francisco, CA	USA	1937	
16	Yangluo	1280	Wuhan	China	2007	
17	Höga Kusten	1210	Kramfors	Sweden	1997	
18	Longjiang	1196	Yunnan	China	2015	
19	Jinshajiang Taku	1190	Yunnan	China	2020	
20	Aizhai	1176	Jishou	China	2012	

25.4 LONGEST SPAN CABLE STAYED BRIDGES

The structural, aesthetic and economical advantages of cable stayed bridges are well established by pioneers like Leonhardt, who is credited to have designed and built the first cable stayed bridge in 1952 across the river Rhine in Dusseldorf. The first modern cable stayed bridge being the Stromsund Bridge in Sweeden designed by Dischinger and constructed during the year 1955. A comparative analysis indicates that a suspension bridge requires a stiffening deck girder with a flexural stiffness which must be about ten times larger than that required for a cable stayed bridge covering the same span. Also the quantity of steel required for the suspension bridge is nearly twice that required for the cable stayed bridge for spans exceeding 1000 m.

The present trend is to opt for cable stayed type in preference to the other types, especially for long span bridges. Economic studies of bridges constructed in Russia clearly indicate the cost effectiveness of cable stayed bridges for long spans. The ranking list of the world's longest cable stayed bridges is shown in Table 25.2 in which spans from 400m to 1100 m are covered. It is significant to note that China is credited to have almost 50 percent of the long span cable stayed bridges in the list.

Table 25.2 List of Longest Span Cable Stayed Bridges in the World

No.	Bridge	Span [m]	Location	Country	Year	Notes
1	Russky	1104	Vladivostok	Russia	2012	
2	Hutong	1092	Jiangsu	China	2018	
3	Sutong	1088	Suzhou-Nantong	China	2008	
4	Stonecutters	1018	Hong Kong	China	2009	
5	Edong	926	Hubei	China	2009	
6	Tatara	890	Onomichi-Imabari	Japan	1999	
7	Pont de Normandie	856	Le Havre	France	1995	
8	Jiujiang	818	Hubei	China	2012	
9	Jingsha	816	Hubei	China	2009	
10	Wuhu-2	806	Anhui	China	2017	
11	Skarsund	530		Norway	1991	
12	Vidyasagar Sethu	457	Kolkota	India	2010	
13	Saint – Nazaire	404		France	1975	
14	Jogighopa	286	Assam	India	1995	
15	Akkar	152	Sikkim	India	1995	

25.5 LONGEST SPAN STEEL ARCHED BRIDGES

Steel arched bridges are ideally suited for spanning deep gorges where suitable rock is available for anchoring the foundations of arch supports. Arched bridges are preferred to cross deep viaducts and ravines in hilly terrain and they are aesthetically superiors to other types of bridges. Table 25.3 is a compilation of the prominent long span arched bridges in the world. The span lengths of the top ten arched bridges shown in the table is in the range of 460 to 550 m.

The disadvantage of steel arched bridges being the maintenance cost to prevent rusting of the steel elements due to adverse environmental conditions. A majority of the top ten steel arched bridges are located in China.

25.6 LONGEST SPAN STEEL TRUSSED BRIDGES

The earliest long span bridges were built using steel trusses in the nineteenth century. Steel trusses were invariably used for the stiffening girders of suspension and cable stayed bridges. The earliest steel trussed girder used for the Firth of Forth suspension bridges dates back to the year 1890. Steel trussed bridges were common used in the 19th century for crossings of wide rivers in Railways. Various types of steel trusses like Warren girders, N and K-type have been widely used for railway bridges in India and other countries. Table 25.4 shows the longest top ten trussed bridges in the world for spans in the range of 400 to 500 m.

Table 25.3 List of Longest Span Steel Arch Bridges in the World

No.	Bridge	Span [m]	Location	Country	Year	Notes
1	Chaotianmen	552	Chongqing	China	2009	
2	Lupu	550	Shanghai	China	2003	
3	New River Gorge	518	Fayetteville, WV	USA	1977	
4	Bosideng	513	Sichuan	China	2012	
5	Bayonne	504	New York, NY	USA	1931	
6	Sydney Harbour	503	Sydney	Australia	1932	
7	Xiangxi	498	Xiangxizhen	China	2017	
8	Chenab	467	Katra	India	2019	
9	Daduhe Tianwan	466	Sichuan	China	2017	
10	Wushan	460	Chongqing	China	2005	

Table 25.4 List of Longest Span Steel Trussed Bridges in the World

No.	Bridge	Span [m]	Location	Country	Year	Notes
1	Pont de Quebec	549	Quebec City	Canada	1917	
2	Firth of Forth	521	Edinburgh	UK	1890	2 spans
3	Minato	510	Osaka	Japan	1974	
4	Commodore Barry	501	Chester, PA	USA	1974	
5	Greater New Orleans-1	480	New Orleans, LA	USA	1958	
6	Greater New Orleans-2	480	New Orleans, LA	USA	1988	
7	Howrah	457	Calcutta	India	1943	
8	Veterans Memorial	445	Gramercy, LA	USA	1995	
9	Tokyo Bay	440	Tokyo	Japan	2010	
10	Transbay	427	San Francisco, CA	USA	1936	[1]

25.7 LONGEST SPAN CONCRETE ARCH BRIDGES

Reinforced concrete arched bridges have been widely used for both road and railway crossings in various countries around the world. Elegant arch bridges were built during the early part of the twentieth century. Concrete being strong in compression, the arch configuration is ideally suited to transmit the dead and live loads from the deck to the arch supports through compressive forces. Various types of arched bridges like, bow string, open spandrel and balanced cantilever types have been widely used for road and railway bridge crossings.

Table 25.5 shows the top ten longest span concrete arched bridges around the world covering the span range from 300 to 400 m. China is credited to possess the maximum number long span Concrete arched bridges in the world.

Table 25.5 List of Longest Span Concrete Arch Bridges in the World

No.	Bridge	Span [m]	Location	Country	Year	Notes
1	Beipanjiang Qinglong	445	Guizhou	China	2016	
2	Wanxian	425	Wanzhou	China	1997	
3	Nanpanjiang	416	Yunnan	China	2015	
4	Krk-1 (east span)	390	Krk Island	Croatia	1980	
5	Almonte	384	Caceres	Spain	2015	
6	Yelanghe	370	Guizhou	China	2017	
7	Jialing	364	Sichuan	China	2012	
8	Jiangjiehe	330	Weng'an	China	1995	
9	Tajo Railway	324	Caceres	Spain	2015	
10	Hoover Dam Bypass	323	Boulder City, NV	USA	2010	

25.8 LONGEST SPAN PRESTRESSED CONCRETE BRIDGES

Ever since the development of the revolutionary building material like Prestressed concrete in 1940 by Eugene Freyssinet in France, significant progress has been achieved in the design and construction of long span bridges around the world. The material is ideally suited for the production of precast elements with superior quality and precision paving the way for use in long span bridges. Rapid developments in construction techniques[2] like cantilever, staging and progressive placement construction methods have been widely employed for the construction of long span prestressed concrete bridges. Also prestressed concrete is ideally suited for use in the deck girders of cable stayed bridges generally adopted for long spans. Table 25.6 lists the top ten longest prestressed concrete bridges around the world with China leading the world with maximum number of long span bridges.

Table 25.6 List of Longest Span Prestressed Concrete Bridges in the World

No.	Bridge	Span [m]	Location	Country	Year	Notes
1	Shibanpo	330	Chongqing	China	2006	
2	Stolmasundet	301	Austevoll	Norway	1998	
3	Raftsundet	298	Lofoten	Norway	1998	
4	Sundoy	298	Leirfjord	Norway	2003	
5	Beipanjiang Shuipan	290	Guizhou	China	2013	
6	Sandsfjord	290	Rogaland	Norway	2015	
7	Humen-2	270	Guangdong	China	1997	
8	Sutong-2	268	Suzhou-Nantong	China	2009	
9	Honghe	265	Yuanjiang	China	2003	
10	Gateway-1	260	Brisbane	Australia	1986	

25.9 LONGEST SPAN STEEL BOX/PLATE GIRDER BRIDGES

During the early part of the 20th century, steel plate girder bridges were commonly used for railway bridges around the world. Later box type plate girders were assembled for long span bridges. The main advantage of the steel plate girder bridges being the faster construction in comparison with the concrete bridges. The structural elements of the plate girders can be easily assembled and fabricated at site and the construction time required for the bridge can be significantly reduced. However the main disadvantage being the increased cost of maintenance to protect the structural elements form deterioration due to adverse environmental conditions. Steel box girder type bridges have been used for the span range of 200 to 300 m according to the top ten steel box/plate girder bridges compiled in Table 25.7.

Table 25.7 List of Longest Span Steel Box/Plate Girder Bridges in the World

No.	Bridge	Span [m]	Location	Country	Year	Notes
1	Ponte Costa e Silva	300	Rio de Janeiro	Brazil	1974	
2	Neckartalbrücke-1	263	Weitingen	Germany	1978	
3	Sava-1	261	Belgrade	Serbia	1956	
4	Ponte de Vitoria-3	260	Espirito Santo	Brazil	1989	
5	Zoobrücke	259	Cologne	Germany	1966	
6	Sava-2 (Gazelle)	250	Belgrade	Serbia	1970	
7	Kaita	250	Hiroshima	Japan	1991	
8	Namihaya (Shirinashi)	250	Osaka	Japan	1994	
9	Auckland Harbour (widening)	244	Auckland	New Zealand	1969	
10	Trans-Tokyo Bay	240	Kawasaki-Kisarazu	Japan	1997	2 spans

25.10 GENERAL REMARKS

The prominent examples of long span bridges compiled above are by no means exhaustive and only bridges built in various countries around the world. Most of the long span bridges are located in developed countries like U.S.A, Germany, France, Norway and Japan. However many long span bridges have been built during the post war years in developing countries like India, China, Argentina and Mexico. During the prewar years, long spans were generally negotiated using steel suspension bridges.

Developments in the field of materials technology, analysis of complex structures under different types of loads, construction techniques like cantilever construction together with the experience gained in maintenance of steel and concrete structures under severe exposure conditions have conclusively proved the advantages of prestressed concrete over steel. As a consequence of low maintenance costs, ease of construction, durability, structural integrity under dead, live and wind loads, cable stayed bridges in conjunction with prestressed concrete decks are preferred to steel suspension bridges for long spans. The reader may refer to the excellent publications by Shirley-Smith[3], Virola[4], Plowden[5], Subba Rao[6], Leonhardt[7], Gerwick[8], Muller[9], Lin *et al*[10] and others for detailed information regarding analysis, design and construction of prestressed concrete bridges.

At the dawn of the 21st century, India has embarked on a gigantic highway development project with the planning of the Golden Quadrilateral connecting the capital cities of various states spread between the north-south and east-west corridors. This massive highway project involves the crossing of number of major rivers and valley's using suitable bridges. In this context, the use of prestressed concrete for the construction of innumerable number of bridges ensures both economy and durability.

REFERENCES

1. Johnson Victor, D., Essentials of Bridge Engineering (Fifth Edition), Oxford & IBH Publishing Co. Pvt. Ltd., New Delhi, 2001, pp. 1–9.
2. Raina, V.K., Concrete Bridge Practice, Analysis, design & Economics, Tata McGraw-Hill Publishng Co. Ltd., New Delhi, 1991, pp. 3–700.
3. Shirley-Smith, H., World's Great Bridges, English language Book Society, London, 1964, pp. 250.
4. Virola, J., The World's Greatest Bridges, Proceedings of the American Society of Civil Engineers, Vol. 38, No.10, Oct. 1968, pp.52–55.
5. Plowden, D., Bridges The Spans of America, The Viking Press, New York, 1974, pp. 328.
6. Subba Rao, T.N., Long Span Prestressed Concrete Bridges in India, Seminar on Problems of Prestressing, Indian National Group of the IABSE, Preliminary Publication, Madras, Jan/Feb 1970, pp. 1–113 to 1–130.
7. Leonhardt, F., New Trends in design and Construction of Long Span Bridges and Viaducts (Skew, Flat slabs, Torsion Box), Preliminary Publication of Eighth Congress of IABSE, New York, 1968.
8. Gerwick, B.C. (Jr)., Precast Segmental Construction for Long Span Bridges, Civil Engineering, New York, January, 1964.
9. Muller, J., Long Span Precast Prestressed Concrete Bridges Built in Cantilever, First International Symposium on Concrete Bridge Design, ACI Publication SP-23 American Concrete Institute, Detroit, 1969, pp. 705–740.
10. Lin, T.Y and Gerwick, B.C., Design of Long Span Concrete Bridges With Special reference to Prestressing, Precasting, Erection, Structural Behaviour and Economics, First International Symposium on Concrete Bridge Design, ACI Publication SP-23 American Concrete Institute, Detroit, 1969, pp. 693–704.

Subject Index

Author Index